VOLUME FOUR HUNDRED AND THIRTY-FIVE

METHODS IN
ENZYMOLOGY

Oxygen Biology and Hypoxia

METHODS IN ENZYMOLOGY

Editors-in-Chief

JOHN N. ABELSON AND MELVIN I. SIMON

Division of Biology
California Institute of Technology
Pasadena, California

Founding Editors

SIDNEY P. COLOWICK AND NATHAN O. KAPLAN

VOLUME FOUR HUNDRED AND THIRTY-FIVE

Methods in ENZYMOLOGY

Oxygen Biology and Hypoxia

EDITED BY

HELMUT SIES
Heinrich-Heine-University Düsseldorf
Institute for Biochemistry and Molecular Biology I
Düsseldorf, Germany

BERNHARD BRÜNE
Johann Wolfgang Goethe-University
Institute for Biochemistry I/Pathobiochemistry (ZAFES)
Frankfurt, Germany

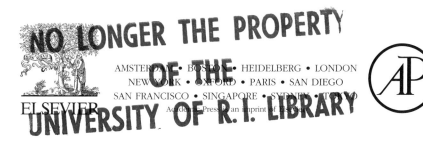

AMSTERDAM • BOSTON • HEIDELBERG • LONDON
NEW YORK • OXFORD • PARIS • SAN DIEGO
SAN FRANCISCO • SINGAPORE • SYDNEY • TOKYO
Academic Press is an imprint of Elsevier

ELSEVIER

Academic Press is an imprint of Elsevier
525 B Street, Suite 1900, San Diego, California 92101-4495, USA
84 Theobald's Road, London WC1X 8RR, UK

For information on all Elsevier Academic Press publications
visit our Web site at www.books.elsevier.com

ISBN: 978-0-12-373970-4

PRINTED IN THE UNITED STATES OF AMERICA
07   08   09   10   9   8   7   6   5   4   3   2   1

Contents

Contributors   *xiii*
Preface   *xxi*
Volumes in Series   *xxiii*

Section I. Hypoxia-Inducible Factor   1

1. Hypoxia-Inducible Factors P^{ER}/ARNT/S^{IM} Domains: Structure and Function   3

Thomas H. Scheuermann, Jinsong Yang, Lei Zhang, Kevin H. Gardner, and Richard K. Bruick

  1. Introduction   4
  2. Delineation of the HIF PAS Domains   5
  3. Expression and Characterization of HIF PAS Domains   7
  4. Assessing PAS Domain Protein–Protein Interactions   11
  5. Discussion   18
  Acknowledgments   20
  References   21

2. Hypoxia-Inducible Factor Prolyl-Hydroxylase: Purification and Assays of PHD2   25

Kirsty S. Hewitson, Christopher J. Schofield, and Peter J. Ratcliffe

  1. Introduction   26
  2. Preparation of Purified PHD2 from a Bacterial Source   28
  3. Assaying of PHD2 Activity   29
  4. Indirect Measurements of PHD2 Activity   29
  5. Direct Measurements of PHD2 Hydroxylation Activity   35
  6. Binding Assays   37
  7. Comparison of Assay Formats   38
  References   39

3. Determination and Modulation of Prolyl-4-Hydroxylase Domain Oxygen Sensor Activity   43

Renato Wirthner, Kuppusamy Balamurugan, Daniel P. Stiehl, Sandra Barth, Patrick Spielmann, Felix Oehme, Ingo Flamme, Dörthe M. Katschinski, Roland H. Wenger, and Gieri Camenisch

1. Introduction   44
2. Production of Functionally Active PHDs   48
3. Determination of PHD by VHL Binding to Peptides Derived from the HIF-1α ODD Domain   48
4. Determination of Prolyl-4-Hydroxylation by Oxidative Decarboxylation of 2-Oxoglutarate   51
5. Crude Tissue Extracts are not a Suitable Source of PHD Activity for the 2-Oxoglutarate Conversion Assay   53
6. Thin Layer Chromatography to Assess the Purity of [5-^{14}C]2-Oxoglutarate   53
7. Application of the 2-Oxoglutarate Conversion Assay to Protein Targets   55
8. Conclusions   55
Acknowledgments   57
References   57

4. Characterization of Ankyrin Repeat–Containing Proteins as Substrates of the Asparaginyl Hydroxylase Factor Inhibiting Hypoxia-Inducible Transcription Factor   61

Sarah Linke, Rachel J. Hampton-Smith, and Daniel J. Peet

1. Introduction   62
2. Experimental Techniques   65
3. Discussion/Conclusion   82
Acknowledgments   83
References   83

5. Transgenic Models to Understand Hypoxia-Inducible Factor Function   87

Andrew Doedens and Randall S. Johnson

1. Introduction   88
2. Hypoxia Response Pathway Genes and Development   90
3. HIF in Physiology   93
4. HIF Function in Tumor Biology   98
5. Summary   100
References   101

6. The Silencing Approach of the Hypoxia-Signaling Pathway   **107**

Edurne Berra and Jacques Pouysségur

1.	A Brief History of RNAi	108
2.	The Hypoxia-Signaling Pathway	109
3.	HIF-α Stability	110
4.	HIF Activity	114
5.	HIF-1/HIF-2 Target Gene Specificity	116
6.	RNAi as a New Potential Therapeutic Strategy	116
	Acknowledgments	118
	References	118

7. Cellular and Developmental Adaptations to Hypoxia:
   A *Drosophila* Perspective   **123**

Nuria Magdalena Romero, Andrés Dekanty, and Pablo Wappner

1.	Introduction	124
2.	*Drosophila melanogaster* as a Model System to Study Physiological Responses to Hypoxia	124
3.	Experimental Advantages of the Model System	125
4.	The *Drosophila* Respiratory System	126
5.	Occurrence of a *Drosophila* System Homologous to Mammalian HIF	128
6.	Regulation of Sima by Oxygen Levels	131
7.	Role of Sima and Fatiga in *Drosophila* Development	132
8.	Hypoxia-Inducible Genes and the Adaptation of *Drosophila* to Oxygen Starvation	134
9.	Regulation of Sima by the PI3K and TOR Pathways	134
10.	Role of the HIF System in Growth Control and Cell Size Determination	136
11.	Concluding Remarks	138
	Acknowledgments	139
	References	139

Section II. Erythropoietin   145

**8. Constitutively Overexpressed Erythropoietin Reduces Infarct Size
   in a Mouse Model of Permanent Coronary Artery Ligation**   **147**

Giovanni G. Camici, Thomas Stallmach, Matthias Hermann, Rutger Hassink,
Peter Doevendans, Beat Grenacher, Alain Hirschy, Johannes Vogel,
Thomas F. Lüscher, Frank Ruschitzka, and Max Gassmann

1.	Introduction	148
2.	Material and Methods	149
3.	Results	151
4.	Discussion	153

Acknowledgments   154
References   154

9. **Use of Gene-Manipulated Mice in the Study of *Erythropoietin*
   Gene Expression**   **157**

   Norio Suzuki, Naoshi Obara, and Masayuki Yamamoto

   1. Introduction   158
   2. Materials   161
   3. Methods and Results   163
   4. Conclusion   173
   Acknowledgments   173
   References   174

10. **Control of Erythropoietin Gene Expression and its
    Use in Medicine**   **179**

    Wolfgang Jelkmann

    1. Introduction   180
    2. Native EPO Gene Expression and its Pharmacologic Stimulation   180
    3. EPO Gene Transfer   184
    4. Recombinant EPOS   187
    5. Conclusions   190
    References   191

11. **Role of Hypoxia-Inducible Factor-2α in Endothelial
    Development and Hematopoiesis**   **199**

    Osamu Ohneda, Masumi Nagano, and Yoshiaki Fujii-Kuriyama

    1. Introduction   200
    2. Vasculogenesis/Angiogenesis and HIFs   200
    3. HIF-1β/ARNT Null Mice   204
    4. Neovascularization and HIFs   205
    5. Hematopoiesis and HIFs   211
    6. Conclusion   214
    References   214

Section III. Hypoxia and Adaptation   **219**

12. **Organ Protection by Hypoxia and Hypoxia-Inducible Factors**   **221**

    Wanja M. Bernhardt, Christina Warnecke, Carsten Willam, Tetsuhiro Tanaka,
    Michael S. Wiesener, and Kai-Uwe Eckardt

    1. Introduction   222
    2. From Ischemic to Hypoxic Preconditioning   222

3. Hypoxia-Inducible Transcription Factors   223
4. Strategies to Activate HIF and HIF Target Genes   227
5. Hypoxic Preconditioning and HIF   231
6. HIF in Chronic Hypoxic/Ischemic Diseases   236
7. Conclusions and Perspectives   237
References   238

13. **Hypoxia and Regulation of Messenger RNA Translation**   **247**

Marianne Koritzinsky and Bradly G. Wouters

1. Introduction   248
2. Changes in Global mRNA Translation During Hypoxia   249
3. Molecular Mechanisms that Regulate mRNA Translation During Hypoxia   251
4. Methods Employed to Study mRNA Translation During Hypoxia   256
5. Protocols   268
References   271

14. **Hypoxia and the Unfolded Protein Response**   **275**

Constantinos Koumenis, Meixia Bi, Jiangbin Ye, Douglas Feldman,
and Albert C. Koong

1. Introduction   276
2. Methods Employed in Detecting Hypoxic Induction of ER Stress   282
Acknowledgments   289
References   290

Section IV. Hypoxia and Tumor Biology   295

15. **Tumor Hypoxia in Cancer Therapy**   **297**

J. Martin Brown

1. Hypoxia in Human Tumors   298
2. The Dynamic Nature of Hypoxia in Tumors   300
3. Consequences of Tumor Hypoxia for Cancer Treatment   300
4. Size of the Oxygen Effect with Radiation   302
5. The Influence of Tumor Hypoxia on Cancer Treatment
   by Radiotherapy   303
6. Influence of Tumor Hypoxia on Response to Chemotherapy   307
7. Exploiting Hypoxia in Cancer Treatment   308
References   315

16. **HIF Gene Expression in Cancer Therapy**   **323**

Denise A. Chan, Adam J. Krieg, Sandra Turcotte, and Amato J. Giaccia

1. Introduction   324
2. Experimental Procedures   326

3. Conclusions   337
Acknowledgments   337
References   338

17. Analysis of Hypoxia-Inducible Factor-1α Expression and its Effects on Invasion and Metastasis   **347**

Balaji Krishnamachary and Gregg L. Semenza

1. Introduction   347
2. Protocol 1: HIF-1α Immunohistochemistry   349
3. Protocol 2: Invasion Assay   350
4. Protocol 3: Transepithelial Resistance Measurement of
   Cell–Cell Adhesion   351
5. Protocol 4: Analysis of mRNA Expression by qRT-PCR   351
References   352

18. Macrophage Migration Inhibitory Factor Manipulation and Evaluation in Tumoral Hypoxic Adaptation   **355**

Millicent Winner, Lin Leng, Wayne Zundel, and Robert A. Mitchell

1. Introduction   356
2. Modulation of MIF Levels by Targeted shRNAs and Assessment of
   Knockdown Efficiency   357
3. Analysis of MIF-Dependent CSN5 and COP9 Signalosome Function   362
4. Determination of Tumor-Associated MIF Expression and MIF
   Polymorphic Disparity   364
5. Conclusions   367
References   367

19. The Von Hippel-Lindau Tumor Suppressor Protein: An Update   **371**

William G. Kaelin, Jr.

1. Introduction   372
2. Regulation of Epithelial Differentiation by pVHL   373
3. Crosstalk Between c-Met and VHL   374
4. Regulation of Neuronal Apoptosis by pVHL   375
5. Possible Links Between p53 and pVHL   376
6. Regulation of pVHL by Phosphorylation   376
7. Polyubiquitylation of pVHL   377
8. Mouse Models for Studying pVHL Function   377
References   378

20. Hypoxia-Inducible Factor 1 Inhibitors   385

Giovanni Melillo

1. Introduction   386
2. Cell-Based High Throughput Screens   387
3. Cell-Free Assays   393
4. Bioassay-Directed Isolation of Natural Product HIF-1 Inhibitors   398
5. Conclusions   399
Acknowledgments   399
References   400

Section V. Hypoxia and Inflammatory Mediators   403

21. Regulation of Hypoxia-Inducible Factors During Inflammation   405

Stilla Frede, Utta Berchner-Pfannschmidt, and Joachim Fandrey

1. Introduction   406
2. Regulation of HIF at the Transcriptional Level   408
3. Regulation of HIF at the Translational Level   410
4. Regulation of HIF-1α at the Posttranslational Level   411
5. Regulation of HIF-1 Activity   413
6. Perspectives   414
7. Conclusions   414
Acknowledgments   415
References   415

22. Superoxide and Derived Reactive Oxygen Species in the Regulation of Hypoxia-Inducible Factors   421

Agnes Görlach and Thomas Kietzmann

1. Introduction   422
2. Reactive Oxygen Species Act as Signaling Molecules   423
3. HIFs are Sensitive to Oxygen   424
4. Reactive Oxygen Species Modulate HIF   425
5. How are HIFs Regulated by Reactive Oxygen Species?   427
6. Summary   431
7. Methods   431
8. The Cytochrome C Reduction Assay for Detection of Extracellular Reactive Oxygen Species   431
9. Chemiluminescence Assay for Detection of Extracellular Reactive Oxygen Species   432
10. Measuring Intracellular Production of Reactive Oxygen Species using Fluorescent Dyes   433

11. Detection of Reactive Oxygen Species by Electron Paramagnetic
    Resonance   436
Acknowledgments   438
References   438

23. **Genetics of Mitochondrial Electron Transport Chain in Regulating Oxygen Sensing**   **447**

Eric L. Bell and Navdeep S. Chandel

1. Introduction   448
2. Detecting HIF-1α Protein Levels   449
3. Detecting Intracellular ROS Levels   452
4. Method 1: Examining Hypoxic Stabilization of HIF-1α Protein in Cells
   Containing RNAI against the Rieske Fe-S Protein   454
5. Method 2: Examining the Role of ROS Generated from Mitochondrial
   Electron Transport in Hypoxic Stabilization of HIF-1α Protein   458
6. Concluding Remarks   459
Acknowledgments   459
References   460

24. **Hypoxia-Inducible Factor-1α Under the Control of Nitric Oxide**   **463**

Bernhard Brüne and Jie Zhou

1. HIF-1 and Oxygen Sensing   464
2. Nitric Oxide: A Multifunctional Messenger   465
3. Accumulation of HIF-1α and Activation of HIF-1 by NO   467
4. Superoxide Stabilizes HIF-1α but Antagonizes NO Actions   470
5. Hypoxic Signal Transmission is Antagonized by NO   472
6. Summary and Conclusions   473
Acknowledgments   475
References   475

25. **Hypoxic Regulation of NF-κB Signaling**   **479**

Eoin P. Cummins, Katrina M. Comerford, Carsten Scholz, Ulrike Bruning,
and Cormac T. Taylor

1. Background   480
2. Treatment Protocols for Cellular Hypoxia Studies   481
3. Measurement of NF-κB Activity in Cultured Cells   484
4. Summary/Conclusions   491
References   491

Author Index   *493*
Subject Index   *535*

Contributors

Kuppusamy Balamurugan
Institute of Physiology and Zürich Center for Integrative Human Physiology (ZIHP), University of Zürich, Zürich, Switzerland

Sandra Barth
Institute of Physiology and Zürich Center for Integrative Human Physiology (ZIHP), University of Zürich, Zürich, Switzerland

Eric L. Bell
Department of Medicine, Northwestern University Medical School, Chicago, Illinois

Utta Berchner-Pfannschmidt
Institut für Physiologie, Universität Duisburg-Essen, Essen, Germany

Wanja M. Bernhardt
Department of Nephrology and Hypertension, Friedrich-Alexander University, Erlangen, Nürnberg, Germany

Edurne Berra
CICbioGUNE Technology Park of Bizkaia, Derio, Spain

Meixia Bi
Department of Radiation Oncology, University of Pennsylvania School of Medicine, Philadelphia, Pennsylvania

J. Martin Brown
Division of Radiation and Cancer Biology, Department of Radiation Oncology, Stanford University School of Medicine, Stanford, California

Bernhard Brüne
Institute of Biochemistry I/Pathobiochemistry (ZAFES), Johann Wolfgang Goethe-University, Frankfurt, Germany

Richard K. Bruick
Department of Biochemistry, University of Texas Southwestern Medical Center, Dallas, Texas

Ulrike Bruning
University of Lübeck, Lübeck, Germany

Gieri Camenisch
Institute of Physiology and Zürich Center for Integrative Human Physiology (ZIHP), University of Zürich, Zürich, Switzerland

Giovanni G. Camici
Zürich Center for Integrative Human Physiology (ZIHP), University of Zürich, Zürich, Switzerland, and Cardiology and Cardiovascular Research and Institute of Physiology, University of Zürich, Zürich, Switzerland

Denise A. Chan
Department of Radiation Oncology, Stanford University School of Medicine, Stanford, California

Navdeep S. Chandel
Department of Cell and Molecular Biology and Department of Medicine, Northwestern University Medical School, Chicago, Illinois

Katrina M. Comerford
UCD Conway Institute, University College Dublin, Dublin, Ireland

Eoin P. Cummins
UCD Conway Institute, University College Dublin, Dublin, Ireland

Andrés Dekanty
Instituto Leloir, Patricias Argentinas, Buenos Aires, Argentina

Andrew Doedens
Molecular Biology Section, Division of Biological Sciences, University of California San Diego, La Jolla, California

Peter Doevendans
Department of Cardiothoracic Surgery, Heart Lung Center, Utrecht, The Netherlands

Kai-Uwe Eckardt
Department of Nephrology and Hypertension, Friedrich-Alexander University, Erlangen, Nürnberg, Germany

Joachim Fandrey
Institut für Physiologie, Universität Duisburg-Essen, Essen, Germany

Douglas Feldman
Department of Radiation Oncology, Stanford University School of Medicine, Stanford, California

Ingo Flamme
Institute for Cardiovascular Research, Bayer HealthCare, Wuppertal, Germany

Stilla Frede
Institut für Physiologie, Universität Duisburg-Essen, Essen, Germany

Yoshiaki Fujii-Kuriyama
Center for TARA, University of Tsukuba, Tsukuba, Japan, and SORST, Japan
Science Technology Agency, Kawaguchi, Japan

Kevin H. Gardner
Department of Biochemistry, University of Texas Southwestern Medical Center,
Dallas, Texas

Max Gassmann
Institute of Veterinary Physiology, Vetsuisse Faculty, and Zürich Center for Integrative Human Physiology (ZIHP), University of Zürich, Zürich, Switzerland

Amato J. Giaccia
Department of Radiation Oncology, Stanford University School of Medicine,
Stanford, California

Agnes Görlach
Experimental Pediatric Cardiology, German Heart Center Munich, Munich,
Germany

Beat Grenacher
Institute of Veterinary Physiology, Vetsuisse Faculty, and Zürich Center for Integrative Human Physiology (ZIHP), University of Zürich, Zürich, Switzerland

Rachel J. Hampton-Smith
School of Molecular and Biomedical Science and The ARC Special Research
Centre for the Molecular Genetics of Development, University of Adelaide,
Adelaide, Australia

Rutger Hassink
Department of Cardiothoracic Surgery, Heart Lung Center, Utrecht, The
Netherlands

Matthias Hermann
Cardiology and Cardiovascular Research and Institute of Physiology and Zürich
Center for Integrative Human Physiology (ZIHP), University of Zürich, Zürich,
Switzerland

Kirsty S. Hewitson
Department of Chemistry, Chemistry Research Laboratory, University of Oxford,
Oxford, United Kingdom

Alain Hirschy
Institute of Cell Biology, Swiss Federal Institute of Technology Zürich, Zürich,
Switzerland

Wolfgang Jelkmann
Institute of Physiology, University of Lübeck, Lübeck, Germany

Randall S. Johnson
Molecular Biology Section, Division of Biological Sciences, University of California San Diego, La Jolla, California

William G. Kaelin, Jr.
Howard Hughes Medical Institute, Harvard, Medical School, Boston, Massachusetts

Dörthe M. Katschinski
Department of Heart and Circulatory Physiology, Center of Physiology and Pathophysiology, Georg-August University Göttingen, Göttingen, Germany

Thomas Kietzmann
Faculty of Chemistry, Department of Biochemistry, University of Kaiserslautern, Kaiserslautern, Germany

Albert C. Koong
Department of Radiation Oncology, Stanford University School of Medicine, Stanford, California

Marianne Koritzinsky
Department of Radiation Oncology (Maastro Lab), GROW Research Institute, Maastricht University, Maastricht, The Netherlands

Constantinos Koumenis
Department of Radiation Oncology, University of Pennsylvania School of Medicine, Philadelphia, Pennsylvania

Adam J. Krieg
Department of Radiation Oncology, Stanford University School of Medicine, Stanford, California

Balaji Krishnamachary
Vascular Biology Program, Institute for Cell Engineering, Departments of Pediatrics, Medicine, Oncology, and Radiation Oncology, and McKusick-Nathans Institute of Genetic Medicine, The Johns Hopkins University School of Medicine, Baltimore, Maryland

Lin Leng
Yale University, New Haven, Connecticut

Sarah Linke
School of Molecular and Biomedical Science and The ARC Special Research Centre for the Molecular Genetics of Development, University of Adelaide, Adelaide, Australia

Thomas F. Lüscher
Cardiology and Cardiovascular Research and Institute of Physiology, and Zürich Center for Integrative Human Physiology (ZIHP), University of Zürich, Zürich, Switzerland

Giovanni Melillo
Developmental Therapeutics Program, SAIC Frederick, Inc., National Cancer Institute at Frederick, Frederick, Maryland

Robert A. Mitchell
Molecular Targets Program, JG Brown Cancer Center, University of Louisville, Louisville, Kentucky

Masumi Nagano
Department of Regenerative Medicine and Stem Cell Biology, University of Tsukuba, Tsukuba, Japan, and SORST, Japan Science Technology Agency, Kawaguchi, Japan

Naoshi Obara
ERATO Environmental Response Project, Japan Science and Technology Agency, Center for Tsukuba Advanced Research Alliance, University of Tsukuba, Tsukuba, Ibaraki, Japan

Felix Oehme
Institute for Cardiovascular Research, Bayer HealthCare, Wuppertal, Germany

Osamu Ohneda
Department of Regenerative Medicine and Stem Cell Biology, University of Tsukuba, Tsukuba, Japan, and SORST, Japan Science Technology Agency, Kawaguchi, Japan

Daniel J. Peet
School of Molecular and Biomedical Science and The ARC Special Research Centre for the Molecular Genetics of Development, University of Adelaide, Adelaide, Australia

Jacques Pouysségur
Centre Antoine Lacassagne, Institute of Signaling, Developmental Biology and Cancer Research, University of Nice, Nice, France

Peter J. Ratcliffe
Henry Wellcome Building for Molecular Physiology, University of Oxford, Oxford, United Kingdom

Nuria Magdalena Romero
Instituto Leloir, Patricias Argentinas, Buenos Aires, Argentina

Frank Ruschitzka
Cardiology and Cardiovascular Research and Institute of Physiology, and Zürich Center for Integrative Human Physiology (ZIHP), University of Zürich, Zürich, Switzerland

Thomas H. Scheuermann
Department of Biochemistry, University of Texas Southwestern Medical Center, Dallas, Texas

Christopher J. Schofield
Department of Chemistry, Chemistry Research Laboratory, University of Oxford, Oxford, United Kingdom

Carsten Scholz
University of Lübeck, Lübeck, Germany

Gregg L. Semenza
Vascular Biology Program, Institute for Cell Engineering, Departments of Pediatrics, Medicine, Oncology, and Radiation Oncology, and McKusick-Nathans Institute of Genetic Medicine, The Johns Hopkins University School of Medicine, Baltimore, Maryland

Patrick Spielmann
Institute of Physiology and Zürich Center for Integrative Human Physiology (ZIHP), University of Zürich, Zürich, Switzerland

Thomas Stallmach
Institute of Pathology, University Hospital Zürich, and Zürich Center for Integrative Human Physiology (ZIHP), University of Zürich, Zürich, Switzerland

Daniel P. Stiehl
Institute of Physiology and Zürich Center for Integrative Human Physiology (ZIHP), University of Zürich, Zürich, Switzerland

Norio Suzuki
ERATO Environmental Response Project, Japan Science and Technology Agency, Center for Tsukuba Advanced Research Alliance, University of Tsukuba, Tsukuba, Ibaraki, Japan

Tetsuhiro Tanaka
Department of Nephrology and Hypertension, Friedrich-Alexander University, Erlangen, Nürnberg, Germany

Cormac T. Taylor
UCD Conway Institute, University College Dublin, Dublin, Ireland

Sandra Turcotte
Department of Radiation Oncology, Stanford University School of Medicine, Stanford, California

Johannes Vogel
Institute of Veterinary Physiology, Vetsuisse Faculty University of Zürich, Zürich, Switzerland, and Zürich Center for Integrative Human Physiology (ZIHP), University of Zürich, Zürich, Switzerland

Pablo Wappner
Instituto Leloir and FBMC, University of Buenos Aires, Patricias Argentinas, Buenos Aires, Argentina

Christina Warnecke
Department of Nephrology and Hypertension, Friedrich-Alexander University, Erlangen, Nürnberg, Germany

Roland H. Wenger
Institute of Physiology and Zürich Center for Integrative Human Physiology (ZIHP), University of Zürich, Zürich, Switzerland

Michael S. Wiesener
Department of Nephrology and Hypertension, Friedrich-Alexander University, Erlangen, Nürnberg, Germany

Carsten Willam
Department of Nephrology and Hypertension, Friedrich-Alexander University, Erlangen, Nürnberg, Germany

Millicent Winner
Molecular Targets Program, JG Brown Cancer Center, University of Louisville, Louisville, Kentucky

Renato Wirthner
Institute of Physiology and Zürich Center for Integrative Human Physiology (ZIHP), University of Zürich, Zürich, Switzerland

Bradly G. Wouters
Department of Radiation Oncology (Maastro Lab), GROW Research Institute, Maastricht University, Maastricht, The Netherlands

Masayuki Yamamoto
ERATO Environmental Response Project, Japan Science and Technology Agency, Center for Tsukuba Advanced Research Alliance, University of Tsukuba, Tsukuba, Ibaraki, Japan, and Department of Medical Biochemistry, Tohoku University Graduate School of Medicine, Sendai, Japan

Jinsong Yang
Department of Biochemistry, University of Texas Southwestern Medical Center, Dallas, Texas

Jiangbin Ye
Department of Radiation Oncology, University of Pennsylvania School of Medicine, Philadelphia, Pennsylvania

Lei Zhang
Department of Biochemistry, University of Texas Southwestern Medical Center, Dallas, Texas

Jie Zhou
Institute of Biochemistry I/ZAFES, Johann Wolfgang Goethe-University, Frankfurt, Germany

Wayne Zundel
Molecular Targets Program, JG Brown Cancer Center, University of Louisville, Louisville, Kentucky

PREFACE

The ability to sense and respond to changes in oxygen partial pressure represents a fundamental property for assuring balanced cellular oxygen supply. The ubiquitous pathway of the hypoxic induction of the transcription factor "hypoxia inducible factor" (HIF) is pivotal. Stability and activity of HIF are regulated as a function of oxygen but also in response to hormones, growth factors, oncogenes and as a result of redox-changes related to the production of superoxide or nitric oxide. HIF, in turn, coordinates gene and protein expression profiles that promote compensatory shifts in anaerobic metabolism, oxygen delivery and adaptive processes to coordinate systemic, local and cellular responses to hypoxia.

Research on hypoxia currently focuses on how cells sense changes in oxygen availability to coordinate genomic and proteomic responses, how HIF affects normal development and physiology, how hypoxia can lead to diseases in close association with human pathologies such as inflammation or cancer, and which of the targets can be approached for potential therapeutic interventions. Although the discovery of molecular details on HIF stability regulation has revealed an unexpectedly direct connection between molecular oxygen and responses to hypoxia, major problems in the field include understanding multiple intracellular changes that occur upon oxygen deprivation. This volume of Methods in Enzymology brings together expertise from various disciplines to shed light on recent developments, methodological approaches and biomedical aspects in the field of Oxygen Biology and Hypoxia.

We thank the authors for their excellent contributions and for advice, and we thank Kirstin Schäfer, Marlies Scholtes and Cindy Minor for their valuable help.

HELMUT SIES AND BERNHARD BRÜNE

METHODS IN ENZYMOLOGY

VOLUME I. Preparation and Assay of Enzymes
Edited by SIDNEY P. COLOWICK AND NATHAN O. KAPLAN

VOLUME II. Preparation and Assay of Enzymes
Edited by SIDNEY P. COLOWICK AND NATHAN O. KAPLAN

VOLUME III. Preparation and Assay of Substrates
Edited by SIDNEY P. COLOWICK AND NATHAN O. KAPLAN

VOLUME IV. Special Techniques for the Enzymologist
Edited by SIDNEY P. COLOWICK AND NATHAN O. KAPLAN

VOLUME V. Preparation and Assay of Enzymes
Edited by SIDNEY P. COLOWICK AND NATHAN O. KAPLAN

VOLUME VI. Preparation and Assay of Enzymes *(Continued)*
Preparation and Assay of Substrates
Special Techniques
Edited by SIDNEY P. COLOWICK AND NATHAN O. KAPLAN

VOLUME VII. Cumulative Subject Index
Edited by SIDNEY P. COLOWICK AND NATHAN O. KAPLAN

VOLUME VIII. Complex Carbohydrates
Edited by ELIZABETH F. NEUFELD AND VICTOR GINSBURG

VOLUME IX. Carbohydrate Metabolism
Edited by WILLIS A. WOOD

VOLUME X. Oxidation and Phosphorylation
Edited by RONALD W. ESTABROOK AND MAYNARD E. PULLMAN

VOLUME XI. Enzyme Structure
Edited by C. H. W. HIRS

VOLUME XII. Nucleic Acids (Parts A and B)
Edited by LAWRENCE GROSSMAN AND KIVIE MOLDAVE

VOLUME XIII. Citric Acid Cycle
Edited by J. M. LOWENSTEIN

VOLUME XIV. Lipids
Edited by J. M. LOWENSTEIN

VOLUME XV. Steroids and Terpenoids
Edited by RAYMOND B. CLAYTON

VOLUME XVI. Fast Reactions
Edited by KENNETH KUSTIN

VOLUME XVII. Metabolism of Amino Acids and Amines (Parts A and B)
Edited by HERBERT TABOR AND CELIA WHITE TABOR

VOLUME XVIII. Vitamins and Coenzymes (Parts A, B, and C)
Edited by DONALD B. MCCORMICK AND LEMUEL D. WRIGHT

VOLUME XIX. Proteolytic Enzymes
Edited by GERTRUDE E. PERLMANN AND LASZLO LORAND

VOLUME XX. Nucleic Acids and Protein Synthesis (Part C)
Edited by KIVIE MOLDAVE AND LAWRENCE GROSSMAN

VOLUME XXI. Nucleic Acids (Part D)
Edited by LAWRENCE GROSSMAN AND KIVIE MOLDAVE

VOLUME XXII. Enzyme Purification and Related Techniques
Edited by WILLIAM B. JAKOBY

VOLUME XXIII. Photosynthesis (Part A)
Edited by ANTHONY SAN PIETRO

VOLUME XXIV. Photosynthesis and Nitrogen Fixation (Part B)
Edited by ANTHONY SAN PIETRO

VOLUME XXV. Enzyme Structure (Part B)
Edited by C. H. W. HIRS AND SERGE N. TIMASHEFF

VOLUME XXVI. Enzyme Structure (Part C)
Edited by C. H. W. HIRS AND SERGE N. TIMASHEFF

VOLUME XXVII. Enzyme Structure (Part D)
Edited by C. H. W. HIRS AND SERGE N. TIMASHEFF

VOLUME XXVIII. Complex Carbohydrates (Part B)
Edited by VICTOR GINSBURG

VOLUME XXIX. Nucleic Acids and Protein Synthesis (Part E)
Edited by LAWRENCE GROSSMAN AND KIVIE MOLDAVE

VOLUME XXX. Nucleic Acids and Protein Synthesis (Part F)
Edited by KIVIE MOLDAVE AND LAWRENCE GROSSMAN

VOLUME XXXI. Biomembranes (Part A)
Edited by SIDNEY FLEISCHER AND LESTER PACKER

VOLUME XXXII. Biomembranes (Part B)
Edited by SIDNEY FLEISCHER AND LESTER PACKER

VOLUME XXXIII. Cumulative Subject Index Volumes I–XXX
Edited by MARTHA G. DENNIS AND EDWARD A. DENNIS

VOLUME XXXIV. Affinity Techniques (Enzyme Purification: Part B)
Edited by WILLIAM B. JAKOBY AND MEIR WILCHEK

VOLUME XXXV. Lipids (Part B)
Edited by JOHN M. LOWENSTEIN

VOLUME XXXVI. Hormone Action (Part A: Steroid Hormones)
Edited by BERT W. O'MALLEY AND JOEL G. HARDMAN

VOLUME XXXVII. Hormone Action (Part B: Peptide Hormones)
Edited by BERT W. O'MALLEY AND JOEL G. HARDMAN

VOLUME XXXVIII. Hormone Action (Part C: Cyclic Nucleotides)
Edited by JOEL G. HARDMAN AND BERT W. O'MALLEY

VOLUME XXXIX. Hormone Action (Part D: Isolated Cells, Tissues,
and Organ Systems)
Edited by JOEL G. HARDMAN AND BERT W. O'MALLEY

VOLUME XL. Hormone Action (Part E: Nuclear Structure and Function)
Edited by BERT W. O'MALLEY AND JOEL G. HARDMAN

VOLUME XLI. Carbohydrate Metabolism (Part B)
Edited by W. A. WOOD

VOLUME XLII. Carbohydrate Metabolism (Part C)
Edited by W. A. WOOD

VOLUME XLIII. Antibiotics
Edited by JOHN H. HASH

VOLUME XLIV. Immobilized Enzymes
Edited by KLAUS MOSBACH

VOLUME XLV. Proteolytic Enzymes (Part B)
Edited by LASZLO LORAND

VOLUME XLVI. Affinity Labeling
Edited by WILLIAM B. JAKOBY AND MEIR WILCHEK

VOLUME XLVII. Enzyme Structure (Part E)
Edited by C. H. W. HIRS AND SERGE N. TIMASHEFF

VOLUME XLVIII. Enzyme Structure (Part F)
Edited by C. H. W. HIRS AND SERGE N. TIMASHEFF

VOLUME XLIX. Enzyme Structure (Part G)
Edited by C. H. W. HIRS AND SERGE N. TIMASHEFF

VOLUME L. Complex Carbohydrates (Part C)
Edited by VICTOR GINSBURG

VOLUME LI. Purine and Pyrimidine Nucleotide Metabolism
Edited by PATRICIA A. HOFFEE AND MARY ELLEN JONES

VOLUME LII. Biomembranes (Part C: Biological Oxidations)
Edited by SIDNEY FLEISCHER AND LESTER PACKER

VOLUME LIII. Biomembranes (Part D: Biological Oxidations)
Edited by SIDNEY FLEISCHER AND LESTER PACKER

VOLUME LIV. Biomembranes (Part E: Biological Oxidations)
Edited by SIDNEY FLEISCHER AND LESTER PACKER

VOLUME LV. Biomembranes (Part F: Bioenergetics)
Edited by SIDNEY FLEISCHER AND LESTER PACKER

VOLUME LVI. Biomembranes (Part G: Bioenergetics)
Edited by SIDNEY FLEISCHER AND LESTER PACKER

VOLUME LVII. Bioluminescence and Chemiluminescence
Edited by MARLENE A. DELUCA

VOLUME LVIII. Cell Culture
Edited by WILLIAM B. JAKOBY AND IRA PASTAN

VOLUME LIX. Nucleic Acids and Protein Synthesis (Part G)
Edited by KIVIE MOLDAVE AND LAWRENCE GROSSMAN

VOLUME LX. Nucleic Acids and Protein Synthesis (Part H)
Edited by KIVIE MOLDAVE AND LAWRENCE GROSSMAN

VOLUME 61. Enzyme Structure (Part H)
Edited by C. H. W. HIRS AND SERGE N. TIMASHEFF

VOLUME 62. Vitamins and Coenzymes (Part D)
Edited by DONALD B. MCCORMICK AND LEMUEL D. WRIGHT

VOLUME 63. Enzyme Kinetics and Mechanism (Part A: Initial Rate and
Inhibitor Methods)
Edited by DANIEL L. PURICH

VOLUME 64. Enzyme Kinetics and Mechanism
(Part B: Isotopic Probes and Complex Enzyme Systems)
Edited by DANIEL L. PURICH

VOLUME 65. Nucleic Acids (Part I)
Edited by LAWRENCE GROSSMAN AND KIVIE MOLDAVE

VOLUME 66. Vitamins and Coenzymes (Part E)
Edited by DONALD B. MCCORMICK AND LEMUEL D. WRIGHT

VOLUME 67. Vitamins and Coenzymes (Part F)
Edited by DONALD B. MCCORMICK AND LEMUEL D. WRIGHT

VOLUME 68. Recombinant DNA
Edited by RAY WU

VOLUME 69. Photosynthesis and Nitrogen Fixation (Part C)
Edited by ANTHONY SAN PIETRO

VOLUME 70. Immunochemical Techniques (Part A)
Edited by HELEN VAN VUNAKIS AND JOHN J. LANGONE

VOLUME 71. Lipids (Part C)
Edited by JOHN M. LOWENSTEIN

VOLUME 72. Lipids (Part D)
Edited by JOHN M. LOWENSTEIN

VOLUME 73. Immunochemical Techniques (Part B)
Edited by JOHN J. LANGONE AND HELEN VAN VUNAKIS

VOLUME 74. Immunochemical Techniques (Part C)
Edited by JOHN J. LANGONE AND HELEN VAN VUNAKIS

VOLUME 75. Cumulative Subject Index Volumes XXXI, XXXII, XXXIV–LX
Edited by EDWARD A. DENNIS AND MARTHA G. DENNIS

VOLUME 76. Hemoglobins
Edited by ERALDO ANTONINI, LUIGI ROSSI-BERNARDI, AND EMILIA CHIANCONE

VOLUME 77. Detoxication and Drug Metabolism
Edited by WILLIAM B. JAKOBY

VOLUME 78. Interferons (Part A)
Edited by SIDNEY PESTKA

VOLUME 79. Interferons (Part B)
Edited by SIDNEY PESTKA

VOLUME 80. Proteolytic Enzymes (Part C)
Edited by LASZLO LORAND

VOLUME 81. Biomembranes (Part H: Visual Pigments and Purple Membranes, I)
Edited by LESTER PACKER

VOLUME 82. Structural and Contractile Proteins (Part A: Extracellular Matrix)
Edited by LEON W. CUNNINGHAM AND DIXIE W. FREDERIKSEN

VOLUME 83. Complex Carbohydrates (Part D)
Edited by VICTOR GINSBURG

VOLUME 84. Immunochemical Techniques (Part D: Selected Immunoassays)
Edited by JOHN J. LANGONE AND HELEN VAN VUNAKIS

VOLUME 85. Structural and Contractile Proteins (Part B: The Contractile Apparatus and the Cytoskeleton)
Edited by DIXIE W. FREDERIKSEN AND LEON W. CUNNINGHAM

VOLUME 86. Prostaglandins and Arachidonate Metabolites
Edited by WILLIAM E. M. LANDS AND WILLIAM L. SMITH

VOLUME 87. Enzyme Kinetics and Mechanism (Part C: Intermediates, Stereo-chemistry, and Rate Studies)
Edited by DANIEL L. PURICH

VOLUME 88. Biomembranes (Part I: Visual Pigments and Purple Membranes, II)
Edited by LESTER PACKER

VOLUME 89. Carbohydrate Metabolism (Part D)
Edited by WILLIS A. WOOD

VOLUME 90. Carbohydrate Metabolism (Part E)
Edited by WILLIS A. WOOD

VOLUME 91. Enzyme Structure (Part I)
Edited by C. H. W. HIRS AND SERGE N. TIMASHEFF

VOLUME 92. Immunochemical Techniques (Part E: Monoclonal Antibodies and
General Immunoassay Methods)
Edited by JOHN J. LANGONE AND HELEN VAN VUNAKIS

VOLUME 93. Immunochemical Techniques (Part F: Conventional Antibodies, Fc
Receptors, and Cytotoxicity)
Edited by JOHN J. LANGONE AND HELEN VAN VUNAKIS

VOLUME 94. Polyamines
Edited by HERBERT TABOR AND CELIA WHITE TABOR

VOLUME 95. Cumulative Subject Index Volumes 61–74, 76–80
Edited by EDWARD A. DENNIS AND MARTHA G. DENNIS

VOLUME 96. Biomembranes [Part J: Membrane Biogenesis: Assembly and
Targeting (General Methods; Eukaryotes)]
Edited by SIDNEY FLEISCHER AND BECCA FLEISCHER

VOLUME 97. Biomembranes [Part K: Membrane Biogenesis: Assembly and
Targeting (Prokaryotes, Mitochondria, and Chloroplasts)]
Edited by SIDNEY FLEISCHER AND BECCA FLEISCHER

VOLUME 98. Biomembranes (Part L: Membrane Biogenesis: Processing
and Recycling)
Edited by SIDNEY FLEISCHER AND BECCA FLEISCHER

VOLUME 99. Hormone Action (Part F: Protein Kinases)
Edited by JACKIE D. CORBIN AND JOEL G. HARDMAN

VOLUME 100. Recombinant DNA (Part B)
Edited by RAY WU, LAWRENCE GROSSMAN, AND KIVIE MOLDAVE

VOLUME 101. Recombinant DNA (Part C)
Edited by RAY WU, LAWRENCE GROSSMAN, AND KIVIE MOLDAVE

VOLUME 102. Hormone Action (Part G: Calmodulin and
Calcium-Binding Proteins)
Edited by ANTHONY R. MEANS AND BERT W. O'MALLEY

VOLUME 103. Hormone Action (Part H: Neuroendocrine Peptides)
Edited by P. MICHAEL CONN

VOLUME 104. Enzyme Purification and Related Techniques (Part C)
Edited by WILLIAM B. JAKOBY

VOLUME 105. Oxygen Radicals in Biological Systems
Edited by LESTER PACKER

VOLUME 106. Posttranslational Modifications (Part A)
Edited by FINN WOLD AND KIVIE MOLDAVE

VOLUME 107. Posttranslational Modifications (Part B)
Edited by FINN WOLD AND KIVIE MOLDAVE

VOLUME 108. Immunochemical Techniques (Part G: Separation and Characterization of Lymphoid Cells)
Edited by GIOVANNI DI SABATO, JOHN J. LANGONE, AND HELEN VAN VUNAKIS

VOLUME 109. Hormone Action (Part I: Peptide Hormones)
Edited by LUTZ BIRNBAUMER AND BERT W. O'MALLEY

VOLUME 110. Steroids and Isoprenoids (Part A)
Edited by JOHN H. LAW AND HANS C. RILLING

VOLUME 111. Steroids and Isoprenoids (Part B)
Edited by JOHN H. LAW AND HANS C. RILLING

VOLUME 112. Drug and Enzyme Targeting (Part A)
Edited by KENNETH J. WIDDER AND RALPH GREEN

VOLUME 113. Glutamate, Glutamine, Glutathione, and Related Compounds
Edited by ALTON MEISTER

VOLUME 114. Diffraction Methods for Biological Macromolecules (Part A)
Edited by HAROLD W. WYCKOFF, C. H. W. HIRS, AND SERGE N. TIMASHEFF

VOLUME 115. Diffraction Methods for Biological Macromolecules (Part B)
Edited by HAROLD W. WYCKOFF, C. H. W. HIRS, AND SERGE N. TIMASHEFF

VOLUME 116. Immunochemical Techniques
(Part H: Effectors and Mediators of Lymphoid Cell Functions)
Edited by GIOVANNI DI SABATO, JOHN J. LANGONE, AND HELEN VAN VUNAKIS

VOLUME 117. Enzyme Structure (Part J)
Edited by C. H. W. HIRS AND SERGE N. TIMASHEFF

VOLUME 118. Plant Molecular Biology
Edited by ARTHUR WEISSBACH AND HERBERT WEISSBACH

VOLUME 119. Interferons (Part C)
Edited by SIDNEY PESTKA

VOLUME 120. Cumulative Subject Index Volumes 81–94, 96–101

VOLUME 121. Immunochemical Techniques (Part I: Hybridoma Technology and Monoclonal Antibodies)
Edited by JOHN J. LANGONE AND HELEN VAN VUNAKIS

VOLUME 122. Vitamins and Coenzymes (Part G)
Edited by FRANK CHYTIL AND DONALD B. MCCORMICK

VOLUME 123. Vitamins and Coenzymes (Part H)
Edited by FRANK CHYTIL AND DONALD B. MCCORMICK

VOLUME 124. Hormone Action (Part J: Neuroendocrine Peptides)
Edited by P. MICHAEL CONN

VOLUME 125. Biomembranes (Part M: Transport in Bacteria, Mitochondria, and
Chloroplasts: General Approaches and Transport Systems)
Edited by SIDNEY FLEISCHER AND BECCA FLEISCHER

VOLUME 126. Biomembranes (Part N: Transport in Bacteria, Mitochondria, and
Chloroplasts: Protonmotive Force)
Edited by SIDNEY FLEISCHER AND BECCA FLEISCHER

VOLUME 127. Biomembranes (Part O: Protons and Water: Structure
and Translocation)
Edited by LESTER PACKER

VOLUME 128. Plasma Lipoproteins (Part A: Preparation, Structure, and
Molecular Biology)
Edited by JERE P. SEGREST AND JOHN J. ALBERS

VOLUME 129. Plasma Lipoproteins (Part B: Characterization, Cell Biology,
and Metabolism)
Edited by JOHN J. ALBERS AND JERE P. SEGREST

VOLUME 130. Enzyme Structure (Part K)
Edited by C. H. W. HIRS AND SERGE N. TIMASHEFF

VOLUME 131. Enzyme Structure (Part L)
Edited by C. H. W. HIRS AND SERGE N. TIMASHEFF

VOLUME 132. Immunochemical Techniques (Part J: Phagocytosis and
Cell-Mediated Cytotoxicity)
Edited by GIOVANNI DI SABATO AND JOHANNES EVERSE

VOLUME 133. Bioluminescence and Chemiluminescence (Part B)
Edited by MARLENE DELUCA AND WILLIAM D. MCELROY

VOLUME 134. Structural and Contractile Proteins (Part C: The Contractile
Apparatus and the Cytoskeleton)
Edited by RICHARD B. VALLEE

VOLUME 135. Immobilized Enzymes and Cells (Part B)
Edited by KLAUS MOSBACH

VOLUME 136. Immobilized Enzymes and Cells (Part C)
Edited by KLAUS MOSBACH

VOLUME 137. Immobilized Enzymes and Cells (Part D)
Edited by KLAUS MOSBACH

VOLUME 138. Complex Carbohydrates (Part E)
Edited by VICTOR GINSBURG

VOLUME 139. Cellular Regulators (Part A: Calcium- and Calmodulin-Binding Proteins)
Edited by ANTHONY R. MEANS AND P. MICHAEL CONN

VOLUME 140. Cumulative Subject Index Volumes 102–119, 121–134

VOLUME 141. Cellular Regulators (Part B: Calcium and Lipids)
Edited by P. MICHAEL CONN AND ANTHONY R. MEANS

VOLUME 142. Metabolism of Aromatic Amino Acids and Amines
Edited by SEYMOUR KAUFMAN

VOLUME 143. Sulfur and Sulfur Amino Acids
Edited by WILLIAM B. JAKOBY AND OWEN GRIFFITH

VOLUME 144. Structural and Contractile Proteins (Part D: Extracellular Matrix)
Edited by LEON W. CUNNINGHAM

VOLUME 145. Structural and Contractile Proteins (Part E: Extracellular Matrix)
Edited by LEON W. CUNNINGHAM

VOLUME 146. Peptide Growth Factors (Part A)
Edited by DAVID BARNES AND DAVID A. SIRBASKU

VOLUME 147. Peptide Growth Factors (Part B)
Edited by DAVID BARNES AND DAVID A. SIRBASKU

VOLUME 148. Plant Cell Membranes
Edited by LESTER PACKER AND ROLAND DOUCE

VOLUME 149. Drug and Enzyme Targeting (Part B)
Edited by RALPH GREEN AND KENNETH J. WIDDER

VOLUME 150. Immunochemical Techniques (Part K: *In Vitro* Models of B and T Cell Functions and Lymphoid Cell Receptors)
Edited by GIOVANNI DI SABATO

VOLUME 151. Molecular Genetics of Mammalian Cells
Edited by MICHAEL M. GOTTESMAN

VOLUME 152. Guide to Molecular Cloning Techniques
Edited by SHELBY L. BERGER AND ALAN R. KIMMEL

VOLUME 153. Recombinant DNA (Part D)
Edited by RAY WU AND LAWRENCE GROSSMAN

VOLUME 154. Recombinant DNA (Part E)
Edited by RAY WU AND LAWRENCE GROSSMAN

VOLUME 155. Recombinant DNA (Part F)
Edited by RAY WU

VOLUME 156. Biomembranes (Part P: ATP-Driven Pumps and Related Transport: The Na, K-Pump)
Edited by SIDNEY FLEISCHER AND BECCA FLEISCHER

VOLUME 157. Biomembranes (Part Q: ATP-Driven Pumps and Related Transport: Calcium, Proton, and Potassium Pumps)
Edited by SIDNEY FLEISCHER AND BECCA FLEISCHER

VOLUME 158. Metalloproteins (Part A)
Edited by JAMES F. RIORDAN AND BERT L. VALLEE

VOLUME 159. Initiation and Termination of Cyclic Nucleotide Action
Edited by JACKIE D. CORBIN AND ROGER A. JOHNSON

VOLUME 160. Biomass (Part A: Cellulose and Hemicellulose)
Edited by WILLIS A. WOOD AND SCOTT T. KELLOGG

VOLUME 161. Biomass (Part B: Lignin, Pectin, and Chitin)
Edited by WILLIS A. WOOD AND SCOTT T. KELLOGG

VOLUME 162. Immunochemical Techniques (Part L: Chemotaxis and Inflammation)
Edited by GIOVANNI DI SABATO

VOLUME 163. Immunochemical Techniques (Part M: Chemotaxis and Inflammation)
Edited by GIOVANNI DI SABATO

VOLUME 164. Ribosomes
Edited by HARRY F. NOLLER, JR., AND KIVIE MOLDAVE

VOLUME 165. Microbial Toxins: Tools for Enzymology
Edited by SIDNEY HARSHMAN

VOLUME 166. Branched-Chain Amino Acids
Edited by ROBERT HARRIS AND JOHN R. SOKATCH

VOLUME 167. Cyanobacteria
Edited by LESTER PACKER AND ALEXANDER N. GLAZER

VOLUME 168. Hormone Action (Part K: Neuroendocrine Peptides)
Edited by P. MICHAEL CONN

VOLUME 169. Platelets: Receptors, Adhesion, Secretion (Part A)
Edited by JACEK HAWIGER

VOLUME 170. Nucleosomes
Edited by PAUL M. WASSARMAN AND ROGER D. KORNBERG

VOLUME 171. Biomembranes (Part R: Transport Theory: Cells and Model Membranes)
Edited by SIDNEY FLEISCHER AND BECCA FLEISCHER

VOLUME 172. Biomembranes (Part S: Transport: Membrane Isolation and Characterization)
Edited by SIDNEY FLEISCHER AND BECCA FLEISCHER

VOLUME 173. Biomembranes [Part T: Cellular and Subcellular Transport: Eukaryotic (Nonepithelial) Cells]
Edited by SIDNEY FLEISCHER AND BECCA FLEISCHER

VOLUME 174. Biomembranes [Part U: Cellular and Subcellular Transport: Eukaryotic (Nonepithelial) Cells]
Edited by SIDNEY FLEISCHER AND BECCA FLEISCHER

VOLUME 175. Cumulative Subject Index Volumes 135–139, 141–167

VOLUME 176. Nuclear Magnetic Resonance (Part A: Spectral Techniques and Dynamics)
Edited by NORMAN J. OPPENHEIMER AND THOMAS L. JAMES

VOLUME 177. Nuclear Magnetic Resonance (Part B: Structure and Mechanism)
Edited by NORMAN J. OPPENHEIMER AND THOMAS L. JAMES

VOLUME 178. Antibodies, Antigens, and Molecular Mimicry
Edited by JOHN J. LANGONE

VOLUME 179. Complex Carbohydrates (Part F)
Edited by VICTOR GINSBURG

VOLUME 180. RNA Processing (Part A: General Methods)
Edited by JAMES E. DAHLBERG AND JOHN N. ABELSON

VOLUME 181. RNA Processing (Part B: Specific Methods)
Edited by JAMES E. DAHLBERG AND JOHN N. ABELSON

VOLUME 182. Guide to Protein Purification
Edited by MURRAY P. DEUTSCHER

VOLUME 183. Molecular Evolution: Computer Analysis of Protein and Nucleic Acid Sequences
Edited by RUSSELL F. DOOLITTLE

VOLUME 184. Avidin-Biotin Technology
Edited by MEIR WILCHEK AND EDWARD A. BAYER

VOLUME 185. Gene Expression Technology
Edited by DAVID V. GOEDDEL

VOLUME 186. Oxygen Radicals in Biological Systems (Part B: Oxygen Radicals and Antioxidants)
Edited by LESTER PACKER AND ALEXANDER N. GLAZER

VOLUME 187. Arachidonate Related Lipid Mediators
Edited by ROBERT C. MURPHY AND FRANK A. FITZPATRICK

VOLUME 188. Hydrocarbons and Methylotrophy
Edited by MARY E. LIDSTROM

VOLUME 189. Retinoids (Part A: Molecular and Metabolic Aspects)
Edited by LESTER PACKER

VOLUME 190. Retinoids (Part B: Cell Differentiation and Clinical Applications)
Edited by LESTER PACKER

VOLUME 191. Biomembranes (Part V: Cellular and Subcellular Transport:
Epithelial Cells)
Edited by SIDNEY FLEISCHER AND BECCA FLEISCHER

VOLUME 192. Biomembranes (Part W: Cellular and Subcellular Transport:
Epithelial Cells)
Edited by SIDNEY FLEISCHER AND BECCA FLEISCHER

VOLUME 193. Mass Spectrometry
Edited by JAMES A. MCCLOSKEY

VOLUME 194. Guide to Yeast Genetics and Molecular Biology
Edited by CHRISTINE GUTHRIE AND GERALD R. FINK

VOLUME 195. Adenylyl Cyclase, G Proteins, and Guanylyl Cyclase
Edited by ROGER A. JOHNSON AND JACKIE D. CORBIN

VOLUME 196. Molecular Motors and the Cytoskeleton
Edited by RICHARD B. VALLEE

VOLUME 197. Phospholipases
Edited by EDWARD A. DENNIS

VOLUME 198. Peptide Growth Factors (Part C)
Edited by DAVID BARNES, J. P. MATHER, AND GORDON H. SATO

VOLUME 199. Cumulative Subject Index Volumes 168–174, 176–194

VOLUME 200. Protein Phosphorylation (Part A: Protein Kinases: Assays,
Purification, Antibodies, Functional Analysis, Cloning, and Expression)
Edited by TONY HUNTER AND BARTHOLOMEW M. SEFTON

VOLUME 201. Protein Phosphorylation (Part B: Analysis of Protein
Phosphorylation, Protein Kinase Inhibitors, and Protein Phosphatases)
Edited by TONY HUNTER AND BARTHOLOMEW M. SEFTON

VOLUME 202. Molecular Design and Modeling: Concepts and Applications
(Part A: Proteins, Peptides, and Enzymes)
Edited by JOHN J. LANGONE

VOLUME 203. Molecular Design and Modeling: Concepts and Applications
(Part B: Antibodies and Antigens, Nucleic Acids, Polysaccharides, and Drugs)
Edited by JOHN J. LANGONE

VOLUME 204. Bacterial Genetic Systems
Edited by JEFFREY H. MILLER

VOLUME 205. Metallobiochemistry (Part B: Metallothionein and
Related Molecules)
Edited by JAMES F. RIORDAN AND BERT L. VALLEE

VOLUME 206. Cytochrome P450
Edited by MICHAEL R. WATERMAN AND ERIC F. JOHNSON

VOLUME 207. Ion Channels
Edited by BERNARDO RUDY AND LINDA E. IVERSON

VOLUME 208. Protein–DNA Interactions
Edited by ROBERT T. SAUER

VOLUME 209. Phospholipid Biosynthesis
Edited by EDWARD A. DENNIS AND DENNIS E. VANCE

VOLUME 210. Numerical Computer Methods
Edited by LUDWIG BRAND AND MICHAEL L. JOHNSON

VOLUME 211. DNA Structures (Part A: Synthesis and Physical Analysis of DNA)
Edited by DAVID M. J. LILLEY AND JAMES E. DAHLBERG

VOLUME 212. DNA Structures (Part B: Chemical and Electrophoretic
Analysis of DNA)
Edited by DAVID M. J. LILLEY AND JAMES E. DAHLBERG

VOLUME 213. Carotenoids (Part A: Chemistry, Separation, Quantitation,
and Antioxidation)
Edited by LESTER PACKER

VOLUME 214. Carotenoids (Part B: Metabolism, Genetics, and Biosynthesis)
Edited by LESTER PACKER

VOLUME 215. Platelets: Receptors, Adhesion, Secretion (Part B)
Edited by JACEK J. HAWIGER

VOLUME 216. Recombinant DNA (Part G)
Edited by RAY WU

VOLUME 217. Recombinant DNA (Part H)
Edited by RAY WU

VOLUME 218. Recombinant DNA (Part I)
Edited by RAY WU

VOLUME 219. Reconstitution of Intracellular Transport
Edited by JAMES E. ROTHMAN

VOLUME 220. Membrane Fusion Techniques (Part A)
Edited by NEJAT DÜZGÜNEŞ

VOLUME 221. Membrane Fusion Techniques (Part B)
Edited by NEJAT DÜZGÜNEŞ

VOLUME 222. Proteolytic Enzymes in Coagulation, Fibrinolysis, and Complement
Activation (Part A: Mammalian Blood Coagulation Factors and Inhibitors)
Edited by LASZLO LORAND AND KENNETH G. MANN

VOLUME 223. Proteolytic Enzymes in Coagulation, Fibrinolysis, and Complement Activation (Part B: Complement Activation, Fibrinolysis, and Nonmammalian Blood Coagulation Factors)
Edited by LASZLO LORAND AND KENNETH G. MANN

VOLUME 224. Molecular Evolution: Producing the Biochemical Data
Edited by ELIZABETH ANNE ZIMMER, THOMAS J. WHITE, REBECCA L. CANN, AND ALLAN C. WILSON

VOLUME 225. Guide to Techniques in Mouse Development
Edited by PAUL M. WASSARMAN AND MELVIN L. DEPAMPHILIS

VOLUME 226. Metallobiochemistry (Part C: Spectroscopic and Physical Methods for Probing Metal Ion Environments in Metalloenzymes and Metalloproteins)
Edited by JAMES F. RIORDAN AND BERT L. VALLEE

VOLUME 227. Metallobiochemistry (Part D: Physical and Spectroscopic Methods for Probing Metal Ion Environments in Metalloproteins)
Edited by JAMES F. RIORDAN AND BERT L. VALLEE

VOLUME 228. Aqueous Two-Phase Systems
Edited by HARRY WALTER AND GÖTE JOHANSSON

VOLUME 229. Cumulative Subject Index Volumes 195–198, 200–227

VOLUME 230. Guide to Techniques in Glycobiology
Edited by WILLIAM J. LENNARZ AND GERALD W. HART

VOLUME 231. Hemoglobins (Part B: Biochemical and Analytical Methods)
Edited by JOHANNES EVERSE, KIM D. VANDEGRIFF, AND ROBERT M. WINSLOW

VOLUME 232. Hemoglobins (Part C: Biophysical Methods)
Edited by JOHANNES EVERSE, KIM D. VANDEGRIFF, AND ROBERT M. WINSLOW

VOLUME 233. Oxygen Radicals in Biological Systems (Part C)
Edited by LESTER PACKER

VOLUME 234. Oxygen Radicals in Biological Systems (Part D)
Edited by LESTER PACKER

VOLUME 235. Bacterial Pathogenesis (Part A: Identification and Regulation of Virulence Factors)
Edited by VIRGINIA L. CLARK AND PATRIK M. BAVOIL

VOLUME 236. Bacterial Pathogenesis (Part B: Integration of Pathogenic Bacteria with Host Cells)
Edited by VIRGINIA L. CLARK AND PATRIK M. BAVOIL

VOLUME 237. Heterotrimeric G Proteins
Edited by RAVI IYENGAR

VOLUME 238. Heterotrimeric G-Protein Effectors
Edited by RAVI IYENGAR

VOLUME 239. Nuclear Magnetic Resonance (Part C)
Edited by THOMAS L. JAMES AND NORMAN J. OPPENHEIMER

VOLUME 240. Numerical Computer Methods (Part B)
Edited by MICHAEL L. JOHNSON AND LUDWIG BRAND

VOLUME 241. Retroviral Proteases
Edited by LAWRENCE C. KUO AND JULES A. SHAFER

VOLUME 242. Neoglycoconjugates (Part A)
Edited by Y. C. LEE AND REIKO T. LEE

VOLUME 243. Inorganic Microbial Sulfur Metabolism
Edited by HARRY D. PECK, JR., AND JEAN LEGALL

VOLUME 244. Proteolytic Enzymes: Serine and Cysteine Peptidases
Edited by ALAN J. BARRETT

VOLUME 245. Extracellular Matrix Components
Edited by E. RUOSLAHTI AND E. ENGVALL

VOLUME 246. Biochemical Spectroscopy
Edited by KENNETH SAUER

VOLUME 247. Neoglycoconjugates (Part B: Biomedical Applications)
Edited by Y. C. LEE AND REIKO T. LEE

VOLUME 248. Proteolytic Enzymes: Aspartic and Metallo Peptidases
Edited by ALAN J. BARRETT

VOLUME 249. Enzyme Kinetics and Mechanism (Part D: Developments in Enzyme Dynamics)
Edited by DANIEL L. PURICH

VOLUME 250. Lipid Modifications of Proteins
Edited by PATRICK J. CASEY AND JANICE E. BUSS

VOLUME 251. Biothiols (Part A: Monothiols and Dithiols, Protein Thiols, and Thiyl Radicals)
Edited by LESTER PACKER

VOLUME 252. Biothiols (Part B: Glutathione and Thioredoxin; Thiols in Signal Transduction and Gene Regulation)
Edited by LESTER PACKER

VOLUME 253. Adhesion of Microbial Pathogens
Edited by RON J. DOYLE AND ITZHAK OFEK

VOLUME 254. Oncogene Techniques
Edited by PETER K. VOGT AND INDER M. VERMA

VOLUME 255. Small GTPases and Their Regulators (Part A: Ras Family)
Edited by W. E. BALCH, CHANNING J. DER, AND ALAN HALL

VOLUME 256. Small GTPases and Their Regulators (Part B: Rho Family)
Edited by W. E. BALCH, CHANNING J. DER, AND ALAN HALL

VOLUME 257. Small GTPases and Their Regulators (Part C: Proteins Involved in Transport)
Edited by W. E. BALCH, CHANNING J. DER, AND ALAN HALL

VOLUME 258. Redox-Active Amino Acids in Biology
Edited by JUDITH P. KLINMAN

VOLUME 259. Energetics of Biological Macromolecules
Edited by MICHAEL L. JOHNSON AND GARY K. ACKERS

VOLUME 260. Mitochondrial Biogenesis and Genetics (Part A)
Edited by GIUSEPPE M. ATTARDI AND ANNE CHOMYN

VOLUME 261. Nuclear Magnetic Resonance and Nucleic Acids
Edited by THOMAS L. JAMES

VOLUME 262. DNA Replication
Edited by JUDITH L. CAMPBELL

VOLUME 263. Plasma Lipoproteins (Part C: Quantitation)
Edited by WILLIAM A. BRADLEY, SANDRA H. GIANTURCO, AND JERE P. SEGREST

VOLUME 264. Mitochondrial Biogenesis and Genetics (Part B)
Edited by GIUSEPPE M. ATTARDI AND ANNE CHOMYN

VOLUME 265. Cumulative Subject Index Volumes 228, 230–262

VOLUME 266. Computer Methods for Macromolecular Sequence Analysis
Edited by RUSSELL F. DOOLITTLE

VOLUME 267. Combinatorial Chemistry
Edited by JOHN N. ABELSON

VOLUME 268. Nitric Oxide (Part A: Sources and Detection of NO; NO Synthase)
Edited by LESTER PACKER

VOLUME 269. Nitric Oxide (Part B: Physiological and Pathological Processes)
Edited by LESTER PACKER

VOLUME 270. High Resolution Separation and Analysis of Biological Macromolecules (Part A: Fundamentals)
Edited by BARRY L. KARGER AND WILLIAM S. HANCOCK

VOLUME 271. High Resolution Separation and Analysis of Biological Macromolecules (Part B: Applications)
Edited by BARRY L. KARGER AND WILLIAM S. HANCOCK

VOLUME 272. Cytochrome P450 (Part B)
Edited by ERIC F. JOHNSON AND MICHAEL R. WATERMAN

VOLUME 273. RNA Polymerase and Associated Factors (Part A)
Edited by SANKAR ADHYA

VOLUME 274. RNA Polymerase and Associated Factors (Part B)
Edited by SANKAR ADHYA

VOLUME 275. Viral Polymerases and Related Proteins
Edited by LAWRENCE C. KUO, DAVID B. OLSEN, AND STEVEN S. CARROLL

VOLUME 276. Macromolecular Crystallography (Part A)
Edited by CHARLES W. CARTER, JR., AND ROBERT M. SWEET

VOLUME 277. Macromolecular Crystallography (Part B)
Edited by CHARLES W. CARTER, JR., AND ROBERT M. SWEET

VOLUME 278. Fluorescence Spectroscopy
Edited by LUDWIG BRAND AND MICHAEL L. JOHNSON

VOLUME 279. Vitamins and Coenzymes (Part I)
Edited by DONALD B. MCCORMICK, JOHN W. SUTTIE, AND CONRAD WAGNER

VOLUME 280. Vitamins and Coenzymes (Part J)
Edited by DONALD B. MCCORMICK, JOHN W. SUTTIE, AND CONRAD WAGNER

VOLUME 281. Vitamins and Coenzymes (Part K)
Edited by DONALD B. MCCORMICK, JOHN W. SUTTIE, AND CONRAD WAGNER

VOLUME 282. Vitamins and Coenzymes (Part L)
Edited by DONALD B. MCCORMICK, JOHN W. SUTTIE, AND CONRAD WAGNER

VOLUME 283. Cell Cycle Control
Edited by WILLIAM G. DUNPHY

VOLUME 284. Lipases (Part A: Biotechnology)
Edited by BYRON RUBIN AND EDWARD A. DENNIS

VOLUME 285. Cumulative Subject Index Volumes 263, 264, 266–284, 286–289

VOLUME 286. Lipases (Part B: Enzyme Characterization and Utilization)
Edited by BYRON RUBIN AND EDWARD A. DENNIS

VOLUME 287. Chemokines
Edited by RICHARD HORUK

VOLUME 288. Chemokine Receptors
Edited by RICHARD HORUK

VOLUME 289. Solid Phase Peptide Synthesis
Edited by GREGG B. FIELDS

VOLUME 290. Molecular Chaperones
Edited by GEORGE H. LORIMER AND THOMAS BALDWIN

VOLUME 291. Caged Compounds
Edited by GERARD MARRIOTT

VOLUME 292. ABC Transporters: Biochemical, Cellular, and Molecular Aspects
Edited by SURESH V. AMBUDKAR AND MICHAEL M. GOTTESMAN

VOLUME 293. Ion Channels (Part B)
Edited by P. MICHAEL CONN

VOLUME 294. Ion Channels (Part C)
Edited by P. MICHAEL CONN

VOLUME 295. Energetics of Biological Macromolecules (Part B)
Edited by GARY K. ACKERS AND MICHAEL L. JOHNSON

VOLUME 296. Neurotransmitter Transporters
Edited by SUSAN G. AMARA

VOLUME 297. Photosynthesis: Molecular Biology of Energy Capture
Edited by LEE MCINTOSH

VOLUME 298. Molecular Motors and the Cytoskeleton (Part B)
Edited by RICHARD B. VALLEE

VOLUME 299. Oxidants and Antioxidants (Part A)
Edited by LESTER PACKER

VOLUME 300. Oxidants and Antioxidants (Part B)
Edited by LESTER PACKER

VOLUME 301. Nitric Oxide: Biological and Antioxidant Activities (Part C)
Edited by LESTER PACKER

VOLUME 302. Green Fluorescent Protein
Edited by P. MICHAEL CONN

VOLUME 303. cDNA Preparation and Display
Edited by SHERMAN M. WEISSMAN

VOLUME 304. Chromatin
Edited by PAUL M. WASSARMAN AND ALAN P. WOLFFE

VOLUME 305. Bioluminescence and Chemiluminescence (Part C)
Edited by THOMAS O. BALDWIN AND MIRIAM M. ZIEGLER

VOLUME 306. Expression of Recombinant Genes in Eukaryotic Systems
Edited by JOSEPH C. GLORIOSO AND MARTIN C. SCHMIDT

VOLUME 307. Confocal Microscopy
Edited by P. MICHAEL CONN

VOLUME 308. Enzyme Kinetics and Mechanism (Part E: Energetics of Enzyme Catalysis)
Edited by DANIEL L. PURICH AND VERN L. SCHRAMM

VOLUME 309. Amyloid, Prions, and Other Protein Aggregates
Edited by RONALD WETZEL

VOLUME 310. Biofilms
Edited by RON J. DOYLE

VOLUME 311. Sphingolipid Metabolism and Cell Signaling (Part A)
Edited by ALFRED H. MERRILL, JR., AND YUSUF A. HANNUN

VOLUME 312. Sphingolipid Metabolism and Cell Signaling (Part B)
Edited by ALFRED H. MERRILL, JR., AND YUSUF A. HANNUN

VOLUME 313. Antisense Technology (Part A: General Methods, Methods of Delivery, and RNA Studies)
Edited by M. IAN PHILLIPS

VOLUME 314. Antisense Technology (Part B: Applications)
Edited by M. IAN PHILLIPS

VOLUME 315. Vertebrate Phototransduction and the Visual Cycle (Part A)
Edited by KRZYSZTOF PALCZEWSKI

VOLUME 316. Vertebrate Phototransduction and the Visual Cycle (Part B)
Edited by KRZYSZTOF PALCZEWSKI

VOLUME 317. RNA–Ligand Interactions (Part A: Structural Biology Methods)
Edited by DANIEL W. CELANDER AND JOHN N. ABELSON

VOLUME 318. RNA–Ligand Interactions (Part B: Molecular Biology Methods)
Edited by DANIEL W. CELANDER AND JOHN N. ABELSON

VOLUME 319. Singlet Oxygen, UV-A, and Ozone
Edited by LESTER PACKER AND HELMUT SIES

VOLUME 320. Cumulative Subject Index Volumes 290–319

VOLUME 321. Numerical Computer Methods (Part C)
Edited by MICHAEL L. JOHNSON AND LUDWIG BRAND

VOLUME 322. Apoptosis
Edited by JOHN C. REED

VOLUME 323. Energetics of Biological Macromolecules (Part C)
Edited by MICHAEL L. JOHNSON AND GARY K. ACKERS

VOLUME 324. Branched-Chain Amino Acids (Part B)
Edited by ROBERT A. HARRIS AND JOHN R. SOKATCH

VOLUME 325. Regulators and Effectors of Small GTPases (Part D: Rho Family)
Edited by W. E. BALCH, CHANNING J. DER, AND ALAN HALL

VOLUME 326. Applications of Chimeric Genes and Hybrid Proteins (Part A: Gene Expression and Protein Purification)
Edited by JEREMY THORNER, SCOTT D. EMR, AND JOHN N. ABELSON

VOLUME 327. Applications of Chimeric Genes and Hybrid Proteins (Part B: Cell Biology and Physiology)
Edited by JEREMY THORNER, SCOTT D. EMR, AND JOHN N. ABELSON

VOLUME 328. Applications of Chimeric Genes and Hybrid Proteins (Part C: Protein–Protein Interactions and Genomics)
Edited by JEREMY THORNER, SCOTT D. EMR, AND JOHN N. ABELSON

VOLUME 329. Regulators and Effectors of Small GTPases (Part E: GTPases Involved in Vesicular Traffic)
Edited by W. E. BALCH, CHANNING J. DER, AND ALAN HALL

VOLUME 330. Hyperthermophilic Enzymes (Part A)
Edited by MICHAEL W. W. ADAMS AND ROBERT M. KELLY

VOLUME 331. Hyperthermophilic Enzymes (Part B)
Edited by MICHAEL W. W. ADAMS AND ROBERT M. KELLY

VOLUME 332. Regulators and Effectors of Small GTPases (Part F: Ras Family I)
Edited by W. E. BALCH, CHANNING J. DER, AND ALAN HALL

VOLUME 333. Regulators and Effectors of Small GTPases (Part G: Ras Family II)
Edited by W. E. BALCH, CHANNING J. DER, AND ALAN HALL

VOLUME 334. Hyperthermophilic Enzymes (Part C)
Edited by MICHAEL W. W. ADAMS AND ROBERT M. KELLY

VOLUME 335. Flavonoids and Other Polyphenols
Edited by LESTER PACKER

VOLUME 336. Microbial Growth in Biofilms (Part A: Developmental and Molecular Biological Aspects)
Edited by RON J. DOYLE

VOLUME 337. Microbial Growth in Biofilms (Part B: Special Environments and Physicochemical Aspects)
Edited by RON J. DOYLE

VOLUME 338. Nuclear Magnetic Resonance of Biological Macromolecules (Part A)
Edited by THOMAS L. JAMES, VOLKER DÖTSCH, AND ULI SCHMITZ

VOLUME 339. Nuclear Magnetic Resonance of Biological Macromolecules (Part B)
Edited by THOMAS L. JAMES, VOLKER DÖTSCH, AND ULI SCHMITZ

VOLUME 340. Drug–Nucleic Acid Interactions
Edited by JONATHAN B. CHAIRES AND MICHAEL J. WARING

VOLUME 341. Ribonucleases (Part A)
Edited by ALLEN W. NICHOLSON

VOLUME 342. Ribonucleases (Part B)
Edited by ALLEN W. NICHOLSON

VOLUME 343. G Protein Pathways (Part A: Receptors)
Edited by RAVI IYENGAR AND JOHN D. HILDEBRANDT

VOLUME 344. G Protein Pathways (Part B: G Proteins and Their Regulators)
Edited by RAVI IYENGAR AND JOHN D. HILDEBRANDT

VOLUME 345. G Protein Pathways (Part C: Effector Mechanisms)
Edited by RAVI IYENGAR AND JOHN D. HILDEBRANDT

VOLUME 346. Gene Therapy Methods
Edited by M. IAN PHILLIPS

VOLUME 347. Protein Sensors and Reactive Oxygen Species (Part A: Selenoproteins and Thioredoxin)
Edited by HELMUT SIES AND LESTER PACKER

VOLUME 348. Protein Sensors and Reactive Oxygen Species (Part B: Thiol Enzymes and Proteins)
Edited by HELMUT SIES AND LESTER PACKER

VOLUME 349. Superoxide Dismutase
Edited by LESTER PACKER

VOLUME 350. Guide to Yeast Genetics and Molecular and Cell Biology (Part B)
Edited by CHRISTINE GUTHRIE AND GERALD R. FINK

VOLUME 351. Guide to Yeast Genetics and Molecular and Cell Biology (Part C)
Edited by CHRISTINE GUTHRIE AND GERALD R. FINK

VOLUME 352. Redox Cell Biology and Genetics (Part A)
Edited by CHANDAN K. SEN AND LESTER PACKER

VOLUME 353. Redox Cell Biology and Genetics (Part B)
Edited by CHANDAN K. SEN AND LESTER PACKER

VOLUME 354. Enzyme Kinetics and Mechanisms (Part F: Detection and Characterization of Enzyme Reaction Intermediates)
Edited by DANIEL L. PURICH

VOLUME 355. Cumulative Subject Index Volumes 321–354

VOLUME 356. Laser Capture Microscopy and Microdissection
Edited by P. MICHAEL CONN

VOLUME 357. Cytochrome P450, Part C
Edited by ERIC F. JOHNSON AND MICHAEL R. WATERMAN

VOLUME 358. Bacterial Pathogenesis (Part C: Identification, Regulation, and Function of Virulence Factors)
Edited by VIRGINIA L. CLARK AND PATRIK M. BAVOIL

VOLUME 359. Nitric Oxide (Part D)
Edited by ENRIQUE CADENAS AND LESTER PACKER

VOLUME 360. Biophotonics (Part A)
Edited by GERARD MARRIOTT AND IAN PARKER

VOLUME 361. Biophotonics (Part B)
Edited by GERARD MARRIOTT AND IAN PARKER

VOLUME 362. Recognition of Carbohydrates in Biological Systems (Part A)
Edited by YUAN C. LEE AND REIKO T. LEE

VOLUME 363. Recognition of Carbohydrates in Biological Systems (Part B)
Edited by YUAN C. LEE AND REIKO T. LEE

VOLUME 364. Nuclear Receptors
Edited by DAVID W. RUSSELL AND DAVID J. MANGELSDORF

VOLUME 365. Differentiation of Embryonic Stem Cells
Edited by PAUL M. WASSAUMAN AND GORDON M. KELLER

VOLUME 366. Protein Phosphatases
Edited by SUSANNE KLUMPP AND JOSEF KRIEGLSTEIN

VOLUME 367. Liposomes (Part A)
Edited by NEJAT DÜZGÜNEŞ

VOLUME 368. Macromolecular Crystallography (Part C)
Edited by CHARLES W. CARTER, JR., AND ROBERT M. SWEET

VOLUME 369. Combinational Chemistry (Part B)
Edited by GUILLERMO A. MORALES AND BARRY A. BUNIN

VOLUME 370. RNA Polymerases and Associated Factors (Part C)
Edited by SANKAR L. ADHYA AND SUSAN GARGES

VOLUME 371. RNA Polymerases and Associated Factors (Part D)
Edited by SANKAR L. ADHYA AND SUSAN GARGES

VOLUME 372. Liposomes (Part B)
Edited by NEJAT DÜZGÜNEŞ

VOLUME 373. Liposomes (Part C)
Edited by NEJAT DÜZGÜNEŞ

VOLUME 374. Macromolecular Crystallography (Part D)
Edited by CHARLES W. CARTER, JR., AND ROBERT W. SWEET

VOLUME 375. Chromatin and Chromatin Remodeling Enzymes (Part A)
Edited by C. DAVID ALLIS AND CARL WU

VOLUME 376. Chromatin and Chromatin Remodeling Enzymes (Part B)
Edited by C. DAVID ALLIS AND CARL WU

VOLUME 377. Chromatin and Chromatin Remodeling Enzymes (Part C)
Edited by C. DAVID ALLIS AND CARL WU

VOLUME 378. Quinones and Quinone Enzymes (Part A)
Edited by HELMUT SIES AND LESTER PACKER

VOLUME 379. Energetics of Biological Macromolecules (Part D)
Edited by JO M. HOLT, MICHAEL L. JOHNSON, AND GARY K. ACKERS

VOLUME 380. Energetics of Biological Macromolecules (Part E)
Edited by JO M. HOLT, MICHAEL L. JOHNSON, AND GARY K. ACKERS

VOLUME 381. Oxygen Sensing
Edited by CHANDAN K. SEN AND GREGG L. SEMENZA

VOLUME 382. Quinones and Quinone Enzymes (Part B)
Edited by HELMUT SIES AND LESTER PACKER

VOLUME 383. Numerical Computer Methods (Part D)
Edited by LUDWIG BRAND AND MICHAEL L. JOHNSON

VOLUME 384. Numerical Computer Methods (Part E)
Edited by LUDWIG BRAND AND MICHAEL L. JOHNSON

VOLUME 385. Imaging in Biological Research (Part A)
Edited by P. MICHAEL CONN

VOLUME 386. Imaging in Biological Research (Part B)
Edited by P. MICHAEL CONN

VOLUME 387. Liposomes (Part D)
Edited by NEJAT DÜZGÜNEŞ

VOLUME 388. Protein Engineering
Edited by DAN E. ROBERTSON AND JOSEPH P. NOEL

VOLUME 389. Regulators of G-Protein Signaling (Part A)
Edited by DAVID P. SIDEROVSKI

VOLUME 390. Regulators of G-Protein Signaling (Part B)
Edited by DAVID P. SIDEROVSKI

VOLUME 391. Liposomes (Part E)
Edited by NEJAT DÜZGÜNEŞ

VOLUME 392. RNA Interference
Edited by ENGELKE ROSSI

VOLUME 393. Circadian Rhythms
Edited by MICHAEL W. YOUNG

VOLUME 394. Nuclear Magnetic Resonance of Biological Macromolecules (Part C)
Edited by THOMAS L. JAMES

VOLUME 395. Producing the Biochemical Data (Part B)
Edited by ELIZABETH A. ZIMMER AND ERIC H. ROALSON

VOLUME 396. Nitric Oxide (Part E)
Edited by LESTER PACKER AND ENRIQUE CADENAS

VOLUME 397. Environmental Microbiology
Edited by JARED R. LEADBETTER

VOLUME 398. Ubiquitin and Protein Degradation (Part A)
Edited by RAYMOND J. DESHAIES

VOLUME 399. Ubiquitin and Protein Degradation (Part B)
Edited by RAYMOND J. DESHAIES

VOLUME 400. Phase II Conjugation Enzymes and Transport Systems
Edited by HELMUT SIES AND LESTER PACKER

VOLUME 401. Glutathione Transferases and Gamma Glutamyl Transpeptidases
Edited by HELMUT SIES AND LESTER PACKER

VOLUME 402. Biological Mass Spectrometry
Edited by A. L. BURLINGAME

VOLUME 403. GTPases Regulating Membrane Targeting and Fusion
Edited by WILLIAM E. BALCH, CHANNING J. DER, AND ALAN HALL

VOLUME 404. GTPases Regulating Membrane Dynamics
Edited by WILLIAM E. BALCH, CHANNING J. DER, AND ALAN HALL

VOLUME 405. Mass Spectrometry: Modified Proteins and Glycoconjugates
Edited by A. L. BURLINGAME

VOLUME 406. Regulators and Effectors of Small GTPases: Rho Family
Edited by WILLIAM E. BALCH, CHANNING J. DER, AND ALAN HALL

VOLUME 407. Regulators and Effectors of Small GTPases: Ras Family
Edited by WILLIAM E. BALCH, CHANNING J. DER, AND ALAN HALL

VOLUME 408. DNA Repair (Part A)
Edited by JUDITH L. CAMPBELL AND PAUL MODRICH

VOLUME 409. DNA Repair (Part B)
Edited by JUDITH L. CAMPBELL AND PAUL MODRICH

VOLUME 410. DNA Microarrays (Part A: Array Platforms and Web-Bench Protocols)
Edited by ALAN KIMMEL AND BRIAN OLIVER

VOLUME 411. DNA Microarrays (Part B: Databases and Statistics)
Edited by ALAN KIMMEL AND BRIAN OLIVER

VOLUME 412. Amyloid, Prions, and Other Protein Aggregates (Part B)
Edited by INDU KHETERPAL AND RONALD WETZEL

VOLUME 413. Amyloid, Prions, and Other Protein Aggregates (Part C)
Edited by INDU KHETERPAL AND RONALD WETZEL

VOLUME 414. Measuring Biological Responses with Automated Microscopy
Edited by JAMES INGLESE

VOLUME 415. Glycobiology
Edited by MINORU FUKUDA

VOLUME 416. Glycomics
Edited by MINORU FUKUDA

VOLUME 417. Functional Glycomics
Edited by MINORU FUKUDA

VOLUME 418. Embryonic Stem Cells
Edited by IRINA KLIMANSKAYA AND ROBERT LANZA

VOLUME 419. Adult Stem Cells
Edited by IRINA KLIMANSKAYA AND ROBERT LANZA

VOLUME 420. Stem Cell Tools and Other Experimental Protocols
Edited by IRINA KLIMANSKAYA AND ROBERT LANZA

VOLUME 421. Advanced Bacterial Genetics: Use of Transposons and Phage for
Genomic Engineering
Edited by KELLY T. HUGHES

VOLUME 422. Two-Component Signaling Systems, Part A
Edited by MELVIN I. SIMON, BRIAN R. CRANE, AND ALEXANDRINE CRANE

VOLUME 423. Two-Component Signaling Systems, Part B
Edited by MELVIN I. SIMON, BRIAN R. CRANE, AND ALEXANDRINE CRANE

VOLUME 424. RNA Editing
Edited by JONATHA M. GOTT

VOLUME 425. RNA Modification
Edited by JONATHA M. GOTT

VOLUME 426. Integrins
Edited by DAVID CHERESH

VOLUME 427. MicroRNA Methods
Edited by JOHN J. ROSSI

VOLUME 428. Osmosensing and Osmosignaling
Edited by HELMUT SIES AND DIETER HAUSSINGER

VOLUME 429. Translation Initiation: Extract Systems and Molecular Genetics
Edited by JON LORSCH

VOLUME 430. Translation Initiation: Reconstituted Systems and Biophysical
Methods
Edited by JON LORSCH

VOLUME 431. Translation Initiation: Cell Biology, High-Throughput and
Chemical-Based Approaches
Edited by JON LORSCH

VOLUME 432. Lipidomics and Bioactive Lipids: Mass-Spectrometry–Based Lipid
Analysis
Edited by H. ALEX BROWN

VOLUME 433. Lipidomics and Bioactive Lipids: Specialized Analytical Methods and
Lipids in Disease
Edited by H. ALEX BROWN

VOLUME 434. Lipidomics and Bioactive Lipids: Lipids and Cell Signaling
Edited by H. ALEX BROWN

VOLUME 435. Oxygen Biology and Hypoxia
Edited by HELMUT SIES AND BERNHARD BRÜNE

VOLUME 436. Globins and Other Nitric Oxide–Reactive Protiens, (Part A)
(in preparation)
Edited by ROBERT K. POOLE

VOLUME 437. Globins and Other Nitric Oxide–Reactive Protiens, (Part B)
(in preparation)
Edited by ROBERT K. POOLE

VOLUME 438. Small GTPases in Diseases, Part A (in preparation)
Edited by WILLIAM E. BALCH, CHANNING J. DER, AND ALAN HALL

HYPOXIA-INDUCIBLE FACTOR

CHAPTER ONE

Hypoxia-Inducible Factors Per/ARNT/Sim Domains: Structure and Function

Thomas H. Scheuermann, Jinsong Yang, Lei Zhang, Kevin H. Gardner, *and* Richard K. Bruick

Contents

1. Introduction	4
2. Delineation of the HIF PAS Domains	5
3. Expression and Characterization of HIF PAS Domains	7
3.1. Recombinant protein expression	7
3.2. Limited proteolysis	8
3.3. Solution NMR spectroscopy	10
4. Assessing PAS Domain Protein–Protein Interactions	11
4.1. NMR spectroscopy	11
4.2. Reporter genes of HIF function	13
4.3. Preparation of nuclear lysates	15
4.4. Coimmunoprecipitation of HIF-α and HIF-β	15
4.5. Electrophoretic mobility shift assay	17
5. Discussion	18
Acknowledgments	20
References	21

Abstract

Hypoxia-inducible factors (HIFs) are key transcriptional regulators of genes involved in cellular adaptation to reduced oxygen availability through effects on anaerobic metabolism, oxygen delivery, angiogenesis, and cellular survival and proliferation. As such, HIFs contribute to the pathogenesis of diseases in which oxygen availability is compromised, notably ischemia and tumorigenesis. Though tremendous progress has been made in elucidating the mechanisms underlying O_2-dependent regulation of HIF by Fe(II)- and 2-oxoglutarate-dependent dioxygenases, HIF induction can be uncoupled from these modes of regulation in diseases such as cancer. Consequently, renewed interest has developed in understanding the structure/function relationships of individual Per/ARNT/Sim (PAS) domains that are important for maintaining transcriptionally active HIF

Department of Biochemistry, University of Texas Southwestern Medical Center, Dallas, Texas

Methods in Enzymology, Volume 435
ISSN 0076-6879, DOI: 10.1016/S0076-6879(07)35001-5

complexes, regardless of the manner by which HIF is induced. This review highlights strategies for the biophysical and biochemical characterization of the PAS domains found within both HIF subunits and provides a platform for future efforts to exploit these domains in therapeutic settings.

1. INTRODUCTION

Mammalian cells have a remarkable capacity to adapt to changes in their environment. One such pathway receiving considerable attention (Kaufman *et al.*, 2004) is the hypoxic response pathway, which is induced in all mammalian cells in response to chronic O_2 deprivation and is critical for many physiological and pathophysiological processes (Semenza, 2000). This response is based on changes in gene expression mediated through HIFs, heterodimers composed of HIF-α and HIF-β subunits (Wang *et al.*, 1995). Oxygen-dependent regulation of HIF appears to be primarily mediated through posttranslational modifications of the HIF-α subunit (reviewed in Dann and Bruick, 2005). Under normoxic conditions, Fe(II)- and 2-oxoglutarate-dependent dioxygenases (Bruick and McKnight, 2001; Epstein *et al.*, 2001; Hewitson *et al.*, 2002; Ivan *et al.*, 2002; Lando *et al.*, 2002b) employ O_2 to hydroxylate proline residues within the HIF-α oxygen-dependent degradation domain and an asparagine residue within the HIF-α C-terminal transactivation domain. These modifications promote HIF-α degradation by the proteasome (Ivan *et al.*, 2001; Jaakkola *et al.*, 2001) and interfere with coactivator recruitment (Lando *et al.*, 2002a), respectively. Under low oxygen conditions, hydroxylation does not occur, and the HIF-α subunit binds to the O_2-insensitive HIF-β subunit (also known as the aryl hydrocarbon nuclear receptor translocator, ARNT) and subsequently induces expression of genes that promote cellular adaptation to hypoxia (for examples, see Semenza, 2003).

HIF can also be induced through O_2-independent mechanisms, as revealed by genetic alterations that activate oncogenic-signaling pathways or inactivate tumor suppressor genes. Perhaps the best-studied examples are mutations that inactivate pVHL, the recognition subunit of the E3 ubiquitin ligase complex, which targets HIF-α for degradation under normoxia (Maxwell *et al.*, 1999). Such mutations result in constitutive stabilization of the HIF-α subunit and contribute to the pathogenesis of a number of tumors (Maxwell, 2005). In contrast, activating mutations in several signaling pathways can induce HIF under normoxic conditions by increasing the rate of HIF-1α protein synthesis (Giaccia *et al.*, 2003; Semenza, 2003). The ability of O_2-dependent and independent mechanisms to induce HIF has prompted renewed interest in characterizing other domains relevant to HIF function, regardless of the manner in which it is induced.

In addition to the O_2-sensitive regions of HIF-α, both HIF subunits contain several O_2-independent elements, including two PAS domains. The PAS domain family derives its name from three of the first proteins noted to contain this motif: *period* (Per; Huang *et al.*, 1993), *aryl hydrocarbon receptor nuclear translocator* (ARNT; Hoffman *et al.*, 1991), and *single-minded* (Sim; Nambu *et al.*, 1991). These domains are found in a wide range of proteins, ranging from transcription factors to enzymes to ion channels (Taylor and Zhulin, 1999), where they typically mediate protein–protein interactions. While HIF PAS domains have been reported to contribute to HIF heterodimer formation (Chapman-Smith *et al.*, 2004; Jiang *et al.*, 1996), they also have been implicated as mediators of nuclear localization (Kallio *et al.*, 1998) and HIF stabilization via heat shock protein 90 (Hsp90) association (Gradin *et al.*, 2002; Isaacs *et al.*, 2004; Kallio *et al.*, 1998; Katschinski *et al.*, 2004; Minet *et al.*, 1999). While these studies provided some initial insight into the importance of PAS domains for HIF function, the interpretation of much of these data is complicated by the absence of structural information on PAS domains at the time and the incomplete biochemical characterization of reagents used in these studies. For example, the use of limited sequence data to define the boundaries of low homology PAS domains can result in the generation of inappropriate truncations within a structural domain, introducing unfolded regions within a protein instead of cleanly excising a single domain. These concerns may be compounded by over-reliance on exogenous protein overexpression (resulting in non-physiological interactions in the cell culture model system) or through the use of uncharacterized recombinant proteins that may not be properly folded. Here we describe efforts to avoid these issues in our studies of HIF PAS domains by using a combination of methods to characterize their structure and function, integrating data from *in vitro* and in-cell studies of the HIF heterodimer.

2. DELINEATION OF THE HIF PAS DOMAINS

Since its initial description in 1993 (Huang *et al.*, 1993), the concept of a "PAS domain" has undergone several revisions as additional sequence and structural information have become available. As originally detailed by sequence comparisons among the Per, ARNT, Sim, and aryl hydrocarbon receptor (AHR) proteins, the PAS domain was thought to be roughly 270 amino acids in length and to contain two copies of a tandem repeat sequence. Subsequent characterization of a larger number of PAS-containing proteins revealed that these proteins actually contain two separate PAS domains (designated PAS-A and PAS-B). This led to several suggestions in which PAS domains actually contain two sequence homology elements, a 50- to 60-residue N-terminal S1 box (Zhulin *et al.*, 1997) or PAS (Ponting and Aravind, 1997) motif that is frequently accompanied by a C-terminal 45- to 50-residue S2 box or PAS C-terminal (PAC) motif. Efforts to identify PAS-containing sequences are complicated by poor sequence similarity and the variable length of loops connecting several conserved secondary structure elements, particularly in the S2/PAC motif. These sources of primary structure diversity are likely of functional importance, perhaps reflecting the broad range of PAS domain interactions with proteins, cofactors, and ligands (Pellequer *et al.*, 1999).

The tertiary structures of several PAS domains reveal a conserved structural unit comprised of a mixed α/β fold in which several α-helices pack against a common face of a five-stranded antiparallel β-sheet (Fig. 1.1) (Amezcua *et al.*, 2002; Card *et al.*, 2005; Erbel *et al.*, 2003; Morais Cabral *et al.*, 1998). These structural studies confirmed earlier suspicions that the S1/PAS and S2/PAC motifs are integral to the PAS domain tertiary structure, with the former encompassing the Aβ-Fα secondary structure elements and the latter containing the last three β-strands of the PAS β-sheet (Hefti *et al.*, 2004) (see Fig. 1.1). Unfortunately, the PAS/PAC nomenclature is still occasionally encountered and has led to studies of 70- to 80-residue PAS domain fragments that undoubtedly lack the last three strands of the PAS β-sheet. As a general rule, minimal PAS domains that lack extensive loop structures contain approximately 100 amino acid residues, and we suggest that shorter constructs are likely missing elements required for a stably folded, biologically relevant protein. It should be noted that some PAS structures display additional structural elements N- or

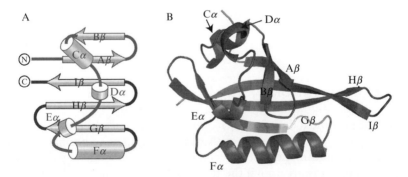

Figure 1.1   Per/ARNT/Sim (PAS) domain topology and structure. (A) Schematic of a typical arrangement of secondary structure elements in PAS domains. (B) Ribbon diagram of the solution structure of the hypoxia-inducible factors (HIF)-β PAS-B domain (Card *et al.*, 2005). In both panels, the PAS motif/S1 box (red) and PAC motif/S2 box (blue) sequence homology elements are indicated, and secondary structure elements are labeled using the nomenclature initially established for FixL (Gong *et al.*, 1998). (See color insert.)

C-terminal to the core PAS fold, and that these non-canonical structural features can associate with the core α/β fold in structurally and functionally essential ways (e.g., Per [Yildiz *et al.*, 2005], photoactive yellow protein [PYP] [Craven *et al.*, 2000], and phototropin1 [phot1] kinase [Harper *et al.*, 2003, 2004]).

We have routinely used a combination of computational and biophysical tools detailed elsewhere (Card and Gardner, 2005) to aid in the identification of PAS domain boundaries that provide well-folded proteins amenable to biophysical and functional characterization. To identify preliminary domain boundaries, the secondary structure of a PAS-containing protein can be predicted with the web-accessible program JPred (Cuff *et al.*, 1998). These predictions are then compared with known PAS tertiary structures, paying particular attention to identify any potential secondary structures adjacent to the minimal α/β PAS fold. Using this information, we design a series of protein constructs with variable N- and C-termini, each of which are evaluated for recombinant expression levels, solubility, structural integrity, and oligomeric state. Such analyses performed on the PAS domains from the three human HIF-α subunits, HIF-1α, HIF-2α, and HIF-3α (Ema *et al.*, 1997; Gu *et al.*, 1998; Tian *et al.*, 1997), and HIF-β have led to the predicted sequence boundaries defining each HIF PAS domain shown in Fig. 1.2.

3. Expression and Characterization of HIF PAS Domains

3.1. Recombinant protein expression

Numerous PAS domains have proven amenable to heterologous expression in *Escherichia coli*, which, as of January 2007, has provided the quantities of protein required to determine the high-resolution structures of at least 16 different domains or their complexes. In our work, we employ a parallel cloning approach with a series of bacterial expression plasmids (Sheffield *et al.*, 1999), each of which provides expression of a different N-terminal fusion protein tag while using a common multiple cloning site. We have extended this system, generating expression plasmids with N-terminal Gβ1 or His$_6$-Gβ1 tags (Amezcua *et al.*, 2002; Harper *et al.*, 2003). The latter tag affords the benefits of affinity chromatography while avoiding the poor expression and/or solubility demonstrated by some His$_6$-tagged proteins (Hammarstrom *et al.*, 2002). While this approach is sufficient for many PAS domains, allowing us to generate large amounts ($>$50 mg/l) of folded protein in certain cases, more challenging expression targets may demand alternative expression strategies. Such problems may sometimes be addressed by the use of different bacterial expression vectors or host cell lines to address

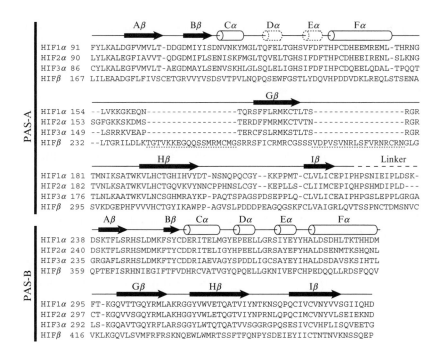

Figure 1.2 Sequence and structural elements of the hypoxia-inducible factors (HIF) Per/ARNT/Sim (PAS) domains. Multiple sequence alignments of PAS-A and -B fragments from human HIF-β (P27540) and three human HIF-α isoforms (Q16665, Q99814, and Q9Y2N7) are shown. The secondary structures plotted above the PAS-B sequences are those observed in the structure of HIF-2α PAS-B (Erbel *et al.*, 2003). For the HIF PAS-A domain, the plotted secondary structures are those predicted by use of the JPred algorithm (Cuff *et al.*, 1998) in combination with manual placement of the D- and E-helices and the F-helix boundaries. All accession numbers given correspond to the sequences deposited in the UniProt Knowledgebase.

proteolysis, redox, codon bias, or toxicity issues. In addition, PAS domains that are poorly soluble and that demonstrate large unstructured regions, such as the 18 and 23 kDa PAS-A domains of HIF-α and HIF-β proteins, may benefit from coexpression approaches, as has been reported for full-length bHLH-PAS-A and -B fragments of the HIF heterodimer (Chachami *et al.*, 2005; Chapman-Smith *et al.*, 2004; Kikuchi *et al.*, 2003).

3.2. Limited proteolysis

The structural integrity of recombinantly expressed proteins is of utmost concern and must be experimentally validated. Using relatively dilute (10–100 μM) protein solutions, techniques such as size exclusion chromatography and circular dichroism spectroscopy can quickly ascertain that the recombinant protein adopts a compact, folded structure. Limited proteolysis experiments can complement these methods by assessing structural integrity, working under the assumption that most potential proteolysis sites are inaccessible in a well-folded protein (Cohen *et al.*, 1995). For limited proteolysis experiments, we generally prepare separate reactions containing a 100:1 w/w ratio of protein to trypsin, chymotrypsin, or V-8 proteases, which cleave proteins after basic, aromatic, or acidic residues, respectively. For small PAS domains on the order of 15 kDa molecular mass, reactions typically consist of 1 mg of purified PAS domain and 10 μg of protease in a 500-μl reaction volume. Following protease addition to the protein solution, we withdraw 20-μl aliquots at time points ranging from 5 min to 16 h and quench the reaction by boiling in the presence of an equivalent volume of denaturing buffer. Samples are subsequently analyzed by sodium dodecyl phosphate–polyacrylamide gel electrophoresis (SDS-PAGE). To precisely determine protease-accessible sites in the primary structure, electrospray ionization (ESI) mass spectrometry–compatible samples may be prepared by quenching with a 1% final concentration of trifluoroacetic acid (in a chemical safety hood), freezing or holding quenched samples on ice, and purifying peptide/protein components by reverse phase high–performance liquid chromatography (HPLC). Given accurate masses of proteolytic fragments, we routinely use the MS-Digest component of the ProteinProspector suite of proteomic tools (http://prospector.ucsf.edu) to identify cleavage sites in the primary structure.

It must be emphasized that limited proteolysis experiments do not directly assay the folded state of a protein, but rather reveal the accessibility of potential protease sites. The most easily interpreted data is collected from protein samples representing the extremes of completely disordered or compact, well-folded polypeptides. The former sample would be completely proteolyzed during initial time points, while the latter sample is expected to show little to no degradation over the course of the experiment. There exists a potentially confusing middle ground where well-folded protein domains demonstrate degradation on account of one or more extended loops. For example, the HIF-2α PAS-B domain has a single, highly-accessible trypsin site within the extended loop connecting the Hβ and Iβ strands (HI loop; see Fig. 1.1) that leads to accumulation of a large protein fragment followed by complete degradation over several hours. Likewise, limited chymotrypsin proteolysis of a PAS-A fragment of PAS kinase (Amezcua *et al.*, 2002) revealed the Gβ/Hβ loop to be similarly protease-accessible. Without high-resolution structures, these data might be misinterpreted as indicating domain boundaries at sites that actually reside within well-folded domains. Attempts to trim these two PAS domains to more minimal amino acid sequences that express only the large proteolytic fragment would excise residues composing internal strand(s) of the PAS β-sheet and almost certainly disrupt the PAS fold. Therefore, one must interpret such data in light of known PAS domain structures and discriminate between the rapid proteolysis of entirely unstructured proteins and the slower degradation of folded proteins.

3.3. Solution NMR spectroscopy

Optimal domain boundaries may also be refined by analysis of two-dimensional (2D) ^{15}N/^{1}H heteronuclear spin quantum coherence (HSQC) spectra (Kay *et al.*, 1992), a solution nuclear magnetic resonance (NMR) experiment commonly used to study protein structure and dynamics. Collected from uniformly ^{15}N-isotopically enriched protein, either purified or in clarified bacterial lysates (Gronenborn and Clore, 1996), a single ^{15}N/^{1}H HSQC experiment can qualitatively describe the extent to which a protein is structurally well-ordered. In these spectra, each cross-peak corresponds to a specific pair of proton and nitrogen nuclei separated by a single bond, as found in amide groups in the protein backbone. The chemical shifts of these peaks are sensitive to their local magnetic environment, which will significantly change depending on the state of protein folding near each site. Unstructured proteins, where amide groups largely interact with solvent, typically exhibit ^{15}N/^{1}H HSQC spectra with most peaks clustered in the middle of the proton dimension ($\sim$7.9–8.5 ppm) (Fig. 1.3A). In contrast, spectra from well-folded proteins have peaks dispersed over a broader range of ^{1}H frequencies ($\sim$5.5–11 ppm) and demonstrate uniform peak intensities and line widths (see Fig. 1.3B). Comparing these spectra collected from protein constructs with different domain boundaries may identify a minimal, well-folded protein construct. Shorter protein constructs that truncate amino acids required for folding will exhibit ^{15}N/^{1}H HSQC spectra with poor amide proton chemical shift dispersion. Larger constructs that merely add unstructured amino acids to the folded domain provide spectra with narrow and intense peaks in the ^{1}H $\sim$7.9- to 8.5-ppm range in addition to the well-dispersed spectral peaks corresponding to the folded domain. Various NMR-based measurements of rotational correlation time or translational diffusion coefficients may also be used to confirm the oligomeric state of recombinant protein fragments (Altieri *et al.*, 1995; Kay *et al.*, 1989), potentially in combination with analytical ultracentrifugation or size exclusion chromatography methods.

By these approaches, boundaries have been identified for the C-terminal PAS (PAS-B) domains of HIF-1α (residues 238–349), HIF-2α (residues 240–350) and HIF-β (residues 350–470) and subsequently used for the solution structure determinations of the isolated HIF-2α and HIF-β domains (Card *et al.*, 2005; Erbel *et al.*, 2003) and to calculate an NMR data-guided model of an HIF-2α:HIF-β PAS-B heterodimer (Card *et al.*, 2005).

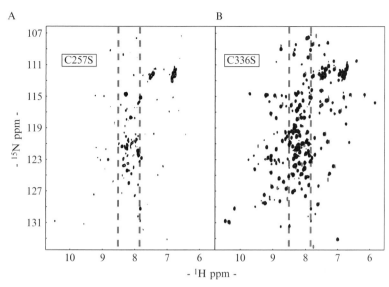

Figure 1.3 Use of nuclear magnetic resonance (NMR) spectroscopy to validate the structures of recombinant proteins. Displayed are $^{15}N/^{1}H$ heteronuclear spin quantum coherence (HSQC) spectra collected from two different Gβ1-HIF-2α PAS-B fusion proteins bearing single amino acid substitutions in the Per/ARNT/Sim (PAS) domain. (A) Hypoxia-inducible factors (HIF)-2α PAS-B (C257S). In this spectrum, the majority of peaks are found within the ^{1}H ∼7.9- to 8.5-ppm (denoted by blue lines) boundaries associated with poorly folded or disordered protein elements. The poor signal intensity of this spectrum also implies that this construct is aggregation prone, a function of NMR spectroscopy's sensitivity to macromolecular size. (B) HIF-2α PAS-B (C336S). A $^{15}N/^{1}H$ HSQC spectrum of this construct indicates that it is a well-folded protein, as demonstrated by the presence of a large number of well-dispersed peaks showing similar line widths. The spectrum collected from the Gβ1-fused mutant PAS domain is given in black, while a reference Gβ1-only spectrum is given in red. This example underscores the importance of confirming the folded state of recombinantly expressed proteins. While neither of these conservative substitutions was expected to disrupt the HIF-2α PAS-B structure, qualitative analysis of these spectra provided early insight regarding the viability of these reagents for subsequent studies. (See color insert.)

4. Assessing PAS Domain Protein–Protein Interactions

4.1. NMR spectroscopy

Investigations of many PAS-containing proteins suggest that they function as protein–protein interaction domains, often using the solvent-exposed β-sheet surface as an interaction interface. Crystal structures of the isolated PAS domains from *E. coli* Dos heme domain EcDOS (Fig. 1.4A) (Kurokawa

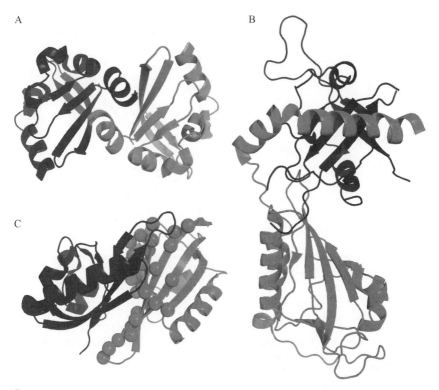

Figure 1.4 Use of the Per/ARNT/Sim (PAS) β-sheet as a dimerization interface. (A) The crystallographic structure of the EcDOS (heme-regulated phosphodiesterase from *Escherichia coli*) PAS domain (Kurokawa *et al.*, 2004; Park *et al.*, 2004) (bound heme omitted for clarity), in which the β-sheet does not directly act as a dimerization surface, but instead hosts interactions with a non-canonical N-terminal helix that serves as part of the interface. (B) The crystallographic structure of tandem PAS domains from the Per protein (Yildiz *et al.*, 2005) reveals a non-canonical helix from the C-terminal PAS domain of one subunit (gray) interacting extensively with the β-sheet of the N-terminal PAS domain (black) from the other subunit. (C) The NMR data-guided model of the hypoxia-inducible factors (HIF)-2α PAS-B (gray) and HIF-β (black) heterodimer, in which the two domains adopt an antiparallel relative orientation with respect to each other. HIF-2α PAS-B amino acids making direct contact with the HIF-β PAS-B domain are indicated by spheres.

et al., 2004; Park *et al.*, 2004) and a tandem PAS fragment from Per (see Fig. 1.4B) (Yildiz *et al.*, 2005) reveal that the solvent-exposed surface of the PAS β-sheet is often used as a homodimerization interface. Photo-sensory PAS domains likewise employ the β-sheet as an intramolecular interaction interface that can be regulated by internally bound cofactors (Craven *et al.*, 2000; Harper *et al.*, 2003). Combined with the established role of PAS domains in protein–protein interactions, it was clear from the initial identification of HIF that these domains could play a potential role in heterodimer assembly (Wang and Semenza, 1995).

To characterize these interactions, we have used solution NMR spectroscopy to take advantage of its strengths in characterizing protein–protein interactions *in vitro*, particularly in its ability to quantitatively evaluate relatively weak complexes (low μM–low mM dissociation constants), which may not be accessible to other techniques, and to provide atomic resolution information. For the HIF-2α:HIF-β PAS-B complex, $^{15}N/^{1}H$ HSQC spectra were collected from samples containing 200 μM ^{15}N-isotopically enriched HIF-2α PAS-B and 0 to 800 μM HIF-β PAS-B with natural abundance ^{15}N (0.37%) (Erbel *et al.*, 2003). Taking advantage of the tendency of NMR peaks to broaden and become less intense with increasing molecular size, diminishing ^{15}N HIF-2α PAS-B peak heights were plotted as a function of HIF-β PAS-B concentration and fit to a model of single-site binding, providing a $K_D = 30$ μM for the two isolated domains. Other $^{15}N/^{1}H$ HSQC peaks shifted position during the course of the titration on account of these nuclei experiencing different magnetic environments in the monomeric and heterodimeric states. Having assigned the chemical shifts of monomeric HIF-2α PAS-B nuclei as a prelude to structure determination, it was possible to correlate changes in cross peak locations to sites that were perturbed upon complex formation. These chemical shift changes mapped to the solvent-exposed β-sheet surface of HIF-2α PAS-B. The reverse titration experiment with unlabeled HIF-2α PAS-B titrated into ^{15}N-HIF-β PAS-B revealed that HIF-β similarly uses its β-sheet as an interaction interface. These findings were supported by analysis of PAS mutants characterized *in vitro* and in the context of living cells (Erbel *et al.*, 2003; Yang *et al.*, 2005). Together, these data were used to calculate the mixed NMR-docking structure of the HIF-2α PAS B:HIF-β PAS B heterodimer (see Fig. 1.4C) (Card *et al.*, 2005). This model shows the two domains packing together via their solvent-exposed sheets, arranged in an antiparallel manner with the N- and C-termini of each domain on opposite sides of the complex. This model was subsequently validated by obtaining long range ($\sim$25 Å) distance restraints using paramagnetic relaxation enhancements to measure intersubunit distances between a nitroxide spin label covalently attached to one subunit of the heterodimeric complex and NMR-active nuclei in the other partner (Card *et al.*, 2005). These data were fully consistent with the antiparallel orientation identified by modeling.

4.2. Reporter genes of HIF function

To functionally validate such models, one can employ a combination of domain deletions and point mutations to probe the involvement of domains and their surfaces in protein–protein interactions *in vitro* and within living cells. Choices underlying the nature and location of mutations benefit from the availability of structural models and are dictated by the particular question(s) being asked. In every case, consideration should be given to the consequences of deletions/mutations with respect to the potential for impinging upon adjacent functional elements or unintentionally compromising global folding. Techniques previously described are useful for assessing the effects of mutations *in vitro* on domain folding/integrity and on protein–protein interactions using recombinant protein preparations. As with any such model, it is important to correlate effects observed in the context of isolated domains with effects on full-length transcription factors in functional settings.

To this end, a number of functional assays for HIF function can be performed. For example, HIF activity can be reported as a function of target gene induction. Several suitable assays are available, employing surrogate reporter constructs (such as luciferase) under the control of HIF response elements (for example, see Bruick, 2000) or the examination of endogenous HIF target genes by Northern blot analysis or quantitative reverse transcriptase polymerase chain reaction (RT-PCR). Interpretation of such experiments must be taken with care, as HIF activity can be influenced by a number of PAS-independent regulatory mechanisms. Furthermore, exogenous overexpression of HIF subunits containing mutations or deletions may overwhelm relevant endogenous regulatory factors leading to artifactual behaviors. Fortunately, a number of cell lines exist in which one of the two HIF subunits has been inactivated (Hoffman *et al.*, 1991; Wood *et al.*, 1998). Such cell lines are well-suited for examining the activity of transiently- or stably-transfected HIF constructs expressed at physiologically relevant levels in a background absent of endogenous HIF.

A variety of mechanisms exist by which mutations to HIF subunits can lead to a decrease in transcriptional activity in cultured cells including, but not limited to, decreased protein accumulation, improper localization, impaired coactivator recruitment, heterodimer disruption, or decreased ability to bind DNA. While many assays are available to assess each of these functions, we have found two to be particularly useful; coimmuno-precipitation (co-IP) assays, which monitor HIF heterodimerization, and electrophoretic mobility shift assay (EMSA), which monitors HIF DNA binding activity, are described here. In our experience, the effects of PAS domain mutations on the function of full-length HIF proteins in these assays strongly correlates with the effects of those same mutations on individual PAS domains (Yang *et al.*, 2005). Though these assays can be performed with recombinantly expressed fragments of HIF, we typically use HIF-containing nuclear lysates, presuming that HIF PAS domains are most likely to be folded properly in this context.

4.3. Preparation of nuclear lysates

Following a modified version of protocols described in other texts (Wang and Semenza, 1996), cells are grown to near confluency in complete medium in 150-mm plates. For HeLa cells, typically 5×10^6 cells are inoculated/plate and grown for 2 days at 37° in an incubator maintained under ambient atmosphere plus 5% CO_2. To induce HIF accumulation, cells are then incubated under hypoxic conditions (1% O_2, 5% CO_2, balance N_2) in a 37° incubator for 12 h. If a hypoxic incubator is not available, HIF accumulation can be induced upon addition of 100 μM desferrioxamine mesylate, an iron chelator that blocks prolyl hydroxylase-mediated degradation of HIF-α under normoxic conditions. Following HIF induction, the cells are washed with cold phosphate buffered saline (PBS), followed by addition of cold PBS plus 1 mM Na_3VO_4, after which cells are harvested by scraping. After pelleting the cells for 2 min at 3500g and carefully removing all the PBS by aspiration, the cell pellet is resuspended in 1 ml buffer containing 20 mM Hepes (pH 7.5), 0.1 mM EDTA, 5 mM NaF, and 10 μM Na_2MoO_4 and incubated on ice for 10 min. Nonidet P-40 is added to a final concentration of 0.5%, followed by vigorous vortexing of the sample for 10 s. To recover the cell nuclei, the sample is pelleted at 13,000 rpm in a microfuge for 1 min at 4°, after which the supernatant is removed by aspiration.

To extract nuclear proteins, resuspend the pelleted nuclei in extraction buffer (420 mM KCl, 20 mM Tris-Cl (pH 7.5), 1.5 mM $MgCl_2$, 20% glycerol, 2 mM DTT, 1 mM Na_3VO_4, 0.3 mM PMSF; add 1:200 protease inhibitor cocktail [PIC; P8340, Sigma-Aldrich, St. Louis, MO] immediately prior to addition). For HeLa cells, 400 μl extraction buffer is added for every 150-mm plate of cells harvested. Incubate for 30 min at 4° with slow mixing. Pellet the insoluble debris at 13,000 rpm in a refrigerated microfuge for 10 min at 4° and recover the supernatant. The nuclear extract can generally be stored at −80° for several months. Prior to storage or usage, the nuclear lysate can be dialyzed into a low salt buffer. However, dialysis is often unnecessary for the assays described here given that the final protein concentration of the lysate is ~1.5 to 2.5 μg protein/μl, as is typically observed when using HeLa cells. Yields will vary among different cell lines.

4.4. Coimmunoprecipitation of HIF-α and HIF-β

Co-IP assays are particularly useful for assessing HIF heterodimerization *in vitro*. Briefly, an antibody specific for one of the HIF subunits is used to immunoprecipitate (IP) that protein from a sample. The IP'ed material is then resolved by SDS-PAGE, and amounts of the IP'ed protein and co-IP'ed–associated proteins can be ascertained by Western blot analysis using appropriate antibodies. However, success of this approach requires antibodies capable of immunoprecipitating a protein without disrupting its association with other factors. We tested several commercially available and custom antibodies raised against each HIF subunit for their ability to IP the HIF heterodimer. Most commercial antibodies we tested were unable to IP an intact HIF heterodimer, though we did find that rabbit polyclonal antibodies raised against human HIF-β (ARNT) residues 1 to 140 or various human HIF-2α epitopes in the C-terminal of the PAS domains worked for this application (Dan Peet, personal communication). Alternatively, one can transfect cells with HIF-α or -β expression constructs in which the HIF subunits are tagged with epitopes (i.e., FLAG or myc) for which commercial antibodies are available. However, care must be taken when using such tagged proteins to correlate HIF activity with heterodimerization via mutational analysis, as the presence of multiple C-terminal epitope tags can independently compromise HIF activity observed upon transfection (J. Garcia, personal communication).

Exact conditions for the co-IP assay will vary depending on the antibodies and cell lines used and should be optimized on a case-by-case basis. As a starting point, we recommend diluting 50 to 100 μg of total protein from the nuclear lysate to a final volume of 400 μl with IP buffer (150 mM KCl, 10 mM Tris-Cl [pH 7.5], 2 mM EDTA [ethylenediamine tetraacetic acid], 0.1% Triton X-100, 0.5 mM PMSF [phenylmethylsulfony fluoride], 1% skim milk powder). Add 15 to 30 μl of a 50% protein G agarose slurry (Cat# 1 243 233; Roche Applied Science, Indianapolis, IN) pre-equilibrated with IP buffer and an experimentally determined amount of the IP antibody. Protein A–agarose beads also can be used, though we have found that protein G beads give more consistent results. The reaction mixture is then incubated for 2 h at 4° with gentle rotation. Antibody-bound beads are pelleted for 1 min at 2000 rpm in a microfuge, and the supernatant is removed by aspiration. The beads are then washed once with 1.25 ml of IP buffer lacking PMSF. To elute proteins, 50 μl of 2× protein loading buffer (125 mM Tris-Cl [pH 6.8], 4% SDS, 20% glycerol, 10% 2-mercaptoethanol) is added to the beads, followed by boiling the sample for 5 min. After a brief centrifugation to pellet the beads, the soluble sample can be resolved in a 10% SDS-PAGE gel.

Proteins from the SDS-PAGE gel are electro-transferred to Hybond-C Extra nitrocellulose (Amersham, Piscataway, NJ) for Western blot analysis. As we typically employ rabbit polyclonal antibodies for immunoprecipitation, mouse monoclonal antibodies are used for Western blot analysis. Several commercially available antibodies perform well in Westerns, including an anti-HIF-1α mouse monoclonal antibody (Cat# 610959; BD Biosciences, San Jose, CA), an anti-EPAS ([endothelial PAS domain protein] HIF-2α) mouse monoclonal antibody (EP190b, NB 100–132; Novus Biological, Littleton, CO), and an anti-ARNT (HIF-1β) mouse monoclonal antibody (H1beta234, NB 100–124; Novus Biological).

Peroxidase-conjugated AffiniPure donkey anti-mouse immunoglobulin (Ig) G (H + L) from Jackson ImmunoResearch (West Grove, PA; Cat# 715–035–150) is used as the secondary antibody because it displays minimal species cross-reactivity with the large amount of rabbit heavy chain present on the blot from the IP antibody. Together, these precautions minimize background signal on the blot following addition of the chemiluminescent peroxidase substrate (Cat# 34080; Pierce, Rockford, IL).

4.5. Electrophoretic mobility shift assay

Complementing the ability of co-IP assays to probe protein–protein interactions in HIF, EMSA assays can be used to assess the DNA-binding activity. Because assembly of the HIF heterodimer is a prerequisite for DNA binding, these assays can also be used to indirectly probe protein–protein interactions between the HIF subunits. Briefly, a radiolabeled double-stranded DNA probe encompassing an HIF binding site is incubated with protein. The binding of HIF to the DNA probe is reflected as a shift in mobility of the HIF:DNA complex with respect to the unbound DNA in a nondenaturing PAGE gel, as observed by autoradiography. The sequence of DNA probes is typically derived from hypoxia-response elements (HREs) of endogenous HIF promoters. To detect HIF-1 and -2 DNA binding, we typically employ a sequence from the vascular endothelial growth factor (VEGF) promoter, 5′-TACACAGTGCA-TACGTGGGTTTCCACAGGTCGTCT-3′ (Ema et al., 1997). Other sequences, such as the W18 probe derived from the erythropoietin (EPO) promoter (5′-GATCGCCCTACGTGCTGTCTCA-3′) (Wang and Semenza, 1995), work better for HIF-1 than HIF-2. To prepare radiolabeled double-stranded DNA probes, equal amounts of these oligonucleotides and their complements are annealed in STE buffer (10 mM Tris, pH 8.0, 50 mM NaCl, 1 mM EDTA) by heating to 95° followed by slow cooling to room temperature over ∼45 min. The DNA is radiolabeled upon incubation with T4 DNA polynucleotide kinase and [γ-^{32}P] ATP (adenosine triphosphate) and purified by extraction in phenol/chloroform/isoamyl alcohol (25:24:1) followed by desalting using a ProbeQuant G-50 Micro column (Amersham, Piscataway, NJ). Approximately 0.2 pmol of the radiolabeled probe is incubated with 10 μg of nuclear lysate in a 20-μl reaction mixture containing 10 mM Tris-Cl (pH 7.5), 1 mM MgCl$_2$, 1 mM EDTA, 2 mM DTT (dithiothreitol), 1:200 PIC, and 0.05 μg/μl herring sperm DNA (Invitrogen, Carlsbad, CA) for 20 min at 4° room temperature. The samples are then resolved in a 5% nondenaturing polyacrylamide gel. To verify that changes in probe mobility are due to binding by HIF, antibodies to one of the subunits can be added to the reaction mixture to form a larger complex with a greater mobility shift (supershift).

The co-IP and EMSA assays can be performed with recombinant protein preparations, though the latter requires the presence of the bHLH domains for DNA-binding; however, we caution that it is important to thoroughly characterize recombinant PAS domain-containing proteins with respect to folding. Careful examination of several HIF-1α and -2α PAS-A–containing constructs expressed in *E. coli* has revealed that HIF-α PAS-A domains, while soluble, are often unfolded (unpublished data). If the HIF PAS domains do act cooperatively, as discussed later, an unfolded PAS-A domain may complicate interpretation of PAS-A and PAS-B function *in vitro*. Tellingly, we have found that consequences of deletions/mutations to recombinant protein preparations in the co-IP and EMSA assays often do not correlate well with the effects of those same deletions/mutations assayed in the context of HIF proteins residing in nuclear lysates. Because data obtained with HIF from nuclear lysates better predict functional consequences in intact cells, one should proceed with caution when using poorly characterized recombinant protein preparations that may not accurately recapitulate biologically meaningful interactions.

5. Discussion

Over the past decade, much has been learned regarding the structure and function of various PAS domains. Structural studies of PAS-containing proteins have typically emphasized isolated PAS domains for a number of reasons, apart from a limited number of structures that detail PAS:PAS or PAS:protein interactions. In contrast, *in vitro* and cellular function studies have focused on intact proteins or larger constructs. By using approaches such as those already described, a framework for understanding how PAS domains contribute to HIF function is emerging. These approaches rely on integrated studies of the molecular mechanisms of PAS function, relating structural observations to *in vitro* properties to behaviors in the context of multi-domain proteins within intact cells.

The solvent-exposed surface of the central β-sheet appears to be frequently used by PAS domains in other proteins to mediate protein–protein interactions. Indeed, the PAS domains within HIF-α and -β serve this function, providing interactions between subunits that are critical for the formation of active heterodimers *in vitro* and in living cells (Erbel *et al.*, 2003; Yang *et al.*, 2005). Structural studies of PAS domains from HIF-α and -β indicate that these interactions involve residues on an exposed β-sheet surface, commonly used among many PAS domains from different proteins (Card *et al.*, 2005; Erbel *et al.*, 2003). Interestingly, the data reveal an antiparallel arrangement of the HIF PAS-B domains, likely maintained by electrostatic complementarity between the subunits (Erbel *et al.*, 2003).

A variety of models are compatible with this antiparallel arrangement maintained within an HIF heterodimer, particularly given the flexibility afforded by the linkers between the two PAS domains in each subunit (see Fig. 1.2). Furthermore, based on precedents observed in recent structural data of complexes involving multiple PAS domains from the Per and LuxQ proteins (Neiditch *et al.*, 2006; Yildiz *et al.*, 2005), there may in fact be a wide range of intra- and intermolecular interactions among PAS domains in the HIF-α and HIF-β subunits. We view it as increasingly likely that all four PAS domains form a cooperative, intermolecular assembly (Fig. 1.5) necessary for HIF heterodimerization. This model is compatible with recent fluorescence resonant energy transfer (FRET) measurements collected from HIF proteins in living cells (Wotzlaw *et al.*, 2006), suggesting a compact tertiary structure for the intact heterodimer. This model has important implications as well for PAS surfaces that mediate interactions between HIF and other proteins, such as coactivators and corepressors (Beischlag *et al.*, 2004; Sadek *et al.*, 2000; To *et al.*, 2006).

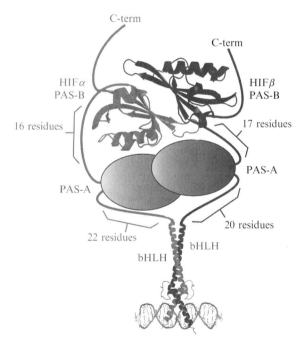

Figure 1.5 Schematic model of hypoxia-inducible factors (HIF) domain interactions within the DNA-bound HIF heterodimer. Shown are the bHLH:DNA interactions, represented by the crystal structure of DNA-bound Max protein (Ferre-D'Amare *et al.*, 1993), and the antiparallel heterodimer of HIF-α (gray) and HIF-β (black) Per/ARNT/ Sim (PAS)-B domains. As the structures of the HIF PAS-A domains have yet to be determined, they are represented schematically. (Reprinted from Card *et al.*, 2005, with permission.)

Lastly, given the absolute dependence on these PAS domain interactions for HIF function, it is appropriate to consider strategies that would target these domains for therapeutic purposes. For example, increased levels of HIF have been associated with tumor aggressiveness, resistance to conventional therapies, and mortality for a variety of human cancers through the induction of genes that promote metabolic adaptation, angiogenesis, changes in cell proliferation/survival, and cellular invasion and metastasis (Semenza, 2003). While a number of agents have been shown to antagonize HIF (Giaccia *et al.*, 2003; Semenza, 2003), almost all are believed to act indirectly. As a consequence, their utility is often dependent on the mechanism of HIF induction while their therapeutic effects mediated through HIF can be difficult to distinguish from effects on other downstream pathways.

Inspiration for a possible route to regulate PAS function in HIF comes from the observations that protein–protein interactions of a wide range of PAS domains are regulated by environmental stimuli, including light, oxygen, and redox state (Taylor and Zhulin, 1999). Sensitization to such diverse stimuli is achieved via small organic cofactors bound within the PAS domain, using changes in the configuration or occupancy of these ligands to alter protein structure and dynamics in ways that modulate function. Examples of this regulation are provided by photosensory PAS domains, which use light-driven bond formation or isomerization to break interactions between the sensory PAS domain and an effector helix (Harper *et al.*, 2003; Rubinstenn *et al.*, 1998). Such observations have led us to speculate that small organic compounds might be used to analogously disrupt an HIF heterodimeric complex. Although high-resolution HIF PAS-B structures have not revealed any internally-bound cofactors or preformed ligand-binding pockets, prior data from other PAS domains (e.g., hPASK [Amezcua *et al.*, 2002]) indicate that this does not preclude ligand binding. Though protein–protein interactions are notoriously difficult to directly disrupt by small molecules, one can imagine that HIF PAS domains might bind small molecules within their cores, leading to conformational changes that propagate to the domain surface and disrupt association with other protein domains. Again, an experimental framework that relates domain structure to *in vitro* activity to *in vivo* function is crucial in identifying and evaluating potential PAS ligands (Park *et al.*, 2006).

ACKNOWLEDGMENTS

This work was supported by grants from the NIH (CA115962 to R.K.B.; CA90601 to K.H.G.; CA95471 to K.H.G. and R.K.B.) and the Robert A. Welch Foundation (I-1568 to R.K.B.). T.H.S. was supported by the American Cancer Society High Plains Division–North Texas Postdoctoral Fellowship. R.K.B. is the Michael L. Rosenberg Scholar in Medical Research and is supported by a Career Award in the Biomedical Sciences from the Burroughs Wellcome Fund.

REFERENCES

Altieri, A. S., Hinton, D. P., and Byrd, R. A. (1995). Association of biomolecular systems via pulse field gradient NMR self-diffusion measurements. *J. Am. Chem. Soc.* **117**, 7566–7567.

Amezcua, C. A., Harper, S. M., Rutter, J., and Gardner, K. H. (2002). Structure and interactions of PAS kinase N-terminal PAS domain: Model for intramolecular kinase regulation. *Structure* **10**, 1349–1361.

Beischlag, T. V., Taylor, R. T., Rose, D. W., Yoon, D., Chen, Y., Lee, W. H., Rosenfeld, M. G., and Hankinson, O. (2004). Recruitment of thyroid hormone receptor/retinoblastoma-interacting protein 230 by the aryl hydrocarbon receptor nuclear translocator is required for the transcriptional response to both dioxin and hypoxia. *J. Biol. Chem.* **279**, 54620–54628.

Bruick, R. K. (2000). Expression of the gene encoding the proapoptotic Nip3 protein is induced by hypoxia. *Proc. Natl. Acad. Sci. USA* **97**, 9082–9087.

Bruick, R. K., and McKnight, S. L. (2001). A conserved family of prolyl-4-hydroxylases that modify HIF. *Science* **294**, 1337–1340.

Card, P. B., and Gardner, K. H. (2005). Identification and optimization of protein domains for NMR studies. *Meth. Enz.* **394**, 3–16.

Card, P. B., Erbel, P. J., and Gardner, K. H. (2005). Structural basis of ARNT PAS-B dimerization: Use of a common β-sheet interface for hetero- and homodimerization. *J. Mol. Biol.* **353**, 664–677.

Chachami, G., Paraskeva, E., Georgatsou, E., Bonanou, S., and Simos, G. (2005). Bacterially produced human HIF-1α is competent for heterodimerization and specific DNA-binding. *Biochem. Biophys. Res. Commun.* **331**, 464–470.

Chapman-Smith, A., Lutwyche, J. K., and Whitelaw, M. L. (2004). Contribution of the Per/Arnt/Sim (PAS) domains to DNA binding by the basic helix-loop-helix PAS transcriptional regulators. *J. Biol. Chem.* **279**, 5353–5362.

Cohen, S. L., Ferre-D'Amare, A. R., Burley, S. K., and Chait, B. T. (1995). Probing the solution structure of the DNA-binding protein Max by a combination of proteolysis and mass spectrometry. *Protein Sci.* **4**, 1088–1099.

Craven, C. J., Derix, N. M., Hendriks, J., Boelens, R., Hellingwerf, K. J., and Kaptein, R. (2000). Probing the nature of the blue-shifted intermediate of photoactive yellow protein in solution by NMR: Hydrogen-deuterium exchange data and pH studies. *Biochemistry* **39**, 14392–14399.

Cuff, J. A., Clamp, M. E., Siddiqui, A. S., Finlay, M., and Barton, G. J. (1998). JPred: A consensus secondary structure prediction server. *Bioinformatics* **14**, 892–893.

Dann, C. E., III, and Bruick, R. K. (2005). Dioxygenases as O_2-dependent regulators of the hypoxic response pathway. *Biochem. Biophys. Res. Commun.* **338**, 639–647.

Ema, M., Taya, S., Yokotani, N., Sogawa, K., Matsuda, Y., and Fujii-Kuriyama, Y. (1997). A novel bHLH-PAS factor with close sequence similarity to hypoxia-inducible factor 1alpha regulates the VEGF expression and is potentially involved in lung and vascular development. *Proc. Natl. Acad. Sci. USA* **94**, 4273–4278.

Epstein, A. C., Gleadle, J. M., McNeill, L. A., Hewitson, K. S., O'Rourke, J., Mole, D. R., Mukherji, M., Metzen, E., Wilson, M. I., Dhanda, A., Tian, Y. M., Masson, N., *et al.* (2001). *C. elegans* EGL-9 and mammalian homologs define a family of dioxygenases that regulate HIF by prolyl hydroxylation. *Cell* **107**, 43–54.

Erbel, P. J., Card, P. B., Karakuzu, O., Bruick, R. K., and Gardner, K. H. (2003). Structural basis for PAS domain heterodimerization in the basic helix–loop–helix-PAS transcription factor hypoxia-inducible factor. *Proc. Natl. Acad. Sci. USA* **100**, 15504–15509.

Ferre-D'Amare, A. R., Prendergast, G. C., Ziff, E. B., and Burley, S. K. (1993). Recognition by Max of its cognate DNA through a dimeric b/HLH/Z domain. *Nature* **363**, 38–45.

Giaccia, A., Siim, B. G., and Johnson, R. S. (2003). HIF-1 as a target for drug development. *Nat. Rev. Drug Discov.* **2,** 803–811.

Gong, W., Hao, B., Mansy, S. S., Gonzalez, G., Gilles-Gonzalez, M. A., and Chan, M. K. (1998). Structure of a biological oxygen sensor: A new mechanism for heme-driven signal transduction. *Proc. Natl. Acad. Sci. USA* **95,** 15177–15182.

Gradin, K., Takasaki, C., Fujii-Kuriyama, Y., and Sogawa, K. (2002). The transcriptional activation function of the HIF-like factor requires phosphorylation at a conserved threonine. *J. Biol. Chem.* **277,** 23508–23514.

Gronenborn, A. M., and Clore, G. M. (1996). Rapid screening for structural integrity of expressed proteins by heteronuclear NMR spectroscopy. *Protein Sci.* **5,** 174–177.

Gu, Y. Z., Moran, S. M., Hogenesch, J. B., Wartman, L., and Bradfield, C. A. (1998). Molecular characterization and chromosomal localization of a third α-class hypoxia inducible factor subunit, HIF3alpha. *Gene Expression* **7,** 205–213.

Hammarstrom, M., Hellgren, N., van Den Berg, S., Berglund, H., and Hard, T. (2002). Rapid screening for improved solubility of small human proteins produced as fusion proteins in *Escherichia coli. Protein Sci.* **11,** 313–321.

Harper, S. M., Neil, L. C., and Gardner, K. H. (2003). Structural basis of a phototropin light switch. *Science* **301,** 1541–1544.

Harper, S. M., Christie, J. M., and Gardner, K. H. (2004). Disruption of the LOV-Jalpha helix interaction activates phototropin kinase activity. *Biochemistry* **43,** 16184–16192.

Hefti, M. H., Francoijs, K. J., de Vries, S. C., Dixon, R., and Vervoort, J. (2004). The PAS fold. A redefinition of the PAS domain based upon structural prediction. *Eur. J. Biochem.* **271,** 1198–1208.

Hewitson, K. S., McNeill, L. A., Riordan, M. V., Tian, Y. M., Bullock, A. N., Welford, R. W., Elkins, J. M., Oldham, N. J., Bhattacharya, S., Gleadle, J. M., Ratcliffe, P. J., Pugh, C. W., and Schofield, C. J. (2002). Hypoxia-inducible factor (HIF) asparagine hydroxylase is identical to factor inhibiting HIF (FIH) and is related to the cupin structural family. *J. Biol. Chem.* **277,** 26351–26355.

Hoffman, E. C., Reyes, H., Chu, F. F., Sander, F., Conley, L. H., Brooks, B. A., and Hankinson, O. (1991). Cloning of a factor required for activity of the Ah (dioxin) receptor. *Science* **252,** 954–958.

Huang, Z. J., Edery, I., and Rosbash, M. (1993). PAS is a dimerization domain common to *Drosophila* period and several transcription factors. *Nature* **364,** 259–262.

Isaacs, J. S., Jung, Y. J., and Neckers, L. (2004). Aryl hydrocarbon nuclear translocator (ARNT) promotes oxygen-independent stabilization of hypoxia-inducible factor-1α by modulating an Hsp90-dependent regulatory pathway. *J. Biol. Chem.* **279,** 16128–16135.

Ivan, M., Kondo, K., Yang, H., Kim, W., Valiando, J., Ohh, M., Salic, A., Asara, J. M., Lane, W. S., and Kaelin, W. G., Jr. (2001). HIF alpha targeted for VHL-mediated destruction by proline hydroxylation: Implications for O_2 sensing. *Science* **292,** 464–468.

Ivan, M., Haberberger, T., Gervasi, D. C., Michelson, K. S., Gunzler, V., Kondo, K., Yang, H., Sorokina, I., Conaway, R. C., Conaway, J. W., and Kaelin, W. G., Jr. (2002). Biochemical purification and pharmacological inhibition of a mammalian prolyl hydroxylase acting on hypoxia-inducible factor. *Proc. Natl. Acad. Sci. USA* **99,** 13459–13464.

Jaakkola, P., Mole, D. R., Tian, Y. M., Wilson, M. I., Gielbert, J., Gaskell, S. J., Kriegsheim, A., Hebestreit, H. F., Mukherji, M., Schofield, C. J., Maxwell, P. H., Pugh, C. W., and Ratcliffe, P. J. (2001). Targeting of HIF-α to the von Hippel-Lindau ubiquitylation complex by O_2-regulated prolyl hydroxylation. *Science* **292,** 468–472.

Jiang, B. H., Rue, E., Wang, G. L., Roe, R., and Semenza, G. L. (1996). Dimerization, DNA binding, and transactivation properties of hypoxia-inducible factor 1. *J. Biol. Chem.* **271,** 17771–17778.

Kallio, P. J., Okamoto, K., O'Brien, S., Carrero, P., Makino, Y., Tanaka, H., and Poellinger, L. (1998). Signal transduction in hypoxic cells: Inducible nuclear translocation and recruitment of the CBP/p300 coactivator by the hypoxia-inducible factor-1α. *EMBO J.* **17**, 6573–6586.

Katschinski, D. M., Le, L., Schindler, S. G., Thomas, T., Voss, A. K., and Wenger, R. H. (2004). Interaction of the PAS B domain with HSP90 accelerates hypoxia-inducible factor-1α stabilization. *Cell Physiol. Biochem.* **14**, 351–360.

Kaufman, B., Scharf, O., Arbeit, J., Ashcroft, M., Brown, J. M., Bruick, R. K., Chapman, J. D., Evans, S. M., Giaccia, A. J., Harris, A. L., Huang, E., Johnson, R., *et al.* (2004). Proceedings of the Oxygen Homeostasis/Hypoxia Meeting. *Can. Res.* **64**, 3350–3356.

Kay, L. E., Torchia, D. A., and Bax, A. (1989). Backbone dynamics of proteins as studied by ^{15}N inverse detected heteronuclear NMR spectroscopy: Application to staphylococcal nuclease. *Biochemistry* **28**, 8972–8979.

Kay, L. E., Keifer, P., and Saarinen, T. (1992). Pure absorption gradient enhanced heteronuclear single quantum correlation spectroscopy with improved sensitivity. *J. Am. Chem. Soc.* **114**, 10663–10665.

Kikuchi, Y., Ohsawa, S., Mimura, J., Ema, M., Takasaki, C., Sogawa, K., and Fujii-Kuriyama, Y. (2003). Heterodimers of bHLH-PAS protein fragments derived from AhR, AhRR, and Arnt prepared by co-expression in *Escherichia coli*: Characterization of their DNA binding activity and preparation of a DNA complex. *J. Biochem.* **134**, 83–90.

Kurokawa, H., Lee, D. S., Watanabe, M., Sagami, I., Mikami, B., Raman, C. S., and Shimizu, T. (2004). A redox-controlled molecular switch revealed by the crystal structure of a bacterial heme PAS sensor. *J. Biol. Chem.* **279**, 20186–20193.

Lando, D., Peet, D. J., Whelan, D. A., Gorman, J. J., and Whitelaw, M. L. (2002a). Asparagine hydroxylation of the HIF transactivation domain: a hypoxic switch. *Science* **295**, 858–861.

Lando, D., Peet, D. J., Gorman, J. J., Whelan, D. A., Whitelaw, M. L., and Bruick, R. K. (2002b). FIH-1 is an asparaginyl hydroxylase enzyme that regulates the transcriptional activity of hypoxia-inducible factor. *Genes Dev.* **16**, 1466–1471.

Maxwell, P. H. (2005). The HIF pathway in cancer. *Semin. Cell Dev. Biol.* **16**, 523–530.

Maxwell, P. H., Wiesener, M. S., Chang, G. W., Clifford, S. C., Vaux, E. C., Cockman, M. E., Wykoff, C. C., Pugh, C. W., Maher, E. R., and Ratcliffe, P. J. (1999). The tumour suppressor protein VHL targets hypoxia-inducible factors for oxygen-dependent proteolysis. *Nature* **399**, 271–275.

Minet, E., Mottet, D., Michel, G., Roland, I., Raes, M., Remacle, J., and Michiels, C. (1999). Hypoxia-induced activation of HIF-1: Role of HIF-1alpha-Hsp90 interaction. *FEBS Lett.* **460**, 251–256.

Morais Cabral, J. H., Lee, A., Cohen, S. L., Chait, B. T., Li, M., and Mackinnon, R. (1998). Crystal structure and functional analysis of the HERG potassium channel N terminus: A eukaryotic PAS domain. *Cell* **95**, 649–655.

Nambu, J. R., Lewis, J. O., Wharton, K. A., Jr., and Crews, S. T. (1991). The *Drosophila* single-minded gene encodes a helix-loop-helix protein that acts as a master regulator of CNS midline development. *Cell* **67**, 1157–1167.

Neiditch, M. B., Federle, M. J., Pompeani, A. J., Kelly, R. C., Swem, D. L., Jeffrey, P. D., Bassler, B. L., and Hughson, F. M. (2006). Ligand-induced asymmetry in histidine sensor kinase complex regulates quorum sensing. *Cell* **126**, 1095–1108.

Park, H., Suquet, C., Satterlee, J. D., and Kang, C. (2004). Insights into signal transduction involving PAS domain oxygen-sensing heme proteins from the X-ray crystal structure of *Escherichia coli* Dos heme domain (Ec DosH). *Biochemistry* **43**, 2738–2746.

Park, E. J., Kong, D., Fisher, R., Cardellina, J., Shoemaker, R. H., and Melillo, G. (2006). Targeting the PAS-A domain of HIF-1α for development of small molecule inhibitors of HIF-1. *Cell Cycle* **5**, 1847–1853.

Pellequer, J. L., Brudler, R., and Getzoff, E. D. (1999). Biological sensors: More than one way to sense oxygen. *Curr. Biol.* **9,** R416–R418.

Ponting, C. P., and Aravind, L. (1997). PAS: A multifunctional domain family comes to light. *Curr. Biol.* **7,** R674–R677.

Rubinstenn, G., Vuister, G. W., Mulder, F. A. A., Düx, P. E., Boelens, R., Hellingwerf, K. J., and Kaptein, R. (1998). Structural and dynamic changes of photoactive yellow protein during its photocycle in solution. *Nat. Structual Biol.* **5,** 568–570.

Sadek, C. M., Jalaguier, S., Feeney, E. P., Aitola, M., Damdimopoulos, A. E., Pelto-Huikko, M., and Gustafsson, J. A. (2000). Isolation and characterization of AINT: A novel ARNT interacting protein expressed during murine embryonic development. *Mech. Dev.* **97,** 13–26.

Semenza, G. L. (2000). HIF-1 and human disease: One highly involved factor. *Genes Dev.* **14,** 1983–1991.

Semenza, G. L. (2003). Targeting HIF-1 for cancer therapy. *Nat. Rev. Cancer* **3,** 721–732.

Sheffield, P., Garrard, S., and Derewenda, Z. (1999). Overcoming expression and purification problems of RhoGDI using a family of "parallel" expression vectors. *Protein Expr. Purif.* **15,** 34–39.

Taylor, B. L., and Zhulin, I. B. (1999). PAS domains: Internal sensors of oxygen, redox potential, and light. *Microbiol. Mol. Biol. Rev.* **63,** 479–506.

Tian, H., McKnight, S. L., and Russell, D. W. (1997). Endothelial PAS domain protein 1 (EPAS1), a transcription factor selectively expressed in endothelial cells. *Genes Dev.* **11,** 72–82.

To, K. K.-W., Sedelnikova, O. A., Samons, M., Bonner, W. M., and Huang, L. E. (2006). The phosphorylation status of PAS-B distinguishes HIF-1α from HIF-2α in NBS1 repression. *EMBO J.* **25,** 4784–4794.

Wang, G. L., and Semenza, G. L. (1995). Purification and characterization of hypoxia-inducible factor 1. *J. Biol. Chem.* **270,** 1230–1237.

Wang, G. L., and Semenza, G. L. (1996). Identification and characterization of transcription factors from mammalian cells. *Meth. Mol. Genet.* **8,** 298–305.

Wang, G. L., Jiang, B. H., Rue, E. A., and Semenza, G. L. (1995). Hypoxia-inducible factor 1 is a basic-helix-loop-helix-PAS heterodimer regulated by cellular O_2 tension. *Proc. Natl. Acad. Sci. USA* **92,** 5510–5514.

Wood, S. M., Wiesener, M. S., Yeates, K. M., Okada, N., Pugh, C. W., Maxwell, P. H., and Ratcliffe, P. J. (1998). Selection and analysis of a mutant cell line defective in the hypoxia-inducible factor-1 α-subunit (HIF-1alpha). Characterization of HIF-1α-dependent and -independent hypoxia-inducible gene expression. *J. Biol. Chem.* **273,** 8360–8368.

Wotzlaw, C., Otto, T., Berchner-Pfannschmidt, U., Metzen, E., Acker, H., and Fandrey, J. (2006). Optical analysis of the HIF-1 complex in living cells by FRET and FRAP. *FASEB J.* **21,** 1–8.

Yang, J., Zhang, L., Erbel, P. J., Gardner, K. H., Ding, K., Garcia, J. A., and Bruick, R. K. (2005). Functions of the Per/ARNT/Sim domains of the hypoxia-inducible factor. *J. Biol. Chem.* **280,** 36047–36054.

Yildiz, Ö., Doi, M., Yujnovsky, I., Cardone, L., Berndt, A., Hennig, S., Schulze, S., Urbanke, C., Sassone-Corse, P., and Wolf, E. (2005). Crystal structure and interactions of the PAS repeat region of the *Drosophila* clock protein PERIOD. *Mol. Cell* **17,** 69–82.

Zhulin, I. B., Taylor, B. L., and Dixon, R. (1997). PAS domain S-boxes in archaea, bacteria and sensors for oxygen and redox. *Trends Biochem. Sci.* **22,** 331–333.

CHAPTER TWO

HYPOXIA-INDUCIBLE FACTOR PROLYL-HYDROXYLASE: PURIFICATION AND ASSAYS OF PHD2

Kirsty S. Hewitson,* Christopher J. Schofield,* *and* Peter J. Ratcliffe[†]

Contents

1. Introduction   26
2. Preparation of Purified PHD2 from a Bacterial Source   28
3. Assaying of PHD2 Activity   29
4. Indirect Measurements of PHD2 Activity   29
    4.1. 1-[^{14}C]-CO$_2$ capture assay   30
    4.2. Fluorescence derivatization of 2OG   32
    4.3. Oxygen consumption assay   33
    4.4. 1-[^{14}C]succinate quantification   35
5. Direct Measurements of PHD2 Hydroxylation Activity   35
    5.1. LC/MS identification of hydroxylated HIF-1α 556 to 574   35
    5.2. pVHL capture assay   36
6. Binding Assays   37
7. Comparison of Assay Formats   38
References   39

Abstract

The adaptation of animals to oxygen availability is mediated by a transcription factor termed hypoxia-inducible factor (HIF). HIF is an alpha (α)/beta (β) hetero-dimer that binds hypoxia response elements (HREs) of target genes, including some of medicinal importance, such as erythropoietin (EPO) and vascular endothelial growth factor (VEGF). While the concentration of the HIF-β subunit, a constitutive nuclear protein, does not vary with oxygen availability, the abundance and activity of the HIF-α subunits are tightly regulated via oxygen-dependent modification of specific residues. Hydroxylation of prolyl residues (Pro402 and Pro564 in HIF-1α) promotes interaction with the von Hippel-Lindau

* Department of Chemistry, Chemistry Research Laboratory, University of Oxford, Oxford, United Kingdom
† Henry Wellcome Building for Molecular Physiology, University of Oxford, Oxford, United Kingdom

Methods in Enzymology, Volume 435
ISSN 0076-6879, DOI: 10.1016/S0076-6879(07)35002-7

E3 ubiquitin ligase and, consequently, proteolytic destruction by the ubiquitin-proteasome pathway. This prolyl hydroxylation is catalyzed by the prolyl-hydroxylase domain (PHD) containing enzymes for which three isozymes have been identified in humans (1–3). Additionally, asparaginyl hydroxylation (Asn803 in HIF-1α) by factor-inhibiting HIF (FIH) ablates interaction of the HIF-α subunit with the coactivator p300, providing an alternative mechanism for down-regulation of HIF-dependent genes. Under hypoxic conditions, when oxygen-mediated regulation of the α-subunits is curtailed or minimized, dimerization of the α- and β-subunits occurs with subsequent target gene upregulation.

Therapeutic activation of HIF signaling has been suggested as a potential treatment for numerous conditions, including ischemia, stroke, heart attack, inflammation, and wounding. One possible route to achieve this is via inhibition of the HIF hydroxylases. This chapter details methods for the purification and assaying of PHD2, the most abundant PHD and the most important in setting steady-state levels of HIF-α. Assays are described that measure the activity of PHD2 via direct and indirect means. Furthermore, conditions for the screening of small molecules against PHD2 are described.

1. INTRODUCTION

The maintenance of oxygen homeostasis is a fundamental physiological challenge. Recent studies have revealed that a substantial component of the cell's total complement of expressed genes (in the range of hundreds or thousands of transcripts) are modulated by changes in oxygen availability and that the majority of these are responding directly or indirectly to novel signal pathways that govern the activity of HIF by post-translational hydroxylation of specific amino acid residues. As previously stated, HIF is an α/β heterodimer that binds HREs in the *cis*-acting regulatory sequences of target genes that encode molecules involved in both systemic responses to hypoxia, such as enhanced erythropoiesis and angiogenesis, and cellular responses to hypoxia, such as alterations in energy metabolism and cell motility, differentiation, and survival decisions (Pugh and Ratcliffe, 2003; Semenza, 2000; Wenger, 2002). Regulation of HIF activity is provided by its α-subunits (Ivan *et al.*, 2001; Jaakkola *et al.*, 2001; Lando *et al.*, 2002; Masson *et al.*, 2001; Yu *et al.*, 2001), which are highly inducible in hypoxic cells and appear specific to the HIF system. The β-subunits (also known as the aryl hydrocarbon nuclear translocator, ARNT) are constitutive nuclear proteins that mediate a number of different transcriptional responses in combination with other dimerization partners.

HIF-α is controlled by at least two steps involving different types of hydroxylation. Thus, prolyl hydroxylation of specific residues within a central degradation domain controls interaction with the von Hippel-Lindau E3 ubiquitin ligase and proteolytic destruction by the ubiquitin-proteasome pathway (Ivan *et al.*, 2001; Jaakkola *et al.*, 2001; Yu *et al.*, 2001). Asparaginyl hydroxylation at a specific residue in the C-terminal transcriptional activation domain controls coactivator recruitment and, hence, the transcriptional activity of HIF-α polypeptides that escape proteolytic destruction (Lando *et al.*, 2002). This chapter will focus on HIF prolyl hydroxylation, which was the first oxygen-regulated step to be discovered and which exerts dominant control over HIF activity through oxygen-dependent destruction of HIF-α.

HIF prolyl hydroxylation occurs at two residues that are tightly conserved within the closely related mammalian HIF-1α and HIF-2α isoforms and at a single conserved residue in mammalian HIF-3α. This process is conserved in invertebrates; and in flies and worms there appears to be a single site of prolyl hydroxylation in a single HIF-α homologue (Bacon *et al.*, 1998; Epstein *et al.*, 2001; Jiang *et al.*, 2001; Lavista-Llanos *et al.*, 2002; Wappner *et al.*, 2003). The reactions are catalyzed by a series of dioxygenases belonging to the Fe(II)- and 2-oxoglutarate–dependent oxygenase superfamily (Clifton *et al.*, 2006; Costas *et al.*, 2004; Hausinger, 2004; Hewitson *et al.*, 2005), which are represented in mammalian cells by three closely related enzymes (PHD 1, 2, and 3, also identified as HPH1–3 and EGLN1–3) (Bruick and McKnight, 2001; Epstein *et al.*, 2001). The absolute requirement for molecular oxygen as cosubstrate confers oxygen sensitivity and, in hypoxia, hydroxylation is suppressed, allowing HIF-α to escape destruction and form a transcriptional complex. PHD1, -2, and -3 share a highly conserved catalytic domain in which the active Fe(II) center is coordinated by a 2-histidine-1 carboxylate motif (HxD ...H) aligned on the second and seventh strands of the eight-stranded β-barrel jelly-roll conformation that is common to the catalytic core of this family of enzymes (Epstein *et al.*, 2001; McDonough *et al.*, 2006).

All three enzymes are physiological regulators of the HIF system, though they do display preferential activity for different HIF prolyl hydroxylation sites. Thus, PHD1 and -2 show activity for both of the hydroxylation sites in HIF-1α and HIF-2α, whereas PHD3 is selective for the more C-terminal of the two sites (Appelhoff *et al.*, 2004; Hirsila *et al.*, 2003). Cellular abundance, however, varies substantially, and PHD1 and -3 show marked tissue-specific restriction of expression (Appelhoff *et al.*, 2004; Lieb *et al.*, 2002; Willam *et al.*, 2006). In most cells studied to date, PHD2 is the most abundant enzyme, while, in oxygenated cells, it is the most important in setting steady-state levels of HIF-α and hence the activity of the system (Appelhoff *et al.*, 2004; Berra *et al.*, 2003). For this reason, we have focused our development of the assay methods later described on PHD2. The assays are of use in the biochemical and physiological characterization of the enzymes as cellular oxygen sensors and in the development and assessment of inhibitors as a therapeutic approach to ischemic/hypoxic disease.

2. PREPARATION OF PURIFIED PHD2 FROM A BACTERIAL SOURCE

Our laboratory uses the procedure described in this section for the preparation of recombinant PHD2 as a source of material for biochemical characterization studies (McNeill et al., 2005b), inhibitor screens (Hewitson et al., 2007), and crystallization (McDonough et al., 2006). Attempts to produce and purify soluble, full-length PHD2 to homogeneity in "milligram" quantities for the these uses, either with (e.g., maltose binding protein [MBP], His, intein-based protein-splicing vectors, glutathione S-transferase [GST]) or without affinity tags in bacterial cultures, were problematic. Sequence comparisons of PHD2 with a bacterial 2-oxyglutarate (2OG)–dependent oxygenase deacetoxycephalosporin C synthase (DAOCS) (Lloyd et al., 1999), for which crystal structures have been reported (Valegård et al., 1998), identified amino acids 181 to 426 of PHD2 as being the putative PHD2 catalytic domain. An N-terminally hexahistidine-tagged fusion protein comprising these residues (PHD2$_{181-426}$) was produced using the pET28a(+) vector (Novagen, Darmstadt, Germany) to yield approximately 5% of total soluble protein when expressed in Escherichia coli BL21(DE3). Following initial growth at 37° in 2TY medium, protein expression is induced with isopropyl β-D-thiogalactoside (0.5 mM) when the OD$_{600}$ reaches ~0.8 to 1.0. Growth is then continued at 37° for a further 3 to 4 h before harvesting by centrifugation at 14,000 rpm. Cell pellets have been stored for at least 6 mo at −80° without any apparent deleterious effects on catalytic activity of purified PHD2$_{181-426}$.

Affinity purification of PHD2$_{181-426}$ can be achieved using the N-terminal hexahistidine tag (McNeill et al., 2005b). His-Bind ^{TM} resin (Novagen, Darmstadt, Germany) is commonly used in our laboratory for such purification and utilizes Ni^{2+} ions held by tridentate chelation to the resin by iminodiacetic acid (IDA). A typical purification involves the resuspension of 20 g of PHD2$_{181-426}$/BL21(DE3) in 100-ml binding buffer (40 mM Tris-HCl, 0.5 M NaCl, 5 mM imidazole, pH 7.9) with subsequent lysis by sonication. After centrifugation at 14,000 rpm (20 min), the cleared lysate is applied to a column (10 ml) containing Ni-IDA His-Bind resin pre-equilibrated with binding buffer. Once the lysate is loaded, a further 10 column volumes of binding buffer is flowed through the column to remove any unbound protein. After washing with six column volumes of wash buffer (40 mM Tris-HCl, 0.5 M NaCl, 60 mM imidazole, pH 7.9), the PHD2$_{181-426}$ is eluted with six column volumes of elute buffer (40 mM Tris-HCl, 0.5 M NaCl, 1 M imidazole, pH 7.9). At this stage, the PHD2$_{181-426}$ can be stored frozen as the His-tagged protein, once desalted (note that, if the protein is stored without the removal of imidazole, it may precipitate upon thawing), or the His tag can be cleaved by treatment with thrombin (Novagen, Darmstadt, Germany). Efficient cleavage of the His-tag is possible with 1 U of thrombin to 20 mg of PHD2$_{181-426}$ following a 16-h incubation at 4°. As a final step, separation of the His-tag from PHD2$_{181-426}$ is achieved by using a 300-ml Superdex S75 gel filtration column, which also serves to desalt the protein. The resulting PHD2$_{181-426}$ is greater than 95% pure by sodium dodecyl sulfate polyacrylamide gel electrophoresis (SDS-PAGE) analysis and has been stored for at least 6 mo at −80° with no resulting loss in enzymatic activity. Typical yields of protein are 30 to 40 mg of PHD2$_{181-426}$ from 20 g of cells. As an alternative to Ni-IDA resin, Ni-NTA and TALON^{TM} Resin (BD Biosciences, San Jose, CA; where cobalt [II] is substituted for nickel [II]) have also been used successfully with similar protocols to those previously detailed to purify PHD2$_{181-426}$. Although not observed for PHD2$_{181-426}$, care must be taken with the use of metal affinity chromatography and the 2OG-dependent oxygenases. Since these enzymes bind iron at the active site, as required for catalysis, any leaching of metal from the affinity column and subsequent coordination at the active site can lead to a reduction in enzymatic activity (Searls *et al.*, 2005).

3. Assaying of PHD2 Activity

A variety of techniques can be used to assay PHD2 activity *in vitro*. Those detailed here are procedures used routinely in our laboratories that yield reproducible results. A brief comparison of the relative merits and disadvantages of each of the assay formats is given at the end of this section.

The assays can themselves, broadly speaking, be classified into two types: those that are generic to the Fe(II) and 2OG-dependent oxygenase family (of which PHD2 is a member) (Bruick and McKnight, 2001; Epstein *et al.*, 2001) and those that are specific to the PHDs. While soluble PHD2$_{181-426}$ is currently the only readily available PHD isoform reported to be prepared in milligram quantities, these techniques could be applied to all PHD family members.

4. Indirect Measurements of PHD2 Activity

PHD2 is a member of the Fe(II) and 2OG-dependent oxygenase superfamily in which almost all members appear to follow the generalized reaction scheme shown in Scheme 2.1 (Costas *et al.*, 2004; Hausinger, 2004; Hewitson *et al.*, 2005).

Substrate + O_2 + 2OG $\longrightarrow$ Oxidised + CO_2 + succinate

(HIF-α)   Product
(hydroxylated HIF-α)

Scheme 2.1 Generalized reaction scheme for the 2oG-dependent oxygenase family.

Oxidative decarboxylation of 2OG to give CO_2 and succinate accompanies hydroxylation of HIF-α substrate by PHD2. Indirect measurement of PHD2 activity can therefore be achieved by quantification of the amount of 2OG or dioxygen consumed or by the amount of CO_2 or succinate formed. It should be noted that, for the 2OG-dependent oxygenase family, uncoupling of 2OG decarboxylation from substrate formation can occur (i.e., turnover of 2OG does not always correlate to prime substrate hydroxylation) (Costas *et al.*, 2004; Hausinger, 2004; Hewitson *et al.*, 2005). Care must be taken in the following procedures to control for such a possibility.

4.1. 1-[^{14}C]-CO_2 capture assay

The activity of PHD2 can be assayed by measuring the release of 1-[^{14}C]-CO_2 from 1-[^{14}C]-2OG (PerkinElmer, Waltham, MA) (Epstein *et al.*, 2001). This measurement is a well-established assay technique for 2OG-dependent oxygenases and has been used successfully in our laboratory for a range of other enzymes from this family, including FIH (Hewitson *et al.*, 2002), AlkB (Welford *et al.*, 2003), phytanoyl CoA hydroxylase (PAHX) (Mukherji *et al.*, 2001), and DAOCS (Lloyd *et al.*, 1999). It has also been used by additional groups for other 2OG-dependent oxygenases, notably the procollagen hydroxylating enzymes, for which the assay was originally developed (Myllyharju and Kivirikko, 1997).

4.1.1. Reagents
A standard PHD2 assay contains the following components in 50 mM Tris-HCl, pH 7.5:

5.7 μM PHD2$_{181-426}$
57 μM HIF-α substrate (discussed later)
160 μM 2OG (5% [^{14}C]-2OG, 95% [^{12}C]-2OG)
80 μM Fe(II)
4 mM ascorbate
1 mM dithiothreitol (DTT)
0.6 mg/ml catalase

All incubations are carried out in a final volume of 100 μl. Care must be taken in preparing the Fe(II) solution since aerobic oxidation to Fe(III) readily occurs. To minimize this, a 500-mM Fe(II) stock is first prepared in 20 mM HCl, which is then subsequently diluted with Milli-Q water (Synthesis^{TM}, Millipore, Billerica, MA) to 4 mM. Fe(II) sulphate is normally used, but it can be replaced by Fe(II) chloride without adverse effect. Several different sources of the HIF-α substrate have been utilized, but the most widely used is a 19mer peptide, HIF-1α 556 to 574 (DLDLEMLAPYIPMDDDFQL, Peptide Protein Research Ltd, Fareham, UK). Peptides of alternative HIF-α isoforms have also been used successfully with this assay, as have recombinant preparations of HIF-α protein (e.g., HIF-1α 530–698). The DTT, ascorbate, and 2OG solutions should be prepared fresh each time in 50 mM Tris-HCl, pH 7.5 to avoid inconsistent results. All the aforementioned components are known to degrade with time and repeated freeze-thawing. These considerations also apply for the additional assays detailed later and shall not be repeated again within this manuscript. The 2OG solution comprises 5% [^{14}C]-2OG and 95% [^{12}C]-2OG. The radiolabeled 2OG is used as a tool to monitor the reaction since, ultimately, [^{14}C]-CO_2 is captured and quantified by scintillation. Samples to be assayed are at least duplicated.

4.1.2. Protocol

The reaction vessel of choice is a 5-ml polycarbonate tube that can be obtained from numerous commercial suppliers (e.g., Thermo Fisher Scientific, Waltham, MA; International Scientific Supplies Ltd., Bradford, UK). A master mix is freshly made that contains the 2OG, Fe(II), ascorbate, DTT, and any additional Tris-HCl buffer required to reach the final 100-μl volume. This mix is placed as a spot at the bottom of the 5-ml tube. Additional separate spots containing the PHD2$_{181-426}$ and HIF-α substrate are also placed at the bottom of the tube. Care must be taken to ensure that no mixing of these spots ensues before the samples are placed in the incubator; otherwise, reaction will occur prematurely in an uncontrolled manner. PHD2$_{181-426}$ is capable of hydroxylating HIF-α at room temperature, albeit at a slower rate than at 37°. A 0.5-ml Eppendorf tube (with the lid removed) containing 200 μl of hyamine hydroxide (MP Biomedicals, Solon, OH) is then carefully placed inside the 5-ml tube, which is then sealed by the use of a rubber septum. The hyamine hydroxide is used to trap any CO_2 released from the 2OG following decarboxylation. Reaction is initiated by shaking the tubes at 200 rpm in an orbital shaker at 37°. After 30 min, the reaction is quenched by the addition of 200 μl of methanol through the rubber septum with a needle and syringe. Samples are subsequently incubated on ice for a further 30 min to enable collection of any released CO_2. In order to quantify CO_2 release, the 0.5-ml Eppendorf tube containing the hyamine hydroxide is carefully removed with tweezers and wiped with a tissue to remove any $[^{14}C]$-2OG that may be adhered to the outside of the Eppendorf vial. The entire Eppendorf vial (containing the hyamine hydroxide) is then placed in a scintillation vial, OptiPhase Safe scintillation fluid is added, and the amount of $[^{14}C]$-CO_2 is quantified with the use of a scintillation counter.

4.1.3. Inhibitor testing

The discussed protocol can be modified readily to screen small molecules or other inhibitors of the enzyme (Hewitson *et al.*, 2007). PHD2$_{181-426}$ can tolerate a dimethyl sulphoxide (DMSO) concentration of 10%, with all substrates tested under the previously mentioned assay conditions. Results from our laboratory actually show that this DMSO concentration is weakly stimulatory to PHD2$_{181-426}$ activity. Compounds to be assayed are dissolved to the required concentration in DMSO (if they are not soluble in aqueous solution) and included with the "master mix" spot previously described. The assay then proceeds as detailed. Methanol can also be used as an alternative small molecule solvent to 10% final concentration with PHD2$_{181-426}$ with no adverse effect.

4.2. Fluorescence derivatization of 2OG

An alternative assay developed in this laboratory (McNeill *et al.*, 2005a) for measuring PHD2 activity is based upon the derivatization of 2OG with *o*-phenylenediamine (OPD; Acros Organics, Geel, Belgium) to produce a fluorescent adduct. This methodology was originally used for the identification of different 2-oxo acids in mixtures and required high performance liquid chromatography (HPLC) separation of the subsequent OPD derivatives (Muhling *et al.*, 2003; Singh *et al.*, 1993). However, for assays of PHD2 (and other 2OG-dependent oxygenases), the only 2-oxo acid present in the reaction mixture is 2OG itself, and, therefore, chromatography separation is not required. This assay has also been used successfully with the 2OG-dependent oxygenases FIH (McNeill *et al.*, 2005a) and PAHX (unpublished results).

4.2.1. Reagents

A standard PHD2 assay combines the following components in 50 mM HEPES, pH 7.0:

4 μM PHD2$_{181-426}$
60 μM HIF-α substrate (discussed later)
300 μM 2OG
150 μM Fe(II)
1 mM DTT
0.6 mg/ml catalase

All concentrations described are final in a 50-μl volume. Both peptide and recombinant protein preparations as HIF-α substrate have been used successfully with this assay. Ascorbate, a stimulatory cofactor for the PHDs (Hewitson et al., 2007; Hirsila et al., 2003), must be omitted from the list of reagents for use in this assay due to its reactivity with OPD. OPD is recrystallized from petroleum ether (120–140°) before use. OPD for use in the assay must be white and flaky. Any color still present after recrystallization will adversely affect the results.

4.2.2. Protocol

PHD2$_{181-426}$ activity is measured by mixing the DTT, 2OG, catalase, substrate, and buffer to a final volume of 44 μl and incubating for 5 min at 37°. Simultaneously, the PHD2$_{181-426}$ is mixed with the Fe(II) solution (final volume 6 μl) and kept at room temperature for 3 min. Reaction is initiated by addition of the PHD2:Fe(II) solution to the assay mix and is stopped by the addition of 100 μl of 0.5-M HCl after 12 min at 37°. Derivatization is achieved by the addition of 50 μl 10-mg/ml OPD in 0.5-M HCl with subsequent heating at 95° for 10 min. Following centrifugation for 5 min, 50 μl of the supernatant is removed and made basic with 30 μl of the 1.25-M NaOH. Reaction of the OPD with the 2-oxo acid moiety of 2OG gives rise to a fluorescent product (3-[2-carboxyethyl]-2 [1H]-quinoxalinone). The resulting fluorescence is measured with a Novostar (BMG LABTECH, Offenburg, Germany) with the excitation filter at 340 nm and the emission filter at 420 nm. As with the 1-[^{14}C]-CO$_2$ assay, inhibitors can be assayed by inclusion in the reaction mix with DMSO/MeOH to a final concentration of 10%.

4.3. Oxygen consumption assay

A further assay for PHD2 involves the continuous measurement of oxygen consumption (Ehrismann et al., 2007). This assay has also been successfully used in this laboratory with other members of the 2OG-dependent oxygenase family, including FIH (Ehrismann et al., 2007), PAHX (unpublished results), and taurine dioxygenase (TauD., unpublished results).

4.3.1. Reagents

A standard PHD2 assay combines the following components in 50 mM Tris-HCl, pH 7.5:

50 μM PHD2$_{181-426}$
50 to 250 μM HIF-α substrate (discussed later)
750 μM 2OG
50 μM Fe(II)

All concentrations described are final in a 200-μl volume with assays typically carried out at either 25 or 37°. Both peptide and recombinant protein preparations as HIF-α substrate have been used successfully with this assay. The addition of DTT and catalase made no significant difference to initial rate measurements and, consequently, were omitted from the final reaction conditions. The inclusion of ascorbate stimulates oxygen consumption in a manner independent of PHD2. Since ascorbate is a reducing agent, molecular oxygen is likely reduced to water in a reaction that is promoted by the presence of transition metals, such as Fe(III) (Xu and Jordan, 1990). Ascorbate is thus excluded from those reagents suitable for use with this assay format.

4.3.2. Protocol

It is essential that the assays are performed under conditions of reduced light due to interference with the fiberoptic probe. Rubber septum-sealed (Wilmad-Labglass, Buena, NJ; NMR rubber caps) reaction vials (Supelco, Bellefont, PA; 0.35-mm HPLC vial glass inserts) are used to allow buffer exchange with the appropriate oxygen gas mixture (to give the required oxygen concentration) without contamination by atmospheric oxygen. The reaction vial is placed in a water bath (water-jacketed holder taken from a Clark-type oxygen electrode from Qubit Systems, Kingston, Ontario), and a FOXY AL-300 probe (Ocean Optics, Dunedin, FL; no additional silicone coating was used, in order to improve response time) is inserted through the vial into a solution containing the 2OG. PHD2 reconstituted with Fe(II) is then injected using a Hamilton syringe into this 2OG solution, which has been equilibrated to the required oxygen concentration. This mixture is allowed to equilibrate for 1 to 2 min, and the reaction is initiated by the injection of appropriate amounts of substrate. Oxygen levels are monitored using a Fiberoptic Oxygen Sensor System (FOXY, Ocean Optics, Dunedin, FL) previously calibrated with both oxygen-saturated and oxygen-depleted (by addition of crystals of Na_2SO_3) aqueous solutions. The initial rate (5–20% conversion) is plotted against the substrate concentration, and the Michaelis-Menton equation is fitted directly to the data using a primary Hanes plot or Systat SigmaPlot 2000. Values are typically quoted as a mean of at least three to five independent measurements. Different oxygen concentrations are obtained by bubbling the 2OG and Tris-HCl buffer solution with either oxygen or nitrogen gas via a Hamilton syringe.

Results with this assay demonstrated the preference of PHD2$_{181-426}$ for longer substrate fragments. K_m values of 21.6 and 1.8 μM were obtained for HIF-1α 556 to 574 and HIF-1α 530 to 698, respectively (Ehrismann *et al.*, 2007). Consequently, a range of substrate concentrations (see "Reagents," Section 4.3.1) dependent upon the identity of the substrate are recommended for use with this assay.

Since both DMSO and methanol are incompatible with the FOXY AL-300 probe, inhibitor screening is limited to those soluble in aqueous solution.

4.4. 1-[¹⁴C]succinate quantification

Although not an assay routinely used in our laboratories, PHD2 activity can also be measured by quantifying the amount of 1-[^{14}C]succinate produced from 5-[^{14}C]-2OG decarboxylation (Cunliffe *et al.*, 1986). This assay requires HPLC (equipped with an online radiochemical detector) separation of the generated succinate from any remaining unreacted 2OG .

5. DIRECT MEASUREMENTS OF PHD2 HYDROXYLATION ACTIVITY

All of the previously described assays rely upon indirect measurement of PHD2 activity, namely quantification of co-substrate/product concentration. The following two assays measure the amount of hydroxylated HIF-α produced after reaction of substrate with PHD2.

5.1. LC/MS identification of hydroxylated HIF-1α 556 to 574

The molecular weight of HIF-1α 556 to 574 (HIF-1α 19mer) is 2254.5 Da and, as such, is amenable to mass spectrometric analyses (Hewitson *et al.*, 2007; McNeill *et al.*, 2005b). Reactions, conditions, and components used with PHD2$_{181-426}$ and this substrate are identical to those described for the 1-[^{14}C]-CO$_2$ capture assay. However, the reaction is quenched by immediate freezing at −80° rather than the addition of methanol, since this organic molecule is not compatible with the chromatography column used for separation purposes. Following quenching, the samples can be stored at −80° until analysis.

A Micromass (now Waters) Q-TOFmicro (quadrupole time-of-flight) mass spectrometer (MS) coupled to an Agilent 1100 capillary liquid chromatography (LC) system with autosampler is used for peptide observation, while a Phenomenex Jupiter C4 column (150 × 4.6 mm, 5-μm particle size) pre-equilibrated with water/0.1% formic acid/5% acetonitrile is used with a gradient of acetonitrile/0.1% formic acid for separation of peptide from the other assay components. The modified and unmodified HIF-1α 19mer peptides can be separated by this method, although long gradients are required for baseline separation, and the +16Da shift is easily seen in the mass spectrum. For both the modified and unmodified HIF-1α 19mers, a potassium adduct can be observed and is associated with a mass shift of $+39$ Da on the respective hydroxylated and non-hydroxylated peptide masses. Reactions using different assay components (e.g., $\pm$ ascorbate) and inhibitor testing can all be compared to a standard reaction mixture to determine the relative effects of such variations. The position of hydroxylation can be verified by tandem mass spectrometry (MS/MS) analyses if required; the stereochemistry of hydroxylation (*trans*-hydroxyproline) has been defined by nuclear magnetic resonance (NMR) analyses.

5.2. pVHL capture assay

The pVHL assay (Tuckerman *et al.*, 2004) is based upon the tight interaction between pVHL (von Hippel-Lindau protein) and the hydroxylated Pro564 of HIF-1α (Epstein *et al.*, 2001; Ivan *et al.*, 2001; Jaakkola *et al.*, 2001). Studies have demonstrated that an HIF-1α 19mer peptide (HIF-1α 556–574) is of sufficient length for recognition by pVHL and the PHDs (Hirsila *et al.*, 2003; Hon *et al.*, 2002). The pVHL capture assay is of sufficient sensitivity to allow the measurement of PHD activity in crude cell extracts and recombinant protein preparations. Since mammalian extracts can be used with this procedure, it has been possible to assay all of the full-length PHD enzymes with this format (Tuckerman *et al.*, 2004).

5.2.1. Reagents

A standard PHD assay combines the following components in $100 \text{ m}M$ Tris-HCL, pH 7.5:

PHD2 extract (discussed later)
$1 \ \mu M$ biotinylated-HIF-1α 19mer peptide (HIF-1α 556–574)
$2 \text{ m}M$ 2OG
$50 \ \mu M$ Fe(II)
$2 \text{ m}M$ ascorbate
$1 \text{ m}M$ DTT
0.3 mg/ml catalase
1 mg/ml bovine serum albumin (BSA)
Stop buffer: $150 \text{ m}M$ NaCl, $20 \text{ m}M$ Tris-HCl, pH 7.5, 0.5% Igepal, $300 \ \mu M$ desferrioxamine mesylate (DFO), $200 \ \mu M$ ethylenediamine tetraacetic acid (EDTA)
Capture buffer: $150 \text{ m}M$ NaCl, $20 \text{ m}M$ Tris-HCl, pH 7.5, 0.5% Igepal, $100 \ \mu M$ DFO

All concentrations described are final, typically in a 50-μl volume. [^{35}S]-Methionine-labeled proteins are produced by *in vitro* transcription/ translation (IVTT) reactions using TNT Quick-Coupled Rabbit Reticulocyte Lysate (Promega, Madison, WI).

The system is calibrated by mixing known amounts of hydroxylated and non-hydroxylated biotinylated HIF-1α 19mer peptide with [^{35}S]pVHL.

The signal obtained from captured [^{35}S]-pVHL can then be correlated with the known amounts of hydroxylated HIF-1α 19mer peptide. Linear calibration curves are obtained and can be used for subsequent calculation.

Determination of PHD2 concentration in lysates is via use of purified FLAG-PHD2 from Sf9 cells. Monoclonal antibodies to PHD2 are used to calibrate the immunoblot signal with this purified material and then compared with the signal obtained using a PHD2 lysate. For PHD1 and -3 however, this procedure cannot be followed, since production of highly purified enzyme has so far proved problematic in this laboratory. Instead, [^{35}S]methionine preparations of PHD1 and -3 are compared to known amounts of PHD2 to determine the absolute concentration of each. Using this methodology, the concentration of very small quantities of PHD enzyme can be determined (as low as 4 fmol).

5.2.2. Protocol

A reaction mix is made that contains the biotinylated-HIF-1α 19mer peptide, 2OG, Fe(II), ascorbate, DTT, BSA, catalase, and any additional Tris-HCl buffer required to reach the final 50-μl volume. Reactions are initiated by the addition of PHD enzyme followed by incubation at 37° for 6 min. Initial reaction rates are linear over this time period. The reaction is quenched by dilution in an equal volume of ice-cold stop buffer. The DFO and EDTA present in this solution are known iron chelators, and, hence, further reaction of the PHDs is prevented, since iron is a crucial cofactor for these enzymes. Biotinylated-HIF-1α 19mer peptide (5 μl of the stopped reaction) is captured using 2.7×10^6 streptavidin–coated magnetic beads (Dynabeads M-280 Streptavidin, Dynal Biotech, Carlsbad, CA) in capture buffer. Following incubation on ice for 30 min, the stopped reaction mix is removed and the beads washed once with 500 μl of capture buffer. The peptide-coated beads are then incubated with excess [^{35}S]pVHL at 4° for 60 min. Following two washes (750 μl capture buffer), the captured [^{35}S]pVHL is quantified by SDS-PAGE and autoradiography using a phosphorimager.

Currently, only water-soluble inhibitors have been screened with this assay technique made by dissolving in 100 mM Tris-HCl, pH 7.5 and inclusion with the main reaction mix. The DMSO/MeOH tolerance with this assay format is currently unknown.

6. BINDING ASSAYS

As an alternative to catalysis-based assays, we have developed a binding assay (Ehrismann *et al.*, 2007) that utilizes the technique of surface plasmon resonance (SPR) to investigate the binding properties of different HIF-α substrates to PHD2$_{181-426}$. Both His$_6$-HIF-1α 344 to 503 (NODD) and

His_6-HIF-1α 530 to 698 (CODD) can be covalently coupled to a CM5 sensor chip using the BIACore (Piscataway, NJ) Amine Coupling kit at pH 4.5 for such a purpose. Binding studies on a BIAcore 2000^{TM} sensor are initiated by the injection of 15 μl of PHD2$_{181-426}$ at 50 μl/min in 10 mM HEPES, pH 7.4, 150 mM NaCl, 0.005% surfactant-P20, over both control and experimental cells. The temperature is held constant at 25°. Experiments with PHD2$_{181-426}$ have been performed over a range of concentrations to allow binding data analysis.

7. COMPARISON OF ASSAY FORMATS

The assays can essentially be divided into those that directly measure HIF-1α hydroxylation (LC/MS and pVHL capture assay), those that use indirect means to quantify PHD2 activity (1-[^{14}C]-CO$_2$ capture, fluorescence derivatization, oxygen consumption assay), and binding assays (SPR). While the first is normally the preferred option for assays aimed at investigating substrate specificity or inhibition studies, limitations do exist with this type of assay (described later); hence, indirect measurements of PHD2 activity are still routinely used. A main drawback with the use of indirect measurements is the potential for (partial) uncoupling of 2OG oxidation from hydroxylation with this family of enzymes. Under appropriate circumstances (e.g., misfolded protein, a suboptimal substrate) the 2OG-dependent oxygenases can catalyze the oxygen-dependent turnover of 2OG that is not directly linked to prime substrate hydroxylation (Costas *et al.*, 2004; Hausinger, 2004; Hewitson *et al.*, 2005). It may appear that the enzyme is displaying activity, but, in fact, each cycle is nonproductive with respect to hydroxylation. This uncoupled turnover can, however, be partly controlled for in the previously described assay formats by the use of a "no substrate control." Confirmation of product formation (e.g., by MS analyses) is usually required, even if controls are used.

The 1-[^{14}C]-CO$_2$ and pVHL capture assays are well-established means for assaying the PHD enzymes and have been used by a variety of groups successfully. In both cases, radiochemicals are employed, which may limit use in some laboratories. The pVHL assay is particularly useful in that, not only can small quantities of PHD be assayed (4–10 fmols in 10–25 μl), but non-purified material (i.e., lysates) can also be used. However, this assay is limited by the number of samples that can be processed in a given time (~10–20 per day). Note also that peptide-derived substrates are required when the assay is used in fully quantitative format, and, to allow calibration of the system, both the hydroxylated and non-hydroxylated peptides must be synthesized. This assay is, however, potentially suitable for modification to enable its use in a high-throughput format.

All other assay formats require the use of purified components, both in terms of the enzyme and the substrate. High background readings in 2OG turnover, 2-oxo acid content, and oxygen consumption may result from use of lysates with the $1\text{-}[^{14}C]\text{-}CO_2$ capture, fluorescence derivatization, and oxygen consumption assay.

Ascorbate is incompatible with two of the current assay formats: the fluorescence derivatization and oxygen consumption assays. PHD2 requires ascorbate (Hewitson *et al.*, 2007; Hirsila *et al.*, 2003) to stimulate optimal *in vitro* activity as observed with several other members of the 2OG-dependent oxygenase family (e.g., procollagen prolyl hydroxylase [Myllyla *et al.*, 1984]). It must be assumed that, with these assay techniques, the activity of PHD2 is less than maximal, which could potentially affect the measurement of kinetic parameters. It is also why, in the case of the oxygen consumption assay, such high (millimolar) concentrations of PHD2 and HIF-1α are required with the current sensors.

In terms of screening for inhibitors with purified enzymes, the $1\text{-}[^{14}C]\text{-}CO_2$ capture assay can be routinely used. This assay is less reliable with partially purified material due to the potential presence of interfering enzymes that utilize 2OG as a substrate. Thirty inhibitors at a single point concentration can be screened in a day, which is higher throughput than can be achieved with most of the other reported techniques. Furthermore, small molecules containing oxo-acid functionality cannot be used with the fluorescence derivatization assay, while both the oxygen consumption and pVHL capture assays are currently limited to those molecules soluble in aqueous solution.

The LC/MS assay has proved to be extremely useful in terms of validating peptides as substrates for PHD2. However, this technique can be time-consuming when considering the long (>30 min) run times required per sample. Inhibitor studies using this technique have yet to be reported, but should be a possibility with improving MS techniques.

In conclusion, still no definitive, readily-usable assay has been described for the PHDs nor, in fact, for the 2OG-dependent oxygenase family as a whole. A useful advance would be the development of a generic continuous assay employing routine spectroscopic methodology and without the use of coupled enzymes. Despite the lack of such an assay, in the case of the PHDs, careful consideration of the assay formats described here should allow for the selection of the technique most appropriate for the required use.

REFERENCES

Appelhoff, R. J., Tian, Y. M., Raval, R. R., Turley, H., Harris, A. L., Pugh, C. W., Ratcliffe, P. J., and Gleadle, J. M. (2004). Differential function of the prolyl hydroxylases

PHD1, PHD2, and PHD3 in the regulation of hypoxia-inducible factor. *J. Biol. Chem.* **279,** 38458–38465.

Bacon, N. C. M., Wappner, P., O'Rourke, J. F., Bartlett, S. M., Shilo, B., Pugh, C. W., and Ratcliffe, P. J. (1998). Regulation of the *Drosophila* bHLH-PAS protein Sima by hypoxia: Functional evidence for homology with mammalian HIF-1 α. *Biochem. Biophys. Res. Commun.* **249,** 811–816.

Berra, E., Benizri, E., Ginouves, A., Volmat, V., Roux, D., and Pouyssegur, J. (2003). HIF prolyl-hydroxylase 2 is the key oxygen sensor setting low steady-state levels of HIF-1 α in normoxia. *EMBO J.* **22,** 4082–4090.

Bruick, R. K., and McKnight, S. L. (2001). A conserved family of prolyl-4-hydroxylases that modify HIF. *Science* **294,** 1337–1340.

Clifton, I., McDonough, M. A., Ehrismann, D., Kershaw, N. J., Granatino, N., and Schofield, C. J. (2006). Structural studies on 2-oxoglutarate oxygenases and related double-stranded β-helix fold proteins. *J. Inorg. Biochem.* **100,** 644–669.

Costas, M., Mehn, M. P., Jensen, M. P., and Que, L., Jr. (2004). Dioxygen activation at mononuclear nonheme iron active sites: Enzymes, models, and intermediates. *Chem. Rev.* **104,** 939–986.

Cunliffe, C. J., Franklin, T. J., and Gaskell, R. M. (1986). Assay of prolyl 4-hydroxylase by the chromatographic determination of C-14 succinic acid on ion-exchange minicolumns. *Biochem. J.* **240,** 617–619.

Ehrismann, D., Flashman, E., Genn, D. N., Mathioudakis, N., Hewitson, K. S., Ratcliffe, P. J., and Schofield, C. J. (2007). Studies on the activity of the hypoxia-inducible-factor hydroxylases using an oxygen consumption assay. *Biochem. J.* **401,** 227–234.

Epstein, A. C. R., Gleadle, J. M., McNeill, L. A., Hewitson, K. S., O'Rourke, J., Mole, D. R., Mukherji, M., Metzen, E., Wilson, M. I., Dhanda, A., Tian, Y. M., Masson, N., *et al.* (2001). *C. elegans* EGL-9 and mammalian homologs define a family of dioxygenases that regulate HIF by prolyl hydroxylation. *Cell* **107,** 43–54.

Hausinger, R. P. (2004). Fe(II)/α-ketoglutarate-dependent hydroxylases and related enzymes. *Crit. Rev. Biochem. Mol. Biol.* **39,** 21–68.

Hewitson, K. S., Granatino, N., Welford, R. W. D., McDonough, M. A., and Schofield, C. J. (2005). Oxidation by 2-oxoglutarate oxygenases: Non-haem iron systems in catalysis and signalling. *Philos. Transact. A Math. Phys. Eng. Sci.* **363,** 807–828.

Hewitson, K. S., Lienard, B. M. R., McDonough, M. A., Clifton, I. J., Butler, D., Soares, A. S., Oldham, N. J., McNeill, L. A., and Schofield, C. J. (2007). Structural and mechanistic studies on the inhibition of the hypoxia-inducible transcription factor hydroxylases by tricarboxylic acid cycle intermediates. *J. Biol. Chem.* **282,** 3293–3301.

Hewitson, K. S., McNeill, L. A., Riordan, M. V., Tian, Y. M., Bullock, A. N., Welford, R. W., Elkins, J. M., Oldham, N. J., Bhattacharya, S., Gleadle, J. M., Ratcliffe, P. J., Pugh, C. W., *et al.* (2002). Hypoxia-inducible factor (HIF) asparagine hydroxylase is identical to factor inhibiting HIF (FIH) and is related to the cupin structural family. *J. Biol. Chem.* **277,** 26351–26355.

Hirsila, M., Koivunen, P., Gunzler, V., Kivirikko, K. I., and Myllyharju, J. (2003). Characterisation of the human prolyl 4-hydroxylases that modify the hypoxia-inducible factor HIF. *J. Biol. Chem.* **278,** 30772–30780.

Hon, W. C., Wilson, M. I., Harlos, K., Claridge, T. D. W., Schofield, C. J., Pugh, C. W., Maxwell, P. H., Ratcliffe, P. J., Stuart, D. I., and Jones, E. Y. (2002). Structural basis for the recognition of hydroxyproline in HIF-1 α by pVHL. *Nature* **417,** 975–978.

Ivan, M., Kondo, K., Yang, H. F., Kim, W., Valiando, J., Ohh, M., Salic, A., Asara, J. M., Lane, W. S., and Kaelin, W. G. (2001). HIF α targeted for VHL-mediated destruction by proline hydroxylation: Implications for O_2 sensing. *Science* **292,** 464–468.

Jaakkola, P., Mole, D. R., Tian, Y. M., Wilson, M. I., Gielbert, J., Gaskell, S. J., von Kriegsheim, A., Hebestreit, H. F., Mukherji, M., Schofield, C. J., Maxwell, P. H., Pugh, C. W., et al. (2001). Targeting of HIF-α to the von Hippel-Lindau ubiquitylation complex by O_2-regulated prolyl hydroxylation. Science 292, 468–472.

Jiang, H. Q., Guo, R., and Powell-Coffman, J. A. (2001). The Caenorhabditis elegans hif-1 gene encodes a bHLH-PAS protein that is required for adaptation to hypoxia. Proc. Natl. Acad. Sci. USA 98, 7916–7921.

Lando, D., Peet, D. J., Whelan, D. A., Gorman, J. J., and Whitelaw, M. L. (2002). Asparagine hydroxylation of the HIF transactivation domain: A hypoxic switch. Science 295, 858–861.

Lavista-Llanos, S., Centanin, L., Irisarri, M., Russo, D. M., Gleadle, J. M., Bocca, S. N., Muzzopappa, M., Ratcliffe, P. J., and Wappner, P. (2002). Control of the hypoxic response in Drosophila melanogaster by the basic helix-loop-helix PAS protein similar. Mol. Cell. Biol. 22, 6842–6853.

Lieb, M. E., Menzies, K., Moschella, M. C., Ni, R., and Taubman, M. B. (2002). Mammalian EGLN genes have distinct patterns of mRNA expression and regulation. Biochem. Cell. Biol. 80, 421–426.

Lloyd, M. D., Lee, H. J., Harlos, K., Zhang, Z. H., Baldwin, J. E., Schofield, C. J., Charnock, J. M., Garner, C. D., Hara, T., van Scheltinga, A. C. T., Valegard, K., Viklund, J. A. C., et al. (1999). Studies on the active site of deacetoxycephalosporin C synthase. J. Mol. Biol. 287, 943–960.

Masson, N., Willam, C., Maxwell, P. H., Pugh, C. W., and Ratcliffe, P. J. (2001). Independent function of two destruction domains in hypoxia-inducible factor-α chains activated by prolyl hydroxylation. EMBO J. 20, 5197–5206.

McDonough, M. A., Li, V., Flashman, E., Chowdhury, R., Mohr, C., Lienard, B. M. R., Zondlo, J., Oldham, N. J., Clifton, I. J., Lewis, J., McNeill, L. A., Kurzeja, R. J. M., et al. (2006). Cellular oxygen sensing: Crystal structure of hypoxia-inducible factor prolyl hydroxylase (PHD2). Proc. Natl. Acad. Sci. USA 103, 9814–9819.

McNeill, L. A., Bethge, L., Hewitson, K. S., and Schofield, C. J. (2005a). A fluorescence-based assay for 2-oxoglutarate-dependent oxygenases. Anal. Biochem. 336, 125–131.

McNeill, L. A., Flashman, E., Buck, M. R. G., Hewitson, K. S., Clifton, I. J., Jeschke, G., Claridge, T. D. W., Ehrismann, D., Oldham, N. J., and Schofield, C. J. (2005b). Hypoxia-inducible factor prolyl-hydroxylase 2 has a high affinity for ferrous iron and 2-oxoglutarate. Mol. Biosyst. 1, 321–324.

Muhling, J., Fuchs, M., Campos, M. E., Gonter, J., Engel, J. M., Sablotzki, A., Menges, T., Weiss, S., Dehne, M. G., Krull, M., and Hempelmann, G. (2003). Quantitative determination of free intracellular α-keto acids in neutrophils. J. Chromatogr. B Analyt. Technol. Biomed. Life Sci. 789, 383–392.

Mukherji, M., Chien, W., Kershaw, N. J., Clifton, I. J., Schofield, C. J., Wierzbicki, A. S., and Lloyd, M. D. (2001). Structure-function analysis of phytanoyl-CoA 2-hydroxylase mutations causing Refsum's disease. Hum. Mol. Genet. 10, 1971–1982.

Myllyharju, J., and Kivirikko, K. I. (1997). Characterization of the iron- and 2-oxoglutarate-binding sites of human prolyl-4-hydroxylase. EMBO J. 16, 1173–1180.

Myllyla, R., Majamaa, K., Gunzler, V., Hanauskeabel, H. M., and Kivirikko, K. I. (1984). Ascorbate is consumed stoichiometrically in the uncoupled reactions catalyzed by prolyl 4-hydroxylase and lysyl hydroxylase. J. Biol. Chem. 259, 5403–5405.

Pugh, C. W., and Ratcliffe, P. J. (2003). Regulation of angiogenesis by hypoxia: Role of the HIF system. Nat. Med. 9, 677–684.

Searls, T., Butler, D., Chien, W., Mukherji, M., Lloyd, M. D., and Schofield, C. J. (2005). Studies on the specificity of unprocessed and mature forms of phytanoyl-CoA 2-hydroxylase and mutation of the iron binding ligands. J. Lipid Res. 46, 1660–1667.

Semenza, G. L. (2000). HIF-1 and human disease: One highly involved factor. *Genes Dev.* **14,** 1983–1991.

Singh, B. K., Szamosi, I., and Shaner, D. (1993). A high-performance liquid-chromatography assay for threonine serine dehydratase. *Anal. Biochem.* **208,** 260–263.

Tuckerman, J. R., Zhao, Y. G., Hewitson, K. S., Tian, Y. M., Pugh, C. W., Ratcliffe, P. J., and Mole, D. R. (2004). Determination and comparison of specific activity of the HIF-prolyl hydroxylases. *FEBS Lett.* **576,** 145–150.

Valegård, K., van Scheltinga, A. C., Lloyd, M. D., Hara, T., Ramaswamy, S., Perrakis, A., Thompson, A., Lee, H. J., Baldwin, J. E., Schofield, C. J., Hajdu, J., and Andersson, I. (1998). Structure of a cephalosporin synthase. *Nature* **394,** 805–809.

Wappner, P., Irisarri, M., Lavista-Llanos, S., Mondotte, J. A., and Centanin, L. (2003). The hypoxic response in *Drosophila* depends on the bHLH-PAS protein similar and the prolyl-4-hydroxylase, fatiga, that operates as an oxygen sensor. *Dev. Biol.* **259,** 520–524.

Welford, R. W. D., Schlemminger, I., McNeill, L. A., Hewitson, K. S., and Schofield, C. J. (2003). The selectivity and inhibition of AlkB. *J. Biol. Chem.* **278,** 10157–10161.

Wenger, R. H. (2002). Cellular adaptation to hypoxia: O_2-sensing protein hydroxylases, hypoxia-inducible transcription factors, and O_2-regulated gene expression. *FASEB J.* **16,** 1151–1162.

Willam, C., Maxwell, P. H., Nichol, L., Lygate, C., Tian, Y.-M., Bernhardt, W., Wiesener, M., Ratcliffe, P. J., Eckardt, K. U., and Pugh, C. W. (2006). HIF prolyl hydroxylases in the rat; organ distribution and changes in expression following hypoxia and coronary artery ligation. *J. Mol. Cell. Cardiol.* **41,** 68–77.

Xu, J. H., and Jordan, R. B. (1990). Kinetics and mechanism of the reaction of aqueous iron(Iii) with ascorbic-acid. *Inorg. Chem.* **29,** 4180–4184.

Yu, F., White, S. B., Zhao, Q., and Lee, F. S. (2001). HIF-1 α binding to VHL is regulated by stimulus-sensitive proline hydroxylation. *Proc. Natl. Acad. Sci. USA* **98,** 9630–9635.

CHAPTER THREE

DETERMINATION AND MODULATION OF PROLYL-4-HYDROXYLASE DOMAIN OXYGEN SENSOR ACTIVITY

Renato Wirthner,* Kuppusamy Balamurugan,* Daniel P. Stiehl,*
Sandra Barth,* Patrick Spielmann,* Felix Oehme,[†] Ingo Flamme,[†]
Dörthe M. Katschinski,[‡] Roland H. Wenger,* *and*
Gieri Camenisch*

Contents

1. Introduction   44
2. Production of Functionally Active PHDs   48
3. Determination of PHD Activity by VHL Binding to Peptides Derived
   from the HIF-1α ODD Domain   48
4. Determination of Prolyl-4-Hydroxylation by Oxidative
   Decarboxylation of 2-Oxoglutarate   51
5. Crude Tissue Extracts are not a Suitable Source of PHD Activity for
   the 2-Oxoglutarate Conversion Assay   53
6. Thin Layer Chromatography to Assess the Purity of
   [5-^{14}C]2-Oxoglutarate   53
7. Application of the 2-Oxoglutarate Conversion Assay to
   Protein Targets   55
8. Conclusions   55
Acknowledgments   57
References   57

Abstract

The prolyl-4-hydroxylase domain (PHD) oxygen sensor proteins hydroxylate hypoxia-inducible transcription factor (HIF)-alpha (α) subunits, leading to their subsequent ubiquitinylation and degradation. Since oxygen is a necessary

* Institute of Physiology and Zürich Center for Integrative Human Physiology (ZIHP), University of Zürich, Zürich, Switzerland
† Institute for Cardiovascular Research, Bayer HealthCare, Wuppertal, Germany
‡ Department of Heart and Circulatory Physiology, Center of Physiology and Pathophysiology, Georg-August University Göttingen, Göttingen, Germany

Methods in Enzymology, Volume 43
ISSN 0076-6879, DOI: 10.1016/S0076-6879(07)35003-9

cosubstrate, a reduction in oxygen availability (hypoxia) decreases PHD activity and, subsequently, HIF-α hydroxylation. Non-hydroxylated HIF-α cannot be bound by the ubiquitin ligase von Hippel-Lindau tumor suppressor protein (pVHL), and HIF-α proteins thus become stabilized. HIF-α then heterodimerizes with HIF-beta (β) to form the functionally active HIF transcription factor complex, which targets approximately 200 genes involved in adaptation to hypoxia. The three HIF-α PHDs are of a different nature compared with the prototype collagen prolyl-4-hydroxylase, which hydroxylates a mass protein rather than a rare transcription factor. Thus, novel assays had to be developed to express and purify functionally active PHDs and to measure PHD activity *in vitro*. A need also exists for such assays to functionally distinguish the three different PHDs in terms of substrate specificity and drug function. We provide a detailed description of the expression and purification of the PHDs as well as of an HIF-α–dependent and a HIF-α–independent PHD assay.

1. INTRODUCTION

Cells sense changes in environmental oxygen availability by a group of enzymes that directly control the cellular response to low oxygen by destabilizing HIF-α subunits, the master transcriptional regulators of the hypoxic response. These oxygen-sensing enzymes have alternatively been termed PHD, HIF prolyl hydroxylase (HPH), or egg-laying defective nine homolog (EGLN). Up to date, three family members are known: PHD1/HPH3/EGLN2, PHD2/HPH2/EGLN1, and PHD3/HPH1/EGLN3 (Bruick, 2000; Epstein *et al.*, 2001; Ivan *et al.*, 2002). PHDs hydroxylate HIF-1α and HIF-2α at two distinct proline residues within the HIF-α oxygen-dependent degradation (ODD) domain (Fig. 3.1A). Under normoxic conditions, prolyl-4-hydroxylase allows binding of pVHL, leading to polyubiquitinylation and proteasomal destruction (Schofield and Ratcliffe, 2004; Wenger, 2002). Under hypoxic conditions, prolyl-4-hydroxylase is reduced, and HIF-1α and HIF-2α become stabilized, heterodimerize with the constitutively expressed HIF-1β subunit aryl hydrocarbon receptor nuclear translocator (ARNT), and regulate the expression of a large number of effector genes involved in adaptation to low oxygen (Wenger *et al.*, 2005). In addition, factor-inhibiting HIF (FIH) hydroxylates an asparagine residue within the C-terminal transactivation domain. Oxygen-dependent asparagine hydroxylation blocks the recruitment of the CREB-binding protein (CBP)/p300 transcriptional coactivators and thereby regulates the transcriptional activity of HIFs (Hewitson *et al.*, 2002; Lando *et al.*, 2002; Mahon *et al.*, 2001).

Prolyl-4-hydroxylase domain proteins do not represent static oxygen sensor molecules, but rather are highly regulated themselves. Importantly, PHD2 and PHD3, but not PHD1, have been reported to be hypoxically induced at both the messenger RNA (mRNA) and protein levels (Epstein

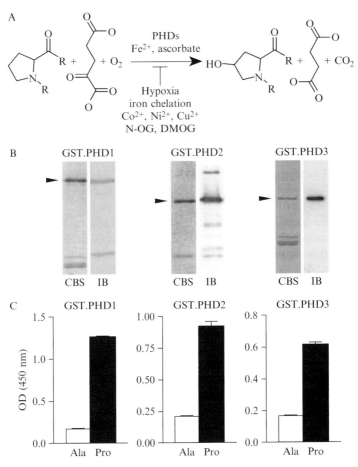

Figure 3.1 (A) Reaction mechanism of prolyl-4-hydroxylase domain (PHD)-mediated oxygen-dependent prolyl-4-hydroxylation by oxidative decarboxylation of the cosubstrate 2-oxoglutarate. Inhibitors of this reaction are indicated. N-OG, N-oxalylglycine; DMOG, dimethyloxalylglycine. (B) Purification of glutathione S-transferase (GST)-tagged PHD1, PHD2, and PHD3 expressed in baculovirus-infected *Spodoptera frugiperda* (Sf) 9 insect cells by glutathione affinity chromatography. The purified GST-PHD proteins were analyzed by sodium dodecyl sulfate polyacrylamide gel electrophoresis (SDS-PAGE) and Coomassie blue staining (CBS) or immuno-blotting (IB) using anti-GST antibodies for the detection of GST-PHD1 and GST-PHD2 or anti-PHD3 antibodies for the detection of PHD3. (C) Hydroxylation activities of GST-PHD preparations determined by a von Hippel-Lindau tumor suppressor protein (pVHL)/elongin B/elongin C (VBC)-binding assay, as described in the text. Wild-type (Pro) and P564A mutant (Ala) hypoxia-inducible transcription factor (HIF)-1α oxygen-dependent degradation (ODD) domain-derived peptides were used as hydroxylation substrates.

et al., 2001). Accordingly, elevated PHD2 and PHD3 levels have been demonstrated in a broad panel of established cancer cell lines (Appelhoff *et al.*, 2004). Functional hypoxia response elements (HREs) are located in the promoter region of the human *PHD2* gene as well as in the first intron of the human *PHD3* gene, demonstrating that *PHD2* and *PHD3* are HIF target genes themselves (Metzen *et al.*, 2005; Pescador *et al.*, 2005). Because the essential cofactor oxygen is basically lacking under hypoxic conditions, the HIF-dependent hypoxic increase in PHD abundance is somehow paradoxical; it has been suggested that PHD induction plays a role in accelerating the termination of the HIF response following reoxygenation (Appelhoff *et al.*, 2004; Aprelikova *et al.*, 2004; Epstein *et al.*, 2001; Marxsen *et al.*, 2004). Indeed, biochemical *in vitro* studies revealed K_m values of purified PHDs for oxygen close to the oxygen partial pressure (pO_2) in air, suggesting that the kinetics of specific HIF-α hydroxylation under hypoxic conditions are rather slow (Ehrismann *et al.*, 2006; Hirsilä *et al.*, 2003). Of note, these K_m values are critically dependent on the length of the target peptide (Koivunen *et al.*, 2006). However, tissues *in situ* have to deal with a great variability of generally very low pO_2 values, even when the inspiratory pO_2 is considered to be "normoxic." We recently showed that HIF-dependent regulation of PHD levels adapts the PHD–HIF oxygen sensing system to a given tissue pO_2 rather than simply accelerating HIF-α destruction following reoxygenation (Stiehl *et al.*, 2006). Such a self-regulatory loop might define a tissue-specific threshold for HIF-α activation as a function of local pO_2.

In addition to transcriptional regulation, PHD levels are also regulated by protein–protein interactions: the ubiquitin ligase Siah2 regulates PHD1 and PHD3, but not PHD2, protein stability (Nakayama *et al.*, 2004); PHD3, but not PHD1 or PHD2, appears to be a substrate for the TRiC chaperonin (Masson *et al.*, 2004); OS-9 apparently is simultaneously interacting with both HIF-α and PHD2 or PHD3, but not PHD1, thereby enhancing HIF-α hydroxylation and degradation (Baek *et al.*, 2005); and Morg1 might provide the molecular scaffold for HIF-α interaction specifically with PHD3 (Hopfer *et al.*, 2006). On the other hand, PHD2 has been shown to inhibit the transactivation function of HIF-1α in VHL-deficient cells (To and Huang, 2005), a process that might involve PHD2-dependent recruitment of ING4, a likely component of a chromatin-remodeling complex (Ozer *et al.*, 2005).

These findings suggest two additional layers in oxygen signaling: first, abundance and function of PHDs actually can be regulated; and second, the three different PHDs are regulated in nonidentical ways, further supporting their non-redundant roles in oxygen sensing. In fact, while all three PHDs can hydroxylate HIF-α with similar efficiency *in vitro*, PHD2 has been suggested to play the main role for normoxic HIF-α turnover in cultured cells (Berra *et al.*, 2003). Consistent with these findings, PHD2, but neither PHD1 nor PHD3, knockout mice die during embryonic development (Takeda *et al.*, 2006). Interestingly, a family with erythrocytosis due to a mutation in the gene encoding PHD2 has recently been reported (Percy et al., 2006). The three PHDs are expressed in most organs; however, strikingly high levels of PHD3 and PHD1 mRNA are expressed in the heart and testis, respectively (Stiehl et al., 2006; Willam et al., 2006).

PHD function can also be regulated by a number of endogenous small molecules as well as by a number of clinically relevant drugs; ascorbate (Knowles et al., 2003), transition metals (Hirsilä et al., 2005; Martin et al., 2005), Krebs cycle intermediates (Dalgard et al., 2004; Lu et al., 2005; Selak et al., 2005), and reactive oxygen species (ROS), including NO (Berchner-Pfannschmidt et al., 2006; Gerald et al., 2004; Metzen et al., 2003), have been shown to influence or completely block the activity of the PHDs. Thus, molecular cross-talks appear to exist between oxygen homeostasis and transition metal homeostasis as well as cellular metabolism. However, it is not completely understood how various transition metals and ascorbate interfere with PHD and FIH function. The transition metals do not simply replace or oxidize the ferrous iron in the active center of the hydroxylases (Hirsilä et al., 2005; Martin et al., 2005); at least cobalt and nickel might deplete intracellular ascorbate (Karaczyn et al., 2006; Salnikow et al., 2004). As known for collagen hydroxylation, ascorbate is essential in reducing ferric iron in the active center of PHDs and FIH, which occurs when oxidative decarboxylation of 2-oxoglutarate is uncoupled from target hydroxylation (McNeill et al., 2005; Myllylä et al., 1984).

In addition to kinase signaling pathways, numerous reports appeared about an involvement of ROS in HIF-α protein stabilization by either growth stimuli or hypoxia. Inconclusive data were obtained about the source(s) of the ROS and whether hypoxia leads to an increase or decrease in ROS levels (Wenger, 2000). However, ROS do have the potential to interfere with the complex process of protein hydroxylation. Indeed, the increase in ROS in $junD^{-/-}$ cells leads to a decrease in PHD activity and hence to HIF-1 α accumulation (Gerald et al., 2004). Apart from direct interference of ROS with the active center of oxygen sensing protein hydroxylases, the redirection of oxygen from mitochondria towards the protein hydroxylases might contribute to these effects (Hagen et al., 2003; Wenger, 2006).

Due to the large therapeutic potential of PHD inhibitors in the treatment of anemic and ischemic diseases, several attempts to develop PHD antagonists are in progress. Currently, N-oxyalylglycine (N–OG) and its cell-permeable derivative dimethyloxalylglycine (DMOG) are the commercially available PHD inhibitors of choice for experimental purposes. Apart from these cosubstrate competitive inhibitors, iron chelators are very well known to induce HIF by PHD inhibition. These include deferoxamine (DFX) and ciclopirox olamine (CPX), two clinically relevant hydroxamic acid iron chelators (Linden et al., 2003). In order to functionally investigate the potency and mechanisms of available and future PHD inhibitors, a need exists for reliable methods to generate PHD proteins and to determine their activity.

2. PRODUCTION OF FUNCTIONALLY ACTIVE PHDs

Since bacterially expressed PHDs retain only little hydroxylase activity, PHDs were expressed and purified from baculovirally infected insect cells. To facilitate purification, the PHDs were tagged with either glutathione S-transferase (GST), maltose-binding protein (MBP), or His_6. According to our experience, best results were obtained with GST-tagged PHD proteins. Thus, GST-tagged expression vectors were prepared by LR clonase-mediated homologous recombination of the corresponding pENTR vector containing PHD1, PHD2, or PHD3 complementary DNA (cDNA) inserts, with the pDEST20 vector (Invitrogen, Carlsbad, CA). Recombined plasmids were transfected into the *Escherichia coli* strain DH10BAC, and the resulting bacmid plasmids were used to generate baculovirus stocks according to the manufacturer's instructions (Invitrogen). *Spodoptera frugiperda* (Sf) 9 cells were infected with baculovirus and cultured in Grace's insect medium (Invitrogen) at $27°$ in a humidified incubator. Infected cells were grown for 80 to 110 h, collected by centrifugation at $700 \times g$ for 10 min at $4°$, and washed with ice-cold phosphate buffered saline (PBS). Lysis was performed on ice for 10 min with 0.1% NP-40 in a buffer containing 10 mM Tris-HCl, pH 7.4, 100 mM NaCl, 100 mM glycine, 100 μM dithiothreitol (DTT), and ethylenediaminetetraacetic acid (EDTA)-free complete protease inhibitor cocktail (Roche Applied Science, Indianapolis, IN). Crude lysate was cleared by centrifugation at $20,000 \times g$ for 20 min at $4°$, and the supernatant was incubated with glutathione-Sepharose beads (previously washed with PBS) for at least 2 h at $4°$ with gentle agitation. After washing three times with PBS, the protein was eluted with 15 mM reduced glutathione (GSH) in 50 mM Tris-HCl, pH 8.0, and 2 μM $FeSO_4$. Eluted proteins were supplemented with 5% glycerol and stored in aliquots at $-80°$. Purity was checked by sodium dodecyl sulfate polyacrylamide gel electrophoresis (SDS-PAGE) followed by Coomassie staining or immunoblotting (see Fig. 3.1B). Activity was routinely determined by the VHL binding assay described later (see Fig. 3.1C).

3. DETERMINATION OF PHD ACTIVITY BY VHL BINDING TO PEPTIDES DERIVED FROM THE HIF-1α ODD DOMAIN

Enzymatic activities of recombinant PHD proteins were determined by an *in vitro* hydroxylation assay based on a 96-well format, as described previously (Oehme *et al.*, 2004). Briefly, biotinylated synthetic peptides derived from human HIF-1α amino acids 556 to 574 (biotin-DLDLEALAPYIPADDDFQL), either wild-type, P564A mutant, or hydroxylated (Hyp564), were bound to NeutrAvidin-coated 96-well plates (Pierce, Rockford, IL). All methionine residues were mutated to alanine residues in these peptides (M561A, M568A). Hydroxylase reactions using purified recombinant GST-tagged PHDs or cellular extracts were carried out for 1 h at room temperature. A polycistronic bacterial expression vector for His_6-tagged and thioredoxin-tagged pVHL/elongin B/elongin C (VBC) complex (kindly provided by S. Tan, Pennsylvania State University, University Park, PA) was used to express VBC in *E. coli* strain BL21AI, followed by purification using nickel affinity chromatography (GE Healthcare, Buckinghamshire, UK), followed by anion exchange chromatography using a Hi Trap Q FF column (GE Healthcare) and gel filtration Hi Prep 26/10 desalting column; GE Healthcare). The VBC complex was allowed to bind to the hydroxylated peptides for 15 min, anti-thioredoxin antibodies were added for 30 min, and horseradish peroxidase-coupled anti-rabbit antibodies (Sigma, St. Louis, MO) were added for 30 min. Bound VBC complex was detected using the 3,3′,5,5′-tetramethylbenzidine (TMB) substrate kit (Pierce). The peroxidase reaction was stopped by adding one volume 2 M H_2SO_4, and absorbance was determined at 450 nm in a microplate photometer. This assay is routinely used to determine the activity of the three GST-tagged PHD enzymes purified from Sf 9 insect cells (see Fig. 3.1C). Interassay comparability was guaranteed by calibration of each experiment to an internal standard curve using increasing fractions of synthetic hydroxyproline–containing peptides (Fig. 3.2A). When performed with pre-equilibrated solutions in a hypoxic glove box (Inviv02 400; Ruskinn Technology, Pencoed, UK), this assay demonstrates that PHD2 and PHD3 hydroxylase activities are functional over a wide range of physiologically relevant oxygen concentrations (see Fig. 3.2B). Furthermore, this assay is strictly dependent on the presence of P564 within the substrate peptide, and the iron chelator DFX as well as the substrate analog N-OG inhibit PHD-dependent prolyl-4-hydroxylation (see Fig. 3.2C).

Apart from purified recombinant PHDs, crude cellular extracts also are suitable sources of PHD activity for the VBC-binding assay. Therefore, Hep3B cells were lysed by dounce homogenization in 100 mM Tris-HCl, pH 7.5, 1.5 mM $MgCl_2$, 8.75% glycerol, 0.01% Tween20, and EDTA-free complete protease inhibitor cocktail (Roche Applied Science, Indianapolis, IN). Cell lysates were cleared by centrifugation at 20,000×g for 30 min at 4°. These extracts stimulated VBC binding to wild-type, but not P564A mutant, HIF-1αODD-derived peptides (see Fig. 3.2D). The specificity of the PHD activity was further demonstrated by inhibition with 2 mM N-OG and by culturing the Hep3B cells under hypoxic conditions (0.4% O_2 for 16 h). Consistent with the known hypoxic induction of PHD2 and PHD3, extracts derived from hypoxic Hep3B cells showed an increased PHD activity (see Fig. 3.2D). Similar results were obtained with HeLa cells (data not shown).

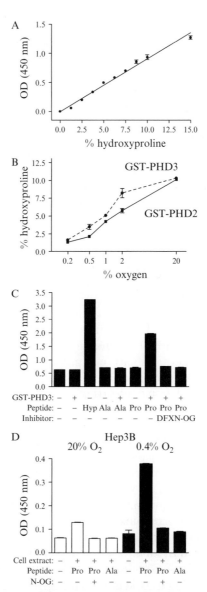

Figure 3.2 (A) Von Hippel-Lindau tumor suppressor protein (pVHL)/elongin B/elongin C (VBC) complex binding to prolyl-4-hydroxylated hypoxia-inducible tran-scription factor (HIF)-1α oxygen-dependent degradation (ODD) domain-derived peptides. Biotinylated peptides containing increasing proportions of synthetically gen-erated hydroxyproline peptides were allowed to bind to NeutrAvidin-coated 96-well enzyme-linked immunosorbent assay (ELISA) plates. Immunodetection of bound VBC complex was performed, as described in the text. (B) Oxygen-dependent activity of recombinant glutathione S-transferase (GST) prolyl-4-hydroxylase domain (PHD) 2 and GST-PHD3. Purified PHD proteins were used to hydroxylate HIF-1α ODD

4. DETERMINATION OF PROLYL-4-HYDROXYLATION BY OXIDATIVE DECARBOXYLATION OF 2-OXOGLUTARATE

While the VHL binding assay is well suited to investigate PHD function, it is strictly dependent on VBC binding and hence unlikely to investigate the hydroxylation of potential novel non–HIF-α targets. To overcome this problem, a 2-oxoglutarate to succinate conversion assay, originally developed to study collagen hydroxylation (Kaule and Günzler, 1990), was adapted to HIF-α hydroxylation. There are basically two possibilities to label the 2-oxoglutarate cosubstrate: when [5-^{14}C]2-oxoglutarate is used, the resulting [^{14}C]succinate can be measured (Kaule and Günzler, 1990); when [1-^{14}C]2-oxoglutarate is used, the resulting [^{14}C]CO$_2$ must be captured and quantified (Hirsilä *et al.*, 2003). Because it is technically less demanding, the former technique is used in our laboratory. This assay was first applied for PHD measurements by Frelin and coworkers using "light mitochondrial rat kidney fractions" (D'Angelo *et al.*, 2003a,b). We measured oxidative decarboxylation of 2-oxoglutarate by purified PHDs because kidney extracts prepared according to Frelin and coworkers solely showed high background activities, but did not contain measurable quantities of specific PHD activity (described later). Peptide (5–50 μM) or protein substrates (5 μM) were incubated with 0.1 μg GST-tagged PHD enzymes in 100 μl 50 mM Tris-HCl, pH 7.4, 10 μM 2-oxoglutarate, 2 mM ascorbate, 100 μM DTT, 1 mg/ml BSA, 0.6 mg/ml catalase, 5 μM freshly prepared FeSO$_4$, and 50,000 to 100,000 dpm [5-^{14}C] 2-oxoglutarate with a specific activity of 50 mCi/mmol (Hartmann-Analytic, Braunschweig, Germany). The 2-oxoglutarate concentration can be varied from 10 to 100 μM, depending on the peptide/protein target concentration and the desired specific radioactivity. Generally, a two-fold molar excess over the target concentration was used. Hydroxylation reactions were carried out for 24 h at 37°. To separate 2-oxoglutarate from succinate, 25 μl 20 mM succinate and 20 mM 2-oxoglutarate were added and mixed by vortexing. Subsequently, 25 μl 0.16 M 2,4-dinitrophenyl hydrazine in 30% HClO$_4$ were added. After incubation at room temperature for 30 min, 50 μl of 1M 2-oxoglutarate was added to remove residual 2,4-dinitrophenyl hydrazine, and the reaction was

---

domain-derived peptides under the indicated oxygen concentrations (%vol/vol in the gas phase). The extent of prolyl-4-hydroxylation was quantitated, as previously described. (C) PHD3 and the wild-type peptide substrate (Pro) are required for VBC binding. The lack of peptide (−), a P564A mutation (Ala), iron chelation by deferroxamine (DFX), or N-oxalylglycine (N-OG) all completely inhibited VBC binding. (A–C) Shown are mean values ± SEM of triplicates. (D) Hypoxia stimulates PHD activity in Hep3B cells. Crude cellular lysates were used to hydroxylate HIF-1α ODD domain-derived peptides, as previously described. Shown are mean values ± SEM of triplicates normalized to protein content.

allowed to proceed for another 30 min at room temperature. Following centrifugation at 20,000×*g* for 15 min at 4°, 150 µl supernatant was carefully removed, mixed with 3 ml scintillation cocktail (PerkinElmer, Waltham, MA), and the amount of [^{14}C]succinate was determined by liquid scintillation counting in a *β*-counter (TRI-CARB 2900TR, Packard).

As depicted in Fig. 3.3A, recombinant GST-PHD2 shows a linear relationship between the availability of an HIF-1α ODD domain-derived synthetic peptide substrate and [^{14}C]succinate production in the range from 8- to 50-µ*M* peptide. Below 5 µ*M*, however, only background values were obtained. This assay is strictly dependent on the presence of 1 m*M* ascorbate as well as a peptide substrate containing the wild-type P564 (see Fig. 3.3B).

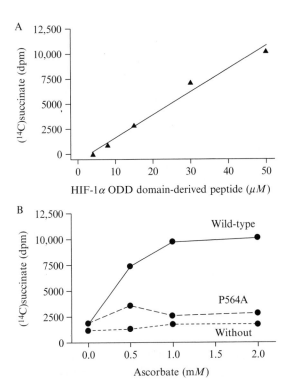

Figure 3.3 (A) Determination of prolyl-4-hydroxylation by oxidative decarboxylation of [5-^{14}C]2-oxoglutarate. Recombinant glutathione S-transferase (GST) prolyl-4-hydroxylase domain (PHD) 2 was incubated with increasing concentrations of a hypoxia-inducible transcription factor (HIF)-1α oxygen-dependent degradation (ODD) domain-derived peptide, as indicated. PHD-dependent generation of [^{14}C] succinate was determined, as described in the text. Background activity was determined in control reactions without substrate peptide and the obtained value subtracted from the peptide-containing reactions. (B) Recombinant GST-PHD2 was incubated with increasing concentrations of ascorbate (as indicated) and either a wild-type, P564A mutant, or no peptide substrate.

The lack of a substrate peptide or the mutation of the critical proline (P564A) results in background 2-oxoglutarate conversion.

5. CRUDE TISSUE EXTRACTS ARE NOT A SUITABLE SOURCE OF PHD ACTIVITY FOR THE 2-OXOGLUTARATE CONVERSION ASSAY

Earlier reports suggested the use of "light mitochondrial rat kidney fractions" as a source for PHD activity in this type of assay (D'Angelo *et al.*, 2003a,b). We therefore prepared rat kidney extracts according to this protocol and tested them for PHD activity. As shown in Fig. 3.4A, 2-oxoglutarate conversion was indeed stimulated by these extracts. However, a very high background activity was also observed in the absence of the HIF-1α ODD domain-derived peptide, and the presence of the wild-type peptide stimulated 2-oxoglutarate conversion only moderately. To ensure PHD-dependent 2-oxoglutarate conversion, ample amounts of the known PHD inhibitors DFX and N-OG (see Fig. 3.2C) or $CuCl_2$ (Martin *et al.*, 2005) were added to the reaction mix. Surprisingly, none of these PHD inhibitors blocked 2-oxoglutarate conversion, whether the substrate peptide was present or not (see Fig. 3.4A, left panel). Moreover, the presence of a P564A mutant peptide stimulated 2-oxoglutarate conversion as well as the wild-type peptide (see Fig. 3.4A, right panel). The P564A mutant peptide cannot be hydroxylated and shows no VBC complex binding after incubation with purified PHDs (see Figs. 3.2C and 3.3B). Therefore, these experiments demonstrate that "light mitochondrial rat kidney fractions" do not represent a source of specific PHD activity, but rather contain high levels of enzymes not specifically metabolizing 2-oxoglutarate. A similar lack of specific (i.e., PHD-dependent) 2-oxoglutarate conversion was observed with crude extracts derived from HeLa, Hep3B, and Sf 9 cells (data not shown).

6. THIN LAYER CHROMATOGRAPHY TO ASSESS THE PURITY OF [5-¹⁴C]2-OXOGLUTARATE

A major problem in the optimization of the 2-oxoglutarate conversion method is the high proportion of background activity when non-purified PHD preparations are used as a source of the enzymatic activity. Indeed, cellular extracts usually contain too high background activities to be suitable as PHD sources, even when the PHDs are exogenously overexpressed. Another background-causing problem was the quality of the radioactively labelled [5-¹⁴C]2-oxoglutarate. To analyze the [5-¹⁴C]2-oxoglutarate preparations for impurities, it was diluted in 1.5 mM unlabeled

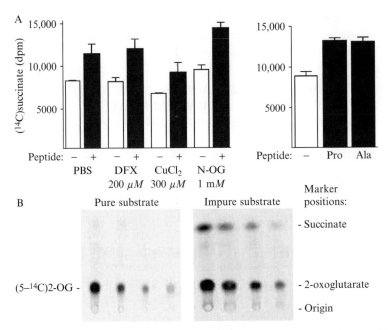

Figure 3.4 (A) Lack of specific 2-oxoglutarate-to-succinate conversion by "light mitochondrial rat kidney fractions." (Left panel) Peptide-stimulated activity in these extracts could not be inhibited by DFX, CuCl₂, or N-OG. (Right panel) A P564A mutant (Ala) hypoxia-inducible transcription factor (HIF)-1α oxygen-dependent degradation (ODD) domain-derived peptide stimulated 2-oxoglutarate conversion and the wild-type (Pro) peptide. (B) Quality control of [5-^{14}C]2-oxoglutarate by thin layer chromatography. The amount of [5-^{14}C]2-oxoglutarate spotted corresponded to 30, 15, 7.5, and 3.8 kdpm, respectively (from left to right). Two [5-^{14}C]2-oxoglutarate batches of different quality are shown. Pure 2-oxoglutarate and succinate served as migration markers (positions indicated in right margin).

2-oxoglutarate and spotted onto thin layer chromatography (TLC) plates (Silica gel 60 F254, Merck, Whitehouse Station, NJ). Following drying, the TLC plate was placed in a chromatography chamber containing a 120:70:15 mixture of diethyl ether/hexane/formic acid. When the eluent reached the top of the TLC plate, it was dried and the radioactivity quantitated by phosphorimaging (BioRad, Hercules, CA). Unlabeled 2-oxoglutarate and succinate were used as standards. They were visualized as bright yellow spots by immersing the TLC plates in a 0.04% bromcresol purple solution in a 1:1 mixture of ethanol and water (adjusted to pH 10.0 with NaOH) and dried with a hairdryer. R_f values for succinate and 2-oxoglutarate were 0.45 and 0.12, respectively. Considerable differences exist in the quality of the available [5-^{14}C]2-oxoglutarate preparations (see Fig. 3.4B). While some batches were of acceptable purity (Fig. 3.4B, left panel), others contained up to 30% of an impurity that co-migrated with the succinate standard (Fig. 3.4B, right panel). Regarding the relatively low specific PHD activities and the rather high unspecific background 2-oxoglutarate conversion, such high impurities are not acceptable in this type of assay.

7. APPLICATION OF THE 2-OXOGLUTARATE CONVERSION ASSAY TO PROTEIN TARGETS

In order to be useful for putative novel PHD substrate proteins without prior knowledge of the actual target prolyl residue, the 2-oxoglutarate conversion assay needs to work also with proteins rather than only with synthetic peptides. To demonstrate the feasibility of this approach, wild-type GST-HIF-2αODD (aa 404–569, MW = 46 kDa) and P405A/P531A double-mutant GST-HIF-2αODD protein fragments were expressed in *E. coli* BL21AI by induction with 0.2% arabinose for 4 h at 37°. After harvesting by centrifugation, the bacteria were lysed with a high-pressure cell disrupter (Basic-Z, Constant Systems Ltd, Sanford, NC) in the presence of EDTA-free complete protease inhibitor cocktail (Roche Applied Science, Indianapolis, IN). Purification of the GST-tagged protein was carried out using affinity chromatography on glutathione Sepharose beads (GE Healthcare, Buckinghamshire, UK), as previously described (see Fig. 3.1B). Purity of the recombinant protein fragments was checked by SDS-PAGE followed by Coomassie staining or immunoblotting (Fig. 3.5A). When compared with an HIF-1αODD–derived wild-type peptide, equimolar concentrations of the HIF-2αODD protein fragment also stimulated PHD1-dependent 2-oxoglutarate conversion, albeit to a somewhat lower extent (see Fig. 3.5B).

8. CONCLUSIONS

The PHD-dependent hydroxylation assay presented here works independently of already known peptide sequences or VHL binding. It should thus be possible to use this assay for the investigation of putative novel, non–HIF-α PHD substrates as well. In addition, even if not shown here, the same assay should principally be applicable to novel FIH substrates and might become useful for screening for novel hydroxylation targets. Whereas no non–HIF-α PHD hydroxylation targets have been reported thus far, FIH has recently been shown to hydroxylate a number of proteins containing

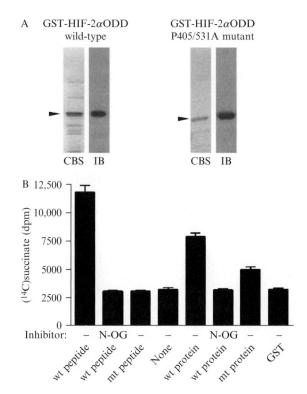

Figure 3.5 (A) Purification of wild-type and P405/531A double-mutant glutathione S-transferase (GST) hypoxia-inducible transcription factor (HIF)-2α oxygen-dependent degradation (ODD human HIF-2α amino acids 404–569) protein fragments expressed in bacteria and purified by glutathione affinity chromatography. The purified proteins were analyzed by sodium dodecyl sulfate polyacrylamide gel electrophoresis (SDS-PAGE) and Coomassie blue staining (CBS) or immunoblotting (IB) using anti-GST antibodies. (B) Determination of prolyl-4-hydroxylation by oxidative decarboxylation of [5-^{14}C]2-oxoglutarate. Recombinant GST-PHD1–dependent generation of [^{14}C] succinate was promoted by HIF-1α ODD-derived wild-type (wt), but not by P564A mutant (mt) peptides (human HIF-1α amino acids 556–574). Generation of [^{14}C]succinate was also promoted by wild-type GST-HIF-2α ODD protein, but not by GST alone and to a much lower extent by the GST-HIF-2α ODD P405/531A double mutant. Equimolar concentrations (5 μM) of the substrate peptides and protein fragments were used. Where indicated, the reactions could be blocked by 5 mM N-oxalylglycine (N-OG). Shown are mean values ± SEM of three independent experiments performed in duplicates.

ankyrin repeats, including NF-κB and IκBα (Cockman *et al.*, 2006). We expect a similar widening of the spectrum of PHD targets. In addition, PHD-dependent hydroxylation assays will be required to study novel drugs that modulate PHD activity and hence will become important for the treatment of anemic and ischemic disease.

ACKNOWLEDGMENTS

The authors wish to thank S. Tan and W. G. Kaelin, Jr. for gifts of plasmids. This work was supported by grants from the Forschungskredit of the University of Zürich (to D.P.S. and G.C.), the *Sassella Stiftung* (to D.P.S. and G.C.), the *Olga Mayenfisch Stiftung* (to G.C.), the Krebsliga des Kantons Zürich (to R.H.W.), the 6th Framework Programme of the European Commission/SBF (EUROXY LSHC-CT-2003–502932/SBF Nr. 03.0647–2 to R. H. W.), and the Swiss National Science Foundation (3100A0–104219 to R. H. W. and G. C.).

REFERENCES

Appelhoff, R. J., Tian, Y. M., Raval, R. R., Turley, H., Harris, A. L., Pugh, C. W., Ratcliffe, P. J., and Gleadle, J. M. (2004). Differential function of the prolyl hydroxylases PHD1, PHD2, and PHD3 in the regulation of hypoxia-inducible factor. *J. Biol. Chem.* **279,** 38458–38465.

Aprelikova, O., Chandramouli, G. V., Wood, M., Vasselli, J. R., Riss, J., Maranchie, J. K., Linehan, W. M., and Barrett, J. C. (2004). Regulation of HIF prolyl hydroxylases by hypoxia-inducible factors. *J. Cell. Biochem.* **92,** 491–501.

Baek, J. H., Mahon, P. C., Oh, J., Kelly, B., Krishnamachary, B., Pearson, M., Chan, D. A., Giaccia, A. J., and Semenza, G. L. (2005). OS-9 interacts with hypoxia-inducible factor 1α and prolyl hydroxylases to promote oxygen-dependent degradation of HIF-1α. *Mol. Cell* **17,** 503–512.

Berchner-Pfannschmidt, U., Yamac, H., Trinidad, B., and Fandrey, J. (2006). Nitric oxide modulates oxygen sensing by HIF-1 dependent induction of prolyl hydroxylase 2. *J. Biol. Chem.* **282,** 1788–1796.

Berra, E., Benizri, E., Ginouves, A., Volmat, V., Roux, D., and Pouysségur, J. (2003). HIF prolyl-hydroxylase 2 is the key oxygen sensor setting low steady-state levels of HIF-1α in normoxia. *EMBO J.* **22,** 4082–4090.

Bruick, R. K. (2000). Expression of the gene encoding the proapoptotic Nip3 protein is induced by hypoxia. *Proc. Natl. Acad. Sci. USA* **97,** 9082–9087.

Cockman, M. E., Lancaster, D. E., Stolze, I. P., Hewitson, K. S., McDonough, M. A., Coleman, M. L., Coles, C. H., Yu, X., Hay, R. T., Ley, S. C., Pugh, C. W., Oldham, N. J., et al. (2006). Posttranslational hydroxylation of ankyrin repeats in IκB proteins by the hypoxia-inducible factor (HIF) asparaginyl hydroxylase, factor inhibiting HIF (FIH). *Proc. Natl. Acad. Sci. USA* **103,** 14767–14772.

D'Angelo, G., Duplan, E., Vigne, P., and Frelin, C. (2003a). Cyclosporin A prevents the hypoxic adaptation by activating hypoxia-inducible factor-1α Pro-564 hydroxylation. *J. Biol. Chem.* **278,** 15406–15411.

D'Angelo, G., Duplan, E., Boyer, N., Vigne, P., and Frelin, C. (2003b). Hypoxia up-regulates prolyl hydroxylase activity: A feedback mechanism that limits HIF-1 responses during reoxygenation. *J. Biol. Chem.* **278,** 38183–38187.

Dalgard, C. L., Lu, H., Mohyeldin, A., and Verma, A. (2004). Endogenous 2-oxoacids differentially regulate expression of oxygen sensors. *Biochem. J.* **380,** 419–424.

Ehrismann, D., Flashman, E., Genn, D. N., Mathioudakis, N., Hewitson, K. S., Ratcliffe, P. J., and Schofield, C. J. (2006). Studies on the activity of the hypoxia-inducible factor hydroxylases using an oxygen consumption assay. *Biochem. J.* **401,** 227–234.

Epstein, A. C., Gleadle, J. M., McNeill, L. A., Hewitson, K. S., O'Rourke, J., Mole, D. R., Mukherji, M., Metzen, E., Wilson, M. I., Dhanda, A., Tian, Y. M., Masson, Y. M., et al.

(2001). *C. elegans* EGL-9 and mammalian homologs define a family of dioxygenases that regulate HIF by prolyl hydroxylation. *Cell* **107**, 43–54.

Gerald, D., Berra, E., Frapart, Y. M., Chan, D. A., Giaccia, A. J., Mansuy, D., Pouysségur, J., Yaniv, M., and Mechta-Grigoriou, F. (2004). JunD reduces tumor angiogenesis by protecting cells from oxidative stress. *Cell* **118**, 781–794.

Hagen, T., Taylor, C. T., Lam, F., and Moncada, S. (2003). Redistribution of intracellular oxygen in hypoxia by nitric oxide: Effect on HIF1α. *Science* **302**, 1975–1978.

Hewitson, K. S., McNeill, L. A., Riordan, M. V., Tian, Y. M., Bullock, A. N., Welford, R. W., Elkins, J. M., Oldham, N. J., Bhattacharya, S., Gleadle, J. M., Ratcliffe, P. J., Pugh, C. W., *et al.* (2002). Hypoxia-inducible factor (HIF) asparagine hydroxylase is identical to factor inhibiting HIF (FIH) and is related to the cupin structural family. *J. Biol. Chem.* **277**, 26351–26355.

Hirsilä, M., Koivunen, P., Günzler, V., Kivirikko, K. I., and Myllyharju, J. (2003). Characterization of the human prolyl 4-hydroxylases that modify the hypoxia-inducible factor. *J. Biol. Chem.* **278**, 30772–30780.

Hirsilä, M., Koivunen, P., Xu, L., Seeley, T., Kivirikko, K. I., and Myllyharju, J. (2005). Effect of desferrioxamine and metals on the hydroxylases in the oxygen sensing pathway. *FASEB J.* **19**, 1308–1310.

Hopfer, U., Hopfer, H., Jablonski, K., Stahl, R. A., and Wolf, G. (2006). The novel WD-repeat protein Morg1 acts as a molecular scaffold for hypoxia-inducible factor prolyl hydroxylase 3 (PHD3). *J. Biol. Chem.* **281**, 8645–8655.

Ivan, M., Haberberger, T., Gervasi, D. C., Michelson, K. S., Günzler, V., Kondo, K., Yang, H., Sorokina, I., Conaway, R. C., Conaway, J. W., and Kaelin, W. G., Jr. (2002). Biochemical purification and pharmacological inhibition of a mammalian prolyl hydroxylase acting on hypoxia-inducible factor. *Proc. Natl. Acad. Sci. USA* **99**, 13459–13464.

Karaczyn, A., Ivanov, S., Reynolds, M., Zhitkovich, A., Kasprzak, K. S., and Salnikow, K. (2006). Ascorbate depletion mediates up-regulation of hypoxia-associated proteins by cell density and nickel. *J. Cell. Biochem.* **97**, 1025–1035.

Kaule, G., and Günzler, V. (1990). Assay for 2-oxoglutarate decarboxylating enzymes based on the determination of [1-^{14}C]succinate: Application to prolyl 4-hydroxylase. *Anal. Biochem.* **184**, 291–297.

Knowles, H. J., Raval, R. R., Harris, A. L., and Ratcliffe, P. J. (2003). Effect of ascorbate on the activity of hypoxia-inducible factor in cancer cells. *Cancer Res.* **63**, 1764–1768.

Koivunen, P., Hirsilä, M., Kivirikko, K. I., and Myllyharju, J. (2006). The length of peptide substrates has a marked effect on hydroxylation by the hypoxia-inducible factor prolyl 4-hydroxylases. *J. Biol. Chem.* **281**, 28712–28720.

Lando, D., Peet, D. J., Gorman, J. J., Whelan, D. A., Whitelaw, M. L., and Bruick, R. K. (2002). FIH-1 is an asparaginyl hydroxylase enzyme that regulates the transcriptional activity of hypoxia-inducible factor. *Genes Dev.* **16**, 1466–1471.

Linden, T., Katschinski, D. M., Eckhardt, K., Scheid, A., Pagel, H., and Wenger, R. H. (2003). The antimycotic ciclopirox olamine induces HIF-1α stability, VEGF expression, and angiogenesis. *FASEB J.* **17**, 761–763.

Lu, H., Dalgard, C. L., Mohyeldin, A., McFate, T., Tait, A. S., and Verma, A. (2005). Reversible inactivation of HIF-1 prolyl hydroxylases allows cell metabolism to control basal HIF-1. *J. Biol. Chem.* **280**, 41928–41939.

Mahon, P. C., Hirota, K., and Semenza, G. L. (2001). FIH-1: A novel protein that interacts with HIF-1α and VHL to mediate repression of HIF-1 transcriptional activity. *Genes Dev.* **15**, 2675–2686.

Martin, F., Linden, T., Katschinski, D. M., Oehme, F., Flamme, I., Mukhopadhyay, C. K., Eckhardt, K., Tröger, J., Barth, S., Camenisch, G., and Wenger, R. H. (2005). Copper-dependent activation of hypoxia-inducible factor (HIF)-1: Implications for ceruloplasmin regulation. *Blood* **105**, 4613–4619.

Marxsen, J. H., Stengel, P., Doege, K., Heikkinen, P., Jokilehto, T., Wagner, T., Jelkmann, W., Jaakkola, P., and Metzen, E. (2004). Hypoxia-inducible factor-1 (HIF-1) promotes its degradation by induction of HIF-α-prolyl-4-hydroxylases. *Biochem. J.* **381**, 761–767.

Masson, N., Appelhoff, R. J., Tuckerman, J. R., Tian, Y. M., Demol, H., Puype, M., Vandekerckhove, J., Ratcliffe, P. J., and Pugh, C. W. (2004). The HIF prolyl hydroxylase PHD3 is a potential substrate of the TRiC chaperonin. *FEBS Lett.* **570**, 166–170.

McNeill, L. A., Flashman, E., Buck, M. R., Hewitson, K. S., Clifton, I. J., Jeschke, G., Claridge, T. D., Ehrismann, D., Oldham, N. J., and Schofield, C. J. (2005). Hypoxia-inducible factor prolyl hydroxylase 2 has a high affinity for ferrous iron and 2-oxoglutarate. *Mol. Biosyst.* **1**, 321–324.

Metzen, E., Zhou, J., Jelkmann, W., Fandrey, J., and Brüne, B. (2003). Nitric oxide impairs normoxic degradation of HIF-1α by inhibition of prolyl hydroxylases. *Mol. Biol. Cell* **14**, 3470–3481.

Metzen, E., Stiehl, D. P., Doege, K., Marxsen, J. H., Hellwig-Bürgel, T., and Jelkmann, W. (2005). Regulation of the prolyl hydroxylase domain protein 2 (*phd2/egln-1*) gene: Identification of a functional hypoxia-responsive element. *Biochem. J.* **387**, 711–717.

Myllylä, R., Majamaa, K., Günzler, V., Hanauske-Abel, H. M., and Kivirikko, K. I. (1984). Ascorbate is consumed stoichiometrically in the uncoupled reactions catalyzed by prolyl 4-hydroxylase and lysyl hydroxylase. *J. Biol. Chem.* **259**, 5403–5405.

Nakayama, K., Frew, I. J., Hagensen, M., Skals, M., Habelhah, H., Bhoumik, A., Kadoya, T., Erdjument-Bromage, H., Tempst, P., Frappell, P. B., Bowtell, D. D., Ronai, Z., *et al.* (2004). Siah2 regulates stability of prolyl-hydroxylases, controls HIF1α abundance, and modulates physiological responses to hypoxia. *Cell* **117**, 941–952.

Oehme, F., Jonghaus, W., Narouz-Ott, L., Huetter, J., and Flamme, I. (2004). A nonradioactive 96-well plate assay for the detection of hypoxia-inducible factor prolyl hydroxylase activity. *Anal. Biochem.* **330**, 74–80.

Ozer, A., Wu, L. C., and Bruick, R. K. (2005). The candidate tumor suppressor ING4 represses activation of the hypoxia inducible factor (HIF). *Proc. Natl. Acad. Sci. USA* **102**, 7481–7486.

Percy, M. J., Zhao, Q., Flores, A., Harrison, C., Lappin, T. R., Maxwell, P. H., McMullin, M. F., and Lee, F. S. (2006). A family with erythrocytosis establishes a role for prolyl hydroxylase domain protein 2 in oxygen homeostasis. *Proc. Natl. Acad. Sci. USA* **103**, 654–659.

Pescador, N., Cuevas, Y., Naranjo, S., Alcaide, M., Villar, D., Landázuri, M. O., and Del Peso, L. (2005). Identification of a functional hypoxia-responsive element that regulates the expression of the *egl nine* homologue 3 (*egln3/phd3*) gene. *Biochem. J.* **390**, 189–197.

Salnikow, K., Donald, S. P., Bruick, R. K., Zhitkovich, A., Phang, J. M., and Kasprzak, K. S. (2004). Depletion of intracellular ascorbate by the carcinogenic metals nickel and cobalt results in the induction of hypoxic stress. *J. Biol. Chem.* **279**, 40337–40344.

Schofield, C. J., and Ratcliffe, P. J. (2004). Oxygen sensing by HIF hydroxylases. *Nat. Rev. Mol. Cell Biol.* **5**, 343–354.

Selak, M. A., Armour, S. M., MacKenzie, E. D., Boulahbel, H., Watson, D. G., Mansfield, K. D., Pan, Y., Simon, M. C., Thompson, C. B., and Gottlieb, E. (2005). Succinate links TCA cycle dysfunction to oncogenesis by inhibiting HIF-α prolyl hydroxylase. *Cancer Cell* **7**, 77–85.

Stiehl, D. P., Wirthner, R., Köditz, J., Spielmann, P., Camenisch, G., and Wenger, R. H. (2006). Increased prolyl 4-hydroxylase domain proteins compensate for decreased oxygen levels. Evidence for an autoregulatory oxygen-sensing system. *J. Biol. Chem.* **281**, 23482–23491.

Takeda, K., Ho, V., Takeda, H., Duan, L. J., Nagy, A., and Fong, G. H. (2006). Placental but not heart defect is associated with elevated HIFα levels in mice lacking prolyl hydroxylase domain protein 2. *Mol. Cell. Biol.* **26,** 8336–8346.

To, K. K., and Huang, L. E. (2005). Suppression of hypoxia-inducible factor 1α (HIF-1α) transcriptional activity by the HIF prolyl hydroxylase EGLN1. *J. Biol. Chem.* **280,** 38102–38107.

Wenger, R. H. (2000). Mammalian oxygen sensing, signalling and gene regulation. *J. Exp. Biol.* **203,** 1253–1263.

Wenger, R. H. (2002). Cellular adaptation to hypoxia: O_2-sensing protein hydroxylases, hypoxia-inducible transcription factors, and O_2-regulated gene expression. *FASEB J.* **16,** 1151–1162.

Wenger, R. H. (2006). Mitochondria: Oxygen sinks rather than sensors? *Med. Hypotheses* **66,** 380–383.

Wenger, R. H., Stiehl, D. P., and Camenisch, G. (2005). Integration of oxygen signaling at the consensus HRE. *Sci. STKE* **2005,** re12.

Willam, C., Maxwell, P. H., Nichols, L., Lygate, C., Tian, Y. M., Bernhardt, W., Wiesener, M., Ratcliffe, P. J., Eckardt, K. U., and Pugh, C. W. (2006). HIF prolyl hydroxylases in the rat; organ distribution and changes in expression following hypoxia and coronary artery ligation. *J. Mol. Cell. Cardiol.* **41,** 68–77.

CHAPTER FOUR

CHARACTERIZATION OF ANKYRIN REPEAT–CONTAINING PROTEINS AS SUBSTRATES OF THE ASPARAGINYL HYDROXYLASE FACTOR INHIBITING HYPOXIA-INDUCIBLE TRANSCRIPTION FACTOR

Sarah Linke, Rachel J. Hampton-Smith, *and* Daniel J. Peet

Contents

1. Introduction	62
2. Experimental Techniques	65
2.1. Production of FIH-1	65
2.2. Protein quantitation	68
2.3. Production of substrates	69
2.4. CO_2 capture assays	71
2.5. Interaction assays	76
2.6. *In vitro* pull-down assay	76
2.7. Co-immunoprecipitation assay	79
3. Discussion/Conclusion	82
Acknowledgments	83
References	83

Abstract

The hypoxia-inducible transcription factors (HIFs) are essential mediators of the genomic response to oxygen deficiency (hypoxia) in multicellular organisms. The HIFs are regulated by four oxygen-sensitive hydroxylases—three prolyl hydroxylases and one asparaginyl hydroxylase. These hydroxylases are all members of the 2-oxoglutarate (2OG)–dependent dioxygenase superfamily and convey changes in cellular oxygen concentration to the HIF-alpha (α) subunit, leading to potent accumulation and activity in hypoxia versus

School of Molecular and Biomedical Science and The ARC Special Research Centre for the Molecular Genetics of Development, University of Adelaide, Adelaide, Australia

Methods in Enzymology, Volume 435
ISSN 0076-6879, DOI: 10.1016/S0076-6879(07)35004-0

degradation and repression in normoxia. HIF-α asparaginyl hydroxylation is catalyzed by factor-inhibiting HIF-1 (FIH-1) and directly regulates the transcription activity of the HIF-α proteins. Recent work has demonstrated that, in addition to hydroxylating HIF-α, FIH-1 can also hydroxylate the ankyrin domains of a wide range of proteins. This paper presents *in vitro* and cell-based techniques for the preliminary characterization of ankyrin domain–containing proteins as FIH-1 substrates and interacting proteins. Strategies are presented for the expression and purification of FIH-1 from mammalian or bacterial cells. Similar to the HIF-α proteins, the ankyrin-containing substrates are examined as purified proteins expressed in bacteria and overexpressed in mammalian cells or in the form of synthetic peptides. Specific conditions for the efficient expression of ankyrin-containing proteins compared with the HIF-α substrates in *Escherichia coli* are detailed. Hydroxylation is rapidly inferred, utilizing the described *in vitro* CO_2 capture assay. Finally, substrate and non-substrate interactions are examined using *in vitro* affinity pull-down assays and mammalian cell-based co-immunoprecipitation assays. Together, these methods are rapid and well suited to the preliminary characterization of potential substrates of the therapeutically relevant oxygen-sensing enzyme FIH-1.

1. INTRODUCTION

In Chapter 1, Richard Bruick and his coauthors describe the essential HIFs, which orchestrate a major genomic response to hypoxia in multicellular organisms. These transcription factors are α/beta (β)-heterodimers that bind DNA and activate transcription of over 70 target genes during cellular hypoxia (Semenza, 2003).

During normoxia, very little fully-active HIF assembles because the α subunit (HIF-α) is degraded and repressed through hydroxylation performed by a family of O_2, 2OG, and Fe^{2+} dependent dioxygenases—the HIF hydroxylases. Three prolyl hydroxylases (PHD1–3) hydroxylate two prolyl residues within the central oxygen-dependent degradation domain (ODD) of HIF-α, triggering ubiquitination and rapid degradation (see preceding chapter). The HIF-α subunits are also hydroxylated on a conserved asparaginyl residue within the C-terminal transactivation domain (CAD). This is performed by an asparaginyl hydroxylase initially cloned as FIH-1 (Hewitson *et al.*, 2002; Lando *et al.*, 2002a,b; Mahon *et al.*, 2001), the subject of this chapter. This single hydroxylation inhibits binding of the obligate CREB-binding protein (CBP)/p300 coactivator proteins to the HIF-α CAD, thus repressing transcriptional activity.

Both classes of HIF hydroxylases are sensitive to physiological oxygen fluctuations, as their apparent K_ms for oxygen *in vitro* (90–300 μM) (Ehrismann *et al.*, 2007; Hirsila *et al.*, 2003; Koivunen *et al.*, 2004) are above normal physiological oxygen concentrations. Hydroxylation of

HIF-α in cells is also limiting (Stolze et al., 2004). Therefore, they are postulated to be direct oxygen sensors, with hydroxylation of the HIF-α substrates decreasing in hypoxia when oxygen levels are limiting, resulting in stabilized and active HIF proteins for robust gene induction.

FIH-1 is currently one of just two asparaginyl hydroxylases known in mammals, the other being aspartyl (asparaginyl) β-hydroxylase (BAH), which is also a member of the 2OG-dependent dioxygenase superfamily (Jia et al., 1992; Wang et al., 1991). However, although FIH-1 and BAH hydroxylate asparaginyl residues, they are not closely related within this superfamily and are functionally distinct, exhibiting different subcellular location, substrate specificity, and stereospecificity.

BAH is a transmembrane enzyme, predicted to reside in the endoplasmic reticulum (ER) where it targets epidermal growth factor (EGF)-like domains within transmembrane or secreted substrates (such as blood coagulation factors and the Notch ligand, Jagged) (Stenflo et al., 2000; Wouters et al., 2005). Although no direct structural information exists, BAH is predicted to adopt a double-stranded β-helix (DSBH) fold characteristic of this family of proteins (Clifton et al., 2006; Lancaster et al., 2004a). BAH targets the Asn/Asp β-carbon and generates the erythro-stereoisomer. Functionally, hydroxylation of substrates by BAH has not been tied directly to cellular effects; however, roles are implied in cell motility (Lahousse et al., 2006) and malignancy (Dinchuk et al., 2002).

The crystal structure of FIH-1 has been reported by three groups, showing that it possesses the DSBH and contains the characteristic iron-binding 2-His-1-carboxylate triad (Dann et al., 2002; Elkins et al., 2003; Lee et al., 2003). FIH-1 exists as a dimer via an extensive C-terminal interface, which is required for efficient substrate hydroxylation (Lancaster et al., 2004b). The DSBH of FIH-1 is related to the cupin superfamily of enzymes and to the JmjC-domain (Hewitson et al., 2002), first identified in the jumonji transcription family. The initial primary sequence homology between FIH-1 and the JmjC domain showed that many other JmjC proteins also have conserved potential Fe^{2+} and 2-OG binding residues, suggesting a catalytic role for other JmjC proteins (Hewitson et al., 2002; Takeuchi et al., 2006). Many JmjC proteins have since been shown to demethylate mono-, di-, or trimethyl-lysine residues within histones by hydroxylation of the methyl carbon (Takeuchi et al., 2006; Tian and Fang, 2007; Trewick et al., 2005). Thus far, FIH-1 has not been shown to demethylate any substrates, and it has not been shown to efficiently hydroxylate aspartyl residues. For HIF-α at least, FIH-1 targets the asparaginyl β-carbon to create the threo-stereoisomer (McNeill et al., 2002; Peet et al., 2004).

The crystal structure of the HIF-α CAD/FIH-1 complex demonstrates an "induced fit" binding process where FIH-1 and HIF-α CAD undergo structural changes upon binding (Elkins et al., 2003). These changes are particularly marked for the HIF-α CAD, which is unstructured in isolation

(Dames *et al.*, 2002). The role of individual amino acids in substrate recognition is not well characterized (Linke *et al.*, 2004), although longer peptide substrates are clearly preferred (Ehrismann *et al.*, 2007; Koivunen *et al.*, 2004), and current data imply that the overall context of the asparagine is a major determinant of recognition rather than a small "consensus" sequence of particular invariant residues.

Our knowledge of FIH-1 substrate repertoire has been expanded recently by Cockman *et al.* (2006), who employed a yeast two-hybrid screen with FIH-1 as bait to identify novel substrates and used mass spectrometry to demonstrate hydroxylation of the interacting NF-κB family members, p105 and IκBα. In mammalian cells, FIH-1 hydroxylates p105 once and IκBα twice on asparagines analogously positioned within the ankyrin repeats. Furthermore, peptides from homologous sites of other ankyrin proteins (FEM-1β, p19-INK4d, GABP-β, Tankyrase-1, gankyrin, MYPT1, myotrophin, fetal globin-increasing factor [FGIF], and ILK-1) were found to promote decarboxylation of 2OG by FIH-1, suggesting that they too can be hydroxylated. With a similar decarboxylation assay ($^{14}CO_2$ capture assay, described later in this chapter), we have independently found FIH-1–mediated activity with the ankyrin repeat domain of gankyrin, which is folded according to our nuclear magnetic resonance (NMR) analyses, and on the FGIF ankyrin repeat region, and we have also defined two asparaginyl hydroxylations within the Notch receptor ankyrin domain by mass spectrometry (I. Murchland, R. Hampton Smith, S. Linke, D. Peet, unpublished data). These results clearly demonstrate that ankyrin repeats comprise a new set of FIH-1 substrates.

The ankyrin repeat is a relatively common, conserved motif found within thousands of proteins, each containing between 1 and 34 repeats, though the majority of ankyrin proteins contain less than seven repeats (Li *et al.*, 2006). The secondary structure of the 33-residue ankyrin motif entails two antiparallel α-helices separated by a short loop. The helix-loop-helix is followed by a longer loop, positioned perpendicular to the helices with a β-hairpin turn connecting adjacent repeats. The targeted asparaginyl residue is located in the longer loop and is semi-conserved in the ankyrin consensus (Mosavi *et al.*, 2002). The ankyrin structure is distinct from the induced fold of the HIF-α CAD bound to FIH-1 (Elkins *et al.*, 2003), thus highlighting the ability of FIH-1 to hydroxylate structurally diverse substrates.

FIH-1 is a therapeutically relevant protein that is straightforward to express, purify, and characterize using various *in vitro* assays. Hydroxylation can be conveniently inferred using *in vitro* assays that measure CO_2 release or cosubstrate consumption and then confirmed with mass spectrometry (detailed in McNeill *et al.*, 2005 and Peet *et al.*, 2004). These techniques have been crucial in the discovery and characterization of FIH-1 substrates, the dependence on oxygen, and a new understanding of cellular oxygen sensing. FIH-1 also has an unexpectedly high affinity for some of its substrates, particularly those containing ankyrin repeats. Thus, it is amenable to techniques such as co-immunoprecipitation, affinity pull-down, and yeast two-hybrid assays for measuring substrate/non-substrate binding (Cockman *et al.*, 2006; Mahon *et al.*, 2001). The characterization of specific sites of hydroxylation by mass spectrometry and amino acid analysis has been presented previously by others (Peet *et al.*, 2004).

This chapter covers the use of *in vitro* hydroxylation and interaction assays to characterize FIH-1, with a focus on the recently discovered ankyrin repeat–containing proteins as substrates. The techniques described utilize affinity-tagged or untagged FIH-1 purified from overexpressing bacteria or mammalian cells. Advantages and limitations of each method for examining binding and hydroxylation rates by FIH-1 are outlined, with a comparison of strategies suitable for ankyrin versus HIF substrates.

2. Experimental Techniques

2.1. Production of FIH-1

FIH-1 is commonly expressed with an N-terminal maltose-binding protein (MBP) affinity tag in *E. coli* and purified via amylose affinity chromatography, obtaining yields of approximately 5 mg/l of culture (Peet *et al.*, 2004). Alternatively, MBP-FIH-1 protein can be purified from stably overexpressing cultured mammalian cells, such as human embryonic kidney (HEK) 293T cells, in which FIH-1 might undergo folding and posttranslational modifications similar to endogenous FIH-1. Purification with a similar amylose affinity chromatography procedure generates a lower yield (typically 0.01 mg from $5 \times 175 \text{ cm}^2$ flasks of HEK 293T cells), although the enzyme has approximately 10-fold greater specific activity than that produced in *E. coli*. Figure 4.1 shows a comparison of MBP-FIH-1 purified from HEK-293T cells versus those from *E. coli*. Although their respective activities appear similar in the *in vitro* CO_2 capture assay (see Fig. 4.1B), it should be noted that 10-fold more MBP-FIH-1 from *E. coli* than from HEK-293T cells is used in this experiment. This may reflect a higher proportion of folded enzyme and/or mammalian cell-specific posttranslational modifications. Therefore, for a kinetic characterization that best reflects the *in vivo* status of FIH-1, expression in mammalian cells is preferred, though this expression is offset by the added expense and maintenance required for mammalian tissue culture and relatively low yields.

For both bacterial- and mammalian-expressed MBP-FIH-1, the MBP tag can be proteolytically cleaved to yield functional untagged FIH-1. The pMBP vector utilized here encodes a tobacco etch virus (TEV) protease site between the MBP tag and FIH-1, thus enabling efficient removal of the tag.

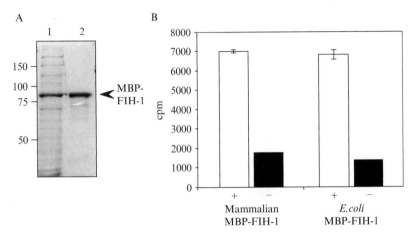

Figure 4.1 Maltose-binding protein factor-inhibiting HIF-1 (MBP-FIH-1), purified from either human embryonic kidney (HEK) 293T or *Escherichia coli* cells, is active in an *in vitro* $^{14}CO_2$ capture assay. (A) MBP-FIH-1 can be purified from HEK 293T cells or *E. coli* cells via amylose affinity chromatography. Analysis by sodium dodecyl sulfate polyacrylamide gel electrophoresis (SDS-PAGE) and Coomassie staining shows 0.25 μg of MBP-FIH-1 derived from HEK 293T cells (lane 1), and *E. coli* cells (lane 2). (B) MBP-FIH-1 purified from HEK 293T cells has 10-fold higher specific activity compared with MBP-FIH-1 purified from *E. coli* cells in an *in vitro* $^{14}CO_2$ capture assay. This experiment was performed using 0.05 μg of MBP-FIH-1 from mammalian (HEK-293T) cells and 0.5 μg from *E. coli*. Data are expressed as mean counts per minute (cpm) ± range ($n = 2$). Background release of $^{14}CO_2$ in the absence of FIH-1 is also presented ($n = 1$). Representative of more than three independent experiments.

Interestingly, cleavage around this site also occurs nonspecifically at low levels during sample preparation. In practice, we find that the MBP tag alone dissociates rapidly from the amylose agarose resin compared with MBP-FIH-1, perhaps because FIH-1 is dimeric. Thus, an additional purification step is required to efficiently separate the MBP tag from the cleaved FIH-1. Therefore, to generate purified FIH-1 without a tag, we recently constructed an expression vector with a thioredoxin-6 histidine (Trx-6H) tag separated by a TEV protease site (F. Whelan, unpublished data). This tag is used to express Trx-6H-FIH-1 in *E. coli* followed by purification using Ni-affinity chromatography, as later discussed in this chapter for the FIH-1 substrates. This method typically provides yields of 5 mg/litre of culture with a purity similar to MBP-FIH-1. Cleavage of the protein using 6H-TEV protease permits purification using Ni-affinity resin to remove both the Trx-6H tag and the protease. Importantly, the MBP or Trx-6H tags do not significantly influence the *in vitro* activity or the interactions of FIH-1 with proteins examined thus far. Therefore, depending on the intended use of the purified FIH-1, either tag can be used to aid expression and purification of functional protein.

2.1.1. Protocols

The expression and purification of MBP-FIH-1 has been previously described, but in less detail than provided here (Lando *et al.*, 2002b; Peet *et al.*, 2004). A 10-ml overnight culture of BL21 DE3 *E. coli* (in Luria Broth [LB] with 100 μg/ml ampicillin [Amp]) transformed with pMBP-FIH-1 is used to inoculate 500 ml of LB/Amp, which is shaken at 37° to OD_{600} 0.6 to 0.8, then induced with 0.2 mM isopropyl-β-D-galactopyranoside (IPTG) and grown for a further 5 h at 30°. The cells are pelleted, resuspended in 30 ml lysis buffer (150 mM NaCl, 20 mM Tris pH 8, 1 mM PMSF) at 4°, and lysed by two passes through a french press. Lysate is then clarified by centrifugation (14,000g at 4° for 45 min), and the supernatant is incubated with 1 ml of amylose agarose (Scientifix, Victoria, Australia) for 1 h at 4° with gentle rocking before being transferred to a chromatography column. The final yield may be limited by the capacity of the resin; hence, yields can often be increased by the addition of more resin, though this can compromise purity. The amylose agarose resin is then washed with 50 ml of lysis buffer followed by a further 50 ml of lysis buffer without phenylmethanesulphonyl fluoride (PMSF), and the MBP-FIH-1 is eluted with 2.5 ml of 10 mM maltose, 150 mM NaCl, and 20 mM Tris-HCl, pH 8.0. Extended incubation in elution buffer can significantly increase the final yield. The final MBP-FIH-1 sample can be buffer-exchanged to remove maltose. However, maltose does not impair enzyme activity *in vitro*.

Expression and purification of Trx-6H-TEV–tagged FIH-1 employs the pET-32a-TEV-FIH-1 vector. This vector was generated by replacement of the enterokinase site in the pET32a vector (Novagen, Darmstadt, Germany; Merck, Whitehouse Station, NJ) with a TEV protease site by cloning in the following oligonucleotide using Kpn1/Nco1 restriction digestion: Upper 5′-cgaaaacctgtacttccagggcgc-3′, lower 5′-catggcgccctggaagtacaggttttcggtac-3′. FIH-1 was then cloned into the pET32a-TEV vector from pMBP-FIH-1 by Nco/Xho1 restriction digestion.

Trx-6H-FIH-1 protein is induced as per MBP-FIH-1 (previously discussed) and purified by Ni-affinity chromatography (as detailed later for FIH-1 substrates). The 6H-TEV (Invitrogen, Carlsbad, CA) protease is used according to the manufacturer's instructions and is separated from the liberated FIH-1 by adding Ni-IDA resin (Scientifix, Victoria, Australia), which removes the 6H-protease and the Trx-6H tag from the FIH-1 solution.

Expression and purification of MBP-FIH-1 from HEK-293T cells is achieved by generating a polyclonal stable cell line overexpressing MBP-FIH-1 from the pEF-IRES-puro6 vector. To generate this expression vector, the NcoI, BamHI, and NdeI sites were removed from pEF-IRES-puro (Hobbs *et al.*, 1998) by sequentially digesting, filling in with DNA Polymerase I to generate blunt ends, and relegating at each site. Then, the XbaI/NotI fragment was excised from the polylinker and replaced with

annealed oligonucleotides (5′-ctagaccatggagatctcatatggatccactagtatcgatgata-
tcgagctgc-3′, 5′-ggccgcgagctcgatatcatcgatactagtggatccatatgagatctccatggt-3′)
to generate pEF-IRES-puro5. A modified version called pEF-IRES-puro6
lacking the NcoI site in the polylinker was generated by digesting pEF-IRES-
puro5—first with NcoI, then ExonucleaseVII—to generate blunt ends before
relegation. Next, the sequence encoding MBP-FIH-1 was amplified by
polymerase chain reaction (PCR) from pMBP-FIH-1 (Lando *et al.*, 2002b),
verified by sequencing, and cloned into the NheI/Not sites of pEF-IRES-
puro6 to generate pEF-MBP-FIH-1-IRES-puro. The MBP-FIH-1
expressed from this vector is identical in primary sequence to that generated
from the pMBP-FIH-1 in bacteria.

The stable 293T cell line is generated by transfection of cells grown in
Dulbecco's modified eagle media (DMEM), 10% FCS at 37° in 5% CO_2,
with the pEF-MBP-FIH-1-IRES-puro vector using Lipofectamine2000
(Invitrogen, Carlsbad, CA), followed by selection with puromycin at
10 μg/ml for 2 wk. The polyclonal cell pool, which expresses MBP-FIH-1
at levels approximately 10^3-fold greater than endogenous FIH-1, is used
for the large-scale expression and purification of MBP-FIH-1 as follows.
A total of five 175-cm^2 flasks of near-confluent MBP-FIH-1 293T cells are
washed in ice-cold phosphate buffer solution (PBS), lysed for 30 min at 4° in
5 ml of whole cell extract buffer (WCEB: 20 mM HEPES, pH 7, 420 mM
NaCl, 0.35 mM EDTA, 1.5 mM $MgCl_2$, 25 % glycerol, 0.5 % igepal, 1 mM
PMSF, 1 mM DTT, 2 μg/ml aprotinin, 4 μg/ml bestatin, 1 μg/ml pepstatin, 5 μg/ml leupeptin), and centrifuged at 14,000g for 30 min at 4°. The
supernatant-containing soluble protein is removed, the volume made up to
10 ml with 20 mM Tris-HCl, pH 8.0, and 1 mM PMSF, and combined
with 200 μl of amylose agarose resin (Scientifix, Victoria, Australia) for 1 h
at 4°. The supernatant is discarded and the slurry washed with 5 ml of
150-mM NaCl, 20 mM Tris, pH 8.0, and 1 mM PMSF followed by 5 ml
of the same buffer without PMSF. MBP-FIH-1 is then eluted with 1 ml of
elution buffer (150 mM NaCl, 20 mM Tris, pH 8.0, 10 mM maltose), after
which it can be buffer-exchanged to remove maltose. Glycerol can also be
added to a final concentration of 20%, allowing the enzyme to be stored at
-20° for extended periods of time with minimal loss of activity. This method
yields approximately 0.01 mg of 60% pure MBP-FIH-1 (see Fig. 4.1A).

2.2. Protein quantitation

The enzyme and substrate concentrations are estimated with extinction coefficients calculated using the complete amino acid sequences via http://ca.
expasy.org/tools/protparam.html and measuring absorbances at 280 nm. The
concentrations are then adjusted where required after comparing samples by
sodium dodecyl sulfate polyacrylamide gel electrophoresis (SDS-PAGE)
and Coomassie blue staining in order to consider purity, which is typically

around 90% for the higher yield, bacterially expressed enzymes and substrates and around 60% for the lower yield mammalian expressed enzymes.

2.3. Production of substrates

Most characterization of FIH-1 has been performed using HIF-α substrates. However, the recent discovery of ankyrin repeat proteins as a new class of substrates has required the adaptation of previous methods for substrate production. One of the most interesting of these new substrates is the Notch protein, which is hydroxylated at two sites within the seven ankyrin repeats. These repeats are located within the intracellular domain (ICD), which is C-terminal to adjacent recombination signal sequence–binding protein for Jkappa genes (RBP-Jκ)–associated molecule (RAM) domain. The ICD is liberated from the full-length transmembrane receptor upon ligand binding and is responsible for signal transduction. Unlike the ankyrin repeats, which fold autonomously, the RAM domain is inherently unstructured and clearly flexible, much like the HIF-α CAD, as it participates in "induced fit" interactions (Nam *et al.*, 2003). However, in contrast to the HIF-α CAD, the RAM domain alone appears unable to interact with or be hydroxylated by FIH-1, though it may contribute to the binding between FIH-1 and the ankyrin repeat region.

For *in vitro* analyses, synthetic peptides, partial proteins, or full-length proteins can be used. Partial protein sequences encoding whole structural domains are often more efficiently expressed, soluble, and folded in *E. coli* than full-length proteins and facilitate the identification of hydroxylated residues due to their smaller size. In addition, since they are commonly folded into their native structure, they provide data more relevant to the hydroxylation of the full-length native protein than do synthetic peptides, which commonly lack significant secondary structure. However, it is convenient to obtain synthetic peptides in a purified form, negating the need for cloning and protein expression and purification. Overall, due to the soluble, folded nature of the ankyrin repeat domains purified from *E. coli* compared to their respective full-length proteins (including Notch, gankyrin, and FGIF) and their relative stability, they have been the substrates of choice for *in vitro* hydroxylation assays.

Analyses of potential and known substrates reveal a general trend in protein preparation, with distinct conditions required for the expression of the ankyrin repeat proteins compared with unstructured protein domains, such as the HIF-α CAD and Notch-RAM. The latter are more straightforward to express with an N-terminal Trx-6H tag in *E. coli* followed by Ni-affinity purification. In contrast, ankyrin repeat domains do not express efficiently unless cells are grown in conditions that maximize retention of the plasmid and minimize leaky expression of the fusion protein prior to induction. Specifically, this involves the use of carbenicillin rather than ampicillin for selection of transformed cells, the inclusion of glucose in culture medium to fully repress expression of the inducible proteins, and removal of starter culture supernatant containing secreted β-lactamase prior to inoculation of large scale culture. Using these conditions, high yields are obtained (20 mg/l of culture) of soluble, 90% pure ankyrin repeat domain proteins. Despite these measures, full-length Trx-6H-Notch ICD is expressed/purified as a highly degraded species, implying that sequences C-terminal to the ankyrin domain are unstable in *E. coli*, given that the full ICD can be expressed efficiently in mammalian cells. The presence of protease inhibitors and performing all purification steps at 4° or lower has little impact on the production of degradation products, suggesting that the degradation occurs within the *E. coli* and not during purification.

2.3.1. Protocols

The pET32a-HIF-α CAD (residues 737–826 human HIF-1α) vector has been previously described by others (Linke *et al.*, 2004), as has the abbreviated method for inducing and purifying this protein (Lando *et al.*, 2002a). Briefly, 10-ml overnight LB/amp cultures of BL21 DE3 *E. coli* transformed with pET32a-HIF-α CAD are used to inoculate a large scale 500-ml LB/amp culture, which is incubated with shaking at 37°. The culture is induced at OD_{600} of 0.6 to 0.8 with 1 mM IPTG, incubated with shaking at 37° for 2 to 5 h followed by lysis via french press (two passes) in 30 ml of lysis buffer (500 mM NaCl, 20 mM Tris-HCl, pH 8.0, 5 mM imidazole, 1 mM PMSF, 0.5 mM DTT). The lysate is then clarified by centrifugation (14,000g at 4° for 45 min) and the supernatant incubated with 1 ml of Ni-IDA agarose (Scientifix, Victoria, Australia) for 1 h at 4° with gentle rocking before transfer to a chromatography column. The bound Ni-IDA resin is then washed with 50 ml of lysis buffer followed by 50 ml of 500 mM NaCl, 20 mM, Tris-HCl (pH 8.0), 10 mM imidazole, and Trx-6H-HIF-α CAD eluted with 2.5 ml of 500 mM NaCl, 20 mM Tris-HCl (pH 8.0), 250 mM imidazole, followed by desalting (sephadex 25 PD-10 column, GE Healthcare, Buckinghamshire, UK) into 150 mM NaCl, 20 mM Tris-HCl, pH 8.0.

The mouse Notch 1 constructs are as follows: pET32a Notch 1 ICD, which encodes Trx-6H-myc, fused to the full-length mouse Notch ICD (1753–2537), was generated by cloning myc-tagged mNotch 1 ICD (Gustafsson *et al.*, 2005) into the BamH1/Sal1 site in pET-32a; pET32-RAM-ankyrin (ANK) (1753–2027), which encodes Trx-6H, fused to the myc-tagged mNotch 1 RAM domain and ankyrin repeats 1 to 4.5 with a C-terminal 6H tag and was generated by deleting the NotI/NotI fragment from pET32-Notch ICD (one NotI site is in mNotch 1 ICD, the other in pET-32a) and relegating the vector; pET32-ANK (1847–2027), which encodes Trx-6H, fused to the mNotch1 ankyrin repeats 1 to 4.5 with a C-terminal 6H tag and was generated by deleting the NcoI/NcoI fragment containing the myc-RAM domain from pET32-RAM-ANK (1753-2027; one NcoI site is in mNotch 1 ICD, the other in the MCS of pET-32a), and relegating the vector; and pET32-RAM, which encodes Trx-6H fused to the mNotch 1myc-RAM, was generated by cloning the Nco1/Nco1 fragment from pET32-RAM-ANK (1753-2027) into the Nco1-digested pET-32a vector.

Expression of Trx-6H-RAM is straightforward and robust and is performed as per Trx-6H-HIF-α CAD. Efficient induction of the ankyrin repeat domain Trx-6H proteins requires the following conditions: the 30-ml overnight starter cultures are grown at 37° in LB containing 100 μg/ml carbenicillin and 2% glucose. The following morning, the culture is centrifuged (4000g for 5 min), and the supernatant containing secreted β-lactamase is discarded. The cell pellet is resuspended in 10-ml LB and used to inoculate a 500-ml LB, 100-μg/ml carbenicillin culture, which is grown to OD$_{600}$ of 0.4 at 37°. This culture is then supplemented with extra carbenicillin (100 μg/ml) to compensate for consumption by the β-lactamase and induced with 1 mM IPTG (37° for 2 h) and proteins purified as per Trx-6H-HIF-α CAD.

2.4. CO$_2$ capture assays

FIH-1 is predicted to catalyze hydroxylation via an ordered substrate-binding pathway whereby 2OG binds the Fe^{2+}-FIH-1 complex first followed by the peptide substrate, which causes a water molecule coordinated to the Fe^{2+} center to be displaced by dioxygen (Hausinger, 2004; Price et al., 2005; Schofield and Ratcliffe, 2005). Reaction of dioxygen with Fe^{2+} leads to decarboxylation of the 2OG, releasing CO$_2$ and leaving succinate and a Fe^{IV}=O species, which oxidizes the asparagine at the β-carbon.

Some enzymes of this class catalyze uncoupled decarboxylation of 2OG, for which it is reasoned that the ascorbate cofactor becomes a surrogate acceptor of the second oxygen atom (Myllylä et al., 1984). We do not find FIH-1 to perform significant uncoupled decarboxylation of 2OG, and related work demonstrates only low levels of uncoupled activity (Koivunen et al., 2004; McNeill et al., 2005). However, FIH-1 does require ascorbate for high activity (Koivunen et al., 2004), probably to keep Fe^{2+} reduced, as is rationalized for other enzymes of the superfamily (reviewed in Schofield and Ratcliffe, 2005).

To infer hydroxylation, the ^{14}CO$_2$ capture assay is routinely used, which quantifies radio-labelled carbon dioxide released from [1-^{14}C]-2OG stoichiometrically to hydroxylation of the peptide. This method is relatively straightforward to set up, simple to perform, provides rapid results, and is very effective for small-scale preliminary screening of soluble potential FIH-1 substrates. For substrates with relatively high solubility and K$_m$, such as the Trx-6H HIF-α CAD substrates, this assay can be used to calculate apparent

V_{max} and K_m constants. However, this method does have significant limitations. First, it is an indirect assay for hydroxylation and does not distinguish substrate hydroxylation from substrate-promoted uncoupled decarboxylation. Therefore, asparaginyl hydroxylation of potential substrates must be formally verified using mass spectrometry (Peet *et al.*, 2004). Second, the method demands considerable effort if meaningful kinetic data are sought in order to avoid the kinetic artifacts detailed later in this chapter. Kinetic constants often cannot be accurately obtained due to limited sensitivity of the assay and limited solubility of some substrates. Useful alternative methods exist, though each has its own limitations. For example, measurement of 2OG consumption via reaction with *o*-phenylenediamine (OPD) to generate a fluorescent derivative avoids the use of radioactive substrates and is well suited to high-throughput screens; however, it is less suited for precise kinetic information (McNeill *et al.*, 2005). Similarly, direct measurement of oxygen consumption using a fiberoptic sensor is a better alternative for obtaining kinetic data, though it is compromised by the necessary exclusion of ascorbate and is technically more difficult (Ehrismann *et al.*, 2007).

The $^{14}CO_2$ capture assay outlined in this chapter is based on methods developed for related enzymes (Rhoads and Udenfriend, 1968; Stenflo *et al.*, 1989; Zhang *et al.*, 1999), optimized using the Trx-6H-HIF-α CAD as substrate, and adapted for use with the ankyrin repeat-containing proteins. The Trx-6H-HIF-α CAD is amenable to kinetic analyses with this assay in terms of having an apparent K_m that is within the sensitivity range of the assay and solubility sufficient to reach a V_{max}. However, with ankyrin repeat-containing substrates, the assay described later is adequate for identifying potential substrates but is unsuitable for kinetic analyses due to the lower K_m (below the sensitivity of this assay). However, this assay can be adapted to increase the sensitivity (Koivunen *et al.*, 2006), thus enabling the kinetic characterization of these substrates. This assay is also amenable to high-throughput screening in a 96-well format (Zhang *et al.*, 1999). Finally, these assays are suitable for screening potential FIH-1 antagonists and agonists for their ability to modulate FIH-1–mediated hydroxylation of specific substrates.

2.4.1. Assay conditions

In standard assay conditions, FIH-1 activity is linear with time, for up to 90 min; thus, we routinely perform assays over 20 min. All cofactors are provided in excess except 2OG, which is added at 40 μM as [1-^{14}C]-2-oxoglutaric acid with a specific activity of approximately 1000 MBq/mmol (PerkinElmer, Waltham, MA). Other groups employing the $^{14}CO_2$ capture assay find the K_m for 2OG to be 10 (Hewitson *et al.*, 2002) or 25 μM (Koivunen *et al.*, 2004). However, as 2OG concentration is increased beyond 40 μM, the activity may be reduced depending on buffer conditions; hence, a final concentration of 40 μM is used. For the other cofactors, ascorbic acid is included at 4 mM and iron as $FeSO_4$ at 1.5 mM. The first reported K_ms for these cofactors are 260 μM (ascorbate) and 0.5 μM (Fe^{2+}) (Koivunen et al., 2004), and we find similar values with our assay. A cofactor stock solution is used, including $FeSO_4$, ascorbate, dithiothreitol (DTT), and bovine serum albumin (BSA) in Tris–HCl buffer at pH 7.0 or 7.5. In general, the protein substrates are more stable and less prone to precipitation at pH 7.5, though, at this pH, the excess Fe forms a noticeable brown precipitate. This effect does not appear to influence the assay, presumably because the Fe^{2+} is present in vast excess. However, the formation of precipitate can be prevented by using less $FeSO_4$ (e.g., 100–500 μM). For the HIF-α CAD proteins, a pH of 7.0 results in higher activity and more consistent results than pH 7.5. However, with the ankyrin repeat–containing substrates, a pH of 7.5 is preferred as, at high concentrations, these proteins can precipitate at pH 7.0, and an $FeSO_4$ concentration of 100 μM is used. To prevent inappropriate disulphide bond formation for enzyme and substrates, 0.5 mM DTT is included, and 0.1 % BSA is included as nonspecific protein to aid stability of the purified proteins and possibly to absorb free radicals generated by Fenton chemistry.

Although the assay described in this chapter is performed under normal atmospheric conditions (21% O_2), by varying the levels of oxygen, the K_m in vitro can be experimentally determined. Such experiments can provide invaluable information about the ability of FIH-1 to respond to physiologically relevant changes in oxygen levels with different protein substrates. However, these experiments require the use of workstations where the atmospheric levels of oxygen can be regulated for the setting up of the individual assays and the pre-equilibration of all reagents under defined oxygen levels.

$^{14}CO_2$ is captured using filters suspended above the reaction and soaked in 30 mM $Ca(OH)_2$ solution, which will react with CO_2 to form $CaCO_3$ precipitate. In practice, the $Ca(OH)_2$ purchased (Asia Pacific Specialty Chemicals, New South Wales, Australia) contains some $CaCO_3$ due to reaction with CO_2 in air; thus, the solution used to soak the filters is a suspension. This solution is adequate as the $Ca(OH)_2$ in the filter is in large excess of what is required to capture all the $^{14}CO_2$ label, as can be verified by checking with excess FIH-1 and substrate. For this reason, manually-cut filters that fit into the top of the tubes can tolerate imprecise sizing between samples.

Pilot reactions are performed to ensure FIH-1 is limiting and the incubation time short enough to make certain that activity is linear over the incubation time (FIH-1 has limited stability under these conditions). The assay is somewhat tolerant to modulation of the cofactor/buffer conditions if required for solubilizing a substrate. We have successfully used dimethyl sulfoxide (DMSO), glycerol, triton X, Tris buffer between pH 7.0 and 8.0, $FeSO_4$ between 100 μM and 1.5 mM, and ascorbate between 400 μM and 4 mM without observing major changes in activity.

2.4.2. Protocol

Each 40-μl reaction is set up individually (1 min between samples) in screw-capped 1.5-ml polypropylene tubes (an airtight seal is essential), including replicates for each sample. First, substrate and enzyme are combined (30-μl total volume in 150 mM NaCl, 20 mM Tris pH 7–8, typically 10–100 μM substrate, 5–100 nM FIH-1), then 5 μl of freshly mixed cofactor solution is added (from concentrated frozen stocks of each cofactor) and mixed thoroughly, followed by initiation of reactions with 5 μl of [1-^{14}C]-2OG. The filter (Schleicher and Schuell gel blot paper), cut to size to fit while suspended above the reaction in the polypropylene tube, is soaked in fresh 30-mM Ca(OH)$_2$ solution, blotted briefly, and immediately placed in the tube above the reactions; the tube is then sealed. Each reaction is timed individually and terminated by removal of filters, which are dried at room temperature for 60 min, followed by application of 130 μl of scintillation fluid (ultimaGOLD XR, PerkinElmer, Waltham, MA) and scintillation counting.

2.4.3. Background calculations

When mixed with cofactors, 2OG spontaneously liberates small amounts of $^{14}CO_2$ in the absence of any FIH-1 or substrate. Addition of FIH-1 does not significantly enhance this liberation, nor does addition of FIH-1 plus a mutant substrate that is unable to be hydroxylated (e.g., HIF-1α CAD N803A). Thus, the "background" $^{14}CO_2$ release, determined for each assay in the absence of FIH-1, is relatively constant, measuring approximately 800 counts per minute (cpm) (varies between 300 and 2000 cpm depending on buffer conditions) from a total of 60,000 cpm (i.e., 0.5–3 %).

2.4.4. Limitations for kinetics

For direct kinetic comparisons, it is best to compare substrates purified under similar conditions and analyzed in the same assay. Apparent substrate K_m is assessed by measuring rate reductions due to varying one substrate concentration, ensuring that less than 10% of the substrate is consumed (or up to 40% if corrected for substrate consumption). This quantitation assumes stoichiometric decarboxylation of 2OG. Verification of hydroxylation should be confirmed by mass spectrometry. Pilot experiments should be performed to test that hydroxylation is proceeding linearly over the course of a standard assay for the lowest substrate concentrations. These pilots are also necessary because precise quantification of protein substrate is difficult, and we have encountered situations with Notch proteins where hydroxylation rates are linear with time (or enzyme concentration) using high-substrate concentrations, but rates with low substrate concentrations are not, presumably because all substrate is consumed. Perhaps the substrate concentration is consistently overestimated by our methods or, alternatively, not all may be folded to a structurally-accessible form for hydroxylation. Regardless, we ensure that a reduction in released $^{14}CO_2$ due to lowering the substrate concentration reflects a genuine rate change rather than serving as an artifact caused by complete consumption of substrate.

Analogous problems occur when using high-substrate concentrations because many proteins and peptides (both bacterially expressed and synthetic peptides) exhibit poor solubility and have a tendency to visibly aggregate in the reaction conditions. Thus, an artifactual maximum hydroxylation rate can occur when the peptide is not completely soluble. Microscopic aggregation can be checked using centrifugation or microfiltration. In addition, as reported by McNeill et al. (2005), we find that the hydroxylation rate reduces from its near-maximum with higher substrate concentrations. This may reflect direct inhibition of FIH-1 by blocking 2OG binding (McNeill et al., 2005), which, for productive hydroxylation, must occur prior to binding of the prime substrate (Elkins et al., 2003).

2.4.5. Sensitivity

Sensitivity limitations prevent determination of substrate K_m constants below 1 to 5 μM because significant activity above background risks the consumption of too much substrate in the reaction. Highly soluble substrates with a K_m of 25 to 100 μM are most amenable to kinetic analyses, such as many synthetic ankyrin and non-ankyrin–derived peptides and the bacterially expressed and purified HIF-α CAD (K_m of 40 μM) (Linke et al., 2004). By contrast, various bacterially expressed, purified ankyrin repeat–containing proteins bind FIH-1 orders of magnitude more efficiently than the HIF-α CAD. Accordingly, we have not been able to measure K_m constants using the $^{14}CO_2$ capture assay, and thus, for initial screening for major changes in FIH-1 binding caused by amino acid substitutions, wild-type and mutant ankyrin constructs are compared using a simple and rapid affinity pull-down method (described later). On the other hand, for certain short peptides with a low affinity for FIH-1, poor solubility prevents estimation of V_{max} and thus of K_m. If kinetic characterization is sought for proteins with a low K_m, a higher sensitivity assay should be performed, such as the modified version of the CO_2 capture assay reported by Koivunen et al. (2006), for which a K_m of 10 nM can be measured.

It is important to note that this assay is an in vitro assay and therefore omits essential cellular support factors and does not supply cofactors at physiological levels. Thus, the physiological relevance of the apparent K_m and V_{max} constants obtained by this method needs to be carefully interpreted. To validate genuine changes in hydroxylation observed in vitro in more physiologically relevant systems, cell-based techniques can be used. For example, HIF-α CAD proteins less efficiently hydroxylated in vitro demonstrate a resistance to normoxic repression of a reporter gene in mammalian cells (Lando et al., 2002a,b; Linke et al., 2004; Peet and Linke, 2006). Similarly, the activity of the Notch–ICD can be assessed using Notch–responsive reporter genes in mammalian cells (Gustafsson *et al.*, 2005).

2.5. Interaction assays

The affinity of FIH-1 for the Notch ankyrin domain is much greater than for the HIF-α CAD and is too high to enable measurement of substrate K_m with the CO_2 capture assay. One strategy to assess relative affinities between FIH-1 and different high affinity substrates (or interacting proteins) is to use an interaction assay, exploiting an affinity or epitope tag on either FIH-1 or the interacting protein. Such experiments not only provide valuable information about relative affinities for different substrates, but, combined with site-directed mutagenesis, can also demonstrate which regions or even which specific amino acids contribute to these interactions. Furthermore, these interaction assays can also be exploited to characterize the ability of FIH-1 inhibitors to disrupt interactions between FIH-1 and specific substrates.

We routinely use two different assays to characterize the interaction between FIH-1 and ankyrin proteins. The first is an *in vitro* pull–down assay using purified FIH-1 and substrate proteins. This assay utilizes the Trx–6H or MBP tags on the substrate or FIH-1 for affinity purification under native conditions. Advantages include the use of the same proteins and affinity resins used in the *in vitro* CO_2 capture assays and a defined system with each component at known concentrations. Importantly, this assay demonstrates a direct protein–protein interaction when only enzyme and substrate are included. This method can also be adapted to use endogenous FIH-1 from cell extracts rather than purified protein; however, the substrates are bacterially expressed and may be differentially folded or posttranslationally modified compared with endogenous substrate. Therefore, the second method is a cell-based co-immunoprecipitation that utilizes an epitope tag on overexpressed substrate protein. Expression in mammalian cells supports protein folding and modification reminiscent of endogenous proteins. This method is significantly more sensitive than the *in vitro* pull–down, presumably due to the higher proportion of folded, functional protein produced within the mammalian cells, and is suitable for proteins that are not expressed or soluble within *E. coli*. However, this method utilizes protein overexpression, and the relative concentrations of each component are poorly defined.

2.6. *In vitro* pull-down assay

Sequence analysis of mNotch1 identified a number of conserved residues of particular interest: one conserved asparagine at position 1945 and two highly conserved leucine residues at positions 1937 and 1938. To investigate the importance of these amino acids in Notch1-binding to FIH-1, a N1945A single mutant and an LL1937/38AA double mutant were generated and compared to wild-type mNotch1 1847 to 2027 using *in vitro* pull-down assays. These experiments utilize bacterially expressed and purified Trx-6H-mNotch1 1847 to 2027 wild-type, N1945A, or LL1937/38AA mutant proteins, which are bound to Ni-affinity resin, mixed with FIH-1 purified from bacteria, and washed. The bound FIH-1 is detected by Western blotting. As a negative control, the Trx-6H-mNotch1-RAM is used, as it does not appear to interact efficiently with or be hydroxylated by FIH-1. Alternatively, purified Trx-6H can be used as a negative control.

Using this method, the N1945A and the LL1937/38AA mutant are observed interacting with FIH-1 with a much lower affinity than the wild-type Notch protein (Fig. 4.2B). The role of the LL motif is likely to be two-fold. First, it is predicted to contact FIH-1 because the analogously positioned residues in the HIF-α CAD form part of the FIH-1 binding site (L795/T796 hHIF-1α) (Elkins *et al.*, 2003), and mutation (T796D) or phosphorylation at T796 reduces binding/hydroxylation between HIF-α CAD and FIH-1 (Lancaster *et al.*, 2004b; Peet and Linke, 2006). Second, the LL is invariant in the ankyrin consensus (Li *et al.*, 2006), and, structurally, this invariance is rationalized by the leucine residues contributing hydrophobic interactions that stabilize the single ankyrin fold and its interaction with neighboring repeats.

Similar experiments can be performed using either endogenous (see Fig. 4.2C) or overexpressed FIH-1 from mammalian cells rather than purified protein. These proteins, specifically the endogenous FIH-1, should be properly folded and posttranslationally modified, and the whole cell extracts should contain any accessory factors that also contribute to this interaction in a physiological environment.

This pull-down assay is a rapid and relatively simple way of characterizing interactions between FIH-1 and ankyrin repeat–containing proteins, pinpointing key substrate residues. However, this assay is only semi-quantitative, does not interrogate on/off rates, and can be followed up with more in-depth analytical binding assays, such as BiaCore (Ehrismann *et al.*, 2007).

2.6.1. Protocol

Mutant Notch constructs were generated via the Quickchange protocol (Stratagene) according to the manufacturer's directions. The pET32a-ANK 1847 to 2027 template DNA was used together with the following primer sets: N1945A upper 5′ gtgcagatgccgccatccaggacaac 3′, lower 5′ gttgtcctgga-tggcggcatctgcac 3′ LL1937/1938AA, upper 5′ gctgcaaagcgcgccgccgaggc-cagtgcag 3′, and lower 5′ctgcactggcctcggcggcgcgctttgcagc 3′. Resins are prepared by binding approximately 2 mg of Trx-6H–tagged bait proteins, including a negative control (e.g., Trx-6H alone or fused to a non-interacting protein), with 200 μl of Ni-IDA resin in a total volume of

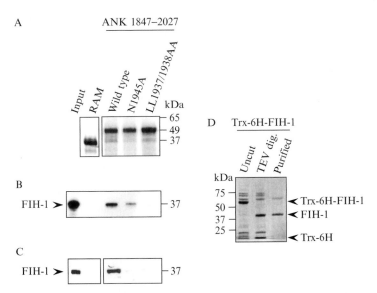

Figure 4.2 The interaction between factor-inhibiting HIF-1 (FIH-1) and mNotch 1 (1847–2027) is disrupted by alanine mutagenesis of a conserved asparagine (N1945) or a conserved dileucine motif (LL1937/1938). (A) Trx-6H-mNotch 1 (1847 to 2027) wild-type, N1945A, or LL1937/1938AA mutant proteins (or Trx-6H-myc-RAM as a negative control) were expressed and purified from *Escherichia coli* by nickel affinity chromatography and equal amounts bound to nickel-resin and checked with sodium dodecyl sulfate polyacrylamide gel electrophoresis (SDS-PAGE) and Coomassie staining. (B) Equal volumes of the resins were mixed with purified FIH-1 and washed, and the bound FIH-1 was eluted and visualized by anti-FIH-1 immunoblotting (50% input is shown). (C) The same proteins were used to bind FIH-1 from HEK 293T cell lysate (10% input is shown). It should be noted that all samples are present upon the same gel; the separation derives from omission of irrelevant intervening samples. (D) FIH-1 purified from *E. coli* cells as Trx-6H-FIH-1, cleaved with TEV protease, and checked with SDS-PAGE and Coomassie staining. Representative of more than three independent experiments using purified tagged or untagged FIH-1 or FIH-1 from lysate, all of which give similar results.

1.2 ml of 20 mM Tris, pH 8.0, and 150 mM NaCl at 4° for 1 h. This combination is then washed with 3 × 1 ml of 20 mM Tris, pH 8.0, and 150 mM NaCl, and a portion is analyzed by SDS-PAGE to ensure they contain equivalent amounts of bound bait protein. The bait-bound resin can be stored for up to 2 mo at 4° (addition of 1 mM PMSF and 0.01% sodium azide to the wash buffer extends the stability of these proteins). Twenty microliters of the bait-bound resin is then used to capture approximately 5 to 20 μg of bacterially expressed and purified FIH-1 with or without an affinity tag (in 20 mM Tris, pH 8, and 150 mM NaCl) and made up to a final volume of 200 μl with WCEB. This combination is bound for 1 h at 4°, the resin is washed with 3 × 1 ml of 20 mM Tris, pH 7.5, 150 mM NaCl, 10 % glycerol, 1% NP40, and bound protein is eluted with 30 μl of 250 mM imidazole, 500 mM NaCl, and 20 mM Tris, pH 8.0. Fifteen microliters of eluted FIH-1 is separated by SDS-PAGE and analyzed by Western blotting using an antibody to FIH-1 (100–428 A2, Novus Biologicals, Littleton, CO) used at 1/1000 in PBS 0.1% Tween.

Whole cell extracts containing endogenous FIH-1 or overexpressed MBP-FIH-1 can be used in place of purified FIH-1 (as previously described for purification of MBP-FIH-1 from HEK-293T cells). In this case, approximately 200 μl of whole cell extract is added to 20 μl of the resin-bound bait and treated as previously stated.

2.7. Co-immunoprecipitation assay

The *in vitro* pull-down assays described earlier are reliant upon successful expression of substrate/interacting proteins in bacteria and solubility of the proteins in the relevant buffer conditions used in the assays. As mentioned previously, for some ankyrin repeat–containing proteins, such as the Notch ICD, it is difficult to obtain a high enough yield of intact protein. Co-immunoprecipitation assays provide a means to minimize these problems, as the FIH-1 and bait proteins utilized in these assays are expressed in mammalian cells. As a result, the proteins undergo folding and posttranslational modifications similar to endogenous protein, thereby promoting solubility. Subsequent analysis of protein complex formation is more physiologically relevant. An additional advantage is the ability to examine complex formation between endogenous proteins. In all cell types tested thus far (Bracken *et al.*, 2006; Cockman *et al.*, 2006; Soilleux *et al.*, 2005; Stolze *et al.*, 2004), FIH-1 protein has been readily detectable by Western blotting, such that endogenous levels of expression are high enough to be amenable to co-immunoprecipitation assays. If endogenous levels of a potential interacting partner are also sufficient, as has been demonstrated for p105 and IκBα (Cockman *et al.*, 2006), the use of artificial affinity tags in the assay can be avoided altogether. However, for simple analysis of potential FIH-1 substrates/binding partners, overexpression of bait proteins in HEK 293T cells with a 6*myc* epitope tag followed by immunoprecipitation with 9E10 anti-*myc* hybridoma supernatant has proved to be a very useful technique. Similar experiments using other affinity tags and cells types have indicated that the quality of FIH-1 interactions with ankyrin repeat proteins is consistently robust (Cockman *et al.*, 2006).

Much like the *in vitro* pull-down assays, analyses of the co-immuno-precipitation data are restricted to semi-quantitative comparisons of binding affinities between different substrates/interacting proteins. Indeed, overexpression of the bait proteins within mammalian cells complicates comparison as it precludes the use of equimolar or physiological amounts of each protein. Despite its limitations, the co-immunoprecipitation assay is not overly labor intensive, as the extra maintenance required for mammalian cell culture is balanced by the small quantity of protein utilized by the assay, which does not require purification. Overall, the opportunity to investigate FIH-1 interactions in a cell-based context is valuable, and provides important information to guide further study.

As an example, to verify candidate FIH-1 interacting proteins isolated by yeast two-hybrid assay, including numerous ankyrin repeat–containing proteins, co-immunoprecipitation assays were utilized. FGIF, which contains three ankyrin repeats, is one such candidate. To confirm a specific interaction, full-length mouse FGIF (aa 1-238) was expressed with a *6myc* epitope tag in HEK-293T cells, which were untreated or treated with dimethyloxalyl glycine (DMOG, an FIH-1 inhibitor) or vehicle (dimethylsulfoxide [DMSO]). DMOG is metabolized by cells to the 2OG analogue N-oxalylglycine (N-OG), which inhibits the 2OG-dependent enzyme activity. The inclusion of DMOG in this co-immunoprecipitation experiment is reasoned to "stall" the interaction between enzyme and substrate and thus makes the interaction more stable. *Myc*-tagged FGIF was then co-immunoprecipitated from cell extracts using the *myc*-specific 9E10 monoclonal antibody (Fig. 4.3). The ribosomal protein S2, previously shown to have no affinity for FIH-1 (data not shown), was used as a nonspecific control, and the endogenous FIH-1 that co-immunoprecipitated with FGIF was detected by Western blot using a polyclonal anti-FIH-1–specific antibody. The Western blots of the input (see Fig. 4.3A) show similar amounts of FIH-1 and either *6myc*-FGIF or S2 protein in each sample. The *6myc*-FGIF and S2 protein were efficiently immunoprecipitated (see Fig. 4.3B, lower panel), but only FGIF co-immunoprecipitated FIH-1, and this partnership was unchanged by treatment with DMOG or DMSO (Fig. 4.3B, upper panel). Hence, these results clearly demonstrate a relatively strong interaction between FGIF and FIH-1.

2.7.1. Protocol

pEF-*6myc*-FGIF (1-238)-BOS-CS was generated through PCR amplification from a mouse 10.5 dpc embryonic complimentary DNA (cDNA) library using primers (upper 5'-cggtcgacaagagtagtaaaaatgg-3', lower 5'-ttgcggccgctgtattacctgtgtcc-3') and cloning of the PCR product into pEF-6myc-BOS-CS using SalI/NotI restriction enzymes. pEF-*6myc*-rpS2-BOS-CS was generated in a two-step process: ribosomal protein S2 cDNA was excised from the pPC86 vector (Invitrogen, Carlsbad, CA) using SalI/NotI restriction enzymes and ligated into pBS(KS) (Stratagene, Cedar Creek, TX), which had been modified with an N-terminal *6myc* tag. *6myc*-rpS2 was then subcloned into the pEF-BOS-CS vector (Goldman *et al.*, 1996) using KpnI/NotI enzymes. To generate pBS(KS)-6myc, a *6myc* tag was amplified by PCR and cloned KpnI/NotI into pBS(KS). pEF-*6myc*-BOS-CS was generated through KpnI/NotI excision of the *6myc* tag from pBS(KS) and ligation into pEF-BOS-CS.

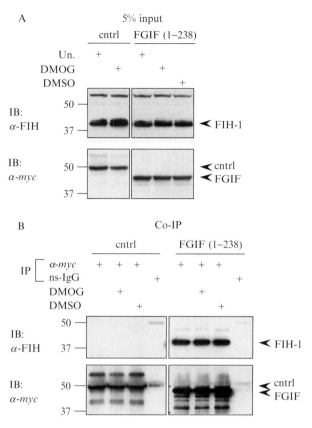

Figure 4.3 The ankyrin repeat protein fetal globin-inducing factor (FGIF) and factor-inhibiting HIF-1 (FIH-1) interact in co-immunoprecipitation assays. pEF-6*myc*-FGIF-BOS and pEF-6myc-rpS2-BOS plasmids were transfected into human embryonic kidney (HEK) 293T cells, and the cells were left untreated or treated with dimethyloxalyl glycine (DMOG) or vehicle (dimethylsulfoxide glycine [DMSO]) prior to generation of whole cell extracts. (A) Five percent of the untreated or DMOG/DMSO-treated extracts used in the immunoprecipitates (IPs) was analyzed by immunoblotting to ensure similar input. (B) The extracts were immunoprecipitated with indicated antibodies and the co-immunoprecipitated FIH-1 analyzed by immunoblotting. Representative of two independent experiments. IB, immunoblot; cntrl, control.

To generate cell extracts, 50% confluent 175-cm^2 flasks of HEK-293T cells (cultured in DMEM with 10% FCS at 37° in 5% CO$_2$) were transfected with 15 μg per flask of pEF-6-*myc*-FGIF-BOS-CS or a control construct (pEF-6-*myc*-rpS2-BOS-CS) using Lipofectamine2000 (Invitrogen, Carlsbad, CA). Five hours after transfection, cells were treated with 1 mM DMOG from a 1000 × stock or vehicle (DMSO) or left untreated for a further 16 h. Whole cell extracts were then prepared (see production of MBP-FIH-1 in HEK 293T cells, shown previously) and stored at −80°.

Anti-*myc* 9E10 hybridoma supernatant (500 μl) or a non-specific mouse IgG as a control are applied to 30 μl of protein G sepharose (PGS) resin (Invitrogen, Carlsbad, CA), pre-equilibrated in IP-binding buffer (10 mM Tris, pH 7.5, 0.1% Triton X-100, 150 mM KCl, 2 mM EDTA, 1 mM PMSF), and incubated overnight (or for 2 h if more convenient) at 4°. Due to the variability of antibody concentration in different hybridoma supernatants, some optimization may be required to determine the volume to saturate the PGS. Next, 300 μl of immunoprecipitate (IP)-binding buffer and 90 μl (approximately 500 to 700 μg total protein) of whole cell extract are added to the antibody-bound PGS resin and incubated for 2 h at 4°. The resin is then washed twice with 500 μl IP-binding buffer. Bound complexes are eluted by heating the Sepharose for 5 min at 95° in 40 μl of elution buffer (10 mM Tris, pH 7.5, 0.5 % SDS, 50 mM β-mercaptoethanol). Alternatively, the Sepharose can be incubated at 55° for 5 min or the concentration of SDS or β-mercaptoethanol in the elution buffer decreased to reduce co-elution of antibodies. The results are then analyzed by Western blotting using a polyclonal antibody against FIH-1 (previously described in Linke *et al.* [2004], used at 1/250 in PBS 0.1% Tween with 2% milk) or the 9E10 hybridoma supernatant (used at 1/4 in PBS 0.1% Tween).

3. DISCUSSION/CONCLUSION

Since the discovery of FIH-1 as an asparaginyl hydroxylase, it has been the subject of intense study as a mediator of the cellular hypoxic response via the HIF-α proteins, as a primary cellular oxygen sensor, and as an attractive therapeutic target. The techniques for *in vitro* and cell-based analyses described herein have been utilized extensively together with mass spectrometry to not only characterize the role of FIH-1 and hydroxylation on HIF signaling, but also to identify new substrates. The recent discoveries of ankyrin domain–containing proteins as FIH-1 substrates, such as IκBα, p105, and Notch, have created a significant amount of interest. The characterization of these new substrates—in particular, their relative affinities for FIH-1—will be crucial for our understanding of the physiological role of FIH-1, and more substrates are likely to be identified in the future.

As our knowledge of the repertoire of FIH-1 substrates expands, so do the implications for therapeutic strategies directly targeting this enzyme. Hence, for a detailed understanding of the physiological implications of these findings, these *in vitro* techniques will need to be complemented with more cellular and whole animal studies. Together, they should provide not only a broader understanding of the role of FIH-1 in normal physiology and disease, but also result in the development of techniques amenable to the characterization of other known peptidyl hydroxylases, such as the PHDs.

ACKNOWLEDGMENTS

We would like to thank Fiona Whelan for the construction and provision of the pET-32a-TEV vector and Maria Gustafsson for assistance with the generation of the Notch constructs.

REFERENCES

Bracken, C. P., Fedele, A. O., Linke, S., Balrak, W., Lisy, K., Whitelaw, M. L., and Peet, D. J. (2006). Cell-specific regulation of hypoxia-inducible factor (HIF)-1α and HIF-2α stabilization and transactivation in a graded oxygen environment. *J. Biol. Chem.* **281,** 22575–22585.

Clifton, I. J., McDonough, M. A., Ehrismann, D., Kershaw, N. J., Granatino, N., and Schofield, C. J. (2006). Structural studies on 2-oxoglutarate oxygenases and related double-stranded β-helix fold proteins. *J. Inorg. Biochem.* **100,** 644–669.

Cockman, M. E., Lancaster, D. E., Stolze, I. P., Hewitson, K. S., McDonough, M. A., Coleman, M. L., Coles, C. H., Yu, X., Hay, R. T., Ley, S. C., Pugh, C. W., Oldham, N. J., *et al.* (2006). Posttranslational hydroxylation of ankyrin repeats in IκB proteins by the hypoxia-inducible factor (HIF) asparaginyl hydroxylase, factor inhibiting HIF (FIH). *Proc. Natl. Acad. Sci. USA* **103,** 14767–14772.

Dames, S. A., Martinez-Yamout, M., De Guzman, R. N., Dyson, H. J., and Wright, P. E. (2002). Structural basis for HIF-1α/CBP recognition in the cellular hypoxic response. *Proc. Natl. Acad. Sci. USA* **99,** 5271–5276.

Dann, C. E., 3rd, Bruick, R. K., and Deisenhofer, J. (2002). Structure of factor-inhibiting hypoxia-inducible factor 1: An asparaginyl hydroxylase involved in the hypoxic response pathway. *Proc. Natl. Acad. Sci. USA* **99,** 15351–15356.

Dinchuk, J. E., Focht, R. J., Kelley, J. A., Henderson, N. L., Zolotarjova, N. I., Wynn, R., Neff, N. T., Link, J., Huber, R. M., Burn, T. C., Rupar, M. J., Cunningham, M. R., *et al.* (2002). Absence of posttranslational aspartyl β-hydroxylation of epidermal growth factor domains in mice leads to developmental defects and an increased incidence of intestinal neoplasia. *J. Biol. Chem.* **277,** 12970–12977.

Ehrismann, D., Flashman, E., Genn, D. N., Mathioudakis, N., Hewitson, K. S., Ratcliffe, P. J., and Schofield, C. J. (2007). Studies on the activity of the hypoxia-inducible factor hydroxylases using an oxygen consumption assay. *Biochem. J.* **401,** 227–234.

Elkins, J. M., Hewitson, K. S., McNeill, L. A., Seibel, J. F., Schlemminger, I., Pugh, C. W., Ratcliffe, P. J., and Schofield, C. J. (2003). Structure of factor-inhibiting hypoxia-inducible factor (HIF) reveals mechanism of oxidative modification of HIF-1α. *J. Biol. Chem.* **278,** 1802–1806.

Goldman, L. A., Cutrone, E. C., Kotenko, S. V., Krause, C. D., and Langer, J. A. (1996). Modifications of vectors pEF-BOS, pcDNA1 and pcDNA3 result in improved convenience and expression. *Biotechniques* **21,** 1013–1015.

Gustafsson, M. V., Zheng, X., Pereira, T., Gradin, K., Jin, S., Lundkvist, J., Ruas, J. L., Poellinger, L., Lendahl, U., and Bondesson, M. (2005). Hypoxia requires notch signaling to maintain the undifferentiated cell state. *Dev. Cell.* **9,** 617–628.

Hausinger, R. P. (2004). FeII/α-ketoglutarate-dependent hydroxylases and related enzymes. *Crit. Rev. Biochem. Mol. Biol.* **39,** 21–68.

Hewitson, K. S., McNeill, L. A., Riordan, M. V., Tian, Y. M., Bullock, A. N., Welford, R. W., Elkins, J. M., Oldham, N. J., Bhattacharya, S., Gleadle, J. M., Ratcliffe, P. J., Pugh, C. W., *et al.* (2002). Hypoxia-inducible factor (HIF) asparagine hydroxylase is identical to factor inhibiting HIF (FIH) and is related to the cupin structural family. *J. Biol. Chem.* **277**, 26351–26355.

Hirsila, M., Koivunen, P., Gunzler, V., Kivirikko, K. I., and Myllyharju, J. (2003). Characterization of the human prolyl 4-hydroxylases that modify the hypoxia-inducible factor. *J. Biol. Chem.* **278**, 30772–30780.

Hobbs, S., Jitrapakdee, S., and Wallace, J. C. (1998). Development of a bicistronic vector driven by the human polypeptide chain elongation factor 1α promoter for creation of stable mammalian cell lines that express very high levels of recombinant proteins. *Biochem. Biophys. Res. Commun.* **252**, 368–372.

Jia, S., VanDusen, W. J., Diehl, R. E., Kohl, N. E., Dixon, R. A., Elliston, K. O., Stern, A. M., and Friedman, P. A. (1992). cDNA cloning and expression of bovine aspartyl (asparaginyl) β-hydroxylase. *J. Biol. Chem.* **267**, 14322–14327.

Koivunen, P., Hirsila, M., Kivirikko, K. I., and Myllyharju, J. (2006). The length of peptide substrates has a marked effect on hydroxylation by the hypoxia-inducible factor prolyl 4-hydroxylases. *J. Biol. Chem.* **281**, 28712–28720.

Koivunen, P., Hirsila, M., Gunzler, V., Kivirikko, K. I., and Myllyharju, J. (2004). Catalytic properties of the asparaginyl hydroxylase (FIH) in the oxygen sensing pathway are distinct from those of its prolyl 4-hydroxylases. *J. Biol. Chem.* **279**, 9899–9904.

Lahousse, S. A., Carter, J. J., Xu, X. J., Wands, J. R., and de la Monte, S. M. (2006). Differential growth factor regulation of aspartyl-(asparaginyl)-β-hydroxylase family genes in SH-Sy5y human neuroblastoma cells. *B.M.C. Cell Biol.* **7**, 41.

Lancaster, D. E., McDonough, M. A., and Schofield, C. J. (2004a). Factor inhibiting hypoxia-inducible factor (FIH) and other asparaginyl hydroxylases. *Biochem. Soc. Trans.* **32**, 943–945.

Lancaster, D. E., McNeill, L. A., McDonough, M. A., Aplin, R. T., Hewitson, K. S., Pugh, C. W., Ratcliffe, P. J., and Schofield, C. J. (2004b). Disruption of dimerization and substrate phosphorylation inhibit factor inhibiting hypoxia-inducible factor (FIH) activity. *Biochem. J.* **383**, 429–437.

Lando, D., Peet, D. J., Whelan, D. A., Gorman, J. J., and Whitelaw, M. L. (2002a). Asparagine hydroxylation of the HIF transactivation domain a hypoxic switch. *Science* **295**, 858–861.

Lando, D., Peet, D. J., Gorman, J. J., Whelan, D. A., Whitelaw, M. L., and Bruick, R. K. (2002b). FIH-1 is an asparaginyl hydroxylase enzyme that regulates the transcriptional activity of hypoxia-inducible factor. *Genes Dev.* **16**, 1466–1471.

Lee, C., Kim, S. J., Jeong, D. G., Lee, S. M., and Ryu, S. E. (2003). Structure of human FIH-1 reveals a unique active site pocket and interaction sites for HIF-1 and von Hippel-Lindau. *J. Biol. Chem.* **278**, 7558–7563.

Li, J., Mahajan, A., and Tsai, M. D. (2006). Ankyrin repeat: A unique motif mediating protein-protein interactions. *Biochemistry* **45**, 15168–15178.

Linke, S., Stojkoski, C., Kewley, R. J., Booker, G. W., Whitelaw, M. L., and Peet, D. J. (2004). Substrate requirements of the oxygen-sensing asparaginyl hydroxylase factor-inhibiting hypoxia-inducible factor. *J. Biol. Chem.* **279**, 14391–14397.

Mahon, P. C., Hirota, K., and Semenza, G. L. (2001). FIH-1: A novel protein that interacts with HIF-1α and VHL to mediate repression of HIF-1 transcriptional activity. *Genes Dev.* **15**, 2675–2686.

McNeill, L. A., Bethge, L., Hewitson, K. S., and Schofield, C. J. (2005). A fluorescence-based assay for 2-oxoglutarate-dependent oxygenases. *Anal. Biochem.* **336**, 125–131.

McNeill, L. A., Hewitson, K. S., Claridge, T. D., Seibel, J. F., Horsfall, L. E., and Schofield, C. J. (2002). Hypoxia-inducible factor asparaginyl hydroxylase (FIH-1) catalyses hydroxylation at the beta-carbon of asparagine-803. *Biochem. J.* **367**, 571–575.

Mosavi, L. K., Minor, D. L., Jr., and Peng, Z. Y. (2002). Consensus-derived structural determinants of the ankyrin repeat motif. *Proc. Natl. Acad. Sci. USA* **99**, 16029–16034.

Myllylä, R., Majamaa, K., Gunzler, V., Hanauske-Abel, H. M., and Kivirikko, K. I. (1984). Ascorbate is consumed stoichiometrically in the uncoupled reactions catalyzed by prolyl 4-hydroxylase and lysyl hydroxylase. *J. Biol. Chem.* **259,** 5403–5405.

Nam, Y., Weng, A. P., Aster, J. C., and Blacklow, S. C. (2003). Structural requirements for assembly of the CSL.intracellular Notch1.Mastermind-like 1 transcriptional activation complex. *J. Biol. Chem.* **278,** 21232–21239.

Peet, D., and Linke, S. (2006). Regulation of HIF: Asparaginyl hydroxylation. *Novartis Found. Symp.* **272,** 37–49.

Peet, D. J., Lando, D., Whelan, D. A., Whitelaw, M. L., and Gorman, J. J. (2004). Oxygen-dependent asparagine hydroxylation. *Methods Enzymol.* **381,** 467–487.

Price, J. C., Barr, E. W., Hoffart, L. M., Krebs, C., and Bollinger, J. M., Jr. (2005). Kinetic dissection of the catalytic mechanism of taurine:α-ketoglutarate dioxygenase (TauD) from *Escherichia coli*. *Biochemistry* **44,** 8138–8147.

Rhoads, R. E., and Udenfriend, S. (1968). Decarboxylation of α-ketoglutarate coupled to collagen proline hydroxylase. *Proc. Natl. Acad. Sci. USA* **60,** 1473–1478.

Schofield, C. J., and Ratcliffe, P. J. (2005). Signalling hypoxia by HIF hydroxylases. *Biochem. Biophys. Res. Commun.* **338,** 617–626.

Semenza, G. L. (2003). Targeting HIF-1 for cancer therapy. *Nat. Rev. Cancer* **3,** 721–732.

Soilleux, E. J., Turley, H., Tian, Y. M., Pugh, C. W., Gatter, K. C., and Harris, A. L. (2005). Use of novel monoclonal antibodies to determine the expression and distribution of the hypoxia regulatory factors PHD-1, PHD-2, PHD-3 and FIH in normal and neoplastic human tissues. *Histopathology* **47,** 602–610.

Stenflo, J., Stenberg, Y., and Muranyi, A. (2000). Calcium-binding EGF-like modules in coagulation proteinases: Function of the calcium ion in module interactions. *Biochim. Biophys. Acta* **1477,** 51–63.

Stenflo, J., Holme, E., Lindstedt, S., Chandramouli, N., Huang, L. H., Tam, J. P., and Merrifield, R. B. (1989). Hydroxylation of aspartic acid in domains homologous to the epidermal growth factor precursor is catalyzed by a 2-oxoglutarate-dependent dioxygenase. *Proc. Natl. Acad. Sci. USA* **86,** 444–447.

Stolze, I. P., Tian, Y. M., Appelhoff, R. J., Turley, H., Wykoff, C. C., Gleadle, J. M., and Ratcliffe, P. J. (2004). Genetic analysis of the role of the asparaginyl hydroxylase factor inhibiting hypoxia-inducible factor (HIF) in regulating HIF transcriptional target genes. *J. Biol. Chem.* **279,** 42719–42725.

Takeuchi, T., Watanabe, Y., Takano-Shimizu, T., and Kondo, S. (2006). Roles of jumonji and jumonji family genes in chromatin regulation and development. *Dev. Dyn.* **235,** 2449–2459.

Tian, X., and Fang, J. (2007). Current perspectives on histone demethylases. *Acta Biochim. Biophys. Sin. (Shanghai)* **39,** 81–88.

Trewick, S. C., McLaughlin, P. J., and Allshire, R. C. (2005). Methylation: Lost in hydroxylation? *EMBO Rep.* **6,** 315–320.

Wang, Q. P., VanDusen, W. J., Petroski, C. J., Garsky, V. M., Stern, A. M., and Friedman, P. A. (1991). Bovine liver aspartyl beta-hydroxylase. Purification and characterization. *J. Biol. Chem.* **266,** 14004–14010.

Wouters, M. A., Rigoutsos, I., Chu, C. K., Feng, L. L., Sparrow, D. B., and Dunwoodie, S. L. (2005). Evolution of distinct EGF domains with specific functions. *Protein Sci.* **14,** 1091–1103.

Zhang, J. H., Qi, R. C., Chen, T., Chung, T. D., Stern, A. M., Hollis, G. F., Copeland, R. A., and Oldenburg, K. R. (1999). Development of a carbon dioxide-capture assay in microtiter plate for aspartyl-beta-hydroxylase. *Anal. Biochem.* **271,** 137–142.

CHAPTER FIVE

TRANSGENIC MODELS TO UNDERSTAND HYPOXIA-INDUCIBLE FACTOR FUNCTION

Andrew Doedens *and* Randall S. Johnson

Contents

1. Introduction	88
2. Hypoxia Response Pathway Genes and Development	90
2.1. HIF-α subunits	90
2.2. HIF-β/ARNT subunits	91
2.3. pVHL	92
2.4. PHDs	92
2.5. VEGF	93
3. HIF in Physiology	93
3.1. HIF-1α heterozygotes	93
3.2. Conditional knockouts using the loxP/cre recombinase system	94
3.3. Liver	94
3.4. Myeloid lineage	94
3.5. Lymphocytes	95
3.6. Hematopoiesis	95
3.7. Skeletal muscle	95
3.8. Heart and cardiovascular system	96
3.9. Mammary gland	96
3.10. Colon	96
3.11. Chondrocytes	96
3.12. Skin	97
3.13. Motor neurons and brain	97
4. HIF Function in Tumor Biology	98
4.1. First insights—xenografts	98
4.2. Astrocytoma	99
4.3. Mammary carcinogenesis	99
4.4. pVHL and VHL disease	99
5. Summary	100
References	101

Molecular Biology Section, Division of Biological Sciences, University of California San Diego, La Jolla, California

Methods in Enzymology, Volume 435
ISSN 0076-6879, DOI: 10.1016/S0076-6879(07)35005-2

Abstract

The hypoxia-inducible factor (HIF) is a heterodimeric basic helix-loop-helix (bHLH) transcription factor that controls the mammalian cellular transcriptional response to low oxygen tension by up-regulating genes including glycolytic enzymes and angiogenic factors, such as the vascular endothelial growth factor (VEGF). Under normal oxygen tensions, the pathway is negatively regulated by posttranslational proteasomal degradation of HIF-alpha (α) subunits in a pathway requiring prolyl-hydroxylase domain (PHD) containing enzyme modification followed by von-Hippel Lindau (VHL) tumor suppressor polyubiquitination (pVHL). Murine knockouts of HIF, pVHL, PHD, and VEGF have demonstrated the essential role of these hypoxic response pathway proteins in development. Conditional deletion of these genes in a wide range of tissues has further shown that ablation or overexpression of the pathway has profound *in vivo* effects, with important implications for physiology, pathology, and tumor biology. This review aims to summarize the insights garnered from key murine knockouts and transgenics involving components of the HIF hypoxia response pathway.

1. INTRODUCTION

Hypoxia-inducible factor is a heterodimeric transcription factor of the bHLH Per-ARNT-Sim (bHLH-PAS) superfamily that regulates over 100 genes involved in glycolysis, metabolism, cell cycle, apoptosis, and angiogenesis (Semenza, 2001). Formation of a heterodimer composed of an HIF-α subunit (HIF-1α, HIF-2α, or HIF-3α) and HIF-beta (β) subunit (typically ARNT1) is required before translocation from the cytoplasm to the nucleus and subsequent binding to hypoxic response elements (HREs) and transcriptional transactivation (Beck *et al.*, 1991; Pugh *et al.*, 1991; Semenza and Wang, 1992; Semenza *et al.*, 1994; Wang *et al.*, 1995). Although both α and β subunits are constitutively expressed, under normoxic conditions one or more PHD containing enzymes modify the α subunit, which allows it to then be bound and polyubiquitinated by the E3 ubiquitin ligase VHL made up of pVHL, Cullin2, Elongin B, Elongin C, and Rbx1 (Kaelin, 2002). Polyubiquitinated α subunits are quickly degraded by the cellular proteasome apparatus, thus blocking HIF transcription under normoxia in most cell types. During cellular exposure to hypoxia, low oxygen tensions limit the activity of the PHD enzymes, leaving α subunits unhydroxylated and unrecognizable by the VHL complex. Accumulation of α subunits in the cytoplasm allows for the aforementioned heterodimeric nuclear translocation and transactivation (Bruick and McKnight, 2001; Epstein *et al.*, 2001; Ivan *et al.*, 2001; Jaakkola *et al.*, 2001). A further level of regulation of the HIF transcriptional machinery is exerted by the factor-inhibiting HIF (FIH). FIH

Table 5.1 Selected milestones and transgenic mice in 15 years of hypoxia-inducible factor research

Year	Event	References
1992	HRE identified and HIF-1α first described	Wang et al., 1995 Wang and Semenza, 1995
1995	HIF-1α purified and characterized	Wang et al., 1995 Wang and Semenza, 1995
1996	VEGF knockout	Carmeliet et al., 1996 Ferrara et al., 1996
1997	ARNT1 knockout	Kozak et al., 1997 Maltepe et al., 1997
1997	pVHL knockout	Gnarra et al., 1997
1997	HIF-2α published	Ema et al., 1997 Flamme et al., 1997 Tian et al., 1997
1998	HIF-3α published	Gu et al., 1998
1998	HIF-1α knockout	Iyer et al., 1998 Ryan et al., 1998
1998	HIF-2α knockout	Tian et al., 1998
2001	ARNT2 knockout	Keith et al., 2001
2001	HIF-1α ODD mutant in skin	Elson et al., 2001
2001	FIH published	Mahon et al., 2001
2001	HIF regulation by prolyl hydroxylases published	Bruick and McKnight, 2001 Epstein et al., 2001 Ivan et al., 2001 Jaakkola et al., 2001
2006	ODD-luciferase mouse	Safran et al., 2006
2006	HIF-1α and HIF-2α proline → alanine mutants	Kim et al., 2006
2006	PHD1, -2, and -3 knockouts described	Takeda et al., 2006

hydroxylates HIF-α subunits under normoxia and mild hypoxia, but not in more pronounced hypoxia. Hydroxylation by FIH inhibits the binding of transcriptional coactivator CREB-binding protein (CBP)/p300 at the HRE, which limits HIF transactivation potential (Hewitson et al., 2002; Mahon et al., 2001). In this way, the HIF system can modulate transcription from HRE-containing genes sensitively over a wide range of oxygen tensions, from mild to severe hypoxia.

Transgenic mice (Table 5.1) have been used to unequivocally determine the requirement of a gene for development and to test the role of a gene in all the complexity of an *in vivo* system—both in normal physiology as well as tumor biology. This review aims to sum up the current state of knowledge of the requirement of HIF and related genes for development, and their role in physiology as determined *in vivo* by transgenic models, to discuss selected publications demonstrating the role of HIF family proteins in *in vivo* tumor biology.

2. Hypoxia Response Pathway Genes and Development

Knockout strategies have been employed in the mouse to determine the role of many of the proteins involved in orchestrating the hypoxic response. Homologous recombination strategies that remove an essential part of the gene or the entire gene (i.e., straight knockout or global knockout) have revealed that many of these proteins are essential for development and thus viability.

2.1. HIF-α subunits

HIF-1α is the most widely expressed α subunit and is expressed in nearly all tissues (Wiener *et al.*, 1996). Genetic ablation of HIF-1α in the mouse results in embryonic lethality at mid-gestation. Embryos exhibited a lack of vascularization, reduction in number of somites, neural fold defects, decreased size, and increased tissue hypoxia (Iyer *et al.*, 1998; Ryan *et al.*, 1998). Further characterization of HIF-1α knockout mice using whole embryo culture revealed defects in ventricle development, pharyngeal arch growth, and cephalic vascularization. Hyperoxia extended survival, but did not result in full rescue (Compernolle *et al.*, 2003).

HIF-2α displays a somewhat more restricted expression pattern, appearing in endothelium, kidney, liver, lung, and brain (Tian *et al.*, 1997). HIF-2α has been germline-deleted by several independent groups. Unlike HIF-1α, HIF-2α seems to be sensitive to the presence of modifier alleles in the distinct mouse backgrounds used to generate the knockouts. As a result, different phenotypes have been reported by the independent groups. Given that manipulating the strain background of HIF-2α null mice can even lead to viable mice (Scortegagna *et al.*, 2003b; see "Hematopoiesis," Section 3.6), perhaps it is not surprising that different strain backgrounds can result in unique developmental failures in HIF-2α null mice.

The first group to report on deletion of HIF-2α observed embryonic lethality at mid-gestation due to bradycardia as a result of a defect in catecholamine synthesis. Viable mice could be generated by administering pregnant females a catecholamine precursor, dioleoyl phosphatidylserine

(DOPS) (Tian *et al.*, 1998). HIF-2α targeting by an independent group (Peng *et al.*, 2000) resulted in embryonic lethality between E9.5 and 13.5. However, significant changes in heart rate or catecholamine levels between genotypes were not observed. Although vasculogenesis took place, post-vasculogenesis changes required for the formation of the adult vasculature did not take place in the null. The authors also discussed important differences in the mouse background of the mice generated in their lab versus the null mice generated by Tian *et al.* (1998) and suggest that strain differences are behind what would appear to be divergent results.

A third group has reported on the role of HIF-2α and the control it exerts over VEGF in lung development (Compernolle *et al.*, 2002). Although half of HIF-2α null mice died *in utero* after E13.5, surviving mice died within 2 to 3 days after birth of respiratory distress syndrome caused by failure of pneumocytes to produce surfactant. Mice could be rescued by stimulation of VEGFR2 by antibody or VEGF itself, which restored the surfactant production.

No published information about a HIF-3α knockout in mice is available at the time of this review.

2.2. HIF-β/ARNT subunits

Another approach to studying HIF function is to genetically manipulate the obligate partners of the α subunits, the ARNT proteins. HIF-1β/ARNT1 null embryos do not survive past 10.5 days of gestation, with embryos smaller than wild-type exhibiting defective angiogenesis in the yolk sac and branchial arches (Maltepe *et al.*, 1997). Another team of investigators reported similar embryonic lethality along with neural tube closure defects, placental hemorrhage, stunted forebrain growth, and visceral arch anomalies. This team faulted embryonic lethality on the failure to vascularize the placenta and to form the labyrinthine spongiotrophoblast, and found no defects in yolk sac circulation (Kozak *et al.*, 1997).

Although ARNT1 is the most widely expressed and prototypical binding partner of HIF-α subunits (Wiener *et al.*, 1996), ARNT2, which is expressed in the brain, liver, and kidney, can also form functional heterodimers with the α subunits and thus may be important to hypoxic response in these organs and in the central nervous system (CNS) in general (Maltepe *et al.*, 2000). Genetic ablation in the mouse results in perinatal lethality with impaired hypothalamic development. The authors noted the similarity between this phenotype and the knockout of Single Minded 1 (Sim1), a bHLH-PAS protein involved in neuronal development that binds with ARNT proteins to form active transcriptional complexes (Keith *et al.*, 2001). This finding highlights the importance of considering all bHLH-PAS family members that rely on ARNT dimerization when interpreting experimental results using ARNT family knockouts. Depending on cell type, these partners potentially include hypoxia response HIF-α subunits, the xenobiotic-sensitive aryl hydrocarbon receptor (AhR), proteins involved in the circadian rhythm such as CLOCK and NPAS2, and the previously mentioned Sim1.

Other ARNT proteins (ARNT3 or -4) could potentially bind HIF-α subunits, but have not been reported to be important regulators of the hypoxic response *in vivo*. ARNT3/MOP3/bMAL1/ARNTL1 is thought not to participate in the hypoxic response (Cowden and Simon, 2002), despite the fact that it can bind HIF-α subunits in yeast two–hybrid assays (Hogenesch *et al.*, 1998). Rather, both ARNT3 and ARNT4/bMAL2/ ARNTL2 are thought to be critical to circadian rhythm regulation in neuronal tissues as binding partners of CLOCK and NPAS2 (Bunger *et al.*, 2000).

2.3. pVHL

The negative regulator of the HIF-α subunits, pVHL, which, as a member of the VHL protein complex, recognizes hydroxylated HIF-α subunits, is also essential for embryonic viability. Ablation of the pVHL component of the complex by homologous recombination results in embryonic lethality between days 10.5 and 12.5 as a result of placental dysgenesis and hemorrhage (Gnarra *et al.*, 1997). The authors also noted that no abnormalities in embryonic development were detected until approximately day 9.5, and that heterozygous null mice were phenotypically normal. Deletion of pVHL in endothelial cells using pVHL-floxed mice crossed to Tie2-cre recombinase also results in embryonic lethality similar to the global pVHL knockout. Histological staining revealed a defect in fibronectin deposition, suggesting the importance of this molecule in successful vasculogenesis in the developing embryo (Ohh *et al.*, 1998; Tang *et al.*, 2006).

2.4. PHDs

Prolyl-hydroxylase domain–containing enzymes play a key role in HIF regulation by modification of α subunits when oxygen is available, leading to recognition by VHL, polyubiquitination, and proteasomal degradation. Interfering with PHD activity is often used to chemically induce HIF activity *in vitro*. Three PHD enzymes exist—PHD1, 2, and 3; however, the relative importance and redundancy of the enzymes in the context of HIF regulation is an important question. Genetic ablation of each of these enzymes has been carried out in mice (Takeda *et al.*, 2006). The team found that PHD1$^{-/-}$ and PHD3$^{-/-}$ null mice were viable and appeared healthy. PHD2$^{-/-}$ embryos died *in utero* between E12.5 and 14.5. Defects in heart and placental development were observed. Placental defects included a reduction in labyrinthine branching morphogenesis, population of the labyrinth with spongiotrophoblasts, and aberrant localization of trophoblast giant cells. Although the placental defects were related to HIF-α subunit levels, PHD2$^{-/-}$ mice did not show an increase of HIF-α subunits in the heart. As the authors point out, the PHD enzymes likely have other roles *in vivo* apart from HIF-α regulation, and these roles may be responsible for the observed defects in the heart.

2.5. VEGF

The vascular endothelial growth factor A (VEGF-A; referred to in this review as VEGF), is an important target of HIF in development and cancer. In development, VEGF is essential, as even a heterozygous null was embryonic-lethal at E11 to 12 (Carmeliet *et al.*, 1996; Ferrara *et al.*, 1996). Further studies using conditional knockouts and an engineered VEGF receptor protein that binds free VEGF in young mice showed VEGF to be important in viability, growth, and development of organs, especially the liver. VEGF continued to be important until about 4 weeks of age, when the animals lost their dependence on VEGF (Gerber *et al.*, 1999).

3. HIF in Physiology

3.1. HIF-1α heterozygotes

Although VEGF heterozygous knockouts are embryonic lethal, heterozygous HIF-1α and HIF-2α null mice are viable and appear normal. One approach to studying HIF function involves comparing heterozygous global knockout mice with wild-type controls. Several groups have used this approach to study the effects of intermittent and long-term hypoxia on whole animal physiology (Brusselmans *et al.*, 2003; Peng *et al.*, 2006; Semenza, 2006; Yu *et al.*, 1999) as well as to study cerebral hypoxia in a model of stroke (see Section 3.13). Along with pulmonary vascular remodeling, rodents exposed to chronic hypoxia are known to develop pulmonary hypertension and right ventricular hypertrophy, the combination of which can lead to death. Heterozygosity of HIF-1α delayed pulmonary hypertension in mice exposed to chronic hypoxia (Yu *et al.*, 1999). However, heterozygosity of HIF-2α eliminated pulmonary hypertension and ventricular hypertrophy (Brusselmans *et al.*, 2003). The authors found increases in endothelin-1 and plasma catecholamine levels in wild-type versus HIF-2α heterozygotes, which suggest they may be key players in HIF-dependent hypoxia-induced pulmonary remodeling.

3.2. Conditional knockouts using the loxP/cre recombinase system

Although regulated HIF function is clearly required for embryonic development, the advent of conditional knockouts has allowed the study of function of hypoxia response genes in adult mice. As a first step towards a conditional knockout approach, investigators typically "flox" essential exons and/or the promoters of target genes by inserting loxP recombination sites in flanking introns or noncoding regions. Mice with "floxed" genes, that is, genes with essential sequences flanked by loxP sites, are normal and display no phenotypes as compared to wild-type controls. When floxed mice are genetically crossed with transgenic mice carrying one copy of the cre recombinase driven by a tissue-specific promoter, the enzyme can act upon the flox sites and excise the target sequence, thus generating a knockout of the target gene, but only after the tissue-specific cre recombinase has been expressed (Orban *et al.*, 1992; Sauer, 1998). Using this approach, many HIF family members and related proteins have been studied in viable adult mice to determine their role in mammalian physiology.

3.3. Liver

Von-Hippel Lindau tumor suppressor ablation via liver-specific Albumin-cre recombinase results in hepatocellular steatosis along with vascular tumors and local angiogenesis (Haase *et al.*, 2001). Glycogen granules were also noted in VHL-null livers, and the mechanism appears to be HIF-dependent upregulation of liver glucose uptake coupled with impaired ability to metabolize the increased sugar (Park *et al.*, 2007). Another study conducted with HIF-1α and HIF-2α transgenes that could not be recognized by VHL recapitulated the major findings from the Albumin-cre study, as well as noting increased hepatocyte proliferation *in vivo*. These findings underscore the role of HIF-1α and HIF-2α as the responsible genes for the phenotypes observed in previous liver pVHL knockout studies (Kim *et al.*, 2006). This study also found that, in the liver, although HIF-1α and 2α shared many targets, the target genes upregulated by the two α factors were not identical, whereas in the skin expression of an HIF-2α subunit that cannot be recognized by VHL essentially phenocopied VHL loss.

3.4. Myeloid lineage

Deletion of HIF-1α in myeloid cells led to an unexpected suppression of macrophage and neutrophil function, including reduced ATP levels, cytokine production, inflammatory response, and ability to kill phagocytosed bacteria (Cramer *et al.*, 2003). Myeloid cell development and cell count was unaffected by loss of HIF-1α. In addition, deletion of VHL in myeloid cells had the opposite effect, potentiating the inflammatory capacity of myeloid cells. The relevance of these *in vitro* effects were confirmed using *in vivo* mouse models of acute inflammation and arthritis. Further studies of the role of HIF-1α in bacterial killing detailed the lack of key bactericidal products in HIF null cells, including granule proteases, antimicrobial peptides, and nitric oxide produced by HIF target inducible nitric oxide synthase (iNOS). *In vivo* studies demonstrated that the null mice were unable to prevent the spread of a bacterial infection (Peyssonnaux *et al.*, 2005).

3.5. Lymphocytes

Using a T cell-specific cre recombinase, Lck-cre, VHL loss in thymocytes resulted in an atrophic, hemorrhagic thymus. Increased caspase-8–dependent apoptosis was noted *in vivo* and *in vitro*, and double knockout studies showed the effect to be HIF-1α–dependent and Bcl-2 and Bcl-xL independent (Biju *et al.*, 2004). Further characterization of this mutant pinned the T-cell activation defect on the overexpression of the HIF target calcium transporter SERCA2, which decreased cytoplasmic Ca^{2+} concentrations after T-cell receptor (TCR) ligation, thus blunting TCR signaling (Neumann *et al.*, 2005).

3.6. Hematopoiesis

As previously mentioned, HIF-2α null mice have been reported to be embryonic/perinatal lethal; however, crossing the null heterozygous alleles on the two inbred mouse backgrounds of 129 × C57 results in about 20% survival (Scortegagna *et al.*, 2003a). Characterization of this surviving portion revealed cardiac hypertrophy, hepatomegaly, increased superoxide levels, and a deficiency in hematopoiesis. These mice could be partially rescued by administration of a superoxide dismutase mimetic, MnTBAP (Scortegagna *et al.*, 2003a, 2005). Deletion of HIF-2α approximately 1 wk after birth using a tamoxifen-inducible cre system results in anemic mice, while deletion of HIF-1α does not affect hematocrit levels (Gruber *et al.*, 2007).

3.7. Skeletal muscle

During exercise, skeletal muscle experiences dramatic changes in oxygen tension (Richardson *et al.*, 1995). Conditional deletion of HIF-1α in skeletal muscle resulted in mice that had increased exercise endurance, but at the cost of increased muscle damage after repeated exercise bouts. The increased endurance was a result of increased mitochondrial activity and decreased lactate production in HIF-1α null animals, both of which point towards increased fatty acid oxidation in the knockout (Mason *et al.*, 2004).

3.8. Heart and cardiovascular system

Loss of HIF-1α in the heart in a ventricle-specific knockout results in contractile dysfunction and significant alterations in calcium flux (Huang et al., 2004). A hypovascularity of the hearts of these animals also occurred, indicating a role for HIF-1α in maintenance of the cardiac vasculature.

Loss of HIF in endothelial cells had significant effects on a number of important vascular parameters—wound healing is retarded in these mice, as is the neovascularization of tumors (Tang et al., 2004). Endothelium lacking HIF-1α exhibits a significant deficit in the response to exogenous VEGF, which is correlated with defects in VEGF-mediated stabilization of the VEGF receptor VEGFR2 in these mice.

3.9. Mammary gland

The microenvironment of the mammary gland undergoes rapid and drastic change during pregnancy and lactation. Proliferation of the epithelium and later secretion of milk could place excessive demands on the vascular bed that was established during relative quiescence of the resting virgin mammary gland. Pregnancy and lactation studies of HIF-1α conditional knockouts in the mammary epithelium of mice demonstrated failure of differentiation and lipid secretion, which resulted in abnormal milk composition. The vascular density between wild-type and null mammary glands was unchanged. Apart from demonstrating the requirement of HIF-1α function in lactation, these results, taken together with results from other experiments (Ryan et al., 2000), highlight the complexity of HIF function above and beyond partial control of VEGF and angiogenesis (Seagroves et al., 2003).

3.10. Colon

Investigators studying hapten-induced colitis noted hypoxia and stabilization of HIF-1α in the colonic epithelium. To test the role of HIF in colitis, they then used an intestinal epithelial specific cre to delete either HIF-1α or pVHL from this tissue. After induction of colitis in these models, the investigators observed a protective function of increased HIF levels, from decreased mortality to diminished weight loss. Mechanistically, decreased barrier function as a result of downregulation of a variety of genes, including CD73, was noted in HIF-1α null animals, whereas increased barrier function was observed in the pVHL nulls (Karhausen et al., 2004).

3.11. Chondrocytes

Chondrocytes, found in growth plates of developing bones, exist in an avascular zone of low oxygen tension and thus could be expected to rely on hypoxia-dependent pathways. Conditional deletion of HIF-1α in this tissue released a hypoxia-induced cell cycle arrest, and, as a result, cells in the hypoxic central region of the growth plate proceeded into untenable cell division, which led to apoptosis. Also noted was an HIF-1α–independent upregulation of VEGF, resulting in angiogenesis in the hypoxic region of the plate (Schipani *et al.*, 2001).

3.12. Skin

Constitutive activation of HIF-1α in the epidermis resulted in drastic increases in dermal capillary density and VEGF expression (Elson *et al.*, 2001). Despite these increases, the authors noted no inflammation, edema, or vascular leakage—an unexpected result given the role of VEGF in increasing vascular permeability. Indeed, in an earlier study with VEGF 120-kDa isoform coupled to a K6 promoter, some progeny died exhibiting increased vascularization and edema, which disrupted the skin architecture (Larcher *et al.*, 1998). The loss of HIF-1α in the epidermis results in normal mice when left unchallenged (Boutin and Johnson, unpublished). Loss of pVHL, similar to the HIF-1α constitutively active model, leads to robust expansion both in vessel size and vascular density; however, in this transgenic system, the effect is coupled with increased vascular permeability (Boutin and Johnson, submitted). Epidermal pVHL null mice or stable HIF-2α–expressing mice also exhibit runting, partial alopecia, and erythema (Kim *et al.*, 2006).

3.13. Motor neurons and brain

Genetic ablation of the HRE sequences from the VEGF promoter resulted in motor neuron degeneration in mice similar to human amyotrophic lateral sclerosis (ALS) (Oosthuyse *et al.*, 2001). Noteworthy in the transgenic mice was the lack of effect on baseline VEGF production from muscle, heart, and fibroblasts, but a 40% reduction from neurons. Mice born with homozygous deletion of the HRE in the VEGF promoter suffered 60% mortality perinatally. The surviving fraction was normal until about 5 months of age, when they developed motor neuron dysfunction as demonstrated by decreased grooming, inability to turn over, and behavior when picked up by the tail. Histological characterization of the muscle revealed atrophy, which appeared to be the result of denervation. The investigators proposed decreased neural perfusion in the transgenics or perhaps a neuroprotective effect of VEGF-165 binding to KDR as likely mechanisms behind the phenotype.

In order to study HIF-1α function in whole body hypoxia and stroke, a brain-specific knockout of HIF-1α was generated by crossing HIF floxed mice to the calcium/calmodulin-dependent kinase cre or by using mice heterozygous for the global knockout of HIF-1α. Neuronal hypoxia was induced by placing mice in a 7% oxygen atmosphere for 1 h, and stroke was modeled by occluding one carotid artery for 75 min. Subsequent analysis of tissue sections from the brain for apoptosis and necrosis revealed that heterozygous null or neuron-specific HIF-1α null mice were protected from hypoxia-induced cell death in both neuronal hypoxia and stroke models. The investigators fault HIF-1α–dependent upregulation of proapoptotic genes and conclude that the loss of HIF-1α is neuroprotective in both hypoxic and ischemic episodes (Helton *et al.*, 2005).

▶ 4. HIF Function in Tumor Biology

This portion of the review will focus on murine VHL and HIF-α and-β transgenics in tumor biology. Given the abundance of reviews available dealing with VEGF function *in vivo*, reports relating to VEGF will be included only in the context of HIF and pVHL models.

4.1. First insights—xenografts

The ability of HIF to upregulate glycolysis and potentially confer a survival advantage in areas of poor oxygenation, coupled with its targeting of angiogenic factors such as VEGF, would suggest a protumorigenic role in cancer. Indeed, subcutaneous injection of HIF-1α null embryonic stem cells resulted in tumors with decreased tumor mass, microvessel density, and expression of VEGF (Ryan *et al.*, 1998). An independent group who also generated HIF-1α null embryonic stem cells also found changes in vascular density, with endothelial cell (EC) cords being more dense and medium-sized vessels and vessel beds less dense in HIF null. This group reported increased tumor mass as a result of decreased apoptosis in HIF-1α null cells (Carmeliet *et al.*, 1998). Further characterization of the role of HIF-1α in tumor growth by injection of mouse embryonic fibroblasts showed a decrease in tumor mass in the null. Interestingly, despite HIF-dependent changes in VEGF expression, no changes were observed in vascular density in these *HRAS*–transformed cell lines (Ryan *et al.*, 2000). Recent small interfering RNA (siRNA) experiments have recapitulated those basic findings already determined to be genetically related to HIF and VEGF as being pro-tumor factors in most scenarios of carcinogenesis (Detwiller *et al.*, 2005; Jensen *et al.*, 2006; Li *et al.*, 2005; Zhang *et al.*, 2004); however, other studies highlight the need to understand how the opposing HIF-mediated effects of increased glycolytic energy production and angiogenesis versus induction of proapoptotic genes and cell cycle inhibitors balance out in various tumor settings (Acker *et al.*, 2005).

4.2. Astrocytoma

Experiments with HIF-1α null cells in an astrocytoma model determined that the site of injection was critical in determining how the WT and HIF-1α null cells behaved. In the poorly vascularized subcutaneous environment, HIF-1α null cells were at a disadvantage; however, the same null cells implanted into the well-vascularized brain had a growth advantage (Blouw *et al.*, 2003). Characterization of the vessel density showed the expected decreased vessel density in null subcutaneous tumors. Unexpectedly, the null tumors grown in the brain had increased vascular density. Furthermore, the null cells were more invasive than WT controls. Parallel experiments conducted with VEGF null cells exhibited slower growth in both subcutaneous and brain microenvironments. These results demonstrate how the tumor microenvironment has a powerful role in determining the net effect of HIF-1α on tumor growth, angiogenesis, and invasiveness.

4.3. Mammary carcinogenesis

The viral oncogene polyoma middle T antigen controlled by the murine mammary tumor virus promoter (MMTV-PyMT) is a well characterized and widely used transgenic mouse model of breast cancer and metastasis (Lin *et al.*, 2003). Female mice carrying one copy of the transgene develop mammary gland hyperplasia, carcinoma, and, finally, pulmonary metastases over the course of approximately 6 months. Mice carrying this transgene were crossed into HIF WT or mammary epithelium conditional null. HIF-1α mammary null mice displayed increased latency to the first palpable tumor along with decreased vascular density. Pulmonary metastases were halved in an HIF-1α–dependent manner. *In vitro* experiments revealed null cells had impaired random migration and directed chemotaxis compared to WT controls when cultured in hypoxia (Liao *et al.*, 2007).

4.4. pVHL and VHL disease

Silencing or mutating the VHL tumor suppressor predisposes humans to a specific set of cancers, including sporadic clear cell renal carcinoma, hemangioblastoma, and pheochromocytoma (Kaelin, 2007). In order to establish the fidelity of the pVHL null as a mouse model of human VHL disease and to determine if pVHL deficiency alone can result in symptoms resembling VHL disease, an important question to ask is whether murine VHL heterozygotes are susceptible to equivalents of the same types of cancer as human VHL disease carriers.

Indeed, mice with only one functional copy of pVHL are prone to developing vascular tumors as they age. The incidence of cavernous hemangiomas was ~50% at 3 to 12 months of age, increasing to 90% as the mice aged from 12 to 17 months. In contrast to the high prevalence of vascular tumors, precancerous kidney lesions were rare—only 1 of 30 heterozygote mice had a renal cyst (Haase *et al.*, 2001). Given the shorter life of mice as compared to humans, perhaps deletion of both pVHL alleles was necessary to observe a higher percentage of renal cysts in the mouse model. Further experiments performed in this area, using Phosphoenolpyruvate carboxykinase (PEPCK)-cre recombinase to knockout pVHL in the kidney and partially in the liver, demonstrated, apart from increased erythropoietin (EPO) production and polycythemia, a higher incidence of macroscopic and microscopic renal cysts that required ARNT but not HIF-1α (Rankin *et al.*, 2006).

Another approach to gain insight into pVHL function is the use of xeno-grafts. Deletion of pVHL in fibroblasts and injection subcutaneously into mice yielded fibrosarcomas with increased vasculature, but with a slower growth rate. Increased levels of the HIF targets, cyclin-dependent kinase inhibitors p21 and p27, were faulted for the decreased growth (Mack *et al.*, 2005).

5. SUMMARY

In the 15 years that have elapsed since discovery of the first HRE in the EPO 3′ enhancer (Beck *et al.*, 1991; Pugh *et al.*, 1991; Semenza and Wang, 1992), research focusing on the hypoxia response pathway has made significant progress in development, physiology, and cancer biology. In general, genes of the hypoxic response pathway are essential for development, as demonstrated by the requirement for HIF-1α, HIF-2α, ARNT1, ARNT2, pVHL, and VEGF for embryonic viability (Ferrara *et al.*, 1996; Gnarra *et al.*, 1997; Iyer *et al.*, 1998; Keith *et al.*, 2001; Kozak *et al.*, 1997; Ryan *et al.*, 1998; Tian *et al.*, 1998). Hypoxia plays a role in physiology and pathology in adult mammals, from the chondrocytes in the growth plates of developing bone to neurons in stroke and cerebral ischemia to myeloid cells in areas of inflammation (Cramer and Johnson, 2003; Cramer *et al.*, 2003; Helton *et al.*, 2005; Schipani *et al.*, 2001). Indeed, in all of these microenvironments, HIF has been found to play a role in physiology and pathogenesis, confirming the relevance of the hypoxic response beyond development. Most reports indicate a pro-tumor role for HIF in the rapidly growing neoplasm, with its requirement for neovascularization and survival under low oxygen tensions (Jensen *et al.*, 2006; Ryan *et al.*, 1998, 2000; Zhang *et al.*, 2004). That said, many important questions still remain in the relatively new area of investigation of the mammalian hypoxic response *in vivo*.

REFERENCES

Acker, T., Diez-Juan, A., Aragones, J., Tjwa, M., Brusselmans, K., Moons, L., Fukumura, D., Moreno-Murciano, M. P., Herbert, J. M., Burger, A., Riedel, J., Evert, G., *et al.* (2005). Genetic evidence for a tumor suppressor role of HIF-2α. *Cancer Cell* **8,** 131–141.

Beck, I., Ramirez, S., Weinmann, R., and Caro, J. (1991). Enhancer element at the 3'-flanking region controls transcriptional response to hypoxia in the human erythropoietin gene. *J. Biol. Chem.* **266,** 15563–15566.

Biju, M. P., Neumann, A. K., Bensinger, S. J., Johnson, R. S., Turka, L. A., and Haase, V. H. (2004). Vhlh gene deletion induces *Hif-1*-mediated cell death in thymocytes. *Mol. Cell. Biol.* **24,** 9038–9047.

Blouw, B., Song, H., Tihan, T., Bosze, J., Ferrara, N., Gerber, H. P., Johnson, R. S., and Bergers, G. (2003). The hypoxic response of tumors is dependent on their microenvironment. *Cancer Cell* **4,** 133–146.

Bruick, R. K., and McKnight, S. L. (2001). A conserved family of prolyl-4-hydroxylases that modify HIF. *Science* **294,** 1337–1340.

Brusselmans, K., Compernolle, V., Tjwa, M., Wiesener, M. S., Maxwell, P. H., Collen, D., and Carmeliet, P. (2003). Heterozygous deficiency of hypoxia-inducible factor-2α protects mice against pulmonary hypertension and right ventricular dysfunction during prolonged hypoxia. *J. Clin. Invest.* **111,** 1519–1527.

Bunger, M. K., Wilsbacher, L. D., Moran, S. M., Clendenin, C., Radcliffe, L. A., Hogenesch, J. B., Simon, M. C., Takahashi, J. S., and Bradfield, C. A. (2000). Mop3 is an essential component of the master circadian pacemaker in mammals. *Cell* **103,** 1009–1017.

Carmeliet, P., Ferreira, V., Breier, G., Pollefeyt, S., Kieckens, L., Gertsenstein, M., Fahrig, M., Vandenhoeck, A., Harpal, K., Eberhardt, C., Declercq, C, Pawling, J., *et al.* (1996). Abnormal blood vessel development and lethality in embryos lacking a single VEGF allele. *Nature* **380,** 435–439.

Carmeliet, P., Dor, Y., Herbert, J. M., Fukumura, D., Brusselmans, K., Dewerchin, M., Neeman, M., Bono, F., Abramovitch, R., Maxwell, P., Koch, C. J., Ratcliffe, P., *et al.* (1998). Role of HIF-1α in hypoxia-mediated apoptosis, cell proliferation and tumour angiogenesis. *Nature* **394,** 485–490.

Compernolle, V., Brusselmans, K., Franco, D., Moorman, A., Dewerchin, M., Collen, D., and Carmeliet, P. (2003). Cardia bifida, defective heart development and abnormal neural crest migration in embryos lacking hypoxia-inducible factor-1α. *Cardiovasc. Res.* **60,** 569–579.

Compernolle, V., Brusselmans, K., Acker, T., Hoet, P., Tjwa, M., Beck, H., Plaisance, S., Dor, Y., Keshet, E., Lupu, F., Nemery, B., Dewerchin, M., *et al.* (2002). Loss of HIF-2α and inhibition of VEGF impair fetal lung maturation, whereas treatment with VEGF prevents fatal respiratory distress in premature mice. *Nat. Med.* **8,** 702–710.

Cowden, K. D., and Simon, M. C. (2002). The bHLH/PAS factor MOP3 does not participate in hypoxia responses. *Biochem. Biophys. Res. Commun.* **290,** 1228–1236.

Cramer, T., and Johnson, R. S. (2003). A novel role for the hypoxia inducible transcription factor HIF-1α: Critical regulation of inflammatory cell function. *Cell Cycle* **2,** 192–193.

Cramer, T., Yamanishi, Y., Clausen, B. E., Förster, I., Pawlinski, R., Mackman, N., Haase, V. H., Jaenisch, R., Corr, M., Nizet, V., Firestein, G. S., Gerber, H. P., *et al.* (2003). HIF-1α is essential for myeloid cell-mediated inflammation. *Cell* **112,** 645–657.

Detwiller, K. Y., Fernando, N. T., Segal, N. H., Ryeom, S. W., D'Amore, P. A., and Yoon, S. S. (2005). Analysis of hypoxia-related gene expression in sarcomas and effect of hypoxia on RNA interference of vascular endothelial cell growth factor A. *Cancer Res.* **65,** 5881–5889.

Elson, D. A., Thurston, G., Huang, L. E., Ginzinger, D. G., McDonald, D. M., Johnson, R. S., and Arbeit, J. M. (2001). Induction of hypervascularity without leakage or inflammation in transgenic mice overexpressing hypoxia-inducible factor-1α. *Genes Dev.* **15,** 2520–2532.

Ema, M., Taya, S., Yokotani, N., Sogawa, K., Matsuda, Y., and Fujii-Kuriyama, Y. (1997). A novel bHLH-PAS factor with close sequence similarity to hypoxia-inducible factor 1α regulates the VEGF expression and is potentially involved in lung and vascular development. *Proc. Natl. Acad. Sci. USA* **94,** 4273–4278.

Epstein, A. C., Gleadle, J. M., McNeill, L. A., Hewitson, K. S., O'Rourke, J., Mole, D. R., Mukherji, M., Metzen, E., Wilson, M. I., Dhanda, A., Tian, Y. M., Masson, N., *et al.* (2001). *C. elegans* EGL-9 and mammalian homologs define a family of dioxygenases that regulate HIF by prolyl hydroxylation. *Cell* **107,** 43–54.

Ferrara, N., Carver-Moore, K., Chen, H., Dowd, M., Lu, L., O'Shea, K. S., Powell-Braxton, L., Hillan, K. J., and Moore, M. W. (1996). Heterozygous embryonic lethality induced by targeted inactivation of the VEGF gene. *Nature* **380,** 439–442.

Flamme, I., Frohlich, T., von Reutern, M., Kappel, A., Damert, A., and Risau, W. (1997). HRF, a putative basic helix-loop-helix-PAS-domain transcription factor is closely related to hypoxia-inducible factor-1 α and developmentally expressed in blood vessels. *Mech. Dev.* **63,** 51–60.

Gerber, H. P., Hillan, K. J., Ryan, A. M., Kowalski, J., Keller, G. A., Rangell, L., Wright, B. D., Radtke, F., Aguet, M., and Ferrara, N. (1999). VEGF is required for growth and survival in neonatal mice. *Development* **126,** 1149–1159.

Gnarra, J. R., Ward, J. M., Porter, F. D., Wagner, J. R., Devor, D. E., Grinberg, A., Emmert-Buck, M. R., Westphal, H., Klausner, R. D., and Linehan, W. M. (1997). Defective placental vasculogenesis causes embryonic lethality in VHL-deficient mice. *Proc. Natl. Acad. Sci. USA* **94,** 9102–9107.

Gruber, M., Hu, C. J., Johnson, R. S., Brown, E. J., Keith, B., and Simon, M. C. (2007). Acute postnatal ablation of *Hif-2α* results in anemia. *Proc. Natl. Acad. Sci. USA* **104,** 2301–2306.

Gu, Y. Z., Moran, S. M., Hogenesch, J. B., Wartman, L., and Bradfield, C. A. (1998). Molecular characterization and chromosomal localization of a third α-class hypoxia inducible factor subunit, HIF3α. *Gene Expr.* **7,** 205–213.

Haase, V. H., Glickman, J. N., Socolovsky, M., and Jaenisch, R. (2001). Vascular tumors in livers with targeted inactivation of the von Hippel-Lindau tumor suppressor. *Proc. Natl. Acad. Sci. USA* **98,** 1583–1588.

Helton, R., Cui, J., Scheel, J. R., Ellison, J. A., Ames, C., Gibson, C., Blouw, B., Ouyang, L., Dragatsis, I., Zeitlin, S., Johnson, R. S., Lipson, S. A., *et al.* (2005). Brain-specific knock-out of hypoxia-inducible factor-1α reduces rather than increases hypoxic-ischemic damage. *J. Neurosci.* **25,** 4099–4107.

Hewitson, K. S., McNeill, L. A., Riordan, M. V., Tian, Y. M., Bullock, A. N., Welford, R. W., Elkins, J. M., Oldham, N. J., Bhattacharya, S., Gleadle, J. M., Ratcliffe, P. J., Pugh, C. W., *et al.* (2002). Hypoxia-inducible factor (HIF) asparagine hydroxylase is identical to factor inhibiting HIF (FIH) and is related to the cupin structural family. *J. Biol. Chem.* **277,** 26351–26355.

Hogenesch, J. B., Gu, Y. Z., Jain, S., and Bradfield, C. A. (1998). The basic-helix-loop-helix-PAS orphan MOP3 forms transcriptionally active complexes with circadian and hypoxia factors. *Proc. Natl. Acad. Sci. USA* **95,** 5474–5479.

Huang, Y., Hickey, R. P., Yeh, J. L., Liu, D., Dadak, A., Young, L. H., Johnson, R. S., and Giordano, F. J. (2004). Cardiac myocyte-specific HIF-1α deletion alters vascularization, energy availability, calcium flux, and contractility in the normoxic heart. *FASEB J.* **18,** 1138–1140.

Ivan, M., Kondo, K., Yang, H., Kim, W., Valiando, J., Ohh, M., Salic, A., Asara, J. M., Lane, W. S., and Kaelin, W. G., Jr. (2001). HIFα targeted for VHL-mediated destruction by proline hydroxylation: Implications for O_2 sensing. *Science* **292,** 464–468.

Iyer, N. V., Kotch, L. E., Agani, F., Leung, S. W., Laughner, E., Wenger, R. H., Gassmann, M., Gearhart, J. D., Lawler, A. M., Yu, A. Y., and Semenza, G. L. (1998). Cellular and developmental control of O_2 homeostasis by hypoxia-inducible factor 1 α. *Genes Dev.* **12,** 149–162.

Jaakkola, P., Mole, D. R., Tian, Y. M., Wilson, M. I., Gielbert, J., Gaskell, S. J., Kriegsheim, A., Hebestreit, H. F., Mukherji, M., Schofield, C. J., Maxwell, P. H., Pugh, C. W., *et al.* (2001). Targeting of HIF-α to the von Hippel-Lindau ubiquitylation complex by O_2-regulated prolyl hydroxylation. *Science* **292,** 468–472.

Jensen, R. L., Ragel, B. T., Whang, K., and Gillespie, D. (2006). Inhibition of hypoxia inducible factor-1α (HIF-1α) decreases vascular endothelial growth factor (VEGF) secretion and tumor growth in malignant gliomas. *J. Neurooncol.* **78,** 233–247.

Kaelin, W. G., Jr. (2002). Molecular basis of the VHL hereditary cancer syndrome. *Nat. Rev. Cancer* **2,** 673–682.

Kaelin, W. G., Jr. (2007). The von Hippel-Lindau tumor suppressor protein and clear cell renal carcinoma. *Clin. Cancer Res.* **13,** 680s–684s.

Karhausen, J., Furuta, G. T., Tomaszewski, J. E., Johnson, R. S., Colgan, S. P., and Haase, V. H. (2004). Epithelial hypoxia-inducible factor-1 is protective in murine experimental colitis. *J. Clin. Invest.* **114,** 1098–1106.

Keith, B., Adelman, D. M., and Simon, M. C. (2001). Targeted mutation of the murine arylhydrocarbon receptor nuclear translocator 2 (Arnt2) gene reveals partial redundancy with. *Arnt. Proc. Natl. Acad. Sci. USA* **98,** 6692–6697.

Kim, W. Y., Safran, M., Buckley, M. R., Ebert, B. L., Glickman, J., Bosenberg, M., Regan, M., and Kaelin, W. G., Jr. (2006). Failure to prolyl hydroxylate hypoxia-inducible factor α phenocopies VHL inactivation *in vivo*. *EMBO J.* **25,** 4650–4662.

Kozak, K. R., Abbott, B., and Hankinson, O. (1997). ARNT-deficient mice and placental differentiation. *Dev. Biol.* **191,** 297–305.

Larcher, F., Murillas, R., Bolontrade, M., Conti, C. J., and Jorcano, J. L. (1998). VEGF/VPF overexpression in skin of transgenic mice induces angiogenesis, vascular hyperpermeability and accelerated tumor development. *Oncogene* **17,** 303–311.

Li, L., Lin, X., Staver, M., Shoemaker, A., Semizarov, D., Fesik, S. W., and Shen, Y. (2005). Evaluating hypoxia-inducible factor-1α as a cancer therapeutic target via inducible RNA interference *in vivo*. *Cancer Res.* **65,** 7249–7258.

Liao, D., Corle, C., Seagroves, T. N., and Johnson, R. S. (2007). Hypoxia-inducible factor-1α is a key regulator of metastasis in a transgenic model of cancer initiation and progression. *Cancer Res.* **67,** 563–572.

Lin, E. Y., Jones, J. G., Li, P., Zhu, L., Whitney, K. D., Muller, W. J., and Pollard, J. W. (2003). Progression to malignancy in the polyoma middle T oncoprotein mouse breast cancer model provides a reliable model for human diseases. *Am. J. Pathol.* **163,** 2113–2126.

Mack, F. A., Patel, J. H., Biju, M. P., Haase, V. H., and Simon, M. C. (2005). Decreased growth of Vhl$^{-/-}$ fibrosarcomas is associated with elevated levels of cyclin kinase inhibitors p21 and p27. *Mol. Cell Biol.* **25,** 4565–4578.

Mahon, P. C., Hirota, K., and Semenza, G. L. (2001). FIH-1: A novel protein that interacts with HIF-1α and VHL to mediate repression of HIF-1 transcriptional activity. *Genes Dev.* **15,** 2675–2686.

Maltepe, E., Schmidt, J. V., Baunoch, D., Bradfield, C. A., and Simon, M. C. (1997). Abnormal angiogenesis and responses to glucose and oxygen deprivation in mice lacking the protein ARNT. *Nature* **386,** 403–407.

Maltepe, E., Keith, B., Arsham, A. M., Brorson, J. R., and Simon, M. C. (2000). The role of ARNT2 in tumor angiogenesis and the neural response to hypoxia. *Biochem. Biophys. Res. Commun.* **273,** 231–238.

Mason, S. D., Howlett, R. A., Kim, M. J., Olfert, I. M., Hogan, M. C., McNulty, W., Hickey, R. P., Wagner, P. D., Kahn, C. R., Giordano, F. J., and Johnson, R. S. (2004). Loss of skeletal muscle HIF-1α results in altered exercise endurance. *PLoS Biol.* **2,** e288.

Neumann, A. K., Yang, J., Biju, M. P., Joseph, S. K., Johnson, R. S., Haase, V. H., Freedman, B. D., and Turka, L. A. (2005). Hypoxia inducible factor 1 α regulates T cell receptor signal transduction. *Proc. Natl. Acad. Sci. USA* **102,** 17071–17076.

Ohh, M., Yauch, R. L., Lonergan, K. M., Whaley, J. M., Stemmer-Rachamimov, A. O., Louis, D. N., Gavin, B. J., Kley, N., Kaelin, W. G., Jr., and Iliopoulos, O. (1998). The von Hippel-Lindau tumor suppressor protein is required for proper assembly of an extracellular fibronectin matrix. *Mol. Cell* **1,** 959–968.

Oosthuyse, B., Moons, L., Storkebaum, E., Beck, H., Nuyens, D., Brusselmans, K., Van Dorpe, J., Hellings, P., Gorselink, M., Heymans, S., Theilmeier, G., Dewerchin, M., *et al.* (2001). Deletion of the hypoxia-response element in the vascular endothelial growth factor promoter causes motor neuron degeneration. *Nat. Genet.* **28,** 131–138.

Orban, P. C., Chui, D., and Marth, J. D. (1992). Tissue- and site-specific DNA recombination in transgenic mice. *Proc. Natl. Acad. Sci. USA* **89,** 6861–6865.

Park, S. K., Haase, V. H., and Johnson, R. S. (2007). Von Hippel Lindau tumor suppressor regulates hepatic glucose metabolism by controlling expression of glucose transporter 2 and glucose 6-phosphatase. *Int. J. Oncol.* **30,** 341–348.

Peng, J., Zhang, L., Drysdale, L., and Fong, G. H. (2000). The transcription factor EPAS-1/hypoxia-inducible factor 2α plays an important role in vascular remodeling. *Proc. Natl. Acad. Sci. USA* **97,** 8386–8391.

Peng, Y. J., Yuan, G., Ramakrishnan, D., Sharma, S. D., Bosch-Marce, M., Kumar, G. K., Semenza, G. L., and Prabhakar, N. R. (2006). Heterozygous HIF-1α deficiency impairs carotid body-mediated systemic responses and reactive oxygen species generation in mice exposed to intermittent hypoxia. *J. Physiol.* **577,** 705–716.

Peyssonnaux, C., Datta, V., Cramer, T., Doedens, A., Theodorakis, E. A., Gallo, R. L., Hurtado-Ziola, N., Nizet, V., and Johnson, R. S. (2005). HIF-1α expression regulates the bactericidal capacity of phagocytes. *J. Clin. Invest.* **115,** 1806–1815.

Pugh, C. W., Tan, C. C., Jones, R. W., and Ratcliffe, P. J. (1991). Functional analysis of an oxygen-regulated transcriptional enhancer lying 3′ to the mouse erythropoietin gene. *Proc. Natl. Acad. Sci. USA* **88,** 10553–10557.

Rankin, E. B., Tomaszewski, J. E., and Haase, V. H. (2006). Renal cyst development in mice with conditional inactivation of the von Hippel-Lindau tumor suppressor. *Cancer Res.* **66,** 2576–2583.

Richardson, R. S., Noyszewski, E. A., Kendrick, K. F., Leigh, J. S., and Wagner, P. D. (1995). Myoglobin O_2 desaturation during exercise. Evidence of limited O_2 transport. *J. Clin. Invest.* **96,** 1916–1926.

Ryan, H. E., Lo, J., and Johnson, R. S. (1998). HIF-1α is required for solid tumor formation and embryonic vascularization. *EMBO J.* **17,** 3005–3015.

Ryan, H. E., Poloni, M., McNulty, W., Elson, D., Gassmann, M., Arbeit, J. M., and Johnson, R. S. (2000). Hypoxia-inducible factor-1α is a positive factor in solid tumor growth. *Cancer Res.* **60,** 4010–4015.

Safran, M., Kim, W. Y., O'Connell, F., Flippin, L., Gunzler, V., Horner, J. W., Depinho, R. A., and Kaelin, W. G., Jr. (2006). Mouse model for noninvasive imaging of HIF prolyl hydroxylase activity: Assessment of an oral agent that stimulates erythropoietin production. *Proc. Natl. Acad. Sci. USA* **103,** 105–110.

Sauer, B. (1998). Inducible gene targeting in mice using the Cre/lox system. *Methods* **14,** 381–392.

Schipani, E., Ryan, H. E., Didrickson, S., Kobayashi, T., Knight, M., and Johnson, R. S. (2001). Hypoxia in cartilage: HIF-1α is essential for chondrocyte growth arrest and survival. *Genes Dev.* **15,** 2865–2876.

Scortegagna, M., Morris, M. A., Oktay, Y., Bennett, M., and Garcia, J. A. (2003a). The HIF family member EPAS1/HIF-2α is required for normal hematopoiesis in mice. *Blood* **102,** 1634–1640.

Scortegagna, M., Ding, K., Oktay, Y., Gaur, A., Thurmond, F., Yan, L. J., Marck, B. T., Matsumoto, A. M., Shelton, J. M., Richardson, J. A., Bennett, M. J., Garcia, J. A., *et al.* (2003b). Multiple organ pathology, metabolic abnormalities and impaired homeostasis of reactive oxygen species in Epas1$^{-/-}$ mice. *Nat. Genet.* **35,** 331–340.

Scortegagna, M., Ding, K., Zhang, Q., Oktay, Y., Bennett, M. J., Bennett, M., Shelton, J. M., Richardson, J. A., Moe, O., and Garcia, J. A. (2005). HIF-2α regulates murine hematopoietic development in an erythropoietin-dependent manner. *Blood* **105,** 3133–3140.

Seagroves, T. N., Hadsell, D., McManaman, J., Palmer, C., Liao, D., McNulty, W., Welm, B., Wagner, K. U., Neville, M., and Johnson, R. S. (2003). HIF1α is a critical regulator of secretory differentiation and activation, but not vascular expansion, in the mouse mammary gland. *Development* **130,** 1713–1724.

Semenza, G. L. (2001). Hypoxia-inducible factor 1: Oxygen homeostasis and disease pathophysiology. *Trends Mol. Med.* **7,** 345–350.

Semenza, G. L. (2006). Regulation of physiological responses to continuous and intermittent hypoxia by hypoxia-inducible factor 1. *Exp. Physiol.* **91,** 803–806.

Semenza, G. L., and Wang, G. L. (1992). A nuclear factor induced by hypoxia via *de novo* protein synthesis binds to the human erythropoietin gene enhancer at a site required for transcriptional activation. *Mol. Cell Biol.* **12,** 5447–5454.

Semenza, G. L., Roth, P. H., Fang, H. M., and Wang, G. L. (1994). Transcriptional regulation of genes encoding glycolytic enzymes by hypoxia-inducible factor 1. *J. Biol. Chem.* **269,** 23757–23763.

Takeda, K., Ho, V. C., Takeda, H., Duan, L. J., Nagy, A., and Fong, G. H. (2006). Placental but not heart defects are associated with elevated hypoxia-inducible factor α levels in mice lacking prolyl hydroxylase domain protein 2. *Mol. Cell Biol.* **26,** 8336–8346.

Tang, N., Mack, F., Haase, V. H., Simon, M. C., and Johnson, R. S. (2006). pVHL function is essential for endothelial extracellular matrix deposition. *Mol. Cell Biol.* **26,** 2519–2530.

Tang, N., Wang, L., Esko, J., Giordano, F. J., Huang, Y., Gerber, H. P., Ferrara, N., and Johnson, R. S. (2004). Loss of HIF-1α in endothelial cells disrupts a hypoxia-driven VEGF autocrine loop necessary for tumorigenesis. *Cancer Cell* **6,** 485–495.

Tian, H., McKnight, S. L., and Russell, D. W. (1997). Endothelial PAS domain protein 1 (EPAS1), a transcription factor selectively expressed in endothelial cells. *Genes Dev.* **11,** 72–82.

Tian, H., Hammer, R. E., Matsumoto, A. M., Russell, D. W., and McKnight, S. L. (1998). The hypoxia-responsive transcription factor EPAS1 is essential for catecholamine homeostasis and protection against heart failure during embryonic development. *Genes Dev.* **12,** 3320–3324.

Wang, G. L., and Semenza, G. L. (1995). Purification and characterization of hypoxia-inducible factor 1. *J. Biol. Chem.* **270,** 1230–1237.

Wang, G. L., Jiang, B. H., Rue, E. A., and Semenza, G. L. (1995). Hypoxia-inducible factor 1 is a basic-helix-loop-helix-PAS heterodimer regulated by cellular O_2 tension. *Proc. Natl. Acad. Sci. USA* **92,** 5510–5514.

Wiener, C. M., Booth, G., and Semenza, G. L. (1996). *In vivo* expression of mRNAs encoding hypoxia-inducible factor 1. *Biochem. Biophys. Res. Commun.* **225,** 485–488.

Yu, A. Y., Shimoda, L. A., Iyer, N. V., Huso, D. L., Sun, X., McWilliams, R., Beaty, T., Sham, J. S., Wiener, C. M., Sylvester, J. T., and Semenza, G. L. (1999). Impaired physiological responses to chronic hypoxia in mice partially deficient for hypoxia-inducible factor 1α. *J. Clin. Invest.* **103,** 691–696.

Zhang, Q., Zhang, Z. F., Rao, J. Y., Sato, J. D., Brown, J., Messadi, D. V., and Le, A. D. (2004). Treatment with siRNA and antisense oligonucleotides targeted to HIF-1α induced apoptosis in human tongue squamous cell carcinomas. *Int. J. Cancer* **111,** 849–857.

CHAPTER SIX

THE SILENCING APPROACH OF THE HYPOXIA-SIGNALING PATHWAY

Edurne Berra* *and* Jacques Pouysségur[†]

Contents

1. A Brief History of RNAi	108
2. The Hypoxia-Signaling Pathway	109
3. HIF-α Stability	110
3.1. The family of the HIF prolyl-hydroxylases	110
3.2. Acetylation/deacetylation (ARD1/HDAC1)	114
4. HIF Activity	114
4.1. The factor-inhibiting HIF	114
5. HIF-1/HIF-2 Target Gene Specificity	116
6. RNAi as a New Potential Therapeutic Strategy	116
Acknowledgments	118
References	118

Abstract

Gene silencing by RNA interference (RNAi) is ushering biological research into a new age, providing an extremely powerful tool for the analysis of loss-of-function phenotypes in vertebrates where alternative approaches are very often arduous or even ineffective. In this review, we will highlight the different RNAi approaches that have been undertaken to evaluate the functional role of several components of the hypoxia-signaling cascade, particularly the various oxygen sensors that control the expression and activity of the hypoxia-inducible factor (HIF). Indeed, this transcription factor lies at the heart of the pathway triggered in response to low O_2 availability to assure O_2 homeostasis, which is crucial for survival.

* CICbioGUNE Technology Park of Bizkaia, Derio, Spain
† Centre Antoine Lacassagne, Institute of Signaling, Developmental Biology and Cancer Research, University of Nice, Nice, France

Methods in Enzymology, Volume 435
ISSN 0076-6879, DOI: 10.1016/S0076-6879(07)35006-4

1. A Brief History of RNAi

In 1998, two American scientists, A. Fire and C. Mello, working with *Caenorhabditis elegans* discovered gene silencing, or RNAi, an evolutionary, conserved gene regulatory mechanism triggered by double-stranded RNA (dsRNA) (Fire *et al.*, 1998). These dsRNAs activate complex biochemical machinery, which is capable of inducing the degradation of the messenger RNA (mRNA) molecules containing homolog sequence to the dsRNA. Therefore, target mRNA is sequence-specifically depleted, resulting in knockdown of the expression of the corresponding gene (Fig. 6.1). This discovery provided a rational explanation for the previously reported silencing of endogenous genes in petunias, *Neurospora crasa*, and *C. elegans*

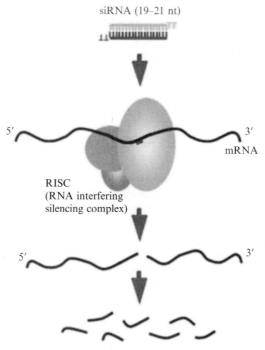

Figure 6.1   Schema of the evolutionary conserved RNA interference (RNAi) process. Chemically synthesized, enzymatically produced, or shRNAs and shRNA-miRs– derived small interfering RNAs (siRNAs) trigger RNAi silencing. One strain of the siRNA (those containing the 5′ extremity are more thermolabile) is incorporated into a protein complex called RNA interfering silencing complex (RISC). This complex "guides" the siRNA to the target messenger RNA (mRNA), then binds to and cleaves it in the middle of the complementary region. Following this first cleavage, several nonspecific nucleases are able to achieve complete mRNA degradation leading to gene silencing. (See color insert.)

by co-suppression, quelling, and sense mRNA, respectively (Guo and Kemphues, 1995; Napoli *et al.*, 1990; Romano and Macino, 1992).

Two years later, two independent groups of biochemists found that 21- to 23-nt dsRNAs always co-purified with RNAi, suggesting that dsRNA was processed into shorter intermediates, small interfering RNAs (siRNAs) capable of binding and cleaving the target mRNA (Hammond *et al.*, 2000; Zamore *et al.*, 2000).

Tuschl and colleagues definitively identified the siRNAs as the effector intermediates triggering RNAi pathway (Elbashir *et al.*, 2001), while Hannon's group identified Dicer, a ribonuclease III family (RNase III) as the key processing enzyme (Bernstein *et al.*, 2001).

Up to this point, the RNAi was limited to flies, worms, and plants since long dsRNA into mammalian cell elicits activation of the interferon response leading to inhibition of protein translation and causing mRNA degradation in a sequence-independent manner (Manche *et al.*, 1992; Minks *et al.*, 1979). However, the finding of Tuschl and colleagues that siRNAs trigger gene silencing heralded the use of RNAi in mammals. These same authors also showed that the RNAi pathway was functional in cultured mammalian cells (Elbashir *et al.*, 2001).

The siRNAs that trigger specific silencing by RNAi in mammalian cells come in many types: chemical synthetic siRNAs (19–21 bp in length with symmetric 2-nt $3'$ overhangs), their enzymatically produced cousins (esiRNAs), and short hairpin RNAs (shRNAs and shRNA-miRs), which can be expressed from an RNA polymerase III (RNA pol III) or an RNA pol II promoter and within several types of expression cassettes (Brummelkamp *et al.*, 2002; Myers *et al.*, 2003; Silva *et al.*, 2005). The use of chemically synthesized siRNAs has some advantage, such as the constant quality of reagents; however, it also has disadvantages, which include: (1) the transient inhibition of target genes, since siRNAs are short-lived and become diluted as cells divide; (2) the low-efficiency delivery to primary cells; and (3) the high cost. To deal with these limitations, shRNA- or shRNA-miR– (based on the endogenous miR-30 expression) producing vectors have been developed. Such vectors provide a persistent source of siRNAs that can mediate constitutive or conditional gene silencing following stable integration of the vector into the genome of the cell. Furthermore, the use of retroviral, lentiviral, or adenoviral vectors provides the delivery into a broad range of cell types.

2. THE HYPOXIA-SIGNALING PATHWAY

The ability of organisms to maintain O_2 homeostasis is crucial for survival. Almost all organisms respond to a low O_2 availability by activating the evolutionary, conserved, hypoxia-signaling cascade. The transcription factor HIF, a topic that is addressed in detail elsewhere in this volume, lies at the heart of this pathway. In this chapter, we discuss the siRNA approaches undertaken during the past years to study the hypoxic pathway and, most particularly, the regulatory mechanisms controlling HIF–alpha (α), the rate-limiting subunit of HIF. The stability and/or activity of HIF-α are tightly modulated by the cellular O_2 concentration.

3. HIF-α STABILITY

3.1. The family of the HIF prolyl-hydroxylases

In well-oxygenated cells, HIF-α is an extremely short-lived protein (half life less than 5 min), whereas reduced O_2 availability induces HIF-α accumulation by relaxing its ubiquitin-proteasome degradation. The mechanism targeting HIF-α for degradation remained obscure until two independent laboratories reported that O_2-dependent hydroxylation (a posttranslational modification that is inherently O_2-dependent) of two proline residues (Pro^{402} and Pro^{564} in the human HIF-1α) within the oxygen–dependent degradation domain (ODD) triggers HIF-α through the proteasome (Ivan et al., 2001; Jaakkola et al., 2001). Following these initial reports, the enzymes catalyzing the hydroxylation reaction were identified (Bruick and McKnight, 2001; Epstein et al., 2001).

Bruick and McKnight queried the GenBank database for sequences related to (at this time) the best characterized prolyl-4-hydroxylases: the catalytic α subunit of the collagen-modifying prolyl-4-hydroxylases. Of the several families of putative prolyl-hydroxylases in the database, five contained human homologs and were further investigated. By using a number of different in vitro techniques, HIF prolyl-hydroxylase 1 (HPH-1; later referred to as PHD3) was shown as the best candidate enzyme-specifying HIF-1α prolyl hydroxylation. To test whether HPH functioned in the hypoxic pathway, the authors incubated the KC167 cell line derived from Drosophila embryos with double-stranded RNAs corresponding to the sequence of the unique HPH gene in Drosophila melanogaster (dmHPH) and looked for the expression of the lactate dehydrogenase (dmLDH, a well-known HIF-dependent gene). As expected, RNAi-mediated silencing of dmHPH mRNA resulted in an increase in dmLDH mRNA levels upon normoxia.

Simultaneously, Epstein et al. (2001) identified egg-laying abnormal-9 (EGL-9) as the HIF prolyl-hydroxylase in C. elegans and a family of three genes in mammals that were homologous to egl-9 (phd1, phd2, and phd3). The three PHDs have been shown to hydroxylate the key proline residues in vitro. Each PHD isoform differs in its subcellular localization and the relative abundance of its mRNA; however, all three show a ubiquitous pattern of expression (for a review, see Berra *et al.* [2006]). The contribution of each PHD isoform to the regulation of HIF-α stability is dependent on its relative abundance, which supports the HIF-α hydroxylation as a non-equilibrium reaction (Appelhoff *et al.*, 2004). In spite of the *in vitro* studies and in order to evaluate the role of the three mammalian *egl-9* orthologies *in cellulo*, we specifically ablated each PHD by exploiting the siRNA approach (Berra *et al.*, 2003). We showed that PHD2 plays a dominant role and is the rate-limiting oxygen-sensor that sets the low, steady-state levels of HIF-1α in well-oxygenated cells (Fig. 6.2). Indeed, we reported that specific silencing of PHD2 is sufficient in stabilizing HIF-1α in all of the human cells we analyzed in normoxia, protecting HIF-1α degradation following re-oxygenation of hypoxia-stressed cells and triggering HIF-1α nuclear translocation and HIF-dependent transcriptional activation. Interestingly, among the mammalian enzymes, PHD2, but not PHD1 or PHD3, contains an N-terminal "MYND" domain that is conserved in EGL-9, which suggests that PHD2 is most closely related to the *C. elegans* gene product (Taylor, 2001).

Despite the explosion that occurred in the RNAi field, a lot remains to be understood about the RNAi process as it occurs naturally. Therefore, it is important to keep in mind that RNAi is still an immature tool, and, in the meantime, it is crucial to proceed cautiously. Although siRNAs are designed to be fully complementary to a unique mRNA transcript, it is known that some siRNAs may inadvertently target unrelated genes with only partial sequence-complementarity (off-target effects)

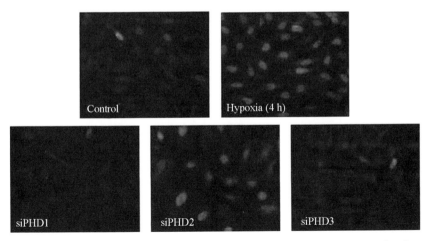

Figure 6.2 Prolyl hydroxylase domain (PHD) 2 is the oxygen sensor setting low, steady-state levels of hypoxia-inducible transcription factor (HIF)-1α in normoxia. HeLa cells were transfected with the corresponding small interfering RNAs (siRNAs; 20 nM), and 48 h later, the expression of HIF-1α was analyzed by immunofluorescence. Cells were incubated in hypoxia (1% O_2 for 4 h) as a positive control. (See color insert.)

(Jackson *et al.*, 2003). We were concerned about this inherent problem, and, to confirm the validity of the observed phenotype, we carried out many specificity controls. Using an irrelevant siRNA-targeting dmHIF (Sima) as a negative control, we showed that two (even three, in the case of *phd2*) distinct siRNAs targeted to different regions each and independently knocked down the corresponding *phd* mRNA specifically and exerted the same phenotypic effect (Fig. 6.3A). Several siRNA sequence selection algorithms and guidelines have been developed in recent years to increase the likelihood of identifying effective and specific siRNAs (for a review, see Pei and Tuschl [2006]); however, for this study, siRNA design relied on rules based on conventional molecular biology oligos.

Saturation of the RNAi machinery and thereby inhibition of endogenous miRNA function is another potential source of nonspecific effects (Persengiev *et al.*, 2004). The lowest effective amount of the siPHD2 that gave rise to HIF-1α accumulation was 0.5 nM, while the maximun specific RNAi response was achieved at 200 nM (see Fig. 6.3B).

Furthermore, we also performed the gold standard control for specificity in RNAi experiments: rescue (for a review, see Cullen [2006]). Taking advantage of the redundancy of the genetic code, we rescued the expression of PHD2 by reintroducing a complementary DNA (cDNA)—PHD2[siRNAmut]—resistant to the siRNA used. Restoration of the PHD2 expression rescued the wild-type phenotype (see Fig. 6.3C). Thus, HIF-1α is no more stabilized by the siPHD2, showing once again the specificity of the phenotype we observed.

In contrast to acute hypoxia and accordingly to two independent reports, we have shown that the HIF-α levels vanish during chronic hypoxic stress *in cellulo* and in mice (Ginouvès *et al.* personal communication; Stiehl *et al.*, 2006; Wang *et al.*, 1995). By using *in vitro* and *in vivo* siRNAs, the three PHDs (PHD1, PHD2, and PHD3) emerge as the key enzymes at the basis of this desensitization mechanism. The expression of PHD2 and PHD3 is induced by hypoxia (Epstein *et al.*, 2001). Furthermore, functional hypoxia–response elements (HREs) are located in the promoter of *phd2* as well as within the first intron of *phd3,* and the RNAi approach has demonstrated that PHD2 is an HIF-1–dependent gene, whereas PHD3 is dependent on HIF-1 and HIF-2 transcriptional complexes (Berra *et al.*, 2003; del Peso *et al.*, 2003; Metzen *et al.*, 2005; Pescador *et al.*, 2005). Accumulation of PHD2 and PHD3, the hypoxia-inducible isoforms, could account, at least partially, for the feedback mechanism, as proposed by Stiehl *et al.*, (2006). However, PDK1 (pyruvate dehydrogenase kinase), an HIF-1–dependent gene product, recently has been reported as a natural inhibitor of mitochondrial activity in hypoxia (Kim *et al.*, 2006; Papandreou *et al.*, 2006). Based on these results, we propose that chronic hypoxia, by inhibiting mitochondrial respiration, progressively increases O_2 availability for PHDs, leading to their over-activation and thus triggering HIF-α "desensitization."

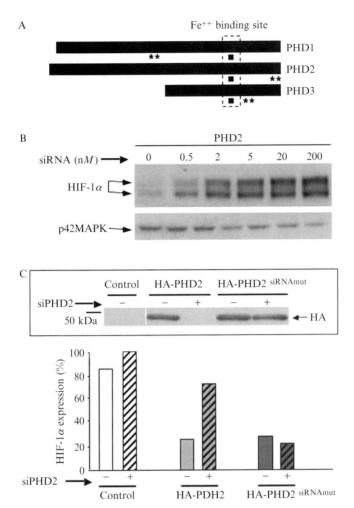

Figure 6.3 Validation of the RNA interference (RNAi) approach. (A) Specificity of the silencing phenotype. Filled squares and double asterisks show localization of the prolyl hydroxylase domain (PHD) region targeted by the two independent sets of small interfering RNAs (siRNAs). (B) Dose-dependent effect of the siPHD2. HeLa cells were transfected with increasing doses of the siPHD2, and the expression of hypoxia-inducible transcription factor (HIF)-1α was analyzed by Western blotting. The anti-p42MAPK antibody was used as a loading control. (C) PHD2^{siRNAmut} overexpression rescues HIF-1α degradation. HeLa cells were transfected with the indicated plasmids in the absence or the presence of the siPHD2 and incubated in hypoxia (1% O_2) for 1 h. HIF-1α expression was analyzed per immunofluorescence, and the histogram corresponded to the percent of positive cells for each condition. Inset shows the expression levels of the overexpressed PHD2 constructs.

Related to the PHDs, Semenza's group identified OS-9 as a protein interacting with HIF-α and PHDs to promote O_2-dependent degradation (Baek *et al.*, 2005). They demonstrate that OS-9 loss of function by RNAi increases HIF-α protein levels and hence HIF-mediated transcription. COP9 signalosome subunit 5 (CSN5) also appears to control HIF-1α stability by inhibiting HIF-1α hydroxylation on P^{564} (Bemis *et al.*, 2004).

3.2. Acetylation/deacetylation (ARD1/HDAC1)

Acetylation of one lysine residue (K532 of the human HIF-1α) within the ODD by an acetyltransferase called Arrest-defective-1 (ARD1) was reported to destabilize HIF-1α by enhancing its interaction with the von Hippel-Lindau protein (pVHL)-containing E3 ubiquitin ligase complex (Jeong *et al.*, 2002). However, contradictory data bring into question the foundations of this mechanism; silencing of ARD1 by using siRNAs (Bilton *et al.*, 2005) or shRNAs delivered by a retroviral vector (Fisher *et al.*, 2005) had no impact on HIF-α protein levels. Nonetheless, Yoo *et al.* (2006) have shown that silencing of metastasis-associated protein-1 (MTA-1) enhances HIF-1α binding to pVHL, PHD2, and thus protein instability, even in severe hypoxia (0.1 %), by a mechanism implicating the histone deacetylase 1 (HDAC 1). Furthermore, MTA-1 has no impact on a mutant HIF-1α^{K532R}. This report reinforces the potential role of HIF-α acetylation and deacetylation in the control of this subunit.

Mechanisms other than HIF-α posttranscriptional modifications appear to be implicated in the control of the protein stability. Looking for HIF-1α interacting protein, Semenza's group has recently identified the receptor of activated protein kinase C (RACK1) (Liu *et al.*, 2007). By overexpressing and silencing of RACK1, the authors present evidence that RACK1 is a critical component of a PHD's/pVHL's independent mechanism for regulating HIF-1α stability through competition with Hsp90 and recruitment of the elongin B/C E3 ubiquitin ligase complex.

4. HIF ACTIVITY

4.1. The factor-inhibiting HIF

A second posttranscriptional hydroxylation has been shown to affect the hypoxia-signaling pathway: asparagine hydroxylation by factor-inhibiting HIF (FIH) (Lando *et al.*, 2002). Instead of the instability conferred by PHDs, the O_2-dependent hydroxylation of one asparagine residue (Asn^{803} in the human HIF-1α) within the COOH-terminal transcriptional activation domain (C-TAD) inactivates HIF. Indeed, FIH represses transcriptional activity by preventing the binding of the transcriptional activator CREB-binding protein (CBP)/p300 to the C-TAD (Mahon *et al.*, 2001). Ratcliffe and colleagues were the first to evaluate the role of FIH in regulating endogenous HIF-dependent gene expression (Stolze *et al.*, 2004). By using two different siRNAs, they showed that the low levels of HIF-1α in normoxia are potentially active and that this transcriptional activity is normally suppressed by FIH. We confirmed these results by co-silencing of PHD2 and FIH using two parallel approaches: a very sensitive reporter plasmid (p3HRE-Δptk-LUC) and endogenous HIF-dependent genes (Dayan *et al.*, 2006) (Fig. 6.4). Moreover, given that HIF-1α possesses an NH_2-terminal TAD (N-TAD) in addition to the C-TAD, we hypothesized that FIH, by discriminating both TADs, might regulate a spectrum of HIF-dependent genes. To address this issue (based on the pTER plasmid developed by Agami's group [van de Wetering *et al.*, 2003]), we engineered a cell line to silence FIH in an inducible manner, particularly in response to tetracycline. Our process showed that theory was indeed the case: FIH specifically controls the expression of a set of HIF-dependent genes.

Datta *et al.* (2004) reported that PKCδ plays an important role in the control of HIF-1 transcriptional activity. By using an siRNA directed against PKCδ, they showed that this protein modulates the association of HIF-α with p300. Inhibition of PKCδ increases FIH mRNA levels in 786–0 and A498 cells (two pVHL-deficient renal cell carcinoma [RCC] cell lines) as well as in ASPC-1 (human pancreatic adenocarcinoma) and HT1080 (fibrosarcoma cell line). Inhibitor of growth family, member 4 (ING4), a likely component of a chromatin-remodeling complex, has

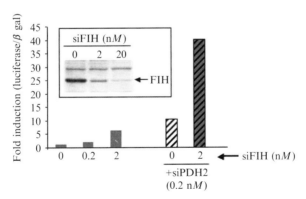

Figure 6.4 Synergistic effect of prolyl hydroxylase domain (PHD) 2 and factor-inhibiting hypoxia-inducible transcription factor (FIH). HeLa cells were transfected with increasing doses of the small interfering FIG (siFIH) in the absence or the presence of the siPHD2 at 0.2 nM and the reporter vector (p3HRE-Δptk-LUC) containing three copies of the hypoxia response element (HRE) from the erythropoietin gene. In addition, a β-galactosidase vector was co-transfected to normalize for transfection efficiency. Inset shows dose-dependent silencing of FIH by Western blotting.

also been shown to repress HIF activation (Ozer *et al.*, 2005). Surprisingly, rather than affecting stability of HIF-α, through the recruitment of PHDs, ING4 mediates HIF activity.

5. HIF-1/HIF-2 Target Gene Specificity

Before the discovery of the RNAi, the relative contribution of HIF-1α and HIF-2α to the hypoxia-signaling pathway was an open question in the field. Indeed, knockout models suggested non-redundant functions, whereas overexpression approaches showed no target gene specificity. The use of RNAi precludes severe perturbations of the signaling cascades arising from the transient or stable expression of HIF-α isoform. Hence, the silencing approach allows for the proper address of this controversial issue. Looking for the expression of a panel of well-known hypoxia-inducible genes, Harris' group was the first to show that HIF-1α is the isoform with the broadest impact on hypoxic gene regulation (Sowter *et al.*, 2003). Similar results were obtained in two independent studies (Aprelikova *et al.*, 2006; Warnecke *et al.*, 2004). Furthermore, by silencing ELK-1 (the most often represented member of ETS), Aprelikova *et al.* (2006) presented evidence of a role of this family of transcription factors in the HIF-2 target gene specificity, suggesting that binding to specific factors might discriminate the choice of target genes by HIF-1 or HIF-2.

Looking for HIF-2–specific target genes, Shibasaki's group performed a yeast two-hybrid analysis and identified the tumor suppressor Int6 as a novel regulator of HIF-2α (Chen *et al.*, 2007). Int6 protein knockdown by siRNA vectors increased endogenous HIF-2α expression, whereas it has no effect on HIF-1α. Furthermore, Int6-mediated HIF-2α degradation was found to be both hypoxia- and pVHL-independent.

6. RNAi as a New Potential Therapeutic Strategy

Based on the highly specific and efficient silencing of a target gene, RNAi has been devoted as a powerful tool for reverse and forward genetic studies. More recently, RNAi emerged as a very promising technique not only for drug discovery and target validation in cell culture but also as a novel therapeutic strategy. Preclinical studies have provided investigators with promising data, and the expectations for the implementation of such technology in the clinical setting are growing.

Many diseases could benefit from such a therapeutic approach. This idea is especially true for those occurring from the expression of mutant genes, aberrant splicing isoforms, or overexpressed genes leading to gain of function effects, like cancers, neurodegenerative diseases, or viral infections. Hence, it is tempting to hope that this new technology, will challenge those diseases that fail conventional therapeutics, by specific silencing of key pathogenic genes. More specifically, therapeutic targets previously deemed "undruggable" by small molecules are now within reach of RNA-based therapy.

However, limitations of transfer vectors may turn out to be the limiting step in the development of RNAi-based therapeutic strategies. Localized delivery of siRNAs or shRNAs to specific organs or tissues has been successful in an increasing number of experimental models; however, this approach is limited to sites that are readily accessible (for a review, see Snove and Rossi [2006]). For systemic use, the biggest challenge in the utilization of RNAi-based therapies is the efficiency of delivery. The first attempts of *in vivo* siRNA delivery were carried out using the hydrodynamic technique of administering naked siRNAs in a large volume of a physiological solution under high pressure into the tail vein of mice (McCaffrey *et al.*, 2002). Although effective in rodents, hydrodynamic delivery is unlikely to be applicable to human therapy. Several methods for systemic delivery are currently under investigation for naked and chemically modified siRNAs or DNA-based systems (shRNAs), which, as previously mentioned, offer the advantage of constitutive expression of siRNAs with a potentially unlimited duration of gene silencing and allow the incorporation of regulatory elements to the promoter region, resulting in tissue-specific silencing. These methods include viral delivery (retrovirus, lentivirus, adenovirus, adeno-associated virus, or baculovirus), envelopes using liposomes and other nanoparticles (especially the cationic polymer polyethylenimine [PEI]) as simple packages and in combination with specific homing signals designed to direct the preferential uptake by a specific target tissue or a group of target cells, or even bacteria-mediated RNAi production.

Regarding the specific targeting of the hypoxia-signaling pathway using the RNAi-based therapeutic approach, most of the studies are related to cancer therapies. Indeed, intratumoral injection of an adenovirus encoding the HIF-targeted siRNA has a significant effect on tumor growth when combined with ionizing radiation (Zhang *et al.*, 2004). RNAi of the urokinase plasminogen activator receptor (uPAR), an HIF-dependent gene, reduces glioma cell invasion and angiogenesis *in vitro* and *in vivo*. Furthermore, intratumoral injections of plasmid vectors expressing shRNAs targeting uPAR resulted in the regression of preestablished intracranial tumors (Gondi *et al.*, 2004). Nevertheless, in addition to cancer, hypoxia is implicated in several pathological processes where the RNAi-based therapeutic approach might be implemented as well. The subretina injection of siRNAs directed against vascular endothelial growth factor (VEGF) to ameliorate macular degeneration symptoms is one example of the successful experience of therapeutic effects mediated by RNAi (Reich *et al.*, 2003).

Moreover, these results prompted three independent biotech companies to initiate phase I clinical trials that are currently underway.

In summary, even if solving the problem of the delivery will be another big step, the pharmacological revolution that RNAi offers is so great that the efforts spent are worth it, making it reasonable to remain optimistic. Indeed, functional systemic delivery of siRNAs has been reported in a nonhuman primate without any evidence of adverse effects (Zimmermann *et al.*, 2006).

ACKNOWLEDGMENTS

We thank Pouysségur's laboratory. This work has been partially supported by the Centre National de la Recherche Scientifique (CNRS), Ministère de l'Education, de la Recherche et de la Technologie, Ligue Nationale Contre le Cancer (Equipe labellisée), Cancéropôle PACA, and Association pour la Recherche sur le Cancer (Subvention n° 3693). E.B. is currently funded by the ETORTEK Research Program 05–07 (Department of Industry, Tourism, and Trade of the Government of the Autonomous Community of the Basque Country) and the BIZKAIA XEDE Program from Bizkaia County. We apologize to our colleagues whose work could not be cited due to space limitations.

REFERENCES

Appelhoff, R. J., Tian, Y. M., Raval, R. R., Turley, H., Harris, A. L., Pugh, C. W., Ratcliffe, P. J., and Gleadle, J. M. (2004). Differential function of the prolyl hydroxylases PHD1, PHD2, and PHD3 in the regulation of hypoxia-inducible factor. *J. Biol. Chem.* **279**, 38458–38465.

Aprelikova, O., Wood, M., Tackett, S., Chandramouli, G. V., and Barrett, J. C. (2006). Role of ETS transcription factors in the hypoxia-inducible factor-2 target gene selection. *Cancer Res.* **66**, 5641–5647.

Baek, J. H., Mahon, P. C., Oh, J., Kelly, B., Krishnamachary, B., Pearson, M., Chan, D. A., Giaccia, A. J., and Semenza, G. L. (2005). OS-9 interacts with hypoxia-inducible factor 1alpha and prolyl hydroxylases to promote oxygen-dependent degradation of HIF-1alpha. *Mol. Cell.* **17**, 503–512.

Bemis, L., Chan, D. A., Finkielstein, C. V., Qi, L., Sutphin, P. D., Chen, X., Stenmark, K., Giaccia, A. J., and Zundel, W. (2004). Distinct aerobic and hypoxic mechanisms of HIF-alpha regulation by CSN5. *Genes Dev.* **18**, 739–744.

Bernstein, E., Caudy, A. A., Hammond, S. M., and Hannon, G. J. (2001). Role for a bidentate ribonuclease in the initiation step of RNA interference. *Nature* **409**, 363–366.

Berra, E., Benizri, E., Ginouves, A., Volmat, V., Roux, D., and Pouyssegur, J. (2003). HIF prolyl-hydroxylase 2 is the key oxygen sensor setting low steady-state levels of HIF-1alpha in normoxia. *EMBO J.* **22**, 4082–4090.

Berra, E., Ginouves, A., and Pouyssegur, J. (2006). The hypoxia-inducible-factor hydroxylases bring fresh air into hypoxia signalling. *EMBO Rep.* **7**, 41–45.

Bilton, R., Mazure, N., Trottier, E., Hattab, M., Dery, M. A., Richard, D. E., Pouyssegur, J., and Brahimi-Horn, M. C. (2005). Arrest-defective-1 protein, an acetyl-transferase, does not alter stability of hypoxia-inducible factor (HIF)-1alpha and is not induced by hypoxia or HIF. *J. Biol. Chem.* **280**, 31132–31140.

Bruick, R. K., and McKnight, S. L. (2001). A conserved family of prolyl-4-hydroxylases that modify HIF. *Science* **294**, 1337–1340.

Brummelkamp, T. R., Bernards, R., and Agami, R. (2002). A system for stable expression of short interfering RNAs in mammalian cells. *Science* **296,** 550–553.

Chen, L., Uchida, K., Endler, A., and Shibasaki, F. (2007). Mammalian Tumor Suppressor Int6 Specifically Targets Hypoxia Inducible Factor 2alpha for Degradation by Hypoxia- and pVHL-independent Regulation. *J. Biol. Chem.* **282,** 12707–12716.

Cullen, B. R. (2006). Enhancing and confirming the specificity of RNAi experiments. *Nat. Methods* **3,** 677–681.

Datta, K., Li, J., Bhattacharya, R., Gasparian, L., Wang, E., and Mukhopadhyay, D. (2004). Protein kinase C zeta transactivates hypoxia-inducible factor alpha by promoting its association with p300 in renal cancer. *Cancer Res.* **64,** 456–462.

Dayan, F., Roux, D., Brahimi-Horn, M. C., Pouyssegur, J., and Mazure, N. M. (2006). The oxygen sensor factor-inhibiting hypoxia-inducible factor-1 controls expression of distinct genes through the bifunctional transcriptional character of hypoxia-inducible factor-1alpha. *Cancer Res.* **66,** 3688–3698.

del Peso, L., Castellanos, M. C., Temes, E., Martin-Puig, S., Cuevas, Y., Olmos, G., and Landazuri, M. O. (2003). The von Hippel Lindau/hypoxia-inducible factor (HIF) pathway regulates the transcription of the HIF-proline hydroxylase genes in response to low oxygen. *J. Biol. Chem.* **278,** 48690–48695.

Elbashir, S. M., Harborth, J., Lendeckel, W., Yalcin, A., Weber, K., and Tuschl, T. (2001). Duplexes of 21-nucleotide RNAs mediate RNA interference in cultured mammalian cells. *Nature* **411,** 494–498.

Epstein, A. C., Gleadle, J. M., McNeill, L. A., Hewitson, K. S., O'Rourke, J., Mole, D. R., Mukherji, M., Metzen, E., Wilson, M. I., Dhanda, A., Tian, Y. M., and Masson, N., *et al.* (2001). C. elegans EGL-9 and mammalian homologs define a family of dioxygenases that regulate HIF by prolyl hydroxylation. *Cell* **107,** 43–54.

Fire, A., Xu, S., Montgomery, M. K., Kostas, S. A., Driver, S. E., and Mello, C. C. (1998). Potent and specific genetic interference by double-stranded RNA in *Caenorhabditis elegans. Nature* **391,** 806–811.

Fisher, T. S., Etages, S. D., Hayes, L., Crimin, K., and Li, B. (2005). Analysis of ARD1 function in hypoxia response using retroviral RNA interference. *J. Biol. Chem.* **280,** 17749–17757.

Gondi, C. S., Lakka, S. S., Dinh, D. H., Olivero, W. C., Gujrati, M., and Rao, J. S. (2004). RNAi-mediated inhibition of cathepsin B and uPAR leads to decreased cell invasion, angiogenesis and tumor growth in gliomas. *Oncogene* **23,** 8486–8496.

Guo, S., and Kemphues, K. J. (1995). par-1, a gene required for establishing polarity in *C. elegans* embryos, encodes a putative Ser/Thr kinase that is asymmetrically distributed. *Cell* **81,** 611–620.

Hammond, S. M., Bernstein, E., Beach, D., and Hannon, G. J. (2000). An RNA-directed nuclease mediates post-transcriptional gene silencing in *Drosophila* cells. *Nature* **404,** 293–296.

Ivan, M., Kondo, K., Yang, H., Kim, W., Valiando, J., Ohh, M., Salic, A., Asara, J. M., Lane, W. S., and Kaelin, W. G., Jr. (2001). HIF-alpha targeted for VHL-mediated destruction by proline hydroxylation: Implications for O_2 sensing. *Science* **292,** 464–468.

Jaakkola, P., Mole, D. R., Tian, Y. M., Wilson, M. I., Gielbert, J., Gaskell, S. J., Kriegsheim, A., Hebestreit, H. F., Mukherji, M., Schofield, C. J., Maxwell, P. H., and Pugh, C. W., *et al.* (2001). Targeting of HIF-alpha to the von Hippel-Lindau ubiquitylation complex by O_2-regulated prolyl hydroxylation. *Science* **292,** 468–472.

Jackson, A. L., Bartz, S. R., Schelter, J., Kobayashi, S. V., Burchard, J., Mao, M., Li, B., Cavet, G., and Linsley, P. S. (2003). Expression profiling reveals off-target gene regulation by RNAi. *Nat. Biotechnol.* **21,** 635–637.

Jeong, J. W., Bae, M. K., Ahn, M. Y., Kim, S. H., Sohn, T. K., Bae, M. H., Yoo, M. A., Song, E. J., Lee, K. J., and Kim, K. W. (2002). Regulation and destabilization of HIF-1alpha by ARD1-mediated acetylation. *Cell* **111,** 709–720.

Kim, J. W., Tchernyshyov, I., Semenza, G. L., and Dang, C. V. (2006). HIF-1-mediated expression of pyruvate dehydrogenase kinase: A metabolic switch required for cellular adaptation to hypoxia. *Cell Metab.* **3,** 177–185.

Lando, D., Peet, D. J., Gorman, J. J., Whelan, D. A., Whitelaw, M. L., and Bruick, R. K. (2002). FIH-1 is an asparaginyl hydroxylase enzyme that regulates the transcriptional activity of hypoxia-inducible factor. *Genes Dev.* **16,** 1466–1471.

Liu, Y. V., Baek, J. H., Zhang, H., Diez, R., Cole, R. N., and Semenza, G. L. (2007). RACK1 competes with HSP90 for binding to HIF-1alpha and is required for O(2)-independent and HSP90 inhibitor-induced degradation of HIF-1alpha. *Mol. Cell.* **25,** 207–217.

Mahon, P. C., Hirota, K., and Semenza, G. L. (2001). FIH-1: A novel protein that interacts with HIF-1alpha and VHL to mediate repression of HIF-1 transcriptional activity. *Genes Dev.* **15,** 2675–2686.

Manche, L., Green, S. R., Schmedt, C., and Mathews, M. B. (1992). Interactions between double-stranded RNA regulators and the protein kinase DAI. *Mol. Cell Biol.* **12,** 5238–5248.

McCaffrey, A. P., Meuse, L., Pham, T. T., Conklin, D. S., Hannon, G. J., and Kay, M. A. (2002). RNA interference in adult mice. *Nature* **418,** 38–39.

Metzen, E., Stiehl, D. P., Doege, K., Marxsen, J. H., Hellwig-Burgel, T., and Jelkmann, W. (2005). Regulation of the prolyl hydroxylase domain protein 2 (phd2/egln-1) gene: Identification of a functional hypoxia-responsive element. *Biochem. J.* **387,** 711–717.

Minks, M. A., West, D. K., Benvin, S., and Baglioni, C. (1979). Structural requirements of double-stranded RNA for the activation of 2′,5′-oligo(A) polymerase and protein kinase of interferon-treated HeLa cells. *J. Biol. Chem.* **254,** 10180–10183.

Myers, J. W., Jones, J. T., Meyer, T., and Ferrell, J. E., Jr. (2003). Recombinant Dicer efficiently converts large dsRNAs into siRNAs suitable for gene silencing. *Nat. Biotechnol.* **21,** 324–328.

Napoli, C., Lemieux, C., and Jorgensen, R. (1990). Introduction of a Chimeric Chalcone Synthase Gene into Petunia Results in Reversible Co-Suppression of Homologous Genes in trans. *Plant Cell.* **2,** 279–289.

Ozer, A., Wu, L. C., and Bruick, R. K. (2005). The candidate tumor suppressor ING4 represses activation of the hypoxia inducible factor (HIF). *Proc. Natl. Acad. Sci. USA* **102,** 7481–7486.

Papandreou, I., Cairns, R. A., Fontana, L., Lim, A. L., and Denko, N. C. (2006). HIF-1 mediates adaptation to hypoxia by actively downregulating mitochondrial oxygen consumption. *Cell Metab.* **3,** 187–197.

Pei, Y., and Tuschl, T. (2006). On the art of identifying effective and specific siRNAs. *Nat. Methods* **3,** 670–676.

Persengiev, S. P., Zhu, X., and Green, M. R. (2004). Nonspecific, concentration-dependent stimulation and repression of mammalian gene expression by small interfering RNAs (siRNAs). *RNA* **10,** 12–18.

Pescador, N., Cuevas, Y., Naranjo, S., Alcaide, M., Villar, D., Landazuri, M. O., and del Peso, L. (2005). Identification of a functional hypoxia-responsive element that regulates the expression of the egl nine homologue 3 (egln3/phd3) gene. *Biochem. J.* **390,** 189–197.

Reich, S. J., Fosnot, J., Kuroki, A., Tang, W., Yang, X., Maguire, A. M., Bennett, J., and Tolentino, M. J. (2003). Small interfering RNA (siRNA) targeting VEGF effectively inhibits ocular neovascularization in a mouse model. *Mol. Vis.* **9,** 210–216.

Romano, N., and Macino, G. (1992). Quelling: Transient inactivation of gene expression in *Neurospora crassa* by transformation with homologous sequences. *Mol. Microbiol.* **6,** 3343–3353.

Silva, J. M., Li, M. Z., Chang, K., Ge, W., Golding, M. C., Rickles, R. J., Siolas, D., Hu, G., Paddison, P. J., Schlabach, M. R., Sheth, N., and Bradshaw, J., et al. (2005). Second-generation shRNA libraries covering the mouse and human genomes. *Nat. Genet.* **37**, 1281–1288.

Snove, O., Jr., and Rossi, J. J. (2006). Expressing short hairpin RNAs *in vivo*. *Nat. Methods* **3**, 689–695.

Sowter, H. M., Raval, R. R., Moore, J. W., Ratcliffe, P. J., and Harris, A. L. (2003). Predominant role of hypoxia-inducible transcription factor (Hif)-1alpha versus Hif-2alpha in regulation of the transcriptional response to hypoxia. *Cancer Res.* **63**, 6130–6134.

Stiehl, D. P., Wirthner, R., Koditz, J., Spielmann, P., Camenisch, G., and Wenger, R. H. (2006). Increased prolyl 4-hydroxylase domain proteins compensate for decreased oxygen levels. Evidence for an autoregulatory oxygen-sensing system. *J. Biol. Chem.* **281**, 23482–23491.

Stolze, I. P., Tian, Y. M., Appelhoff, R. J., Turley, H., Wykoff, C. C., Gleadle, J. M., and Ratcliffe, P. J. (2004). Genetic analysis of the role of the asparaginyl hydroxylase factor inhibiting hypoxia-inducible factor (HIF) in regulating HIF transcriptional target genes. *J. Biol. Chem.* **279**, 42719–42725.

Taylor, M. S. (2001). Characterization and comparative analysis of the EGLN gene family. *Gene* **275**, 125–132.

van de Wetering, M., Oving, I., Muncan, V., Pon Fong, M. T., Brantjes, H., van Leenen, D., Holstege, F. C., Brummelkamp, T. R., Agami, R., and Clevers, H. (2003). Specific inhibition of gene expression using a stably integrated, inducible small-interfering-RNA vector. *EMBO Rep.* **4**, 609–615.

Wang, G. L., Jiang, B. H., Rue, E. A., and Semenza, G. L. (1995). Hypoxia-inducible factor 1 is a basic-helix-loop-helix-PAS heterodimer regulated by cellular O_2 tension. *Proc. Natl. Acad. Sci. USA* **92**, 5510–5514.

Warnecke, C., Zaborowska, Z., Kurreck, J., Erdmann, V. A., Frei, U., Wiesener, M., and Eckardt, K. U. (2004). Differentiating the functional role of hypoxia-inducible factor (HIF)-1alpha and HIF-2alpha (EPAS-1) by the use of RNA interference: Erythropoietin is a HIF-2alpha target gene in Hep3B and Kelly cells. *FASEB J.* **18**, 1462–1464.

Yoo, Y. G., Kong, G., and Lee, M. O. (2006). Metastasis-associated protein 1 enhances stability of hypoxia-inducible factor-1alpha protein by recruiting histone deacetylase 1. *EMBO J.* **25**, 1231–1241.

Zamore, P. D., Tuschl, T., Sharp, P. A., and Bartel, D. P. (2000). RNAi: Double-stranded RNA directs the ATP-dependent cleavage of mRNA at 21 to 23 nucleotide intervals. *Cell* **101**, 25–33.

Zhang, X., Kon, T., Wang, H., Li, F., Huang, Q., Rabbani, Z. N., Kirkpatrick, J. P., Vujaskovic, Z., Dewhirst, M. W., and Li, C. Y. (2004). Enhancement of hypoxia-induced tumor cell death *in vitro* and radiation therapy *in vivo* by use of small interfering RNA targeted to hypoxia-inducible factor-1alpha. *Cancer Res.* **64**, 8139–8142.

Zimmermann, T. S., Lee, A. C., Akinc, A., Bramlage, B., Bumcrot, D., Fedoruk, M. N., Harborth, J., Heyes, J. A., Jeffs, L. B., John, M., Judge, A. D., and Lam, K., et al. (2006). RNAi-mediated gene silencing in non-human primates. *Nature* **441**, 111–114.

CHAPTER SEVEN

Cellular and Developmental Adaptations to Hypoxia: A Drosophila Perspective

Nuria Magdalena Romero,* Andrés Dekanty,* and Pablo Wappner[†]

Contents

1. Introduction	124
2. Drosophila melanogaster as a Model System to Study Physiological Responses to Hypoxia	124
3. Experimental Advantages of the Model System	125
4. The Drosophila Respiratory System	126
5. Occurrence of a Drosophila System Homologous to Mammalian HIF	128
6. Regulation of Sima by Oxygen Levels	131
7. Role of Sima and Fatiga in Drosophila Development	132
8. Hypoxia-Inducible Genes and the Adaptation of Drosophila to Oxygen Starvation	134
9. Regulation of Sima by the PI3K and TOR Pathways	134
10. Role of the HIF System in Growth Control and Cell Size Determination	136
11. Concluding Remarks	138
Acknowledgments	139
References	139

Abstract

The fruit fly Drosophila melanogaster, a widely utilized genetic model, is highly resistant to oxygen starvation and is beginning to be used for studying physiological, developmental, and cellular adaptations to hypoxia. The Drosophila respiratory (tracheal) system has features in common with the mammalian circulatory system so that an angiogenesis-like response occurs upon exposure of Drosophila larvae to hypoxia. A hypoxia-responsive system homologous to

* Instituto Leloir, Patricias Argentinas, Buenos Aires, Argentina
† Instituto Leloir and FBMC, University of Buenos Aires, Patricias Argentinas, Buenos Aires, Argentina

Methods in Enzymology, Volume 435
ISSN 0076-6879, DOI: 10.1016/S0076-6879(07)35007-6

mammalian hypoxia-inducible factor (HIF) has been described in the fruit fly, where Fatiga is a *Drosophila* oxygen-dependent HIF prolyl hydroxylase, and the basic helix-loop-helix Per/ARNT/Sim (bHLH-PAS) proteins Sima and Tango are, respectively, the *Drosophila* homologues of mammalian HIF-alpha (α) and HIF-beta (β). Tango is constitutively expressed regardless of oxygen tension and, like in mammalian cells, Sima is controlled at the level of protein degradation and subcellular localization. Sima is critically required for development in hypoxia, but, unlike mammalian model systems, it is dispensable for development in normoxia. In contrast, *fatiga* mutant alleles are all lethal; however, strikingly, viability to adulthood is restored in *fatiga sima* double mutants, although these double mutants are not entirely normal, suggesting that Fatiga has Sima-independent functions in fly development. Studies in cell culture and *in vivo* have revealed that Sima is activated by the insulin receptor (InR) and target-of-rapamycin (TOR) pathways. Paradoxically, Sima is a negative regulator of growth. This suggests that Sima is engaged in a negative feedback loop that limits growth upon stimulation of InR/TOR pathways.

1. INTRODUCTION

This chapter intends to provide an overview of the recent progress attained in the field of hypoxia and HIF biology in the fruit fly *D. melanogaster*. We will begin with a brief description of the model, its life cycle, its advantages as a genetic system, and a summary of some of the useful genetic methods available in this species. We will continue with a description of the *Drosophila* respiratory system and its oxygen-dependent plasticity, which shares many features with angiogenesis of vertebrate organisms. We will then summarize the current knowledge on the *Drosophila* HIF system, including the cellular mechanisms of oxygen-dependent regulation as well as a brief description of the target genes that mediate adaptation to oxygen starvation. We will finish by discussing the regulation that the InR and TOR pathways exert on the *Drosophila* HIF system and the current knowledge about the role of the hypoxia-responsive machinery in growth control and cell size determination.

2. *DROSOPHILA MELANOGASTER* AS A MODEL SYSTEM TO STUDY PHYSIOLOGICAL RESPONSES TO HYPOXIA

The fruit fly *D. melanogaster* is a cosmopolitan species with striking capacity to colonize a wide array of different habitats and environmental conditions (Berrigan and Partridge, 1997; Dillon and Frazier, 2006; Gibert *et al.*, 2001). It belongs to the so-called group of holometabolous insects in which the general body plan undergoes a dramatic reorganization in the larva-to–pupa transition. The duration of the entire life cycle is about 12 days at 25°

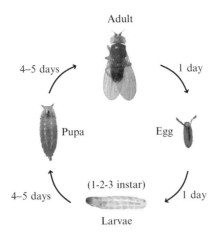

Adult

4–5 days

1 day

Pupa

Egg

4–5 days

(1-2-3 instar)

1 day

Larvae

Figure 7.1 The life cycle of *Drosophila melanogaster*. The duration of the entire *Drosophila* life cycle is about 12 days at 25°. Adult females lay eggs; embryogenesis lasts 24 h and takes place within the eggshell. The three larvae instars take 4 to 5 days; the third instar larva gives rise to pupae that undergo metamorphosis to become an adult that attains sexual maturity within 24 h.

(Fig. 7.1) so that generations can be reared within a short time period (Ashburner *et al.*, 2004). Females lay eggs into fermenting fruits, and, after completion of embryonic development, a first instar larva hatches from the eggshell. The resulting larva feeds very actively and increases its weight by several folds, while it molts twice to a second and a third larval instar. After attaining a critical weight, third instar larvae stop feeding, become immobile, and encapsulate in the pupal case, where they undergo a 4-day metamorphosis. During this period, the larval tissues are degraded to their basic components, and the entire body is rebuilt to give rise to a pupa that undergoes one additional cuticle molt to become a pharate adult. Finally, a fully formed adult emerges from the pupal case and attains sexual maturity within 24 h.

In nature, *Drosophila* larvae live mostly in fermenting fruits and feed with fruit pulp and yeast that usually grows therein. Thus, in its normal habitat, the larvae compete with microorganisms for limited amounts of oxygen. Therefore, *Drosophila* larvae are permanently exposed to a hypoxic microenvironment, anticipating a well-developed cellular machinery that responds to oxygen starvation (Gorr *et al.*, 2006).

3. EXPERIMENTAL ADVANTAGES OF THE MODEL SYSTEM

Genetic studies in *D. melanogaster* started about 100 years ago with the pioneering work of Thomas Hunt Morgan in 1909. Since then, genetic work in the fruit fly has become increasingly intense, and thousands of mutations have been isolated, targeting a large portion of the genes of the fruit fly. This ample repertoire of mutations has been gradually enriched with sophisticated genetic tools that facilitated manipulations (Greenspan, 1997). The availability of molecular biology techniques in the early 1980s provoked a true revolution in the *Drosophila* field. *Drosophila* genetic mobile elements were isolated, molecularly characterized, and converted into gene-delivering vectors (Cooley *et al.*, 1988; Rubin and Spradling, 1982; Spradling and Rubin, 1982). In just a few years, transformation of *Drosophila* embryos and generation of transgenic fly lines became a routine practice in every *Drosophila* laboratory.

Progress in *Drosophila* gene technology has accelerated dramatically over the last decade and, nowadays, the *Drosophila* genome is fully sequenced (Adams *et al.*, 2000), mutations targeting most of the genes are available, genes can be readily overexpressed in almost any desired pattern (limited just by the availability of known promoters) (Brand and Perrimon, 1993), and gene expression can be conditionally silenced in spatially and temporally restricted patterns by delivering RNA interference (RNAi) (Carthew, 2001), by inducing loss-of-function mitotic clones (Chou and Perrimon, 1992), or by overexpressing dominant-negative constructs *in vivo* (Wilk *et al.*, 1996).

4. THE *DROSOPHILA* RESPIRATORY SYSTEM

The circulatory system of insects is primitive and basically composed of a single dorsal vessel that plays the role of a primitive heart. Insect blood (hemolymph) does not circulate through veins and arteries but, rather, fills the entire body cavity, delivering nutrients to organs and tissues throughout the body. Gas transport (i.e., oxygen delivery and carbon dioxide release) does not depend on such a primitive circulatory system. Instead, gases are delivered directly to organs and tissues of the organism through the respiratory system named *tracheal system* (Ghabrial *et al.*, 2003). The air enters the insect body through orifices called spiracles, which are directly connected to a complex network of ramified epithelial-like tubes, the *tracheae* (Samakovlis *et al.*, 1996a), which provide oxygen to every cell or tissue in the organism (Fig. 7.2). As we will discuss later, the *Drosophila* tracheal network shares many cellular features with the mammalian circulatory system. Development of the *Drosophila* tracheal system begins at mid-embryogenesis when 10 clusters of approximately 80 cells at each side of the embryo begin to express the transcription factors Tracheless (Isaac and Andrew, 1996; Wilk *et al.*, 1996) and Drifter/Ventral veinless (Anderson *et al.*, 1995; Llimargas and Casanova, 1997), which promote differentiation of ectodermal cells into a tracheal cell fate. Following differentiation, these cell clusters, called tracheal placodes, invaginate to form tracheal pits (Llimargas and Casanova, 1999),

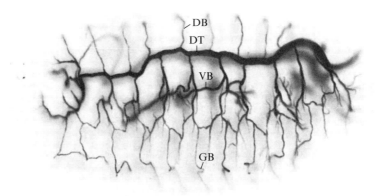

Figure 7.2 The *Drosophila* respiratory (tracheal) system. A stage 17 *Drosophila* embryo is stained with a 2A12 monoclonal antibody to visualize the tracheal tree, which is a continuous network of ramified tubes that conduces air to all tissues and cells in the body. Some of the main branches of the tracheal system are marked: DB, dorsal branch; DT, dorsal trunk; GB, ganglionic branch; VB, visceral branch.

and, immediately afterwards, discrete groups of cells in every cluster start migrating in highly stereotyped directions to give rise to the various tracheal branches (Llimargas, 2000; Wappner *et al.*, 1997). As embryogenesis proceeds, tracheal branches from adjacent or contralateral clusters get in close proximity and fuse (Jiang and Crews, 2003; Samakovlis *et al.*, 1996b; Tanaka-Matakatsu *et al.*, 1996), generating a continuous tubular network—the tracheal tree (see Fig. 7.2). By the end of embryogenesis, one single cell at the tip of each branch (Llimargas, 1999), the terminal cell, emits subcellular processes that will provide air to all cells in target tissues (Guillemin *et al.*, 1996). Noteworthy, from the moment tracheal pits have been formed, the entire process of tracheal development depends exclusively on tracheal cell migration in the complete absence of cell divisions. Thus, tracheal development begins with 20 tracheal placodes of about 80 cells each (total: ∼1600 cells) and ends up with the same number of cells, forming a complex tubular network that supplies air to all the cells and tissues in the body.

This sophisticated process of guided cell migration depends mainly on the chemotactic activity of the fly fibroblast growth factor (FGF) homologue Branchless (Bnl) (Sutherland *et al.*, 1996), which is expressed in target tissues outside the tracheae, attracting the extension of tracheal branches. Bnl binds the FGF receptor homologue Breathless (Btl) (Dossenbach *et al.*, 2001; Klambt *et al.*, 1992; Reichman-Fried and Shilo, 1995; Reichman-Fried *et al.*, 1994), which is expressed in tracheal cells and relays the signal intracellularly, provoking modifications in the cytoskeleton that induce changes of cell shape that result in guided cell migration. Thus, the expression pattern of Bnl in target tissues predicts the direction of tracheal cell migration and consequent branch extension. This expression pattern is very dynamic during tracheogenesis; once a leading cell of a given branch has reached a Bnl-positive cluster in the target tissue, *bnl* expression is switched-off, and the gene is turned on again a little further on the track of the growing branch. Thus, throughout the process of tracheal development, *bnl* expression is turned on and off many times along the path of migrating tracheal cells (Ribeiro *et al.*, 2002, 2003; Sutherland *et al.*, 1996). This tubulogenic process has features in common with mammalian vasculogenesis, where migration of blood vessel primordia is guided by the expression of different isoforms of the vascular endothelial growth factor (VEGF) that target receptors on the plasma membrane of epithelial cells (Metzger and Krasnow, 1999; Wappner and Ratcliffe, 2001).

By the end of embryogenesis, the stereotypic phase of tracheal development has been completed, and later, in larval stages, terminal tracheal branches are plastic and have the capacity to sprout out projections towards oxygen-starved areas in the surrounding tissues, very much like angiogenesis in mammals (Jarecki *et al.*, 1999). This hypoxia-dependent response of tracheal terminal branches is also mediated by Bnl, which is induced in oxygen-starved target tissues, and its receptor, Btl, expressed in tracheal cells. This hypoxia-dependent behavior of the *Drosophila* tracheal system is remarkably similar to mammalian angiogenesis (Metzger and Krasnow, 1999; Wappner and Ratcliffe, 2001), where VEGF is induced in oxygen-starved cells, promoting the outgrowth of new blood capillaries that provide additional oxygenation to hypoxic target tissues.

5. Occurrence of a *Drosophila* System Homologous to Mammalian HIF

The occurrence in *Drosophila* of a system homologous to mammalian HIF was first inferred by electromobility shift assays (EMSA), in which nuclear extracts prepared from *Drosophila* S2 cells were incubated with oligonucleotides derived from enhancers of mammalian genes that are induced in hypoxia. In these conditions, hypoxia-inducible complexes were formed, suggesting the occurrence of an endogenous *Drosophila* nuclear protein that can bind HIF consensus motifs on the DNA (Nagao *et al.*, 1996). Almost simultaneously, a ubiquitously expressed *Drosophila* gene encoding a 1505 amino acid basic helix-loop-helix Per/ARNT/Sim (bHLH-PAS) transcription factor, closely related to mammalian HIF-α, was cloned and named *similar* (sima) (Fig. 7.3) (Nambu *et al.*, 1996). Sima is remarkably bigger than all mammalian HIF-α proteins described so far, exhibiting a molecular weight of approximately 180 kDa; it displays a 45% amino acid identity with human HIF-1α in the PAS domain

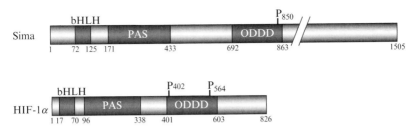

Figure 7.3   Schematic representation of the *Drosophila* hypoxia-inducible factor (HIF) homologue Sima and mammalian HIF-1α proteins. The basic-helix-loop-helix (bHLH), Per/ARNT/Sim (PAS), and oxygen-dependent degradation (ODDD) domains are shown. Note that the Sima prolyl 850 residue is the substrate of the *Drosophila* HIF prolyl hydroxylase, Fatiga.

(the highest among *Drosophila* bHLH-PAS proteins) and 63% in the bHLH domain. The first experimental evidence that Sima could be a *Drosophila* HIF-α functional homologue was reported soon afterwards. Sima protein is expressed in *Drosophila* S2 cells at low levels, but when cell cultures are exposed to severe hypoxia, Sima protein levels rise dramatically (Bacon *et al.*, 1998). Paralleling the regulation of mammalian HIF-α proteins (Huang *et al.*, 1998; Jiang *et al.*, 1997; Maxwell *et al.*, 1999; Pugh *et al.*, 1997), Sima regulation occurs at the level of protein stability, and a transferable central domain of the protein is responsible for oxygen-dependent proteasomal degradation (Bacon *et al.*, 1998).

The *Drosophila* homologue of HIF-β/ARNT was identified at approximately the same time and, in addition to forming a heterodimer with Sima, was found to be a common partner for several different *Drosophila* bHLH-PAS proteins, as occurs with HIF-β/ARNT in mammalian cells (Ma and Haddad, 1999; Ohshiro and Saigo, 1997; Sonnenfeld *et al.*, 1997; Zelzer *et al.*, 1997). As expected, Sima and Tango interact physically through their HLH motifs and PAS domains (Sonnenfeld *et al.*, 1997), and functional studies in cell culture and in developing embryos confirmed that Sima and Tango are absolutely required for inducing a transcriptional response to hypoxia (Bruick and McKnight, 2001; Centanin *et al.*, 2005; Dekanty *et al.*, 2005; Gorr *et al.*, 2004; Lavista-Llanos *et al.*, 2002). Expression of Tango protein is ubiquitous in all tissues of the fruit fly throughout development. Interestingly, Tango is primarily localized in the cytoplasm of all cells in the embryo, unless an α-subunit partner, such as Trachealess (involved in tracheal development) (Wilk *et al.*, 1996) or Single minded (involved in glial cell differentiation in the embryonic nervous system) (Nambu *et al.*, 1991), is coexpressed in the same cell. When α- and β-subunits are expressed in the same cell, they localize in the nucleus and can readily induce transcription of target genes (Ward *et al.*, 1998). A possible role of Tango in subcellular localization of Sima has not been studied so far. Given that subcellular localization of Sima seems itself to depend on a complex cellular machinery controlled by oxygen tension (discussed later), the participation of Tango in this regulation might be more complex than in the subcellular localization of other bHLH-PAS protein partners. As with other aspects of HIF cell biology, *Drosophila* genetics might help to understand the role of Tango/HIF-β in the regulation of Sima/HIF-α subcellular localization.

Unlike Tango, Sima protein levels are far too low to be detectable in normoxic embryos by immunofluorescence. Only upon exposure to severe hypoxia can the Sima protein be observed in the nuclei of cells of the tracheal system (Lavista-Llanos *et al.*, 2002). Transgenic fly lines bearing transcriptional *lacZ* or green fluorescent protein (GFP) reporters that are specifically induced in hypoxia have been developed and, consistent with the expression pattern of Sima protein, are induced with maximal sensitivity in cells of the tracheal system (Fig. 7.4). This high sensitivity to hypoxia displayed by the tracheal cells is maintained throughout the life cycle of the fruit fly. Remarkably, all the rest of the tissues in the organism are also able to accumulate Sima protein and induce hypoxia-responsive reporters, though induction occurs at stronger hypoxic conditions (Lavista-Llanos *et al.*, 2002). The existence of enhanced responses to hypoxia in tracheal cells is of interest and raises the question as to the identity of tracheal endogenous inducible genes. These observations pose an interesting paradox as, in the currently accepted model of tracheal adaptation to hypoxia, Bnl upregulation in non-tracheal cells is the key determinant of oxygen-dependent tracheal plasticity (Jarecki *et al.*, 1999). Therefore, the physiological significance of the observation that Sima-dependent gene induction

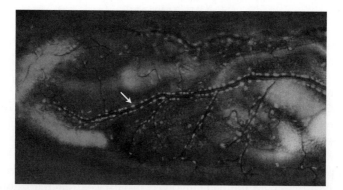

Figure 7.4   The transcriptional response to hypoxia is highly sensitive in cells of the tracheal system. *Drosophila* transgenic larvae bearing a hypoxia-inducible reporter exposed to 5% O_2 express the reporter mainly in the cells of the respiratory (tracheal) system, so that all nuclei of tracheal cells are decorated with green fluorescent protein (GFP; arrow) (for details, see Lavista-Llanos *et al.* [2002]). Reproduced with permission from Gorr *et al.* (2006).

is particularly sensitive in tracheal cells is unclear. The *Drosophila* system offers an opportunity to apply genetic tools to investigate this point, which might also contribute to better understanding the molecular basis of angiogenesis.

6. REGULATION OF SIMA BY OXYGEN LEVELS

As in mammalian cells, the *Drosophila* HIF-α protein Sima is primarily regulated by oxygen at the level of protein degradation (Gorr *et al.*, 2004). An oxygen-dependent degradation domain (ODDD) encompassing amino acids 692 to 863 has been identified (Lavista-Llanos *et al.*, 2002), and, remarkably, this domain contains a prolyl residue (P850) (Arquier *et al.*, 2006; Jaakkola *et al.*, 2001) (see Fig. 7.3), which appears to be the substrate of a *Drosophila* HIF prolyl hydroxylase that operates as an oxygen sensor. Consistent with this, an open-reading frame encoding a *Drosophila* gene highly homologous to mammalian prolyl hydroxylase domains (PHDs) was discovered (Bruick and McKnight, 2001; Epstein *et al.*, 2001; Lavista-Llanos *et al.*, 2002) and named *fatiga* (Centanin *et al.*, 2005). RNAi-mediated silencing of this gene provokes constitutive accumulation of Sima protein both in normoxic cell cultures (Bruick and McKnight, 2001) and in *Drosophila* embryos (Centanin *et al.*, 2005), and, as expected, constitutive accumulation of Sima led to upregulation of genes that are typically induced in hypoxia. These results could be mimicked in various *fatiga* loss-of-function alleles; they all display higher-than-normal Sima protein levels accompanied by normoxic induction of hypoxia-responsive transgenic reporters (Centanin *et al.*, 2005).

In addition to being regulated by the prolyl hydroxylase Fatiga at the level of protein stability, experiments carried out in embryos revealed that Sima subcellular localization depends on oxygen tension as well. Studies of Sima subcellular localization have been carried out by overexpressing Sima in transgenic embryos, thereby overriding the rate of protein degradation. In this experimental setting, Sima is primarily cytoplasmic in normoxia and accumulates in the nuclear compartment upon exposure to hypoxia (Lavista-Llanos *et al.*, 2002). However, this is not an all-or-none response. Detailed studies in normoxia and graded hypoxia revealed that regulation of Sima subcellular localization is dose-dependent and modulated by developmental parameters (Dekanty *et al.*, 2005) (Fig. 7.5). Whereas in normoxic early embryos, Sima is localized exclusively in the cytoplasm, by the end of embryogenesis, a significant proportion of normoxic embryos show an even distribution of Sima within the cell, and a lower proportion of individuals exhibit Sima localized in the nuclear compartment. When challenged with increasingly stronger hypoxic stimuli, developing embryos have a higher proportion of Sima protein localized in the nucleus, becoming totally

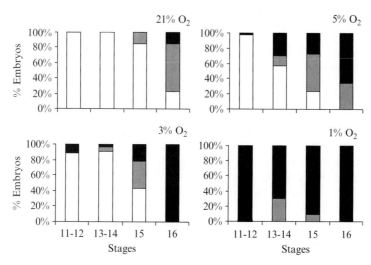

Figure 7.5 Sima subcellular localization depends on oxygen concentrations in a dose-dependent manner and is modulated by developmental parameters. White: cytoplasmic location; grey: ubiquitous; black: nuclear. Numbers refer to embryonic stages. Reproduced from Dekanty *et al.* (2005).

nuclear in embryos exposed to 1% O_2. Thus, Sima becomes increasingly nuclear as hypoxia becomes more severe and predominantly cytoplasmic in conditions of abundant oxygen supply (see Fig. 7.5). Contrary to some initial predictions, nuclear localization seems to be the "default state" of Sima subcellular localization, since deletion of the ODDD renders Sima constitutively nuclear, regardless of oxygen tension (Lavista-Llanos *et al.*, 2002). In mammalian cells, similar regulation of HIF-α subcellular localization seems to occur; as in von Hippel-Lindau (VHL)-lacking cells, HIF-α is constitutively localized in the nucleus, suggesting that the ODDD is involved in regulation of subcellular localization (Groulx and Lee, 2002). The molecular mechanism by which the ODDD mediates this regulation is unclear; two models in principle could account for the observations: the ODDD is necessary for cytoplasmic retention in normoxia or the ODDD is required for active nuclear export. Genetic experiments in *Drosophila* might help to understand this unresolved issue of HIF regulation.

7. ROLE OF SIMA AND FATIGA IN *DROSOPHILA* DEVELOPMENT

Analyses of "knockout" mouse strains have revealed that mammalian HIF proteins have essential functions in embryonic and postembryonic development. They have been shown to participate in the formation of the embryonic heart, vasculature, brain, cartilages, and the placenta in adult females (Adelman *et al.*, 2000; Covello and Simon, 2004; Giaccia *et al.*, 2004; Iyer *et al.*, 1998; Pfander *et al.*, 2004; Tomita *et al.*, 2003), suggesting that local oxygen tension might play a role in these developmental processes. Unexpectedly, the *Drosophila* homozygous Sima loss-of-function mutants are fully viable in normoxia, but fail to develop in mild hypoxic conditions (Centanin *et al.*, 2005), indicating that Sima is necessary for development in hypoxia but not in normoxia. In contrast, *fatiga* mutant alleles provoke lethality at different stages of the life cycle in normoxia (none of them can attain the adult stage), implying that Fatiga is critically required for normal development. Strikingly, *fatiga sima* double mutants are viable, attaining the adult stage in normoxia (not in hypoxia) (Centanin *et al.*, 2005) (Fig. 7.6), suggesting that the most fundamental functions of Fatiga/PHD in *Drosophila* development involve downregulation of Sima protein levels.

Noteworthy, *fatiga sima* double mutants are not entirely normal, as they show defects in ovary and wing development. This observation suggests that Fatiga is apparently involved in patterning these organs in a Sima-independent fashion. This conclusion is clearly of interest, since alternative target molecules for HIF prolyl hydroxylases have not been identified as yet. Forthcoming studies in the field of *Drosophila* developmental genetics might help in identifying these elusive target molecules of HIF prolyl hydroxylases.

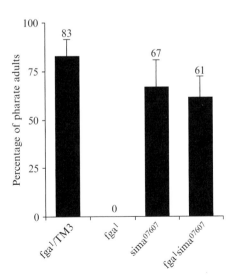

Figure 7.6 A mutation in *sima* gene reverts lethality of *fatiga* mutants. *fatiga* heterozygous individuals (*fga¹*/TM3) are viable to adulthood, but *fatiga* homozygotes (*fga¹*) are lethal; lethality is fully reverted in *fatiga sima* double mutants (*fga¹ sima^{07607}*). Reproduced from Centanin *et al.* (2005).

8. HYPOXIA-INDUCIBLE GENES AND THE ADAPTATION OF *DROSOPHILA* TO OXYGEN STARVATION

Studies of oxygen-regulated genes have been carried out in *Drosophila*, and genetic screens based on behavioral responses to hypoxia have been successful in identifying loci that are relevant for adaptation to low oxygen conditions (Haddad, 1998; Haddad *et al.*, 1997). *Drosophila* is extremely tolerant of oxygen starvation and can survive in hypoxia for long periods of time (Haddad, 2006; Vigne and Frelin, 2006). For instance, flies that are challenged with 0.5% O_2 for more than 6 h do not die, although they enter a state of stupor in which they do not move or respond to stimuli (Liu *et al.*, 2006). A few minutes after reoxygenation, flies wake-up, recover their usual locomotor activity, flying capacity and normal behavior, and fertility is not significantly affected. Screens for genes that participate in the adaptation of *Drosophila* to such extreme hypoxic conditions have led to the isolation of loci that provoke a lengthened or shortened waking period after hypoxia-induced stupor. For instance, a mutation affecting the trehalose phosphate synthase gene was shown to increase the post-stupor recovering period; it was proposed that the disaccharide trehalose prevents protein denaturation in the central nervous system, thereby improving the outcome upon oxygen starvation (Chen *et al.*, 2002).

More recently, a genome-wide expression screen was performed in *Drosophila* adults by comparing mRNA expression levels of each of the genes of the transcriptome in different conditions of oxygen deprivation (Liu *et al.*, 2006). Different sets of genes could be defined according to their expression profile in different hypoxic conditions. Some of the transcripts are induced in mild hypoxia; others in stronger hypoxia; a third subgroup of transcripts is induced after an acute, but not a chronic exposure to hypoxia, and a set of transcripts is induced only if exposure to hypoxia has a certain minimal duration. The whole set of genes induced in hypoxic conditions is functionally diverse, reflecting the plethora of molecular and cellular changes that occur in an oxygen-starved organism. Interestingly, transcription factors that mediate responses to various types of (non-hypoxic) stresses are upregulated in low oxygen, suggesting that the physiological changes that occur in hypoxia activate multiple stress-responsive pathways simultaneously (Liu *et al.*, 2006).

9. REGULATION OF SIMA BY THE PI3K AND TOR PATHWAYS

In addition to oxygen-dependent mechanisms that control the abundance or activity of HIF proteins, non-hypoxic stimuli, such as nitric oxide, growth factors, hormones, and cytokines, also play a role in mammalian

HIF regulation (Conrad *et al.*, 1999; Feldser *et al.*, 1999; Fukuda *et al.*, 2002; Kasuno *et al.*, 2004; Richard *et al.*, 1999, 2000; Zhong *et al.*, 2000). For instance, insulin or insulin growth factors (IGFs) can increase HIF-α protein levels, triggering the induction of hypoxia-responsive genes (Kietzmann *et al.*, 2003; Zelzer *et al.*, 1998). The effect of these growth factors seems to depend mostly on the phosphoinositide-3-kinase (PI3K) signaling pathway (Roth *et al.*, 2004; Treins *et al.*, 2002; Zundel *et al.*, 2000), although the mitogen-activated protein kinase (MAPK) pathway seems to contribute to HIF activation as well. *Drosophila* insulin-like peptides (DILPs) displaying high-sequence identity with mammalian insulin are secreted by discrete groups of neurosecretory cells in the brain and target a unique *Drosophila* InR homologue that activates a conserved downstream kinase cascade that includes PI3K and the protein kinase B (PKB), also called AKT (Fernandez *et al.*, 1995; Ikeya *et al.*, 2002; Rulifson *et al.*, 2002). Like in mammalian cells, the activity of PI3K is antagonized by the phosphatase protein and tensin homolog (PTEN) (Goberdhan *et al.*, 1999; Huang *et al.*, 1999), and activation of the InR pathway brings about the activation of the TOR and phosphorylation of downstream effectors (Miron *et al.*, 2003; Oldham and Hafen, 2003; Oldham *et al.*, 2000). It has been recently demonstrated that InR and TOR pathways respond to the nutrition state of the organism, regulating larval growth (Hafen, 2004; Kim and Rulifson, 2004; Zhang *et al.*, 2000). When nutrients are abundant, InR and TOR pathways are fully active and promote growth, and, conversely, in conditions of nutrient deprivation, the activity of these pathways is reduced, leading to growth inhibition. The effect of TOR on cell growth depends at least in part on the activation of S6K, a kinase that phosphorylates the ribosomal protein S6, promoting an increase of protein translation and the inactivation of eIF4E-binding protein (4E-BP), a translation initiation inhibitor (Neufeld, 2004; Rintelen *et al.*, 2001).

Regulation of HIF by the InR and TOR pathways seems to be another conserved feature of the *Drosophila* hypoxia-responsive pathway (Dekanty *et al.*, 2005). Insulin can trigger the expression of a Sima-Tango–dependent luciferase reporter in *Drosophila* S2 cells to levels that are comparable with those observed upon exposure to extreme hypoxia. Induction of the luciferase reporter is paralleled by upregulation of Sima-endogenous target genes, such as *lactate dehydrogenase-A* (*ldh-A*), confirming the physiological relevance of this insulin-induced response. Pharmacological and RNAi-mediated silencing experiments revealed that Sima-dependent transcription upon induction with insulin depends on the PI3K and TOR pathways (Dekanty *et al.*, 2005). Genetically-induced over-activation of these pathways in living embryos also provokes the upregulation of Sima-dependent transcription, as revealed by the expression of a lacZ hypoxia-inducible reporter in transgenic embryos. Detailed studies carried out in cell culture and *in vivo* showed that an accumulation of Sima protein and an increase of its nuclear localization account for Sima-dependent gene induction upon activation of PI3K and TOR pathways (Dekanty *et al.*, 2005).

 ## 10. ROLE OF THE HIF SYSTEM IN GROWTH CONTROL AND CELL SIZE DETERMINATION

From nematodes to humans, the PI3K and TOR pathways play a cardinal role in growth control and cell size determination (Oldham and Hafen, 2003). Do Sima and Fatiga participate in this regulation? It has long been appreciated that insects exposed to hypoxia grow at reduced rates and remain smaller than controls kept in normoxia (Frazier *et al.*, 2001; Peck and Maddrell, 2005), but the cellular basis of this phenomenon remains largely unresolved. We have recently observed that *fatiga* mutant pupae are strikingly smaller and their growth rate is reduced compared with that of their wild-type siblings (Centanin *et al.*, 2005) (Fig. 7.7A). Consistent with this, a recent study showed that cells in *fatiga* loss–of–function mitotic clones in the larval fat body, an organ analogous to the mammalian liver, are clearly smaller than wild-type cells of the same organ (Frei and Edgar, 2004). The same study showed that, conversely, overexpression of *fatiga* in wing imaginal discs (the primordia of adult wings) was sufficient to increase cell size. Thus, Fatiga seems to be required for normal growth and cell size determination. Given that Fatiga is a negative regulator of Sima/HIF-α, it was relevant to answer whether or not over-accumulation of Sima can account for cell size reduction in *fatiga* mutant cells. This appears indeed to be the case, since *sima fatiga* double mutant pupae have a normal size (see Fig. 7.7A), and normal growth rate is also restored in these double mutants (Centanin *et al.*, 2005). Consistent with this, experiments involving overexpression of Sima in cells of the fat body, an organ composed of endo-replicative cells, were strikingly smaller than wild-type control cells (see Fig. 7.7B), suggesting that Sima is a cell-autonomous negative regulator of growth (Centanin *et al.*, 2005).

These results pose an apparent paradox, since: (1) activation of InR and TOR pathways induce growth; (2) activation of InR and TOR pathways induce Sima-dependent transcription; and (3) Sima is a negative regulator of growth. A model accounting for these data might involve a negative feedback loop, where InR/TOR pathways promote growth but also activate Sima, which in turn downregulates InR/TOR signaling, thereby limiting growth. A recent genetic screen aimed to identify suppressors of the InR/TOR network has led to the discovery of a novel *Drosophila* gene, *Scylla*, a negative regulator of these pathways (Reiling and Hafen, 2004). *Scylla* is evolutionarily conserved—the mammalian orthologue is called REDD1 (Brugarolas *et al.*, 2004)—and induced by Sima/HIF upon exposure to

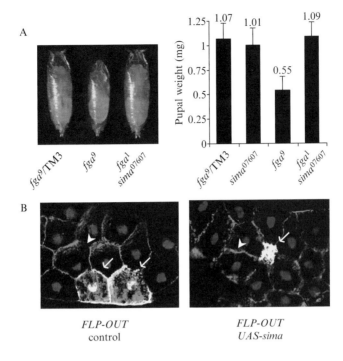

Figure 7.7 Fatiga and Sima are involved in growth control and cell size determination. (A) *fatiga* homozygous mutants (*fga⁹*) are remarkably smaller than their heterozygous siblings (*fga⁹*/TM3), but normal size is restored in *fatiga sima* double mutants (*fga¹ sima^07607*). (B) Random expression of Sima in cells of the larval fat body provokes striking reduction of cell size. (Left panel) Control larvae in which green fluorescent protein (GFP) has been expressed in random cells of the fat body (arrow); these cells have the same size as the cells that do not express GFP (arrowhead). (Right panel) Sima was expressed in random cells of the fat body that also express GFP as a marker (arrow); these cells are remarkably smaller than control neighboring cells that do not express Sima protein (arrowhead). Reproduced from Centanin *et al.* (2005).

hypoxia. Thus, it seems likely that a negative feedback loop involving *Scylla*/REDD1, and perhaps other Sima/HIF-inducible negative effectors of InR/TOR signaling, accounts for size and growth-rate reduction in hypoxia (Fig. 7.8).

Yet, it cannot be ruled-out that Fatiga plays a Sima-independent role in *Drosophila* growth regulation. In a genetic screen for modifiers of CycD/Cdk4-induced overgrowth in eye imaginal discs, a mutation in the *fatiga* gene emerged as a dominant suppressor (Frei and Edgar, 2004). As this effect is not suppressed in *tango* mutant clones in the eye, it seems unlikely that the suppression of growth mediated by the *fatiga* mutant in this tissue might involve upregulation of Sima/Tango. Thus, a likely scenario is that growth

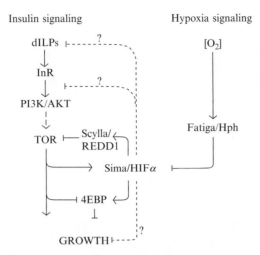

Figure 7.8 Model for insulin receptor (InR)/target-of-rapamycin (TOR) signaling through Sima and the role of Sima in growth control. *Drosophila* insulin-like peptides (DILPs) bind InR, activating the PI3K/AKT and TOR pathways, which in turn promote growth. TOR activates Sima-dependent transcription that in turn induces *Scylla/REDD1*, a negative regulator of TOR signaling, eIF4E-BP, whose induction inhibits growth and, possibly, other hypothetical genes that mediate downregulation of InR/TOR (dashed lines with question marks). Through such a mechanism, Sima would be engaged in a negative feedback loop that limits overgrowth induced by InR/TOR signaling. Modified from Dekanty *et al.* (2005).

impairment in *fatiga* mutant endoreplicative cells is due to over-accumulation of Sima, while growth impairment in *fatiga* mutant cells in the eye (non-endoreplicative) involves regulation of a putative HIF-independent pathway (Frei and Edgar, 2004).

11. CONCLUDING REMARKS

A system homologous to mammalian HIF is largely conserved in *Drosophila melanogaster*, a genetically tractable organism with advantages as an *in vivo* model system for cell and developmental biology studies. Initial analysis of the regulation of Sima/HIF-α has confirmed that all the basic features of mammalian HIF biology are largely maintained in *Drosophila*, suggesting that any progress in understanding hypoxic responses in this species can probably be extrapolated to human HIF. The availability of a wide array of mutants as well as the simplicity of the methods used for gene silencing, transgenesis, and overexpression studies provide an ideal frame-work for tracking HIF regulation *in vivo* and for investigating the cellular basis of poorly understood aspects of HIF biology (e.g., oxygen-dependent control of subcellular localization). Mammalian HIF proteins are regulated by several different signaling transduction pathways, but the biochemical and molecular mechanisms involved in this regulation remain largely unclear. The genetic tools available in *Drosophila* might help to shed light on these poorly understood processes and contribute in better defining at what level and by which mechanisms the "hard wired" genetic networks controlling development interface with oxygen-sensing cellular machineries.

ACKNOWLEDGMENTS

N. M. R. is a fellow of CONICET, A. D. is a fellow of the ANPCyT; P. W. is a career investigator of CONICET. Howard Hughes Medical Institute Grant #55005973; ANPCyT #01–10839; UBA #X-147.

REFERENCES

Adams, M. D., Celniker, S. E., Holt, R. A., Evans, C. A., Gocayne, J. D., Amanatides, P. G., Scherer, S. E., Li, P. W., Hoskins, R. A., Galle, R. F., George, R. A., Lewis, S. E., *et al.* (2000). The genome sequence of *Drosophila melanogaster. Science* **287**, 2185–2195.

Adelman, D. M., Gertsenstein, M., Nagy, A., Simon, M. C., and Maltepe, E. (2000). Placental cell fates are regulated *in vivo* by HIF-mediated hypoxia responses. *Genes Dev.* **14**, 3191–3203.

Anderson, M. G., Perkins, G. L., Chittick, P., Shrigley, R. J., and Johnson, W. A. (1995). Drifter, a *Drosophila* Pou-domain transcription factor, is required for correct differentiation and migration of tracheal cells and midline glia. *Genes Dev.* **9**, 123–137.

Arquier, N., Vigne, P., Duplan, E., Hsu, T., Therond, P. P., Frelin, C., and D'Angelo, G. (2006). Analysis of the hypoxia-sensing pathway in *Drosophila melanogaster. Biochem. J.* **393**, 471–480.

Ashburner, M., Golic, K. G., and Hawley, R. S. (2004). "*Drosophila*: A Laboratory Handbook" edn. 2. Cold Spring Harbor Laboratory Press, Cold Spring Harbor.

Bacon, N. C., Wappner, P., O'Rourke, J. F., Bartlett, S. M., Shilo, B., Pugh, C. W., and Ratcliffe, P. J. (1998). Regulation of the *Drosophila* bHLH-PAS protein Sima by hypoxia: Functional evidence for homology with mammalian HIF-1α. *Biochem. Biophys. Res. Commun.* **249**, 811–816.

Berrigan, D., and Partridge, L. (1997). Influence of temperature and activity on the metabolic rate of adult *Drosophila melanogaster. Comp. Biochem. Physiol. A Physiol.* **118**, 1301–1307.

Brand, A. H., and Perrimon, N. (1993). Targeted gene expression as means of altering cell fates and generating dominant phenotypes. *Development* **118**, 401–415.

Brugarolas, J., Lei, K., Hurley, R. L., Manning, B. D., Reiling, J. H., Hafen, E., Witters, L. A., Ellisen, L. W., and Kaelin, W. G., Jr. (2004). Regulation of mTOR function in response to hypoxia by REDD1 and the TSC1/TSC2 tumor suppressor complex. *Genes Dev.* **18**, 2893–2904.

Bruick, R. K., and McKnight, S. L. (2001). A conserved family of prolyl-4-hydroxylases that modify HIF. *Science* **294**, 1337–1340.

Carthew, R. W. (2001). Gene silencing by double-stranded RNA. *Curr. Opin. Cell Biol.* **13,** 244–248.

Centanin, L., Ratcliffe, P. J., and Wappner, P. (2005). Reversion of lethality and growth defects in Fatiga oxygen-sensor mutant flies by loss of hypoxia-inducible factor-α/Sima. *EMBO Rep.* **6,** 1070–1075.

Chen, Q., Ma, E., Behar, K. L., Xu, T., and Haddad, G. G. (2002). Role of trehalose phosphate synthase in anoxia tolerance and development in *Drosophila melanogaster*. *J. Biol. Chem.* **277,** 3274–3279.

Chou, T. B., and Perrimon, N. (1992). Use of a yeast site-specific recombinase to produce female germline chimeras in *Drosophila*. *Genetics* **131,** 643–653.

Conrad, P. W., Rust, R. T., Han, J., Millhorn, D. E., and Beitner-Johnson, D. (1999). Selective activation of p38α and p38γ by hypoxia. Role in regulation of cyclin D1 by hypoxia in PC12 cells. *J. Biol. Chem.* **274,** 23570–23576.

Cooley, L., Kelley, R., and Spradling, A. (1988). Insertional mutagenesis of the *Drosophila* genome with single P elements. *Science* **239,** 1121–1128.

Covello, K. L., and Simon, M. C. (2004). HIFs, hypoxia, and vascular development. *Curr. Top. Dev. Biol.* **62,** 37–54.

Dekanty, A., Lavista-Llanos, S., Irisarri, M., Oldham, S., and Wappner, P. (2005). The insulin-PI3K/TOR pathway induces a HIF-dependent transcriptional response in *Drosophila* by promoting nuclear localization of HIF-α/Sima. *J. Cell Sci.* **118,** 5431–5441.

Dillon, M. E., and Frazier, M. R. (2006). *Drosophila melanogaster* locomotion in cold thin air. *J. Exp. Biol.* **209,** 364–371.

Dossenbach, C., Rock, S., and Affolter, M. (2001). Specificity of FGF signaling in cell migration in *Drosophila*. *Development* **128,** 4563–4572.

Epstein, A. C., Gleadle, J. M., McNeill, L. A., Hewitson, K. S., O'Rourke, J., Mole, D. R., Mukherji, M., Metzen, E., Wilson, M. I., Dhanda, A., Tian, Y. M., Masson, N., *et al.* (2001). *C. elegans* EGL-9 and mammalian homologs define a family of dioxygenases that regulate HIF by prolyl hydroxylation. *Cell* **107,** 43–54.

Feldser, D., Agani, F., Iyer, N. V., Pak, B., Ferreira, G., and Semenza, G. L. (1999). Reciprocal positive regulation of hypoxia-inducible factor 1α and insulin-like growth factor 2. *Cancer Res.* **59,** 3915–3918.

Fernandez, R., Tabarini, D., Azpiazu, N., Frasch, M., and Schlessinger, J. (1995). The *Drosophila* insulin receptor homolog: A gene essential for embryonic development encodes two receptor isoforms with different signaling potential. *EMBO J.* **14,** 3373–3384.

Frazier, M. R., Woods, H. A., and Harrison, J. F. (2001). Interactive effects of rearing temperature and oxygen on the development of *Drosophila melanogaster*. *Physiol. Biochem. Zool.* **74,** 641–650.

Frei, C., and Edgar, B. A. (2004). *Drosophila* cyclin D/Cdk4 requires HIF-1 prolyl hydroxylase to drive cell growth. *Dev. Cell* **6,** 241–251.

Fukuda, R., Hirota, K., Fan, F., Jung, Y. D., Ellis, L. M., and Semenza, G. L. (2002). Insulin-like growth factor 1 induces hypoxia-inducible factor 1-mediated vascular endothelial growth factor expression, which is dependent on MAP kinase and phosphatidylinositol 3-kinase signaling in colon cancer cells. *J. Biol. Chem.* **277,** 38205–38211.

Ghabrial, A., Luschnig, S., Metzstein, M. M., and Krasnow, M. A. (2003). Branching morphogenesis of the *Drosophila* tracheal system. *Annu. Rev. Cell Dev. Biol.* **19,** 623–647.

Giaccia, A. J., Simon, M. C., and Johnson, R. (2004). The biology of hypoxia: The role of oxygen sensing in development, normal function, and disease. *Genes Dev.* **18,** 2183–2194.

Gibert, P., Huey, R. B., and Gilchrist, G. W. (2001). Locomotor performance of *Drosophila melanogaster*: Interactions among developmental and adult temperatures, age, and geography. *Evolution Int. J. Org. Evolution* **55,** 205–209.

Goberdhan, D. C., Paricio, N., Goodman, E. C., Mlodzik, M., and Wilson, C. (1999). *Drosophila* tumor suppressor PTEN controls cell size and number by antagonizing the Chico/PI3-kinase signaling pathway. *Genes Dev.* **13,** 3244–3258.

Gorr, T. A., Gassmann, M., and Wappner, P. (2006). Sensing and responding to hypoxia via HIF in model invertebrates. *J. Insect Physiol.* **52,** 349–364.

Gorr, T. A., Tomita, T., Wappner, P., and Bunn, H. F. (2004). Regulation of *Drosophila* hypoxia-inducible factor (HIF) activity in SL2 cells: Identification of a hypoxia-induced variant isoform of the HIFα homolog gene similar. *J. Biol. Chem.* **279,** 36048–36058.

Greenspan, R. J. (1997). "Fly Pushing: The Theory and Practice of *Drosophila* Genetics." Cold Spring Harbor Laboratory Press, Cold Spring Harbor.

Groulx, I., and Lee, S. (2002). Oxygen-dependent ubiquitination and degradation of hypoxia-inducible factor requires nuclear-cytoplasmic trafficking of the von Hippel-Lindau tumor suppressor protein. *Mol. Cell Biol.* **22,** 5319–5336.

Guillemin, K., Groppe, J., Ducker, K., Treisman, R., Hafen, E., Affolter, M., and Krasnow, M. A. (1996). The pruned gene encodes the *Drosophila* serum response factor and regulates cytoplasmic outgrowth during terminal branching of the tracheal system. *Development* **122,** 1353–1362.

Haddad, G. G. (1998). Mechanisms of anoxia tolerance. A novel approach using a *Drosophila* model system. *Adv. Exp. Med. Biol.* **454,** 273–280.

Haddad, G. G. (2006). Tolerance to low O_2: Lessons from invertebrate genetic models. *Exp. Physiol.* **91,** 277–282.

Haddad, G. G., Sun, Y., Wyman, R. J., and Xu, T. (1997). Genetic basis of tolerance to O_2 deprivation in *Drosophila melanogaster*. *Proc. Natl. Acad. Sci. USA* **94,** 10809–10812.

Hafen, E. (2004). Interplay between growth factor and nutrient signaling: Lessons from *Drosophila* TOR. *Curr. Top. Microbiol. Immunol.* **279,** 153–167.

Huang, H., Potter, C. J., Tao, W., Li, D. M., Brogiolo, W., Hafen, E., Sun, H., and Xu, T. (1999). PTEN affects cell size, cell proliferation and apoptosis during *Drosophila* eye development. *Development* **126,** 5365–5372.

Huang, L. E., Gu, J., Schau, M., and Bunn, H. F. (1998). Regulation of hypoxia-inducible factor 1α is mediated by an O_2-dependent degradation domain via the ubiquitin-proteasome pathway. *Proc. Natl. Acad. Sci. USA* **95,** 7987–7992.

Ikeya, T., Galic, M., Belawat, P., Nairz, K., and Hafen, E. (2002). Nutrient-dependent expression of insulin-like peptides from neuroendocrine cells in the CNS contributes to growth regulation in *Drosophila*. *Curr. Biol.* **12,** 1293–1300.

Isaac, D. D., and Andrew, D. J. (1996). Tubulogenesis in *Drosophila*: A requirement for the trachealess gene product. *Genes Dev.* **10,** 103–117.

Iyer, N. V., Kotch, L. E., Agani, F., Leung, S. W., Laughner, E., Wenger, R. H., Gassmann, M., Gearhart, J. D., Lawler, A. M., Yu, A. Y., and Semenza, G. L. (1998). Cellular and developmental control of O_2 homeostasis by hypoxia-inducible factor 1 alpha. *Genes Dev.* **12,** 149–162.

Jaakkola, P., Mole, D. R., Tian, Y. M., Wilson, M. I., Gielbert, J., Gaskell, S. J., Kriegsheim, A. V., Heberstreit, H. F., Mukherji, M., Schofield, C. J., Maxwell, P. H., Pugh, C. W., *et al.* (2001). Targeting of HIF-α to the von Hippel Lindau ubiquitylation complex by O_2-regulatred prolyl hydroxylation. *Science* **292,** 468–472.

Jarecki, J., Johnson, E., and Krasnow, M. (1999). Oxygen regulation of airway branching in *Drosophila* is mediated by Branchless FGF. *Cell* **99,** 211–220.

Jiang, L., and Crews, S. T. (2003). The *Drosophila* dysfusion basic helix-loop-helix (bHLH)-PAS gene controls tracheal fusion and levels of the trachealess bHLH-PAS protein. *Mol. Cell Biol.* **23,** 5625–5637.

Jiang, B. H., Zheng, J. Z., Leung, S. W., Roe, R., and Semenza, G. L. (1997). Transactivation and inhibitory domains of hypoxia-inducible factor 1α. Modulation of transcriptional activity by oxygen tension. *J. Biol. Chem.* **272,** 19253–19260.

Kasuno, K., Takabuchi, S., Fukuda, K., Kizaka-Kondoh, S., Yodoi, J., Adachi, T., Semenza, G. L., and Hirota, K. (2004). Nitric oxide induces hypoxia-inducible factor 1 activation that is dependent on MAPK and phosphatidylinositol 3-kinase signaling. *J. Biol. Chem.* **279,** 2550–2558.

Kietzmann, T., Samoylenko, A., Roth, U., and Jungermann, K. (2003). Hypoxia-inducible factor-1 and hypoxia response elements mediate the induction of plasminogen activator inhibitor-1 gene expression by insulin in primary rat hepatocytes. *Blood* **101,** 907–914.

Kim, S. K., and Rulifson, E. J. (2004). Conserved mechanisms of glucose sensing and regulation by *Drosophila corpora cardiaca* cells. *Nature* **431,** 316–320.

Klambt, C., Glazer, L., and Shilo, B. Z. (1992). Breathless, a *Drosophila* FGF receptor homolog, is essential for migration of tracheal and specific midline glial cells. *Genes Dev.* **6,** 1668–1678.

Lavista-Llanos, S., Centanin, L., Irisarri, M., Russo, D. M., Gleadle, J. M., Bocca, S. N., Muzzopappa, M., Ratcliffe, P. J., and Wappner, P. (2002). Control of the hypoxic response in *Drosophila melanogaster* by the basic helix-loop-helix PAS protein similar. *Mol. Cell Biol.* **22,** 6842–6853.

Liu, G., Roy, J., and Johnson, E. A. (2006). Identification and function of hypoxia-response genes in *Drosophila melanogaster*. *Physiol. Genomics* **25,** 134–141.

Llimargas, M. (1999). The Notch pathway helps to pattern the tips of the *Drosophila* tracheal branches by selecting cell fates. *Development* **126,** 2355–2364.

Llimargas, M. (2000). Wingless and its signaling pathway have common and separable functions during tracheal development. *Development* **127,** 4407–4417.

Llimargas, M., and Casanova, J. (1997). Ventral veinless, a Pou domain transcription factor, regulates different transduction pathways required for tracheal branching in *Drosophila*. *Development* **124,** 3273–3281.

Llimargas, M., and Casanova, J. (1999). EGF signaling regulates cell invagination as well as cell migration during formation of tracheal system in *Drosophila*. *Dev. Genes Evol.* **209,** 174–179.

Ma, E., and Haddad, G. G. (1999). Isolation and characterization of the hypoxia-inducible factor 1β in *Drosophila melanogaster*. *Brain Res. Mol. Brain Res.* **73,** 11–16.

Maxwell, P. H., Wiesener, M. S., Chang, G. W., Clifford, S. C., Vaux, E. C., Cockman, M. E., Wykoff, C. C., Pugh, C. W., Maher, E. R., and Ratcliffe, P. J. (1999). The tumour suppressor protein VHL targets hypoxia-inducible factors for oxygen-dependent proteolysis. *Nature* **399,** 271–275.

Metzger, R. J., and Krasnow, M. A. (1999). Genetic control of branching morphogenesis. *Science* **284,** 1635–1639.

Miron, M., Lasko, P., and Sonenberg, N. (2003). Signaling from Akt to FRAP/TOR targets both 4E-BP and S6K in *Drosophila melanogaster*. *Mol. Cell Biol.* **23,** 9117–9126.

Nagao, M., Ebert, B. L., Ratcliffe, P. J., and Pugh, C. W. (1996). *Drosophila melanogaster* SL2 cells contain a hypoxically inducible DNA binding complex which recognizes mammalian HIF-binding sites. *FEBS Lett.* **387,** 161–166.

Nambu, J. R., Chen, W., Hu, S., and Crews, S. T. (1996). The *Drosophila melanogaster* similar bHLH-PAS gene encodes a protein related to human hypoxia-inducible factor 1 alpha and *Drosophila* single-minded. *Gene* **172,** 249–254.

Nambu, J. R., Lewis, J. O., Wharton, K. A., Jr., and Crews, S. T. (1991). The *Drosophila* single-minded gene encodes a helix-loop-helix protein that acts as a master regulator of CNS midline development. *Cell* **67,** 1157–1167.

Neufeld, T. P. (2004). Genetic analysis of TOR signaling in *Drosophila*. *Curr. Top. Microbiol. Immunol.* **279,** 139–152.

Ohshiro, T., and Saigo, K. (1997). Transcriptional regulation of breathless FGF receptor gene by binding of TRACHEALESS/dARNT heterodimers to three central midline elements in *Drosophila* developing trachea. *Development* **124,** 3975–3986.

Oldham, S., and Hafen, E. (2003). Insulin/IGF and target of rapamycin signaling: A TOR de force in growth control. *Trends Cell. Biol.* **13,** 79–85.

Oldham, S., Montagne, J., Radimerski, T., Thomas, G., and Hafen, E. (2000). Genetic and biochemical characterization of dTOR, the *Drosophila* homolog of the target of rapamycin. *Genes Dev.* **14,** 2689–2694.

Peck, L. S., and Maddrell, S. H. (2005). Limitation of size by hypoxia in the fruit fly *Drosophila melanogaster. J. Exp. Zoolog. A Comp. Exp. Biol.* **303,** 968–975.

Pfander, D., Kobayashi, T., Knight, M. C., Zelzer, E., Chan, D. A., Olsen, B. R., Giaccia, A. J., Johnson, R. S., Haase, V. H., and Schipani, E. (2004). Deletion of Vhlh in chondrocytes reduces cell proliferation and increases matrix deposition during growth plate development. *Development* **131,** 2497–2508.

Pugh, C. W., O'Rourke, J. F., Nagao, M., Gleadle, J. M., and Ratcliffe, P. J. (1997). Activation of hypoxia-inducible factor-1; definition of regulatory domains within the alpha subunit. *J. Biol. Chem.* **272,** 11205–11214.

Reichman-Fried, M., and Shilo, B. Z. (1995). Breathless, a *Drosophila* FGF receptor homolog, is required for the onset of tracheal cell migration and tracheole formation. *Mech. Dev.* **52,** 265–273.

Reichman-Fried, M., Dickson, B., Hafen, E., and Shilo, B. Z. (1994). Elucidation of the role of breathless, a *Drosophila* FGF receptor homolog, in tracheal cell migration. *Genes Dev.* **8,** 428–439.

Reiling, J. H., and Hafen, E. (2004). The hypoxia-induced paralogs Scylla and Charybdis inhibit growth by down-regulating S6K activity upstream of TSC in *Drosophila. Genes Dev.* **18,** 2879–2892.

Ribeiro, C., Ebner, A., and Affolter, M. (2002). *In vivo* imaging reveals different cellular functions for FGF and Dpp signaling in tracheal branching morphogenesis. *Dev. Cell* **2,** 677–683.

Ribeiro, C., Petit, V., and Affolter, M. (2003). Signaling systems, guided cell migration, and organogenesis: Insights from genetic studies in *Drosophila. Dev. Biol.* **260,** 1–8.

Richard, D. E., Berra, E., and Pouyssegur, J. (2000). Nonhypoxic pathway mediates the induction of hypoxia-inducible factor 1alpha in vascular smooth muscle cells. *J. Biol. Chem.* **275,** 26765–26771.

Richard, D. E., Berra, E., Gothie, E., Roux, D., and Pouyssegur, J. (1999). p42/p44 mitogen-activated protein kinases phosphorylate hypoxia-inducible factor 1alpha (HIF-1α) and enhance the transcriptional activity of HIF-1. *J. Biol. Chem.* **274,** 32631–32637.

Rintelen, F., Stocker, H., Thomas, G., and Hafen, E. (2001). PDK1 regulates growth through Akt and S6K in *Drosophila. Proc. Natl. Acad. Sci. USA* **98,** 15020–15025.

Roth, U., Curth, K., Unterman, T. G., and Kietzmann, T. (2004). The transcription factors HIF-1 and HNF-4 and the coactivator p300 are involved in insulin-regulated glucokinase gene expression via the phosphatidylinositol 3-kinase/protein kinase B pathway. *J. Biol. Chem.* **279,** 2623–2631.

Rubin, G. M., and Spradling, A. C. (1982). Genetic transformation of *Drosophila* with transposable element vectors. *Science* **218,** 348–353.

Rulifson, E. J., Kim, S. K., and Nusse, R. (2002). Ablation of insulin-producing neurons in flies: Growth and diabetic phenotypes. *Science* **296,** 1118–1120.

Samakovlis, C., Hacohen, N., Manning, G., Sutherland, D. C., Guillemin, K., and Krasnow, M. A. (1996a). Development of the *Drosophila* tracheal system occurs by a series of morphologically distinct but genetically coupled branching events. *Development* **122,** 1395–1407.

Samakovlis, C., Manning, G., Steneberg, P., Hacohen, N., Cantera, R., and Krasnow, M. A. (1996b). Genetic control of epithelial tube fusion during *Drosophila* tracheal development. *Development* **122,** 3531–3536.

Sonnenfeld, M., Ward, M., Nystrom, G., Mosher, J., Stahl, S., and Crews, S. (1997). The *Drosophila tango* gene encodes a bHLH-PAS protein that is orthologous to mammalian Arnt and controls CNS midline and tracheal development. *Development* **124**, 4571–4582.

Spradling, A. C., and Rubin, G. M. (1982). Transposition of cloned P elements into *Drosophila* germ line chromosomes. *Science* **218**, 341–347.

Sutherland, D., Samakovlis, C., and Krasnow, M. A. (1996). Branchless encodes a *Drosophila* FGF homolog that controls tracheal cell migration and the pattern of branching. *Cell* **87**, 1091–1101.

Tanaka-Matakatsu, M., Uemura, T., Oda, H., Takeichi, M., and Hayashi, S. (1996). Cadherin-mediated cell adhesion and cell motility in *Drosophila* trachea regulated by the transcription factor Escargot. *Development* **122**, 3697–3705.

Tomita, S., Ueno, M., Sakamoto, M., Kitahama, Y., Ueki, M., Maekawa, N., Sakamoto, H., Gassmann, M., Kageyama, R., Ueda, N., Gonzalez, F. J., and Takahama, Y. (2003). Defective brain development in mice lacking the HIF-1α gene in neural cells. *Mol. Cell Biol.* **23**, 6739–6749.

Treins, C., Giorgetti-Peraldi, S., Murdaca, J., Semenza, G. L., and Van Obberghen, E. (2002). Insulin stimulates hypoxia-inducible factor 1 through a phosphatidylinositol 3-kinase/target of rapamycin-dependent signaling pathway. *J. Biol. Chem.* **277**, 27975–27981.

Vigne, P., and Frelin, C. (2006). A low protein diet increases the hypoxic tolerance in *Drosophila*. *PLoS ONE* **1**, e56.

Wappner, P., Gabay, L., and Shilo, B. Z. (1997). Interactions between the EGF receptor and DPP pathways establish distinct cell fates in the tracheal placodes. *Development* **124**, 4707–4716.

Wappner, P., and Ratcliffe, P. J. (2001). Development of branched structures and the cellular response to hypoxia: An evolutionary perspective. *In* "Genetic models in cardio-respiratory biology" (G. Haddad, ed.), pp. 91–138. Informa Healthcare, New York, NY.

Ward, M. P., Mosher, J. T., and Crews, S. T. (1998). Regulation of bHLH-PAS protein subcellular localization during *Drosophila* embryogenesis. *Development* **125**, 1599–1608.

Wilk, R., Weizman, I., and Shilo, B. Z. (1996). Trachealess encodes a bHLH-PAS protein that is an inducer of tracheal cell fates in *Drosophila*. *Genes Dev.* **10**, 93–102.

Zelzer, E., Wappner, P., and Shilo, B. Z. (1997). The PAS domain confers target gene specificity of *Drosophila* bHLH/PAS proteins. *Genes Dev.* **11**, 2079–2089.

Zelzer, E., Levy, Y., Kahana, C., Shilo, B. Z., Rubinstein, M., and Cohen, B. (1998). Insulin induces transcription of target genes through the hypoxia-inducible factor HIF-1alpha/ARNT. *EMBO J.* **17**, 5085–5094.

Zhang, H., Stallock, J. P., Ng, J. C., Reinhard, C., and Neufeld, T. P. (2000). Regulation of cellular growth by the *Drosophila* target of rapamycin dTOR. *Genes Dev.* **14**, 2712–2724.

Zhong, H., Chiles, K., Feldser, D., Laughner, E., Hanrahan, C., Georgescu, M. M., Simons, J. W., and Semenza, G. L. (2000). Modulation of hypoxia-inducible factor 1alpha expression by the epidermal growth factor/phosphatidylinositol 3-kinase/PTEN/AKT/FRAP pathway in human prostate cancer cells: Implications for tumor angiogenesis and therapeutics. *Cancer Res.* **60**, 1541–1545.

Zundel, W., Schindler, C., Haas-Kogan, D., Koong, A., Kaper, F., Chen, E., Gottschalk, A. R., Ryan, H. E., Johnson, R. S., Jefferson, A. B., Stokoe, D., and Giaccia, A. J. (2000). Loss of PTEN facilitates HIF-1-mediated gene expression. *Genes Dev.* **14**, 391–396.

ERYTHROPOIETIN

Constitutively Overexpressed Erythropoietin Reduces Infarct Size in a Mouse Model of Permanent Coronary Artery Ligation

Giovanni G. Camici,*,** Thomas Stallmach,[†],** Matthias Hermann,*,** Rutger Hassink,[‡] Peter Doevendans,[‡] Beat Grenacher,[¶],** Alain Hirschy,[§] Johannes Vogel,[¶],** Thomas F. Lüscher,*,** Frank Ruschitzka,*,** and Max Gassmann[¶],**

Contents

1. Introduction	148
2. Material and Methods	149
2.1. Animals and surgery	149
2.2. Determination of infarct size and immunohistochemical analysis	150
3. Results	151
3.1. EPO plasma levels and EPO-R expression in the myocardium	151
3.2. Cardiac protection by EPO	153
3.3. Infarct size correlates with WT1 staining	153
4. Discussion	153
Acknowledgments	154
References	154

Abstract

In view of the emerging role of recombinant human erythropoietin (rhEPO) as a novel therapeutical approach in myocardial ischemia, we performed the first two-way parallel comparison to test the effects of rhEPO pretreatment

* Cardiology and Cardiovascular Research and Institute of Physiology, University of Zürich, Zürich, Switzerland
† Institute of Pathology, University Hospital Zürich, Zürich, Switzerland
‡ Department of Cardiothoracic Surgery, Heart Lung Center, Utrecht, The Netherlands
§ Institute of Cell Biology, Swiss Federal Institute of Technology Zürich, Zürich, Switzerland
¶ Institute of Veterinary Physiology, Vetsuisse Faculty, University of Zürich, Zürich, Switzerland
** Zürich Center for Integrative Human Physiology (ZIHP), University of Zürich, Zürich, Switzerland

Methods in Enzymology, Volume 435
ISSN 0076-6879, DOI: 10.1016/S0076-6879(07)35008-8

(1000 U/kg, 12 h before surgery) versus EPO transgenic overexpression in a mouse model of myocardial infarction. Unlike EPO transgenic mice who doubled their hematocrit, rhEPO pretreated mice maintained an unaltered hematocrit, thereby offering the possibility to discern erythropoietic-dependent from erythropoietic-independent protective effects of EPO. Animals pretreated with rhEPO as well as EPO transgenic mice underwent permanent left anterior descending (LAD) coronary artery ligation. Resulting infarct size was determined 24 h after LAD ligation by hematoxylin/eosin staining, and morphometrical analysis was performed by computerized planimetry. A large reduction in infarction size was observed in rhEPO-treated mice ($-74\% \pm 14.51$; $P = 0.0002$) and an even more pronounced reduction in the EPO transgenic group ($-87\% \pm 6.31$; $P < 0.0001$) when compared to wild-type controls. Moreover, while searching for novel early ischemic markers, we analyzed expression of hypoxia-sensitive Wilms' tumor suppressor gene (WT1) in infarcted hearts. We found that its expression correlated with the infarct area, thereby providing the first demonstration that WT1 is a useful early marker of myocardial infarction. This study demonstrates for the first time that, despite high hematocrit levels, endogenously overexpressed EPO provides protection against myocardial infarction in a murine model of permanent LAD ligation.

1. INTRODUCTION

The ability to sense fluctuations in oxygen supply is crucial for life. One key response to hypoxic exposure is the transcriptional upregulation of oxygen-dependent genes, such as EPO, which help restore oxygen homeostasis (Schofield and Ratcliffe, 2004) and whose transcription is regulated by hypoxia-inducible factor-1 (HIF-1). Synthesis of EPO is elevated following HIF-1 stabilization under hypoxic conditions (Fandrey *et al.*, 2006). Apart from its erythropoietic function, EPO has been shown to have protective effects in cerebral, myocardial, and renal ischemia (Digicaylioglu and Lipton, 2001; Moon *et al.*, 2003; Vesey *et al.*, 2004; Yang *et al.*, 2003). To analyze the cardioprotective effects of EPO in more detail, we recently established a transgenic mouse line (termed tg6) that over-expresses human EPO in a constitutive and oxygenindependent manner (Ruschitzka *et al.*, 2000). We previously showed that tg6 mice cope well with excessive erythrocytosis by NO-mediated vasodilatation and enhance red blood cell flexibility (Bogdanova *et al.*, 2007; Vogel *et al.*, 2003). Moreover, they do not suffer from cardiovascular or thromboembolic complications (Wagner *et al.*, 2001; Wiessner *et al.*, 2001). In addition to regulating transcription of EPO, HIF-1 also regulates expression of several other genes, one of which is the transcription factor WT1, which is critical for correct embryonic development of the kidneys and the heart (Wagner *et al.*, 2003). The product of WT1 gene belongs to the steadily growing list of transcription factors whose crucial role in cardiac development have been recently recognized (Scholz and Kirschner, 2005). WT1 knockout embryos develop an extremely thin myocardium, often consisting of no more than a single layer of cells. Furthermore, WT1-deficient mice exhibit disruption of the epicardium that lacks much of the cranial surface of the ventricles.

In the present study, we made use of tg6 mice to perform the first parallel comparison testing the effects of rhEPO pretreatment versus chronically over-expressed EPO in a mouse model of permanent LAD coronary artery ligation. Considering that the WT1 gene is an HIF-1 target, we predicted a rapid induction of WT1 upon ligation-induced ischemia, and, in view of the scarcity of early histological markers of post-ischemic fibrosis, we tested whether WT1 represents a valid and specific alternative.

2. Material and Methods

2.1. Animals and surgery

Wild-type (wt) C57BL/6 (Charles River Laboratories, Wilmington, MA) and EPO-overexpressing transgenic tg6 male mice termed tg6 (Ruschitzka *et al.*, 2000) were used for all experiments. All mice were age- and weight-matched (age, 8 wk; weight, 25 ± 3 g). Animals were divided into three groups: tg6, EPO pretreated (wt + EPO), and wild-type mice (wt). EPO mice were given an rhEPO (Roche, Basel, Switzerland) intraperitoneal (IP) injection (1000 U/kg) 12 h before intervention. Each group was successively subdivided in LAD-ligated ($n = 20$) and sham-operated ($n = 10$) animals, resulting in a total of six sets of animals. Following anesthesia by IP injection (90 mg/kg ketamine from Bimeda-MTC, Cambridge, Ontario, Canada; 10 mg/kg xylazine from Streuli Pharma, Uznach, Switzerland), mice were subjected to LAD ligation. Mice were placed in a supine position with paws taped to the operating table. A 5-O ligature was placed behind the front lower incisors and pulled tight so that the neck was in a straight and slightly extended position. The endotracheal respirator tube was made of polyethylene percutaneous endoscopic jejunostomy (PEJ) tubing size 90, one end of which was beveled at a 45° angle. To place the endotracheal tube, the tongue was gently pulled out, and the beveled end of the catheter was inserted through the larynx and into the trachea with care not to injure the trachea or other structures in the pharyngeal region. The tubing was inserted approximately 5 mm from the larynx and taped in place to prevent dislodgment. Ventilation was provided by a rodent ventilator (Harvard, S. Natick, MA). The PE-90 mouse endotracheal tube was inserted loosely into the PE-160 connection to the ventilator. This connection allowed the volume-cycled ventilator to provide sufficient volume (2–4 ml) to the lungs. Normal chest expansion was noted as seen with a conscious mouse. Oxygen

(100%) was provided to the inflow of the ventilator. Dissection was performed using an operating microscope (Olympus, Volketswil, Switzerland), and the chest was opened by a lateral cut with tenotomy scissors along the right or left side of the sternum, cutting through the ribs to approximately midsternum. The chest walls were retracted by use of 5-O or S-O silk or monofilament suture. Slight rotation of the animal to the right oriented the heart to better expose the left ventricle (LV). The left auricle was slightly retracted, exposing the entire left main coronary artery system. Ligation proceeded with a 7-O silk suture passed with a tapered needle underneath the LAD branch of the left coronary artery 1 to 3 mm from the tip of the normally positioned left auricle. In the mouse heart, the cardiac venous network was clearly visible on the heart surface, and no veins appeared to be occluded. Clear visibility of the LAD coronary artery was achieved by surgical lights, right lateral orientation of the heart, and a dissection microscope. The chest wall was then closed by a 5-O Ticron blue polyester fiber suture with one layer through the chest wall and muscle and a second layer through the skin and subcutaneous material. On completion of the surgical procedure, mice were removed from the respirator, and the respirator tube was withdrawn. Postoperatively, mice were kept warm by a heat lamp and treated subcutaneously (s.c.) with analgesic (2.5 mg/kg buprenorphine from Essex Chemie AG, Luzern, Switzerland). Twenty-four hours after the operation, mice were euthanized, and organs were collected. Hematocrit and plasma EPO levels were determined, as described (Ruschitzka *et al.*, 2000). All animal experiments were performed following the Swiss Animal Protection Laws.

2.2. Determination of infarct size and immunohistochemical analysis

Formaline (4%)-fixed hearts were embedded in paraffin, and transverse 5-μm-thick sections were cut serially and stained with hematoxylin/eosin (H&E). Morphometrical analysis was performed by computerized planimetry using an Axioplan microscope and NIH *Image* software. Infarct size was expressed as percentage of infarcted tissue relative to the total ventricular area. For statistical analysis, unpaired Student's t-test and/or analysis of variance (ANOVA) were used, as appropriate, and a P value of less than 0.05 was considered significant. For immunohistochemical analysis, 4% formaline-fixed hearts were mounted on paraffin blocks. WT1, EPO receptor (EPO-R), and CD45 staining was performed using mouse monoclonal (DakoCytomation, Zug, Switzerland), rabbit polyclonal, and mouse monoclonal (both from Santa Cruz Biotechnologies, Santa Cruz, CA) antibodies, respectively. All antibodies were diluted in 2% bovine serum albumin.

3. RESULTS

3.1. EPO plasma levels and EPO-R expression in the myocardium

Plasma EPO levels were 22.1 ± 5 U/l, 157 ± 24 U/l, and 259 ± 79 U/l in wt, wt + EPO, and tg6 mice, respectively. wt and wt + EPO mice had normal hematocrit levels of 40 ± 2%, while tg6 mice showed excessive erythrocytosis with hematocrit levels of 80 ± 3.9%. Since EPO affects infarct size *in vivo* (Parsa *et al.*, 2003), we predicted the presence of EPO-R on cardiomyocytes. Immunohistochemical analysis of myocardial sections revealed that EPO-R are expressed throughout mouse myocardium (Fig. 8.1A).

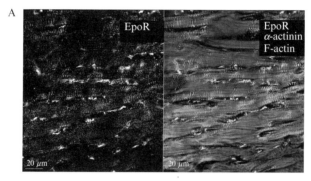

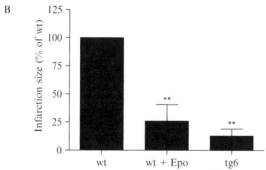

Figure 8.1   (A) Immunohistochemical analysis reveals abundant expression of erythropoietin receptor (EPO-R) in wild-type (wt) mice myocardium (left panel). According to the expression pattern of the cardiomyocyte-specific marker alpha (α)-actinin (right panel, red) and myofibril marker F-actin (right panel, blue), EPO-R (right panel, green) localizes in the membrane of cardiomyocytes. (B) Administration of recombinant EPO (rhEPO) (wt+EPO) as well as constitutively overexpressed EPO (tg6) drastically reduce infarction size upon left anterior descending (LAD) ligation ($n = 8$; $P < 0.05$ for all groups versus wt). Infarct size is represented as percentage of wt mice ± SEM (Standard Error Mean). (See color insert.)

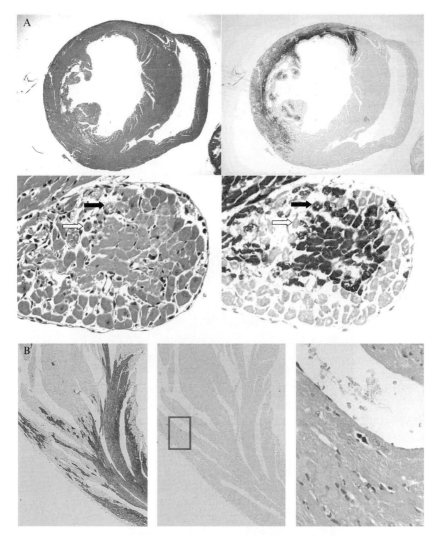

Figure 8.2 (A) Serially cut sections from left anterior descending (LAD) ligated hearts of wild-type (wt) mice were stained with hematoxylin and eosin (H&E) (left panel) and Wilms' tumor suppressor gene (WT1; right panel). Upper and lower panel correspond to a 20- and 60-fold magnification, respectively. Black and white arrows indicate necrotic and viable cardiomyocytes, respectively. (B) To exclude nonspecific binding of WT1 (left panel), parallel sections from infarcted hearts were stained with CD45 antibody. No unspecific staining was observed (middle panel). Magnified (60x) selection (red box) in the right panel shows specific staining of CD45-positive lymphocytes. (See color insert.)

3.2. Cardiac protection by EPO

Twenty-four hours after surgery, infarction size was determined. Sham-operated mice (i.e., thoracotomy only) showed no signs of infarction irrespective of their group, while LAD-ligated mice developed consistently large infarctions. Compared to wt controls, wt + EPO mice showed a 74% (±14.5, $n = 8$) reduction in infarct size ($P = 0.0002$ versus wt; see Fig. 8.1B), while tg6 mice showed an even larger reduction of 87% (±6.3, $n = 8$) compared to wt controls ($P < 0.0001$ versus wt; see Fig. 8.1B).

3.3. Infarct size correlates with WT1 staining

In view of the scarcity of early histological markers of ischemic injury, we tested whether WT1 represents a valid and specific alternative. In a first set of control experiments, we performed WT1 staining on non-ligated hearts and revealed no positive staining in any of the groups (data not shown). Next, we compared LAD-ligated and H&E-stained hearts to the corresponding WT1-stained slides. WT1 staining pattern was identical to that obtained by H&E (Fig. 8.2A): cardiomyocytes showed a disrupted cytoplasmic membrane and no clear nucleic structure (see Fig. 8.2A, black arrows) was necrotic (Vargas *et al.*, 1999) and, as postulated, WT1-positive. In contrast, viable cardiomyocytes (see Fig. 8.2A, white arrows) were negative for WT1 staining. Further, we repeated infarct size measurement using WT1-stained sections and obtained identical data to those obtained with H&E. Unspecific WT1 staining was excluded by performing further immunohistological analysis using different monoclonal antibodies against unrelated proteins, such as CD45. Compared to WT1-stained sections (see Fig. 8.2B, left panel), CD45 resulted in specific staining (see Fig. 8.2B, middle panel). Closer examination of CD45 stained sections (see Fig. 8.2B, right panel) confirmed proper staining by this antibody, as positively stained lymphocytes appeared in the infarcted area. To exclude infarct-unrelated effects of EPO on WT1 expression, we stained hearts from sham-operated wt + EPO and tg6 mice with anti-WT1 antibody. WT1 expression was not detected in either section (data not shown).

4. DISCUSSION

We show for the first time that constitutively EPO-overexpressing tg6 mice are protected against myocardial infarction; thus, we indicate that constitutively elevated (12-fold higher compared to wt) EPO plasma levels precondition the heart to cope very efficiently with ischemic injury despite the existing excessive erythrocytosis. In line with this, a recent report showed that high serum EPO levels are associated with smaller infarct size in patients with acute myocardial infarction (Namiuchi et al., 2005). Taken together, these observations add on to recent considerations of therapeutical administration of novel EPO derivatives that exert tissue-protective but not erythropoietic properties in myocardial ischemia (Leist et al., 2004).

Apart from testing the impact of constitutively overexpressed EPO on myocardial infarction, we made use of our animal model to search for early markers of ischemic heart injury. The fact that WT1 is an oxygen-dependent gene upregulated via instantaneous stabilization of HIF-1 upon hypoxic/ischemic exposure (Jewell et al., 2001) encouraged us to investigate the likelihood of WT1 representing such an early marker. Indeed, we found that WT1 expression occurs only in infarcted hearts, and its staining pattern precisely delineates infarcted areas as demonstrated by comparison with H&E. Our data are partially comparable to those of others who demonstrated WT1 expression in the infarcted rat heart (Wagner et al., 2002). Although species differences certainly play a role, at present, we have no explanation why Wagner et al. observed WT1 expression in the coronary vasculature at the periphery of the infarction, but not, as presented here, in the infarcted myocardium. Based on the congruent overlap between H&E and WT1 staining shown in this work, we propose WT1 as a specific and unique early marker of myocardial infarction that can be easily visualized and accurately quantified.

ACKNOWLEDGMENTS

This work was supported by the Swiss National Science Foundation grants 32–57225.99 and 3100–100214, and by EUROXY (EU grant 502932). The authors would like to thank S. Keller (Vetsuisse Faculty, Zürich, Switzerland) for mouse breeding and H. Scholz (Charité, Berlin, Germany) for discussion and advice on the manuscript.

REFERENCES

Bogdanova, A., Mihov, S., Lutz, H., Saam, B., Gassmann, M., and Vogel, J. (2007). Enhanced erythro-phagocytosis in polycytemic mice overexpressing erythropoietin. *Blood* **110**, 762–769.

Digicaylioglu, M., and Lipton, S. A. (2001). Erythropoietin-mediated neuroprotection involves cross-talk between Jak2 and NF-κB signaling cascades. *Nature* **412**, 641–647.

Fandrey, J., Gorr, T. A., and Gassman, M. (2006). Regulating cellular oxygen sensing by hydroxylation. *Cardiovascular Res.* **71**, 642–651.

Jewell, U. R., Kvietikova, I., Scheid, A., Bauer, C., Wenger, R. H., and Gassmann, M. (2001). Induction of HIF-1α in response to hypoxia is instantaneous. *FASEB J.* **15**, 1312–1314.

Leist, M., Ghezzi, P., Grasso, G., Bianchi, R., Villa, P., Fratelli, M., Savino, C., Bianchi, M., Nielsen, J., Gerwien, J., Kallunki, P., Larsen, A. K., et al. (2004). Derivatives of erythropoietin that are tissue protective but not erythropoietic. *Science* **305**, 239–242.

Moon, C., Krawczyk, M., Ahn, D., Ahmet, I., Paik, D., Lakatta, E. G., and Talan, M. I. (2003). Erythropoietin reduces myocardial infarction and left ventricular functional decline after coronary artery ligation in rats. *Proc. Natl. Acad. Sci. USA* **100,** 11612–11617.

Namiuchi, S., Kagaya, Y., Ohta, J., Shiba, N., Sugi, M., Oikawa, M., Kunii, H., Yamao, H., Komatsu, N., Yui, M., Tada, H., Sakuma, M., *et al.* (2005). High serum erythropoietin level is associated with smaller infarct size in patients with acute myocardial infarction who undergo successful primary percutaneous coronary intervention. *J. Am. Coll. Cardiol.* **45,** 1406–1412.

Parsa, C. J., Matsumoto, A., Kim, J., Riel, R. U., Pascal, L. S., Walton, G. B., Thompson, R. B., Petrofski, J. A., Annex, B. H., Stamler, J. S., and Koch, W. J. (2003). A novel protective effect of erythropoietin in the infarcted heart. *J. Clin. Invest.* **112,** 999–1007.

Ruschitzka, F. T., Wenger, R. H., Stallmach, T., Quaschning, T., de Wit, C., Wagner, K., Labugger, R., Kelm, M., Noll, G., Rulicke, T., Shaw, S., Lindberg, R. L., *et al.* (2000). Nitric oxide prevents cardiovascular disease and determines survival in polyglobulic mice overexpressing erythropoietin. *Proc. Natl. Acad. Sci. USA* **97,** 11609–11613.

Schofield, C. J., and Ratcliffe, P. J. (2004). Oxygen sensing by HIF hydroxylases. *Nat. Rev. Mol. Cell Biol.* **5,** 343–354.

Scholz, H., and Kirschner, K. M. (2005). A role for the Wilms' tumor protein WT1 in organ development. *Physiology (Bethesda)* **20,** 54–59.

Vargas, S. O., Sampson, B. A., and Schoen, F. J. (1999). Pathologic detection of early myocardial infarction: A critical review of the evolution and usefulness of modern techniques. *Mod. Pathol.* **12,** 635–645.

Vesey, D. A., Cheung, C., Pat, B., Endre, Z., Gobe, G., and Johnson, D. W. (2004). Erythropoietin protects against ischaemic acute renal injury. *Nephrol. Dial. Transplant* **19,** 348–355.

Vogel, J., Kiessling, I., Heinicke, K., Stallmach, T., Ossent, P., Vogel, O., Aulmann, M., Frietsch, T., Schmid-Schonbein, H., Kuschinsky, W., and Gassmann, M. (2003). Transgenic mice overexpressing erythropoietin adapt to excessive erythrocytosis by regulating blood viscosity. *Blood* **102,** 2278–2284.

Wagner, K. D., Wagner, N., Bondke, A., Nafz, B., Flemming, B., Theres, H., and Scholz, H. (2002). The Wilms' tumor suppressor WT1 is expressed in the coronary vasculature after myocardial infarction. *FASEB J.* **16,** 1117–1119.

Wagner, K. D., Wagner, N., Wellmann, S., Schley, G., Bondke, A., Theres, H., and Scholz, H. (2003). Oxygen-regulated expression of the Wilms' tumor suppressor WT1 involves hypoxia-inducible factor-1 (HIF-1). *FASEB J.* **17,** 1364–1366.

Wagner, K. F., Katschinski, D. M., Hasegawa, J., Schumacher, D., Meller, B., Gembruch, U., Schramm, U., Jelkmann, W., Gassmann, M., and Fandrey, J. (2001). Chronic inborn erythrocytosis leads to cardiac dysfunction and premature death in mice overexpressing erythropoietin. *Blood* **97,** 536–542.

Wiessner, C., Allegrini, P. R., Ekatodramis, D., Jewell, U. R., Stallmach, T., and Gassmann, M. (2001). Increased cerebral infarct volumes in polyglobulic mice overexpressing erythropoietin. *J. Cereb. Blood Flow Metab.* **21,** 857–864.

Yang, C. W., Li, C., Jung, J. Y., Shin, S. J., Choi, B. S., Lim, S. W., Sun, B. K., Kim, Y. S., Kim, J., Chang, Y. S., and Bang, B. K. (2003). Preconditioning with erythropoietin protects against subsequent ischemia-reperfusion injury in rat kidney. *FASEB J.* **17,** 1754–1755.

CHAPTER NINE

USE OF GENE-MANIPULATED MICE IN THE STUDY OF *ERYTHROPOIETIN* GENE EXPRESSION

Norio Suzuki,* Naoshi Obara,* *and* Masayuki Yamamoto*,†

Contents

1. Introduction	158
2. Materials	161
2.1. Genetic manipulation of mouse lines	161
2.2. Hypoxic chamber	161
2.3. BAC clones	162
2.4. Plasmids and bacterial strains for recombination of BAC clones	162
2.5. Antibodies and immunohistochemistry	162
3. Methods and Results	163
3.1. Real-time and noninvasive monitoring of EPO activity *in vivo*	163
3.2. Analysis of erythropoietic and non-erythropoietic function of EPO–EPOR pathway *in vivo*	165
3.3. Transgenic mouse-expressing GFP under the control of *Epo* gene regulatory region	166
3.4. Regulatory region sufficient for *in vivo Epo* gene expression	168
3.5. Identification of the REP cell	169
3.6. Essential *cis*-elements for cell type–specific and inducible *Epo* gene expression *in vivo*	171
4. Conclusion	173
Acknowledgments	173
References	174

Abstract

Transcriptional regulation of animal genes has been classified into two major categories: tissue-specific and stress-inducible. Erythropoietin (EPO), an erythroid growth factor, plays a central role in the regulation of red blood cell production. In response to hypoxic and/or anemic stresses, *Epo* gene

* ERATO Environmental Response Project, Japan Science and Technology Agency, Center for Tsukuba Advanced Research Alliance, University of Tsukuba, Tsukuba, Ibaraki, Japan
† Department of Medical Biochemistry, Tohoku University Graduate School of Medicine, Sendai, Japan

Methods in Enzymology, Volume 435   © 2007 Elsevier Inc.
ISSN 0076-6879, DOI: 10.1016/S0076-6879(07)35009-X   All rights reserved.

expression is markedly induced in kidney and liver; thus, the *Epo* gene has been used as a model for elucidating stress-inducible gene expression in animals. A key transcriptional regulator of the hypoxia response, hypoxia-inducible transcription factor (HIF), has been identified and cloned through studies on the *Epo* gene. Recently developed gene-modified mouse lines have proven to be a powerful means of exploring the regulatory mechanisms as well as the physiological significance of the tissue-specific and hypoxia-inducible expression of the *Epo* gene. In this chapter, several gene-modified mouse lines related to EPO and the EPO receptor are introduced, with emphasis placed on the examination of *in vivo* EPO activity, EPO function in nonhematopoietic tissues, EPO-producing cells in the kidney, and *cis*-acting regulatory elements for *Epo* gene expression. These *in vivo* studies of the *Epo* gene have allowed for a deeper understanding of transcriptional regulation operated in a tissue-specific and stress-inducible manner.

1. INTRODUCTION

In order to maintain oxygen homeostasis in tissues, our body activates various regulatory reactions upon exposure to hypoxia (Poellinger and Johnson, 2004). It has been known for a while that hypoxia enhances erythropoiesis, and, through the increase of red blood cells, hypoxia increases oxygen supply in the body. This reaction has been referred to as "stress erythropoiesis" (Bailey and Davies, 1997; Koury, 2005). Stress erythropoiesis is one of the most prominent biological responses in our body, and EPO plays a central role in this response (Koury, 2005). Indeed, plasma EPO concentrations increase significantly in response to hypoxia or anemia. EPO subsequently binds to erythropoietin receptor (EPOR) on the surface of erythroid progenitors and transduces anti–apoptotic, proliferative, and cell differentiation signals to the nucleus (Jelkmann, 2004). It is well recognized that recombinant EPO protein is currently the most successful therapeutic agent against anemia (Jelkmann, 2004). In this regard, *Epo* and *EpoR* gene knockout lines of mice have been generated and analyzed in detail, and both lines exhibit severe anemia and die by embryonic day 13.5 (E13.5) (Wu *et al.*, 1995).

Anemic or hypoxic stress increases EPO production by inducing transcription of the *Epo* gene in the kidney and liver. In 1987, Bunn *et al.* examined a number of cell lines derived from kidney or liver to see whether the EPO production could be induced in these cells in response to hypoxia. Two human hepatoma cell lines, Hep3B and HepG2, were found to express EPO in a hypoxia–inducible manner (Goldberg *et al.*, 1987). Since then, these cell lines have been a useful means for studies on *Epo* gene regulation. By exploiting these cell lines, the hypoxia–responsive element (HRE) was identified in the 3′ flanking region of the *Epo* gene (Blanchard *et al.*, 1992; Semenza *et al.*, 1991a), and this HRE contributed to the molecular cloning of HIF (Wang and Semenza, 1995). HIF regulates a group of hypoxia-responsible genes and cellular response to hypoxia (Poellinger and Johnson, 2004).

The major EPO-producing tissue in adult animals is the kidney. Indeed, in adult animals, plasma EPO levels reduce dramatically upon extirpation of the kidney (Koury, 2005; Zanjani *et al.*, 1981). Similarly, human renal failure patients often suffer from EPO-responsive anemia (Jelkmann, 2004). Whereas liver is also known as an EPO-producing tissue, the EPO-producing capacity of the liver is much lower than that of the kidney in adult animals. In the liver, hepatocytes produce EPO. As summarized in Table 9.1, some studies have suggested that, in kidney, peritubular interstitial cells express EPO, while other studies suggest that the tubular cells produce EPO. This discrepancy may be due to the difficulty in detecting EPO expression *in vivo* (Rich, 1991).

Whereas regulatory mechanisms of the *Epo* gene in the hepatocytes have been addressed through taking advantage of the EPO-producing hepatoma cell lines, mechanisms regulating the *Epo* gene expression in the kidney remain to be clarified. No feasible approach has been discovered for kidney because no kidney-derived cell lines have been identified that retain hypoxia-inducible EPO-producing properties. In this regard, transgenic reporter mouse assays emerge as an alternative method to examine hypoxia-inducible and tissue-specific *Epo* gene regulation *in vivo*. A new era of the gene expression analysis arrived with the introduction of bacteria artificial chromosome (BAC) transgenic reporter mice. BACs are large vectors that are easy to manipulate and can incorporate very large genomic fragments (usually around 200 kb) (Lee *et al.*, 2001). A transgenic reporter mouse assay with the *Epo* gene BAC is currently ongoing.

EPO has been reported to protect neural and cardiac cells from ischemic damage (Brines and Cerami, 2006), suggesting that EPO may play a role in preventing cell death in certain nonhematopoietic tissues. However, hypoxic response usually induces EPO and improves anemia. Therefore, upon the assessment of EPO function in nonhematopoietic tissues, it is important to consider the influence of stress erythropoiesis in addition to the local effect of EPO. One genetic approach to elucidate the contribution of EPO in nonhematopoietic tissues is the use of mouse line specifically expressing EPOR in hematopoietic lineage cells but not in other cell lineages. The hematopoietic cell-specific gene regulatory domain of the mouse *Gata1* gene (*G1-HRD*) serves as an excellent means for this purpose (Onodera *et al.*, 1997; Suzuki *et al.*, 2003). The *G1-HRD* faithfully recapitulates the *Gata1* gene expression profile and drives transgenic expression only in hematopoietic cell lineages (Suzuki *et al.*, 2004). Indeed, *G1-HRD–EpoR* transgenic mice have provided data suggesting novel function of EPO *in vivo*. The designing of *G1-HRD–EpoR* transgenic mouse lines, a unique mouse model for examination of nonhematopoietic EPO function (Suzuki *et al.*, 2002), will be described in detail in this manuscript. *G1-HRD–Luciferase* transgenic lines allow for

Table 9.1   Cells reported as renal erythropoietin-producing sites

Site	Cell type	Species	Method	References
Glomerulus	Not defined	Sheep	Immunohistochemistry	Fisher *et al.*, 1965
	Not defined	Dog	Immunohistochemistry	Busuttil *et al.*, 1972
	Not defined	Goat	Primary culture	Burlington *et al.*, 1972
	Epithelial cell	Human	Immunohistochemistry	Mori *et al.*, 1985
Mesangium	Mesangial cell	Rat	Primary culture	Kurtz *et al.*, 1983
Proximal tubule	Tubular cell	Mouse	Tg–lacZ	Loya *et al.*, 1994
	Tubular Cell	Mouse	Immunohistochemistry and *in situ* hybridization	Shanks *et al.*, 1996
	Tubular cell	Mouse	Immunohistochemistry and *in situ* hybridization	Suliman *et al.*, 2004
Peritubule	Endothelial cell	Mouse	*In situ* hybridization	Lacombe *et al.*, 1988
(Interstitial space)	Not defined	Mouse	*In situ* hybridization	Koury *et al.*, 1988
	Endothelial cell	Mouse	Immunohistochemistry	Suzuki and Sasaki, 1990
	Not defined	Mouse	Tg/ISH	Semenza *et al.*, 1991b
	Not defined	Rat	*In situ* hybridization	Schuster *et al.*, 1992
	Fibroblastic cell	Rat	*In situ* hybridization	Bachmann *et al.*, 1993
	Fibroblastic cell	Mouse	Tg/IHC	Maxwell *et al.*, 1993
	Fibroblastic cell	Mouse	Primary culture	Plotkin and Goligorsky, 2006

Tg/IHC, detection of T-antigen by immunohistochemistry (IHC) in transgenic mice expressing SV40 T-antigen under the control of mouse *Epo* gene; Tg/ISH, detection of human erythropoietin (EPO) messenger RNA (mRNA) in human *Epo* transgenic mice by in situ hybridization (ISH); Tg-lacZ, transgenic mice with *lacZ* reporter gene.

noninvasive and real-time monitoring of stress erythropoiesis *in vivo* (Suzuki *et al.*, 2006).

Thus, genetically modified mouse lines have provided a powerful means to explore the regulatory mechanisms as well as the physiological significance of the tissue-specific and hypoxia-inducible expression of the *Epo* gene. In this chapter, several such mouse lines are introduced, placing emphases on the examination of *in vivo* EPO activity, EPO function in nonhematopoietic tissues, EPO-producing cells in the kidney, and the *cis*-acting regulatory elements involved in *Epo* gene expression.

2. MATERIALS

2.1. Genetic manipulation of mouse lines

All mice were strictly kept in specific pathogen-free conditions and were treated according to the regulations of *The Standards for Human Care and Use of Laboratory Animals of the University of Tsukuba*. G1-HRD–Luciferase transgenic mice were used for real-time noninvasive monitoring of EPO activity *in vivo* (Suzuki *et al.*, 2006). *In vivo* bioluminescence imaging was performed with the IVIS imaging system (Xenogen, Alameda, CA). Transgenic mice were anesthetized with isoflurane and injected intraperitoneally with 75 mg/kg of D-luciferin (Xenogen), which is a substrate of the firefly luciferase. Ten minutes after luciferin injection, mice were imaged for 5 to 10 s. Photons emitted from specific regions were quantified with Living Image software (Xenogen).

EPOR-rescued mice exclusively expressing EPOR in hematopoietic cells were generated by crossing *G1-HRD–EpoR* transgenic mice with heterozygous EPOR knockout mice (Suzuki *et al.*, 2002). Whereas EPOR knockout mice die *in utero* due to severe anemia (Kieran *et al.*, 1996), the EPOR-rescued mice escape lethality caused by anemia because of the transgenic expression of EPOR exclusively in hematopoietic cells (Suzuki *et al.*, 2002). The *G1-HRD–GFP* transgenic line of mice is useful for single, cell-based analysis of *Gata1*-expressing cells by detecting the green fluorescence (Suzuki *et al.*, 2003). The *G1-HRD–GFP*, *G1-HRD–Luciferase*, and EPOR-rescued lines of mice are available from RIKEN Bioresource Center for academic research use (RIKEN BRC, Tsukuba, Japan).

2.2. Hypoxic chamber

To expose mice to hypoxia, mice were kept in an isolated chamber (Natsume, Tokyo) with a Zirconia oxygen analyzer (LC-750, Toray Engineering, Tokyo). Oxygen concentrations were kept at 6% in the chamber by control of nitrogen gas input.

2.3. BAC clones

A C57black/6 mouse genomic BAC library (in pBeloBAC11 vector) was screened *in silico* through utilization of the CELERA database (Rockville, MD), and three clones, Epo-60K/BAC (RP27826), Epo-30K/BAC (RP27790), and Epo-22K/BAC (RP27827), were purchased from Incyte Genomics Inc. (St. Louis, MO).

2.4. Plasmids and bacterial strains for recombination of BAC clones

pFNEoba2 was constructed by cloning *FRT-Km^R-FRT* cassette from pICGN21 (Lee *et al.*, 2001) into pBluescript II-SK (Stratagene, La Jolla, CA). Green fluorescent protein (GFP) complementary DNA (cDNA) from pEGFP-N1 (Clontech, Mountain View, CA) and a polyadenylation signal sequence from pSVbeta (*β*)(Clontech) were ligated in frame with a 8.3-kb mouse *Epo* genomic fragment upstream from the second exon (Imagawa *et al.*, 2002), and the fused fragment was integrated into pFNEoba2 with the 4.5-kb genomic fragment between *Sma*I (in the fourth intron) and *Xba*I (in 3′ downstream region) to construct the targeting plasmid. The plasmid was used for homologous recombination in *Escherichia coli* after isolation of insert (Targeting vector, Fig. 9.3A).

The amine-terminal region of the transgenic GFP translates into an additional 35 amino acids corresponding to the EPO amino terminus, which contains the signal peptide sequence for secretion of EPO (1–26 amino acids). However, no secretion or specific subcellular localization of GFP was observed in the transgenic mice or in cultured cells transfected with the transgene. This observation supports the notion that the signal sequence of EPO is insufficient for secretion, and additional regions must be required for EPO secretion (Herrera *et al.*, 2000). We concluded that the GFP fused with the extra 35 amino acids of EPO sequence could be used as a transgenic reporter.

DH10B-derived *E. coli* strain EL250 contains an *araBAD*-promoter driven *flpe* gene in the genome (Lee *et al.*, 2001). Culturing EL250 with 0.1% arabinose (Sigma, St Louis, MO) for 1.5 h enhanced flippase (FLP) recombinase (FLPE)-mediated FLP recognition target (FRT) recombination. EL250 was usually cultured at 32°, but the cells were cultured at 42° to induce competence for FLP recombination. Chloramphenicol (Cm; 25 μg/ml; Wako, Osaka) and kanamycin (Km; 10 μg/ml; Wako) were used as antibiotics for selection of transformants.

2.5. Antibodies and immunohistochemistry

Frozen mouse kidney sections (10-μm thickness) were fixed in 4% paraformaldehyde for 30 min and were incubated with rabbit anti-GFP polyclonal antibody (diluted 1:1000; Molecular Probes, Eugene, OR) at 4° overnight.

After treatment with 3% hydrogen peroxide, sections were incubated with horseradish peroxidase-conjugated anti-rabbit IgG secondary antibody (diluted 1:500; Biosource, Camarillo, CA). Color detection was performed using diaminobenzidine (DAKO, Carpinteria, CA) as a chromogen (brown color staining). Hematoxylin was used for counterstaining. Phycoerythrin (PE)-conjugated anti-PECAM1 (BD PharMingen, San Diego, CA) was diluted 1:100 and incubated with the thin sections for 2 h at room temperature. Fluorescent and bright field images were observed using the BioZero imaging system (Keyence, Tokyo).

3. Methods and Results

3.1. Real-time and noninvasive monitoring of EPO activity *in vivo*

To examine *in vivo* activity of EPO, we usually measure hematological indices, such as hematocrit, hemoglobin concentration, and reticulocyte counts (Koury, 2005). However, an apparent time delay exists between changes in the plasma EPO concentration and changes in these hematopoietic parameters. Since EPO stimulates proliferation and differentiation of erythroid progenitors, we hypothesized that measurement of erythroid progenitor numbers would provide for sensitive monitoring of the EPO activity *in vivo*. GATA1 is a transcription factor responsible for erythroid, megakaryocytic, eosinophilic, and mast cell lineage development (reviewed in Shimizu and Yamamoto, 2005). During erythroid differentiation, GATA1 mRNA expression peaks in the late erythroid progenitor (LEP) stage, which corresponds to colony-forming unit erythroid (CFU-E) cells and proerythroblasts (Suzuki *et al.*, 2003). LEP cells express the *EpoR* gene abundantly, and the cells are highly responsive to EPO stimulation (Suzuki *et al.*, 2002, 2003). Therefore, LEP cells seem to be good target cells for real-time monitoring of EPO activity.

To this end, we established *G1-HRD–Luciferase* transgenic mouse lines (Suzuki *et al.*, 2006) and measured transgenic luciferase expression with an *in vivo* bioluminescent imaging system (IVIS) optical bioluminescent imaging detector (Fig. 9.1A) (Massoud and Gambhir, 2003). In the spleen and bone marrow of *G1-HRD–Luciferase* transgenic mice, a weak IVIS signal was detected. After subcutaneous injection of recombinant human EPO (300 U/kg; kindly supplied by Chugai Pharmaceutical), the IVIS signal was increased in the spleen. The signal intensity was increased continuously for 4 days by daily administration of EPO (see Fig. 9.1B). The IVIS signal started decreasing immediately after stopping the EPO injection, and the level returned to basal levels 4 days after the last injection (see Fig. 9.1B). In contrast, hematocrit began to increase 4 days after the first EPO

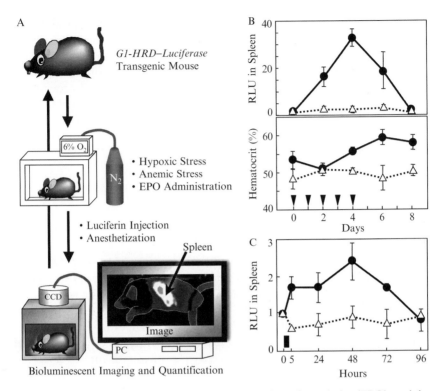

Figure 9.1 Real-time noninvasive monitoring of erythropoietin (EPO) activity. (A) *G1-HRD–Luciferase* transgenic mouse lines and an *in vivo* bioluminescent imaging system (IVIS, Xenogen). Transgenic mice are stimulated with EPO, hypoxia, or anemia to expand the erythroid progenitor population. Bioluminescence is monitored with a charge-coupled device (CCD) camera, and the signal is quantified with Living Image software (Xenogen). Luciferase activity is noninvasively and consecutively measured. The lateral view of transgenic mice shows exclusive luciferase activity in the spleen. (B) Relative light units (RLU) in the spleen (upper) and hematocrit (lower) of the transgenic mice are changed by the EPO administration. *G1-HRD–Luciferase* transgenic mice are treated with 300 U/kg of EPO (solid line) or saline (dotted line) for 5 consecutive days (arrowheads). (C) *G1-HRD–Luciferase* transgenic mice are exposed to 6% oxygen for 4 h (black box), and then RLU of the spleen is measured (solid line). Mean RLU of control normoxic mice is also shown (dotted line). RLU before treatments are set to 1 in each group (B and C).

administration and continued increasing after the last injection. The hematocrit level returned to normal 8 days after the stop of EPO administration (data not shown). The increase in magnitude of the IVIS signal after the EPO injection was larger than the resultant increase of the hematocrit and other hematopoietic parameters (see Fig. 9.1B). These data clearly demonstrate that bioluminescent imaging is a sensitive method for real-time monitoring of the EPO activity *in vivo*.

To test whether this erythropoietic *in vivo* monitoring system allows for real-time detection of endogenous *Epo* gene expression, we measured the IVIS signal after exposing *G1-HRD–Luciferase* mice to hypoxic conditions in an isolated chamber. We applied hypoxic stress by adjusting oxygen levels to 6% for 4 h for mice. This treatment rapidly induced the IVIS signal in the transgenic mouse spleen (see Fig. 9.1C). Plasma EPO concentrations were measured by enzyme-linked immunosorbent assay (ELISA) (Roche, Basel) and increased simultaneously (data not shown). On the contrary, the hematocrit did not significantly change during the experiment (Suzuki *et al.*, 2006). These results demonstrate that stress-induced erythropoiesis can be monitored by utilization of the IVIS signal, which is reflected by the resultant increase of *G1-HRD–Luciferase*–expressing cells.

One of the technical advantages inherent to this noninvasive method is that an individual mouse can be repetitively used for the real-time monitoring (see Fig. 9.1A). We surmise that the increase in the IVIS signal directly reflects the increase of the numbers of the cells expressing the luciferase transgene rather than the transcriptional induction of the transgene in each cell. One immediate application of this mouse system is the assessment of erythropoiesis in the mutant lines of mice harboring defects in erythropoiesis. In addition, a possibility exists that transfusion of human-derived monitoring cells containing integrated *G1-HRD–Luciferase* transgene may help to diagnose the endogenous erythropoietic activity in EPO-treated patients.

3.2. Analysis of erythropoietic and non-erythropoietic function of EPO–EPOR pathway *in vivo*

Non-hematopoietic functions of the EPO–EPOR system have been suggested, and clinical studies related to this issue demonstrate some of the alternative impacts EPO may have in various disease states. For example, EPO improves neuronal diseases, heart failure, and angiogenesis in laboratory animals (Brines and Cerami, 2006). *EpoR* is expressed in many mouse tissues and cell lineages (Suzuki *et al.*, 2002), yet the physiological contribution of the EPO–EPOR system to non-hematopoietic cells has not been fully clarified yet due to the early lethality (E13.5) of EPOR-null mice (Kieran *et al.*, 1996; Wu *et al.*, 1995).

Using *G1-HRD*, transgenic mouse lines expressing GFP or EPOR exclusively in hematopoietic cells have been established (Suzuki *et al.*, 2002). The mouse full-length EPOR cDNA or a *GFP* tag was ligated with *G1-HRD* for generation of *G1-HRD–EpoR* or *G1-HRD–GFP* transgenic mouse lines, respectively. The latter mouse embryos showed hematopoietic expression of GFP (see Fig. 9.1A) and no hematopoietic abnormalities. We then crossed *G1-HRD–EpoR* transgenic mice with *EpoR (+/−)* mice, and EPOR knockout mice with the transgene (EPOR-rescued mice; *EpoR [−/−]::G1-HRD–EpoR* transgene [+]) were generated (Fig. 9.2B). EPOR-rescued

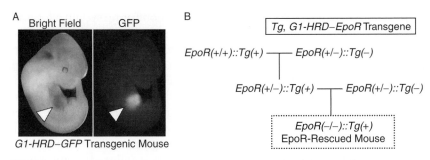

Figure 9.2 Establishment of erythropoietin receptor (EPOR)-rescued mouse expressing EPOR exclusively in hematopoietic cells. (A) Green fluorescent signals from the *G1-HRD–GFP* transgene are exclusively detected in the hematopoietic cells of the fetal liver (arrowheads) at E12.5. (B) Schematic for generating EPOR-rescued mice.

mice showed no abnormalities in development or in erythropoiesis, and hematopoietic tissue-specific expression of EPOR was verified in the EPOR-rescued mice (Suzuki *et al.*, 2002). The rescue of EPOR knockout mouse from lethality by the *G1-HRD–EpoR* transgene clearly demonstrates that the primary function of the EPO–EPOR system *in vivo* is the support of erythropoiesis.

To test whether the EPO–EPOR system plays an important role in non-hematopoietic tissues, the EPOR-rescued mice were pathologically challenged. As a result, novel non-hematopoietic functions of the EPO–EPOR system have been identified. For example, EPO contributes to angiogenesis in the retinopathy of prematurity (Morita *et al.*, 2003) and to the mobilization of the bone marrow–derived endothelial progenitor cells in pulmonary hypertension (Satoh *et al.*, 2006). We also found it critically important to verify whether non-hematopoietic abnormalities in EPOR-mutant embryos were a direct consequence of the loss of EPOR function. In fact, while defects in heart development were reported in the EPOR knockout mouse embryos (Wu *et al.*, 1999), the hematopoietic EPOR-rescued mice had no such abnormalities (Suzuki *et al.*, 2002). The latter result indicates that the EPOR deficiency in cardiac muscle is not the direct cause of the malformation, but indirectly affects the heart development by causing severe anemia. In addition, *in vitro* investigations of non-hematopoietic functions of the EPO–EPOR system are currently ongoing in a number of laboratories utilizing hematopoietic EPOR-rescued mice.

3.3. Transgenic mouse-expressing GFP under the control of *Epo* gene regulatory region

To elucidate regulatory mechanisms governing the *Epo* gene, we performed transgenic mouse reporter gene expression analyses using an 11-kb mouse *Epo* gene-flanking region. However, this 11-kb region did not recapitulate fully the *Epo* gene expression (unpublished data), demonstrating that the regulatory element(s) for *Epo* gene spans much wider genomic regions. Therefore, we decided to utilize the BAC system, which allows for manipulation of much larger fragments of DNA.

Three mouse genomic BAC clones containing 170 to 190 kb of the *Epo* gene and its flanking regions were obtained (see "Materials" section and Table 9.2). The EL250 *E. coli* strain was first transformed with 100 ng of BAC DNA using an electroporator. Transformants were electroporated again with 500 ng of the targeting construct (Fig. 9.3A), and Cm/Km double-resistant colonies containing homologously recombined BAC clones were selected. After checking the BAC DNA recombination by

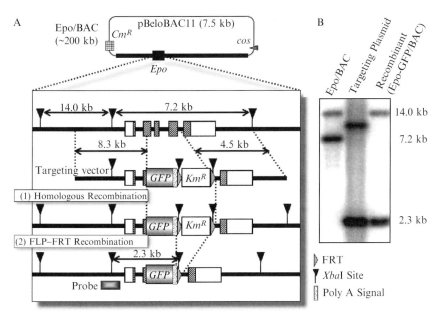

Figure 9.3 Construction of *Epo-GFP* transgenes. (A) Strategy for *GFP* insertion into the coding region of the *Epo* gene in a bacteria artificial chromosome (BAC) clone. The *GFP* gene with a polyadenylation signal sequence and kanamycin resistance (Km^R) gene was replaced with the region spanning second exon to forth intron of the mouse *Epo* gene in the BAC clone by homologous recombination. Boxes indicate exons and shaded boxes indicate the translational region of the *Epo* gene. The Km^R gene cassette is removed by the flippase–FLP recognition target (FLP–FRT) recombination in EL250 strain harboring inducible FLP expression system. pBeloBAC11 has the chloramphenicol-resistance (Cm^R) gene for a bacteria selection marker and the *cos* sequence for linearization by λ-terminase. (B) Southern blotting analysis of the recombinant Epo–green fluorescent protein (GFP) construct. DNA was isolated from the bacteria and hybridized with a probe indicated in (A) after digestion with *Xba*I. Two different-sized (14.0-kb and 7.2-kb) bands are detected in the original Epo/BAC clone by the probe containing an *Xba*I site, but the size of 7.2-kb band shifts to 2.3 kb in the recombinant Epo-GFP construct.

Southern blotting (see Fig. 9.3B), the recombinant bacteria were incubated in L-broth (LB) medium supplemented with 0.1% arabinose for 1.5 h at 32°, and the Km-resistant gene cassette was removed to avoid interference for the target gene activity in the BAC transgene (Lee *et al.*, 2001; Yu *et al.*, 2000). Finally, BAC DNA was purified from the Cm-resistant, Km-sensitive cells using a Nucleobond AX-500 cartridge (Macherey-Nagel, Düren, Germany). Recombination of the BAC-derived Epo-GFP construct was confirmed by Southern blotting. We refer to the three Epo–GFP constructs derived from Epo-60K/BAC and Epo-22K/BAC as wt-Epo–GFP and 22K-Epo–GFP, respectively (see Table 9.2).

For microinjection of BAC-based constructs, purified BAC DNA was linearized by digestion with lambda (λ-terminase and purified by gel electrophoresis (Chrast *et al.*, 1999). Using an AM6000 semiautomatic microinjection system (Leica Microsystems, Wetzlar, Germany), purified transgenes (5 ng/μl) were microinjected into the pronuclei of fertilized eggs from bromodomain factor 1 (BDF1) parents. The resultant *Epo-GFP* transgenic mouse lines are shown in Table 9.2. Low-copy (2 to 4 copies) integration of the transgene into the mouse genome was confirmed by the genomic Southern blotting and by polymerase chain reaction (PCR).

3.4. Regulatory region sufficient for *in vivo Epo* gene expression

Four lines of *wt-Epo–GFP* transgenic mouse harboring a 180-kb genomic region containing the *Epo* gene and flanking regions (upstream 60 kb and downstream 120 kb) were generated (see Table 9.2), and the expression profile of the GFP reporter was compared with that of the endogenous *Epo* gene. Recapitulation of endogenous *Epo* gene expression was relatively normal, with transgenic GFP expression marginal under unstressed conditions. Under severe anemic conditions (i.e., hematocrit less than 15%) generated by retroorbital bleeding four times every 12 h (Suzuki *et al.*, 2002), GFP expression was upregulated and green fluorescence was clearly visible in tissue sections of the kidney and liver. ELISA and quantitative reverse transcription PCR (RT-PCR) analyses (Kobayashi *et al.*, 2002) revealed that plasma EPO concentrations were increased 20- to 50-fold and that EPO mRNA expression was increased 500- to 1000-fold and 20- to 100-fold in the kidney and liver, respectively. Similarly, both endogenous *Epo* gene expression and transgenic GFP expression were induced in the kidney and liver of mildly anemic mice (i.e., hematocrit about 30%), albeit the induction level was much weaker than that in the profound anemia. Hypoxia also stimulated *Epo* gene expression, but the conditions employed (i.e., 6% oxygen for 6 h) resulted in the mild increase of EPO mRNA compared with the profound anemia. The positive signals for GFP expression were observed in three out of four lines of the *wt-Epo–GFP*

Table 9.2 Summary of erythropoietin/bacteria artificial chromosome clones and transgene reporter analyses

BAC clone	Epo-60K/BAC (RP27826)	Epo-30K/BAC (RP27790)	Epo-22K/BAC (RP27827)
Upstream[a]	60 kb	30 kb	22 kb
Downstream[a]	120 kb	135 kb	163 kb
Recombinant Tg	*wt-Epo–GFP*	*30K-Epo–GFP*	*22K-Epo–GFP*
Tg lines, *n*	4	Not determined	4
Expression[b]			
Kidney	3	Not determined	3
Liver	3	Not determined	3
Other tissues	0	Not determined	0

[a] The nucleotide length between the end points (upstream and downstream) and the transcriptional initiation site (see Fig. 9.3A).
[b] Green fluorescent protein (GFP) reporter expression was detected by fluorescent microscopy and immunohistochemistry with anti-GFP antibody. GFP expression was induced in the kidney and liver by anemic or hypoxic stress, but the expression was not detectable in other tissues. BAC, bacteria artificial chromosome; EPO, erythropoietin; Tg, T-antigen.

transgenic mice (see Table 9.2). In the positive three lines, the GFP inducibility appeared similar, but basal expression levels were different from each other, and this difference showed no correlation to the transgene copy numbers. These results indicate that the *wt-Epo–GFP* transgene contains the sufficient regulatory domains for the *Epo* gene expression *in vivo*, but the large genomic construct was still affected by the chromosomal position effect variegation.

We then examined the expression profile of the *22K-Epo–GFP* transgene (upstream 22 kb and downstream 163 kb) and found that it was similar to that of the *wt-Epo–GFP* transgene (see Table 9.2). Taken together, these results led us to conclude that 22-kb upstream and 120-kb downstream regions from the transcription start site of the *Epo* gene are sufficient to recapitulate the *Epo* gene expression *in vivo* in a tissue-specific and hypoxia-inducible manner (see Table 9.2).

3.5. Identification of the REP cell

Taking advantage of the fact that the *wt-Epo–GFP* reporter transgenic mice can fully recapitulate the *Epo* gene expression, we attempted to precisely identify the EPO-producing cells in the liver and kidney. In the liver, GFP expression was detected exclusively in the hepatocytes (data not shown). As described in the "Introduction" section, the localization of EPO-producing cells in the kidney has been controversial (see Table 9.1).

Previous reports claimed that the cells might be proximal tubular cells, glomerular, mesangial, or interstitial cells of the renal cortex.

Fluorescent microscopic analysis of the anemic *wt-Epo–GFP* transgenic mouse kidney revealed that GFP is expressed in the peritubular interstitial cells located in the deep renal outer medulla region (Fig. 9.4A and B). This result supports the idea that EPO expresses in fibroblast-like peritubular interstitial cells (see Table 9.1). We referred to these cells as renal EPO-producing (REP) cells. We executed further characterization of REP cells with anti-GFP antibody and fluorescent microscopy. We especially have succeeded in taking three-dimensional images of REP cells using laser

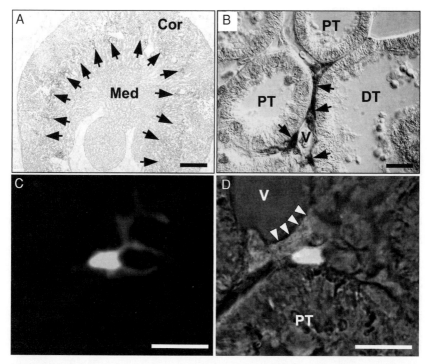

Figure 9.4 The renal erythropoietin (EPO)–producing (REP) cells. (A, B) Anti–green fluorescent protein (GFP) immunohistochemistry of the kidney section from the anemic *wt-Epo–GFP* transgenic mouse. The renal EPO-producing cells are stained black (arrows). The cells are surrounded by the renal tubules (DT, distal tube; PT, proximal tube) in deeper regions of the outer medulla (Med). Cor and V indicate the renal cortex and vessel, respectively. (C, D) Fluorescent image of REP cell in the anemic kidney of *wt-Epo–GFP* transgenic mouse. REP cells display a unique stellar shape, with the projections extending in various directions. (D) Anti-PECAM1-PE immunofluorescence and phase contrast images are merged with the GFP image. Arrowheads indicate a PECAM1-positive vascular endothelial cell associating with REP cells. Scale bars indicate 1 mm (A) and 10 μm (B–D). (See color insert.)

confocal microscopy (data not shown). The REP cells displayed a unique stellar shape, with the projections extending to various directions (see Fig. 9.4B and C). The REP cells are located between the tubular cells and vascular endothelial cells positive for anti-PECAM1 staining (see Fig. 9.4B and D).

To elucidate the *Epo* gene regulatory mechanism, it is helpful to consider physiological implications of the REP cell distribution. Since sodium transport and water reabsorption require a large amount of oxygen, the kidney is the second highest organ in terms of oxygen consumption, only behind the heart (O'Connor, 2006). REP cells are mainly concentrated in the deeper regions of the renal outer medulla where the oxygen concentration is lower (10–15 mm Hg) than in other regions of the kidney (20–60 mm Hg). Indeed, the S3 segments of proximal tubules, which have the highest reabsorption rate, penetrate the deeper regions of the outer medulla (O'Connor, 2006). The REP cells localize between the tubular epithelial cells, vascular endothelial cells (see Fig. 9.4D), and, during the tubular reabsorption process, various molecules passing through or around the REP cells. Therefore, REP cells may be able to sense oxygen concentrations in the urine during reabsorption. Alternatively, some paracrine molecules from the tubular cells may transduce the signal for the inducible production of EPO in REP cells. Actually, hyperactivation of the renin-angiotensin system in transgenic mice caused constitutive overexpression of EPO in the kidney and polycythemia (Kato *et al.*, 2006), suggesting that angiotensin stimulates EPO production in REP cells. These observations suggest that the REP cells may be effective in sensing hypoxia in collaboration with the other cells in the kidney and may generate a response to hypoxia through EPO production.

3.6. Essential *cis*-elements for cell type–specific and inducible *Epo* gene expression *in vivo*

The BAC transgenic system has been a powerful tool in the investigation of *cis*-acting gene regulatory elements. To evaluate the *in vivo* function of *cis*-acting elements on *Epo* gene expression, we introduced point mutations into the regulatory elements of BAC reporter transgene constructs within the homologous arms by PCR-based mutagenesis (Imagawa *et al.*, 2002; Ito *et al.*, 1991), and resulting plasmids were used as a targeting vector (see Fig. 9.3A). After selection of recombinant BAC clones, integration of the point mutations was confirmed via sequencing. Direct mutagenesis was successfully applied for analyses of *cis*-elements near the GFP reporter cassette.

Several regions spanning from 5′ to 3′ of the *Epo* gene show high conservation between mouse and human genomes (Fig. 9.5). Other than the coding exons, these conserved regions are prime candidates for the regulatory domains of the *Epo* gene (Shoemaker and Mitsock, 1986).

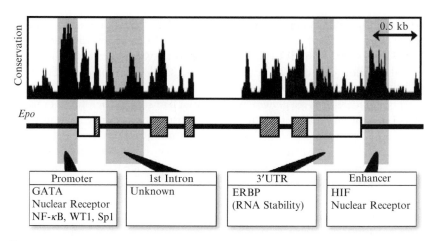

Promoter	1st Intron	3′UTR	Enhancer
GATA Nuclear Receptor NF-κB, WT1, Sp1	Unknown	ERBP (RNA Stability)	HIF Nuclear Receptor

Figure 9.5 Genomic conservation and putative regulatory elements of the *Epo* gene. Blast-like alignment tool (BLAT) analysis (http://genome.ucsc.edu/cgi-bin/hgBlat) of the mouse and human *Epo* gene shows that four highly conserved regions exist (gray boxes). Shaded boxes are the translated regions. The proximal promoter region contains binding motifs for GATA factors, nuclear receptors, NF-κB, WT1, and Sp1. The 3′ proximal region of the *Epo* gene is known as a hypoxia-inducible enhancer recognized by HIF and nuclear receptors. The first intron and 3′ untranslated region (UTR, open box) are also highly conserved. 3′UTR is reported to be involved in mRNA stability control through binding to EPO messenger RNA (mRNA) binding protein (ERBP) (McGary *et al.*, 1997).

Studies using the hepatoma cell lines have identified transcription factor–binding sequences in the 5′ flanking region (e.g., GATA, nuclear receptors, NF-κB, WT1, and Sp1) (Dame *et al.*, 2006; Imagawa *et al.*, 1997; La Ferla *et al.*, 2002). In the 3′ flanking region, a very strong hypoxia-inducible enhancer was identified that was recognized by HIF and nuclear receptors (Stockmann and Fandrey, 2006). The first intron and 3′ untranslated region, both of which contain high homology sequences (see Fig. 9.5), are also considered to be related to *Epo* gene regulation (Imagawa *et al.*, 1991; McGary *et al.*, 1997).

The GATA-binding motif (5′-TGATAA-3′) is located in the position where the TATA box usually resides and acts as a negative regulator of *Epo* gene expression (Imagawa *et al.*, 1997), while another group recently reported that the GATA motif positively regulates the *Epo* gene expression in the liver (Dame *et al.*, 2004). To assess *in vivo* function of the GATA motif, Epo-60K–BAC was recombined, and *Epo-GFP* transgene containing point mutation(s) in the GATA motif was constructed. We prepared two types of GATA mutants (5′-T*T*ATAA-3′ and 5′-T*TT*TAA-3′) and assessed the GFP reporter expression in the transgenic mouse system. To our surprise, GFP expression from both mutant transgenes was detected in many tissues of the transgenic mice, even under unstressed conditions. This constitutive expression is exclusively observed in the epithelial cells, including the renal tubules, hepatic bile duct, thymus, and lung. GFP expression does not change during hypoxic stress, but the anemia-inducible GFP expression is maintained in the hepatocytes and REP cells of the mutant Epo-GFP transgenic mice. We identified that GATA-2 and GATA-3 bind to the GATA motif in the renal tubular cells (data not shown). These results demonstrate that the GATA motif is essential for constitutive repression of ectopic *Epo* gene expression in the epithelial lineage cells, and GATA-based repression is essential for the inducible and cell type–specific expression of the *Epo* gene.

We also examined the *in vivo* function of the *cis*-regulatory motifs located in the 3′ enhancer of the *Epo* gene (Semenza *et al.*, 1991a). We especially focused on the HRE (5′-ACGTG-3′). A mutant *Epo-GFP* transgene containing point mutations in the HRE (5′-A*AAA*G-3′) (Wang and Semenza, 1993) that could prevent HIF transcription factor binding was constructed. Although three out of the five mutant transgenic mouse lines expressed GFP in REP cells under anemic conditions, the GFP expression could not be detected at all in the adult mouse liver, even if bleeding anemia was used to stimulate EPO production. These data demonstrate that the 3′ enhancer is a liver-specific and hypoxia-inducible enhancer, yet is dispensable for renal EPO expression.

4. Conclusion

It has been 20 years since the cloning of the *Epo* gene (Jacobs *et al.*, 1985; Lin *et al.*, 1985), and recombinant human EPO is one of the most successful therapeutic agents in the current medical repertoire (Jelkmann, 2004; Winearls *et al.*, 1986). Whereas many of the aspects of the *Epo* gene have been vigorously analyzed in the basic biology field and have provided important insight into the molecular mechanisms of stress-inducible gene expression (Poellinger and Johnson, 2004), there still remain many unsolved issues related to the function of the *Epo* gene *in vivo*. Unanswered questions include the non-hematopoietic functions of the EPO–EPOR system, tissue-specific expression of the *Epo* gene, and the sensor mechanisms that are stimulated by hypoxia. We propose that the gene-manipulated mouse lines introduced in this chapter will be elegant tools to settle these problems. EPO is one of the most classical growth factors; however, because it is so unique, we still have more to learn from the *Epo* gene.

ACKNOWLEDGMENTS

We thank Drs. Jon Maher, Mikiko Suzuki, and Shigehiko Imagawa (University of Tsukuba) and Dr. Yoshiaki Kondo (Tohoku University) for their help and stimulating discussions. This work is supported in part by the Japan Science and Technology Agency through the ERATO Environmental Response Project.

REFERENCES

Bachmann, S., Le Hir, M., and Eckardt, K. U. (1993). Co-localization of erythropoietin mRNA and ecto-5′-nucleotidase immunoreactivity in peritubular cells of rat renal cortex indicates that fibroblasts produce erythropoietin. *J. Histochem. Cytochem.* **41,** 335–341.

Bailey, D. M., and Davies, B. (1997). Physiological implications of altitude training for endurance performance at sea level: A review. *Br. J. Sports Med.* **31,** 183–190.

Blanchard, K. L., Acquaviva, A. M., Galson, D. L., and Bunn, H. F. (1992). Hypoxic induction of the human erythropoietin gene: Cooperation between the promoter and enhancer, each of which contains steroid receptor response elements. *Mol. Cell. Biol.* **12,** 5373–5385.

Brines, M., and Cerami, A. (2006). Discovering erythropoietin's extra-hematopoietic functions: Biology and clinical promise. *Kidney Int.* **70,** 246–250.

Burlington, H., Cronkite, E. P., Reincke, U., and Zanjani, E. D. (1972). Erythropoietin production in cultures of goat renal glomeruli. *Proc. Natl. Acad. Sci. USA* **69,** 3547–3550.

Busuttil, R. W., Roh, B. L., and Fisher, J. W. (1972). Localization of erythropoietin in the glomerulus of the hypoxic dog kidney using a fluorescent antibody technique. *Acta Haematol.* **47,** 238–242.

Chrast, R., Scott, H. S., and Antonarakis, S. E. (1999). Linearization and purification of BAC DNA for the development of transgenic mice. *Transgenic Res.* **8,** 147–150.

Dame, C., Kirschner, K. M., Bartz, K. V., Wallach, T., Hussels, C. S., and Scholz, H. (2006). Wilms' tumor suppressor, Wt1, is a transcriptional activator of the erythropoietin gene. *Blood* **107,** 4282–4290.

Dame, C., Sola, M. C., Lim, K. C., Leach, K. M., Fandrey, J., Ma, Y., Knopfle, G., Engel, J. D., and Bungert, J. (2004). Hepatic erythropoietin gene regulation by GATA-4. *J. Biol. Chem.* **279,** 2955–2961.

Fisher, J. W., Taylor, E., and Porteous, D. D. (1965). Localization of erythropoietin in glomeruli of sheep kidney by fluorescent antibody technique. *Nature* **205,** 611.

Goldberg, M. A., Glass, G. A., Cunningham, J. M., and Bunn, H. F. (1987). The regulated expression of erythropoietin by two human hepatoma cell lines. *Proc. Natl. Acad. Sci. USA* **84,** 7972–7976.

Herrera, A. M., Musacchio, A., Fernandez, J. R., and Duarte, C. A. (2000). Efficiency of erythropoietin's signal peptide for HIV(MN)-1 gp 120 expression. *Biochem. Biophys. Res. Commun.* **273,** 557–559.

Imagawa, S., Yamamoto, M., and Miura, Y. (1997). Negative regulation of the erythropoietin gene expression by the GATA transcription factors. *Blood* **89,** 1430–1439.

Imagawa, S., Goldberg, M. A., Doweiko, J., and Bunn, H. F. (1991). Regulatory elements of the erythropoietin gene. *Blood* **77,** 278–285.

Imagawa, S., Suzuki, N., Ohmine, K., Obara, N., Mukai, H. Y., Ozawa, K., Yamamoto, M., and Nagasawa, T. (2002). GATA suppresses erythropoietin gene expression through GATA site in mouse erythropoietin gene promoter. *Int. J. Hematol.* **75,** 376–381.

Ito, W., Ishiguro, H., and Kurosawa, Y. (1991). A general method for introducing a series of mutations into cloned DNA using the polymerase chain reaction. *Gene* **102,** 67–70.

Jacobs, K., Shoemaker, C., Rudersdorf, R., Neill, S. D., Kaufman, R. J., Mufson, A., Seehra, J., Jones, S. S., Hewick, R., and Fritsch, E. F. (1985). Isolation and characterization of genomic and cDNA clones of human erythropoietin. *Nature* **6,** 806–810.

Jelkmann, W. (2004). Molecular biology of erythropoietin. *Intern. Med.* **43,** 649–659.

Kato, H., Ishida, J., Imagawa, S., Saito, T., Suzuki, N., Matsuoka, T., Sugaya, T., Tanimoto, K., Yokoo, T., Ohneda, O., Sugiyama, F., Yagami, K., *et al.* (2005). Enhanced erythropoiesis mediated by activation of the renin-angiotensin system via angiotensin II type 1a receptor. *FASEB J.* **19,** 2023–2025.

Kieran, M. W., Perkins, A. C., Orkin, S. H., and Zon, L. I. (1996). Thrombopoietin rescues *in vitro* erythroid colony formation from mouse embryos lacking the erythropoietin receptor. *Proc. Natl. Acad. Sci. USA* **93,** 9126–9131.

Kobayashi, T., Yanase, H., Iwanaga, T., Sasaki, R., and Nagao, M. (2002). Epididymis is a novel site of erythropoietin production in mouse reproductive organs. *Biochem. Biophys. Res. Commun.* **296,** 145–151.

Koury, M. J. (2005). Erythropoietin: T-story of hypoxia and a finely regulated hematopoietic hormone. *Exp. Hematol.* **33,** 1263–1270.

Koury, S. T., Bondurant, M. C., and Koury, M. J. (1988). Localization of erythropoietin-producing cells in murine kidneys by *in situ* hybridization. *Blood* **71,** 524–527.

Kurtz, A., Jelkmann, W., Sinowaitz, F., and Bauer, C. (1983). Renal mesangial cell cultures as a model for study of erythropoietin production. *Proc. Natl. Acad. Sci. USA* **80,** 4008–4011.

Lacombe, C., Da Silva, J. L., Bruneval, P., Fournier, J. G., Wendling, F., Casadevall, N., Camilleri, J. P., Bariety, J., Varet, B., and Tamourin, P. (1988). Peritubular cells are the site of erythropoietin synthesis in murine hypoxic kidney. *J. Clin. Invest.* **81,** 620–623.

La Ferla, K., Reimann, C., Jelkmann, W., and Hellwig-Bürgel, T. (2002). Inhibition of erythropoietin gene expression signaling involves the transcription factors GATA-2 and NF-κB. *FASEB. J.* **16,** 1811–1813.

Lee, E. C., Yu, D., Martinez, J., Tessarollo, L., Swing, D. A., Court, D. L., Jenkins, N. A., and Copeland, N. G. (2001). A highly efficient *Escherichia coli*–based chromosome engineering system adapted for recombinogenic targeting and subcloning of BAC DNA. *Genomics* **73,** 56–65.

Lin, F. K., Suggs, S., Lin, C. H., Browne, J. K., Smalling, R., Egrie, J. C., Chen, K. K., Fox, G. M., Martin, F., Stabinsky, Z., Badrawi, S. M., Lai, P. H., and Goldwasser, E. (1985). Cloning and expression of the human erythropoietin gene. *Proc. Natl. Acad. Sci. USA* **82,** 7580–7584.

Loya, F., Yang, Y., Lin, H., Goldwasser, E., and Albitar, M. (1994). Transgenic mice carrying the erythropoietin gene promoter linked to *lacZ* express the reporter in proximal convoluted tubule cells after hypoxia. *Blood* **84,** 1831–1836.

Massoud, T. F., and Gambhir, S. S. (2003). Molecular imaging in living subjects: Seeing fundamental biological processes in a new light. *Genes Dev.* **17,** 545–580.

Maxwell, P. H., Osmond, M. K., Pugh, C. W., Heryet, A., Nicholls, L. G., Tan, C. C., Doe, B. G., Ferguson, D. J., Johnson, M. H., and Ratcliffe, P. J. (1993). Identification of the renal erythropoietin-producing cells using transgenic mice. *Kidney Int.* **44,** 1149–1162.

McGary, E. C., Rondon, I. J., and Beckman, B. S. (1997). Posttranscriptional regulation of erythropoietin mRNA stability by erythropoietin mRNA-binding protein. *J. Biol. Chem.* **272,** 8628–8634.

Mori, S., Sato, T., Morishita, Y., Saito, K., Urabe, A., Wakabayashi, T., and Takaku, F. (1985). Glomerular epithelium as the main locus of erythropoietin in human kidney. *Jpn. J. Exp. Med.* **55,** 69–70.

Morita, M., Ohneda, O., Yamashita, T., Takahashi, S., Suzuki, N., Nakajima, O., Kawauchi, S., Ema, M., Shibahara, S., Udono, T., Tomita, K., Tamai, M., *et al.* (2003). HLF/HIF-2α is a key factor in retinopathy of prematurity in association with erythropoietin. *EMBO J.* **22,** 1134–1146.

O'Connor, P. M. (2006). Renal oxygen delivery: Matching delivery to metabolic demand. *Clin. Exp. Pharmacol. Physiol.* **33,** 961–967.

Onodera, K., Takahashi, S., Nishimura, S., Ohta, J., Motohashi, H., Yomogida, K., Hayashi, N., Engel, J. D., and Yamamoto, M. (1997). GATA-1 transcription is controlled by distinct regulatory mechanisms during primitive and definitive erythropoiesis. *Proc. Natl. Acad. Sci. USA* **94,** 4487–4492.

Plotkin, M. D., and Goligorsky, M. S. (2006). Mesenchymal cells from adult kidney support angiogenesis and differentiate into multiple interstitial cell types including erythropoietin-producing fibroblasts. *Am. J. Physiol. Renal. Physiol.* **291**, F902–F912.

Poellinger, L., and Johnson, R. S. (2004). HIF-1 and hypoxic response: The plot thickens. *Curr. Opin. Genet. Dev.* **14**, 81–85.

Rich, I. N. (1991). The site of erythropoietin production: Localization of erythropoietin mRNA by radioactive *in situ* hybridization. *Blood* **78**, 2469–2471.

Satoh, K, Kagaya, Y., Nakano, M., Ito, Y., Ohta, J., Tada, H., Karibe, A., Minegishi, N., Suzuki, N., Yamamoto, M., Ono, M., Watanabe, J., *et al.* (2006). Important role of endogenous erythropoietin system in recruitment of endothelial progenitor cells in hypoxia-induced pulmonary hypertension in mice. *Circulation* **113**, 1442–1450.

Schuster, S. J., Koury, S. T., Bohrer, W., and Caro, J. (1992). Cellular sites of renal and extrarenal erythropoietin production in anemic rats. *Br. J. Haematol.* **81**, 153–159.

Semenza, G. L., Nejfelt, M. K., Chi, S. M., and Antonarakis, S. E. (1991a). Hypoxia-inducible nuclear factors bind to an enhancer element located 3′ to the human erythropoietin gene. *Proc. Natl. Acad. Sci. USA* **88**, 5680–5684.

Semenza, G. L., Koury, S. T., Nejfelt, M. K., Gearhart, J. D., and Antonarakis, S. E. (1991b). Cell-type-specific and hypoxia-inducible expression of the human erythropoietin gene in transgenic mice. *Proc. Natl. Acad. Sci. USA* **88**, 8725–8729.

Shanks, J. H., Hill, C. M., Lappin, T. R., and Maxwell, A. P. (1996). Localization of erythropoietin gene expression in proximal renal tubular cells detected by digoxigenin-labelled oligonucleotide probes. *J. Pathol.* **179**, 283–287.

Shimizu, R., and Yamamoto, M. (2005). Gene expression regulation and domain function of hematopoietic GATA factors. *Semin. Cell Dev. Biol.* **16**, 129–136.

Shoemaker, C. B., and Mitsock, L. D. (1986). Murine erythropoietin gene: Cloning, expression, and human gene homology. *Mol. Cell. Biol.* **6**, 849–858.

Stockmann, C., and Fandrey, J. (2006). Hypoxia-induced erythropoietin production: A paradigm for oxygen-regulated gene expression. *Clin. Exp. Pharmacol. Physiol.* **33**, 968–979.

Suliman, H. B., Ali, M., and Piantadosi, C. A. (2004). Superoxide dismutase-3 promotes full expression of the EPO response to hypoxia. *Blood* **104**, 43–50.

Suzuki, M., Ohneda, K., Hosoya-Ohmura, S., Tsukamoto, S., Ohneda, O., Philipsen, S., and Yamamoto, M. (2006). Real-time monitoring of stress erythropoiesis *in vivo* using Gata1 and beta-globin LCR luciferase transgenic mice. *Blood* **108**, 726–733.

Suzuki, N., Imagawa, S., Noguchi, C. T., and Yamamoto, M. (2004). Do *β*-globin, GATA-1, or EPOR regulatory domains specifically mark erythroid progenitors in transgenic reporter mice? *Blood* **104**, 2988–2989.

Suzuki, N., Ohneda, O., Takahashi, S., Higuchi, M., Mukai, H. Y., Nakahata, T., Imagawa, S., and Yamamoto, M. (2002). Erythroid-specific expression of the erythropoietin receptor rescued its null mutant mice from lethality. *Blood* **100**, 2279–2288.

Suzuki, N., Suwabe, N., Ohneda, O., Obara, N., Imagawa, S., Pan, X., Motohashi, H., and Yamamoto, M. (2003). Identification and characterization of two types of erythroid progenitors that express GATA-1 at distinct levels. *Blood* **102**, 3575–3583.

Suzuki, T., and Sasaki, R. (1990). Immunocytochemical demonstration of erythropoietin immunoreactivity in peritubular endothelial cells of anemic mouse kidney. *Arch. Histol. Cytol.* **53**, 121–124.

Wang, G. L., and Semenza, G. L. (1993). Characterization of hypoxia-inducible factor 1 and regulation of DNA binding activity by hypoxia. *J. Biol. Chem.* **268**, 21513–21518.

Wang, G. L., and Semenza, G. L. (1995). Purification and characterization of hypoxia-inducible factor 1. *J. Biol. Chem.* **270**, 1230–1237.

Winearls, C. G., Oliver, D. O., Pippard, M. J., Reid, C., Downing, M. R., and Cotes, P. M. (1986). Effect of human erythropoietin derived from recombinant DNA on the anaemia of patients maintained by chronic haemodialysis. *Lancet* **ii**, 1175–1178.

Wu, H., Liu, X., Jaenisch, R., and Lodish, H. F. (1995). Generation of committed erythroid BFU-E and CFU-E progenitors does not require erythropoietin or the erythropoietin receptor. *Cell* **83,** 59–67.

Wu, H., Lee, S. H., Gao, J., Liu, X., and Iruela-Arispe, M. L. (1999). Inactivation of erythropoietin leads to defects in cardiac morphogenesis. *Development* **126,** 3597–3605.

Yu, D., Ellis, H. M., Lee, E. C., Jenkins, N. A., Copeland, N. G., and Court, D. L. (2000). An efficient recombination system for chromosome engineering in *Escherichia coli. Proc. Natl. Acad. Sci. USA* **97,** 5978–5983.

Zanjani, E. D., Ascensao, J. L., McGlave, P. B., Banisadre, M., and Ash, R. C. (1981). Studies on the liver to kidney switch of erythropoietin production. *J. Clin. Invest.* **67,** 1183–1188.

CHAPTER TEN

CONTROL OF ERYTHROPOIETIN GENE EXPRESSION AND ITS USE IN MEDICINE

Wolfgang Jelkmann

Contents

1. Introduction	180
2. Native EPO Gene Expression and its Pharmacologic Stimulation	180
3. EPO Gene Transfer	184
4. Recombinant EPOS	187
5. Conclusions	190
References	191

Abstract

Erythropoietin (EPO) gene expression is under the control of inhibitory (GATA-2, NF-κB) and stimulatory (hypoxia-inducible transcription factor [HIF]-2, hepatocyte nuclear factor [HNF]-4alpha [α]) transcription factors. EPO deficiency is the main cause of the anemia in chronic kidney disease (CKD) and a contributing factor in the anemias of inflammation and cancer. Small, orally active compounds capable of stimulating endogenous EPO production are in preclinical or clinical trials for treatment of anemia. These agents include stabilizers of the HIFs that bind to the EPO enhancer and GATA inhibitors which prevent GATA from suppressing the EPO promoter. While HIF stabilizing drugs may prove useful as inexpensive second-line choices, at present, their side effects—particularly tumorigenicity—preclude their use as first-choice therapy. As an alternative, EPO gene therapy has been explored in animal studies and in trials on CKD patients. Here, a major problem is immunogenicity of *ex vivo* transfected implanted cells and of the recombinant protein produced after *ex vivo* or *in vivo* EPO complementary DNA (cDNA) transfer. Recombinant human EPO (rhEPO) engineered in Chinese hamster ovary (CHO) cell cultures (epoetin α and epoetin beta [β]) and its hyperglycosylated analogue darbepoetin α are established and safe drugs to avoid allogeneic red blood cell transfusion. Gene-activated EPO (epoetin delta [δ]) from human fibrosarcoma cells (HT-1080) has recently been launched for use in CKD. It is important to know the basics of the technologies, production processes, and structural properties of the novel anti-anemic strategies and drugs.

Institute of Physiology, University of Lübeck, Lübeck, Germany

Methods in Enzymology, Volume 435
ISSN 0076-6879, DOI: 10.1016/S0076-6879(07)35010-6

1. INTRODUCTION

The hormone EPO is essential for red blood cell (RBC) production. Human EPO is an acidic glycoprotein with a molecular mass of 30.4 kDa. Its peptide core consists of a single chain of 165 amino acids with specific binding sites for the EPO receptor (EPO-R). The carbohydrate portion (40% of the molecule) comprises three tetra-antennary N-linked (at Asn[24], Asn[38], and Asn[83]) and one small O-linked (at Ser[126]) glycans. Plasma EPO exhibits several isoforms that differ in glycosylation and biological activity. The N-glycans, which possess terminal sialic acid residues, are of major importance for the *in vivo* biological activity of EPO (Weikert *et al.*, 1999). The primary target cells of EPO are the myeloid colony-forming unit erythroids (CFU-Es). EPO enables the CFU-Es to escape from apoptosis and to generate proerythroblasts and normoblasts. The human EPO-R of hemopoietic cells is a 60-kDa glycoprotein. One EPO molecule binds to two EPO-R molecules, which constitute a homodimer.

The kidneys are the main source of circulating EPO in adults, whereas the liver is the primary production site before birth. Within the kidney, cortical fibroblast-like cells express EPO messenger RNA (mRNA); however, attempts have failed to culture these cells *in vitro* for study of the molecular mechanisms of renal EPO synthesis. Because EPO deficiency leads to anemia, patients with Chronic Kidney disease are commonly substituted with rhEPO (epoetin) or its mutated and hyperglycosylated analogue darbepoetin α (Jelkmann, 2007). In addition, therapy with recombinant erythropoietic proteins may be indicated in cancer patients with symptomatic anemia receiving chemotherapy, in anemic patients with AIDS or autoimmune diseases, and in patients scheduled for major elective surgery. In any case, the primary therapeutic goal is to reduce allogeneic RBC transfusions.

More recent strategies are aimed at developing anti-anemic drugs that stimulate endogenous EPO production, performing EPO gene therapy, and producing rhEPO in human cells by gene activation technology. This chapter describes these approaches that have been enabled by the advanced knowledge of the molecular mechanisms of hypoxia-induced EPO gene expression.

2. NATIVE EPO GENE EXPRESSION AND ITS PHARMACOLOGIC STIMULATION

EPO synthesis increases on tissue hypoxia in the kidneys and, to a minor degree, in a few other organs, such as the liver and the brain (Fandrey, 2004). The tissue O_2-pressure (pO_2) in an organism depends on the blood hemoglobin (Hb) concentration, the arterial pO_2, the O_2-affinity of the blood, and the rate of blood flow. The understanding of the molecular mechanisms in control of EPO gene expression *in vivo* is still insufficient. Most of the information described in this chapter has been based on *in vitro* studies utilizing human hepatoma cells (lines Hep3B and HepG2). Significant differences between renal and hepatic EPO gene expression have already been recognized. In contrast to human hepatoma cell lines, which respond in a graded way to hypoxia, the EPO-expressing cells in hypoxic kidneys *in situ* respond in an all-or-nothing fashion (Koury *et al.*, 1989). The hypoxia response elements (HREs) are located upstream (between 9.5 and 14 kb 5' to the EPO gene) in the kidney, but downstream (within 0.7 kb 3' to the EPO gene) in the liver according to studies in transgenic mice (Kochling *et al.*, 1998). The expression of the EPO gene (chromosomal location: 7q22) is under the control of several transcription factors. GATA-2 and NF-κB act on the 5' promoter and inhibit EPO gene expression (La Ferla *et al.*, 2002). GATA-2 binding to the EPO promoter is reduced under hypoxic conditions. For therapeutic purposes, GATA inhibitors are under development that can be taken orally and prevent GATA-2 from suppressing the EPO promoter (Imagawa *et al.*, 2003; Nakano *et al.*, 2004). GATA inhibitors are non-peptidic organic compounds that increase the acetylation of GATA and enhance DNA-binding of HIFs (Imagawa *et al.*, 2003). The GATA inhibitor K-11706 exerts even stronger effects *in vitro* and in experimental animals than the first-generation compound K-7174 (Nakano *et al.*, 2004). Because GATA inhibitors partially reverse the inhibition of EPO gene expression by proinflammatory cytokines, they may become useful, particularly for treatment of the anemias of inflammatory diseases and cancer.

The hepatic hypoxia-inducible EPO enhancer is a 50-basepair element located 3' of the EPO gene and contains at least two transcription factor binding sites. The proximal site of the EPO enhancer binds the HIFs (Pugh *et al.*, 1991; Semenza *et al.*, 1991), which are composed of an O_2-labile α-subunit and a constitutive β-subunit (Wang *et al.*, 1995). HIF-2 (rather than HIF-1) appears to be the main physiological transcription factor inducing EPO gene expression in hypoxia. HIF-2α is dominantly found in the EPO-producing peritubular fibroblasts in rat kidney while HIF-1α is mainly present in renal tubular cells and has different cell-protective functions (Rosenberger *et al.*, 2002, 2005). EPO mRNA induction is almost completely abolished by HIF-2α RNA interference (RNAi), as shown in studies utilizing human hepatoma (Hep3B), human neuroblastoma (Kelly), or mouse brain cortex astrocytic cell lines (Chavez *et al.*, 2006; Warnecke *et al.*, 2004). Evidence suggests that HIF-2α binds preferentially to the HRE within the native EPO enhancer, whereas HIF-1α may bind to isolated HREs (Rankin *et al.*, 2007). The conditional postnatal knockout of HIF-2α, but not of HIF-1α, results in anemia in mice (Gruber *et al.*, 2007).

Renal (Scortegagna *et al.*, 2005) and hepatic (Rankin *et al.*, 2007) EPO production in response to hypoxia is diminished in HIF-2α knockout mice.

The distal site of the hepatic EPO enhancer consists of a direct repeat of two nuclear hormone receptor half sites for HNF-4α, which cooperates with HIF (Galson *et al.*, 1995). Down-regulation of HNF-4α is considered the main mechanism by which cytokines like interleukin 1 inhibit EPO production in human hepatoma cells (Krajewski *et al.*, 2007). For full hypoxic induction of the EPO gene, the 3′ enhancer and the 5′ promoter need to cooperate physically. It is likely that the transcriptional coactivator p300 bridges HIF, HNF-4α, and the EPO promoter for recruitment of histone-acetyltransferase activity (Sanchez-Elsner *et al.*, 2004; Soutoglou *et al.*, 2001).

EPO gene expression is suppressed in normoxia because HIF-α undergoes prolyl and asparaginyl hydroxylation. Prolyl hydroxylation causes binding of von Hippel–Lindau protein (pVHL) E3 ubiquitin ligase complex to HIF-α, which is then degraded in the proteasome. Asparaginyl hydroxylation of HIF-α reduces its transcriptional activity by preventing binding of the coactivator p300. The three HIF-α prolyl hydroxylases (PHDs) and the one asparaginyl hydroxylase (factor-inhibiting HIF [FIH]) identified in human tissues do not only require O_2 for their catalytic action but also Fe^{2+} and 2-oxoglutarate. Accordingly, HIF-α hydroxylation can be prevented by iron depletion or by the application of 2-oxoglutarate–competing analogues (for further information, see Bruegge *et al.* [2007]).

Iron chelators, such as desferrioxamine (DFO) or epolones, which stabilize HIF-α (Gleadle *et al.*, 1995; Linden *et al.*, 2003; Wang and Semenza, 1993; Wanner *et al.*, 2000), promote EPO expression in cell cultures (Kling *et al.*, 1996; Linden *et al.*, 2003; Wanner *et al.*, 2000) as well as in humans *in vivo* (Kling *et al.*, 1996; Ren *et al.*, 2000). However, iron chelators are not suited for stimulation of RBC production in anemic patients because iron is required for heme synthesis and it is an important constituent of many proteins. Actually, DFO is administered clinically to reduce iron overload in patients who need RBC transfusions on a regular basis. Divalent transition metals, such as Co^{2+} or Ni^{2+}, can induce EPO gene expression under normoxic conditions by displacing Fe^{2+} from the HIF-α hydroxylases (Salnikow *et al.*, 2000; Topol *et al.*, 2006). Co^{2+} and Ni^{2+} may also cause depletion of the ascorbate (Salnikow *et al.*, 2004) that is required to reduce Fe^{3+} to Fe^{2+} within the catalytic site of the HIF-α PHDs and of FIH. Note that ascorbate *per se* does not stimulate EPO production in hypoxia-responsive hepatoma cell culture models (Jelkmann *et al.*, 1997). Apart from inhibiting the HIF-α hydroxylases, Co^{2+} can bind to HIF-2α, thereby preventing its degradation (Yuan *et al.*, 2003). Because cobalt stimulates EPO production and erythropoiesis in a dose-dependent way, the international EPO standard was originally calibrated against cobalt, with

1 unit (U) of EPO producing the same erythropoiesis-stimulating response in experimental animals as 5-μmol cobaltous chloride (for further information, see Jelkmann [2003]). Although definitive information on the health risks of cobalt in humans is not available (Lippi *et al.*, 2005) cobalt therapy has been abandoned for treatment of anemia because of its toxicity.

Distinct 2-oxoglutarate analogues (medical jargon: HIF stabilizers) prevent the degradation of HIF-α and the inactivation of HIF by inhibiting the HIF-α PHDs and FIH (Jaakkola *et al.*, 2001; Maxwell *et al.*, 1999). HIF stabilizers fail to support HIF-α hydroxylation, which requires the concomitant translocation of the second oxygen atom into the carbon-2 of 2-oxoglutarate to form succinate and CO_2. In several HIF stabilizing analogues, the CH_2-group present in position 3 of 2-oxoglutarate is replaced by an NH_2-group. Among the HIF stabilizers are simple molecules, such as N-oxalylglycine, ethyl-3,4-dihydroxybenzoate, or L-mimosine, that are easy to synthesize and are active when taken orally. HIF-α hydroxylase inhibitors for clinical use have been developed primarily by FibroGen (San Francisco, CA). The chemical structure of the compounds presently under study (FG-.....) can be found elsewhere (Klaus *et al.*, 2005). The 2-oxoglutarate analogue FG-4383 stimulates EPO production in normal and in nephrectomized mice (Safran *et al.*, 2006). The superior compound FG-2216 increases plasma EPO levels and reticulocyte numbers in humans, as shown in a phase I study on 54 healthy volunteers (Urquilla *et al.*, 2004), and raises Hb in persons suffering from CKD, as shown in a phase II study on five predialysis patients (Wiecek *et al.*, 2005). It is not known whether the EPO produced by CKD patients treated with HIF stabilizers derives from remnant intact renal tissue or from other organs, such as the liver. This issue will be critical when electing patients who are likely to respond to the therapy with HIF stabilizers. Note that CKD patients often receive intravenous (IV) iron therapy, which will activate the PHDs and FIH and thus counteract the effectiveness of 2-oxoglutarate competitors. HIF stabilizers induce the expression of other genes coding proteins important in erythropoiesis, such as transferrin and ferroportin (Klaus *et al.*, 2005).

Notably, the clinical use of the present 2-oxoglutarate analogues for treatment of anemia will be associated with the expression of many other HIF-dependent genes, which may cause unwanted effects. A major concern is the tumorigeneic potential of the HIF stabilizers because HIFs activate several genes coding proteins (e.g., vascular endothelial growth factor [VEGF]) that promote tumor growth (O'donnell *et al.*, 2006; Welsh *et al.*, 2006). Thus, more specific drugs should be developed. First, it will be important to design cell membrane-permeable HIF stabilizers, which would require lower dosing. For example, the 2-oxoglutarate analogue 2,4-pyridine dicarboxylic (PDC) acid, which is not membrane-permeable, has been successfully esterified to di-*tert*-butyroyl-oxymethyl-PDC (^{t}Bu-2,4 PDC) to increase its potency to stimulate EPO gene expression in cell cultures. ʹBu-2,4 PDC acts as a pro-drug that is intracellularly activated by enzymatic hydrolysis. It is effective at 250-fold lower concentrations compared to PDC (Doege et al., 2006). Second, for clinical use, it is desirable to have 2-oxoglutarate analogues at hand that preferentially stabilize HIF-2α and increase EPO production. Evidence for a hierarchy of HIF-responsive genes can be deduced from the symptoms of Chuvash disease, in which erythrocytosis is the main manifestation, although the mutation of the pVHL gene leads to ubiquituous HIF-α accumulation (Ang et al., 2002). Third, it will be of major importance to develop HIF stabilizers that allow for a tissue-specific delivery of the drugs, thus promoting HIF-2 dependent HRE induction in EPO-expressing cells only. However, even then, important questions as to the safety of the drugs would remain unanswered. For example, EPO is produced in the eye, and increasing this production could worsen diabetic retinopathy because EPO promotes angiogenesis (Watanabe et al., 2005). On the other hand, stimulation of EPO production in the brain, where astrocytes are the main type of cells secreting EPO, may be beneficial in patients with stroke (Ehrenreich et al., 2002) or schizophrenia (Ehrenreich et al., 2006). Compared with EPO (Jumbe, 2002), the small, non-peptidic HIF stabilizers can probably cross the blood–brain barrier more readily.

3. EPO Gene Transfer

In vivo or *ex vivo*, EPO gene therapy is an attractive alternative to the administration of erythropoietic drugs. *In vivo* approaches often utilize replication-deficient recombinant viral vectors to transfer genes like the EPO gene with a high level expression potential. For *ex vivo* transfer, autologous or homologous cells in culture are transferred into the individual after transfection (Naffakh and Danos, 1996).

Effectivity of *in vivo* EPO gene transfer was first shown with one-time administration of AdMLP.Epo-injected intraperitoneally (i.p.) or subcutaneously (s.c.) in cotton rats (Setoguchi et al., 1994). AdMLP.Epo is constructed by deleting the majority of E1 from adenovirus type 5 and replacing E1 with an expression cassette containing the adenovirus type 5 major late promoter (MLP) and the human EPO gene, including the 3ʹ *cis*-acting HRE, to increase transcription under hypoxic conditions. Indeed, a major challenge in human EPO gene therapy is to adapt the expression of the transgene to the physiological needs of erythropoiesis. EPO overexpression would result in erythrocytosis. Binley et al. (2002) have developed an adenoviral vector that expresses murine EPO under the control of a synthetic HRE multimer referred to as OBHRE (Oxford Biomedica HRE), which corrects hematocrit (Hct) in anemic mice to the physiological level, but does not cause erythrocytosis (Binley *et al.*, 2002). Compared with the direct plasmid DNA injection intramuscular (i.m.), electrotransfer by means of controlled electrical field pulses to promote plasmid DNA transfer into myotubes gives rise to much higher and more regular EPO expression (Kreiss *et al.*, 1999). Muscle-targeted electroporation-mediated transfer of plasmid-expressing rat EPO corrects anemia in uremic, subtotally nephrectomized rats (Ito *et al.*, 2001). Sustained EPO production in uremic rats is achieved on i.m. injection of lentiviral vector carrying EPO cDNA (Oh *et al.*, 2006). Since adenoviral i.m. transfer of mouse EPO cDNA under the control of the cytomegalovirus (CMV) promoter into leg muscles of β-thalassemic mice proved to increase β-minor globin synthesis and to correct anemia (Bohl *et al.*, 2000), it was important to show that similar results are obtained on repeated i.m. injections of such plasmids by electrotransfer (Payen *et al.*, 2001). The electroporation approach can also be applied for gene delivery in skin, as shown in a rat study (Maruyama *et al.*, 2001). Kidney-targeted EPO gene transfer is also possible, as demonstrated with rat EPO cDNA inserted into the *Xho*I site of the pCAGGS expression vector and transferred by retrograde renal vein injection in rats (Maruyama *et al.*, 2002). Plasmid DNA expression can be increased when saline solution is replaced by sodium phosphate as the vehicle for plasmid DNA infection (Hartikka *et al.*, 2000). While skeletal muscle generally has been used as the target organ for *in vivo* EPO gene transfer, an alternative method has been developed with adipose tissue for transduction in applying membranotropic polymeric surfactants (Pluronic F88) for adenoviral EPO gene transfer (Mizukami *et al.*, 2006). Adipose tissue has several advantages, including its easy accessibility, the lack of cell divisions, and the possibility to harmlessly remove the transfected tissue when unexpected events occur.

Nonviral (Sebestyen *et al.*, 2007) and adenoviral (Rivera *et al.*, 2005) EPO gene transfer has also been preformed in nonhuman primates. Note that the structure of the N-glycans of EPO generated on homologous *in vivo* adenoviral gene transfer into muscle differs from that of the N-glycans of native EPO, as shown in an isoelectric focusing (IEF) analysis of serum EPO from cynomolgus macaques (Lasne *et al.*, 2004). Moreover, when homologous EPO cDNA was administered to macaques in different ectopic organs (skeletal muscle and lung) via recombinant adeno-associated virus, severe anemia developed in many animals after 1 to 4 months (Chenuaud *et al.*, 2004; Gao *et al.*, 2004). Thus, this kind of genetic intervention can break tolerance to a self-antigen, resulting in autoimmune anemia. These findings are of general importance because they demonstrate that gene transfer studies in inbred rodent models may lead to results that cannot be transferred to humans. Apparently, autoimmune anemias do not occur in rodents on adenoviral EPO gene transfer.

The seminal *ex vivo* EPO gene therapy studies were performed with fibroblasts transformed in mice with mouse EPO cDNA in a Moloney-driven retrovirus vector under control of the phosphoglycerate kinase (PGK) promoter (Naffakh *et al.*, 1995) and in rats with rat vascular smooth muscle cells transformed with an *Eco*RI/*Bam*HI fragment of the rat EPO cDNA inserted into retroviral plasmid LXSN (Osborne *et al.*, 1995). Transplantation of a C_2 myoblast cell clone that stably secretes high levels of human EPO proved to increase Hct in mice with experimental renal failure (Hamamori *et al.*, 1995). To keep tabs on long-term delivery of EPO from transfected myoblasts, a tetracycline-inducible promoter controlling EPO cDNA expression can be introduced into retroviral vectors for transfer (Bohl and Heard, 1997; Bohl *et al.*, 1997). On/off switching of EPO production depends on the addition of the tetracycline derivative doxycycline to the drinking water of the transplanted animals. To achieve oxygen-dependent EPO expression, murine C_2C_{12} fibroblasts have been successfully transfected with a vector containing human EPO cDNA driven by the hypoxia-responsive promoter of the murine PGK gene. Cells transfected in this way produce more EPO on encapsulation and s.c. implantation in hypoxemic than in normoxemic mice (Rinsch *et al.*, 1997). In cell culture, the rate of the expression of the EPO gene in transfected cells can be modulated by iron donors and iron chelators (Dalle *et al.*, 2000), which indicates that EPO gene expression is HIF-dependent.

Another important question relates to the immune response to transplanted autologous, homologous (allogeneic), or xenogeneic cells in *ex vivo* gene therapy approaches. Schneider *et al.* (2001) have shown that the survival of C_2C_{12} mouse fibroblasts can be prolonged from a few days to at least 1 month when the cells are co-transfected with transcription elements coding CTLA4Ig, a soluble factor blocking T-cell activation. In a nonhuman primate model, baboon mesenchymal stem cells transfected with the human EPO gene, and homologously implanted s.c. via an inert immunoisolatory device (TheraCyte, Laguna Hills, CA) produced human EPO in the recipients for 1 to 22 weeks after implantation (Bartholomew *et al.*, 2001). A promising procedure to prevent cell-to-cell contact of implanted *ex vivo* transfected cells with the host is their enclosure in semipermeable polymer microcapsules (Orive *et al.*, 2005; Rinsch *et al.*, 2002). While recognition of the non-self cells by the host is prevented by this procedure, the secretory product may still be immunogenic. Administration of anti-CD4[+] monoclonal antibody beginning at the time of cell implantation can prevent the decrease in Hct due to the production of neutralizing antibodies towards EPO, as shown in a homologous gene transfer study with C_2C_{12} myoblasts genetically engineered to produce mouse EPO (Rinsch *et al.*, 2002).

In view of the studies showing that anti-EPO antibodies develop in nonhuman primates on *in vivo* adenoviral homologous EPO gene transfer (Chenuaud *et al.*, 2004; Gao *et al.*, 2004), an autologous *ex vivo* approach was chosen in the first human EPO gene therapy trial on patients with CKD (Lippin *et al.*, 2005). From each of the 10 patients, an individual dermal core sample was excised and incubated with EPO cDNA inserted into a pAd-lox shuttle vector containing the CMV promoter and simian virus-40 polyA site (Ad-MG/EPO-1; Ad5 E1/E3 deleted) in 24-well plates for transfection. Following several days in culture, the EPO production rate of the dermal cores was determined for prior dosing of the so-called "Biopumps," which were reimplanted under the abdominal skin 9 to 10 days after harvest. Serum EPO levels peaked in most cases on day 3 and then decreased, reaching baseline levels, likely due to immunologic rejection of the transplants. Anti-EPO antibodies were not detected until day 90 following transplantation. The transient EPO increase produced a reticulocytosis, but was not sufficient to raise Hb levels (Lippin *et al.*, 2005). This study indicates that the implantation of genetically modified autologous tissue into human dermis can be used to deliver recombinant human proteins. However, major problems still need to be overcome, namely difficulties in the dosing of the delivered protein and the apparent immunogenicity of the transplants.

While EPO gene therapy appears less promising for anemia treatment, it may become suitable for site-directed local short-term delivery of EPO for tissue protection. Convincing evidence has been provided for the expression of functional EPO-R outside hemopoietic tissues. EPO protects cells from apoptosis due to hypoxia and inflammatory mediators in many organs, including brain, heart, kidney, and liver (Brines and Cerami, 2006; Jelkmann and Wagner, 2004; Sasaki, 2003). Very recently, a hypoxia-inducible luciferase reporter plasmid was constructed with the EPO 3′ untranslated region, which can enhance the stability of EPO mRNA. Luciferase reporter activity increased on injection of the plasmid into injured—but not normal—rat spinal cord (Choi *et al.*, 2007).

4. RECOMBINANT EPOS

Therapeutic human glycoproteins are usually engineered in mammalian host cells because prokaryotes do not glycosylate proteins. Transformed bacteria, such as *Escherichia coli*, are only useful for engineering non-glycosylated recombinant proteins. Yeast and filamentous fungi glycosylate proteins differently from mammalian cells, adding, for example, high mannose-type N-glycans (Sodoyer, 2004). The synthesis of N-glycans in mammalian cells is a complicated process, which starts with the addition of distinct sugar molecules to dolichol in the cytosol, from where the glycans are transferred to the growing polypeptide in the endoplasmic reticulum. After the protein is folded and several sugars are removed from the glycans,

N-acetylglucosamine, galactose, and sialic acid (N-acetylneuraminic acid) are eventually added in the Golgi complex (Helenius and Aebi, 2001). Note, however, that, very recently, a yeast cell line could be "humanized" to produce functional sialylated recombinant EPO by means of knockout of four genes to eliminate yeast-specific glycosylation and introduction of 14 heterologous genes coding human enzymes for N-glycan synthesis (Hamilton *et al.*, 2006). This novel approach is exciting because the manufacture of human glycoproteins in yeast offers several advantages, including the use of chemically defined media for culture and high protein production rates.

The first-generation rhEPO preparations (epoetin α and epoetin β) have been manufactured in cultures of CHO (*Cricetulus griseus*) cells transfected with the human EPO gene or EPO cDNA (Lin *et al.*, 1985). rhEPO (epoetin omega [ω]) engineered in transfected baby hamster kidney (BHK, *Mesocricetus auratus*) cell cultures are used clinically in some countries outside the European Union and Northern America. Recombinant glycoproteins manufactured in heterologous eukaryotic host cells are heterogenous with respect to the composition of their glycans. The main factors determining the type of glycan isoforms are the type of host cell, the culturing conditions, and the purification procedures (Wurm, 2004). For example, the CHO cell "line," which is commonly used for the production of biopharmaceuticals, truly represents a family of derivative immortalized and transfected cell sub-lines with frequent chromosomal aberrations (Derouazi *et al.*, 2006). rhEPOs produced in CHO or other mammalian cell line cultures are not single entities, but are a mixture of EPO isoforms that can be separated by IEF. As shown for CHO cell-derived rhEPO, these isoforms differ from each other by their sialic acid content (Strickland *et al.*, 2004). For the manufacture of clinically used rhEPO, isoforms with a strong *in vivo* biological activity are selectively purified by preparative methods, such as ion exchange chromatography, reverse phase chromatography, and filtration (Strickland *et al.*, 2004). Ongoing research is attempted to improve rhEPO purification by application of differential precipitation, hydrophobic interaction chromatography, tandem exchange chromatographies, and molecular size exclusion chromatography (Carcagno *et al.*, 2006).

All epoetins have an amino acid sequence similar to that of endogenous EPO, while the Greek letters are added to indicate that the products vary in glycosylation pattern (World Health Organization, 2006). The N-glycans of EPO have a major role in secretion, molecular stability, solubility, EPO-R–binding affinity, and elimination (Takeuchi and Kobata, 1991). The introduction of new N-glycans into the EPO backbone increases its *in vivo* survival (Sinclair and Elliott, 2005). The CHO cell–derived hyperglycosylated rhEPO analogue darbepoetin α possesses two extra N-glycans as a result of site-directed mutagenesis for the exchange of five amino acids. Darbepoetin α has a three-fold longer half-life compared to epoetin α and epoetin β, which have a plasma half life of 6 to 8 h (Egrie and Browne, 2001).

The biological activity of 200 U/μg rhEPO peptide core corresponds to that of 1 μg darbepoetin α peptide (Jelkmann, 2002).

Very recently, epoetin δ has been launched for treatment of CKD patients (Kwan and Pratt, 2007; Spinowitz and Pratt, 2006). Epoetin δ is produced in cultured human fibrosarcoma cell cultures (line R223, a derivative of the HT-1080 cell line) by homologous recombination and "gene activation," a process that was developed by Transkaryotic Therapies, Inc. (TKT; Cambridge, MA). Hence, epoetin δ is also called gene-activated EPO (GA-EPO) because the expression of the dormant native human EPO gene is switched on by transfection of the cells with the CMV promoter. In detail, the DNA targeting construct (pREPO22) is composed of seven major sequences: some 3000 nucleotides upstream of the EPO gene, a dihydrofolate reductase (DHFR) gene to provide resistance against methotrexate, the CMV promoter for activation of the endogenous coding regions, a sequence encoding human growth hormone (hGH), the neomycin phosphotransferase (NEO) gene for selection of stably transfected cells, an EPO splice donor site to ensure correct processing of the primary transcript, and another targeting sequence of some 1800 nucleotides upstream of the EPO gene (outlined in Kirin v. Hoechst [2002]). After homologous recombination of the pREPO22 construct and the endogenous DNA, the CMV promoter and the hGH cDNA fragment are integrated into the genomic DNA upstream of the native promoter of the EPO gene. Following several steps of mRNA splicing, the translation results in an EPO precursor whose leader peptide is removed, and recombinant EPO is secreted with an amino acid sequence that is identical to that of endogenous human EPO. With respect to the N-glycans, however, it is unlikely that epoetin δ is identical to endogenous human EPO because the structure of N-glycans is not only species-specific, but also tissue-specific. Earlier studies have shown that rhEPO from a human lymphoblastoid cell line (RPMI 1788) transfected with the human EPO gene possesses N-glycans with unusual characteristics (Cointe et al., 2000). In contrast to CHO or BHK cell–derived rhEPOs, epoetin δ does not possess N-glycolylneuraminic acid (Neu5Gc) because—in contrast to other mammals, including great apes—humans are genetically unable to produce Neu5Gc due to an evolutionary mutation (Tangvoranuntakul et al., 2003). All normal humans have circulating antibodies against Neu5Gc that were raised against the Neu5Gc contained in food. In any case, Neu5Gc present in recombinant proteins is not considered to be of immunogeneic relevance. The prevalence of the so-called Hanganutziu-Deicher antibody, which recognizes Neu5Gc moiety of sialoglycoconjugates, is not increased in patients receiving CHO cell–derived rhEPO (Noguchi et al., 1996). Thus, while the gene activation technology is biotechnologically a fascinating approach, the therapeutic advantages of the use of epoetin δ over that of the established drugs epoetin α, epoetin β, and darbepoetin α still need to be demonstrated.

The "gene activation" process has posed interesting legal questions with respect to granted patents on isolated DNA sequences and the recombinant proteins therefrom (isolating the human EPO gene, introducing it into a cloning vector, and using the insert for engineering recombinant EPO in host cells). For example, with respect to the patents related to rhEPO, a major disputed point was whether a "host cell" must be understood as either a host to an EPO encoding DNA sequence that it does not contain naturally or to a foreign DNA sequence not coding EPO but activating its production. Other litigations addressed "product by product claims" (Kirin v. Hoechst, 2002). The results of such assessments have varied, even from jurisdiction to jurisdiction in Europe and the United States (Basheer, 2005).

 ## 5. CONCLUSIONS

The conventional rhEPO preparations (epoetin α, epoetin β) and their hyperglycosylated analogue (darbepoetin α) have proved to be safe and effective drugs for treatment of anemia, primarily those associated with CKD and cancer in combination with chemotherapy. Epoetin α and epoetin β have been used in clinical settings for two decades (Jelkmann, 2007). The only major disadvantages of the use of recombinant erythropoiesis-stimulating agents are the need for regular parenteral administration and the high costs. It is mainly for these reasons that novel anti-anemic drugs and therapies are explored (Table 10.1).

Table 10.1 Anti-anemic drugs and techniques based on EPO gene expression

GATA inhibitors
HIF stabilizers
Iron chelators
Cobalt
2-Oxoglutarate competing analogues
EPO gene therapy
In vivo
Ex vivo
Recombinant erythropoietic proteins
rhEPOs (Epoetin alpha, Epoetin beta, Epoetin omega, Epoetin delta)
Mutated hyperglycosylated rhEPO (Darbepoetin alpha)
Copy rhEPOs, biosimilars and counterfeit products
Continuous erythropoiesis receptor activator (C.E.R.A.)
EPO fusion proteins

GATA inhibitors and HIF stabilizers are small, orally active chemical compounds that stimulate endogenous EPO production. So far, only HIF stabilizers have been used in humans. The present 2-oxoglutarate competitors affect many genes and tissues apart from stimulating EPO production. Until more specific HIF-2α–stabilizing and EPO gene–activating 2-oxoglutarate analogues are at hand, concern remains about the tumorigeneic effect of this class of drugs. Eventually, a better insight into the molecular mechanisms of the structure and action of the pO$_2$-dependent HIF hydroxylases, transcription factors, and targeted DNA elements may enable it to develop drugs that are at least useful as inexpensive second-line antianemic therapeutics.

Even more doubt exists as to whether EPO gene therapy may reach routine clinical practice. Here, the immunogenicity of transplanted transfected cells and of the secreted transgenic EPO is the major problem. In addition, regulation of the expression of the transgene requires fine adjustment to the red cell demands of the organism. On the other side, there are no obvious advantages of EPO gene therapy over the established therapy of anemias with exogenously administered erythropoiesis-stimulating drugs with respect to versatility, effectiveness, safety, and costs. In addition, several novel erythropoietic drugs with improved pharmacokinetic properties are under development (see Table 10.1), including pegylated epoetins and EPO analogues, EPO fusion proteins, and EPO mimetics (Bunn, 2007; Jelkmann, 2007; MacDougall and Eckardt, 2006).

REFERENCES

Ang, S. O., Chen, H., Hirota, K., Gordeuk, V. R., Jelinek, J., Guan, Y., Liu, E., Sergueeva, A. I., Miasnikova, G. Y., Mole, D., Maxwell, P. H., Stockton, D. W., *et al.* (2002). Disruption of oxygen homeostasis underlies congenital Chuvash polycythemia. *Nat. Genet.* **32,** 614–621.

Bartholomew, A., Patil, S., Mackay, A., Nelson, M., Buyaner, D., Hardy, W., Mosca, J., Sturgeon, C., Siatskas, M., Mahmud, N., Ferrer, K., Deans, R., *et al.* (2001). Baboon mesenchymal stem cells can be genetically modified to secrete human erythropoietin *in vivo*. *Hum. Gene Ther.* **12,** 1527–1541.

Basheer, S. (2005). Block me not: Are patented genes "essential facilities"? *ExpressO Preprint Series* Paper 577.

Binley, K., Askham, Z., Iqball, S., Spearman, H., Martin, L., de Alwis, M., Thrasher, A. J., Ali, R. R., Maxwell, P. H., Kingsman, S., and Naylor, S. (2002). Long-term reversal of chronic anemia using a hypoxia-regulated erythropoietin gene therapy. *Blood* **100,** 2406–2413.

Bohl, D., and Heard, J. M. (1997). Modulation of erythropoietin delivery from engineered muscles in mice. *Hum. Gene Ther.* **8,** 195–204.

Bohl, D., Naffakh, N., and Heard, J. M. (1997). Long-term control of erythropoietin secretion by doxycycline in mice transplanted with engineered primary myoblasts. *Nat. Med.* **3,** 299–305.

Bohl, D., Bosch, A., Cardona, A., Salvetti, A., and Heard, J. M. (2000). Improvement of erythropoiesis in β-thalassemic mice by continuous erythropoietin delivery from muscle. *Blood* **95,** 2793–2798.

Brines, M., and Cerami, A. (2006). Discovering erythropoietin's extra-hematopoietic functions: Biology and clinical promise. *Kidney Int.* **70,** 246–250.

Bruegge, K., Jelkmann, W., and Metzen, E. (2007). Hydroxylation of hypoxia-inducible transcription factors and chemical compounds targeting the HIF-hydroxylases. *Curr. Med. Chem.* **14,** 1853–1862.

Bunn, H. F. (2007). New agents that stimulate erythropoiesis. *Blood* **109,** 868–873.

Carcagno, C. M., Criscuolo, M. E., Melo, M. E., and Vidal, J. A. (2006). Methods of purifying recombinant human erythropoietin from cell culture supernatants No. 09830964; filed on 1999–11–08. US Patent 7012130.

Chavez, J. C., Baranova, O., Lin, J., and Pichiule, P. (2006). The transcriptional activator hypoxia inducible factor 2 (HIF-2/EPAS-1) regulates the oxygen-dependent expression of erythropoietin in cortical astrocytes. *J. Neurosci.* **26,** 9471–9481.

Chenuaud, P., Larcher, T., Rabinowitz, J. E., Provost, N., Cherel, Y., Casadevall, N., Samulski, R. J., and Moullier, P. (2004). Autoimmune anemia in macaques following erythropoietin gene therapy. *Blood* **103,** 3303–3304.

Choi, B. H., Ha, Y., Ahn, C. H., Huang, X., Kim, J. M., Park, S. R., Park, H., Park, H. C., Kim, S. W., and Lee, M. (2007). A hypoxia-inducible gene expression system using erythropoietin $3'$ untranslated region for the gene therapy of rat spinal cord injury. *Neurosci. Lett.* **412,** 118–122.

Cointe, D., Beliard, R., Jorieux, S., Leroy, Y., Glacet, A., Verbert, A., Bourel, D., and Chirat, F. (2000). Unusual N-glycosylation of a recombinant human erythropoietin expressed in a human lymphoblastoid cell line does not alter its biological properties. *Glycobiology* **10,** 511–519.

Dalle, B., Payen, E., and Beuzard, Y. (2000). Modulation of transduced erythropoietin expression by iron. *Exp. Hematol.* **28,** 760–764.

Derouazi, M., Martinet, D., Besuchet, S. N., Flaction, R., Wicht, M., Bertschinger, M., Hacker, D. L., Beckmann, J. S., and Wurm, F. M. (2006). Genetic characterization of CHO production host DG44 and derivative recombinant cell lines. *Biochem. Biophys. Res. Commun.* **340,** 1069–1077.

Doege, K., Jelkmann, W., and Metzen, E. (2006). Synthesis of a new cell permeable HIF hydroxylase inhibitor [abstract 13232]. *Acta Physiol.* **86** (Suppl. 1)**,** 141.

Egrie, J. C., and Browne, J. K. (2001). Development and characterization of novel erythropoiesis stimulating protein (NESP). *Br. J. Cancer* **84,** 3–10.

Ehrenreich, H., Hasselblatt, M., Dembrowski, C., Cepek, L., Lewczuk, P., Stiefel, M., Rustenbeck, H.-H., Breiter, N., Jacob, S., Knerlich, F., Bohn, M., Poser, W., *et al.* (2002). Erythropoietin therapy for acute stroke is both safe and beneficial. *Mol. Med.* **8,** 495–505.

Ehrenreich, H., Hinze-Selch, D., Stawicki, S., Aust, C., Knolle-Veentjer, S., Wilms, S., Heinz, G., Erdag, S., Jahn, H., Degner, D., Ritzen, M., Mohr, A., *et al.* (2006). Improvement of cognitive functions in chronic schizophrenic patients by recombinant human erythropoietin. *Mol. Psychiatry* **12,** 206–220.

Fandrey, J. (2004). Oxygen-dependent and tissue-specific regulation of erythropoietin gene expression. *Am. J. Physiol. Regul. Integr. Comp. Physiol.* **286,** R977–R988.

Galson, D. L., Tsuchiya, T., Tendler, D. S., Huang, L. E., Ren, Y., Ogura, T., and Bunn, H. F. (1995). The orphan receptor hepatic nuclear factor 4 functions as a transcriptional activator for tissue-specific and hypoxia-specific erythropoietin gene expression and is antagonized by EAR3/COUP-TF1. *Mol. Cell. Biol.* **15,** 2135–2144.

Gao, G., Lebherz, C., Weiner, D. J., Grant, R., Calcedo, R., McCullough, B., Bagg, A., Zhang, Y., and Wilson, J. M. (2004). Erythropoietin gene therapy leads to autoimmune anemia in macaques. *Blood* **103,** 3300–3302.

Gleadle, J. M., Ebert, B. L., Firth, J. D., and Ratcliffe, P. J. (1995). Regulation of angiogenic growth factor expression by hypoxia, transition metals, and chelating agents. *Am. J. Physiol.* **268,** C1362–C1368.

Gruber, M., Hu, C. J., Johnson, R. S., Brown, E. J., Keith, B., and Simon, M. C. (2007). Acute postnatal ablation of Hif-2α results in anemia. *Proc. Natl. Acad. Sci. USA* **104,** 2301–2306.

Hamamori, Y., Samal, B., Tian, J., and Kedes, L. (1995). Myoblast transfer of human erythropoietin gene in a mouse model of renal failure. *J. Clin. Invest.* **95,** 1808–1813.

Hamilton, S. R., Davidson, R. C., Sethuraman, N., Nett, J. H., Jiang, Y., Rios, S., Bobrowicz, P., Stadheim, T. A., Li, H., Choi, B. K., Hopkins, D., Wischnewski, H., et al. (2006). Humanization of yeast to produce complex terminally sialylated glycoproteins. *Science* **313,** 1441–1443.

Hartikka, J., Bozoukova, V., Jones, D., Mahajan, R., Wloch, M. K., Sawdey, M., Buchner, C., Sukhu, L., Barnhart, K. M., Abai, A. M., Meek, J., Shen, N., et al. (2000). Sodium phosphate enhances plasmid DNA expression *in vivo. Gene Ther.* **7,** 1171–1182.

Helenius, A., and Aebi, M. (2001). Intracellular functions of N-linked glycans. *Science* **291,** 2364–2369.

Imagawa, S., Nakano, Y., Obara, N., Suzuki, N., Doi, T., Kodama, T., Nagasawa, T., and Yamamoto, M. (2003). A GATA-specific inhibitor (K-7174) rescues anemia induced by IL-1β, TNF-α, or L-NMMA. *FASEB J.* **17,** 1742–1744.

Ito, Y., Ataka, K., Gejyo, F., Higuchi, N., Ito, Y., Hirahra, H., Imazeki, I., Hirata, M., Ichikawa, F., Neichi, T., Kikuchi, H., Sugawa, M., et al. (2001). Long-term production of erythropoietin after electroporation-mediated transfer of plasmid DNA into the muscles of normal and uremic rats. *Gene Ther.* **8,** 461–468.

Jaakkola, P., Mole, D. R., Tian, Y. M., Wilson, M. I., Gielbert, J., Gaskell, S. J., Kriegsheim, A., Hebestreit, H. F., Mukherji, M., Schofield, C. J., Maxwell, P. H., Pugh, C. W., et al. (2001). Targeting of HIF-α to the von Hippel-Lindau ubiquitylation complex by O$_2$-regulated prolyl hydroxylation. *Science* **292,** 468–472.

Jelkmann, W. (2002). The enigma of the metabolic fate of circulating erythropoietin (Epo) in view of the pharmacokinetics of the recombinant drugs rhEpo and NESP. *Eur. J. Haematol.* **69,** 265–274.

Jelkmann, W. (2003). Biochemistry and assays of Epo. *In* "Erythropoietin: Molecular Biology and Clinical Use" (W. Jelkmann, ed.), pp. 35–63. FP Graham Publishing Co., Johnson City, TN.

Jelkmann, W. (2007). Erythropoietin after a century of research: Younger than ever. *Eur. J. Haematol.* **78,** 183–205.

Jelkmann, W., and Wagner, K. (2004). Beneficial and ominous aspects of the pleiotropic action of erythropoietin. *Ann. Hematol.* **83,** 673–686.

Jelkmann, W., Pagel, H., Hellwig, T., and Fandrey, J. (1997). Effects of antioxidant vitamins on renal and hepatic erythropoietin production. *Kidney Int.* **51,** 497–501.

Jumbe, N. L. (2002). Erythropoietic agents as neurotherapeutic agents: What barriers exist? *Oncology* **16,** 91–107.

Kirin v. Hoechst (2002). EWCA Civ. 1096. Accessible atwww.hmcourts-service.gov.uk/judgmentsfiles/j1329/Kirin_v_Hoechst.htm.

Klaus, S. J., Molineaux, C. J., Neff, T. B., Guenzler-Pukall, V., Langsetmo Parobok, I., Seeley, T. W., and Stephenson, R. C. (2005). Enhanced erythropoiesis and iron metabolism No. 861590; filed on 2004–06–06. US Patent 20050020487.

Kling, P. J., Dragsten, P. R., Roberts, R. A., Dos-Santos, B., Brooks, D. J., Hedlund, B. E., and Taetle, R. (1996). Iron deprivation increases erythropoietin production *in vitro,* in normal subjects and patients with malignancy. *Br. J. Haematol.* **95,** 241–248.

Kochling, J., Curtin, P. T., and Madan, A. (1998). Regulation of human erythropoietin gene induction by upstream flanking sequences in transgenic mice. *Br. J. Haematol.* **103,** 960–968.

Koury, S. T., Koury, M. J., Bondurant, M. C., Caro, J., and Graber, S. E. (1989). Quantitation of erythropoietin-producing cells in kidneys of mice by *in situ* hybridization: Correlation with hematocrit, renal erythropoietin mRNA, and serum erythropoietin concentration. *Blood* **74,** 645–651.

Krajewski, J., Batmunkh, C., Jelkmann, W., and Hellwig-Bürgel, T. (2007). Interleukin-1β inhibits the hypoxic inducibility of the erythropoietin enhancer by suppressing hepatocyte nuclear factor-4α. *Cell. Mol. Life Sci.* **64,** 989–998.

Kreiss, P., Bettan, M., Crouzet, J., and Scherman, D. (1999). Erythropoietin secretion and physiological effect in mouse after intramuscular plasmid DNA electrotransfer. *J. Gene Med.* **1,** 245–250.

Kwan, J. T., and Pratt, R. D. (2007). Epoetin delta, erythropoietin produced in a human cell line, in the management of anaemia in predialysis chronic kidney disease patients. *Curr. Med. Res. Opin.* **23,** 307–311.

La Ferla, K., Reimann, C., Jelkmann, W., and Hellwig-Bürgel, T. (2002). Inhibition of erythropoietin gene expression signaling involves the transcription factors GATA-2 and NF-κB. *FASEB J.* **16,** 1811–1813.

Lasne, F., Martin, L., de Ceaurriz, J., Larcher, T., Moullier, P., and Chenuaud, P. (2004). "Genetic Doping" with erythropoietin cDNA in primate muscle is detectable. *Mol. Ther.* **10,** 409–410.

Lin, F. K., Suggs, S., Lin, C. H., Browne, J. K., Smalling, R., Egrie, J. C., Chen, K. K., Fox, G. M., Martin, F., Stabinsky, Z., Badrawi, S. M., Lai, P. H., *et al.* (1985). Cloning and expression of the human erythropoietin gene. *Proc. Natl. Acad Sci. USA* **82,** 7580–7584.

Linden, T., Katschinski, D. M., Eckhardt, K., Scheid, A., Pagel, H., and Wenger, R. H. (2003). The antimycotic ciclopirox olamine induces HIF-1α stability, VEGF expression, and angiogenesis. *FASEB J.* **17,** 761–763.

Lippi, G., Franchini, M., and Guidi, G. C. (2005). Cobalt chloride administration in athletes: A new perspective in blood doping? *Br. J. Sports Med.* **39,** 872–873.

Lippin, Y., Dranitzki-Elhalel, M., Brill-Almon, E., Mei-Zahav, C., Mizrachi, S., Liberman, Y., Iaina, A., Kaplan, E., Podjarny, E., Zeira, E., Harati, M., Casadevall, N., *et al.* (2005). Human erythropoietin gene therapy for patients with chronic renal failure. *Blood* **106,** 2280–2286.

MacDougall, I. C., and Eckardt, K. U. (2006). Novel strategies for stimulating erythropoiesis and potential new treatments for anaemia. *Lancet* **368,** 947–953.

Maruyama, H., Ataka, K., Higuchi, N., Sakamoto, F., Gejyo, F., and Miyazaki, J. (2001). Skin-targeted gene transfer using *in vivo* electroporation. *Gene Ther.* **8,** 1808–1812.

Maruyama, H., Higuchi, N., Nishikawa, Y., Hirahara, H., Iino, N., Kameda, S., Kawachi, H., Yaoita, E., Gejyo, F., and Miyazaki, J. (2002). Kidney-targeted naked DNA transfer by retrograde renal vein injection in rats. *Hum. Gene Ther.* **13,** 455–468.

Maxwell, P. H., Wiesener, M. S., Chang, G. W., Clifford, S. C., Vaux, E. C., Cockman, M. E., Wykoff, C. C., Pugh, C. W., Maher, E. R., and Ratcliffe, P. J. (1999). The tumour suppressor protein VHL targets hypoxia-inducible factors for oxygen-dependent proteolysis. *Nature* **399,** 271–275.

Mizukami, H., Mimuro, J., Ogura, T., Okada, T., Urabe, M., Kume, A., Sakata, Y., and Ozawa, K. (2006). Adipose tissue as a novel target for *in vivo* gene transfer by adeno-associated viral vectors. *Hum. Gene Ther.* **17,** 921–928.

Naffakh, N., and Danos, O. (1996). Gene transfer for erythropoiesis enhancement. *Mol. Med. Today* **2,** 343–348.

Naffakh, N., Henri, A., Villeval, J. L., Rouyer-Fessard, P., Moullier, P., Blumenfeld, N., Danos, O., Vainchenker, W., Heard, J. M., and Beuzard, Y. (1995). Sustained delivery of erythropoietin in mice by genetically modified skin fibroblasts. *Proc. Natl. Acad. Sci. USA* **92,** 3194–3198.

Nakano, Y., Imagawa, S., Matsumoto, K., Stockmann, C., Obara, N., Suzuki, N., Doi, T., Kodama, T., Takahashi, S., Nagasawa, T., and Yamamoto, M. (2004). Oral administration of K-11706 inhibits GATA binding activity, enhances hypoxia-inducible factor 1 binding activity, and restores indicators in an *in vivo* mouse model of anemia of chronic disease. *Blood* **104,** 4300–4307.

Noguchi, A., Mukuria, C. J., Suzuki, E., and Naiki, M. (1996). Failure of human immunoresponse to N-glycolylneuraminic acid epitope contained in recombinant human erythropoietin. *Nephron.* **72,** 599–603.

O'donnell, J. L., Joyce, M. R., Shannon, A. M., Harmey, J., Geraghty, J., and Bouchier-Hayes, D. (2006). Oncological implications of hypoxia inducible factor-1alpha (HIF-1α) expression. *Cancer Treat. Rev.* **32,** 407–416.

Oh, T. K., Quan, G. H., Kim, H. Y., Park, F., and Kim, S. T. (2006). Correction of anemia in uremic rats by intramuscular injection of lentivirus carrying an erythropoietin gene. *Am. J. Nephrol.* **26,** 326–334.

Orive, G., De Castro, M., Ponce, S., Hernandez, R. M., Gascon, A. R., Bosch, M., Alberch, J., and Pedraz, J. L. (2005). Long-term expression of erythropoietin from myoblasts immobilized in biocompatible and neovascularized microcapsules. *Mol. Ther.* **12,** 283–289.

Osborne, W. R., Ramesh, N., Lau, S., Clowes, M. M., Dale, D. C., and Clowes, A. W. (1995). Gene therapy for long-term expression of erythropoietin in rats. *Proc. Natl. Acad. Sci. USA* **92,** 8055–8058.

Payen, E., Bettan, M., Rouyer-Fessard, P., Beuzard, Y., and Scherman, D. (2001). Improvement of mouse β-thalassemia by electrotransfer of erythropoietin cDNA. *Exp. Hematol.* **29,** 295–300.

Pugh, C. W., Tan, C. C., Jones, R. W., and Ratcliffe, P. J. (1991). Functional analysis of an oxygen-regulated transcriptional enhancer lying $3'$ to the mouse erythropoietin gene. *Proc. Natl. Acad. Sci. USA* **88,** 10553–10557.

Rankin, E. B., Biju, M. P., Liu, Q., Unger, T. L., Rha, J., Johnson, R. S., Simon, M. C., Keith, B., and Haase, V. H. (2007). Hypoxia-inducible factor-2 (HIF-2) regulates hepatic erythropoietin *in vivo. J. Clin. Invest.* **117,** 1068–1077.

Ren, X., Dorrington, K. L., Maxwell, P. H., and Robbins, P. A. (2000). Effects of desferrioxamine on serum erythropoietin and ventilatory sensitivity to hypoxia in humans. *J. Appl. Physiol.* **89,** 680–686.

Rinsch, C., Regulier, E., Déglon, N., Dalle, B., Beuzard, Y., and Aebischer, P. (1997). A gene therapy approach to regulated delivery of erythropoietin as a function of oxygen tension. *Hum. Gene Ther.* **8,** 1881–1889.

Rinsch, C., Dupraz, P., Schneider, B. L., Déglon, N., Maxwell, P. H., Ratcliffe, P. J., and Aebischer, P. (2002). Delivery of erythropoietin by encapsulated myoblasts in a genetic model of severe anemia. *Kidney Int.* **62,** 1395–1401.

Rivera, V. M., Gao, G. P., Grant, R. L., Schnell, M. A., Zoltick, P. W., Rozamus, L. W., Clackson, T., and Wilson, J. M. (2005). Long-term pharmacologically regulated expression of erythropoietin in primates following AAV-mediated gene transfer. *Blood* **105,** 1424–1430.

Rosenberger, C., Mandriota, S., Jurgensen, J. S., Wiesener, M. S., Horstrup, J. H., Frei, U., Ratcliffe, P. J., Maxwell, P. H., Bachmann, S., and Eckardt, K. U. (2002). Expression of hypoxia-inducible factor-1α and -2α in hypoxic and ischemic rat kidneys. *J. Am. Soc. Nephrol.* **13,** 1721–1732.

Rosenberger, C., Heyman, S. N., Rosen, S., Shina, A., Goldfarb, M., Griethe, W., Frei, U., Reinke, P., Bachmann, S., and Eckardt, K. U. (2005). Upregulation of HIF in experimental acute renal failure: Evidence for a protective transcriptional response to hypoxia. *Kidney Int.* **67,** 531–542.

Safran, M., Kim, W. Y., O'Connell, F., Flippin, L., Gunzler, V., Horner, J. W., Depinho, R. A., and Kaelin, W. G., Jr. (2006). Mouse model for noninvasive imaging of HIF prolyl hydroxylase activity: Assessment of an oral agent that stimulates erythropoietin production. *Proc. Natl. Acad. Sci. USA* **103,** 105–110.

Salnikow, K., Su, W., Blagosklonny, M. V., and Costa, M. (2000). Carcinogenic metals induce hypoxia-inducible factor-stimulated transcription by reactive oxygen species-independent mechanism. *Cancer Res.* **60,** 3375–3378.

Salnikow, K., Donald, S. P., Bruick, R. K., Zhitkovich, A., Phang, J. M., and Kasprzak, K. S. (2004). Depletion of intracellular ascorbate by the carcinogenic metals nickel and cobalt results in the induction of hypoxic stress. *J. Biol. Chem.* **279,** 40337–40344.

Sanchez-Elsner, T., Ramirez, J. R., Sanz-Rodriguez, F., Varela, E., Bernabeu, C., and Botella, L. M. (2004). A cross-talk between hypoxia and TGF-β orchestrates erythropoietin gene regulation through SP1 and Smads. *J. Mol. Biol.* **336,** 9–24.

Sasaki, R. (2003). Pleiotropic functions of erythropoietin. *Intern. Med.* **42,** 142–149.

Schneider, B. L., Peduto, G., and Aebischer, P. (2001). A self-immunomodulating myoblast cell line for erythropoietin delivery. *Gene Ther.* **8,** 58–66.

Scortegagna, M., Ding, K., Zhang, Q., Oktay, Y., Bennett, M. J., Bennett, M., Shelton, J. M., Richardson, J. A., Moe, O., and Garcia, J. A. (2005). HIF-2α regulates murine hematopoietic development in an erythropoietin-dependent manner. *Blood* **105,** 3133–3140.

Sebestyen, M. G., Hegge, J. O., Noble, M. A., Lewis, D. L., Herweijer, H., and Wolff, J. A. (2007). Progress toward a nonviral gene therapy protocol for the treatment of anemia. *Hum. Gene Ther.* **18,** 269–285.

Semenza, G. L., Nejfelt, M. K., Chi, S. M., and Antonarakis, S. E. (1991). Hypoxia-inducible nuclear factors bind to an enhancer element located 3′ to the human erythropoietin gene. *Proc. Natl. Acad. Sci. USA* **88,** 5680–5684.

Setoguchi, Y., Danel, C., and Crystal, R. G. (1994). Stimulation of erythropoiesis by *in vivo* gene therapy: Physiologic consequences of transfer of the human erythropoietin gene to experimental animals using an adenovirus vector. *Blood* **84,** 2946–2953.

Sinclair, A. M., and Elliott, S. (2005). Glycoengineering: The effect of glycosylation on the properties of therapeutic proteins. *J. Pharm. Sci.* **94,** 1626–1635.

Sodoyer, R. (2004). Expression systems for the production of recombinant pharmaceuticals. *BioDrugs* **18,** 51–62.

Soutoglou, E., Viollet, B., Vaxillaire, M., Yaniv, M., Pontoglio, M., and Talianidis, I. (2001). Transcription factor-dependent regulation of CBP and P/CAF histone acetyltransferase activity. *EMBO J.* **20,** 1984–1992.

Spinowitz, B. S., and Pratt, R. D. (2006). Epoetin delta is effective for the management of anaemia associated with chronic kidney disease. *Curr. Med. Res. Opin.* **22,** 2507–2513.

Strickland, T. W., Byrne, T., and Elliot, S. G. (2004). Erythropoietin isoforms European Patent Office. New European Patent Specification: 90311193.8. EP0428267, Kind Code B2.

Takeuchi, M., and Kobata, A. (1991). Structures and functional roles of the sugar chains of human erythropoietins. *Glycobiology* **1,** 337–346.

Tangvoranuntakul, P., Gagneux, P., Diaz, S., Bardor, M., Varki, N., Varki, A., and Muchmore, E. (2003). Human uptake and incorporation of an immunogenic nonhuman dietary sialic acid. *Proc. Natl. Acad. Sci. USA* **100,** 12045–12050.

Topol, I. A., Nemukhin, A. V., Salnikow, K., Cachau, R. E., Abashkin, Y. G., Kasprzak, K. S., and Burt, S. K. (2006). Quantum chemical modeling of reaction mechanism for 2-oxoglutarate dependent enzymes: Effect of substitution of iron by nickel and cobalt. *J. Phys. Chem. A Mol. Spectrosc. Kinet. Environ. Gen. Theory* **110,** 4223–4228.

Urquilla, P., Fong, A., Oksanen, S., Leigh, S., Turtle, E., Flippin, L., Brenner, M., Muthukrishnan, E., Fourney, P., Lin, A., Yeowell, D., and Molineaux, C. (2004). Upregulation of endogenous EPO in healthy sbjects by inhibition of HIF-PH. *J. Am. Soc. Nephrol.* **15,** 546.

Wang, G. L., and Semenza, G. L. (1993). Desferrioxamine induces erythropoietin gene expression and hypoxia-inducible factor 1 DNA-binding activity: Implications for models of hypoxia signal transduction. *Blood* **82,** 3610–3615.

Wang, G. L., Jiang, B. H., Rue, E. A., and Semenza, G. L. (1995). Hypoxia-inducible factor 1 is a basic helix-loop-helix PAS heterodimer regulated by cellular O_2 tension. *Proc. Natl. Acad. Sci. USA* **92,** 5510–5514.

Wanner, R. M., Spielmann, P., Stroka, D. M., Camenisch, G., Camenisch, I., Scheid, A., Houck, D. R., Bauer, C., Gassmann, M., and Wenger, R. H. (2000). Epolones induce erythropoietin expression via hypoxia-inducible factor-1α activation. *Blood* **96,** 1558–1565.

Warnecke, C., Zaborowska, Z., Kurreck, J., Erdmann, V. A., Frei, U., Wiesener, M., and Eckardt, K. U. (2004). Differentiating the functional role of hypoxia-inducible factor (HIF)-1α and HIF-2α (EPAS-1) by the use of RNA interference: Erythropoietin is a HIF-2α target gene in Hep3B and Kelly cells. *FASEB J.* **18,** 1462–1464.

Watanabe, D., Suzuma, K., Matsui, S., Kurimoto, M., Kiryu, J., Kita, M., Suzuma, I., Ohashi, H., Ojima, T., Murakami, T., Kobayashi, T., Masuda, S., *et al.* (2005). Erythropoietin as a retinal angiogenic factor in proliferative diabetic retinopathy. *N. Engl. J. Med.* **353,** 782–792.

Weikert, S., Papac, D., Briggs, J., Cowfer, D., Tom, S., Gawlitzek, M., Lofgren, J., Mehta, S., Chisholm, V., Modi, N., Eppler, S., Carroll, K., *et al.* (1999). Engineering Chinese hamster ovary cells to maximize sialic acid content of recombinant glycoproteins. *Nat. Biotechnol.* **17,** 1116–1121.

Welsh, S. J., Koh, M. Y., and Powis, G. (2006). The hypoxic inducible stress response as a target for cancer drug discovery. *Semin. Oncol.* **33,** 486–497.

Wiecek, A., Piecha, G., Ignacy, W., Schmidt, R., Neumayer, H. H., Scigalla, P., and Urquilla, P. (2005). Pharmacological stabilization of HIF increases hemoglobin concentration in anemic patients with chronic kidney disease. *Nephrol. Dial. Transplant.* **20** (Suppl. 5), 195.

World Health Organization. (2006). International nonproprietary names (INN) for biological and biotechnological substances. INN Working Document 05.179. Accessible at http://www.who.int/medicines/services/inn/INN_Biorev11-06.pdf.

Wurm, F. M. (2004). Production of recombinant protein therapeutics in cultivated mammalian cells. *Nat. Biotechnol.* **22,** 1393–1398.

Yuan, Y., Hilliard, G., Ferguson, T., and Millhorn, D. E. (2003). Cobalt inhibits the interaction between hypoxia-inducible factor-alpha and von Hippel-Lindau protein by direct binding to hypoxia-inducible factor-alpha. *J. Biol. Chem.* **278,** 15911–15916.

ROLE OF HYPOXIA-INDUCIBLE FACTOR-2α IN ENDOTHELIAL DEVELOPMENT AND HEMATOPOIESIS

Osamu Ohneda,[*,‡] Masumi Nagano,[*,‡] and Yoshiaki Fujii-Kuriyama[†,‡]

Contents

1. Introduction	200
2. Vasculogenesis/Angiogenesis and HIFs	200
2.1. HIF-2α null mice	201
2.2. HIF-2α knockdown mice	201
2.3. HIF-1α null mice	203
2.4. HIF-1α null EC	204
3. HIF-1β/ARNT Null Mice	204
4. Neovascularization and HIFs	205
4.1. HIF-2α knockdown mice	205
4.2. HIF-1α null EC	210
5. Hematopoiesis and HIFs	211
5.1. HIF-2α null mice	211
5.2. HIF-2α knockdown mice	212
5.3. HIF-1α null mice	213
5.4. HIF-1β/ARNT null mice	213
6. Conclusion	214
References	214

Abstract

Endothelial cells (EC) are important components for vessel formation and hematopoiesis. The proliferation and differentiation of EC are performed under the close influence of hypoxia-inducible factors (HIFs), which are master transcription factors that regulate vasculogenesis and angiogenesis in response to hypoxic stimuli. During early development of embryos, EC are directly involved in hematopoiesis and are known to act as stromal cells, which generate a variety

* Department of Regenerative Medicine and Stem Cell Biology, University of Tsukuba, Tsukuba, Japan
† Center for TARA, University of Tsukuba, Tsukuba, Japan
‡ SORST, Japan Science Technology Agency, Kawaguchi, Japan

Methods in Enzymology, Volume 435
ISSN 0076-6879, DOI: 10.1016/S0076-6879(07)35011-8

of regulatory factors, including cytokines and growth factors, and maintain adhesive interactions with the hematopoietic cells essential for their survival and function in the microenvironment.

Mouse gene-targeting technology provides us with the information that HIFs are crucial for the development of not only EC but also hematopoietic cells. Although we have determined some particular roles of HIF in association with neovascularization and hematopoiesis in the experiments using gene knockout mice, many crucial roles of HIFs in these processes still remain to be elucidated. Because of the complexity of vasculo/angiogenesis and hematopoiesis *in vivo*, it is very difficult to analyze distinct involvement of each of the HIFs in the regulation of vessel formation and development of hematopoietic cells.

In this chapter, we review the role of HIFs in neovascularization and hemato-poiesis with special attention to the usefulness of gene knockdown mice of HIF-2alpha (α) to analyze the respective roles of HIFs in these complex processes.

1. INTRODUCTION

HIF-1α was originally identified over 15 years ago as a transcription factor mediating erythropoietin (EPO) induction in response to hypoxia *in vitro*. Since that time, other HIF family members, HIF-2α and HIF-3α, were identified, and a common partner molecule for all HIF family members, HIF-1beta (β) (also known as arylhydrocarbon-receptor nuclear translocator [ARNT]), was also characterized. An increasing number of reports suggest that HIFs are master regulators of oxygen homeostasis, controlling the induction and expression of genes involved in glucose metabolism, erythropoiesis, and vasculo/angiogenesis. These conclusions have been confirmed and expanded upon by studies of genetically disrupted mice lacking HIFs. However, the lack of HIF genes leads to severe developmental defects and embryonic lethality, thus limiting the utility of these mice for the detailed analysis of HIF function in adult mice. Recently developed genetic techniques have allowed the generation of mice with knocked-down HIF gene expression, and these mice will facilitate the acquisition of greater mechanistic insight into HIF gene function.

2. VASCULOGENESIS/ANGIOGENESIS AND HIFs

During embryogenesis, an early vascular network is formed by vasculogenesis through the differentiation of angioblasts or hemangioblasts, and this is subsequently followed by angiogenesis (Flamme *et al.*, 1997; Risau and Flamme, 1995). Angiogenesis encompasses a number of complex processes, including sprouting, bridging, and vascular investment by mural cells. Vascular network maturation occurs in an ordered pattern of vessel growth, organization, and specialization (Cleaver and Melton, 2003), and multiple molecular mechanisms regulate the expression and action of angiogenic factors (Carmeliet, 2000; Risau, 1997).

Vascular endothelial growth factor (VEGF) is a critical component of vasculogenesis and remodeling. VEGF initiates vessel formation and induces vascular network maturation (Ferrara and Davis-Smyth, 1997). Sprouting angiogenesis is facilitated by hypoxia, and a large number of genes associated with this process, including VEGF, VEGF receptor 1 (Flt-1) (Gerber *et al.*, 1997), VEGF receptor 2 (Flk-1) (Kappel, 1999), angiopoietin-1 (Ang-1) (Enholm *et al.*, 1997), angiopoietin-2 (Oh *et al.*, 1999), Tie-2 (Currie *et al.*, 2002), placenta growth factor (Nicosia, 1998), platelet-derived growth factor (PDGF) (Kuwabara *et al.*, 1995), SDF-1 (Ceradini *et al.*, 2004), CXCR4 (Schioppa *et al.*, 2003), and matrix metaloprotenases (MMPs) (Ben-yosef *et al.*, 2002) are upregulated in an HIF-dependent manner under hypoxic conditions.

2.1. HIF-2α null mice

Several different groups have generated mice with targeted disruptions of *HIF-2α*; however, the phenotypes of these mice differ somewhat, possibly due to differences in genetic backgrounds. Peng *et al.* (2000) observed defective vascular remodeling with abnormally fenestrated capillaries and local hemorrhage, while mice generated by other groups exhibited defective fetal catecholamine production (Tian *et al.*, 1998) or altered lung maturation secondary to impaired surfactant secretion by alveolar type 2 cells (Compernolle *et al.*, 2002). The complete absence of HIF-2α led to severe vascular defects in the yolk sac and embryo proper, and mutant embryos died by E12.5 (Peng *et al.*, 2000). The severely disorganized yolk sac vasculature seen in these mice suggests that HIF-2α plays a critical role in controlling vascular remodeling during early embryonic development.

2.2. HIF-2α knockdown mice

The role of HIF-2α in adult mice was extensively investigated using mice in which HIF-2α expression was knocked-down by insertion of a phospho-glycerate kinase (PGK) promoter-neomycin gene cassette sandwiched between two loxP sites into the first exon of the HIF-2α gene encoding 5′-untranslated (UTR) (Morita *et al.*, 2003) (Fig. 11.1). The double poly(A) signal at the end of the inserted *neo* gene leads to reduced expression of the following HIF-2α coding sequence. These mice were born in the expected Mendelian ratios and grew normally, but the expression of HIF-2α messenger RNA (mRNA) was reduced by 20 to 80% compared to wild-type (*wt*) mice when determined by the real time quantitative reverse transcription

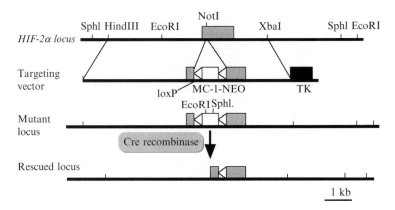

Figure 11.1 Generation of hypoxia-inducible factor (*HIF*)-2α knockdown mice. Depiction of the targeting vector and restriction map around the first exon of *HIF-2α* (closed box). MC-1-NEO (open box) indicates the neomycin resistance (*NEO*) gene sandwiched between loxP sequences, and this was inserted into exon 1 at the *Not*I site. The inserted *NEO* gene was removed by mating *HIF-2α^{kd/kd}* mice with AYU-1-Cre mice that ubiquitously express the cre enzyme to generate *HIF-2α^{lox/lox}* mice.

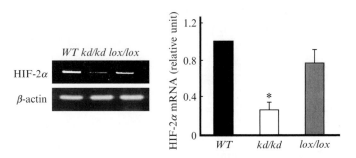

Figure 11.2 Reverse transcription polymerase chain reaction (RT-PCR) analysis of hypoxia-inducible factor (HIF)-2α messenger RNA (mRNA) expression. RNA was prepared from the eyes of wild-type (wt), *HIF-2α^{kd/kd}* (kd/kd), and *HIF-2α^{lox/lox}* (lox/lox) mice and subjected to RT-PCR analysis. The amount of HIF-2α mRNA from the eyes of wt mice was set to 1.

polymerase chain reaction (RT-PCR). Reductions in mRNA varied by organ (data not shown). Additionally, removal of the *neo* gene by expression of Cre recombinase led to normal HIF-2α mRNA expression (Fig. 11.2), clearly indicating that reduced HIF-2α expression is a result of *neo* gene insertion (Niwa *et al.*, 1993). In *HIF-2α^{kd/kd}* mice, no hemorrhaging was observed during embryogenesis, and the heart rate was not significantly effected at E18.5. Thus, two of the phenotypes of HIF-2α^{−/−} mice (Tian *et al.*, 1998) were not macroscopically detected in *HIF-2α^{kd/kd}* mice. However, when mice were immunohistochemically examined by staining for

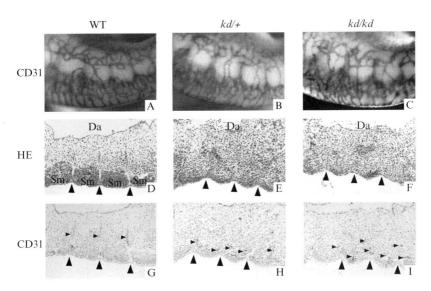

Figure 11.3 Neovascularization in wild-type (*wt*), hypoxia-inducible factor (*HIF*)-2α$^{kd/+}$, HIF-2α$^{kd/kd}$ embryos. Whole mount embryos (A, B, C) (E10.5) and sectioned samples (G, H, I) were immunostained with an anti-CD31 antibody. Blue-color staining represents positive for CD31. Sections were also stained by hematoxylene and eosin (H & E) in D, E, and F. There is altered CD31 staining, vessel formation, and somite structure in the *HIF-2α* mutant embryos. Large arrowheads indicate the intersomite regions, and small arrowheads indicate vessels. Da: dosal aorta, Sm: somite. (See color insert.)

CD31 (an endothelial cell marker), the vascular network in the somites of *HIF-2α$^{kd/kd}$* and *HIF-2α$^{kd/+}$* embryos (E10.5) was more disordered than seen in wt embryos (Fig. 11.3A–C). Additionally, staining of somite sections with hematoxylin and eosin (H & E) (see Fig. 11.3D–F) and anti-CD31 (see Fig. 11.3G–I) clearly demonstrated disordered somites and poorly formed intersomitic arteries (see Figs. 11.3E,F and 12.3H,I) in *HIF-2α$^{kd/+}$* and *HIF-2α$^{kd/kd}$* compared to *wt* embryos (see Fig. 11.3D,G). In contrast, the dorsal aortas of *HIF-2α* mutant mice appeared normal (see Fig. 11.3D–F). These results suggest that normal vessel formation is extremely sensitive to HIF-2α expression levels during embryonic development, but the mechanisms by which this effect is mediated should be investigated in greater detail.

2.3. HIF-1α null mice

The loss of HIF-1α results in early embryonic lethality characterized by defective vascularization of the developing embryos and yolk sacs (Iyer *et al.*, 1998). VEGF is a crucial factor regulating vasculogenesis and angiogenesis, and *VEGF*$^{+/-}$ heterozygous embryos die early during development (E11.5)

(Ferrara *et al.*, 1996). Surprisingly, VEGF expression was elevated in HIF-1$\alpha^{-/-}$ embryos (Kotch *et al.*, 1999), suggesting that factors other than VEGF are responsible for the observed defective EC development. The authors hypothesized that enhanced mesenchymal cell death in HIF-1$\alpha^{-/-}$ embryos may be responsible for the embryonic lethality (Kotch *et al.*, 1999). Consistent with these results, HIF-2α overexpression was not able to rescue cells lacking HIF-1α from hypoxia-induced cell death (Hu *et al.*, 2003), suggesting that HIF-1α and HIF-2α have different downstream target genes (Park *et al.*, 2003).

2.4. HIF-1α null EC

Mice specifically lacking HIF-1α expression in EC were generated by crossing *Tie2-cre* (Kisanuki *et al.* 2001) and *HIF-1$\alpha^{+f/+f}$* mice (Ryan *et al.*, 2000, Tang *et al.*, 2004). Despite the absence of HIF-1α protein expression in EC, these mice developed normally to adulthood, indicating that the primary cause of lethality seen in *HIF-1$\alpha^{-/-}$* embryos is due to the absence of HIF-1α in a non–EC cell population.

EC lines derived from *HIF-1$\alpha^{-/-}$* mice were generated by intercrossing with *p53$^{-/-}$* mice (Tang *et al.*, 2004), and the derived cells grew at a rate similar to *wt* EC under normoxic conditions (20% O_2). However, HIF-1α–deficient EC experience growth arrest after growing for 48 h in hypoxic conditions (0.5% O_2). Additionally, under normoxic conditions, tubular network formation by EC lacking HIF-1α was impaired compared to *wt* EC, and these differences were exacerbated by hypoxia. Hypoxia induces more rapid tube formation in both *wt*- and HIF-1α–deficient EC, suggesting that HIF-2α mediates tubular network formation in response to hypoxia.

3. HIF-1β/ARNT NULL MICE

The arylhydrocarbon-receptor nuclear translocator (ARNT) is a member of the basic helix-loop-helix Per/ARNT/Sim (bHLH-PAS) family and functions as a common partner molecule forming heterodimers with HIFs (Wang *et al.*, 1995). *ARNT$^{-/-}$* embryos died by E10.5, and this lethality was likely due to defective angiogenesis in the yolk sac and branchial arches (Maltepe *et al.*, 1997).

Platelet endothelial cell adhesion molecule (PECAM)-1-positive endothelial cells were found in E9.5 *ARNT$^{-/-}$* embryos, but whole cell lysates from ARNT-deficient embryos contained only half the amount of VEGF as *wt* embryos. Consistent with this, sprouting vessel maturation was severely impaired in *ARNT$^{-/-}$* paraaortic splanchnopleural (PSP) explant cultures (Ramírez-Bergeron *et al.*, 2006). Notably, addition of exogenous

VEGF rescued fine vascular network formation in *ARNT*$^{-/-}$ PSP explants, but Ang-1 had no effect. These data argue that VEGF acts as a prion to or differently from Ang-1. These experiments are complicated by the expanded presence of hematopoietic cells in the PSP explants from *ARNT*$^{-/-}$ mice producing pro-vasculogenic molecules, and the relative importance of these cells has yet to be determined. Under hypoxic conditions, additional VEGF may be needed to support the extensive proliferation and differentiation of EC during embryonic environment, suggesting that embryonic VEGF expression levels may be precisely regulated.

HIF-1α$^{-/-}$ EC derived from the lungs of adult mice exhibit defective proliferation, differentiation, and survival *in vitro* under hypoxic conditions, and there are a number of reasons why differences in EC survival and function could be seen between the *ARNT*$^{-/-}$ PSP explant culture system reflecting embryonic vascular development and *HIF-1α*$^{-/-}$ EC derived from adult animals. Further studies are needed to clarify the relationship between ARNT and HIFs in EC and hematopoietic cell development during embryogenesis and in adult animals.

4. NEOVASCULARIZATION AND HIFS

Angiogenesis, or neovascularization, is the process of generating new blood vessels through extensions of the preexisting vasculature. The principal cells involved in this process are EC, which line all blood vessels and are virtually the sole constituents of capillaries (Folkman and Klagsbrun, 1987). During angiogenesis, EC must first relocate from their stable location by breaking through the basement membrane (Carmeliet *et al.*, 2000; Yancopoulos *et al.*, 2000). Once this is achieved, EC migrate toward angiogenic stimuli released from tumor cells or activated inflammatory cells (Rafii and Lyden, 2003). After migration, EC proliferate to generate a sufficient number of cells to support new vessel formation. Following this process, the newly generated EC network must reorganize into a three-dimensional tubular structure.

4.1. HIF-2α knockdown mice

Retinal neovascularization is the most common cause of retinopathy of prematurity (ROP) (Moss *et al.*, 1994; Prost, 1988), and several observations suggest that ischemia or hypoxia initiates retinal neovascularization by inducing the excessive production of angiogenic factors.

Under hyperoxic conditions in a mouse model of ROP, obliteration of the retinal capillary network occurs (Smith *et al.*, 1994), and, upon a return to room air (P12), the decreased vascular network leads to conditions of relative hypoxia. Subsequently, retinal neovascularization is triggered in 100% of treated animals under these conditions, as evidenced by increases in VEGF mRNA (Pierce *et al.*, 1995).

In this ROP model (Smith *et al.*, 1994) (Fig. 11.4A), HIF-2α was induced within 12 h after a return to normoxia in the eyes of *wt* mice; however, virtually no induction in *HIF-2α$^{kd/kd}$* mice occurred (see Fig. 11.4B). Conversely, HIF-1α expression did not change with relative hypoxia in either mouse genotype (see Fig. 11.4B), and this conflicts with previously published data, for unknown reasons (Miyamoto *et al.*, 2002; Ozaki *et al.*, 1999).

Immediately after shifting from hyperoxic to normoxic conditions, neovascular bud formation was triggered on the retinal ganglion cell layer (GCL) surface of *wt* and mutant mice; however, fewer buds were present in the mutant mice (Fig. 11.5A,B). After 12 h of normoxic conditions, the neovascular buds elongated and infiltrated into the *wt* inner nuclear layer (INL). In the mutant mice, the nascent buds completely disappeared by 12 h (see Fig. 11.5C,D). Relative hypoxia induced extensive neovascularization in the *wt* retinas (Fig. 11.5E), but it was almost completely suppressed on the

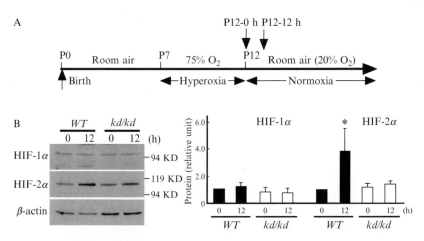

Figure 11.4 Murine model of retinopathy of prematurity (ROP). (A) An outline of the murine model of hyperoxia-normoxia–induced proliferative retinopathy is shown. At P7, wild-type (*wt*) and hypoxia-inducible factor (*HIF*)- 2α$^{kd/kd}$ mice, along with nursing mothers, were exposed to 75% oxygen for 5 days, and at P12, the animals were returned to room air (20% O_2). Following exposure to room air, mice were examined at each time point indicated (0 h or 12 h). (B) Expression of HIF-2α and HIF-1α in retinas following relative hypoxia. Whole cell extracts obtained at the indicated time points (0 h and 12 h) were subjected to sodium dodecyl sulfate polyacrylamide gel electrophoresis (SDS-PAGE) and Western blotting to detect HIF-2α and HIF-1α. β-actin antibody was used as a loading and transfer control. NIH *ImageJ* was used for quantitative analysis. HIF-2α and HIF-1α protein expression at P12–0 h were set to 1. ⋆$P < 0.01$.

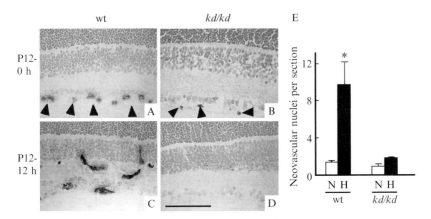

Figure 11.5 Retinal neovascularization in a mouse model of retinopathy of prematurity (ROP). Retinas from wild-type (*wt*) (A, C) and hypoxia-inducible factor (*HIF*)-2α*kd/kd* (B, D) mice at P12–0 h and 12 h after shifting from hyperoxia to normoxia were subjected to immunostaining with anti-CD34 antibody (A–D). Arrowheads indicate neovascular buds detected in the ganglion cell layer (GCL) (A, B). Note the elongation of the neovascular buds and vessel infiltration into the inner nuclear layer (INL) of *wt* mice (C); neovascularizatin was altered in the retina of *HIF-2α kd/kd* mice (D). The number of neovascular nuclei in the retinas of wt and *HIF-2α kd/kd* (*kd/kd*) mice was counted, and the average number of nuclei per section is shown. Bar indicates 100 μm. $^*P < 0.01$. (See color insert.)

inner retinal surface of the mutant mice. These results clearly demonstrate that the development of ROP is dependent upon HIF-2α expression, and HIF-1α does not compensate for its absence in this model.

When the expression profiles of vasculogenic/angiogenic factors were compared between the *wt* and the *HIF-2α kd/kd* mice in the ROP model, EPO mRNA expression was dependent upon HIF-2α expression (Fig. 11.6). In contrast, other factors examined, including VEGF, did not significantly vary between the mouse genotypes, and this data was confirmed on the protein level.

EPO is thought to have roles during both hematopoiesis and angiogenesis (Anagnostou *et al.*, 1990; Ribati *et al.*, 1999), and it is possible that EPO is essential for the development of ROP downstream of HIF-2α. It remains unclear why EPO gene transcription is not activated by HIF-1α following relative hypoxia in this ROP model, but it suggests that target genes may be distinctly induced by each of the HIF family members separately.

VEGF is thought to be the major factor initiating neovascularization in the retina during relative hypoxia (Aiello *et al.*, 1995; Seo *et al.*, 1999). Indeed, VEGF was expressed in the INL and GCL following exposure to relative hypoxia (Fig. 11.7A,B), and its receptor Flk-1 was expressed in elongated retinal EC (see Fig. 11.7C). These data emphasize the importance

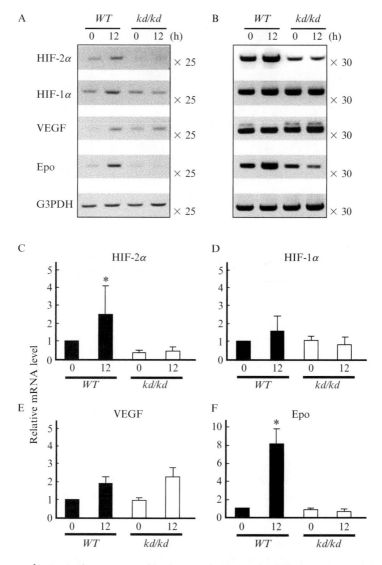

Figure 11.6 Retinal expression of angiogenic factors analyzed by reverse transcription polymerase chain reaction (RT-PCR). Gene expression in wild-type (*wt*) and hypoxia-inducible factor (*HIF*)-2α$^{kd/kd}$ (*kd/kd*) mice was analyzed using specific pairs of primers for HIF-2α, HIF-1α, erythropoietin (EPO), and two forms of vascular endothelial growth factor (VEGF) at the indicated times in the retinopathy of prematurity (ROP) model. The results of the polymerase chain reaction (PCR) analysis shown at 25 (A) and 30 cycles (B) are shown. G3PDH was used as an internal control. Quantitative PCR analysis was performed for HIF-2α (C), HIF-1α (D), vascular endothelial growth factor (VEGF) (E), and EPO (F). Black columns represent *wt* and white columns represent *HIF-2α*$^{kd/kd}$ mice. The relative messenger RNA (mRNA) level in *wt* mice at P12–0 h was set to 1. $^{*}P < 0.01$.

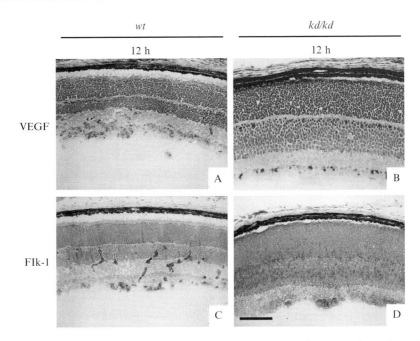

Figure 11.7 Immunohistochemical staining for angiogenic factors in the retinas of wild-type (*wt*) and mutant mice in the retinopathy of prematurity (ROP) model. Frozen retinal sections from *wt* (A, C) and hypoxia-inducible factor (*HIF*)-*2α*$^{kd/kd}$ (*kd/kd*), (B, D) mice were immunostained with anti-vascular endothelial growth factor (VEGF) (A, B), and FlK-1 (C, D) antibodies at P12–12 h time point. Bar indicates 100 *μm*. (See color insert.)

of the VEGF/Flk-1 interaction in inducing proliferation of the retinal vasculature. Although transient neovascular bud formation occurred immediately after the shift to normoxia in *HIF-2α*$^{kd/kd}$ mice, further vascularization was completely inhibited (see Fig. 11.5). Thus, other HIF-2α–inducible factors are necessary for the induction of neovascularization. Exogenous administration of EPO to *HIF2-α*$^{kd/kd}$ mice partially corrected this defect, but there were still significant differences in the number of neovascular buds between *wt* and EPO-treated *HIF2-α*$^{kd/kd}$ mice in this ROP model (Fig. 11.8). EPO may play a protective and/or anti-apoptotic role in EC buds (Carlini *et al.*, 1999) during neovasculogenesis in ROP, and future studies should examine the interplay between EPO and VEGF during neovascularization in response to hypoxia.

Makino *et al.* (2001) recently isolated a dominant-negative regulator of HIFs designated inhibitory PAS (IPAS). IPAS is induced in response to hypoxia and inhibits the activities of HIF-1α and HIF-2α in the corneal epithelium and retinal GCL and INL. Thus, there may be an inhibitory feedback loop consisting of HIFs and IPAS during relative hypoxia in ROP.

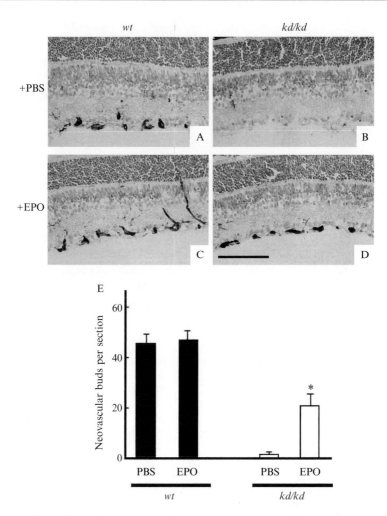

Figure 11.8 Effects of EPO on retinal neovascularization. Wild-type (*wt*) (A, C) and hypoxia-inducible factor (*HIF*)-*2α$^{kd/kd}$* (*kd/kd*) (B, D) mice were examined for the formation of neovascular buds following the administration of phosphate-buffered saline (PBS) (A, B) or EPO (C, D) at P12–24 h. CD34-positive neovascular buds were scored under the microscope (E). *$P < 0.01$. Bar indicates 100 μm. (See color insert.)

4.2. HIF-1α null EC

Mice specifically lacking HIF-1α in EC were generated, and neovascularization in these mice was substantially altered during wound healing and tumorigenesis (Tang *et al.*, 2004). Impaired vessel formation at the wound site significantly delayed wound healing in the *HIF-1α$^{-/-}$* EC mice compared to *wt* mice. In addition, tumor growth was impaired in *HIF-1α$^{-/-}$* EC mice, with tumors reaching only 60% of the weight achieved in *wt* mice.

Pathologically, tumors from *HIF-1α*$^{-/-}$ EC mice exhibited severe central necrosis secondary to a 50% reduction in tumor vessel density. HIF-2α protein expression was not affected by the absence of HIF-1α, suggesting that HIF-2α differentially regulates EC physiology in response to hypoxia *in vivo*.

HIF-1α$^{-/-}$ EC also exhibit defective cell growth and tube formation in response to hypoxia, and this characteristic was thought to arise from the loss of VEGF receptor 2 (VEGFR-2) expression. The altered VEGFR-2 expression seen in *HIF-1α*$^{-/-}$ EC disrupts the VEGF/VEGFR2 autocrine loop (Tang *et al.*, 2004), leading to impaired tumor neovascularization, decreased tumor growth, and necrosis (Ryan *et al.*, 1998).

5. HEMATOPOIESIS AND HIFs

Hematopoietic stem cells and progenitor cells proliferate and differentiate in hematopoietic microenvironments, such as the bone marrow, spleen, fetal liver, aorta-gonad mesonephros, and yolk sac (Sadlon *et al.*, 2004). A variety of pathways regulate hematopoiesis and interactions between hematopoietic and stromal cells.

5.1. HIF-2α null mice

HIF-2α$^{-/-}$ mice on the C57BL/6 genetic background were born with more reduced frequency than expected, and, when *HIF-2α*$^{-/-}$ mice in C57BL/6J background and *HIF-2α*$^{-/-}$ mice in the 129S6/SvEvTac background were crossed, some F1 hybrids survived after birth (Scortegagna *et al.*, 2003b). Although most of the F1 hybrid *HIF-2α*$^{-/-}$ mice died within 2 days after birth, 20% of the expected number of hybrid *HIF-2α*$^{-/-}$ mice survived. The surviving *HIF-2α*$^{-/-}$ mice were small in size and exhibited abnormalities in multiple organ systems, including pancytopenia, hepatomegaly, and cardiac hypertrophy, among others (Scortegagna *et al.*, 2003b).

Notably, the F1 hybrid *HIF-2α*$^{-/-}$ mice had fewer peripheral blood cells of all lineages, but no proportional differences were observed in hematopoietic cell surface markers in the bone marrow, suggesting that the hematopoietic microenvironment in these mice does not support normal cell production (Scortegagna *et al.*, 2003a).

Differences in hematopoiesis could arise from defects in hematopoietic progenitor cells or bone marrow stromal cells, and bone marrow transplantation studies are frequently used to differentiate between these possibilities. When this procedure was performed with these mice, transfer of HIF-2α–deficient bone marrow into *wt* mice supported normal hematopoiesis, suggesting that the absence of HIF-2α from stromal cells was responsible for the observed phenotype. This notion was confirmed by transferring *wt* bone marrow cells into *HIF-2α$^{-/-}$* mice, and, under these conditions, abnormal hematopoiesis was seen. Thus, HIF-2α expression in the stromal cells of the bone marrow is required for normal blood cell development (Scortegagna *et al.*, 2003a).

Further experiments are needed to determine the molecular mechanisms responsible for the altered hematopoietic microenvironment present in the hybrid *HIF-2α$^{-/-}$* mice. It is unclear if the effects of HIF-2α deficiency on the hematopoietic microenvironment are direct or indirect. Additionally, the target molecules downstream of the HIF-2α in the bone marrow are unknown. Finally, these data further confirm the functional non-redundancy of HIF family members and emphasize the need to identify the different regulatory mechanisms of HIF-1α and HIF-2α.

Scorteganga *et al.* (2005) have conducted further studies of the *HIF-2α$^{-/-}$* mice, and they recently reported that altered EPO production and signaling may be responsible for the observed hematopoiesis defect. However, EPO is not known to affect lymphoid or myeloid differentiation, and other pathways involving HIF-2α might affect interactions between hematopoietic cells and bone marrow stromal cells.

Hematopoietic cell proliferation and differentiation depends upon adhesive interactions with bone marrow stromal cells and the regulatory factors that these cells produce, including cytokines and growth factors (Yin and Li, 2006). In addition, adhesive interactions themselves may serve as growth or survival signals, and adhesion could modulate cytokine- or growth factor–dependent signals (Ohneda *et al.*, 2001). Accordingly, these cell–cell and cell–extracellular matrix (ECM) interactions are not only responsible for the localization of hematopoietic cells, but also play an important role in the regulation of hematopoiesis in the hematopoietic microenvironment.

The expression of fibronectin (a component of the ECM) and vascular cell adhesion molecule (VCAM)-1 mRNA were increased in the bone marrow of *HIF-2α$^{-/-}$* mice compared to *wt*; however, urokinase-type plasminogen activator receptor (uPAR) mRNA expression was decreased (Scortegagna *et al.*, 2003a). Further studies are needed to precisely determine the HIF-2α target molecules expressed in bone marrow stromal cells regulating hematopoietic cell development.

5.2. HIF-2α knockdown mice

HIF-2α$^{kd/kd}$ mice on the C57BL/6J genetic background all develop anemia; however, the numbers of white blood cells and platelets in the periphery are comparable to those of wt mice (data not shown). HIF-2α mRNA expression in the hematopoietic organs of *HIF-2α$^{kd/kd}$* mice is decreased (bone marrow: 70% reduction; spleen: 50% reduction), suggesting that there may be a dose effect of HIF-2α expression on hematopoiesis. We are currently investigating the molecular mechanisms underlying the constitutive anemia seen in *HIF-2α^{kd/kd}* mice.

The differences in hematopoiesis between *HIF-2α^{-/-}* and *HIF-2α^{kd/kd}* mice are striking. When considered together, these studies suggest that HIF-2α expression levels differentially affect the development of different cell lineages, and it is likely that different target genes and/or regulatory factors are sensitive to variances in HIF-2α expression levels. Expression profiling using bone marrow stromal and hematopoietic cells derived from these diverse mouse genotypes should illuminate the differences in hematopoiesis and the role of HIF-2α in this process.

5.3. HIF-1α null mice

HIF-1α deficiency is embryonic lethal, and, to study the effects of HIF-1α on lymphopoiesis, a RAG-2–deficient blastocyst complementation system was employed (Chen *et al.*, 1993). Using such an approach, Kojima *et al.* (2002) identified an essential role for HIF-1α in T- and B-cell development. In particular, the loss of HIF-1α led to functional abnormalities in cytotoxic T-lymphocytes, and it completely blocked B-cell development in bone marrow. The absence of highly proliferating B-cell progenitors in bone marrow led to the development of abnormal peritoneal B-1–like lymphocytes with elevated CD45 expression. Additionally, defective B-2 cell maturation was seen with the associated development of autoimmunity.

Myeloid cells derived from *lysomal (lysM)-cre/HIF-1α^{+f/+f}* conditionally null mice had defects in cell aggregation, motility, invasion, and bacterial killing, suggesting that HIF-1α controls the inflammatory response through its regulation of the metabolic switch to glycolysis in association with myeloid cell survival and function (Cramer *et al.*, 2003). Thus, HIF-1α plays a variety of different roles in immune cell development, while HIF-2α affects all hematopoietic cells with an emphasis on cells of the erythroid cell lineage.

5.4. HIF-1β/ARNT null mice

The hematopoietic potential of embryonically lethal mice can be determined by isolating embryoid bodies (EB) and culturing under conditions that give rise to different colonies corresponding to different cell lineages. When EBs were isolated from *ARNT^{-/-}* mice, the number of colonies, including CFU-E, CFU-M, CFU-GEMM, and CFU-GM, decreased compared to those of *wt* (Adelman *et al.*, 1999). VEGF mRNA was induced by 4.3-fold in EBs derived from *wt* mice under hypoxic conditions, but VEGF expression increased by only 2-fold in *ARNT^{-/-}* EBs in response to hypoxia. However, the addition of exogenous VEGF to the *ARNT^{-/-}* EB cultures restored colony formation to wt levels. Multiple combinations

of a number of cytokines failed to rescue hematopoietic colony formation in
$ARNT^{-/-}$ EB, suggesting that the VEGF/Flk-1–signaling pathway is
absolutely essential for the differentiation of EBs into hematopoietic cells.

In the EB culture system, hypoxic conditions promote mesoderm
differentiation, and VEGF is required for the differentiation of hemangioblasts, common precursor cells for EC and hematopoietic cells (Adelman
et al., 1999). The ability of VEGF to rescue colony formation in $ARNT^{-/-}$
EBs suggests that hematopoiesis is not impaired, but the differentiation of
the cellular support network required for the commitment toward hematopoietic cell lineage is disrupted. Thus, consistent with the data from
HIF-2α deficient mice, hypoxia-related gene expression in the surrounding
cells is required for hematopoietic cell commitment and development.

6. CONCLUSION

The regulation of vasculo/angiogenesis and hematopoiesis by HIFs is
critical for normal development under hypoxic conditions. Additionally,
during neovascularization caused by ischemia and cancer, HIFs play central
roles in the induction of early, disordered neovascularization. Despite recent
progress on our understanding of HIF function, the mechanism(s) by which
the HIF proteins affect neovascularization and hematopoiesis remain
unclear. Additionally, current data and models do not explain the differences in HIF-1α and HIF-2α function, and, in particular, unique target
molecules of these proteins are unknown. The next several years should
provide great insight into these questions.

REFERENCES

Adelman, D. M., Maltepe, E., and Simon, M. C. (1999). Multilineage embryonic hematopoiesis requires hypoxic ARNT activity. *Genes & Dev.* **13,** 2478–2483.
Aiello, L. P., Pierce, E. A., Foley, E. D., Takagi, H., Chen, H., Riddle, L., Ferrara, N.,
King, G. L., and Smith, L. E. (1995). Suppression of retinal neovascularization *in vivo* by
inhibition of vascular endothelial growth factor (VEGF) using soluble VEGF-receptor
chimeric proteins. *Proc. Natl. Acad. Sci. USA* **92,** 10457–10461.
Anagnostou, A., Lee, E. S., Kessimian, N., Levinson, R., and Steiner, M. (1990). Erythro-
poietin has a mitogenic and positive chemotactic effect on endothelial cells. *Proc. Natl.
Acad. Sci. USA* **87,** 5978–5982.
Ben-yosef, Y., Lahat, N., Shapiro, S., Bitterman, H., and Miller, A. (2002). Regulation of
endothelial matrix metalloproteinase-2 by hypoxia/reoxygenation. *Circ. Res.* **90,**
784–791.
Carlini, R. G., Alonzo, E. J., Dominguez, J., Blanca, I., Weisinger, J. R., Rothstein, M., and
Bellorin-Font, E. (1999). Effect of recombinant human erythropoietin on endothelial cell
apoptosis. *Kidney Int.* **55,** 546–553.

Carmeliet, P. (2000). Mechanisms of angiogenesis and arteriogenesis. *Nature Med.* **6,** 389–395.

Carmeliet, P., and Jain, R. K. (2000). Angiogenesis in cancer and other diseases. *Nature* **407,** 249–257.

Ceradini, D. J., Kulkarni, A. R., Callaghan, M. J., Tepper, O. M., Bastidas, N., Kleinman, M. E., Capla, J. M., Galiano, R. D., Levine, J. P., and Gurtner, G. C. (2004). Progenitor cell trafficking is regulated by hypoxic gradients through HIF-1 induction of SDF-1. *Nature Med.* **10,** 858–864.

Chen, J., Lansford, R., Stewart, V., Young, F., and Alt, F. W. (1993). RAG-2-deficient blastocyst complementation: An assay of gene function in lymphocyte development. *Proc. Natl. Acad. Sci. USA* **90,** 4528–4532.

Cleaver, O., and Melton, D. A. (2003). Endothelial signaling during development. *Nature Med.* **9,** 661–668.

Compernolle, V., Brusselmans, K., Acker, T., Hoet, P., Tjwa, M., Beck, H., Plaisance, S., Dor, Y., Keshet, E., Lupu, F., Nemery, B., Dewerchin, M., *et al.* (2002). Loss of HIF-2α and inhibition of VEGF impair fetal lung maturation, whereas treatment with VEGF prevent fatal respiratory distress in premature mice. *Nature Med.* **8,** 702–710.

Cramer, T., Yamanishi, Y., Clausen, B. E., Förster, I., Pawlinski, R., Mackman, N., Haase, V. H., Jaenisch, R., Corr, M., Nizet, V., Firestein, G. S., Gerber, H.-P., Ferrara, N., and Johnson, R. S. (2003). HIF-1α is essential for myeloid cell-mediated inflammation. *Cell* **112,** 645–657.

Currie, M. J., Gunningham, S. P., Turner, K., Han, C., Scott, P. A., Robinson, B. A., Chong, W., Harris, A. L., and Fox, S. B. (2002). Expression of the angiopoietins and their receptor Tie2 in human renal clear cell carcinomas; regulation by the von Hippel-Lindau gene and hypoxia. *J. Pathol.* **198,** 502–510.

Enholm, B., Paavonen, K., Ristimäki, A., Kumar, V., Gunji, Y., Klefstrom, J., Kivinen, L., Laiho, M., Olofsson, B., Joukov, V., Eriksson, U., and Alitalo, K. (1997). Comparison of VEGF, VEGF-B, VEGF-C and Ang-1 mRNA regulation by serum, growth factors, oncoproteins and hypoxia. *Oncogene* **14,** 2475–2483.

Ferrara, N., and Davis-Smyth, T. (1997). The biology of vascular endothelial growth factor. *Endocrine Rev.* **18,** 4–25.

Ferrara, N., Carver-Moore, K., Chen, H., Dowd, M., Lu, L., O'Shea, K. S., Powell-Braxton, L., Hillan, K. J., and Moore, M. W. (1996). Heterozygous embryonic lethality induced by targeted inactivation of the VEGF gene. *Nature* **380,** 439–442.

Flamme, I., Frolich, T., and Risau, W. (1997). Molecular mechanisms of vasculogenesis and embryonic angiogenesis. *J. Cell Physiol.* **173,** 206–210.

Folkman, J., and Klagsbrun, M. (1987). Angiogenic factors. *Science* **235,** 442–447.

Gerber, H.-P., Condorelli, F., Park, J., and Ferrara, N. (1997). Differential transcriptional regulation of the two vascular endothelial growth factor receptor genes. *J. Biol. Chem.* **272,** 23659–23667.

Hu, C.-J., Wang, L.-Y., Chodosh, L. A., Keith, B., and Simon, M. C. (2003). Differential roles of hypoxia-inducible factor 1alpha (HIF-1α) and HIF-2α in hypoxic gene regulation. *Mol. Cell. Biol.* **23,** 9361–9374.

Iyer, N. V., Kotch, L. E., Agani, F., Leung, S. W., Laughner, E., Wenger, R. H., Gassmann, M., Gearhart, J. D., Lawler, A. M., Yu, A. Y., and Semenza, G. L. (1998). Cellular and developmental control of O_2 homeostasis by hypoxia-inducible factor 1α. *Genes & Dev.* **12,** 149–162.

Kappel, A., Rönicke, V., Damaert, A., Flamme, I., Risau, W., and Breier, G. (1999). Identification of vascular endothelial growth factor (VEGF) receptor-2 (flk-1) promoter/enhancer sequences sufficient for angioblast and endothelial cell-specific transcription in transgenic mice. *Blood* **93,** 4284–4292.

Kisanuki, Y. Y., Hammer, R. E., Miyazaki, J., Williams, S. C., Richardson, J. A., and Yanagisawa, M. (2001). Tie2-Cre transgenic mice: A new model for endothelial cell-lineage analysis *in vivo*. *Dev. Biol.* **230**, 230–242.

Kojima, H., Gu, H., Nomura, S., Cadwell, C. C., Kobata, T., Carmeliet, P., Semenza, G. L., and Sitkovsky, M. V. (2002). Abnormal B lymphocyte development and autoimmunity in hypoxia-inducible factor 1α-deficient chimeric mice. *Proc. Natl. Acad. Sci. USA* **99**, 2170–2174.

Kotch, L. E., Iyer, N. V., Laughner, E., and Semenza, G. L. (1999). Defective vascularization of HIF-1α-null embryos is not associated with VEGF deficiency but with mesenchymal cell death. *Dev. Biol.* **209**, 254–267.

Kuwabara, K., Ogawa, S., Matsumoto, M., Koga, S., Clauss, M., Pinsky, D. J., Lyn, P., Leavy, J., Witte, L., Joseph-Silverstein, J., Furie, M. B., Torcia, G., *et al.* (1995). Hypoxia-mediated induction of acidic/basic fibroblast growth factor and platelet-derived growth factor in mononuclear phagocytes stimulates growth of hypoxic endothelial cells. *Proc. Natl. Acad. Sci. USA* **92**, 4606–4610.

Makino, Y., Cao, R., Svensson, K., Bertilsson, G., Asman, M., Tanaka, H., Cao, Y., Berkenstam, A., and Poellinger, L. (2001). Inhibitory PAS domain protein is a negative regulator of hypoxia-inducible gene expression. *Nature* **414**, 550–554.

Maltepe, E., Schmidt, J. V., Baunoch, D., Bradfield, C. A., and Simon, M. C. (1997). Abnormal angiogenesis and responses to glucose and oxygen deprivation in mice lacking the protein ARNT. *Nature* **386**, 403–407.

Miyamoto, N., Mandai, M., Takagi, H., Suzuma, I., Suzuma, K., Koyama, S., Otani, A., Oh, H., and Honda, Y. (2002). Contrasting effect of estrogen on VEGF induction under different oxygen status and its role in murine ROP. *Invest. Ophthalmol. Vis. Sci.* **43**, 2007–2014.

Morita, M., Ohneda, O., Yamashita, T., Takahashi, S., Suzuki, N., Nakajima, O., Kawauchi, S., Ema, M., Shibahara, S., Udono, T., Tomita, K., Tamai, M., *et al.* (2003). HLF/HIF-2α is key factor in retinopathy of prematurity in association with erythropoietin. *EMBO J.* **22**, 1134–1146.

Moss, S. E., Klein, R., and Klein, B. E. (1994). Ten-year incidence of visual loss in a diabetic population. *Ophthalmology* **101**, 1061–1070.

Nicosia, R. F. (1998). What is the role of vascular endothelial growth factor-related molecules in tumor angiogenesis? *Am. J. Pathol.* **153**, 11–16.

Niwa, H., Araki, K., Kimura, S., Taniguchi, S., Wakasugi, S., and Yamamura, K. (1993). An efficient gene-trap method using poly A trap vectors and characterization of gene-trap events. *J. Biochem. (Tokyo)* **113**, 343–349.

Oh, H., Takagi, H., Suzuma, K., Otani, A., Matsumura, M., and Honda, Y. (1999). Hypoxia and vascular endothelial growth factor selectively up-regulate angiopoietin-2 in bovine microvascular endothelial cells. *J. Biol. Chem.* **274**, 15732–15739.

Ohneda, O., Ohneda, K., Arai, F., Lee, J., Miyamoto, T., Fukushima, Y., Dowbenko, D., Lasky, L. A., and Suda, T. (2001). ALCAM (CD166): Its role in hematopoietic and endothelial development. *Blood* **98**, 2134–2142.

Ozaki, H., Yu, A. Y., Della, N., Ozaki, K., Luna, J. D., Yamada, H., Hackett, S. F., Okamoto, N., Zack, D. J., Semenza, G. L., and Campochiaro, P. A. (1999). Hypoxia inducible factor-1α is increased in ischemic retina: Temporal and spatial correlation with VEGF expression. *Invest. Ophthalmol. Vis. Sci.* **40**, 182–189.

Park, S.-K., Dadak, A. M., Haase, V. H., Fontana, L., Giaccia, A. J., and Johnson, R. S. (2003). Hypoxia-induced gene expression occurs solely through the action of hypoxia-inducible factor 1α (HIF-1α): Role of cytoplasmic trapping of HIF-2α. *Mol. Cell. Biol.* **23**, 4959–4971.

Peng, J., Zhang, L., Drysdale, L., and Fong, G.-H. (2000). The transcription factor EPAS-1/hypoxia-inducible factor-2α plays an important role in vascular remodeling. *Proc. Natl, Acad. Sci. USA* **97**, 8386–8391.

Pierce, E. A., Avery, R. L., Foley, E. D., Aiello, L. P., and Smith, L. E. (1995). Vascular endothelial growth factor/vascular permeability factor expression in a mouse model of retinal neovascularization. *Proc. Natl. Acad. Sci. USA* **92**, 905–909.

Prost, M. (1988). Experimental studies on the pathogenesis of retinopathy of prematurity. *Br. J. Ophthalmol.* **72**, 363–367.

Rafii, S., and Lyden, D. (2003). Therapeutic stem and progenitor cell transplantation for organ vascularization and regeneration. *Nature Med.* **9**, 702–712.

Ramírez-Bergeron, D. L., Runge, A., Adelman, D. M., Gohil, M., and Simon, M. C. (2006). HIF-dependent hematopoietic factors regulate the development of the embryonic vasculature. *Dev. Cell* **11**, 81–92.

Ribati, D., Presta, M., Vacca, A., Ria, R., Giuliani, R., Dell'Era, P., Nico, B., Roncali, L., and Dammacco, F. (1999). Human erythropoietin induces a pro-angiogenic phenotype in cultured endothelial cells and stimulates neovascularization *in vivo*. *Blood* **93**, 2627–2636.

Risau, W. (1997). Mechanism of angiogenesis. *Nature* **386**, 671–674.

Risau, W., and Flamme, I. (1995). Vasculogenesis. *Ann. Rev. Cell. Dev. Biol.* **11**, 73–91.

Ryan, H. E., Lo, J., and Johnson, R. S. (1998). HIF-1α is required for solid tumor formation and embryonic vascularization. *EMBO J.* **17**, 3005–3015.

Ryan, H. E., Poloni, M., McNulty, W., Elson, D., Gassmann, M., Arbeit, J. M., and Johnson, R. S. (2000). Hypoxia-inducible factor-1α is a positive factor in solid tumor growth. *Cancer Res.* **60**, 4010–4015.

Sadlon, T. J., Lewis, I. D., and D'Andrea, R. J. (2004). BMP4: Its role of development of the hematopoietic system and potential as a hematopoietic growth factor. *Stem Cells* **22**, 457–474.

Schioppa, T., Uranchimeg, B., Saccani, A., Biswas, S. K., Doni, A., Raspisarda, A., Bernasconi, S., Saccani, S., Nebuloni, M., Vago, L., Mantovani, A., Melillo, G., and Sica, A. (2003). Regulation of the chemokine receptor CXCR4 by hypoxia. *J. Exp. Med.* **198**, 1391–1402.

Scortegagna, M., Morris, M. A., Oktay, Y., Bennett, M., and Garci, J. A. (2003a). The HIF family member EPAS1/HIF-2α is required for normal hematopoiesis in mice. *Blood* **102**, 1634–1640.

Scortegagna, M., Ding, K., Oktay, Y., Gaur, A., Thurmond, F., Yan, L.-J., Marck, B. T., Matsumoto, A. M., Shelton, J. M., Richardson, J. A., Bennett, M. J., and Garcia, J. A. (2003b). Multiple organ pathology, metabolic abnormalities and impaired homeostasis of reactive oxygen species in *Epas1*$^{-/-}$ mice. *Nature Genet.* **35**, 331–340.

Scortegagna, M., Ding, K., Zhang, Q., Oktay, Y., Bennett, M. J., Bennett, M., Shelton, J. M., Richardson, J. A., Moe, O., and Garcia, J. A. (2005). HIF-2α regulates murine hematopoietic development in an erythropoietin-dependent manner. *Blood* **105**, 3133–3140.

Seo, M. S., Kwak, N., Ozaki, H., Yamada, H., Okamoto, N., Yamada, E., Fabbro, D., Hofmann, F., Wood, J. M., and Campochiaro, P. A. (1999). Dramatic inhibition of retinal and choroidal neovascularization by oral administration of a kinase inhibitor. *Am. J. Pathol.* **154**, 1743–1753.

Smith, L. E., Wesolowski, E., McLellan, A., Kostyk, S. K., D'Amato, R., Sullivan, R., and D'Amore, P. A. (1994). Oxygen-induced retinopathy in the mouse. *Invest. Ophthalmol. Vis. Sci.* **35**, 101–111.

Tang, N., Wang, L., Esko, J., Giordano, F. J., Huang, Y., Gerber, H.-P., Ferrara, N., and Johnson, R. S. (2004). Loss of HIF-1α in endothelial cells disrupts a hypoxia-driven VEGF autocrine loop necessary for tumorigenesis. *Cancer Cell.* **6**, 485–495.

Tian, H., Hammer, R. E., Matsumoto, A. M., Russell, D. W., and McKnight, S. L. (1998). The hypoxia-responsive transcription factor EPAS-1 is essential for catecholamine homeostasis and protection against heart failure during embryonic development. *Genes & Dev.* **12**, 3320–3324.

Wang, G. L., Jiang, B.-H., Rue, E. A., and Semenza, G. L. (1995). Hypoxia-inducible factor 1 is a basic helix-loop-helix PAS heterodimer regulated by cellular O_2 tension. *Proc. Natl. Acad. USA* **92,** 5510–5514.

Yancopoulos, G. D., Davis, S., Gale, N. W., Rudge, J. S., Wiegand, S. J., and Holash, J. (2000). Vascular-specific growth factors and blood vessel formation. *Nature* **407,** 242–248.

Yin, T., and Li, L. (2006). The stem cell niches in bone. *J. Clin. Invest.* **116,** 1195–1201.

HYPOXIA AND ADAPTATION

ORGAN PROTECTION BY HYPOXIA AND HYPOXIA-INDUCIBLE FACTORS

Wanja M. Bernhardt, Christina Warnecke, Carsten Willam, Tetsuhiro Tanaka, Michael S. Wiesener, *and* Kai-Uwe Eckardt

Contents

1. Introduction	222
2. From Ischemic to Hypoxic Preconditioning	222
3. Hypoxia-Inducible Transcription Factors	223
4. Strategies to Activate HIF and HIF Target Genes	227
4.1. Hypoxic hypoxia and carbon monoxide	227
4.2. Inhibition of HIF prolyl hydroxylases	229
4.3. Additional strategies to activate the HIF pathway	230
5. Hypoxic Preconditioning and HIF	231
5.1. Protective role of HIF target genes	231
5.2. HIF activation and organ protection	232
6. HIF in Chronic Hypoxic/Ischemic Diseases	236
7. Conclusions and Perspectives	237
References	238

Abstract

Since the first description of a protective effect of hypoxic preconditioning in the heart, the principle of reducing tissue injury in response to ischemia by prior exposure to hypoxia was confirmed in a number of cells and organs. However, despite impressive preclinical results, hypoxic preconditioning has so far failed to reach clinical application. Nevertheless, it remains of significant interest to induce genes that are normally activated during hypoxia and ischemia as part of an endogenous escape mechanism prior to or during the early phase of an ischemic insult. This approach has recently been greatly facilitated by the identification of hypoxia-inducible factors (HIFs), transcription factors that operate as a master switch in the cellular response to hypoxia. Far more than 100 target genes are regulated by HIF, including genes such as erythropoietin and hemoxygenase-1, which have been shown to be tissue-protective. The identification of small

Department of Nephrology and Hypertension, Friedrich–Alexander University, Erlangen, Nürnberg, Germany

Methods in Enzymology, Volume 435
ISSN 0076-6879, DOI: 10.1016/S0076-6879(07)35012-X

molecule inhibitors of the oxygen-sensing HIF-prolyl hydroxlases now offers the possibility to mimic the hypoxic response by pharmacological stabilization of HIF in order to achieve organ protection. Oxygen-independent activation of HIF is therefore a promising therapeutic strategy for the prevention of organ injury and failure.

1. INTRODUCTION

Preconditioning can be defined as injury-induced protection from a subsequent, more severe injury. Ischemia and hypoxia are the most broadly used conditions for preconditioning organs in order to protect them against ischemia. Hypoxia is defined as an inadequate supply of oxygen, whereas ischemia is defined as deprivation of oxygen and metabolites, such as glucose, due to compromised blood flow. Oxygen is essential for all metabolic processes, including oxidative phosphorylation for adenosine triphosphate (ATP) formation. A mismatch between oxygen supply and oxygen demand results in hypoxia with deleterious consequences for the affected organs.

A number of clinically relevant disorders are characterized by an impaired blood supply, which is insufficient to maintain adequate oxygen delivery. Among them, coronary artery disease and myocardial infarction are the leading causes of morbidity and mortality in the western countries and will presumably be the major cause of death in the whole world by 2020 (Murray and Lopez, 1997). An induction of adaptive mechanisms, which are able to rapidly increase the hypoxic tolerance of the respective organ, is thus of great interest. In 1986, Murry *et al.* (1986) first described the principle of "ischemic" preconditioning. Four cycles of 5-min ischemia followed by short episodes of reperfusion resulted in a markedly reduced infarct size in a subsequently induced myocardial infarction in dogs. In the following years, the protective potential of this principle has been demonstrated in different experimental animals and organs with the use of different protocols (Bolli, 2007; Bonventre, 2002; Gidday, 2006). A large number of studies have demonstrated that the cellular consequences that are induced by repetitive ischemia provide immediate (early phase) and long-lasting (late phase) protection against a subsequent, more severe ischemic insult.

2. FROM ISCHEMIC TO HYPOXIC PRECONDITIONING

Despite impressive results in preclinical studies, the concept of ischemic preconditioning has not reached routine clinical application, and attempts to apply it in patients led to variable, frequently disappointing results (Schlaifer and Kerensky, 1997). At least three reasons exist for this problem of translation. First, there are obvious limitations in inducing "protective ischemia" in patients. Second, the protective effect appears to be sensitive to details of the experimental design, such as the duration of ischemia and reperfusion periods. Third, the molecular mechanisms, which mediate the ischemic preconditioning, have not been unequivocally identified.

In 1992, Shizukuda *et al.* (1992, 1993) were able to demonstrate equipotent protective effects of hypoxia pretreatment and ischemic preconditioning and introduced the term hypoxic preconditioning. Both the perfusion of dog hearts with severely hypoxic blood (9.2 $\pm$ 0.6 ml O_2/l) and a short pre-insult period of ischemia of 5 min followed by 10 min of reperfusion resulted in a comparable reduction of the size of an infarct subsequently induced by 60 min of coronary artery occlusion (Shizukuda *et al.*, 1993). Consequently, protection mediated by hypoxic pretreatment has been adopted and reported for a number of cell types, tissues, and organs.

Recently, an exciting novel approach applies the concept of tissue protection by inducing part of the biological response to hypoxia. This concept is based on studies on the regulation of the hormone erythropoietin (EPO), which led to the identification of a widespread system of hypoxia-inducible gene expression, which is mediated by the family of HIFs.

3. HYPOXIA-INDUCIBLE TRANSCRIPTION FACTORS

Hypoxia-inducible transcription factors are in a central position in the cellular adaptation to hypoxic environments. HIFs are members of the basic helix-loop-helix Per-ARNT-Sim (bHLH-PAS) protein family and consist of one of three alternative oxygen-regulated alpha-chains (HIF-1α, -2α, and -3α) and a constitutive Beta (β)-chain (HIF-β, aryl hydrocarbon receptor translocator [ARNT]) (Maxwell, 2005; Ratcliffe, 2006). HIF-1α was identified and first characterized by Wang and Semenza in the early 1990s as transcription complex bound to the $3'$ end of the EPO gene (Semenza and Wang, 1992; Wang and Semenza, 1993, 1995). In the following years, the HIF-1α homologue HIF-2α (initially termed endothelial PAS protein-1 [EPAS-1]) has been described and characterized (Wiesener *et al.*, 1998). HIF-1α and -2α share functional and regulatory features. Under normoxia, HIF-α is rapidly degraded and virtually undetectable in most cells due to rapid proteasomal destruction (Fig. 12.1A), whereas, under hypoxic conditions, HIF-α accumulates in the cell and forms heterodimers with HIF-β (see Fig. 12.1B). This complex binds to a specific DNA motif, the hypoxia response element (HRE), and recruits the acetyltransferase p300, which serves as transcriptional activator (Arany *et al.*, 1996) (see Fig. 12.1B). The accumulation of HIF increases progressively with reduction of oxygen concentrations from 20% to below 1% (Jiang *et al.*, 1996; Wiesener *et al.*, 1998). Although structural and functional homology of HIF-1α and -2α suggest redundancy, knockout studies in mice (Iyer *et al.*, 1998; Rankin *et al.*, 2007;

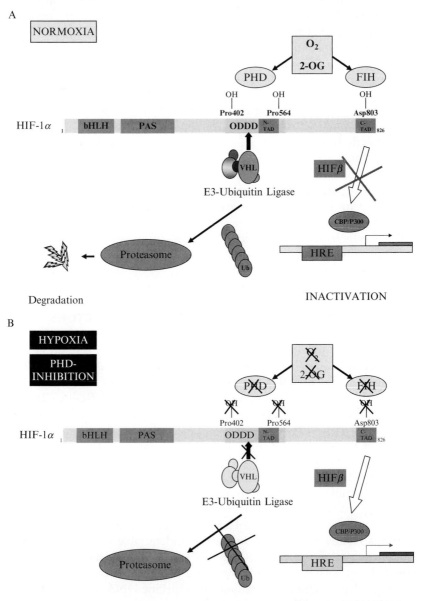

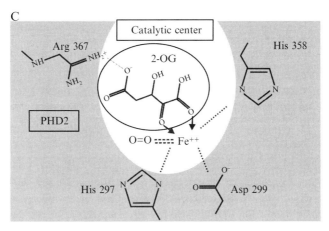

Figure 12.1 (A) Schematic) view of the domain structure and normoxic degradation of hypoxia-inducible transcription factor-1alpha (HIF-1α). HIF-1α consists of the oxygen-dependent degradation domain (ODDD) containing the two prolyl hydroxylation motifs, the basic helix-loop-helix (bHLH) domain, the Per-ARNT-Sim (PAS) domain, the N-terminal transactivation domain (N-TAD), and the C-terminal transactivation domain (C-TAD) with the asparagyl hydroxylation site. At normal oxygen tensions, the oxygen-sensing prolyl hydroxylases (PHD) and asparaginyl hydroxylase (factor-inhibiting HIF [FIH]) are active. These enzymes require molecular oxygen (O_2) and 2-oxoglutarate (2-OG) as co-substrates to hydroxylate the two proline residues (Pro402, Pro564) in the ODDD and the asparaginyl residue (Asp803) in the C-TAD of HIF-1α, respectively. As part of an E3 ubiquitin ligase, the von Hippel-Lindau protein (VHL) binds to the ODDD after prolyl hydroxylation. As a consequence, HIF-1α is ubiquitinylated (Ub) and thus marked for proteasomal destruction. Binding to the transcriptional coactivator, CREB-binding protein (CBP)/p300 is abrogated by hydroxylation of the Asp803. (B) Reduced activity of the PHD results in HIF-1α accumulation. In the absence of oxygen (hypoxia, red) or the presence of hydroxylation inhibitors (green), HIF-1α is stabilized, forms a dimer with HIF-β (ARNT), and binds to the hypoxia-responsive element (HRE) with recruitment of the transcriptional coactivators CBP/p300, thus transactivating its target genes. (C) Schematic representation of the catalytic pocket of human PHD2. PHD2 contains a 2-histidine-1-carboxylate coordination motif, a nonheme-bound Fe^{2+}, dioxygen, and 2-OG. For reasons of simplicity, essential water molecules in the catalytic center and the HIF prolyl residue, which instead binds at the surface of the pocket, are not depicted. Figure adapted from Warnecke *et al.* (2003).

Ryan *et al.*, 1998; Scortegagna *et al.*, 2003a) and differences in the distributional pattern of both isoforms (Philipp *et al.*, 2006b; Rosenberger *et al.*, 2002; Wiesener *et al.*, 2003) indicate different, non-redundant roles. However, the regulatory regions that mediate oxygen-dependent transactivation and/or degradation of the α-chains are conserved between the two HIF-α-isoforms: a C-terminal transactivation domain (C-TAD) and an oxygen-dependent degradation domain (ODD) (shown in Fig. 12.1A for HIF-1α). Under normoxic conditions, the von Hippel-Lindau protein (pVHL) binds to the ODD and acts as part of an E3-ubiquitin ligase complex, which ubiquitinylates HIF-α, thereby marking it for proteasomal degradation (Maxwell *et al.*, 1999) (see Fig. 12.1A). How the cell senses changes in oxygen tensions has long remained elusive. Later observations revealed that the pVHL-dependent destruction of HIF-1α requires 4-hydroxylation of defined proline-residues (Pro402 or Pro564) of the α-chain (Ivan *et al.*, 2002; Jaakkola *et al.*, 2001). In addition, binding of p300 is prevented by the hydroxylation of the asparagine residue in position 803 (Lando *et al.*, 2002). Shortly after this discovery, an enzyme family of dioxygenases that catalyze oxygen-dependent hydroxylation of specific amino acids of the HIF-α chain was identified as being responsible for asparaginyl and prolyl hydroxylation of HIF. Members of this family are termed prolyl-hydroxylase domain (PHD)1 through -3 (or EGLN1 through -3, respectively, according to the *Caenorhabditis elegans* homolog EGL-9 [Bruick and McKnight, 2001; Epstein *et al.*, 2001] and factor-inhibiting HIF (FIH-1) (Hewitson *et al.*, 2002; Lando *et al.*, 2002; reviewed in Masson and Ratcliffe, 2003). These enzymes require molecular oxygen as a substrate and 2-oxoglutarate as a cosubstrate to hydroxylate the specific amino acid residues in a peptide chain, thereby producing carbon dioxide and succinate. The catalytic pocket of these enzymes contains a conserved 2-histidine-1-carboxylate coordination motif for the ligand Fe^{2+} that in turn binds the molecular oxygen and the 2-oxoglutarate (McDonough *et al.*, 2006). The relative contribution of each PHD isoform to HIF-α regulation still remains uncertain.

Nevertheless, the insights into cellular oxygen sensing have already opened new therapeutic avenues. HIF target genes exert protective effects in many forms of systemic or regional hypoxia by facilitating adequate energy supply and exerting survival signals within hours and, in the long term, improving oxygen transport capacity and vascularization.

To date, far more than 100 HIF target genes have been identified (Fig. 12.2, selected HIF target genes), including genes involved in hemato-poiesis (e.g., EPO), iron metabolism (e.g., transferrin, transferrin-receptor), angiogenesis and vascular tone (e.g., vascular endothelial growth factor [VEGF], Flt-1 [VEGF receptor 1], endoglin, plasminogen activator inhibitor-1, adrenomedullin, endothelin-1, heme oxygenase-1, nitric oxide synthase-2), energy metabolism (e.g., glucose transporters 1 and 3, lactate dehydrogenase A), cell proliferation and differentiation (e.g., insulin-like growth factor (IGF) binding proteins 1 and 3, Transforming growth factor [TGF]-β, cyclin G2, opioid growth factor [OGF]-2, caspase 9), pH-regulation (e.g., carbonic anhydrase 9), and matrix metabolism (e.g., collagen prolyl-4-hydroxylase-α1) (Maxwell, 2005; Ratcliffe, 2006; Schofield and Ratcliffe, 2004).

This key role and the specific mode of regulation render the HIF system an attractive target for pharmacological intervention in ischemic diseases (Ratcliffe, 2006).

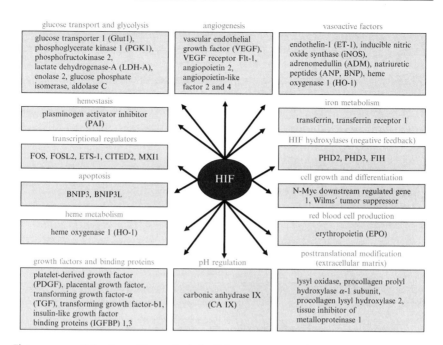

glucose transport and glycolysis

glucose transporter 1 (Glut1), phosphoglycerate kinase 1 (PGK1), phosphofructokinase 2, lactate dehydrogenase-A (LDH-A), enolase 2, glucose phosphate isomerase, aldolase C

angiogenesis

vascular endothelial growth factor (VEGF), VEGF receptor Flt-1, angiopoietin 2, angiopoietin-like factor 2 and 4

vasoactive factors

endothelin-1 (ET-1), inducible nitric oxide synthase (iNOS), adrenomedullin (ADM), natriuretic peptides (ANP, BNP), heme oxygenase 1 (HO-1)

hemostasis

plasminogen activator inhibitor (PAI)

iron metabolism

transferrin, transferrin receptor 1

transcriptional regulators

FOS, FOSL2, ETS-1, CITED2, MXI1

HIF hydroxylases (negative feedback)

PHD2, PHD3, FIH

apoptosis

BNIP3, BNIP3L

cell growth and differentiation

N-Myc downstream regulated gene 1, Wilms' tumor suppressor

heme metabolism

heme oxygenase 1 (HO-1)

red blood cell production

erythropoietin (EPO)

growth factors and binding proteins

platelet-derived growth factor (PDGF), placental growth factor, transforming growth factor-α (TGF), transforming growth factor-b1, insulin-like growth factor binding proteins (IGFBP) 1,3

pH regulation

carbonic anhydrase IX (CA IX)

posttranslational modification (extracellular matrix)

lysyl oxidase, procollagen prolyl hydroxylase α-1 subunit, procollagen lysyl hydroxylase 2, tissue inhibitor of metalloproteinase 1

HIF

Figure 12.2 Selection of hypoxia-inducible transcription factor (HIF) target genes. HIF transactivates a whole array of target genes involved in various adaptive processes in response to hypoxia.

4. Strategies to Activate HIF and HIF Target Genes

4.1. Hypoxic hypoxia and carbon monoxide

The physiological stimulus to activate the HIF system is a lack of oxygen with subsequently reduced activity of the PHDs (see Fig. 12.1B). HIF is ubiquitously expressed, and it appears to be crucial for normal embryonic development and physiological organ function (Cramer et al., 2003; Iyer et al., 1998; Maltepe et al., 1997; Rankin et al., 2007; Ryan et al., 1998; Scortegagna et al., 2003b). Hypoxia leads to a widespread accumulation of the oxygen-regulated subunits HIF-1α and -2α in virtually all organs with subsequent target gene activation (Rosenberger et al., 2002; Stroka et al., 2001; Wiesener et al., 2003). A suitable protocol to stabilize HIF-α in vivo is to create a severely hypoxic environment by exposing rats to 8% O_2 for 1 to 5 h (Rosenberger et al., 2002) or mice to 6% O_2 for 1 to 12 h (Stroka et al., 2001). Usually, oxygen concentrations in the atmosphere below 9% are required to demonstrate HIF-1α induction, whereas others found HIF-1α to be expressed under normoxic conditions (Stroka et al., 2001); in the brain, HIF-1α was already detectable when the animals were exposed to less severe hypoxia (15–18% O_2; [Stroka *et al.*, 2001]). In our experience, 4 to 6 h of 8% O_2 are sufficient for HIF-α accumulation (Fig. 12.3B), while the mortality of animals increases with exposure times exceeding 6 h or oxygen concentrations below 8% O_2. A confounding factor with regard to hypoxic HIF activation *in vivo* may be cardiorespiratory changes of the animals during systemic hypoxia. With ongoing hypoxia, a respiratory alkalosis develops (due to hyperventilation) as well as a reduction in heart rate and blood pressure. It remains unclear to which extent changes in pH might affect HIF activation. The increased production of lactic acid under hypoxic conditions with a subsequent decrease in pH was shown to activate HIF

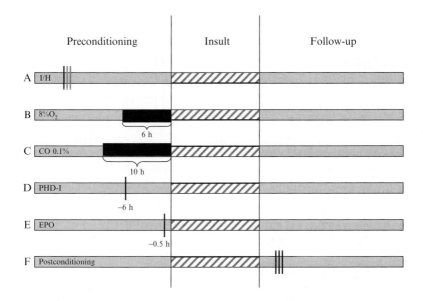

Figure 12.3 Example protocols of preconditioning. After an episode of preconditioning, the respective organ is protected against a subsequent insult. In a follow-up period, organ-function and -morphology is monitored. Example protocols are shown of (A) delayed (late) ischemic or hypoxic preconditioning (about 12 to 24 h prior, a single episode [black] or repetitive [gray] episodes of ischemia and reperfusion should be applied; (B) continuous hypoxia: animals should be exposed to an ongoing episode of hypoxia (e.g., 8% O_2 for 1–6 h) without reoxygenation in order to confer protection; (C) carbon monoxide (functional anemia): continuous exposition of the animals to CO 0.1% for 10 h until the initiation of the insult; (D) HIF prolyl hydroxylase inhibitor: a single intraperitoneal or intravenous dose of a small molecule inhibitor of the HIF PHD (PHD-I) 6 h prior to an insult was shown to be protective against ischemic injury; (E) EPO treatment: high dosages up to 5000 IU/kg b.w. given i.p. or intravenously (i.v.) shortly before onset of ischemia; (F) Post-conditioning: repetitive short episodes of ischemia or hypoxia applied at defined time frames after an ischemic insult or repetitive dosing of a PHD-I after the injury have been demonstrated to be still capable of conferring protection when given after the insult.

by neutralizing the capability of pVHL to degrade HIF (Mekhail *et al.*, 2004). However, *in vitro* experiments from our own group showed inconsistent effects of changes in pH on HIF-α activation (Willam *et al.*, 2006b). Irrespective of the changes in systemic pH, the repression of circulation in animals exposed to hypoxia may lead to hypoperfusion due to functional sympatholysis of organs with potential aggravation of the severity of hypoxia (Dinenno, 2003).

A more stable and better tolerated experimental stimulus of HIF *in vivo* is the induction of functional anemia by carbon monoxide (CO). CO binds with very high affinity to hemoglobin, thus lowering the oxygen-carrying capacity of the blood. Inhalation of CO 0.1% was shown to increase rat plasma EPO levels (Jelkmann and Seidl, 1987). Indeed, CO inhalation leads to a widespread HIF-α accumulation in various organs (Rosenberger *et al.*, 2002; Wiesener *et al.*, 2003) with transactivation of HIF target genes, including EPO, heme-oxygenase-1 (HO-1), and Glut-1 (Bernhardt *et al.*, 2006; Rosenberger *et al.*, 2002). For the purpose of stabilizing HIF and in an effort to confer protection by preconditioning, the exposure of rats to CO 0.1% added to ambient air for 10 h proved useful (Bernhardt *et al.*, 2006) (see Fig. 12.3C).

4.2. Inhibition of HIF prolyl hydroxylases

In addition to hypoxia, inhibitors of the HIF prolyl and asparaginyl hydroxylases, which stabilize HIF and subsequently activate HIF targets, can be used to mimic the hypoxic response (see Fig. 12.3B). In a first step, the PHDs are decarboxylating 2-oxoglutarate to succinate, forming a reactive iron–oxo complex, which subsequently hydroxylates the defined amino acid residue of the peptide substrate. The cofactor ascorbate is only required after repeated reaction cycles and prevents auto-oxidation of the enzyme in the uncoupled reaction.

Another member of the family of 2-oxoglutarate and Fe(II)-dependent dioxygenases is the mammalian procollagen prolyl-4-hydroxylase (P4H). A number of small molecule inhibitors of these P4H have been developed (Baader *et al.*, 1994; Gunzler *et al.*, 1988) for the treatment of fibrotic disorders. Some of these were shown to also inhibit the HIF PHDs at low concentrations (Ivan *et al.*, 2002), thus becoming potentially suitable for HIF stabilization *in vivo*. The majority of these PHD inhibitors (PHD-I) are competitive with regard to 2-oxoglutarate in the catalytic center of the PHD, leading to an oxygen-independent activation of HIF (see Fig. 12.1B,C).

The 2-oxoglutarate analogous N-oxalyl glycine and dimethyloxalglycin (DMOG) were the first substances shown to inhibit PHDs *in vitro* and in worms (Epstein *et al.*, 2001; Jaakkola *et al.*, 2001); however, these compounds are not potent enough to be suitable for clinical use. Further screening for more potent inhibitors of HIF PHD revealed a number of 2-oxoglutarate analogues, such as FG-0041 (Ivan *et al.*, 2002; Nwogu *et al.*,

2001), the plant extract L-mimosine (Warnecke *et al.*, 2003), FG-2216 (Philipp *et al.*, 2006b), or FG-4487 (Bernhardt *et al.*, 2006), all of which have the potential to activate HIF *in vivo*. Repetitive dosing of a small molecular inhibitor of PHDs was reported to confer protection (Nwogu *et al.*, 2001; Philipp *et al.*, 2006b), as was a single dose of PHD-I given 6 h prior to ischemia (see Fig. 12.3D) (Bernhardt *et al.*, 2006; Siddiq *et al.*, 2005). With regard to the effects of PHD-I, PHD2 and 3 and FIH are HIF target genes themselves, and their upregulation under hypoxia results in a negative feedback loop, which limits HIF activation (Aprelikova *et al.*, 2004; D'Angelo *et al.*, 2003; Metzen *et al.*, 2005; Stiehl *et al.*, 2006). PHD2 is almost ubiquitously expressed and was shown to be the enzyme with the strongest impact on HIF stability under normoxic conditions (Berra *et al.*, 2003), whereas PHD3 has the highest amplitude of hypoxic induction (Willam *et al.*, 2006a). Thus, pharmacological PHD inhibitors should inhibit both enzymes in order to stabilize HIF effectively. Further-more, FIH needs to be blocked to achieve maximum effect on HIF-1α, while HIF-2α seems to be relatively resistant to FIH effects (Yan *et al.*, 2007). It should also be taken into account that additional interaction partners of the HIF hydroxylases may exist, such as IκB (Cockman *et al.*, 2006). To understand the whole effect of the inhibition of HIF PHDs, the impact of alternative PHD/FIH functions *in vivo* still needs to be assessed.

4.3. Additional strategies to activate the HIF pathway

The early observation that the iron chelator desferrioxamine (DFO) is able to stabilize HIF (Wang and Semenza, 1993) led to the hypothesis that the cellular oxygen sensor contains a heme molecule. The fact that the oxygen-sensing PHDs contain a non-heme–bound divalent iron (Fe^{2+}) in their catalytic center (see Fig. 12.1C) can now explain these earlier findings. However, in terms of potential application, iron chelators presumably constitute the least HIF-specific activators of this pathway. Since iron is an essential cofactor for many cellular enzymes (e.g., iron is involved in the conversion of ribonucleotides to desoxyribonucleotides and in cell cycle progression), iron chelation affects DNA synthesis and cell growth (Le and Richardson, 2002). To which extent the antiproliferative effects of iron chelators are HIF-dependent is not completely understood.

Also, cobaltous chloride was long suggested to compete for the Fe^{2+} in the catalytic center of PHDs; however, recent evidence revealed that it may instead lead to the depletion of cellular ascorbate and thereby induce oxidant stress and inhibit PHD activity (Salnikow *et al.*, 2004).

Genetic approaches based on adenovirus-mediated expression of stable HIF-α mutants demonstrated the efficacy of HIF overexpression in induc-ing angiogenesis and HIF target genes (Pachori *et al.*, 2004). However, despite the difficulties in cell type–specific targeting and the potential risk of immune complex disease associated with repeated administration of replication-defective adenoviruses, these approaches are under clinical evaluation (http://clinicaltrials.gov/show/NCT00117650).

5. HYPOXIC PRECONDITIONING AND HIF

5.1. Protective role of HIF target genes

Hypoxic preconditioning leads to an upregulation of a number of genes, including the HIF target genes EPO and HO-1. EPO was originally considered a very specific erythropoietic factor. Since then, however, it has been demonstrated to be tissue protective independently from its effects on red cell production. The application of comparatively high doses of EPO already proved to be effective in reducing ischemic insults in the central nervous system, myocardium, retina, and kidney (Bahlmann *et al.*, 2004; Baker, 2005; Grimm *et al.*, 2006; Patel *et al.*, 2004; Sharples *et al.*, 2004, 2006). In addition, it is the only secreted HIF target gene product that is available as a licensed drug with an excellent safety profile for almost two decades (Macdougall and Eckardt, 2006).

Initial experiments used very high doses of EPO (Siren *et al.*, 2001); however, more recently, the protective effects of EPO were still observed in several models of organ injury, using doses of EPO that are not increasing hematocrit levels (Bahlmann *et al.*, 2004). The protective effects have been attributed to the anti-apoptotic, anti-necrotic, and mitogenic properties of EPO, which are mediated by the canonical JAK2 kinase pathway and by PI3 kinase/Akt and mitogen-activated protein (MAP) kinase activation.

Another HIF target gene, HO-1 (also known as heat-shock protein [Hsp] 32), is one of three isoforms that catalyzes the degradation of the heme molecule into Fe^{2+}, biliverdin, and CO. Induction or overexpression of HO-1 confers protection from injury in various models of liver, kidney, and cardiac ischemia (Blydt-Hansen *et al.*, 2003; Kato *et al.*, 2001; Katori *et al.*, 2002; Tullius *et al.*, 2002). The protective effect of HO-1 is conveyed by its products biliverdin, bilirubin (due to their antioxidant properties), and CO (which has been shown to act as an inhibitor of inflammation [Otterbein *et al.*, 2000], platelet aggregator [Brune and Ullrich, 1987], and vasodilator [Ishikawa *et al.*, 2005]).

Although it is possible that the effect of one or a few target gene products may play a dominant role in tissue protection that is exerted by HIF stabilization, activation of this "master switch" as an important physiologic response mechanism is likely to result in more reproducible and widely applicable effects than intervention that is based on a single gene. Importantly, the activation of the whole spectrum of HIF target genes may be synergistic (e.g., increasing glucose uptake into cells and subsequent steps of anaerobic glycolysis, enhanced by transcriptional activation of almost every glycolytic enzyme) (Maxwell, 2005; Wenger, 2002).

5.2. HIF activation and organ protection

Most evidence is available for the role of HIF in neuronal, myocardial, and renal protection; however, data is increasing on HIF-mediated protection of additional organs.

5.2.1. Neuroprotection

The brain was the first organ that was shown to be protected by EPO (Siren et al., 2001), probably the best-examined HIF target gene in terms of brain protection. The dominant effect of EPO is the reduction of cellular apoptosis of the damaged organ. Mechanistically, the EPO-mediated protection of neurons seems to involve a crosstalk between Jak2 and NF-κB signaling cascades (Digicaylioglu and Lipton, 2001). The majority of studies on the role of HIF and hypoxia in brain injury have focused on HIF-1, and, as expected, hypoxic preconditioning has been shown to induce HIF-1 in the brain with subsequent upregulation of HIF target genes (Bergeron et al., 1999). Concordant with findings in other organs (Jurgensen et al., 2004; Rosenberger, 2003), HIF-1α and HIF-target genes are upregulated in the penumbra of brain infarcts (Sharp et al., 2001). Microarray analysis of the neonatal rat brain in the first 24 h following 3 h of hypoxia (8% O_2) showed a number of HIF-α target genes to be upregulated, including VEGF, EPO, Glut-1, adrenomedullin, and t-PA (Bernaudin et al., 2002). Under comparable conditions, HIF-1α and PHD2 protein were shown to be upregulated within the 24-h reoxygenation period after 3 h of hypoxia with a differential expression pattern of HIF-1α and PHD2 (Jones et al., 2006).

Different strategies associated with HIF activation were shown to reduce the cerebral infarction zone in rodents. In the first strategy, which involved hypoxia, various time periods of pre-insult hypoxia (8% O_2 for 1, 3, and 6 h) reduced the infarct size by 30% in a mouse model of permanent focal ischemia, which was associated with HIF-1α and target gene activation (Bernaudin et al., 2002). In the immature rat brain, hypoxic preconditioning (3 h of 8% O_2 24 h prior to injury) reduced infarct size by 96% (Jones and Bergeron, 2001). The second strategy, which involved administration of intraperitoneal cobaltous chloride (60 mg/kg) resulted in upregulation of HIF and conferred significant protection of rat b from ischemia, which was, however, inferior to the level of protection reached with hypoxic preconditioning (Bergeron et al., 2000; Jones and Bergeron, 2001). In a third strategy, which involves iron chelators, intraperitoneal injection of DFO (200 mg/kg) only moderately upregulated HIF-1α protein within 1 to 3 h after injection; however, preconditioning with desferrioxamine 24 h before ischemia still remained protective when compared to controls.

It was less effective, though, than cobaltous chloride or hypoxia (Bergeron et al., 2000). In order to stimulate the HIF system more specifically and independent of oxygen tensions, Siddiq et al. (2005) developed a fourth strategy by first using a small molecule inhibitor of the PHDs named "compound A," a 2-oxoglutarate analogue, at a dosage of 100 mg/kg body weight, applied by oral gavage 6 h prior to middle cerebral artery occlusion of rats. Compound A stabilized HIF-1α and transactivated EPO in the brain of treated rats, leading to a significant reduction of infarct volume (Siddiq et al., 2005).

Mechanistically, it was speculated that HIF may—among many other effects—play a role in protecting the mitochondrial respiratory chain. In an astroglial cell culture, pre-conditional treatment by L-mimosine, cobaltous chloride, or DFO resulted in an HIF-1α-dependent protection against mitochondrial damage of these cells, which was independent of p42/44, NF-κB, and ERK (Yang et al., 2005), but may involve HIF-1α-dependent upregulation of cytochrome P450 2C11 (Liu and Alkayed, 2005).

Although most evidence supports the hypothesis that HIF protects against brain injury, evidence for negative effects of HIF-1α in severe ischemic injury of the mouse brain also have been reported (Helton et al., 2005). In contrast to other studies demonstrating a protective effect of HIF in case of ischemia, mice with late-stage deletion of HIF-1α in the brain subjected to very severe ischemia were protected from hypoxia-induced cell death, indicating that a reduced abundance of HIF-1α could be neuroprotective. Although most hypoxia-responsive genes were unaffected in these HIF-1α–deficient mice, a number of apoptotic genes were specifically downregulated, suggesting that HIF-1α may be proapoptotic and that loss of function of HIF-1α leads to neuroprotection by reduction of apoptosis.

In a specific population of neurons (the retinal photoreceptors), hypoxic preconditioning with upregulation of HIF-1α, VEGF, and EPO protects from light-induced retinal damage (Grimm et al., 2002). Thus, HIF-1α activation by hypoxia (6% O_2 for 6 h followed by 4 h of reoxygenation) and systemic EPO administration (5000 IU intraperitoneally) were shown to be equipotent in mediating protection (Grimm et al., 2002).

5.2.2. Myocardial protection

Functioning HIF is necessary for myocardial integrity. Cardiac myocyte-specific knockout of HIF-1α results in impaired vascularization and physiology of the non-hypoxic heart (Huang et al., 2004). HIF is also upregulated after myocardial infarction, with HIF-1α primarily expressed in the border zone of the infarct and HIF-2α subsequently also in the remaining myocardium (Jurgensen et al., 2004), suggesting that the HIF system plays an important role in restricting ischemic myocardial damage and adaptive remodeling.

The heart was the first organ for which the principle of hypoxic preconditioning was described (Shizukuda *et al.*, 1992, 1993). Hypoxia prior to myocardial infarction reduces the infarct area (Shizukuda *et al.*, 1992, 1993), enhances vascular density, and improves left ventricular contractility (Sasaki *et al.*, 2002). Experiments with heterozygous HIF-1α knockout mice in which the protective effect of a preconditioning protocol was blunted suggest that HIF plays an important role in this adaptation. Wild-type (wt) and heterozygous HIF-1α knockout mice (HIF-1α$^{+/-}$) were exposed to 5 cycles of 6 min of hypoxia (6% O_2) followed by 6 min of reperfusion. After 24 h of recovery (21% O_2), the hearts were isolated (Langendorff preparation) and subjected to 30 min of ischemia followed by 2 h of reperfusion. Preconditioned hearts of wt mice were substantially protected as indicated by a significantly reduced infarct size and higher left ventricular diastolic pressure, whereas no difference was observed between hearts from HIF-1α$^{+/-}$ mice and hearts from animals without hypoxic preconditioning (Cai *et al.*, 2003). While the loss of protection was not associated with different levels of the HIF target genes VEGF, Glut-1, or NOS2, the heterozygous animals lost the capability to transactivate EPO in response to hypoxia (Cai *et al.*, 2003). The protective effect of hypoxic preconditioning could be mimicked by the application of 5000 U/kg EPO to the animals 24 h prior to Langendorff preparation and ischemia (30′)/ reflow (2 h) (Cai *et al.*, 2003), further supporting a critical role of EPO for myocardial protection. However, endogenous EPO production in the myocardium and the expression and function of EPO receptors in the heart remain poorly understood. Other studies have also shown that EPO is regulated by HIF-2α and not by HIF-1α (Rankin *et al.*, 2007; Scortegagna *et al.*, 2003a; Warnecke *et al.*, 2004), which raises additional questions about how HIF-1 effects can be mediated by EPO.

In order to activate the HIF system, desferrioxamine (Philipp *et al.*, 2006a) and cobaltous chloride (Xi *et al.*, 2004) have been used for preconditioning. Both compounds were shown to protect the rodent myocardium via induction of HIF and HIF target genes. Again, the more specific approach of HIF activation is to inhibit the PHDs by 2-oxoglutarate analogues. Using the hearts of female rats, Nwogu *et al.* (2001) aimed to reduce myocardial fibrosis 48 h after myocardial infarction with FG-0041, which was, at that time, assumed to be a specific inhibitor of the procollagen prolyl hydroxylases; it was later found to inhibit the HIF PHDs and FIH as well, leading to stabilization and activation of HIF (Ivan *et al.*, 2002). A first protective effect was already observed after 1 wk, and, after 4 wk, left ventricular function of FG-0041-treated rats was significantly improved compared to controls (Nwogu *et al.*, 2001). Likely, this effect is at least partially due to HIF activation. In further experiments, different PHD-I were used for preconditioning prior to myocardial injury (as compared to early intervention after injury). Intravenous administration of DMOG

(20 mg/kg) 24 h prior to the induction of myocardial infarction in rabbits significantly reduced infarct size and was associated with activation of HO-1 and reduced interleukin-8 production (Ockaili *et al.*, 2005). Philipp *et al.* (2006b) used the more potent PHD-I FG-2216, which stabilizes HIF-α in the heart and other organs. Rats received 30 mg/kg of FG-2216 twice daily per os starting 48 h prior to myocardial infarction and were treated for 9 or 32 days, respectively. One week after myocardial infarction, FG-2216-treated animals had a significantly reduced infarct size, and, after 28 days, FG-2216 treatment led to an improved left ventricular contractility (Philipp *et al.*, 2006b).

An alternative genetic approach to activate the HIF system was used by silencing of PHD2 using small interfering RNA (siRNA). A strong reduction of the cardiac PHD2 messenger RNA (mRNA) expression 24 h after treatment was associated with stabilization of HIF-1α (Natarajan *et al.*, 2006). This PHD2 siRNA transfection of mouse hearts with subsequent upregulation of HIF-1α significantly reduced the infarct size after 30′ ischemia followed by 60 min of reperfusion in comparison to controls. The protective effect was shown to be predominantly inducible nitric oxide synthase (iNOS)–dependent as far as the effect of PHD2 silencing was abolished in iNOS knockout mice (Natarajan *et al.*, 2006).

5.2.3. Nephroprotection

The kidney has a widespread capacity to activate HIF-1α and -2α and HIF target genes in response to hypoxia (Rosenberger *et al.*, 2002), carbon monoxide (Bernhardt *et al.*, 2006; Rosenberger *et al.*, 2002), cobaltous chloride (Rosenberger *et al.*, 2002), and PHD-I (Bernhardt *et al.*, 2006). In several studies, these stimuli were used for preconditioning in order to protect the kidney from acute injury via HIF activation.

Using a rat model of ischemia reperfusion injury, we administered the PHD-I FG-4487 6 h prior to clamping of renal arteries for 40 min (Bernhardt *et al.*, 2006). In a second group of animals, HIF was induced by exposure to 0.1% CO for 10 h. FG-4487 treatment led to HIF-1α protein accumulation in virtually all nephron segments with the exception of the distal convoluted tubule, whereas, after CO exposure, HIF-1α signals were predominant in the proximal tubule, as described previously (Bernhardt *et al.*, 2006; Rosenberger *et al.*, 2002). HIF-2α was observed in interstitial and glomerular cells under hypoxic (0.1% CO) conditions and after pharmacological induction (Bernhardt *et al.*, 2006; Rosenberger *et al.*, 2002), reflecting a clear cell selectivity of the expression pattern. Increases of HO-1 and EPO mRNA levels confirmed the transcriptional activation of HIF-1α and HIF-2α target genes, respectively. The observed increases of serum creatinine and urea after the ischemic injury were significantly decreased by 40 to 60% in the FG-4487– and CO-treated rats after 24 and 72 h. Concordantly, histomorphological scoring revealed a significant reduction in cast formation, the number of apoptotic cells, and the extent of tubular necrosis in the treatment groups, with a trend toward a stronger effect of FG-4487 in the inner stripe of outer medulla, the region with the most pronounced differences in HIF-1α induction between FG-4487 and CO. An intriguing difference was the significant reduction in the number of ED1-positive macrophages in the CO-treated group, which was not reproduced by the PHD-I, suggesting that this effect was due to HIF-independent anti-inflammatory properties of CO (Otterbein *et al.*, 2000).

A previous report has shown similar protective effects using cobaltous chloride. When administered via drinking water for 14 d prior to clamping of renal arteries or only at day −1 and 0, cobalt blunted the increase of serum creatinine levels, improved histomorphological parameters, such as tubular injury, tubular dilation, and cast formation, reduced macrophage infiltration, and prevented peritubular capillary loss (Matsumoto *et al.*, 2003). Paradoxically, cobalt was shown to stabilize HIF in the distal nephron, whereas ischemia-induced renal injury in the renal artery clamp model is mainly confined to the proximal tubules (Rosenberger *et al.*, 2002). In rat models of cisplatin-induced nephrotoxicity, cobalt and hypoxia were also found to ameliorate injury; however, the role of HIF in this setting is, so far, less clear (Tanaka *et al.*, 2005; Wang *et al.*, 2006). Due to its carcinogenic properties, which are probably caused by oxidative DNA damage, the administration of cobalt chloride is limited to preclinical studies.

5.2.4. Other organs

Protection mediated by hypoxic preconditioning has also been demonstrated in a number of other organs, including the lung (Zhang *et al.*, 2004), the intestine (Ceylan *et al.*, 2005), and the liver (Lai *et al.*, 2004); however, the potentially protective role of HIF in these organs is not well defined yet and requires further examination.

In skeletal muscle, HIF-1α is required for metabolism and integrity (Ameln *et al.*, 2005; Dapp *et al.*, 2006; Mason *et al.*, 2004), and hypoxic preconditioning improves contractility and recovery after a subsequent result of single muscle fibers (Kohin *et al.*, 2001). An increase in HIF-1α in mouse skeletal muscle induced by treatment with DMOG was found to lead to increased vessel density after ligation of the femoral artery (Milkiewicz *et al.*, 2004).

6. HIF in Chronic Hypoxic/Ischemic Diseases

To date, little is known about the long-term effects of PHD-I on organ function in models of chronic ischemic disease. The specific requirements for a compound that ameliorates chronic ischemic injury differ markedly from those required for acute protection, and yet, due to the widespread genomic response elicited by HIF activation, PHD-I may fulfill this task. The induction of angiogenic growth factors in order to protect the endothelium and to induce new vessel growth would presumably be the most important part of the HIF response in the chronic setting. HIF-1α induces not only VEGF-A levels, but also expression of the VEGF receptor Flt-1, of angiopoietins and their receptors, and of angiopoietin-like factors (see Fig. 12.2). Some studies demonstrated that HIF activation by PHD-I stimulates angiogenesis within days and weeks (Linden *et al.*, 2003; Warnecke *et al.*, 2003). Administration or overexpression of VEGF-A alone leads to the generation of leaky vessels and edema, whereas the coordinated induction of angiopoietic genes through HIF activation leads to the formation of a functionally competent vascular network (Elson *et al.*, 2001; Pajusola *et al.*, 2005). Besides the angiogenic and vasoactive effects of HIF, the induction of glycolytic genes, HO-1, growth factors, and EPO may further improve cellular energy and oxygen supply, promote cell survival, and reduce oxidant stress and thereby may have a further impact on organ function and tissue integrity.

Some concern exists regarding the side effects of prolonged HIF activation, which could include enhanced fibrosis, since some pro-fibrotic genes were found to be HIF-dependent (Haase, 2006). Furthermore, different levels of target gene expression may have opposite effects, with protection on the one hand (varez Arroyo *et al.*, 2002) and progression of organ failure on the other (Eremina *et al.*, 2003). A further potential risk of the chronic (but not acute) use of PHD-I may lie in the relevant role of HIF for tumor progression. Some forms of congenital erythrocytosis have been linked to chronic over-activation of the HIF system, and affected individuals show no apparently increased incidence of malignant disease (Gordeuk *et al.*, 2004; Pastore *et al.*, 2003), suggesting that HIF stimulation alone may not be sufficient to enhance tumor formation. Nevertheless, the safety of prolonged HIF stabilization will require careful attention.

7. CONCLUSIONS AND PERSPECTIVES

Since the first description of ischemic preconditioning approximately 20 years ago (Murry *et al.*, 1986), the identification of the molecular control of hypoxia-inducible gene transcription mediated by HIF has led to a novel pharmacological approach of tissue protection.

Although the HIF system is only one component of the spectrum of reactions induced by ischemia reperfusion, the animal experiments performed so far point to a dominant and functionally important role. In addition, the available evidence suggests that, although HIF is principally induced in various tissues under conditions of ischemic injury, this response has some time lag and is submaximal so that additional therapeutic activation can lead to an earlier and more robust activation of its target genes, which confers benefit of the intervention as compared to control conditions.

Although protection may be maximal when the intervention can be carried out well before the onset of ischemic injury, or true "preconditioning" (e.g., before elective surgery or organ transplantation), it is likely that benefit can still be achieved in the early phase after an insult (see Fig. 12.3B), which would greatly broaden the therapeutic applicability (e.g., to conditions such as acute myocardial infarction or stroke). The pharmacologic approach of HIF induction with 2-oxoglutarate analogues is easily applicable, and clinical trials in which orally active PHD inhibitors are being used for the treatment of anemia through stimulation of EPO production are already underway in humans (http://clinicaltrials.gov/ct/show/NCT00456053).

One question will be whether the cell specificity of the expression of HIF-α isoforms and the tissue and cellular distribution of PHD1 to 3 and FIH-1 as well as subtle differences between these enzymes will allow us to design inhibitors that lead to a selective activation of certain HIF target genes at specific sites. For example, an inhibitor of FIH-1, which seems not to have an impact on HIF-2α, may thus not activate HIF-2α, and the potential protective effect would be singularly HIF-1α–dependent. Theoretically, PHD-I and/or FIH-I may be tailored for maximal protection under different indications with minimal side effects.

A further question relates to the role of single target genes in the spectrum of the response induced by HIF stabilization and the extent to which individual HIF target genes contribute to the observed protective effect of pharmacological HIF induction. Although in several experimental settings specific genes, such as EPO, HO-1, or iNOS, were found to play important roles, it is more likely that the combined activation of a whole set of target genes exerts more reproducible and stronger effects than that of single genes or their products.

REFERENCES

Ameln, H., Gustafsson, T., Sundberg, C. J., Okamoto, K., Jansson, E., Poellinger, L., and Makino, Y. (2005). Physiological activation of hypoxia inducible factor-1 in human skeletal muscle. *FASEB J.* **19,** 1009–1011.

Aprelikova, O., Chandramouli, G. V., Wood, M., Vasselli, J. R., Riss, J., Maranchie, J. K., Linehan, W. M., and Barrett, J. C. (2004). Regulation of HIF prolyl hydroxylases by hypoxia-inducible factors. *J. Cell Biochem.* **92,** 491–501.

Arany, Z., Huang, L. E., Eckner, R., Bhattacharya, S., Jiang, C., Goldberg, M. A., Bunn, H. F., and Livingston, D. M. (1996). An essential role for p300/CBP in the cellular response to hypoxia. *Proc. Natl. Acad. Sci. USA* **93,** 12969–12973.

Baader, E., Tschank, G., Baringhaus, K. H., Burghard, H., and Gunzler, V. (1994). Inhibition of prolyl 4-hydroxylase by oxalyl amino acid derivatives *in vitro*, in isolated microsomes and in embryonic chicken tissues. *Biochem. J.* **300**(Pt. 2), 525–530.

Bahlmann, F. H., De Groot, K., Spandau, J. M., Landry, A. L., Hertel, B., Duckert, T., Boehm, S. M., Menne, J., Haller, H., and Fliser, D. (2004). Erythropoietin regulates endothelial progenitor cells. *Blood* **103**, 921–926.

Baker, J. E. (2005). Erythropoietin mimics ischemic preconditioning. *Vascul. Pharmacol.* **42**, 233–241.

Bergeron, M., Yu, A. Y., Solway, K. E., Semenza, G. L., and Sharp, F. R. (1999). Induction of hypoxia-inducible factor-1 (HIF-1) and its target genes following focal ischaemia in rat brain. *Eur. J. Neurosci.* **11**, 4159–4170.

Bergeron, M., Gidday, J. M., Yu, A. Y., Semenza, G. L., Ferriero, D. M., and Sharp, F. R. (2000). Role of hypoxia-inducible factor-1 in hypoxia-induced ischemic tolerance in neonatal rat brain. *Ann. Neurol.* **48**, 285–296.

Bernaudin, M., Tang, Y., Reilly, M., Petit, E., and Sharp, F. R. (2002). Brain genomic response following hypoxia and re-oxygenation in the neonatal rat. Identification of genes that might contribute to hypoxia-induced ischemic tolerance. *J. Biol. Chem.* **277**, 39728–39738.

Bernhardt, W. M., Campean, V., Kany, S., Jurgensen, J. S., Weidemann, A., Warnecke, C., Arend, M., Klaus, S., Gunzler, V., Amann, K., Willam, C., and Wiesener, M. S., *et al.* (2006). Preconditional activation of hypoxia-inducible factors ameliorates ischemic acute renal failure. *J. Am. Soc. Nephrol.* **17**, 1970–1978.

Berra, E., Benizri, E., Ginouves, A., Volmat, V., Roux, D., and Pouyssegur, J. (2003). HIF prolyl-hydroxylase 2 is the key oxygen sensor setting low steady-state levels of HIF-1α in normoxia. *EMBO J.* **22**, 4082–4090.

Blydt-Hansen, T. D., Katori, M., Lassman, C., Ke, B., Coito, A. J., Iyer, S., Buelow, R., Ettenger, R., Busuttil, R. W., and Kupiec-Weglinski, J. W. (2003). Gene transfer-induced local heme oxygenase-1 overexpression protects rat kidney transplants from ischemia/reperfusion injury. *J. Am. Soc. Nephrol.* **14**, 745–754.

Bolli, R. (2007). Preconditioning: A paradigm shift in the biology of myocardial ischemia. *Am. J. Physiol Heart Circ. Physiol.* **292**, H19–H27.

Bonventre, J. V. (2002). Kidney ischemic preconditioning. *Curr. Opin. Nephrol. Hypertens.* **11**, 43–48.

Bruick, R. K., and McKnight, S. L. (2001). A conserved family of prolyl-4-hydroxylases that modify HIF. *Science* **294**, 1337–1340.

Brune, B., and Ullrich, V. (1987). Inhibition of platelet aggregation by carbon monoxide is mediated by activation of guanylate cyclase. *Mol. Pharmacol.* **32**, 497–504.

Cai, Z., Manalo, D. J., Wei, G., Rodriguez, E. R., Fox-Talbot, K., Lu, H., Zweier, J. L., and Semenza, G. L. (2003). Hearts from rodents exposed to intermittent hypoxia or erythropoietin are protected against ischemia-reperfusion injury. *Circulation* **108**, 79–85.

Ceylan, H., Yuncu, M., Gurel, A., Armutcu, F., Gergerlioglu, H. S., Bagci, C., and Demiryurek, A. T. (2005). Effects of whole-body hypoxic preconditioning on hypoxia/reoxygenation-induced intestinal injury in newborn rats. *Eur. J. Pediatr. Surg.* **15**, 325–332.

Cockman, M. E., Lancaster, D. E., Stolze, I. P., Hewitson, K. S., McDonough, M. A., Coleman, M. L., Coles, C. H., Yu, X., Hay, R. T., Ley, S. C., Pugh, C. W., Oldham, N. J., *et al.* (2006). Posttranslational hydroxylation of ankyrin repeats in IκB proteins by the hypoxia-inducible factor (HIF) asparaginyl hydroxylase, factor inhibiting HIF (FIH). *Proc. Natl. Acad. Sci. USA* **103**, 14767–14772.

Cramer, T., Yamanishi, Y., Clausen, B. E., Forster, I., Pawlinski, R., Mackman, N., Haase, V. H., Jaenisch, R., Corr, M., Nizet, V., Firestein, G. S., and Gerber, H. P., *et al.* (2003). HIF-1α is essential for myeloid cell-mediated inflammation. *Cell* **112**, 645–657.

D'Angelo, G., Duplan, E., Boyer, N., Vigne, P., and Frelin, C. (2003). Hypoxia upregulates prolyl hydroxylase activity: A feedback mechanism that limits HIF-1 responses during reoxygenation. *J. Biol. Chem.* **278**, 38183–38187.

Dapp, C., Gassmann, M., Hoppeler, H., and Fluck, M. (2006). Hypoxia-induced gene activity in disused oxidative muscle. *Adv. Exp. Med. Biol.* **588**, 171–188.

Digicaylioglu, M., and Lipton, S. A. (2001). Erythropoietin-mediated neuroprotection involves cross-talk between Jak2 and NF-κB signaling cascades. *Nature* **412**, 641–647.

Dinenno, F. A. (2003). Hypoxic regulation of blood flow in humans. Alpha-adrenergic receptors and functional sympatholysis in skeletal muscle. *Adv. Exp. Med. Biol.* **543**, 237–248.

Elson, D. A., Thurston, G., Huang, L. E., Ginzinger, D. G., McDonald, D. M., Johnson, R. S., and Arbeit, J. M. (2001). Induction of hypervascularity without leakage or inflammation in transgenic mice overexpressing hypoxia-inducible factor-1alpha. *Genes Dev.* **15**, 2520–2532.

Epstein, A. C., Gleadle, J. M., McNeill, L. A., Hewitson, K. S., O'Rourke, J., Mole, D. R., Mukherji, M., Metzen, E., Wilson, M. I., Dhanda, A., Tian, Y. M., Masson, N., *et al.* (2001). *C. elegans* EGL-9 and mammalian homologs define a family of dioxygenases that regulate HIF by prolyl hydroxylation. *Cell* **107**, 43–54.

Eremina, V., Sood, M., Haigh, J., Nagy, A., Lajoie, G., Ferrara, N., Gerber, H. P., Kikkawa, Y., Miner, J. H., and Quaggin, S. E. (2003). Glomerular-specific alterations of VEGF-A expression lead to distinct congenital and acquired renal diseases. *J. Clin. Invest.* **111**, 707–716.

Gidday, J. M. (2006). Cerebral preconditioning and ischaemic tolerance. *Nat. Rev. Neurosci.* **7**, 437–448.

Gordeuk, V. R., Sergueeva, A. I., Miasnikova, G. Y., Okhotin, D., Voloshin, Y., Choyke, P. L., Butman, J. A., Jedlickova, K., Prchal, J. T., and Polyakova, L. A. (2004). Congenital disorder of oxygen sensing: Association of the homozygous Chuvash polycythemia VHL mutation with thrombosis and vascular abnormalities but not tumors. *Blood* **103**, 3924–3932.

Grimm, C., Wenzel, A., Acar, N., Keller, S., Seeliger, M., and Gassmann, M. (2006). Hypoxic preconditioning and erythropoietin protect retinal neurons from degeneration. *Adv. Exp. Med. Biol.* **588**, 119–131.

Grimm, C., Wenzel, A., Groszer, M., Mayser, H., Seeliger, M., Samardzija, M., Bauer, C., Gassmann, M., and Reme, C. E. (2002). HIF-1–induced erythropoietin in the hypoxic retina protects against light-induced retinal degeneration. *Nat. Med.* **8**, 718–724.

Gunzler, V., Hanauske-Abel, H. M., Myllyla, R., Kaska, D. D., Hanauske, A., and Kivirikko, K. I. (1988). Syncatalytic inactivation of prolyl 4-hydroxylase by anthracyclines. *Biochem. J.* **251**, 365–372.

Haase, V. H. (2006). Hypoxia-inducible factors in the kidney. *Am. J. Physiol. Renal Physiol.* **291**, F271–F281.

Helton, R., Cui, J., Scheel, J. R., Ellison, J. A., Ames, C., Gibson, C., Blouw, B., Ouyang, L., Dragatsis, I., Zeitlin, S., Johnson, R. S., and Lipton, S. A., *et al.* (2005). Brain-specific knockout of hypoxia-inducible factor-1alpha reduces rather than increases hypoxic-ischemic damage. *J. Neurosci.* **25**, 4099–4107.

Hewitson, K. S., McNeill, L. A., Riordan, M. V., Tian, Y. M., Bullock, A. N., Welford, R. W., Elkins, J. M., Oldham, N. J., Bhattacharya, S., Gleadle, J. M., Ratcliffe, P. J., and Pugh, C. W., *et al.* (2002). Hypoxia-inducible factor (HIF) asparagine hydroxylase is identical to factor inhibiting HIF (FIH) and is related to the cupin structural family. *J. Biol. Chem.* **277**, 26351–26355.

Huang, Y., Hickey, R. P., Yeh, J. L., Liu, D., Dadak, A., Young, L. H., Johnson, R. S., and Giordano, F. J. (2004). Cardiac myocyte-specific HIF-1α deletion alters vascularization, energy availability, calcium flux, and contractility in the normoxic heart. *FASEB J.* **18,** 1138–1140.

Ishikawa, M., Kajimura, M., Adachi, T., Maruyama, K., Makino, N., Goda, N., Yamaguchi, T., Sekizuka, E., and Suematsu, M. (2005). Carbon monoxide from heme oxygenase-2 is a tonic regulator against NO-dependent vasodilatation in the adult rat cerebral microcirculation. *Circ. Res.* **97,** e104–e114.

Ivan, M., Haberberger, T., Gervasi, D. C., Michelson, K. S., Gunzler, V., Kondo, K., Yang, H., Sorokina, I., Conaway, R. C., Conaway, J. W., and Kaelin, W. G., Jr. (2002). Biochemical purification and pharmacological inhibition of a mammalian prolyl hydroxylase acting on hypoxia-inducible factor. *Proc. Natl. Acad. Sci. USA* **99,** 13459–13464.

Iyer, N. V., Kotch, L. E., Agani, F., Leung, S. W., Laughner, E., Wenger, R. H., Gassmann, M., Gearhart, J. D., Lawler, A. M., Yu, A. Y., and Semenza, G. L. (1998). Cellular and developmental control of O_2 homeostasis by hypoxia-inducible factor 1 alpha. *Genes Dev.* **12,** 149–162.

Jaakkola, P., Mole, D. R., Tian, Y. M., Wilson, M. I., Gielbert, J., Gaskell, S. J., Kriegsheim, A., Hebestreit, H. F., Mukherji, M., Schofield, C. J., Maxwell, P. H., and Pugh, C. W., *et al.* (2001). Targeting of HIF-α to the von Hippel-Lindau ubiquitylation complex by O_2-regulated prolyl hydroxylation. *Science* **292,** 468–472.

Jelkmann, W., and Seidl, J. (1987). Dependence of erythropoietin production on blood oxygen affinity and hemoglobin concentration in rats. *Biomed. Biochim. Acta* **46,** S304–S308.

Jiang, B. H., Semenza, G. L., Bauer, C., and Marti, H. H. (1996). Hypoxia-inducible factor 1 levels vary exponentially over a physiologically relevant range of O_2 tension. *Am. J. Physiol.* **271,** C1172–C1180.

Jones, N. M., and Bergeron, M. (2001). Hypoxic preconditioning induces changes in HIF-1 target genes in neonatal rat brain. *J. Cereb. Blood Flow Metab.* **21,** 1105–1114.

Jones, N. M., Lee, E. M., Brown, T. G., Jarrott, B., and Beart, P. M. (2006). Hypoxic preconditioning produces differential expression of hypoxia-inducible factor-1alpha (HIF-1α) and its regulatory enzyme HIF prolyl hydroxylase 2 in neonatal rat brain. *Neurosci. Lett.* **404,** 72–77.

Jurgensen, J. S., Rosenberger, C., Wiesener, M. S., Warnecke, C., Horstrup, J. H., Grafe, M., Philipp, S., Griethe, W., Maxwell, P. H., Frei, U., Bachmann, S., and Willenbrock, R., *et al.* (2004). Persistent induction of HIF-1α and -2α in cardiomyocytes and stromal cells of ischemic myocardium. *FASEB J.* **18,** 1415–1417.

Kato, H., Amersi, F., Buelow, R., Melinek, J., Coito, A. J., Ke, B., Busuttil, R. W., and Kupiec-Weglinski, J. W. (2001). Heme oxygenase-1 overexpression protects rat livers from ischemia/reperfusion injury with extended cold preservation. *Am. J. Transplant.* **1,** 121–128.

Katori, M., Buelow, R., Ke, B., Ma, J., Coito, A. J., Iyer, S., Southard, D., Busuttil, R. W., and Kupiec-Weglinski, J. W. (2002). Heme oxygenase-1 overexpression protects rat hearts from cold ischemia/reperfusion injury via an antiapoptotic pathway. *Transplantation* **73,** 287–292.

Kohin, S., Stary, C. M., Howlett, R. A., and Hogan, M. C. (2001). Preconditioning improves function and recovery of single muscle fibers during severe hypoxia and reoxygenation. *Am. J. Physiol. Cell Physiol.* **281,** C142–C146.

Lai, I. R., Ma, M. C., Chen, C. F., and Chang, K. J. (2004). The protective role of heme oxygenase-1 on the liver after hypoxic preconditioning in rats. *Transplantation* **77,** 1004–1008.

Lando, D., Peet, D. J., Whelan, D. A., Gorman, J. J., and Whitelaw, M. L. (2002). Asparagine hydroxylation of the HIF transactivation domain: A hypoxic switch. *Science* **295,** 858–861.

Le, N. T., and Richardson, D. R. (2002). The role of iron in cell cycle progression and the proliferation of neoplastic cells. *Biochim. Biophys. Acta* **1603,** 31–46.

Linden, T., Katschinski, D. M., Eckhardt, K., Scheid, A., Pagel, H., and Wenger, R. H. (2003). The antimycotic ciclopirox olamine induces HIF-1α stability, VEGF expression, and angiogenesis. *FASEB J.* **17,** 761–763.

Liu, M., and Alkayed, N. J. (2005). Hypoxic preconditioning and tolerance via hypoxia inducible factor (HIF) 1alpha-linked induction of P450 2C11 epoxygenase in astrocytes. *J. Cereb. Blood Flow Metab.* **25,** 939–948.

Macdougall, I. C., and Eckardt, K. U. (2006). Novel strategies for stimulating erythropoiesis and potential new treatments for anaemia. *Lancet* **368,** 947–953.

Maltepe, E., Schmidt, J. V., Baunoch, D., Bradfield, C. A., and Simon, M. C. (1997). Abnormal angiogenesis and responses to glucose and oxygen deprivation in mice lacking the protein ARNT. *Nature* **386,** 403–407.

Mason, S. D., Howlett, R. A., Kim, M. J., Olfert, I. M., Hogan, M. C., McNulty, W., Hickey, R. P., Wagner, P. D., Kahn, C. R., Giordano, F. J., and Johnson, R. S. (2004). Loss of skeletal muscle HIF-1α results in altered exercise endurance. *PLoS Biol.* **2,** e288.

Masson, N., and Ratcliffe, P. J. (2003). HIF prolyl and asparaginyl hydroxylases in the biological response to intracellular O_2 levels. *J. Cell Sci.* **116,** 3041–3049.

Matsumoto, M., Makino, Y., Tanaka, T., Tanaka, H., Ishizaka, N., Noiri, E., Fujita, T., and Nangaku, M. (2003). Induction of renoprotective gene expression by cobalt ameliorates ischemic injury of the kidney in rats. *J. Am. Soc. Nephrol.* **14,** 1825–1832.

Maxwell, P. H. (2005). Hypoxia-inducible factor as a physiological regulator. *Exp. Physiol.* **90,** 791–797.

Maxwell, P. H., Wiesener, M. S., Chang, G. W., Clifford, S. C., Vaux, E. C., Cockman, M. E., Wykoff, C. C., Pugh, C. W., Maher, E. R., and Ratcliffe, P. J. (1999). The tumour suppressor protein VHL targets hypoxia-inducible factors for oxygen-dependent proteolysis. *Nature* **399,** 271–275.

McDonough, M. A., Li, V., Flashman, E., Chowdhury, R., Mohr, C., Lienard, B. M., Zondlo, J., Oldham, N. J., Clifton, I. J., Lewis, J., McNeill, L. A., Kurzeja, R. J., *et al.* (2006). Cellular oxygen sensing: Crystal structure of hypoxia-inducible factor prolyl hydroxylase (PHD2). *Proc. Natl. Acad. Sci. USA* **103,** 9814–9819.

Mekhail, K., Gunaratnam, L., Bonicalzi, M. E., and Lee, S. (2004). HIF activation by pH-dependent nucleolar sequestration of VHL. *Nat. Cell Biol.* **6,** 642–647.

Metzen, E., Stiehl, D. P., Doege, K., Marxsen, J. H., Hellwig-Burgel, T., and Jelkmann, W. (2005). Regulation of the prolyl hydroxylase domain protein 2 (PHD2/EGLN-1) gene: Identification of a functional hypoxia-responsive element. *Biochem. J.* **387,** 711–717.

Milkiewicz, M., Pugh, C. W., and Egginton, S. (2004). Inhibition of endogenous HIF inactivation induces angiogenesis in ischemic skeletal muscles of mice. *J. Physiol.* **560,** 21–26.

Murray, C. J., and Lopez, A. D. (1997). Alternative projections of mortality and disability by cause 1990–2020: Global Burden of Disease Study. *Lancet* **349,** 1498–1504.

Murry, C. E., Jennings, R. B., and Reimer, K. A. (1986). Preconditioning with ischemia: A delay of lethal cell injury in ischemic myocardium. *Circulation* **74,** 1124–1136.

Natarajan, R., Salloum, F. N., Fisher, B. J., Kukreja, R. C., and Fowler, A. A., III (2006). Hypoxia inducible factor-1 activation by prolyl 4-hydroxylase-2 gene silencing attenuates myocardial ischemia reperfusion injury. *Circ. Res.* **98,** 133–140.

Nwogu, J. I., Geenen, D., Bean, M., Brenner, M. C., Huang, X., and Buttrick, P. M. (2001). Inhibition of collagen synthesis with prolyl 4-hydroxylase inhibitor improves left ventricular function and alters the pattern of left ventricular dilatation after myocardial infarction. *Circulation* **104,** 2216–2221.

Ockaili, R., Natarajan, R., Salloum, F., Fisher, B. J., Jones, D., Fowler, A. A., III, and Kukreja, R. C. (2005). HIF-1 activation attenuates postischemic myocardial injury: Role for heme oxygenase-1 in modulating microvascular chemokine generation. *Am. J. Physiol Heart Circ. Physiol.* **289,** H542–H548.

Otterbein, L. E., Bach, F. H., Alam, J., Soares, M., Tao, L. H., Wysk, M., Davis, R. J., Flavell, R. A., and Choi, A. M. (2000). Carbon monoxide has anti-inflammatory effects involving the mitogen-activated protein kinase pathway. *Nat. Med.* **6,** 422–428.

Pachori, A. S., Melo, L. G., Hart, M. L., Noiseux, N., Zhang, L., Morello, F., Solomon, S. D., Stahl, G. L., Pratt, R. E., and Dzau, V. J. (2004). Hypoxia-regulated therapeutic gene as a preemptive treatment strategy against ischemia/reperfusion tissue injury. *Proc. Natl. Acad. Sci. USA* **101,** 12282–12287.

Pajusola, K., Kunnapuu, J., Vuorikoski, S., Soronen, J., Andre, H., Pereira, T., Korpisalo, P., Yla-Herttuala, S., Poellinger, L., and Alitalo, K. (2005). Stabilized HIF-1α is superior to VEGF for angiogenesis in skeletal muscle via adeno-associated virus gene transfer. *FASEB J.* **19,** 1365–1367.

Pastore, Y., Jedlickova, K., Guan, Y., Liu, E., Fahner, J., Hasle, H., Prchal, J. F., and Prchal, J. T. (2003). Mutations of von Hippel-Lindau tumor-suppressor gene and congenital polycythemia. *Am. J. Hum. Genet.* **73,** 412–419.

Patel, N. S., Sharples, E. J., Cuzzocrea, S., Chatterjee, P. K., Britti, D., Yaqoob, M. M., and Thiemermann, C. (2004). Pretreatment with EPO reduces the injury and dysfunction caused by ischemia/reperfusion in the mouse kidney *in vivo*. *Kidney Int.* **66,** 983–989.

Philipp, S., Cui, L., Ludolph, B., Kelm, M., Schulz, R., Cohen, M. V., and Downey, J. M. (2006a). Desferoxamine and ethyl-3,4-dihydroxybenzoate protect myocardium by activating NOS and generating mitochondrial ROS. *Am. J. Physiol Heart Circ. Physiol.* **290,** H450–H457.

Philipp, S., Jurgensen, J. S., Fielitz, J., Bernhardt, W. M., Weidemann, A., Schiche, A., Pilz, B., Dietz, R., Regitz-Zagrosek, V., Eckardt, K. U., and Willenbrock, R. (2006b). Stabilization of hypoxia inducible factor rather than modulation of collagen metabolism improves cardiac function after acute myocardial infarction in rats. *Eur. J. Heart Fail.* **8,** 347–354.

Rankin, E. B., Biju, M. P., Liu, Q., Unger, T. L., Rha, J., Johnson, R. S., Simon, M. C., Keith, B., and Haase, V. H. (2007). Hypoxia-inducible factor-2 (HIF-2) regulates hepatic erythropoietin *in vivo*. *J. Clin. Invest.* **117,** 1068–1077.

Ratcliffe, P. J. (2006). Understanding hypoxia signalling in cells—a new therapeutic opportunity? *Clin. Med.* **6,** 573–578.

Rosenberger, C., Griethe, W., Gruberg, G., Wiesener, M., Frei, U., Bachmann, S., and Eckardt, K. U. (2003). Cellular responses to hypoxia after renal segmented infarction. *Kidney Int.* **64,** 874–886.

Rosenberger, C., Mandriota, S., Jurgensen, J. S., Wiesener, M. S., Horstrup, J. H., Frei, U., Ratcliffe, P. J., Maxwell, P. H., Bachmann, S., and Eckardt, K. U. (2002). Expression of hypoxia-inducible factor-1alpha and -2alpha in hypoxic and ischemic rat kidneys. *J. Am. Soc. Nephrol.* **13,** 1721–1732.

Ryan, H. E., Lo, J., and Johnson, R. S. (1998). HIF-1α is required for solid tumor formation and embryonic vascularization. *EMBO J.* **17,** 3005–3015.

Salnikow, K., Donald, S. P., Bruick, R. K., Zhitkovich, A., Phang, J. M., and Kasprzak, K. S. (2004). Depletion of intracellular ascorbate by the carcinogenic metals nickel and cobalt results in the induction of hypoxic stress. *J. Biol. Chem.* **279,** 40337–40344.

Sasaki, H., Fukuda, S., Otani, H., Zhu, L., Yamaura, G., Engelman, R. M., Das, D. K., and Maulik, N. (2002). Hypoxic preconditioning triggers myocardial angiogenesis: A novel approach to enhance contractile functional reserve in rat with myocardial infarction. *J. Mol. Cell Cardiol.* **34,** 335–348.

Schlaifer, J. D., and Kerensky, R. A. (1997). Ischemic preconditioning: Clinical relevance and investigative studies. *Clin. Cardiol.* **20,** 602–606.

Schofield, C. J., and Ratcliffe, P. J. (2004). Oxygen sensing by HIF hydroxylases. *Nat. Rev. Mol. Cell Biol.* **5,** 343–354.

Scortegagna, M., Morris, M. A., Oktay, Y., Bennett, M., and Garcia, J. A. (2003a). The HIF family member EPAS1/HIF-2α is required for normal hematopoiesis in mice. *Blood* **102,** 1634–1640.

Scortegagna, M., Ding, K., Oktay, Y., Gaur, A., Thurmond, F., Yan, L. J., Marck, B. T., Matsumoto, A. M., Shelton, J. M., Richardson, J. A., Bennett, M. J., and Garcia, J. A. (2003b). Multiple organ pathology, metabolic abnormalities and impaired homeostasis of reactive oxygen species in EPAS1$^{-/-}$ mice. *Nat. Genet.* **35,** 331–340.

Semenza, G. L., and Wang, G. L. (1992). A nuclear factor induced by hypoxia via *de novo* protein synthesis binds to the human erythropoietin gene enhancer at a site required for transcriptional activation. *Mol. Cell Biol.* **12,** 5447–5454.

Sharp, F. R., Bergeron, M., and Bernaudin, M. (2001). Hypoxia-inducible factor in brain. *Adv. Exp. Med. Biol.* **502,** 273–291.

Sharples, E. J., Thiemermann, C., and Yaqoob, M. M. (2006). Novel applications of recombinant erythropoietin. *Curr. Opin. Pharmacol.* **6,** 184–189.

Sharples, E. J., Patel, N., Brown, P., Stewart, K., Mota-Philipe, H., Sheaff, M., Kieswich, J., Allen, D., Harwood, S., Raftery, M., Thiemermann, C., Yaqoob, M. M., *et al.* (2004). Erythropoietin protects the kidney against the injury and dysfunction caused by ischemia-reperfusion. *J. Am. Soc. Nephrol.* **15,** 2115–2124.

Shizukuda, Y., Mallet, R. T., Lee, S. C., and Downey, H. F. (1992). Hypoxic preconditioning of ischaemic canine myocardium. *Cardiovasc. Res.* **26,** 534–542.

Shizukuda, Y., Iwamoto, T., Mallet, R. T., and Downey, H. F. (1993). Hypoxic preconditioning attenuates stunning caused by repeated coronary artery occlusions in dog heart. *Cardiovasc. Res.* **27,** 559–564.

Siddiq, A., Ayoub, I. A., Chavez, J. C., Aminova, L., Shah, S., Lamanna, J. C., Patton, S. M., Connor, J. R., Cherny, R. A., Volitakis, I., Bush, A., Langsetmo, I., *et al.* (2005). HIF prolyl 4-hydroxylase inhibition: A target for neuroprotection in the central nervous system. *J. Biol. Chem.* **280,** 41732–41743.

Siren, A. L., Fratelli, M., Brines, M., Goemans, C., Casagrande, S., Lewczuk, P., Keenan, S., Gleiter, C., Pasquali, C., Capobianco, A., Mennini, T., Heumann, R., *et al.* (2001). Erythropoietin prevents neuronal apoptosis after cerebral ischemia and metabolic stress. *Proc. Natl. Acad. Sci. USA* **98,** 4044–4049.

Stiehl, D. P., Wirthner, R., Koditz, J., Spielmann, P., Camenisch, G., and Wenger, R. H. (2006). Increased prolyl 4-hydroxylase domain proteins compensate for decreased oxygen levels. Evidence for an autoregulatory oxygen-sensing system. *J. Biol. Chem.* **281,** 23482–23491.

Stroka, D. M., Burkhardt, T., Desbaillets, I., Wenger, R. H., Neil, D. A., Bauer, C., Gassmann, M., and Candinas, D. (2001). HIF-1 is expressed in normoxic tissue and displays an organ-specific regulation under systemic hypoxia. *FASEB J.* **15,** 2445–2453.

Tanaka, T., Kojima, I., Ohse, T., Inagi, R., Miyata, T., Ingelfinger, J. R., Fujita, T., and Nangaku, M. (2005). Hypoxia-inducible factor modulates tubular cell survival in cisplatin nephrotoxicity. *Am. J. Physiol. Renal Physiol.* **289,** F1123–F1133.

Tullius, S. G., Nieminen-Kelha, M., Buelow, R., Reutzel-Selke, A., Martins, P. N., Pratschke, J., Bachmann, U., Lehmann, M., Southard, D., Iyer, S., Schmidbauer, G., Sawitzki, B., *et al.* (2002). Inhibition of ischemia/reperfusion injury and chronic graft deterioration by a single-donor treatment with cobalt-protoporphyrin for the induction of heme oxygenase-1. *Transplantation* **74,** 591–598.

varez Arroyo, M. V., Suzuki, Y., Yague, S., Lorz, C., Jimenez, S., Soto, C., Barat, A., Belda, E., Gonzalez-Pacheco, F. R., Deudero, J. J., Castilla, M. A., Egido, J., *et al.* (2002). Role of endogenous vascular endothelial growth factor in tubular cell protection against acute cyclosporine toxicity. *Transplantation* **74,** 1618–1624.

Wang, G. L., and Semenza, G. L. (1993). Desferrioxamine induces erythropoietin gene expression and hypoxia-inducible factor 1 DNA-binding activity: Implications for models of hypoxia signal transduction. *Blood* **82,** 3610–3615.

Wang, G. L., and Semenza, G. L. (1995). Purification and characterization of hypoxia-inducible factor 1. *J. Biol. Chem.* **270,** 1230–1237.

Wang, J., Biju, M. P., Wang, M. H., Haase, V. H., and Dong, Z. (2006). Cytoprotective effects of hypoxia against cisplatin-induced tubular cell apoptosis: Involvement of mitochondrial inhibition and p53 suppression. *J. Am. Soc. Nephrol.* **17,** 1875–1885.

Warnecke, C., Zaborowska, Z., Kurreck, J., Erdmann, V. A., Frei, U., Wiesener, M., and Eckardt, K. U. (2004). Differentiating the functional role of hypoxia-inducible factor (HIF)-1α and HIF-2α (EPAS-1) by the use of RNA interference: Erythropoietin is a HIF-2α target gene in Hep3B and Kelly cells. *FASEB J.* **18,** 1462–1464.

Warnecke, C., Griethe, W., Weidemann, A., Jurgensen, J. S., Willam, C., Bachmann, S., Ivashchenko, Y., Wagner, I., Frei, U., Wiesener, M., and Eckardt, K. U. (2003). Activation of the hypoxia-inducible factor-pathway and stimulation of angiogenesis by application of prolyl hydroxylase inhibitors. *FASEB J.* **17,** 1186–1188.

Wenger, R. H. (2002). Cellular adaptation to hypoxia: O_2-sensing protein hydroxylases, hypoxia-inducible transcription factors, and O_2-regulated gene expression. *FASEB J.* **16,** 1151–1162.

Wiesener, M. S., Turley, H., Allen, W. E., Willam, C., Eckardt, K. U., Talks, K. L., Wood, S. M., Gatter, K. C., Harris, A. L., Pugh, C. W., Ratcliffe, P. J., *et al.* (1998). Induction of endothelial PAS domain protein-1 by hypoxia: Characterization and comparison with hypoxia-inducible factor-1alpha. *Blood* **92,** 2260–2268.

Wiesener, M. S., Jurgensen, J. S., Rosenberger, C., Scholze, C. K., Horstrup, J. H., Warnecke, C., Mandriota, S., Bechmann, I., Frei, U. A., Pugh, C. W., Ratcliffe, P. J., Bachmann, S., *et al.* (2003). Widespread hypoxia-inducible expression of HIF-2α in distinct cell populations of different organs. *FASEB J.* **17,** 271–273.

Willam, C., Warnecke, C., Schefold, J. C., Kugler, J., Koehne, P., Frei, U., Wiesener, M., and Eckardt, K. U. (2006a). Inconsistent effects of acidosis on HIF-α protein and its target genes. *Pflugers Arch.* **451,** 534–543.

Willam, C., Maxwell, P. H., Nichols, L., Lygate, C., Tian, Y. M., Bernhardt, W., Wiesener, M., Ratcliffe, P. J., Eckardt, K. U., and Pugh, C. W. (2006b). HIF prolyl hydroxylases in the rat; organ distribution and changes in expression following hypoxia and coronary artery ligation. *J. Mol. Cell Cardiol.* **41,** 68–77.

Xi, L., Taher, M., Yin, C., Salloum, F., and Kukreja, R. C. (2004). Cobalt chloride induces delayed cardiac preconditioning in mice through selective activation of HIF-1α and AP-1 and iNOS signaling. *Am. J. Physiol. Heart Circ. Physiol.* **287,** H2369–H2375.

Yan, Q., Bartz, S., Mao, M., Li, L., and Kaelin, W. G., Jr. (2007). The hypoxia-inducible factor 2alpha N-terminal and C-terminal transactivation domains cooperate to promote renal tumorigenesis *in vivo*. *Mol. Cell Biol.* **27,** 2092–2102.

Yang, Y. T., Ju, T. C., and Yang, D. I. (2005). Induction of hypoxia inducible factor-1 attenuates metabolic insults induced by 3-nitropropionic acid in rat C6 glioma cells. *J. Neurochem.* **93,** 513–525.

Zhang, S. X., Miller, J. J., Gozal, D., and Wang, Y. (2004). Whole-body hypoxic preconditioning protects mice against acute hypoxia by improving lung function. *J. Appl. Physiol.* **96,** 392–397.

CHAPTER THIRTEEN

HYPOXIA AND REGULATION OF MESSENGER RNA TRANSLATION

Marianne Koritzinsky *and* Bradly G. Wouters

Contents

1. Introduction   248
2. Changes in Global mRNA Translation During Hypoxia   249
   2.1. Kinetics and oxygen dependency   250
   2.2. Influence of genetic background   250
3. Molecular Mechanisms that Regulate mRNA Translation
   During Hypoxia   251
   3.1. Translational regulation during hypoxia by
        eIF2α phosphorylation   252
   3.2. Translational regulation during hypoxia by
        eIF4F complex availability   254
   3.3. Translational regulation by eEF2 phosphorylation   255
4. Methods Employed to Study mRNA Translation During Hypoxia   256
   4.1. Protein synthesis   256
   4.2. The polysome assay   256
   4.3. Enzymatic activity of reporter constructs   267
5. Protocols   268
   5.1. ^{35}S methionine labeling   268
   5.2. Polysome fractionation   269
References   271

Abstract

Poor oxygenation (hypoxia) influences important physiological and pathological conditions, including development, ischemia, stroke, and cancer. The influence of hypoxia is due in large part to changes in gene expression, which occur through changes in transcription and translation. In response to hypoxic conditions, cells reduce their overall rate of messenger RNA (mRNA) translation. However, individual mRNA species are effected to highly varying degrees, with some even translationally stimulated under these conditions. Regulation of

Department of Radiation Oncology (Maastro Lab), GROW Research Institute, Maastricht University, Maastricht, The Netherlands

Methods in Enzymology, Volume 435
ISSN 0076-6879, DOI: 10.1016/S0076-6879(07)35013-1

translation in response to hypoxia thereby gives rise to differential gene expression. The ability of cells to regulate translation during hypoxia is important for their survival. In the first part of this chapter, we review the effects of hypoxia on overall and gene-specific mRNA translation efficiencies and summarize the molecular pathways activated by hypoxia that regulate mRNA translation. In the second part, we describe the methods employed to investigate overall and gene-specific translation, including radioactive metabolic labeling, polysome fractionation, and reporter assays. We have emphasized the qualitative and quantitative analysis of polysome profiles, which can yield interesting information regarding the mechanistic basis for (gene-specific) translational regulation.

1. INTRODUCTION

Hypoxia (low oxygenation) plays an important role in physiological and pathological processes, including development, exercise, altitude adaptation, wound healing, stroke, ischemia, and cancer. This role has stimulated investigations aimed at understanding the activation, regulation, and functional consequences of molecular pathways that respond to hypoxic conditions. This research has resulted in a detailed understanding of the most prominent hypoxia–induced transcriptional regulators, the hypoxia-inducible transcription factor (HIF) family (reviewed in Semenza, 2003). These transcription factors are activated primarily via posttranscriptional mechanisms during hypoxia and drive the expression of a large number of genes whose products are involved in angiogenesis, erythropoiesis, metabolic transition to anaerobic glycolysis, and cell motility and invasion. Although these transcriptional programs are very important, they do not account for all of the known biological consequences to hypoxia. Recently, it has become clear that hypoxia also results in the activation of cellular pathways that regulate mRNA translation. Components of these pathways show unique patterns of dependency on oxygen concentration and duration of hypoxic exposures and may be particularly important for responding to the rapid fluctuations in oxygenation that can occur in the microenvironment of solid tumors (Bennewith and Durand, 2004; Cardenas–Navia et al., 2004). These fluctuations presumably give rise to a need for the activation of pathways that respond faster than those involving transcriptional activation by HIF.

Regulation of translation affects cellular responses to hypoxia through at least two distinct mechanisms. First, inhibition of the overall level of mRNA translation will have a significant impact on energy expenditure since protein synthesis can account for up to 70% of cellular adenosine triphosphate (ATP) consumption (Hochachka et al., 1996). This effect may be especially important during hypoxia due to the lack of oxidative phosphorylation and hence inefficient ATP production. Second, regulation of mRNA translation affects differential protein expression and thereby cell phenotype. This occurs in part because a reduction in the overall rate of mRNA translation will result in the preferential loss of short-lived proteins. Under conditions of translation inhibition, the ratio of stable to labile proteins can increase significantly, and important cellular processes that are dependent on the balance between short-lived and long-lived proteins will thus be affected. An interesting example is the triggering of apoptosis via reaper-mediated translational inhibition, which increases the ratio of long-lived caspases to inhibitors of apoptosis (IAPs) (Holley *et al.*, 2002).

Specific changes in protein expression also result from the fact that individual mRNA transcripts are differentially affected by the pathways that regulate overall levels of mRNA translation. Studies in yeast have shown a lack of correlation between levels of mRNA and the corresponding protein (Gygi *et al.*, 1999), and genome-wide approaches in human cells have demonstrated that the translation efficiency of individual mRNA species vary widely under normal conditions. Importantly, cells can significantly alter their pool of efficiently translated mRNA in response to stress conditions, such as hypoxia (Koritzinsky *et al.*, 2005; Thomas and Johannes, 2007), virus infection (Johannes *et al.*, 1999), or oncogenic activation (Rajasekhar *et al.*, 2003). The realization that mRNA translation regulation vastly influences energy homeostasis and differential gene expression has caused widespread interest in determining its regulation and biological impact during hypoxic conditions.

2. Changes in Global mRNA Translation During Hypoxia

Inhibition of overall protein synthesis, measured by incorporation of radioactively labeled amino acids, was early regarded as a hallmark of hypoxia (Pettersen *et al.*, 1986). In more recent years, it has been demonstrated that hypoxia causes loss of polyribosome complexes (polysomes) consistent with the inhibition of global mRNA translation (Blais *et al.*, 2004; Koritzinsky *et al.*, 2005, 2006, 2007; Lang *et al.*, 2002; Thomas and Johannes, 2007). Hypoxia-induced loss of polysomes is accompanied by an increase in free ribosomal subunits and free or monosome-bound mRNA. The reduction in polysome complexes is therefore not a result of low availability of transcripts or ribosomes, but represents a specific inhibition of the mRNA translation process itself (Koritzinsky *et al.*, 2006).

2.1. Kinetics and oxygen dependency

Inhibition of mRNA translation occurs very rapidly in response to hypoxia (Koritzinsky et al., 2006) far prior to loss of nutrients or ATP (Lefebvre et al., 1993) and has emerged as a highly regulated process. Interestingly, the kinetics of inhibition of mRNA translation reflect a biphasic response. A reduction in polysomes can be detected within minutes of establishing an anoxic atmosphere *in vitro* and reaches a minimum following 1 to 2 h of exposure. This maximal inhibition is followed by a small recovery, although overall levels remain suppressed for more than 16 h of anoxia (Koritzinsky et al., 2006). This biphasic response appears to reflect the consequences of different molecular mechanisms that together regulate overall translation during hypoxia (described later). Importantly, inhibition of mRNA translation is completely reversible upon reoxygenation, even following long-term (24 h) anoxia (Kraggerud et al., 1995; Lang et al., 2002; Pettersen et al., 1986), illustrating that inhibition is not related to cell death or media depletion. Translation efficiency decreases further following even more extended exposures (2–3 days) to anoxia (unpublished observations); how-ever, for such long and toxic exposures, energy and pH homeostasis must also be taken into account as contributing factors.

Changes in the overall rate of mRNA translation at more moderate hypoxic conditions are somewhat less clear. The measurements and quantifi-cation of any biological responses from cell monolayers to more moderate hypoxia is tricky because oxygen availability at the cellular level may differ significantly from that in the atmosphere. This is due to the establishment of steep oxygen gradients in media due to cellular respiration (Pettersen et al., 2007). As a consequence, the pericellular oxygen concentration is highly dependent on cell density. Nevertheless, translation has convincingly been shown also to be repressed in response to more moderate hypoxia, but to a milder degree than during anoxia. Polysome analysis has demonstrated reduced amounts of polysomes following short (2 h) exposures to 2.0% oxygen, with the majority of the dose-response relationship occurring between 0.0% and 0.2% O_2 (Koritzinsky et al., 2007). Kinetic studies have indicated that the polysome fraction remains stably lower ($\sim$85%) than control conditions for 2 to 72 h of 0.2% O_2. A number of reports demonstrate inhibition of protein synthesis measured by incorporation of radioactive amino acids following long exposures ($>$ 24 h) to 0.5 to 1.5% O_2 (Connolly et al., 2006; Lang et al., 2002; Liu et al., 2006; Thomas and Johannes, 2007).

2.2. Influence of genetic background

Inhibition of mRNA translation in response to hypoxia is a general phenomenon that occurs in a broad range of cell lines *in vitro*. A recent study (Koritzinsky et al., 2007) demonstrated a remarkable homogeneity in the loss of polysomes of cells from different origins in response to acute (2 h) hypoxia (0.2% O_2) or anoxia (0.0% O_2). These included immortalized human fibroblasts, five different human cancer cell lines (cervix carcinoma, colorectal adenocarcinoma, prostate carcinoma, mammary adenocarcinoma, and glioblastoma-astrocytoma), and transformed mouse embryo fibroblasts. Prolonged moderate hypoxia (1% O_2) also caused loss of polysomes and inhibition of protein synthesis in cells from prostate adenocarcinoma (Thomas and Johannes, 2007). Likewise, hypoxia and anoxia cause protein synthesis inhibition in human cell lines from kidney (Liu *et al.*, 2006) and lung carcinoma (Koumenis *et al.*, 2002), glioblastoma (Kaper *et al.*, 2006), cervical carcinoma (Kraggerud *et al.*, 1995), and rat hepatocytes (Lefebvre *et al.*, 1993) and cell lines derived from rat renal carcinoma (Liu *et al.*, 2006). The homogenous response of cells from different tissues and species suggests that the ability of cells to regulate translation in response to hypoxia is an important adaptation mechanism that is largely preserved during carcinogenesis.

Although hypoxic inhibition of translation has been demonstrated in a vast number of malignant cancer cell lines, it is likely that tumor-specific changes in translation modify this process in more subtle ways. One study compared inhibition of protein synthesis in response to 24 h of 0.5% O_2 in three different breast cell lines and noted an inverse correlation between the magnitude of the response and malignancy (Connolly *et al.*, 2006). Some of the molecular pathways that might regulate overall translation in response to hypoxia (described later) are frequently mutated in cancer, providing a feasible explanation for this result. These cancer-specific changes may be especially relevant for regulation of gene-specific changes in mRNA translation efficiency.

3. MOLECULAR MECHANISMS THAT REGULATE mRNA TRANSLATION DURING HYPOXIA

Regulation of overall mRNA translation occurs primarily at the level of initiation, although the translation efficiency of some mRNAs is limited by the rate of elongation/termination (reviewed in Hershey and Merrick, 2000). During initiation, the eukaryotic initiation factor 4F (eIF4F) is assembled at the m^7GpppN cap structure of the mRNA (Gingras *et al.*, 1999). eIF4F consists of three subunits: the cap binding protein eIF4E, the scaffolding protein eIF4G, and the ATP-dependent helicase eIF4α. Binding of eIF4F to a transcript facilitates the recruitment of the 43S pre-initiation complex, which consists of the small 40S ribosomal subunit, the initiation factor eIF3, and the ternary complex eIF2-GTP-tRNA^{Met}. The 43S pre-initiation complex scans through the 5′ untranslated region (UTR) of the mRNA until it recognizes the AUG initiation codon. Pairing between the AUG and the anticodon of the methionine-charged transfer RNA (tRNA) triggers hydrolysis of guanosine triphosphate (GTP) and release of eIF2–GDP. Subsequently, the 60S ribosomal subunit is recruited to assemble the complete 80S ribosome, and the elongation phase of translation begins (Hinnebusch, 2000; Raught et al., 2000). Eukaryotic elongation factor eEF1A recruits the subsequent aminoacylated tRNA to the ribosome, which catalyzes the formation of a new peptide bond between the incoming amino acid in the A site and the previous amino acid found in the P site. GTP-bound eEF2 catalyzes the GTP-dependent translocation of the newly added peptidyl tRNA into the P site, allowing for a new round of peptide recruitment (Merrick and Nyborg, 2000). The formation of the eIF4F and eIF2-GTP-tRNA^{Met} complexes form important regulatory control points during the translation initiation process in response to many forms of cell stress (Holcik and Sonenberg, 2005), including hypoxia. In the elongation cycle, evidence also exists that the activity of eEF2 is altered during hypoxia.

3.1. Translational regulation during hypoxia by eIF2α phosphorylation

3.1.1. Global effects on translation

The assembly of the eIF2-GTP-tRNA^{Met} complex depends on the exchange of GDP or GTP catalyzed by eIF2B. Phosphorylation of the serine-51 (S51A) residue of the eIF2alpha (α) subunit inhibits the catalytic activity of eIF2B and thereby prevents translation initiation (Hinnebusch, 2000). Anoxia causes a rapid and robust phosphorylation of eIF2α, which is then followed by a partial dephosphorylation (recovery) at later time points. The overall levels of phosphorylated eIF2α correlate well with the magnitude of translation inhibition as assessed by polysome analysis (Koritzinsky et al., 2006; Koumenis et al., 2002). The transient nature of eIF2α phosphorylation is also observed under other conditions and is due to eIF2α phosphorylation-dependent induction of growth arrest and DNA damage-inducible protein (GADD) 34 (Novoa et al., 2001). This protein associates with and activates an eIF2α phosphatase, thus promoting a feedback recovery from eIF2α-mediated translational inhibition. Although transient, eIF2α phosphorylation is important for hypoxia tolerance. Prevention of eIF2α phosphorylation by genetic manipulation causes decreased survival in vitro and impaired tumor growth accompanied by increased apoptosis in hypoxic tumor areas (Bi et al., 2005; Koritzinsky et al., 2007; Koumenis et al., 2002).

Phosphorylation of eIF2α during hypoxia requires activation of the endoplasmic reticulum (ER) resident kinase (PERK), which functions as part of the unfolded protein response (UPR) (Fig. 13.1) (Bi et al., 2005; Koumenis et al., 2002). The activation of the UPR in response to hypoxia is the subject of another chapter in this book (Chapter 14). Genetic models that prevent eIF2α phosphorylation have convincingly demonstrated that the initial inhibition of translation in response to acute anoxia and hypoxia

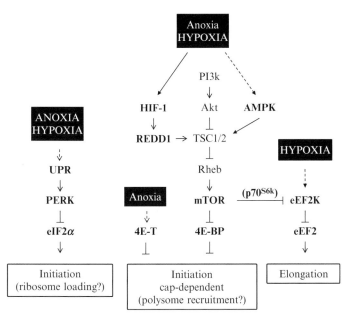

Figure 13.1 Molecular pathways that regulate messenger RNA (mRNA) translation in response to hypoxia/anoxia. Hypoxia and/or anoxia is capitalized for pathways where functional evidence of an effect on overall or gene-specific translation has been demonstrated. Pathway elements in bold type represent proteins where hypoxia-induced changes in expression, posttranslational modification, or activity has been demonstrated. Dashed line: activation mechanism unknown.

(0.2% O_2) depends on eIF2α phosphorylation (Koritzinsky *et al.*, 2006, 2007; Koumenis *et al.*, 2002). These models include overexpression of an eIF2α S51A–dominant negative allele, expression of a constitutively active GADD34 fragment, and mouse embryo fibroblasts (MEFs) with a knock-in mutation for eIF2α, which prevents phosphorylation at the S51A residue. In contrast, mRNA translation is equally inhibited in wild-type (wt) and S51A MEFs following longer anoxic exposures (8–16 h), demonstrating the presence of a separate eIF2α-independent mechanism regulating global translation efficiency following more chronic anoxia.

3.1.2. Gene-specific effects on translation
The importance of eIF2α phosphorylation in promoting hypoxia tolerance may be explained in part by the consequential energy conservation that is achieved through mRNA translation inhibition. However, eIF2α phosphorylation also leads to differential gene expression that may significantly impact cell phenotype. As previously outlined, translation inhibition causes enrichment of stable proteins independent of the molecular mechanism responsible for translational regulation. In addition, individual mRNA

species are differentially affected by translational regulation because they have varying dependencies on translation factors. This specificity is largely conferred by elements in the 5′ and 3′ UTRs of the mRNAs. The best-described example of an mRNA that is differentially regulated directly by eIF2α phosphorylation is that encoding the transcription factor ATF4 (Lu *et al.*, 2004; Vattem and Wek, 2004). ATF4 mRNA contains two conserved upstream open-reading frames (uORFs) within its 5′ UTR that inhibit ATF4 translation during normal conditions. When availability of the ternary complex is high, the ribosome initiates at the first and second uORF, which prevents initiation of the bona fide AUG. However, under conditions where eIF2α is phosphorylated, the scanning ribosome is more likely to bypass the second uORF and reinitiate translation at the correct AUG start codon (Vattem and Wek, 2004). By analyzing the distribution of mRNA within polysomes, we also found that ATF4 translation was rapidly stimulated during hypoxia in a manner dependent on eIF2α phosphorylation (Koritzinsky *et al.*, 2006). Similarly, Blais *et al.* (2004) showed that hypoxia-induced stimulation of ATF4 translation is PERK-dependent and dependent on its 5′ UTR. The fact that ATF4 protects cells against hypoxia (Bi *et al.*, 2005) strongly supports an important role for eIF2α-phosphorylation–regulated protein expression in determining the hypoxic phenotype. Many other mRNA species are also selectively translationally induced by eIF2α phosphorylation during hypoxia, including downstream targets of ATF4, such as C/EBP homology protein CHOP and GADD34 (Bi *et al.*, 2005; Koritzinsky *et al.*, 2006). The mechanism of this selective translation remains unknown, but might be conferred by the presence of uORFs, also in these mRNAs.

3.2. Translational regulation during hypoxia by eIF4F complex availability

Studies have shown that the interaction between eIF4E and eIF4G is disrupted during both anoxia and more moderate hypoxia (Connolly *et al.*, 2006; Koritzinsky *et al.*, 2006). This effect occurs predominantly following prolonged (several hours') exposure and thus correlates with the second phase of translation inhibition during chronic anoxia/hypoxia. eIF4F complex formation is regulated by the eIF4E binding proteins (4E-BP) 1 through 3. When these proteins are hypophosphorylated, they prevent eIF4F complex assembly by competing with eIF4G for binding to eIF4E. Anoxia and hypoxia have both been shown to cause dephosphorylation of 4E-BP1 in concert with decreased activity of its kinase, the mammalian target of rapamycin (mTOR) (see Fig. 13.1) (Arsham *et al.*, 2003; Brugarolas *et al.*, 2004; Connolly *et al.*, 2006; de la Vega *et al.*, 2001; Kaper *et al.*, 2006; Liu *et al.*, 2006; Martin *et al.*, 2000). Hypoxia can inhibit mTOR activity through HIF-dependent and -independent pathways (see Fig. 13.1). Under moderate hypoxia (0.2–0.5% O$_2$), dephosphorylation of

4E-BP1 is functionally important since RNA interference (RNAi)-mediated knockdown of 4E-BP1 results in a reduced ability to inhibit overall protein synthesis (Connolly *et al.*, 2006; Koritzinsky *et al.*, 2007). However, the importance of 4E-BP1 dephosphorylation during anoxia is less clear. Under these conditions, we have found that loss of 4E-BP1 does not affect overall mRNA translation. Furthermore, a discrepancy exists in the kinetics of 4E-BP1 dephosphorylation and loss of eIF4F complex assembly, the latter occurring prior to significant changes in the phosphorylation status of 4E-BP1 (Koritzinsky *et al.*, 2006, 2007). Likewise, Kaper *et al.* (2006) showed that they could prevent anoxia-induced 4E-BP1 hypophosphorylation by overexpression of mTOR, but this did not prevent inhibition of protein synthesis (Kaper *et al.*, 2006).

Irrespective of the role of 4E-BP1 phosphorylation in regulating overall mRNA translation efficiency, it is important to emphasize that regulation of the eIF4F complex formation during hypoxia and anoxia is likely to have important effects on gene-specific mRNA translation and hence cell phenotype. Indeed, we have found that RNAi-mediated knockdown of 4E-BP1 causes changes in gene expression during normoxia and hypoxia (M.G. Magagnin, manuscript in preparation) without significantly influencing hypoxic effects on overall mRNA translation. Other studies have shown that changes in mTOR activity can significantly alter the translation efficiency of a subset of genes without significantly affecting overall translational levels (Grolleau *et al.*, 2002; Rajasekhar *et al.*, 2003). Some hypoxia-induced proteins, including murine HIF-1α and vascular endothelial growth factor (VEGF), have been reported to remain efficiently translated during hypoxia due to the presence of internal ribosomal entry sites (IRES) in their 5′ UTRs (Lang *et al.*, 2002; Stein *et al.*, 1998). These elements bypass the requirement for eIF4F formation at the 5′ UTR cap structure.

3.3. Translational regulation by eEF2 phosphorylation

In the elongation cycle, phosphorylation of eEF2 prevents its interaction with the ribosome and thereby stalls peptide elongation. Phosphorylation of eEF2 has been reported at both early and late time points during moderate hypoxia (0.5–1.5% O_2) (Liu *et al.*, 2006). Expression of the eEF2 kinase (eEF2K) is induced by hypoxia (Connolly *et al.*, 2006), and its activity can be negatively regulated by mTOR-dependent phosphorylation (see Fig. 13.1) (Browne and Proud, 2004; Wang *et al.*, 2001). RNAi-mediated knockdown of eEF2K prevented hypoxia-induced eEF2 phosphorylation accompanied with a decreased ability to inhibit protein synthesis following 24 h of 0.5% O_2 (Connolly *et al.*, 2006). In this study, combined knockdown of 4E-BP1 and eEF2K resulted in almost complete ablation of the hypoxia-induced inhibition of protein synthesis, indicating that these effectors act independently of each other.

4. METHODS EMPLOYED TO STUDY mRNA TRANSLATION DURING HYPOXIA

4.1. Protein synthesis

4.1.1. Global effects

The most commonly implemented approach to assess the overall rate of protein synthesis in living cells is to measure the incorporation of a radioactively labeled amino acid, and ^{35}S-labeled methionine/cysteine is the most common choice. To facilitate efficient uptake, the unlabeled amino acid is removed from the growth media for a short period prior to the labeling. The rate of protein synthesis can be monitored by measuring the amount of radioactive protein that is produced within a given period of time, which can be visualized on a polyacrylamide gel (Fig. 13.2A) or measured by scintillation counting of Trichloroacetic acid (TCA)-precipitate. The main advantage of this method is that it gives a direct measure of the rate of protein synthesis. However, this approach cannot be used to determine whether changes in protein synthesis occur due to reduced mRNA translation or to reduced mRNA transcript or ribosome availability. It is also a concern that this method requires a period of amino acid starvation, which on its own may influence mRNA translation through the same pathways activated by hypoxia. Amino acid starvation and hypoxia cause phosphorylation of eIF2α mediated by the kinases general control amino–acid synthesis 1–like 2 (GCN2) and PERK, respectively (Kimball and Jefferson, 2000; Koumenis *et al.*, 2002).

4.1.2. Gene-specific effects

The rate of *de novo* synthesis of a specific protein can also easily be investigated using this method by including an immunoprecipitation step with a specific antibody prior to gel electrophoresis and autoradiography. Although useful for particular proteins of interest, this approach is limited by its throughput as well as antibody availability.

4.2. The polysome assay

The polysome assay has also been extensively used to investigate translation efficiency during hypoxia. In this assay, the rate of overall translation is assessed by quantifying the level of polysomes—mRNA molecules that are attached to several ribosomes. One important advantage of this assay is that it requires no initial starvation or depletion of the growth media, unlike that required for radioactive labeling. When used to estimate overall translation efficiency, the polysome assay is also more sensitive and can provide additional mechanistic information (discussed later) compared with radioactive amino acid incorporation. When used to measure gene-specific

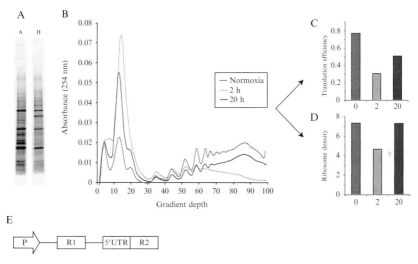

Figure 13.2   (A) Demonstration of changes in *de novo* protein synthesis during hypoxia. Cells were exposed to aerobic or hypoxic (0.0% O₂) (H) conditions for 2 h and incubated with ³⁵S-labeled methionine/cysteine. Proteins were separated on sodium dodecyl sulfate polyacrylamide gel electrophoresis (SDS-PAGE) and incorporated ³⁵S detected by autoradiography. Hypoxia-induced inhibition of protein synthesis is evident by the weaker signal from hypoxia-treated cells. (B) Demonstration of changes in polysomal messenger RNA (mRNA) during hypoxia. HT-29 cells were exposed to 0.0% O₂ for 0, 2, or 20 h. Cell lysates were sedimented through a sucrose gradient, and the optical density (absorbance) at 254 nm recorded as a function of gradient depth. The shallowest peak represents the small (40S) ribosomal subunit, followed by the large (60S) subunit, the monosome (80S), and polysomes, with an increasing number of ribosomes. Hypoxia-induced inhibition of mRNA translation is evidenced by the loss of polysomes in hypoxia-treated cells. (C) Translation efficiency was estimated as the fraction of ribosome RNA (rRNA) participating in polysomes according to the method outlined in Figure 13.3. Acute anoxia inhibits translation followed by a partial recovery after longer exposures. (D) The average number of ribosomes per mRNA in the polysomes was estimated according to the method outlined in Figure 13.3. Acute anoxia results in a decrease in ribosome density, which is completely recovered during more chronic exposures. (E) A bicistronic expression vector with promoter (P), reporter gene 1 and 2 (R1 and R2), and a 5′ UTR of interest. Hairpin structures may be inserted in the linker regions to prevent ribosome scanning.

translation, the advantages extend to the use of primers/probes for mRNA detection rather than antibodies for protein detection, which also facilitates high-throughput (microarray) analysis. The main limitation of this assay lies in the assumption that translation efficiency is reflected solely by the amount of polysomal RNA, a condition that is valid only when changes in translation are mediated through the initiation step.

Polysomes can be separated from single ribosomes (monosomes) and ribosomal subunits by sedimentation of native cell lysate in a (swinging bucket) ultracentrifuge (see Fig. 13.2B). Sedimentation is performed through a sucrose gradient in order to prevent convective mixing during and follow-ing centrifugation. Due to their higher mass, complexes containing many ribosomes will move faster in the gravitational field and end up deeper in the gradient than complexes with few ribosomes. The relative abundance of the separated complexes can be analyzed by recording the optical RNA density (absorbance at 254 nm) as a function of gradient depth. The resulting absor-bance recordings vary qualitatively depending both on biological parameters, such as cell type and growth conditions, and technical parameters, such as gradient composition and centrifugation time. Figure 13.2B shows high resolution profiles displaying unique absorbance peaks reflecting the small (40S) and large (60S) ribosomal subunit, the 80S monosome, as well as individual polysomes containing two to seven ribosomes. The identity of the individual peaks can be verified by isolating the corresponding RNA and protein and demonstrating the presence or absence of ribosomal subunits and translation factors.

Polysome profiles from HT-29 colon carcinoma cells exposed to 0, 2, and 20 h of anoxia (are shown in Fig. 13.2B). Two important observations can be made from a simple qualitative interpretation of such profiles. First, a severe inhibition in overall translation is evident at both 2 and 20 h, as depicted by the relative loss of polysomes and a corresponding increase in ribosomal subunits and single ribosomes (monosomes). Second, hypoxia causes not only an overall reduction in the amount of polysomes, but also leads to changes in the "shape" of their distribution (see Fig. 13.2B). Following 2 h of anoxia, there is a preferential loss of the high molecular weight polysomes representing the most efficiently translated mRNA transcripts with many ribosomes. Thus, at this time point, not only are fewer mRNA transcripts used for translation, those which remain translated are associated with far fewer ribosomes than normal and will thus produce significantly less protein per transcript. In cells exposed to anoxia for 20 h, a partial recovery in the amount of polysomes and a near complete recovery in the shape of the polysome distribution occurs. This indicates that, at this time point, *fewer* mRNA transcripts remain translated than during normoxia (less polysomes); however, the transcripts that *are* being translated contain (on average) the same number of ribosomes as those under aerobic conditions. This example illustrates the fact that polysome profiles can provide information about both mRNA *recruitment* into the translated fraction and *ribosome density* of recruited transcripts. The fact that these characteristics can be observed individually reflects that they are controlled by different molecular mechanisms. Acute anoxia causes a loss in ribosome density, while recruitment is inhibited following chronic anoxia.

4.2.1. Quantitative analysis of polysome profiles

The absorbance recordings obtained in the polysome assay can contribute important quantitative and mechanistic information if analyzed carefully (Fig. 13.3). Components in the gradients and lysis buffer, such as sucrose

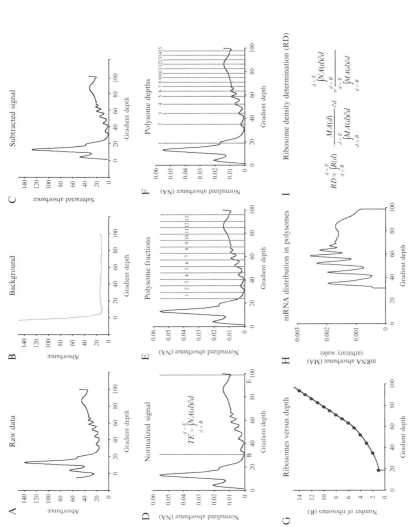

Figure 13.3 (A) A polysome profile derived by recording the absorbance at 254 nm continuously through a sucrose gradient in which the contents of a cell lysate has been separated according to sedimentation speed. (B) Absorbance recorded from a sucrose gradient in which only lysis buffer (without cells) has been layered. Compounds in the lysis buffer and gradient, such as detergent and sucrose, give rise to some background absorbance. (C) To obtain an absorbance profile derived only from the biological material in the cell lysate, the signal in (B) has been subtracted

and detergent, give rise to some absorbance in the 254-nm range. Therefore, one gradient should always be prepared in which only lysis buffer is layered (see Fig. 13.3B). The absorbance of this background gradient can be subtracted from all the others to yield the absorbance of the biological material (see Fig. 13.3C).

Quantifying translation: Fraction of rRNA in polysomes   The principal behind using polysome profiles to quantitatively estimate overall mRNA translation is that the integrated area under the curve in any part of the gradient is proportional to the amount of ribosomal RNA (rRNA) and thus to the number of ribosomes. The possible contribution of other biological components, such as protein, is assumed to be negligible in this absorbance wavelength. It is also assumed that the vastly dominating RNA species is rRNA and that the contribution of tRNA and mRNA can be neglected. This approximation is better for more ribosome-rich complexes deep in the gradient than for poorly translated transcripts attached to one or few ribosomes. Several different approaches to estimate changes in translation can be employed. One option is to simply determine the fraction of ribosomes (estimated by rRNA amount) found within polysomes (i.e., complexes containing two or more ribosomes). This fraction can be calculated simply by integrating the area under the curve in the background subtracted gradient, which contains polysomes, and dividing this value by the total area under the curve (see Fig. 13.3D). Because the value is normalized by the total area, this measurement is not sensitive to differences in the amount of material that is loaded onto the gradient. Since each ribosome within the polysome is producing one protein, this value should correlate well with the amount of overall protein synthesis. Figure 13.2C shows the effects of hypoxia on translation estimated in this way from the polysome profiles shown in Fig. 13.2B. The figure clearly reflects the severe inhibition of

---

from the signal in (A). (D) To correct for loading inaccuracies, the integrated signal in (c) can be normalized. As a measure of translation efficiency (TE), the fractional area under the curve representing polysomes can be calculated by integrating from the beginning to the end of that area. With the approximation that the signal is principally derived from ribosome RNA (rRNA), TE is an estimation of the fraction of ribosomes that are in polysomes. (E) During recording of a polysome profile, small fractions can be collected to investigate the presence of specific messenger RNA (mRNA) or protein. (F) The position of the peaks representing polysomes of different orders is determined. (G) The position of the polysome peaks is used to derive a function describing the ribosome distribution through the sucrose gradient. (H) The ribosome depth relationship in (G) is used to derive the relative distribution of mRNA from the polysomal signal in (D). (I) The average number of ribosomes per mRNA (ribosome density) in any part of the profiles can be shown to equal the integration of (D) divided by the integration of (H). Here, the integration is shown for the whole polysomal area (Beginning to End), but could also, for example, be executed for each collected fraction shown in (E).

mRNA translation that occurs during anoxia as well as the partial recovery following prolonged exposure.

Mechanisms of translational regulation: Ribosomes per mRNA   As previously mentioned, a reduction in overall translation can occur either through a reduction in the number of mRNA transcripts that are translated (referred to here as polysome recruitment) or by a reduction in the rate at which ribosomes are loaded onto a given transcript (ribosome density). A reduction in translation that occurs strictly through changes in recruitment will affect the rRNA area within the polysomes without affecting the shape of the distribution. Conversely, changes in overall translation could theoretically be caused entirely by a reduction in ribosome density, leading to a change in polysome profile shape. Measurement of the rRNA fractional polysome area alone will not indicate to what degree each of these two processes contribute to overall changes. However, the contribution of these two mechanisms can be assessed by further analyzing polysome profiles to directly determine the average ribosome density within the polysomes. We have developed algorithms to estimate this number from polysome profiles. The position of the individual ribosome peaks are recorded (see Fig. 13.3F) and used to derive parameters that describe the number of ribosomes as a continuous function of gradient depth (see Fig. 13.3G). After obtaining this function, it becomes possible to estimate the distribution of mRNA within the polysomes. In the original polysome profiles, the optical density is determined by the amount of rRNA present in ribosomes on the mRNA transcripts. In order to transform the polysome (rRNA signal) into a distribution of mRNA, one must simply divide the signal at any depth by the number of ribosomes per mRNA present at that depth (see Fig. 13.3H). mRNA levels are proportionally higher than the polysome signal in ribosome-poor fractions (viewed through comparison of Fig. 13.3H to 13.3C), as would be expected. Finally, the mRNA distribution (see Fig. 13.3H) and the ribosome distribution (see Fig. 13.3G) provide a basis to estimate the average ribosome density in the polysomes (or any other part of the gradient) (see Fig. 13.3I).

The estimated average number of ribosomes per mRNA transcript from the polysome profiles in Fig. 13.2B are shown in Fig. 13.2D. These data illustrate that the inhibition of translation induced by an acute anoxic exposure of 2 h is accompanied by a significant drop in the average number of ribosomes per mRNA in the polysomal fraction. In contrast, no detectable change is seen in the number of ribosomes per mRNA following chronic anoxia, in spite of sustained inhibition of translation efficiency. As previously outlined, the interpretation of this result is that acute anoxia causes inhibition of ribosome loading (by eIF2α phosphorylation) and that chronic anoxia causes inhibition of mRNA recruitment. Consistent with this interpretation, MEFs expressing a non-phosphorylateable eIF2α allele

(S51A) show a constant number of ribosomes per polysomal mRNA at all time points of hypoxia in spite of a clearly detectable drop in overall translation efficiency (Koritzinsky *et al.*, 2006). The dependency of ribosome density on eIF2α phosphorylation is to be expected since this event causes a reduction in the availability of the ternary complex levels containing the aminoacylated tRNA destined for the initiation codon. Each initiation event requires a new ternary complex to load onto the mRNA, and thus its concentration can often be rate-limiting.

Interestingly, the molecular mechanism(s) that are responsible for the residual translation inhibition during short exposures to anoxia in MEF S51A as well as during prolonged exposures in wt and S51A cells did not cause a drop in ribosome density. Consequently, the drop in translation in these cases appears to result strictly from a reduction in mRNA recruitment to the translation machinery. As discussed earlier, disruption of the eIF4F complex is a likely candidate mechanism for this eIF2α-independent mode of translation inhibition. In this scenario, a drop in eIF4F levels reflects the fact that fewer mRNA transcripts are capable of participating in translation. For a large percentage of transcripts, eIF4E will bind to the cap together with its inhibitor 4E-BP1, and this binding would be expected to prevent translation of that particular transcript. However, a smaller fraction of transcripts may contain eIF4E bound to eIF4G and, for this translation, would proceed with normal efficiency, and thus ribosome density would not be altered.

Changes in ribosome density can also occur under conditions that alter translation elongation or termination. In these cases, ribosome density could increase due to ribosome stalling or slowing. It is unlikely that the general increase in ribosome density over time seen during chronic hypoxic conditions (e.g., compare 2 and 20 h in Fig. 13.2D) is due to this possibility. If this theory were the case, we would have expected to see a similar *increase* in ribosome density in the S51A MEFs during hypoxia. Instead, these cells showed a constant ribosome density during all time points of anoxia. Nevertheless, the possibility that elongation rates affect the shape of the polysome profiles during hypoxia needs to be further investigated, especially in light of the recent data that suggest a role for eEF2 phosphorylation during moderate hypoxia (Connolly *et al.*, 2006).

4.2.2. Gene-specific translation efficiency

The polysome assay also renders analysis of gene-specific translation efficiency possible. RNA from different gradient fractions can be collected (see Fig. 13.3E), isolated, and assayed by standard techniques, such as Northern blotting, ribonuclease (RNase) protection assay, quantitative polymerase chain reaction (PCR), or microarrays. Consequently, it is possible to determine the distribution of any mRNA within the polysome profile. Detection of the transcript of interest in deeper fractions indicates its association with more ribosomes and therefore higher translation efficiency. Detection of (reverse transcribed) mRNA by quantitative PCR (qPCR) or microarrays requires careful attention to RNA quality and to normalization procedures, but facilitates high-throughput quantitative analysis and additional mechanistic insight.

RNA isolation   Individual fractions from the polysome profiles (see Fig. 13.3E) are typically collected in guanidine hydrochloride, which efficiently denatures proteins, including ribonucleases. To obtain pure and high-quality RNA from these fractions, it is subjected to several rounds of precipitation. The lysis buffer and sucrose gradients contain heparin, which is a potent inhibitor of complementary DNA (cDNA) synthesis and PCR reactions and which acts as a carrier in the precipitation of RNA. It is therefore crucial to include a lithium-chloride precipitation to remove heparin. The presence of a PCR inhibitor can often be detected by performing a dilution series over several logs for an abundant mRNA or rRNA. Increased reverse transcriptase (RT) PCR efficiency in higher dilutions is an indication of impurities in RNA/cDNA.

Normalization   In order to compare the amount of mRNA present in the original fractions, it is necessary to correct for differences in the efficiency of RNA isolation and RT efficiency. A relatively easy way of accomplishing this task is to add a known amount of non-endogenous ("spike") RNA to each collected fraction prior to RNA isolation and then normalize the relative abundance of the detected gene of interest to the recovered abundance of the spike. An alternative approach that we have used is to take advantage of the fact that the integrated area under the polysome curve corresponding to a collected fraction is directly proportional to the amount of (18S) rRNA in the fraction. The recovered amount of 18S (as measured by qPCR) divided by the original amount in the fraction (determined by polysome profile integration) is therefore a direct measure of RNA isolation and RT efficiency and can be used for normalization. Either method relies on the assumption that all RNA species are isolated and reverse transcribed with equal efficiency. An advantage of the 18S method is that differences in initial loading of the gradients can also be taken into account by first normalizing the total area under the curve in all gradients.

This method has been used to derive the data presented in Fig. 13.4A. Here, fractions were collected during recording of polysome profiles (see Fig. 13.2B). Following analysis to determine the number of ribosomes as a function of gradient depth (see Fig. 13.3), collected fractions were pooled to yield samples with a defined number of ribosomes per transcript. The relative abundance of β-actin mRNA was determined by qPCR for each sample and normalized by the 18S method previously outlined. β-actin mRNA is very efficiently translated under normoxic conditions, with a

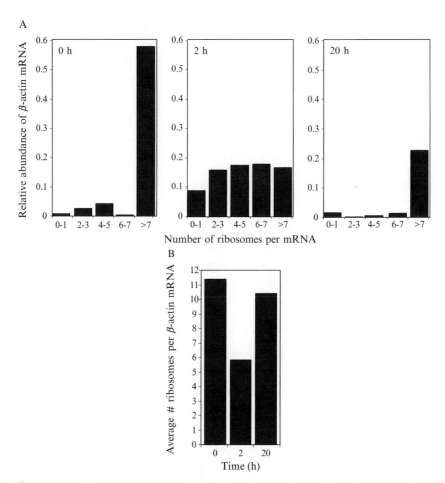

Figure 13.4 (A) Fractions were collected during recording of the polysome profiles shown in Fig. 13.2B and pooled into five samples to yield monosomes/polysomes with 0 to 1, 2 to 3, 4 to 5, 6 to 7, or greater than 7 ribosomes. RNA was isolated and reverse transcribed, and the abundance of β-actin and 18S quantified with qualitative polymerase chain reaction (qPCR). The abundance of β-actin was normalized by F = (18S abundance measured by qPCR/fractional area under the polysome profile), which accounts for differences in RNA isolation and reverse transcription efficiencies. β-actin is efficiently translated during normoxia, followed by translational inhibition after 2 h of anoxia, with a significant recovery after a 20-h exposure. Overall reduction in signal following 20 h is consistent with a loss of cellular β-actin transcript at this time. (B) The ribosome density per fraction was calculated according to the method outlined in Fig. 13.3 and multiplied with the relative β-actin abundance in that fraction. For each time point, the sum of this product over all polysome fractions (i.e., two ribosomes or more) yielded the average number of ribosomes per β-actin transcript *in the polysomes.*

majority of transcripts containing seven or more ribosomes (Fig. 13.4A). Following 2 h of anoxia, β-actin mRNA is redistributed to fractions containing fewer ribosomes, reflecting an inhibition of translation. However, the distribution of β-actin mRNA following 20 h of anoxia resembles that under normoxia, reflecting a recovery of translation efficiency following longer exposures. The kinetic pattern of β-actin translation efficiency hence reflects that of overall translation efficiency with an initial inhibition followed by a partial recovery. Interestingly, many gene transcripts do not follow this "average" kinetic pattern of translation efficiency, which gives rise to time-dependent differential gene expression during hypoxic conditions (Koritzinsky *et al.*, 2006).

Analyzing gene-specific translation: Mechanistic parameters   Similar to what was previously described for quantifying changes in overall translation, several different parameters can be calculated to evaluate changes in gene-specific translation. The translation efficiency of any given gene depends on the relative recruitment of its transcripts to the translational machinery as well as the rate at which ribosomes are loaded onto the transcript (ribosome density).

Recruitment to polysomes: Percentage of translated mRNA   Upon changes in eIF4E availability, certain mRNA transcripts demonstrate a selective ability to compete with limiting amounts of this factor and thus may show a reduced drop in recruitment compared with other genes. As a consequence, changes in translation efficiency arise in part through differential transcript recruitment, and this value can be expressed as the percentage of total mRNA that can be found associated with polysomes. To calculate this value, the abundance of the gene of interest in all polysome fractions must be summed and then divided by the total abundance of the transcript in the cell. Consequently, this value is independent of changes in transcription and represents only translation effects.

Ribosome density: Position within the gradient   Changes in the translation efficiency of a given gene are also mediated by increases or decreases in the rate at which ribosomes are loaded onto the transcript, and this rate can be estimated by evaluating the distribution of mRNA within the polysome profile. Gene transcripts that are found predominantly in the high molecular weight part of the polysome profile are comparatively better translated (contain more ribosomes) than an equivalent amount of mRNA found in the lower molecular weight fractions. This difference can be expressed quantitatively by the average ribosome density for any particular gene. Following determination of the distribution of a specific mRNA within the polysomes, it becomes possible to estimate the average number of ribosomes per transcript for that gene. This can be calculated because the amount of mRNA in each fraction and the position of the fraction within the polysome profile are known. Each fraction in the polysome profile represents mRNAs with a known number of ribosomes (see Fig. 13.3I). In effect, this value describes the average rate at which a given gene transcript in the polysome is converted into protein. From the data presented in Figs. 13.2B and 13.4A, the average number of ribosomes per β-actin mRNA during normoxia was estimated to 11.4 (see Fig. 13.4B). The average number of ribosomes per transcript dropped to 5.8 following 2 h of anoxia before recovering to 10.4 after a 20 h exposure.

Total translation efficiency and protein production   The total translation efficiency of any given gene combines recruitment and ribosome density and reflects directly the rate at which the mRNA is converted to protein. Here, we have termed this value the total translation efficiency, which is calculated simply by multiplying the average *ribosome density* with the *recruitment*. If one also factors in the transcriptional change, it is possible to make an estimate for relative changes in *de novo* protein production.

4.2.3. Genome-wide analysis of gene-specific translation efficiency

Application of polysomal RNA to microarrays can provide a genome-wide perspective of the pool of mRNA that is being translated at any given time (i.e., the *translatome*). The translatome is believed to reflect cell phenotype to a significantly better extent than the transcriptome because it, in essence, corresponds much more closely with protein production. This method thereby becomes an attractive alternative to transcriptomics, which correlates poorly with the proteome, and proteomics, which offers significantly lower sensitivity and genome coverage. In addition, polysomal microarray analysis allows estimation of gene-specific translation efficiency on a genome-wide basis. The amount of an individual mRNA within the polysomes depends both on its abundance and its translation efficiency. By parallel determination of the transcriptome (i.e., relative transcript abundance), one can estimate translation efficiency (TE) for each gene by dividing the relative abundance of the transcript in the polysomes by the overall cellular abundance.

   In general, polysome-bound mRNA/total mRNA is a robust parameter representing translation efficiency (Pradet-Balade *et al.*, 2001), although it may not adequately represent mRNAs that are subject to highly regulated splicing or nucleocytoplasmic transport. To accurately determine translation efficiency, it would be necessary to run microarrays on all fractions throughout the polysome in order to characterize the distribution of each gene transcript. Such a trial has been carried out in a large study of yeast, allowing estimation of translation efficiency and ribosome density for all yeast genes (Kuhn *et al.*, 2001). However, this method is very costly and time-consuming. An alternative approach is to evaluate the levels of transcripts within an arbitrarily defined "well-translated" fraction. One assumes that genes showing an increase in ribosome density in response to some stress will show an increase in this fraction and that genes showing a drop in ribosome density will decrease. Although it is not possible to estimate the total translation efficiency with this approach, one can estimate relative changes in translation and translation efficiency. We have used this approach to screen for changes in translation efficiency in cells exposed to anoxia for different periods of time over a 24 h interval (van den Beucken, manuscript in preparation). The data indicate that a large fraction of the genome exhibits changes in translation that differ significantly from the global average. Genes show individual differences in their susceptibility to translational inhibition as well as unique dependencies on time that may be related to the different molecular mechanisms that are responsible for translational control during hypoxia. The mRNA elements in the 5′ and 3′ UTRs, which confer gene-specific translational control, remain poorly defined; however, these types of genome-wide analyses make it possible to search for common elements in groups of genes that show similar patterns of translational regulation.

4.3. Enzymatic activity of reporter constructs

In order to confirm and examine in greater detail gene-specific translation efficiency, alternatives to the polysome assay can be used. These alternatives include *in vitro* translation assays and quantification of activity from enzymatic reporter constructs. The latter is particularly suited to investigate the impact of regulatory elements in UTRs of mRNAs, as these can be fused in frame with the coding sequence of a reporter gene, such as luciferase. Estimation of changes in translation efficiency that are conferred by the UTR sequence can be made by comparing the relative levels of protein (or activity in the case of luciferase) with mRNA abundance. An interesting extension of this assay is the use of bicistronic vectors to investigate the cap-independent translation efficiency conferred by the 5′ UTR of certain genes (see Fig. 13.2E). These vectors produce a capped transcript with a downstream coding sequence for one reporter gene, followed by the 5′ UTR of a gene of interest fused in frame with a second reporter gene. As ribosomes dissociate from the mRNA following translation of the first reporter, production of the second reporter can only occur from new initiation via a cap-independent mechanism. Such new initiations are believed to demonstrate the presence of an internal ribosomal entry site (IRES) in the intercistronic space. The validity of this interpretation relies on exclusion of the presence of two transcripts producing the two different reporters through the presence of a cryptic promoter, splicing events, or mRNA degradation. The possibility that the ribosome remains associated with the mRNA following translation of the first reporter and reinitiates at the following start-codon should also be excluded. Although conceptually very elegant, this approach remains somewhat controversial due to the issues previously mentioned (Kozak, 2001). Nevertheless, such studies have shown that important hypoxia-responsive genes, such as (murine) HIF-1α and VEGF, contain elements in their 5′ UTRs that ensure efficient translation during hypoxia (Lang *et al.*, 2002; Stein *et al.*, 1998).

5. PROTOCOLS

5.1. ^{35}S methionine labeling

The level of incorporation of radiolabeled amino acids is directly proportional to the length of labeling time for a certain period, after which it reaches a plateau. When all limiting amino acids are consumed, protein synthesis ceases. The length of the linear phase will depend on the amino acid, the cell density, and metabolic activity. It should be ensured that uptake is within the linear phase by labeling for a few different periods of time. The following protocol is adapted for an ~80% confluent monolayer in 35-mm dishes, assuming an immunoprecipitation will be included to measure synthesis of a specific protein. Wash the cells twice with methionine/cysteine/serum-free media and incubate 15 to 30 min to deplete intracellular methionine/cysteine pools. This media must be equilibrated to the oxygen level of interest at 37° prior to use. Aspirate and add 200 μl pulse medium (methionine/cysteine/serum-free medium, to which 50 uCi of ^{35}S-labeled methionine-cysteine has been added). Incubate for 15 min and add 1 ml full media to stop incorporation. Place the dishes on ice and wash twice with ice-cold phosphate-buffered saline (PBS).

5.1.1. Gel electrophoresis

Scrape in 300 μl lysis buffer (20 mM MES; 100 mM NaCl; 30 mM Tris-HCl, pH 7.4; 0.5% Triton X-100; and fresh 1 mM phenylmethanesulphonylfluoride [PMSF], 1 mM ethylenediamine tetraacetic acid [EDTA], 10 μg/ml each of chymostatin, leupeptin, antipain, and pepstatin) and centrifuge 10 min at 16,000g. Use 5 μl to assess changes in overall protein synthesis. The remainder can be used for immunoprecipitation to measure *de novo* synthesis of a specific protein. Add 2× loading buffer (0.5 M Tris-HCl, pH6.9; 10% sodium dodecyl sulfate [SDS], 20% glycerol, 0.01% bromphenol blue) and run on SDS polyacrylamide gel electrophoresis (SDS-PAGE). Place the gel in 0.25% Coomassie brilliant blue in 30% methanol and 10% acetic acid for 5 min on a rocker to stain protein and ensure equal loading. Destain in 30% methanol and 10% acetic acid for 45 min and place the gel in a neutralizing solution of 30% methanol in PBS for 5 min. Transfer the gel to an enhancing solution of 8% sodium salicylate and 30% methanol for 15 min before transferring to a filter paper and covering with plastic wrap. Dry in a gel dryer for 1 h at 80° and expose to a phosphorimager screen for ~1 day (or film for ~1 wk).

5.1.2. Scintillation counting

Incorporation of radioactive amino acids can also be assessed by scintillation counting. Add 0.1 ml lysate to cold 10% w/v trichloroacetic acid (TCA). Incubate on ice for 30 min and pass the suspension through a 2.5-cm glass microfiber filter (GF/C Millipore). Wash the filter twice with 5 ml cold 5-10% trichloroacetic acid and twice with ethanol before letting it air dry. Transfer filters to scintillation vials, add 5 ml scintillation fluid, and measure radioactivity in a scintillation counter.

5.2. Polysome fractionation

5.2.1. Sucrose gradients

Ultracentrifuge tubes for gradients must be thin-walled for later puncture (e.g., Beckman Coulter 331372). Gradient makers can be used to produce 10 ml 20-50% sucrose gradients, where controlled mixing of the heavier and lighter fraction result in a linear gradient. Alternatively, it may be convenient to initially produce step gradients by layering lighter fractions on top of frozen heavier fractions. At the day of the experiment, these step gradients can be thawed and left to diffuse to a smoother gradient for a few (~4) hours at 4°. Buffer: 0.3 M NaCl, 15 mM MgCl$_2$, 15 mM Tris-HCl (pH 7.4), 0.1 mg/ml cycloheximide, and 0.1 mg/ml heparin.

5.2.2. Preparation of lysate

Cell cultures (if monolayers) should be ~80% confluent, yielding ~7 to 20 E6 cells. Increased cell number improves the signal, but decreases resolution. The following protocols assume monolayers on a 15-cm dish. During the last 3 min of treatment, dishes are incubated with 0.1 mg/ml cycloheximide (CHX), which prevents ribosome progression and "fixes" the ribosomes to the mRNA. In the following procedures, all disposables and reagents must be RNase-free and kept cold to prevent RNA degradation. Transfer culture plates to ice and wash twice with 8 ml 0.1 mg/ml CHX in PBS. Scrape cells in 300 μl lysis buffer (0.3 M NaCl, 15 mM MgCl$_2$, 15 mM Tris-HCl [pH 7.4], 0.1 mg/ml cycloheximide, 200 units Superase-In [Ambion, Foster City, CA]). Vortex and pellet nuclei at 2000 g for 5 min. Add heparin (broad-range RNase inhibitor) to the supernatant to a final concentration of 200 μg/ml. Remove debris by centrifugation at 10,000g for 5 min. Transfer 10% of the volume to a new tube and snap-freeze for later isolation of total RNA. Carefully layer the remaining 90% on 20-50% prepared sucrose gradients. Include one gradient for background determination, where an equal volume of lysis buffer and PBS is layered without cell lysate.

5.2.3. Fractionation, detection, and collection

Centrifuge the gradients in a swinging-bucket rotor (e.g., Beckman SW-41Ti) for 90 min at 39,000 rpm and place them on ice. Spectrometers with a low-volume flow-through cell are available that offer an integrated solution to sample extraction, detection, and data collection. Alternatively, a detection set-up can be assembled using a peristaltic pump, gradient holder, ultraviolet (UV) monitor with a low-volume flow cell (e.g., BioRad), chart recorder, or digital data collection unit and a fraction collector. The samples that flow through the set-up must remain laminar, and great care must be taken to avoid air bubbles or turbulence. The sample can enter the spectrometer from the top or bottom of the gradient, but we have found better results with pushing the gradient through the top by pumping a solution of 75% sucrose into the bottom of the gradient tube. Following detection, fractions can be collected in guanidinium hydrochloride (G-HCl) at a final concentration of 5.5 M and frozen at $-80°$ for later RNA extractions. It may be practical to collect small fractions that allow analysis of the polysome profiles before fractions are pooled to represent mRNA with a defined (standard) number of ribosomes. The presence of protein can be detected directly in the samples by Western blotting.

5.2.4. RNA isolation from gradient fractions

Quantitative RT-PCR or microarray analysis requires high-quality RNA, and hence the polysome samples are subjected to several rounds of purification before analysis. To allow for easy normalization (e.g., correction for differences in RNA isolation efficiency and reverse transcription), a standard amount of an exogenous RNA can be added to each of the (pooled) samples before RNA isolation. Precipitate samples with an equal volume of isopropanol at $-80°$ overnight and centrifuge for 20 min at 10,000g. Wash with 85% ethanol, centrifuge again (10 min), and resuspend in 40 μl TE, pH 8. Precipitate by adding 4 μl 3 M NaAcetate (pH 5.3) and 100 μl ethanol. Resuspend in 75 μl water and add 75 μl Tris-buffered phenol:chloroform, mixing well. Incubate 2 to 15 min at room temperature and centrifuge for 15 min at 12,000g. Transfer the aqueous (upper) phase to a new tube and add lithium chloride (LiCl) to a final concentration of 2 M. LiCl precipitation is essential to remove heparin, a potent RT-PCR inhibitor, from the samples. Incubate overnight at $-20°$, centrifuge for 20 min at 12,000g, and wash with 75% ethanol. Resuspend in 20 μl water and precipitate by adding 2 μl 3 M Na Acetate (pH 5.3) and 60 μl 100% ethanol. Centrifuge for 20 min at 12,000g and wash with 75% ethanol. Remove all ethanol with a fine-tipped pipette and let the RNA air dry for 10 min before resuspending in 10 μl water or TE.

REFERENCES

Arsham, A. M., Howell, J. J., and Simon, M. C. (2003). A novel hypoxia-inducible factor-independent hypoxic response regulating mammalian target of rapamycin and its targets. *J. Biol. Chem.* **278,** 29655–29660.

Bennewith, K. L., and Durand, R. E. (2004). Quantifying transient hypoxia in human tumor xenografts by flow cytometry. *Cancer Res.* **64,** 6183–6189.

Bi, M., *et al.* (2005). ER stress-regulated translation increases tolerance to extreme hypoxia and promotes tumor growth. *EMBO J.* **24,** 3470–3481.

Blais, J. D., Filipenko, V., Bi, M., Harding, H. P., Ron, D., Koumenis, C., Wouters, B. G., and Bell, J. C. (2004). Activating transcription factor 4 is translationally regulated by hypoxic stress. *Mol. Cell Biol.* **24,** 7469–7482.

Browne, G. J., and Proud, C. G. (2004). A novel mTOR-regulated phosphorylation site in elongation factor 2 kinase modulates the activity of the kinase and its binding to calmodulin. *Mol. Cell Biol.* **24,** 2986–2997.

Brugarolas, J., Lei, K., Hurley, R. L., Manning, B. D., Reiling, J. H., Hafen, E., Witters, L. A., Ellisen, L. W., and Kaelin, W. G., Jr. (2004). Regulation of mTOR function in response to hypoxia by REDD1 and the TSC1/TSC2 tumor suppressor complex. *Genes Dev.* **18,** 2893–2904.

Cardenas-Navia, L. I., Yu, D., Braun, R. D., Brizel, D. M., Secomb, T. W., and Dewhirst, M. W. (2004). Tumor-dependent kinetics of partial pressure of oxygen fluctuations during air and oxygen breathing. *Cancer Res.* **64,** 6010–6017.

Connolly, E., Braunstein, S., Formenti, S., and Schneider, R. J. (2006). Hypoxia inhibits protein synthesis through a 4E-BP1 and elongation factor 2 kinase pathway controlled by mTOR and uncoupled in breast cancer cells. *Mol. Cell Biol.* **26,** 3955–3965.

Gingras, A. C., Raught, B., and Sonenberg, N. (1999). eIF4 initiation factors: effectors of mRNA recruitment to ribosomes and regulators of translation. *Annu. Rev. Biochem.* **68,** 913–963.

Grolleau, A., Bowman, J., Pradet-Balade, B., Puravs, E., Hanash, S., Garcia-Sanz, J. A., and Beretta, L. (2002). Global and specific translational control by rapamycin in T cells uncovered by microarrays and proteomics. *J. Biol. Chem.* **277,** 22175–22184.

Gygi, S. P., Rochon, Y., Franza, B. R., and Aebersold, R. (1999). Correlation between protein and mRNA abundance in yeast. *Mol. Cell Biol* **19,** 1720–1730.

Hershey, J. W., and Merrick, W. C. (2000). Pathway and mechanism of initiation of protein synthesis. *In* "Translational Control of Gene Expression" (N. Sonenberg, *et al.*, eds.), pp. 33–88. Cold Spring Harbor Laboratory Press, Cold Spring Harbor.

Hinnebusch, A. G. (2000). Mechanism and regulation of initiator methionyl-tRNA binding to ribosomes. *In* "Translational Control of Gene Expression" (N. Sonenberg, *et al.*, eds.), pp. 185–244. Cold Spring Harbor Laboratory Press, Cold Spring Harbor.

Hochachka, P. W., Buck, L. T., Doll, C. J., and Land, S. C. (1996). Unifying theory of hypoxia tolerance: molecular/metabolic defense and rescue mechanisms for surviving oxygen lack. *Proc. Natl. Acad. Sci. USA* **93,** 9493–9498.

Holcik, M., and Sonenberg, N. (2005). Translational control in stress and apoptosis. *Nat. Rev. Mol. Cell Biol.* **6,** 318–327.

Holley, C. L., Olson, M. R., Colon-Ramos, D. A., and Kornbluth, S. (2002). Reaper eliminates IAP proteins through stimulated IAP degradation and generalized translational inhibition. *Nat. Cell Biol.* **4,** 439–444.

Johannes, G., Carter, M. S., Eisen, M. B., Brown, P. O., and Sarnow, P. (1999). Identification of eukaryotic mRNAs that are translated at reduced cap binding complex eIF4F concentrations using a cDNA microarray. *Proc. Natl. Acad. Sci. USA* **96,** 13118–13123.

Kaper, F., Dornhoefer, N., and Giaccia, A. J. (2006). Mutations in the PI3K/PTEN/TSC2 pathway contribute to mammalian target of rapamycin activity and increased translation under hypoxic conditions. *Cancer Res.* **66,** 1561–1569.

Kimball, S. R., and Jefferson, L. S. (2000). Regulation of translation initiation in mammalian cells by amino acids. *In* "Translational Control of Gene Expression" (N. Sonenberg, *et al.*, eds.), pp. 561–579. Cold Spring Harbor Laboratory Press, Cold Spring Harbor.

Koritzinsky, M., *et al.* (2006). Gene expression during acute and prolonged hypoxia is regulated by distinct mechanisms of translational control. *EMBO J.* **25,** 1114–1125.

Koritzinsky, M., Rouschop, K. R., van den Beucken, T., Magagnin, M. G., Savelkouls, K., Lambin, P., and Wouters, B. G. (2007). Phosphorylation of eIF2a is required for mRNA translation inhibition and survival during moderate hypoxia. Radiotherapy and Oncology. In press.

Koritzinsky, M., Seigneuric, R., Magagnin, M. G., Beucken, T., Lambin, P., and Wouters, B. G. (2005). The hypoxic proteome is influenced by gene-specific changes in mRNA translation. *Radiother Oncol.* **76,** 177–186.

Koumenis, C., Naczki, C., Koritzinsky, M., Rastani, S., Diehl, A., Sonenberg, N., Koromilas, A., and Wouters, B. G. (2002). Regulation of Protein Synthesis by Hypoxia via Activation of the Endoplasmic Reticulum Kinase PERK and Phosphorylation of the Translation Initiation Factor eIF2alpha. *Mol. Cell Biol.* **22,** 7405–7416.

Kozak, M. (2001). New ways of initiating translation in eukaryotes? *Mol. Cell Biol.* **21,** 1899–1907.

Kraggerud, S. M., Sandvik, J. A., and Pettersen, E. O. (1995). Regulation of protein synthesis in human cells exposed to extreme hypoxia. *Anticancer Res.* **15,** 683–686.

Kuhn, K. M., DeRisi, J. L., Brown, P. O., and Sarnow, P. (2001). Global and specific translational regulation in the genomic response of Saccharomyces cerevisiae to a rapid transfer from a fermentable to a nonfermentable carbon source. *Mol. Cell Biol.* **21,** 916–927.

Lang, K. J., Kappel, A., and Goodall, G. J. (2002). Hypoxia-inducible Factor-1alpha mRNA Contains an Internal Ribosome Entry Site That Allows Efficient Translation during Normoxia and Hypoxia. *Mol. Biol Cell.* **13,** 1792–1801.

Lefebvre, V. H., Van Steenbrugge, M., Beckers, V., Roberfroid, M., and Buc-Calderon, P. (1993). Adenine nucleotides and inhibition of protein synthesis in isolated hepatocytes incubated under different pO2 levels. *Arch. Biochem. Biophys.* **304,** 322–331.

Liu, L., Cash, T. P., Jones, R. G., Keith, B., Thompson, C. B., and Simon, M. C. (2006). Hypoxia-induced energy stress regulates mRNA translation and cell growth. *Mol. Cell.* **21,** 521–531.

Lu, P. D., Harding, H. P., and Ron, D. (2004). Translation reinitiation at alternative open reading frames regulates gene expression in an integrated stress response. *J. Cell Biol.* **167,** 27–33.

Martin de la Vega, C., Burda, J., Nemethova, M., Quevedo, C., Alcazar, A., Martin, M. E., Danielisova, V., Fando, J. L., and Salinas, M. (2001). Possible mechanisms involved in the down-regulation of translation during transient global ischaemia in the rat brain. *Biochem. J.* **357,** 819–826.

Martin, M. E., Munoz, F. M., Salinas, M., and Fando, J. L. (2000). Ischaemia induces changes in the association of the binding protein 4E- BP1 and eukaryotic initiation factor (eIF) 4G to eIF4E in differentiated PC12 cells. *Biochem. J.* **351**(Pt 2), 327–334.

Merrick, W. C., and Nyborg, J. (2000). The protein biosynthesis elongation cycle. *In* "Translational Control of Gene Expression" (N. Sonenberg, *et al.*, eds.), pp. 89–126. Cold Spring Harbor Laboratory Press, Plainview, NY.

Novoa, I., Zeng, H., Harding, H. P., and Ron, D. (2001). Feedback inhibition of the unfolded protein response by GADD34-mediated dephosphorylation of eIF2alpha. *J. Cell Biol.* **153,** 1011–1022.

Pettersen, E. O., Bjorhovde, I., Sovik, A., Edin, N. F., Zachar, V., Hole, E. O., Sandvik, J. A., and Ebbesen, P. (2007). Response of chronic hypoxic cells to low dose-rate irradiation. *Int. J. Radiat Biol.* **83,** 331–345.

Pettersen, E. O., Juul, N. O., and Ronning, O. W. (1986). Regulation of protein metabolism of human cells during and after acute hypoxia. *Cancer Res.* **46,** 4346–4351.

Pradet-Balade, B., Boulme, F., Mullner, E. W., and Garcia-Sanz, J. A. (2001). Reliability of mRNA profiling: verification for samples with different complexities. *Biotechniques.* **30,** 1352–1357.

Rajasekhar, V. K., Viale, A., Socci, N. D., Wiedmann, M., Hu, X., and Holland, E. C. (2003). Oncogenic Ras and Akt signaling contribute to glioblastoma formation by differential recruitment of existing mRNAs to polysomes. *Mol. Cell.* **12,** 889–901.

Raught, B., Gingras, A.-C., and Sonenberg, N. (2000). Regulation of ribosomal recruitment in eukaryotes. *In* "Translational Control of Gene Expression" (N. Sonenberg, *et al.,* eds.), pp. 245–294. Cold Spring Harbor Laboratory Press, Plainview, NY.

Semenza, G. L. (2003). Targeting HIF-1 for cancer therapy. *Nat. Rev. Cancer.* **3,** 721–732.

Stein, I., Itin, A., Einat, P., Skaliter, R., Grossman, Z., and Keshet, E. (1998). Translation of vascular endothelial growth factor mRNA by internal ribosome entry: Implications for translation under hypoxia. *Mol. Cell Biol.* **18,** 3112–3119.

Thomas, J. D., and Johannes, G. J. (2007). Identification of mRNAs that continue to associate with polysomes during hypoxia. *Rna* .

Vattem, K. M., and Wek, R. C. (2004). Reinitiation involving upstream ORFs regulates ATF4 mRNA translation in mammalian cells. *Proc. Natl. Acad. Sci. USA* **101,** 11269–11274.

Wang, X., Li, W., Williams, M., Terada, N., Alessi, D. R., and Proud, C. G. (2001). Regulation of elongation factor 2 kinase by p90(RSK1) and p70 S6 kinase. *EMBO J.* **20,** 4370–4379.

Hypoxia and the Unfolded Protein Response

Constantinos Koumenis,* Meixia Bi,* Jiangbin Ye,*
Douglas Feldman,† *and* Albert C. Koong†

Contents

1. Introduction	276
1.1. Mechanisms of cellular adaptation to low oxygen environment	276
1.2. Causes of ER stress and UPR	277
1.3. The PERK-eIF2α-ATF4 arm of the UPR	278
1.4. The IRE1-XBP1 arm of the UPR	279
1.5. The ATF6 pathway	280
1.6. Consequences of aberrant UPR induction for tumor formation	281
2. Methods Employed in Detecting Hypoxic Induction of ER Stress	282
2.1. Events proximal to ER stress	282
2.2. Events distal to ER stress	286
Acknowledgments	289
References	290

Abstract

Tumor hypoxia refers to the development of regions within solid tumors in which the oxygen concentration is lower (0–3%) compared to that in most normal tissues (4–9%) (Vaupel and Hockel, 2000). Considerable experimental and clinical evidence exists supporting the notion that hypoxia fundamentally alters the physiology of the tumor towards a more aggressive phenotype (Hockel and Vaupel, 2001). Therefore, delineating the mechanisms by which hypoxia affects tumor physiology at the cellular and molecular levels will be crucial for a better understanding of tumor development and metastasis and for designing better antitumor modalities.

* Department of Radiation Oncology, University of Pennsylvania School of Medicine, Philadelphia, Pennsylvania
† Department of Radiation Oncology, Stanford University School of Medicine, Stanford, California

Methods in Enzymology, Volume 435
ISSN 0076-6879, DOI: 10.1016/S0076-6879(07)35014-3

1. INTRODUCTION

1.1. Mechanisms of cellular adaptation to low oxygen environment

The hypoxic adaptation response can be roughly divided into pathways mediated in large part by the hypoxia-inducible factors (HIF) 1 and 2 or pathways that result in the general slowdown of processes that involve oxygen and energy consumption. These pathways are largely HIF-independent processes (Koumenis and Maxwell, 2006). HIF-1 and HIF-2 regulate the expression of over 100 genes involved in angiogenesis, anaerobic glycolysis, and cell survival, and their coordinated expression results in cellular adaptation to prolonged and acute hypoxia (Ratcliffe *et al.*, 1998; Semenza, 2000). Though the roles of HIF-1 and HIF-2 in cell survival are complex and dependent on the tumor microenvironment, cells deficient in these transcription factors usually display increased susceptibility to proapoptotic stimuli, reduced overall survival under hypoxic conditions, and, in some cases, reduced growth-forming abilities in nude mice (Maxwell *et al.*, 1997; Ryan *et al.*, 2000). Exposure of cells to anoxia also results in a rapid inhibition of replicon initiation, while acute or prolonged hypoxia leads to a reduction in the rates of global protein synthesis, responses that presumably reduce energy demands when oxygen and adenosine triphosphate (ATP) levels are low (Hochachka *et al.*, 1996). This translational inhibition appears to occur in distinct phases and to involve multiple pathways, and the decrease in the rate of translation is postulated to play an important role in cellular adaptation to the new environment of low oxygen and of energy deficiency (Koritzinsky *et al.*, 2006; Koumenis and Wouters, 2006; Liu *et al.*, 2006).

Translational inhibition in response to hypoxia appears to involve several pathways that converge on the translation machinery and, more specifically, on modifying translation initiation (and possible elongation) factors (eIFs). One important translation regulation pathway is initiated in the endoplasmic reticulum (ER), which, upon acute or moderate prolonged hypoxia, activates the ER-resident kinase PERK, which phosphorylates eIF2alpha (α) on ser51, resulting in a general slowdown or inhibition of CAP-dependent translation (Blais *et al.*, 2004; Koritzinsky *et al.*, 2006; Koumenis *et al.*, 2002; Liu *et al.*, 2006). The other major pathway initiates in the cytoplasm and involves the inhibition of the kinase mTOR, resulting in the hypophosphorylation of eIF4E-BP, and a reduced activity of the eIF4G complex (Arsham *et al.*, 2003; Koritzinsky *et al.*, 2006). Despite global inhibition of translation, many protein levels are actually increased under these conditions, including HIF-1α, its downstream gene products, and several other

HIF-1–independent proteins. (For more details and protocols regarding this pathway, see the review by Wouters *et al.* in this issue.) Therefore, in addition to conserving energy in terms of ATP consumption, the general translational inhibition induced by hypoxia may lead to the upregulation of several hypoxia-adaptive proteins. PERK is a critical component of a well-coordinated cellular program activated in response to misfolded or unfolded proteins in the endoplasmic reticulum called the unfolded protein response (UPR); thus, the finding that PERK becomes activated and subsequently phosphorylates eIF2α in response to hypoxia provided a link between hypoxic stress, inhibition of translation, and UPR activation.

1.2. Causes of ER stress and UPR

Several lines of evidence point to a strong relationship between hypoxia and the accumulation of misfolded proteins in the ER, a condition also known as ER stress (Blais *et al.*, 2004; Feldman *et al.*, 2005; Koumenis *et al.*, 2002; Romero-Ramirez *et al.*, 2004). Glucose-regulated protein 78 kDa (GRP78, also known as binding protein [BiP]), an ER chaperone protein, is a key regulator of the UPR. GRP78 is robustly induced in a variety of cell types exposed to severe hypoxia. Under these conditions, GRP78 serves a pro-survival function (Gazit *et al.*, 1999; Koong *et al.*, 1994; Little *et al.*, 1994). In eukaryotic cells, the endoplasmic reticulum serves as the first compartment of the secretory pathway and as a processing station for all secreted and transmembrane proteins. Unlike their counterparts in the cytosol, secreted proteins undergo a series of posttranslational modifications, including glycosylation and disulfide bond formation, which are required for progression to become properly folded proteins. These additional steps increase the likelihood of errors that can cause loss of protein function as well as the potentially toxic accumulation of misfolded proteins in the ER. To ensure the proper folding of ER proteins, cells utilize a dedicated machinery of molecular chaperones and folding enzymes that associate directly with nascent ER polypeptides. ER resident folding enzymes, such as protein disulfide isomerase (PDI) and ERO1, catalyze the formation of disulfide linkages by transiently forming mixed disulfides with their client proteins and acting as an electron relay system for oxidative folding (Tu and Weissman, 2004).

Molecular oxygen is the major electron acceptor at the end of this relay system, providing the driving force for protein folding in the ER (Tu and Weissman, 2002). In this manner, intracellular depletion of oxygen or glucose can result in the accumulation of misfolded proteins in the ER. Because protein folding is an energy-intensive process, the problems associated with the accumulation of misfolded proteins in solid tumors are magnified in hypoxic and hypoglycemic regions where the availability of

ATP may be even more limited. Furthermore, aberrant tumor vasculature formation results in disturbed microcirculation, producing regions of fluctuating oxygen and nutrient availability. In response to this stress, cells activate the UPR, a conserved, adaptive cellular program that counteracts the accumulation of misfolded proteins in the ER. As a homeostatic mechanism, the UPR precisely matches the folding capacity of the ER. With elevated demand, the UPR increases the expression of ER resident molecular chaperones and folding enzymes while decreasing the flux of nascent polypeptides imported into the ER (Mori, 2003; Schroder and Kaufman, 2005). The UPR also greatly expands ER capacity through increased expression of components of the ER-associated protein degradation (ERAD) machinery that mediate the clearance of terminally misfolded proteins (Schroder and Kaufman, 2005).

Activation of the UPR generally serves a protective function in the response to ER stress. However, under some circumstances, prolonged activation of the UPR can lead to cell death through the activation of proapoptotic pathways. A recent study by Rutkowski *et al.* (2006) illustrates the complexities of the cellular response to ER stress. Depending on the severity of ER stress, the balance between expression of proapoptotic and prosurvival genes ultimately determines the cellular fate through posttranscriptional and posttranslational mechanisms. Within solid tumors, multiple microenvironmental factors, such as glucose and oxygen concentration, contribute to ER stress. These elements are continuously in flux as a result of dynamic changes in the delivery and consumption of cellular nutrients. Thus, solid tumors experience both mild and severe ER stress, and understanding the balance between these factors has important implications in developing cancer therapies aimed at modulating the UPR (Koong *et al.*, 2006; Li and Lee, 2006; Ma and Hendershot, 2004).

1.3. The PERK-eIF2α-ATF4 arm of the UPR

Hypoxia was shown to induce a time-dependent and oxygen-dependent induction of eIF2α phosphorylation on ser51, and this modification correlated with a reduction in the rates of protein synthesis and was not observed by other stresses, such as low serum or genotoxic stress (Koritzinsky *et al.*, 2006; Koumenis and Wouters, 2006; Koumenis *et al.*, 2002; Wouters *et al.*, 2005). Cells expressing a trans-dominant–negative mutant allele of eIF2α or cells harboring a knock-in mutation of ser51 to ala exhibited significantly attenuated repression of protein synthesis under hypoxia (Koumenis *et al.*, 2002). The protein kinase responsible for phosphorylation of eIF2α was determined to be PERK, based on the observed hyperphosphorylation of exogenously expressed and tagged PERK protein in cells exposed to hypoxia, and that overexpression of wild-type (wt) PERK increased the levels of phosphorylation of eIF2α. Moreover, unlike PERK$^{+/+}$ mouse embryo fibroblasts (MEFs), PERK$^{-/-}$ MEFs failed to phosphorylate eIF2α in response to treatments with hypoxia or thapsigargin, and expression of a trans-dominant PERK allele with a truncated C-terminus also led to inhibition of eIF2α phosphorylation by hypoxia (Koumenis et al., 2002). Notably, cells with inactivated PERK also exhibited attenuated inhibition of protein synthesis in response to hypoxia (Koumenis et al., 2002; Liu et al., 2006). These cells indicate that PERK is the primary kinase that phosphorylates eIF2α, leading to downregulation of protein synthesis in response to hypoxic stress (Koumenis et al., 2002).

The activation of PERK and phosphorylation of eIF2α serve two major functions in a cell experiencing ER stress. The first is to rapidly downregulate protein synthesis, which in turn reduces the load of misfolded proteins in the ER and leads to lower energy expenditure, since both protein synthesis and protein folding are ATP-requiring processes (Dorner et al., 1990; Harding et al., 2000a; Shi et al., 1998). The second function is to upregulate genes that promote amino acid sufficiency and redox homeostasis (Harding et al., 2003), thereby further promoting cell survival. Some, but not all, of the effects of PERK are mediated by the transcription factor ATF4, which is translationally upregulated by ER stress in an eIF2α phosphorylation-dependent manner (Harding et al., 2000b, 2003). ATF4 activates the induction of downstream UPR genes, but it has also been implicated in pathways activated in response to amino acid deprivation and prooxidant stress (Fawcett et al., 1999; Harding et al., 2003; Lu et al., 2004). This pathway, known as the integrated stress response (ISR) (Ron, 2002), constitutes one arm of the UPR, which promotes cellular adaptation to conditions of ER stress. Hypoxia-induced ATF4 accumulation occurs primarily at the posttranscriptional level because ATF4 messenger RNA (mRNA) levels are not upregulated in hypoxic PERK$^{+/+}$ MEFs, and the data support both translational upregulation of the ATF4 levels and increased stability of the synthesized protein under hypoxia (Ameri et al., 2004; Bi et al., 2005; Blais et al., 2004). Cells lacking ATF4 are quite sensitive to hypoxic stress and expression of the ISR proteins ATF4, and CHOP is upregulated in hypoxic areas of primary human tumors (Bi et al., 2005). The expression of ATF4 was also significantly higher in malignant tissues obtained from patients with brain, breast, cervical, and skin cancers compared to corresponding normal tissues. Collectively, these results point towards a prosurvival, protumorigenic activity of the PERK-eIF2α-ATF4 pathway elicited by hypoxia and (potentially) other stresses within the tumor microenvironment.

1.4. The IRE1-XBP1 arm of the UPR

The ER resident transmembrane protein IRE1α functions as a pivotal UPR signal transducer via its dual cytoplasmic kinase and endoribonuclease domains (Tirasophon et al., 1998). In the absence of ER stress, Grp78 directly associates with IRE1α, suppressing its activation by blocking dimerization. Misfolded proteins competitively titrate away Grp78 as they accumulate in the ER, initiating stress signaling (Yoshida *et al.*, 2000).

Mammalian IRE1α, the major functional homolog of yeast IRE1, excises a 26-nucleotide intron from the mRNA encoding the basic leucine zipper (bZIP) transcription factor XBP-1. This encoding results in a translational frame shift downstream of the splice site to generate XBP-1s, a potent transcriptional activator of UPR targets (Calfon *et al.*, 2002; Lee *et al.*, 1998, 2002; Yoshida *et al.*, 2001). XBP-1s associates with the general transcription factor NF-Y to activate transcription from the ER stress element (ERSE) (Yamamoto *et al.*, 2004; Yoshida *et al.*, 1998). In addition, XBP-1s binds as a homodimer preferentially to the unfolded protein response element (UPRE), a *cis*-regulatory sequence located in the promoter regions of its transactivation target genes (Wang *et al.*, 2000).

ER-to-nucleus signaling via IRE1-mediated mRNA processing of XBP-1 (HAC1 in yeast) is the most distantly conserved component of the UPR, underscoring the critical importance of this pathway for cell survival under ER stress. Comprehensive genomic studies in yeast have demonstrated that IRE1 signaling occurs entirely through the endonuclease-mediated activation of HAC1 (Niwa *et al.*, 2005). However, in higher organisms, IRE1 has acquired additional functions. In metazoans, the target repertoire of the IRE1 nuclease has expanded considerably and includes a pool of mRNAs encoding secreted proteins (Hollien and Weissman, 2006). Activation of mammalian IRE1α by ER stress triggers multiple signaling outputs that extend beyond the endonuclease, including the activation of autophagy and apoptosis pathways through the IRE1α kinase domain and its downstream effectors caspase-12, ASK1, and JNK1 (Ogata *et al.*, 2006; Urano *et al.*, 2000). Thus, IRE1α may participate in both cytoprotective and pro-apoptotic pathways.

1.5. The ATF6 pathway

ATF6 is also an ER transmembrane protein that is constitutively expressed but remains inactive in unstressed cells. ATF6 exists as two isoforms, the 90-kDa ATF6α and the 100-kDa ATF6β. Upon ER stress activation, both isoforms are processed by site-1 protease (S1P) and site-2 protease (S2P), which also cleave the ER–associated transmembrane sterol-response element-binding protein (SREBP) upon cholesterol deprivation to generate 50-kDa cytosolic, a bZIP-containing transcription factor that migrates to the nucleus (Haze *et al.*, 1999; Kaufman, 2002; Okada *et al.*, 2002; Shen *et al.*, 2002). There, both XBP1 and ATF6 regulate the expression of downstream genes by binding to the endoplasmic reticulum stress element (ERSE) and the unfolded protein response element (UPRE) of downstream gene promoters (Kaufman, 2002). Gene expression array analysis using

XBP1$^{-/-}$ MEFs and cells in which ATF6α, as inhibited by RNAi, revealed that XBP1 and ATF6 are responsible for the induction of a distinct subset of UPR genes, with some redundancy in their activities (Lee *et al.*, 2003).

1.6. Consequences of aberrant UPR induction for tumor formation

Several studies have demonstrated that major components of the UPR are required for cellular adaptation to hypoxic stress *in vitro* and tumorigenesis *in vivo*. Genetic deletion of PERK or ATF4 sensitizes cells to apoptosis following exposure to hypoxia (Bi *et al.*, 2005; Koumenis *et al.*, 2002), and Ras-transformed PERK$^{-/-}$ cells or mouse fibroblasts expressing a mutated form of eIF2α that cannot be phosphorylated by PERK are severely impaired for tumor formation *in vivo* (Bi *et al.*, 2005). Although the mechanisms by which an abrogation of the PERK-eIF2α-ATF4 pathway leads to profound inhibition of tumor growth are not currently fully elucidated, two factors that are likely to contribute to this effect include an increased level of *in vivo* apoptosis of cells deficient in PERK or eIF2α phosphorylation, especially in hypoxic areas (Bi *et al.*, 2005), and inhibition of angiogenesis due to abnormal blood vessel formation (Blais *et al.*, 2006).

The IRE1α/XBP-1 branch of the UPR also plays a pivotal role in the control of tumor growth. IRE1α is extensively activated in hypoxic regions of human tumor xenografts throughout tumorigenesis (Feldman *et al.*, 2005), and MEFs genetically deleted for XBP-1 exhibit increased sensitivity to hypoxia and fail to grow as tumors when implanted into immune-deficient mice (Romero-Ramirez *et al.*, 2004), even though they secrete comparable levels of angiogenic and survival factors, such as vascular endothelial growth factor (VEGF) and basic fibroblastic growth factor (bFGF). Similar results were obtained in wt MEFs stably expressing an XBP-1 short hairpin RNA (shRNA) construct. In addition, HT1080 tumor cells stably expressing an XBP-1 shRNA construct also demonstrated impaired tumor growth (Chen *et al.*, 2005). Further evidence demonstrating the importance of the IRE1/XBP-1 pathway in tumor growth comes from the observation that expression of a dominant-negative IRE1α construct compromises the survival of HT1080 fibrosarcoma and MiaPaCa-2 pancreatic carcinoma cells under hypoxia and results in inhibition of tumor growth (Feldman and Koong, manuscript in preparation). The significance of the UPR in the pathophysiology of cancer is reinforced by histochemical analyses of human tissue specimens, which exhibit significantly higher levels of ATF4 and XBP-1s in malignant tumors relative to the corresponding normal tissues (Ameri *et al.*, 2004; Bi *et al.*, 2005; Fujimoto *et al.*, 2003; Shuda *et al.*, 2003).

In addition to solid tumors, XBP-1 also plays a critical role in myeloma cells. Lee *et al.* (2003) demonstrated that PS-341, a proteasome inhibitor, blocks IRE1α-mediated splicing of XBP-1 and stabilizes the unspliced form of the protein, which can function as a dominant-negative inhibitor. Furthermore, myeloma cells rendered functionally deficient in XBP-1 undergo increased apoptosis in response to ER stress. Thus, these results demonstrate the key contribution of the IRE1α/XBP-1 branch of the UPR in mediating survival under hypoxic conditions and regulating tumor growth.

Although it is expected, a definitive role in hypoxia survival and tumorigenesis for ATF6 family members ATF6α and ATF6β has not been reported yet by either biochemical or genetic studies. Nevertheless, an important principle emerges from considering the consequences of eliminating PERK and IRE1α/XBP-1 in tumors. The widespread occurrence of hypoxia in tumors and the severe impairment of growth in UPR-deficient tumors demonstrate that ER stress is inextricably linked to tumorigenesis. Tumor growth may be envisioned as a continuous act of overcoming stress-induced misfolding and proteotoxicity in a process coordinated by the UPR.

2. Methods Employed in Detecting Hypoxic Induction of ER Stress

2.1. Events proximal to ER stress

2.1.1. PERK phosphorylation by supershift in immunoblots and the use of phosphospecific antibody

Activation of PERK involves autophosphorylation, which results in a shift in the apparent molecular weight (MW) on a polyacrylamide gel (Bertolotti *et al.*, 2000; Harding *et al.*, 1999). Mutation of a lysine at position 618 to alanine, which lies close to the catalytic center, reduces the autocatalytic activity of the protein. We therefore used transient transfections of plasmids expressing either wt mouse PERK or the mutant K618A (KA) mouse PERK in A549 cells to examine these effects (Koumenis *et al.*, 2002). As shown in Fig. 14.1A (top panel), under normoxic conditions, wt PERK exhibits a slightly slower electrophoretic mobility compared to the KA PERK, which has been postulated to arise from activation of wt PERK by autophosphorylation (Harding *et al.*, 1999). Such an autophosphorylation activity has been reported when PERK is expressed at high levels, presumably because of oligomerization. Also, overexpression of KA PERK moderately increased the levels of eIF2α phosphorylation due to residual catalytic activity of this mutant (two middle panels). Expression of the wt PERK protein, on the other hand, induced a significantly larger increase in eIF2α phosphorylation. Upon treatments with hypoxia, an even larger shift occurred in the electrophoretic mobility of wt PERK compared to untreated cells, while the KA PERK protein failed to exhibit a similar shift. Hypoxia increased the levels of eIF2α phosphorylation in untransfected cells, as

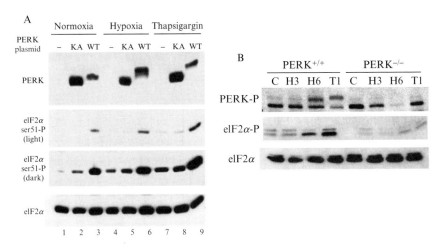

Figure 14.1 (A) Hypoxia induces a shift in the electrophoretic mobility of PERK. A549 cells were transfected with either wild-type (wt)-mouse PERK or K618A (KA)-mouse PERK and exposed to hypoxia or thapsigargin, as indicated. Immunoblotting was performed using a rabbit polyclonal antibody raised against human PERK (top panel) or with the anti-phospho– eukaryotic initiation factor 2alpha (eIF2α) antibody (two middle panels; a light and dark exposure from the same film are shown for easier comparison of levels of phosphorylated eIF2α). (Bottom panel) The same blot was stripped and reblotted with the anti-eIF2α (total eIF2α) antibody. The upper band indicated as P-PERK corresponds to phosphorylated PERK. (B) Extreme hypoxia induces PERK and eIF2α phosphorylation. PERK$^{+/+}$ and PERK$^{-/-}$ mouse embryo fibroblasts (MEFs) were exposed to at most 0.02% O_2 or treated with 1 μM thapsigargin for the times indicated. Immunoblots were performed on lysates with antibodies against phospho-PERK, phosphor-eIF2α, or total eIF2α. Adapted with permission from Koumenis *et al.* (2002) and Bi *et al.* (2005).

expected, while in cells transfected with wt PERK, the increase in eIF2α phosphorylation was greater than that observed in cells transfected with the KA PERK mutant construct or non-transfected cells. Similar results were obtained in cells treated with thapsigargin, an inhibitor of the sarcoplasmic/endoplasmic reticulum calcium ATPase (SERCA) (Ca^{2+}-ATPase) pump and potent ER stressor. These findings suggest that PERK becomes phosphorylated and activated under hypoxia and may be involved in hypoxia-induced eIF2α phosphorylation.

Recently, a phosphospecific antibody that recognizes phosphorylated PERK has become commercially available (Cell Signaling Technologies, Beverly, MA). Our experience with use of this antibody in one of our labs (C. K.) is that it works well in mouse cells and homogenates, but poorly in human cell extracts. We do not know the reason for this preferential activity. To investigate the effects of hypoxia on the PERK activation MEFs from PERK$^{+/+}$ and PERK$^{-/-}$ animals were exposed to extreme hypoxia (= 0.02%) or were treated with thapsigargin, as a positive control.

Consistent with previous findings (Koumenis *et al.*, 2002), hypoxia caused a time-dependent increase in PERK autophosphorylation and eIF2α phosphorylation in PERK$^{+/+}$ MEFs (see Fig. 14.1B). Hypoxia-induced eIF2α phosphorylation was substantially reduced in PERK$^{-/-}$ MEFs below the basal levels observed in PERK$^{+/+}$ MEFs. Cell extracts were immunoblotted with anti-phospho–PERK, anti-phospho–eIF2α, and anti-eIF2α antibodies.

Methods   The wt PERK and KA-PERK plasmids have been previously described (Tu *et al.*, 2000). A549 cells were transfected with 5 μg of wt- or K516-mutant PERK plasmid. Forty-eight hours later, cells were treated with hypoxia or thapsigargin. Following treatments, cells were washed three times with ice-cold phosphate-buffered saline (PBS) and resuspended in PBS with 1% NP-40 containing 2 μg/ml leupeptin, 2 μg/ml pepstatin, 2 μg/ml aprotinin, 2 μg/ml antipain, 1 mM phenylmethanesulphonylfluoride (PMSF), 1 mM Na$_3$VO$_5$, 1 mM NaF, 1 μM microcystin L, and 2 mM ethylenediaminetetraacetic acid (EDTA). Cells were lysed on ice for 15 min and centrifuged for 10 min to separate cell debris. Protein concentrations of each sample were determined by the modified Bradford assay, as recommended by the manufacturer (Biorad, Hercules, CA). Forty to fifty micrograms of whole-cell protein extract were mixed with an equal volume of 2x Laemmli sample buffer (62 mM Tris [pH 6.8], 10% glycerol, 2% sodium dodecyl sulfate [SDS], 5% beta (β-mercaptoethanol, and 0.003% bromophenol blue) and heated at 95° for 5 min. Proteins were resolved on a 10 to 12% SDS-polyacrylamide gel and transferred onto Hybond ECL nitrocellulose membrane (Amersham, Arlington Heights, IL) using a wet transfer system (Bio-Rad, Hercules, CA). Membranes were stained with 0.15% ponceau red (Sigma-Aldrich, St. Louis, MO) to ensure equal loading and transfer and then blocked with 5% (w/v) dried nonfat milk in Tris-buffered saline (TBS) (20 mM Tris-base, 137 mM NaCl, pH 7.6). Immunoblotting for eIF2α was performed using a rabbit polyclonal antibody raised against the phosphorylated form of eIF2α (anti-Pser51) (Cell Signaling Technologies, Beverly, MA) or a rabbit polyclonal antibody recognizing both phosphorylated and unphosphorylated eIF2α (Cell SignalingTechnologies, Beverly, MA) in TBS with 0.1% Tween-20 (TBST). Immunoblotting for PERK was performed using a goat polyclonal antibody (Santa Cruz Biotechnology, Santa Cruz, CA). Following incubation with the primary antibody, the membranes were washed in TBST and incubated with secondary antibody, and immunoreactive bands were visualized using an enhanced chemiluminescence (ECL-Plus) reagent kit according to the manufacturer's recommendations (Amersham, Arlington Heights, IL). Films were exposed to the membranes for varying periods of time and scanned with a personal scanner (Microtec, San Jose, CA). Similar methods for immunoblotting were followed to analyze phosphorylated PERK (see Fig. 14.1B). Immunoblotting was performed using the following antibodies: Rabbit polyclonal raised against the phosphorylated form of eIF2α (anti-Pser51) (Cell Signaling Technologies, Beverly, MA) or a rabbit polyclonal recognizing both phosphorylated and unphosphorylated eIF2α (Cell Signaling Technologies) and a rabbit polyclonal against phosphorylated PERK (Thr 980) (Cell Signaling Technologies).

2.1.2. ATF6 proteolytic cleavage and nuclear translocation of N-terminus

To determine whether hypoxia can activate the ATF6 pathway, we first tested whether hypoxia could induce proteolytic cleavage of ATF6. Thus, we transfected N-terminal FLAG-tagged ATF6 into HeLa cells, and, 48 h later, cells were exposed to hypoxia. Immunoblot analysis indicated that the N-terminal portion of ATF6 was cleaved in a time-dependent manner by hypoxia, reaching maximal levels 16 h after hypoxia (Fig. 14.2A). Confocal microscopy was then used to determine the cellular localization of the N-terminal fragment following exposure to hypoxia. Under normoxic

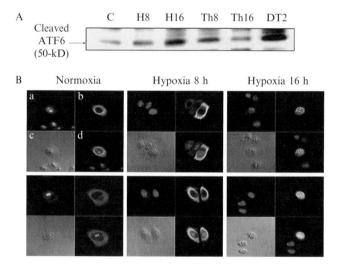

Figure 14.2 (A) Hypoxia induces proteolytic cleavage and activation of ATF6. Production of the 50-kD proteolytic fragment of ATF6 is increased under prolonged hypoxia. HeLa cells were transfected with FLAG-tagged ATF6 and exposed to hypoxia (≤0.02% O₂) for 8 and 16 h, the endoplasmic reticulum (ER) stress activators thapsigargin for 8 or 16 h, and the dithiothreitol (DTT) for 2 h. Cell lysates were immunoblotted with an anti-FLAG antibody. (B) The N-terminal, 50-kD–cleaved ATF6 fragment translocates to the nucleus following prolonged hypoxia. Confocal microscopy was performed on fixed cells using an anti-FLAG fluorescein isothiocyanate (FITC)-labeled antibody (Panel b) and counterstaining nuclei with propidium iodide (Panel a). (Panel c) Differential interference contrast (DIC) microscopy. (Panel d) Combined image of Panels a and b. (See color insert.)

conditions, the N-terminal of ATF6 was found in the cytoplasm, consistent with localization to the ER (see Fig. 14.2B). No significant change was observed after 8 h of hypoxia, but, after 16 h, the majority of the cells displayed nuclear localization of the N-terminal fragment. Hypoxia, thapsigargin, and dithiothreitol (agents that induce ER stress) promote cleavage of ATF6 and production of a 50-kD fragment. HeLa cells were exposed to 0.02% O_2 or treated with 500 nM thapsigargin or 1 mM dithiothreitol for the indicated times before immunoblotting with an antibody specific for cleaved-ATF6.

Methods HeLa cells were transfected with FLAG-tagged ATF6 and exposed to hypoxia (0.02% O_2) for 8 and 16 h. Confocal microscopy was performed on fixed cells using an anti-FLAG fluorescein isothiocyanate-labeled antibody (see Fig. 14.2B, Panel b) and counterstaining nuclei with propidium iodide (see Fig. 14.2B, Panel a). Figure 14.2B, Panel c shows DIC microscopy, and Panel d is the combined image of Panels a and b. For detection of cleaved ATF6 and localization of the N-terminal fragment following ER stress, cells were transfected with 4 μg of an expression plasmid p3XFLAG-CMV 7.1 (generous gift of Dr. Ron Prywes, Columbia University) in which the full-length human ATF6 gene was cloned down-stream of a cytomegalovirus (CMV) promoter and in frame with an N-terminal–encoded tri-repeat of the FLAG epitope. Forty-eight hours after transfection, HeLa cells were treated with indicated stresses, cells were collected in lysis buffer (discussed later), and N-terminal–cleaved ATF6 was detected with a mouse monoclonal anti-FLAG epitope (Sigma, St. Louis, MO). For immunolocalization of ATF6, cells were fixed in a 4% paraformaldehyde solution and blocked with 1% BSA plus 0.1% Triton X in PBS for 30 min. Cells were then incubated with anti-FLAG M2 mono-clonal antibody (Sigma-Aldrich, St. Louis, MO) followed by Alexa Flour 488–conjugated anti-mouse (Molecular Probes, Eugene, OR). Cells were counterstained with DAPI and subjected to confocal microscopy.

2.2. Events distal to ER stress

2.2.1. ATF4 detection by immunohistochemistry in hypoxic areas

To investigate whether the induction of the ISR also occurs in primary human tumors, we analyzed ATF4 levels in human cervical tumor sections obtained from patients that had been injected with the hypoxia-sensitive dye pimonidazole. Expression of ATF4 co-localized with hypoxic areas or areas adjacent to hypoxic regions. ATF4 expression showed stronger association with highly hypoxic areas (pimonidazole-positive staining; see Fig. 14.3).

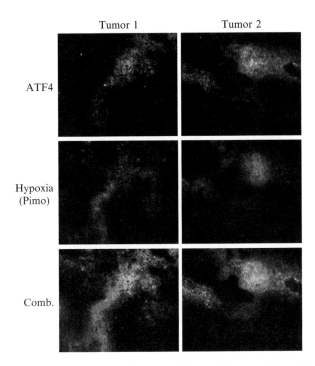

Figure 14.3 Immunofluorescence of ATF4 and hypoxia in two primary human cervical tumors. Sections were stained with an anti-ATF4 (top) polyclonal antibody followed by a fluorescein isothiocyanate (FITC)-conjugated secondary antibody and with a Cy-3–labelled anti-pimonidazole antibody (middle). Bottom panels are combined images. Adapted with permission from Bi *et al.* (2005). (See color insert.)

Methods Human cervical squamous cell carcinoma tumors were obtained at the University of North Carolina at Chapel Hill (UNC-CH) School of Medicine. Patients were injected with pimonidazole prior to surgical resection of tumors. Tumors were flash-frozen in liquid nitrogen, embedded in OCT, and stored at −80°. Frozen human cervical tumor sections (5 μm) were fixed to glass slides in 4% formaldehyde and 95, 75, and 50% ethanol and blocked in 3% BSA in PBS and hypoxia adducts detected with Cy3-conjugated anti-pimonidazole monoclonal antibody (2 μg/ml). Sections were incubated with anti-ATF4 polyclonal antibodies, followed with a FITC-conjugated secondary antibody.

2.2.2. *In vivo* bioluminescence imaging of XBP-1 slicing

To investigate IRE1α signaling in tumors, we employed a chimeric XBP-luciferase reporter in which firefly luciferase is fused downstream of XBP-1 (Fig. 14.4). In cells expressing this reporter, luciferase is translated only when XBP-1 is spliced by IRE1α. Consistent with a previous analysis of

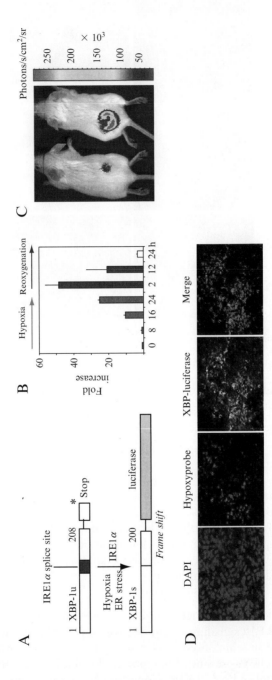

Figure 14.4 Hypoxia activates XBP-1 splicing (A,B). XBP-1 splicing occurs within solid tumors throughout their growth (C,D). (See color insert.)

XBP-1 splicing, the XBP-luciferase reporter was strongly induced after 16 to 24 h incubation in hypoxic conditions ($< 0.1\%$ O_2). IRE1α reporter activity decayed to near-background levels within 24 h after reoxygenation, indicating that cellular luminescence reflects recent or ongoing ER stress (see Fig. 14.4B). HT1080 fibrosarcoma cells stably expressing this reporter were implanted subcutaneously into the hind flanks of immune-compromised severe combined immune deficiency SCID mice and allowed to develop into tumors. IRE1α-dependent luminescence was detected in live tumors by optical bioluminescence imaging (see Fig. 14.4C). Both small (20–50 mm^3) and large, well-established (300–600 mm^3) tumors exhibited strong luminescence. In larger tumors, XBP-1 splicing activity localized to rapidly growing, non-necrotic regions (see Fig. 14.4D). In addition to the detection of bioluminescent activity from subcutaneously implanted tumors, it is also possible to observe activation of this reporter construct in tumors implanted in orthotopic sites (data not shown).

Quantitative analysis of IRE1α luciferase reporter activity as a function of tumor size indicated that, for each three- to four-fold increase in tumor volume, IRE1α signaling increased more than 10-fold. Expression of the XBP-luciferase fusion protein overlapped with acutely hypoxic cells, as judged by positive immunostaining for XBP-luciferase and the hypoxia-specific marker pimonidazole (hypoxyprobe), administered to tumor-bearing mice prior to tumor excision (see Fig. 14.4D). These observations are consistent with results demonstrating a strict requirement for molecular oxygen in the correct folding of ER proteins (Tu *et al.*, 2000). However, XBP-luciferase expression also extended to areas that did not stain strongly for pimonidazole. These findings suggest that additional factors, such as low glucose or low pH, may contribute to activation of ER stress in the tumor microenvironment. Taken together, these findings demonstrate that ER stress and IRE1α activity occur throughout tumorigenesis.

In the bioluminescent imaging experiments previously described, HT1080 fibrosarcoma cells stably expressing the XBP-luciferase reporter were implanted subcutaneously into SCID mice and allowed to develop into tumors. Ten minutes prior to imaging, mice were injected intraperitoneally with D-luciferin (150 mg/kg) solubilized in PBS. Optical biolumi-nescence imaging was performed using the IVIS charged-coupled device camera system (Xenogen, Alameda, CA). Mice were imaged for 1 to 4 min per acquisition scan. Signal intensities were analyzed using Living Image software (Xenogen, Alameda, CA).

ACKNOWLEDGMENTS

The authors would like to acknowledge the following grant support: 1R01CA112108 (A. C. K.), Multiple Myeloma Research Foundation (D. F.), and 1R01 CA94214 (C. K.).

REFERENCES

Ameri, K., Lewis, C. E., Raida, M., Sowter, H., Hai, T., and Harris, A. L. (2004). Anoxic induction of ATF-4 through HIF-1–independent pathways of protein stabilization in human cancer cells. *Blood* **103,** 1876–1882.

Arsham, A. M., Howell, J. J., and Simon, M. C. (2003). A novel hypoxia-inducible factor-independent hypoxic response regulating mammalian target of rapamycin and its targets. *J. Biol. Chem.* **278,** 29655–29660.

Bertolotti, A., Zhang, Y., Hendershot, L. M., Harding, H. P., and Ron, D. (2000). Dynamic interaction of BiP and ER stress transducers in the unfolded-protein response. *Nat. Cell Biol.* **2,** 326–332.

Bi, M., Naczki, C., Koritzinsky, M., Fels, D., Blais, J., Hu, N., Harding, H., Novoa, I., Varia, M., Raleigh, J., Scheuner, D., Kaufman, R. J., *et al.* (2005). ER stress-regulated translation increases tolerance to extreme hypoxia and promotes tumor growth. *EMBO J.* **24,** 3470–3481.

Blais, J. D., Filipenko, V., Bi, M., Harding, H. P., Ron, D., Koumenis, C., Wouters, B. G., and Bell, J. C. (2004). Activating transcription factor 4 is translationally regulated by hypoxic stress. *Mol. Cell Biol.* **24,** 7469–7482.

Blais, J. D., Addison, C. L., Edge, R., Falls, T., Zhao, H., Wary, K., Koumenis, C., Harding, H. P., Ron, D., Holcik, M., and Bell, J. C. (2006). PERK-dependent translational regulation promotes tumor cell adaptation and angiogenesis in response to hypoxic stress. *Mol. Cell Biol.* **26,** 9517–9532.

Calfon, M., Zeng, H., Urano, F., Till, J. H., Hubbard, S. R., Harding, H. P., Clark, S. G., and Ron, D. (2002). IRE1 couples endoplasmic reticulum load to secretory capacity by processing the XBP-1 mRNA. *Nature* **415,** 92–96.

Chen, Y., Feldman, D. E., Deng, C., Brown, J. A., De Giacomo, A. F., Gaw, A. F., Shi, G., Le, Q. T., Brown, J. M., and Koong, A. C. (2005). Identification of mitogen-activated protein kinase signaling pathways that confer resistance to endoplasmic reticulum stress in *Saccharomyces cerevisiae*. *Mol. Cancer Res.* **3,** 669–677.

Dorner, A. J., Wasley, L. C., and Kaufman, R. J. (1990). Protein dissociation from GRP78 and secretion are blocked by depletion of cellular ATP levels. *Proc. Natl. Acad. Sci. USA* **87,** 7429–7432.

Fawcett, T. W., Martindale, J. L., Guyton, K. Z., Hai, T., and Holbrook, N. J. (1999). Complexes containing activating transcription factor (ATF)/cAMP-responsive-element-binding protein (CREB) interact with the CCAAT/enhancer-binding protein (C/EBP)-ATF composite site to regulate Gadd153 expression during the stress response. *Biochem. J.* **339**(Pt 1), 135–141.

Feldman, D. E., Chauhan, V., and Koong, A. C. (2005). The unfolded protein response: A novel component of the hypoxic stress response in tumors. *Mol. Cancer Res.* **3,** 597–605.

Fujimoto, T., Onda, M., Nagai, H., Nagahata, T., Ogawa, K., and Emi, M. (2003). Upregulation and overexpression of human X-box binding protein 1 (hXBP-1) gene in primary breast cancers. *Breast Cancer* **10,** 301–306.

Gazit, G., Hung, G., Chen, X., Anderson, W. F., and Lee, A. S. (1999). Use of the glucose starvation-inducible glucose-regulated protein 78 promoter in suicide gene therapy of murine fibrosarcoma. *Cancer Res.* **59,** 3100–3106.

Harding, H. P., Zhang, Y., and Ron, D. (1999). Protein translation and folding are coupled by an endoplasmic-reticulum-resident kinase. *Nature* **397,** 271–274.

Harding, H. P., Zhang, Y., Bertolotti, A., Zeng, H., and Ron, D. (2000a). PERK is essential for translational regulation and cell survival during the unfolded protein response. *Mol. Cell* **5,** 897–904.

Harding, H. P., Novoa, I., Zhang, Y., Zeng, H., Wek, R., Schapira, M., and Ron, D. (2000b). Regulated translation initiation controls stress-induced gene expression in mammalian cells. *Mol. Cell* **6**, 1099–1108.

Harding, H. P., Zhang, Y., Zeng, H., Novoa, I., Lu, P. D., Calfon, M., Sadri, N., Yun, C., Popko, B., Paules, R., Stojdl, D. F., Bell, J. C., *et al.* (2003). An integrated stress response regulates amino acid metabolism and resistance to oxidative stress. *Mol. Cell* **11**, 619–633.

Haze, K., Yoshida, H., Yanagi, H., Yura, T., and Mori, K. (1999). Mammalian transcription factor ATF6 is synthesized as a transmembrane protein and activated by proteolysis in response to endoplasmic reticulum stress. *Mol. Biol. Cell* **10**, 3787–3799.

Hochachka, P. W., Buck, L. T., Doll, C. J., and Land, S. C. (1996). Unifying theory of hypoxia tolerance: Molecular/metabolic defense and rescue mechanisms for surviving oxygen lack. *Proc. Natl. Acad. Sci. USA* **93**, 9493–9498.

Hockel, M., and Vaupel, P. (2001). Biological consequences of tumor hypoxia. *Semin. Oncol.* **28**, 36–41.

Hollien, J., and Weissman, J. S. (2006). Decay of endoplasmic reticulum-localized mRNAs during the unfolded protein response. *Science* **313**, 104–107.

Kaufman, R. J. (2002). Orchestrating the unfolded protein response in health and disease. *J. Clin. Invest.* **110**, 1389–1398.

Koong, A. C., Chauhan, V., and Romero-Ramirez, L. (2006). Targeting XBP-1 as a novel anti-cancer strategy. *Cancer Biol. Ther.* **5**, 756–759.

Koong, A. C., Chen, E. Y., Lee, A. S., Brown, J. M., and Giaccia, A. J. (1994). Increased cytotoxicity of chronic hypoxic cells by molecular inhibition of GRP78 induction. *Int. J. Radiat. Oncol. Biol. Phys.* **28**, 661–666.

Koritzinsky, M., Magagnin, M. G., van den Beucken, T., Seigneuric, R., Savelkouls, K., Dostie, J., Pyronnet, S., Kaufman, R. J., Weppler, S. A., Voncken, J. W., Lambin, P., Koumenis, C., *et al.* (2006). Gene expression during acute and prolonged hypoxia is regulated by distinct mechanisms of translational control. *EMBO J.* **25**, 1114–1125.

Koumenis, C., and Maxwell, P. H. (2006). Low oxygen stimulates the intellect. Symposium on hypoxia and development, physiology and disease. *EMBO Rep.* **7**, 679–684.

Koumenis, C., and Wouters, B. G. (2006). "Translating" tumor hypoxia: Unfolded protein response (UPR)-dependent and UPR-independent pathways. *Mol. Cancer Res.* **4**, 423–436.

Koumenis, C., Naczki, C., Koritzinsky, M., Rastani, S., Diehl, A., Sonenberg, N., Koromilas, A., and Wouters, B. G. (2002). Regulation of protein synthesis by hypoxia via activation of the endoplasmic reticulum kinase PERK and phosphorylation of the translation initiation factor eIF2α. *Mol. Cell Biol.* **22**, 7405–7416.

Lee, A. H., Iwakoshi, N. N., and Glimcher, L. H. (2003). XBP-1 regulates a subset of endoplasmic reticulum resident chaperone genes in the unfolded protein response. *Mol. Cell Biol.* **23**, 7448–7459.

Lee, A. H., Iwakoshi, N. N., Anderson, K. C., and Glimcher, L. H. (2003). Proteasome inhibitors disrupt the unfolded protein response in myeloma cells. *Proc. Natl. Acad. Sci. USA* **100**, 9946–9951.

Lee, K., Tirasophon, W., Shen, X., Michalak, M., Prywes, R., Okada, T., Yoshida, H., Mori, K., and Kaufman, R. J. (2002). IRE1-mediated unconventional mRNA splicing and S2P-mediated ATF6 cleavage merge to regulate XBP1 in signaling the unfolded protein response. *Genes Dev.* **16**, 452–466.

Li, J., and Lee, A. S. (2006). Stress induction of GRP78/BiP and its role in cancer. *Curr. Mol. Med.* **6**, 45–54.

Little, E., Ramakrishnan, M., Roy, B., Gazit, G., and Lee, A. S. (1994). The glucose-regulated proteins (GRP78 and GRP94): Functions, gene regulation, and applications. *Crit. Rev. Eukaryot. Gene Expr.* **4**, 1–18.

Liu, L., Cash, T. P., Jones, R. G., Keith, B., Thompson, C. B., and Simon, M. C. (2006). Hypoxia-induced energy stress regulates mRNA translation and cell growth. *Mol. Cell* **21**, 521–531.

Lu, P. D., Jousse, C., Marciniak, S. J., Zhang, Y., Novoa, I., Scheuner, D., Kaufman, R. J., Ron, D., and Harding, H. P. (2004). Cytoprotection by preemptive conditional phosphorylation of translation initiation factor 2. *EMBO J.* **23,** 169–179.

Ma, Y., and Hendershot, L. M. (2004). The role of the unfolded protein response in tumour development: Friend or foe? *Nat. Rev. Cancer* **4,** 966–977.

Maxwell, P. H., Dachs, G. U., Gleadle, J. M., Nicholls, L. G., Harris, A. L., Stratford, I. J., Hankinson, O., Pugh, C. W., and Ratcliffe, P. J. (1997). Hypoxia-inducible factor-1 modulates gene expression in solid tumors and influences both angiogenesis and tumor growth. *Proc. Natl. Acad. Sci. USA* **94,** 8104–8109.

Mori, K. (2003). Frame switch splicing and regulated intramembrane proteolysis: Key words to understand the unfolded protein response. *Traffic* **4,** 519–528.

Niwa, M., Patil, C. K., DeRisi, J., and Walter, P. (2005). Genome-scale approaches for discovering novel nonconventional splicing substrates of the Ire1 nuclease. *Genome Bio.* **6,** R3.

Ogata, M., Hino, S. I., Saito, A., Morikawa, K., Kondo, S., Kanemoto, S., Murakami, T., Taniguchi, M., Tanii, I., Yoshinaga, K., Shiosaka, S., Hammarback, J. A., *et al.* (2006). Autophagy is activated for cell survival after ER stress. *Mol. Cell Biol.* **26,** 9220–9231.

Okada, T., Yoshida, H., Akazawa, R., Negishi, M., and Mori, K. (2002). Distinct roles of activating transcription factor 6 (ATF6) and double-stranded RNA-activated protein kinase-like endoplasmic reticulum kinase (PERK) in transcription during the mammalian unfolded protein response. *Biochem. J.* **366,** 585–594.

Ratcliffe, P. J., O'Rourke, J. F., Maxwell, P. H., and Pugh, C. W. (1998). Oxygen sensing, hypoxia-inducible factor-1 and the regulation of mammalian gene expression. *J. Exp. Biol.* **201**(Pt 8), 1153–1162.

Romero-Ramirez, L., Cao, H., Nelson, D., Hammond, E., Lee, A. H., Yoshida, H., Mori, K., Glimcher, L. H., Denko, N. C., Giaccia, A. J., Le, Q. T., and Koong, A. C. (2004). XBP1 is essential for survival under hypoxic conditions and is required for tumor growth. *Cancer Res.* **64,** 5943–5947.

Ron, D. (2002). Translational control in the endoplasmic reticulum stress response. *J. Clin. Invest.* **110,** 1383–1388.

Rutkowski, D. T., Arnold, S. M., Miller, C. N., Wu, J., Li, J., Gunnison, K. M., Mori, K., Sadighi Akha, A. A., Raden, D., and Kaufman, R. J. (2006). Adaptation to ER stress is mediated by differential stabilities of prosurvival and proapoptotic mRNAs and proteins. *PLoS Biol.* **4,** e374.

Ryan, H. E., Poloni, M., McNulty, W., Elson, D., Gassmann, M., Arbeit, J. M., and Johnson, R. S. (2000). Hypoxia-inducible factor-1alpha is a positive factor in solid tumor growth. *Cancer Res.* **60,** 4010–4015.

Schroder, M., and Kaufman, R. J. (2005). ER stress and the unfolded protein response. *Mut. Res.* **569,** 29–63.

Semenza, G. L. (2000). Surviving ischemia: Adaptive responses mediated by hypoxia-inducible factor 1. *J. Clin. Invest.* **106,** 809–812.

Shen, J., Chen, X., Hendershot, L., and Prywes, R. (2002). ER stress regulation of ATF6 localization by dissociation of BiP/GRP78 binding and unmasking of Golgi localization signals. *Dev. Cell* **3,** 99–111.

Shi, Y., Vattem, K. M., Sood, R., An, J., Liang, J., Stramm, L., and Wek, R. C. (1998). Identification and characterization of pancreatic eukaryotic initiation factor 2 alpha-subunit kinase, PEK, involved in translational control. *Mol. Cell Biol.* **18,** 7499–7509.

Shuda, M., Kondoh, N., Imazeki, N., Tanaka, K., Okada, T., Mori, K., Hada, A., Arai, M., Wakatsuki, T., Matsubara, O., Yamamoto, N., and Yamamoto, M. (2003). Activation of the ATF6, XBP1 and grp78 genes in human hepatocellular carcinoma: A possible involvement of the ER stress pathway in hepatocarcinogenesis. *J. Hep.* **38,** 605–614.

Tirasophon, W., Welihinda, A. A., and Kaufman, R. J. (1998). A stress response pathway from the endoplasmic reticulum to the nucleus requires a novel bifunctional protein kinase/endoribonuclease (Ire1p) in mammalian cells. *Genes Dev.* **12,** 1812–1824.

Tu, B. P., and Weissman, J. S. (2002). The FAD- and O(2)-dependent reaction cycle of Ero1-mediated oxidative protein folding in the endoplasmic reticulum. *Mol. Cell* **10,** 983–994.

Tu, B. P., and Weissman, J. S. (2004). Oxidative protein folding in eukaryotes: Mechanisms and consequences. *J. Cell Bio.* **164,** 341–346.

Tu, B. P., Ho-Schleyer, S. C., Travers, K. J., and Weissman, J. S. (2000). Biochemical basis of oxidative protein folding in the endoplasmic reticulum [see comment]. *Science* **290,** 1571–1574.

Urano, F., Wang, X., Bertolotti, A., Zhang, Y., Chung, P., Harding, H. P., and Ron, D. (2000). Coupling of stress in the ER to activation of JNK protein kinases by transmembrane protein kinase IRE1. *Science* **287,** 664–666.

Vaupel, P., and Hockel, M. (2000). Blood supply, oxygenation status and metabolic micromilieu of breast cancers: Characterization and therapeutic relevance [review]. *Int. J. Oncol.* **17,** 869–879.

Wang, Y., Shen, J., Arenzana, N., Tirasophon, W., Kaufman, R. J., and Prywes, R. (2000). Activation of ATF6 and an ATF6 DNA binding site by the endoplasmic reticulum stress response. *J. Bio. Chem.* **275,** 27013–27020.

Wouters, B. G., van den Beucken, T., Magagnin, M. G., Koritzinsky, M., Fels, D., and Koumenis, C. (2005). Control of the hypoxic response through regulation of mRNA translation. *Semin. Cell Dev. Biol.* **16,** 487–501.

Yamamoto, K., Yoshida, H., Kokame, K., Kaufman, R. J., and Mori, K. (2004). Differential contributions of ATF6 and XBP1 to the activation of endoplasmic reticulum stress-responsive *cis*-acting elements ERSE, UPRE and ERSE-II. *J. Biochem.* **136,** 343–350.

Yoshida, H., Haze, K., Yanagi, H., Yura, T., and Mori, K. (1998). Identification of the cis-acting endoplasmic reticulum stress response element responsible for transcriptional induction of mammalian glucose-regulated proteins. Involvement of basic leucine zipper transcription factors. *J. Bio. Chem.* **273,** 33741–33749.

Yoshida, H., Matsui, T., Yamamoto, A., Okada, T., and Mori, K. (2001). XBP1 mRNA is induced by ATF6 and spliced by IRE1 in response to ER stress to produce a highly active transcription factor. *Cell* **107,** 881–891.

Yoshida, H., Okada, T., Haze, K., Yanagi, H., Yura, T., Negishi, M., and Mori, K. (2000). ATF6 activated by proteolysis binds in the presence of NF-Y (CBF) directly to the *cis*-acting element responsible for the mammalian unfolded protein response. *Mol. Cell Biol.* **20,** 6755–6767.

HYPOXIA AND TUMOR BIOLOGY

CHAPTER FIFTEEN

TUMOR HYPOXIA IN CANCER THERAPY

J. Martin Brown

Contents

1. Hypoxia in Human Tumors	298
2. The Dynamic Nature of Hypoxia in Tumors	300
3. Consequences of Tumor Hypoxia for Cancer Treatment	300
4. Size of the Oxygen Effect with Radiation	302
5. The Influence of Tumor Hypoxia on Cancer Treatment by Radiotherapy	303
6. Influence of Tumor Hypoxia on Response to Chemotherapy	307
7. Exploiting Hypoxia in Cancer Treatment	308
7.1. Hypoxic cytotoxins	308
7.2. Hypoxia-selective gene therapy	311
7.3. Targeting HIF-1	312
7.4. Exploiting tumor necrosis with obligate anaerobes	314
References	315

Abstract

Human solid tumors are invariably less well-oxygenated than the normal tissues from which they arose. This so-called tumor hypoxia leads to resistance to radiotherapy and anticancer chemotherapy as well as predisposing for increased tumor metastases. In this chapter, we examine the resistance of tumors to radiotherapy produced by hypoxia and, in particular, address the question of whether this resistance is the result of the physicochemical free radical mechanism that produces resistance to radiation killing of cells *in vitro*. We conclude that a major part of the resistance, though perhaps not all, is the result of the physicochemical free radical mechanism of the oxygen effect in sensitizing cells to ionizing radiation. However, in modeling studies used to evaluate the effect of fractionated irradiation on tumor response, it is essential to consider the fact that the tumor cells are at a wide range of oxygen concentrations, not just at the extremes of oxygenated and hypoxic. Prolonged hypoxia of the tumor tissue also leads to necrosis, and necrotic regions are also characteristic of solid tumors. These two characteristics—hypoxia and

Division of Radiation and Cancer Biology, Department of Radiation Oncology, Stanford University School of Medicine, Stanford, California

Methods in Enzymology, Volume 435
ISSN 0076-6879, DOI: 10.1016/S0076-6879(07)35015-5

necrosis—represent clear differences between tumors and normal tissues and are potentially exploitable in cancer treatment. We discuss strategies for exploiting these differences. One such strategy is to use drugs that are toxic only under hypoxic conditions. The second strategy is to take advantage of the selective induction under hypoxia of the hypoxia-inducible factor (HIF)-1. Gene therapy strategies based on this strategy are in development. Finally, tumor hypoxia can be exploited using live obligate anaerobes that have been genetically engineered to express enzymes that can activate nontoxic prodrugs into toxic chemotherapeutic agents.

1. HYPOXIA IN HUMAN TUMORS

The presence of cells at extremely low levels of oxygen (hereafter termed hypoxic cells) in human tumors was postulated more than 50 years ago by Thomlinson and Gray (1955) based on their observations of the distribution of necrosis relative to blood vessels in human tumors. They observed that a relatively uniform distance exists of approximately 100 to 150 μM from blood vessels to necrosis and calculated that this distance would be approximately the diffusion distance of oxygen through respiring tissue given the oxygen concentration in blood vessels. Thus, they postulated, necrosis was the result of cells being at zero, or close to zero, levels of oxygen for extended periods of time and that the cells adjacent to these necrotic areas would be viable, but at pathologically low levels of oxygen. These observations, combined with the results of studies performed by Gray *et al.* (1953) in the preceding few years, which showed that low levels of oxygenation always protected biological systems from the cytotoxic effects of X-rays, led them to conclude that hypoxic cells in human tumors could be a major limiting factor in the curability of tumors by radiotherapy. This conclusion led to clinical trials with patients undergoing radiotherapy in hyperbaric oxygen chambers to try to force more oxygen into the blood and into the tumor (Churchill-Davidson *et al.*, 1958). However, these trials were not particularly successful (Henk, 1981), in part because it was not known at the time that, in addition to the chronic hypoxia that typically occurs adjacent to necrotis, hypoxia in tumors can also occur by temporary obstruction or variable blood flow in tumor vessels (Brown, 1979; Chaplin *et al.*, 1987) (discussed later), and hyperbaric oxygen would not be expected to eliminate these cells.

Hypoxia-visualizing techniques have since verified the suggestion of Thomlinson and Gray that human solid tumors contain regions of viable hypoxic cells. Markers that are metabolized under hypoxic conditions and then visualized with a fluorescent antibody show low oxygenation levels in tumor cells located distant from blood vessels (Fig. 15.1A). A more

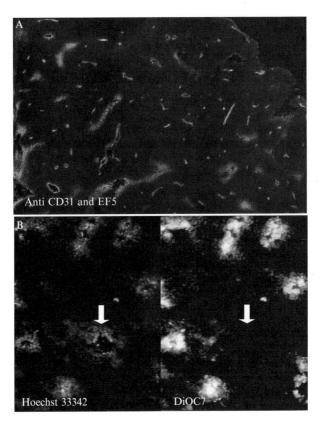

Figure 15.1 Two types of hypoxia affect tumor cells: chronic hypoxia (A), which is a consequence of the fact that the distance between the cells and the vasculature is greater than the diffusion distance of oxygen through respiring tissue, and acute hypoxia (B), which is caused by fluctuating blood flow in the tumor. (A) Chronic hypoxic tumor cells (anti-EF5, red) in this human soft tissue sarcoma are seen in regions distant from blood vessels (anti-CD31, green). Hypoxic cells were labeled with a fluorescent monoclonal antibody against EF5, a nitroimidazole that is metabolized selectively in hypoxic cells and then covalently binds to them. Photograph courtesy of Drs. Cameron Koch and Sydney Evans, University of Philadelphia; adapted from Brown (2002). (B) Acute hypoxia. Visualization of a specific tumor area from a mouse injected with Hoechst 3342 (blue) 20 min prior to injection with DiOC$_7$ (green). Each dye labels the cells immediately surrounding open blood vessels. In these photographs, the two dyes label the same vessels except for the one marked with an arrow, which has shut down in the 20-min period between injection of the first and second dyes. Photographs courtesy of Dr. Andrew Minchinton, BC Cancer Center, Vancouver; adapted from Brown (2002). (See color insert.)

quantitative technique is the polarographic oxygen electrode, a device that measures oxygen tension in tissues. This electrode is mounted on the tip of a needle and advanced rapidly and automatically through the tissue of interest.

The resulting histogram reflects the oxygenation status of the tissues being examined. In normal subcutaneous tissue, the median value for oxygen partial pressure is usually between 40 and 60 mmHg. Tumors, however, show much lower levels of oxygenation. For example, a "typical" tumor might have a median oxygen partial pressure of approximately 10 mmHg. By definition, this amount means that half of the tumor cells have oxygen levels of less than 10 mm Hg and half have oxygen levels of greater than 10 mmHg. In a recent review, we showed that all human tumors so far studied have median oxygen levels lower than their normal tissue of origin, sometimes dramatically so (Brown and Wilson, 2004).

2. THE DYNAMIC NATURE OF HYPOXIA IN TUMORS

Some years ago, we proposed that a second form of hypoxia—acute hypoxia—could occur in tumors as a result of fluctuating blood flow (Brown, 1979). This hypothesis has since been elegantly proven by Chaplin et al. (1986, 1987) and Trotter et al. (1989) by examining tumor blood flow using two fluorescent dyes (Hoechst 33342 and $DiOC_7$) with different excitation and emission properties. Both dyes have a very short half life in blood vessels, but remain bound to cells adjacent to the blood vessel. Blood vessels that can be visualized with the first dye (Hoechst 33342) at one time point were not necessarily present when the second dye ($DiOC_7$) was injected some 20 min later (see Fig. 15.1B). Similarly, and equally often, vessels not perfused by the first dye can sometimes be seen with the second dye. This fluctuation in blood flow suggests that tumor blood vessels are often temporarily occluded, necessarily resulting in hypoxia downstream of the occlusion. However, it is important to realize that acute and chronic hypoxia are really the two ends of a continuous spectrum with fluctuations in blood flow without total occlusion, which are common in both experimental (Kimura et al., 1996; Lanzen et al., 2006) and human tumors (Hill et al., 1996), producing a dynamic situation with fluctuating oxygen diffusion distances in many parts of tumors.

3. CONSEQUENCES OF TUMOR HYPOXIA FOR CANCER TREATMENT

As previously noted, studies performed by Gray et al. (1953) in the early 1950s established that the effects of ionizing radiation on cells and on tissues were markedly reduced in the absence of oxygen. Though it had been known earlier that the absence of oxygen produced radioresistance (Crabtree and Cramer, 1933; Holthusen, 1921; Mottram, 1935;

Petry, 1923), this was thought to be a product of a lower metabolic rate of the cells as opposed to the oxygen tension at the time of irradiation. In fact, rapid–mixing experiments demonstrated that, for radiation sensitization of hypoxic cells, oxygen has to be present at the time of irradiation with no effect, even if the oxygen is present fractions of a second before or after the radiation exposure (Hodgkiss *et al.*, 1987). These studies led to the general hypothesis that oxygen acts at a physicochemical level to "fix" (i.e., make permanent) the radiation damage as a consequence of the high affinity the oxygen molecule has for the unpaired electron on the free radical produced by radiation. Because a number of studies have demonstrated that the DNA is the target of radiation cell killing (Hall, 2000), the overall mechanism of the oxygen effect can be represented by a competition reaction (Fig. 15.2A).

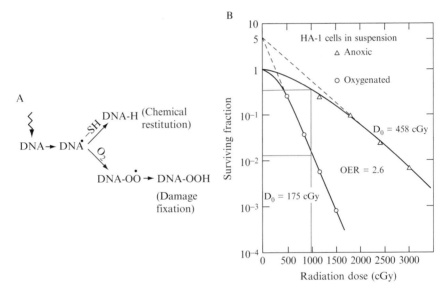

Figure 15.2 The mechanism for the resistance of hypoxic cells to radiation and a typical radiation survival curve of cells irradiated *in vitro* under either hypoxic or aerobic conditions. (A) Radiation produces free radicals in cellular molecules, including DNA, which is the target for radiation killing. Molecular oxygen has a high affinity for this radical and binds covalently to it, producing a peroxy radical, which "fixes" the damage leading to DNA strand breaks and base damage. In the absence of oxygen, cellular nonprotein sulfhydryls (of which glutathione is the most important) can donate an H atom to the radical, thereby restoring it to its undamaged form. (B) Radiation survival curves of Chinese hamster HA-1 cells determined by clonogenic assay. The cells were irradiated under either hypoxic or aerobic conditions and plated for colony formation. All colonies of more than 50 cells after 12 days of incubation at 37° were counted as indicating surviving cells. The red lines show that, at a dose of 1000 cGy (10 Gy), survival under hypoxic conditions is approximately 50% and under aerobic conditions is approximately 1%. (See color insert.)

4. SIZE OF THE OXYGEN EFFECT WITH RADIATION

Figure 15.2B shows typical data for the effect of oxygen on cell killing by radiation of mammalian cells *in vitro*. As can be seen from these data, the effect of oxygen is to reduce the slope (D_0) of the radiation survival curve. The size of this radiation sensitization is usually expressed as the "oxygen enhancement ratio" (OER), defined as the ratio of doses under hypoxic to aerobic conditions to give the same cell survival. Typically, this value for mammalian cells is in the range of 2.5 to 3.0 for a variety of endpoints, including cell killing, mutagenesis, and the induction of chromosome aberrations (Hall, 2000). This sensitization by a factor of 2.5 to 3.0 may not seem large, but what matters in assessing the effects of oxygen on cell killing by a given radiation dose is the ratio of cells killed for a given radiation dose. For example, a dose of 1000 cGy kills approximately 50% of the cells under hypoxic conditions, but close to 99% of the cells under aerobic conditions (see Fig. 15.2B). This result has a very large effect on the sensitivity of cells to radiation and is only matched in size by the effects of inactivating ataxia telangiectasia mutated (ATM) or non-homologous end joining (NHEJ), the most important DNA repair process in mammalian cells (Riballo *et al.*, 2004).

To obtain cells that are fully resistant to radiation under hypoxic conditions (see Fig. 15.2B), investigators typically irradiate cells under near anoxic conditions. This action raises the question of how much oxygen is needed to sensitize anoxic cells to radiation. Data relevant to this question (Fig. 15.3A)

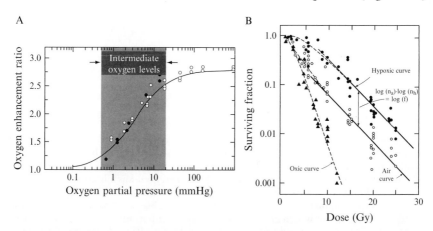

Figure 15.3 Cells in tumors are at both intermediate and full resistance to killing by irradiation. (A) The oxygen enhancement ratio (OER) is shown as a function of the oxygen partial pressure. The published OER values at different oxygen partial pressures are adapted from Koch *et al.* (1984) (●) and Whillans and Hunt (1982) (○). Figure adapted from Wouters and Brown (1997). (B) Paired survival curve of EMT6 tumors under clamped hypoxic or air-breathing conditions showing that approximately 20% of the tumor cells are at oxygen concentrations that produce maximum radioresistance. From Moulder and Rockwell (1987), with permission.

demonstrate that full radiation resistance requires very low oxygen levels of around 0.2 mmHg; however, resistance to radiation begins to develop at much higher oxygen concentrations of approximately 20 mmHg, creating a zone of intermediate oxygen levels between approximately 0.2 and 20 mmHg, at which point radiation resistance is at intermediate levels. The curve in Fig. 15.3A could be conveniently described as the oxygen concentration used to give half-maximum resistance, and this amount is at approximately 3 mmHg for the cells in these studies. This level of oxygen is lower than that typically found in normal tissues, though such values are common in tumors when they have been measured using oxygen electrodes (Brown and Wilson, 2004).

5. THE INFLUENCE OF TUMOR HYPOXIA ON CANCER TREATMENT BY RADIOTHERAPY

The radiation resistance of cells at very low oxygen levels provides a convenient means of assessing whether cells at oxygen levels low enough to produce maximum radiation resistance occur in experimental tumor models. Multiple investigators have performed such experiments using the fact that many experimental tumor models are amenable to clonogenic assay *in vitro*. The experiments consist of irradiating the tumor *in vivo*, removing the tumor after irradiation, producing a single cell suspension, and plating the cells for colony formation *in vitro*. If all of the cells in the tumor were fully oxygenated, one would expect a sensitive survival curve (Oxic curve in Fig. 15.3B), whereas, if all of the cells in the tumor were at low enough oxygen levels to be maximally resistant, one would expect a resistant survival curve (hypoxic curve in Fig. 15.3B). What is typically found is a survival curve intermediates between these two extremes, but with a slope parallel to that of anoxic cells (air curve in Fig. 15.3B). The separation of these two curves is a measure of the proportion of the cells in the tumor at maximum radioresistance. This so-called paired survival curve method and a similar method using tumor growth delay or tumor control have been used by many investigators to determine the proportion of hypoxic cells in experimental tumors, and the results show large variations with values typically in the range of 5 to 50% of the tumor cells (Moulder and Rockwell, 1987).

These studies using excision of tumors and plating of the cells *in vitro* or studies with growth delay and tumor cure have established that relatively high proportions of cells in experimental tumors are at sufficiently low oxygen levels (<0.2 mmHg) to be maximally resistant to killing by radiation. However, this issue is only part of the problem; given that hypoxic cells contribute to the resistance of tumors to radiation, an important question is by what mechanism? Is it by the physicochemical oxygen fixation hypothesis (see Fig. 15.2A), well-established as the mechanism of resistance of hypoxic cells *in vitro*, or is it by other mechanisms, such as the protection of tumor vasculature by presence or upregulation of the HIF-1alpha (α) protein that has been suggested in some studies (Moeller *et al.*, 2004; Williams *et al.*, 2005)? This question is very important because it is through understanding the mechanism of resistance of tumors by hypoxia that strategies to overcome the resistance can be formulated.

Fortunately, studies performed some years ago using nitroimidazole hypoxic cell radiosensitizers, such as misonidazole, have provided insight into the mechanism for the resistance of tumors produced by their hypoxic cells. Nitroimidazole hypoxic cell radiosensitizers are electron–affinic agents that can substitute for oxygen in the physicochemical reaction (see Fig. 15.2A), but do not suffer the problem of diffusion through tumor tissue as they are not involved in normal metabolism, like oxygen. These agents thus penetrate to the hypoxic cells in the tumor. Various investigators have shown that high doses of misonidazole or analogs sensitize tumors to radiation by a large amount (Brown, 1975; Denekamp and Harris, 1975; Sheldon *et al.*, 1974) (Fig. 15.4). In addition, studies performed by Hill *et al.* (1998) have shown that breathing high levels of oxygen for as little as 5 min prior to a large single dose of irradiation is sufficient to radiosensitize transplanted

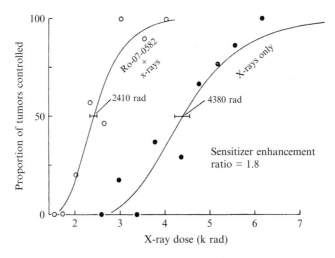

Figure 15.4 Injection of misonidazole (Ro-07-0582) (1 mg/g body weight) 30 min prior to radiation sensitizes C3H mammary tumors to radiation using the tumor control assay at 150 days after irradiation. As misonidazole acts to substitute for oxygen in the physicochemical mechanism of the oxygen effect, this sensitization by a high dose of drug prior to irradiation (but not after) is strong evidence that the radioresistance of this tumor produced by hypoxia is largely, if not entirely, a result of the physicochemical mechanism of the oxygen effect. From Sheldon *et al.* (1974), with permission.

mouse tumors. These studies demonstrate that the resistance of these tumors to large single doses of radiation is largely the result of the physicochemical free radical reaction between the target of radiation damage (DNA) and oxygen.

But what of fractionated irradiation? Most radiotherapy is given in multiple daily doses, usually over 6 wk. These regimes have evolved over the years as those that produce maximum tumor response for tolerable normal tissue damage. For many years, the rationale for this superiority of fractionated radiation over large single doses was not fully understood; however, in the 1960s, the phenomenon of reoxygenation was discovered (Kallman, 1972). This phenomenon describes the kinetics of reestablishment of the pre-irradiation ratio of hypoxic cells to oxygenated cells in the tumor following a dose of radiation. Immediately after a large single dose of radiation, most of the surviving cells in the tumor will be hypoxic (because these are the most resistant cells). However, it was found that several hours later, the equilibrium of aerobic to hypoxic cells that was present prior to irradiation had been reestablished, implying that some of the surviving hypoxic cells had become oxygenated again. Several mechanisms could account for this result, but the most likely is that provided by fluctuating blood flow (Brown, 1979).

Various modeling studies have suggested that the expectation with fractionated irradiation is that, if reoxygenation occurs for all radiation doses, the problem of hypoxic cells in tumors is largely overcome (Denekamp and Joiner, 1982; Hill, 1992). However, this theory does not agree with clinical data for which ample evidence shows that tumor hypoxia is a major contributor to the ability to obtain local control of tumors with radiotherapy (Brizel et al., 1996; Nordsmark et al., 1996, 2005). Does this mean that the resistance of clinical tumors to radiotherapy produced by hypoxia is not a consequence of the physicochemical resistance of hypoxic cells that is the basis of the modeling studies?

The answer to this question is no. It became clear a number of years ago that modeling a tumor composed of fully resistant hypoxic cells and fully oxygenated aerobic cells could be extremely misleading (Wouters and Brown, 1997). Cells within the range of 0.5 to 20 mmHg are at an intermediate radiation resistance (see Fig. 15.3A). Given that this range covers that of median oxygen tensions in human tumors (Brown and Wilson, 2004), it is likely that a large proportion, perhaps the majority of the cells within human tumors, are at intermediate radiosensitivity. When this number is taken into account in modeling studies, it becomes clear that even full reoxygenation between radiation doses does not produce full sensitivity of the tumor cells. The difference between the predicted cell kill from a typical radiotherapy regime with the binary model of aerobic and hypoxic cells and one that includes cells at intermediate oxygenation is substantial (Fig. 15.5). This result

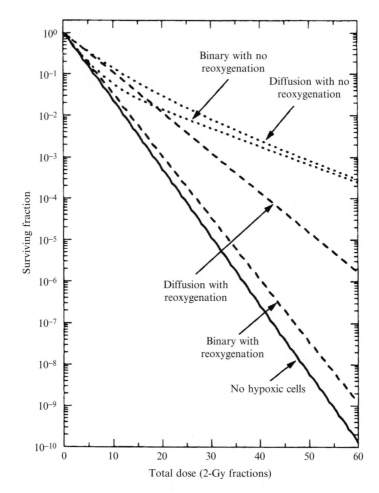

Figure 15.5   The effect of reoxygenation is considerably less in increasing tumor cell kill to a typical radiotherapy fractionation scheme (30 fractions of 2 Gy) if the contribution of cells at intermediate oxygenation is taken into account (diffusion model). The solid line demonstrates the response expected for a tumor containing 100% well-oxygenated (maximally sensitive) cells. Also shown are the responses of hypothetical tumors in which 10% of the cells are radiobiologically hypoxic (maximally resistant) and the remaining 90% either well-oxygenated (binary model) or at oxygen levels determined by a model based on radial diffusion of oxygen from a blood vessel. The expected responses are shown for conditions of full reoxygenation between fractions (dashed lines) or no reoxygenation between fractions (dotted lines). From Wouters and Brown (1997), with permission.

confirms an extensive study of the influence of breathing different levels of oxygen on the control of mouse tumors performed by Suit *et al.* (1988) several years earlier. These investigators showed that, for three different spontaneous mouse tumors, the dose to control the tumors (TCD_{50}) was reduced by a factor of close to 2.0 if the mice breathed pure oxygen at three atmospheres pressure starting 15 min prior to each of five or 15 equal dose fractions. These data demonstrate experimentally that a large fraction, at least of the resistance of these tumors to fractionated irradiation, is the result of the fast free radical mechanism of the oxygen effect.

Thus, the clinical data demonstrating that human tumors, particularly head and neck tumors, that are more hypoxic are more difficult to control by standard radiotherapy is predicted both via experimental data and by modeling studies that include tumor cells at full sensitivity, full resistance, and intermediate resistance with full reoxygenation of the hypoxic cells between each radiation dose. These studies invoke no more than the known resistance of hypoxic cells to radiation by the physicochemical free radical fixation hypothesis. However, this proof does not exclude the possibility that other factors resulting from tumor hypoxia might contribute to the resistance of the tumors. For example, and of particular relevance to the survival of the patient as opposed only to local tumor control, hypoxia predisposes tumors to metastatic spread (Le *et al.*, 2004). Thus, a more hypoxic tumor, in addition to being more difficult to control locally, is also more likely to have spread to distant sites and hence be more difficult to cure. In addition, hypoxia stabilizes HIF-1α, which increases levels of various survival factors, such as vascular endothelial growth factor (VEGF), which can protect against radiation damage to the tumor vasculature and hence protect the tumor (Moeller *et al.*, 2004; Wachsberger *et al.*, 2003). Also, hypoxia activates the unfolded protein response, and proteins in this pathway can protect against hypoxic stress and promote tumor growth (Romero-Ramirez *et al.*, 2004). An important question that remains unanswered is the relative contributions of the physicochemical radiation resistance resulting from hypoxia and other factors, such as those contributed by the tumor vasculature and cytokines induced by hypoxia.

6. INFLUENCE OF TUMOR HYPOXIA ON RESPONSE TO CHEMOTHERAPY

Hypoxic cells are also considered to be resistant to most anticancer drugs for several reasons. First, hypoxic cells are distant from blood vessels and, as a result, are not adequately exposed to some types of anticancer drugs (Durand, 1994; Hicks *et al.*, 2006; Tannock, 1998). Second, cellular proliferation decreases as a function of distance from blood vessels (Tannock, 1968), an effect at least partially due to hypoxia. Third, hypoxia selects for cells that have lost sensitivity to p53-mediated apoptosis (Graeber *et al.*, 1996), which might lessen sensitivity to some anticancer agents. Fourth, the action of some anticancer agents (e.g., bleomycin) resembles that of radiation in that oxygen increases the cytotoxicity of the DNA lesions they cause (Batchelder *et al.*, 1996; Teicher *et al.*, 1981). Fifth, hypoxia upregulates genes involved in drug resistance, including p-glycoprotein (Comerford *et al.*, 2002; Wartenberg *et al.*, 2003). Thus, clear links between hypoxia and intrinsic resistance to chemotherapy are provided by preclinical studies; yet, surprisingly, clinical studies investigating the role of hypoxia in response to chemotherapy have not been reported.

7. Exploiting Hypoxia in Cancer Treatment

So far, we have discussed the fact that hypoxia is an adverse prognostic factor for treatment of tumors by radiotherapy and chemotherapy. However, as hypoxia is a unique feature of tumors, it follows that strategies based on this could be used for selective antitumor efficacy. We next describe several strategies that are currently being explored at the clinical and preclinical levels.

7.1. Hypoxic cytotoxins

Figure 15.6 shows a common mechanism by which a drug could be converted from a nontoxic, so-called "prodrug" into a toxic drug selectively in hypoxic cells. In essence, hypoxic selective cytotoxicity requires one-electron reduction of a relatively nontoxic prodrug to a radical, which then becomes a substrate for back oxidation by oxygen to the original compound. If the radical produced, or downstream products of the radical, is much more toxic than the superoxide generated by redox cycling in oxic cells, then this result gives rise to greater toxicity of the compound under hypoxic conditions. Examples of hypoxia-activated prodrugs in clinical trials are illustrated next.

7.1.1. Tirapazamine
Tirapazamine ([TPZ] 3-amino-1,2,4-benzotriazine-1,4-dioxide, SR4233), a drug discovered some 20 years ago (Zeman *et al.*, 1986), is the first purely hypoxic cytotoxin for which antitumor activity has been demonstrated in clinical trials. Prior to the discovery of this benzotriazine di-N-oxide, two other classes of agents were known that produced some selective killing of hypoxic cells: quinone-containing alkylating agents (of which mitomycin C is the prototype) and nitroaromatic compounds (of which misonidazole and metronidazole are examples). TPZ was a significant advance over the previously known classes because its differential toxicity towards hypoxic cells was considerably larger (Zeman *et al.*, 1986), and combination studies

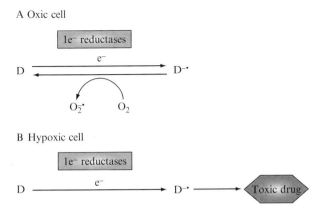

A Oxic cell

B Hypoxic cell

Figure 15.6 Typical mechanism by which prodrugs act as hypoxic-selective cytotoxins. The nontoxic prodrug must be a substrate for intracellular one-electron reductases, such as cytochrome P450 reductase, which adds an electron to the prodrug, thereby converting it to a free radical. In oxic cells, the unpaired electron in the prodrug radical is rapidly transferred to molecular oxygen, forming superoxide and regenerating the initial prodrug. This futile redox cycle prevents buildup of the prodrug radical when O_2 is present. Hypoxia-selective cell killing is achieved if the prodrug radical that accumulates in hypoxic cells is more cytotoxic than the superoxide formed in oxic cells. In principle, the prodrug radical could itself be the cytotoxin; however, more commonly, it undergoes further reactions to form the ultimate toxic species. From Brown and Wilson (2004), with permission. (See color insert.)

with radiation demonstrated its ability to preferentially kill hypoxic cells in transplanted tumors (Brown and Lemmon, 1990).

The mechanism for selective toxicity of TPZ to hypoxic cells follows a general scheme (outlined in Fig. 15.6). TPZ potentiates the antitumor effect of radiation by selectively killing the hypoxic cells in the tumors. As these are the most radiation-resistant cells in tumors, TPZ and radiation act as complementary cytotoxins, each one killing the cells resistant to the other, thereby potentiating the efficacy of radiation on the tumor. TPZ is also very effective in enhancing the anticancer activity of the chemotherapeutic drug cisplatin (Dorie and Brown, 1993), an interaction that again depends on hypoxia (Kovacs et al., 1999), but which results from an increase in cisplatin sensitivity in non-lethally damaged TPZ-treated cells rather than from complementary killing of oxic and hypoxic cells by the two agents (as is the case with radiation). The interaction with cisplatin has been tested in a phase III clinical trial with advanced non–small-cell lung cancer and has been shown to be effective—the addition of TPZ to the standard cisplatin regimen doubled the overall response rate and significantly prolonged survival (von Pawel et al., 2000). TPZ has also been tested in a randomized phase II trial with cisplatin-based chemoradiotherapy of advanced head and neck cancer, and the results of this trial also show improved survival in the TPZ arm (Rischin et al., 2005). A phase III study with cisplatin-based chemoradiotherapy is currently underway.

7.1.2. Banoxantrone (AQ4N)

The anthraquinone AQ4N was designed specifically as a hypoxia-selective cytotoxin. It resembles TPZ in that it is a di-N-oxide but has a distinct mechanism of activation and cytotoxicity. AQ4N is a prodrug of a potent DNA intercalator/topoisomerase poison, AQ4, which is formed by reduction of the two tertiary amine N-oxide groups, which mask DNA binding in the prodrug form (Patterson, 1993). AQ4N is unusual among hypoxia-activated prodrugs in that it is activated by two-electron reduction, which is effected mainly by the CYP3A members of the cytochrome P450 family (Patterson, 2002), which are strongly expressed in some human tumors. Inhibition by oxygen results from competition between O_2 and prodrug for binding at the reduced heme group in the enzyme-active site rather than from redox cycling (Patterson and Murray, 2002). While AQ4 is selective for cycling cells, its long residence time in tissue probably enables it to persist until hypoxic cells come into cycle (Patterson and McKeown, 2000). AQ4N has substantial activity against hypoxic cells in a variety of transplanted tumors (Patterson et al., 2000), and has recently completed phase I and II clinical trials with lymphomas and leukemias and a phase1b/2a in combination with radiotherapy and temozolomide with glioblastoma multiforme is in progress.

7.1.3. PR-104

PR-104 is a dinitrobenzamide mustard (DNBM) developed by Wilson and Denny at the University of Auckland (New Zealand) with considerable advantages over TPZ or AQ4N (Wilson et al., 2005). DNBMs were originally developed for nitroreductase-based gene-directed enzyme prodrug therapy (GDEPT) and were not considered to have sufficient antitumor activity alone. The prototype for this class is the prodrug CB1954, which first came to attention because of its dramatic curative activity against the Walker rat tumor (Cobb et al., 1969). It was subsequently shown to be a bioreductive prodrug, activated within the tumors by rat DT-diaphorase (DTD), which reduces its 4-nitro group to the corresponding hydroxylamine, a potent DNA crosslinking agent (Knox et al., 1988a,b). However, CB1954 is even more efficiently activated by an Escherichia coli nitroreductase (NTR) (Anlezark et al., 1992) and has recently entered clinical trials in the context of GDEPT using a nonreplicating adenoviral vector that expresses NTR (Chung-Faye et al., 2001). CB1954 has weak or little selective toxicity to hypoxic cells, mainly because of its high level of activation by the two-electron reductase DTD, thereby avoiding the oxygen-sensitive one-electron reduction product. Though the related DNBMs have less sensitivity to activation by DTD and therefore have some hypoxia selectivity, this characteristic has not been sufficient for antitumor activity in the absence of an exogenously applied nitroreductase as in GDEPT (Wilson *et al.*, 2002).

However, it was found that the mixed halogen/mesylate (OSO_2CH_3) mustards, of which PR-104 is a member, have higher hypoxic selectivity than the single halide dual mustard. PR-104 is a phosphate ester that is in effect a "pre-prodrug"; systemic or tumor phosphatases generate the corresponding alcohol (prodrugs), which is subsequently activated by reduction of one or more of the nitro groups by nitroreductase, including one-electron reductase, to produce hypoxia-selective cytotoxicity (Wilson *et al.*, 2004).

Bystander killing is a common feature of this class of DNBMs (Wilson *et al.*, 2002), including PR-104 (Wilson *et al.*, 2005). Thus, this compound has an interesting advantage over TPZ in that it is highly effective not only in killing the hypoxic cells in the tumor, but also better oxygenated cells by virtue of the fact that the toxic metabolite formed in hypoxic cells can diffuse to kill nearby aerobic cells. PR-104 is therefore active against tumors alone, not requiring the addition of a second agent (such as radiation) to kill the aerobic cells of the tumor. The drug has completed a multicenter phase I trial and is progressing to phase II trials.

7.2. Hypoxia-selective gene therapy

Because of the lack of sufficient tumor specificity of current gene therapy strategies, essentially all of the protocols currently being investigated in cancer gene therapy now involve local administration of the delivery vectors directly into the tumor, usually by needle injection. This has limited applicability to cancer treatment because metastases from the primary tumor are usually too numerous, inaccessible, or undetected to allow for direct injection. An alternative to direct targeting of tumors is to have the therapeutic gene transcribed or translated by a tumor-specific property so that expression of a particular protein would be tumor specific. One way to do this is to use the fact that the transcription factor HIF-1 is expressed at high levels in most tumors but not generally in normal tissues (Talks *et al.*, 2000; Zhong *et al.*, 1999). HIF-1 is comprised of a dimer of HIF-1α and HIF-1beta (β), and it is the former that is increased in tumor cells both by increased transcription by transformed cells and by stabilization of the protein under hypoxic conditions (Harris, 2002). HIF-1α expression is also associated with poor prognosis and resistance to therapy in head and neck cancer, ovarian cancer, and esophageal cancer (Talks *et al.*, 2000; Zhong *et al.*, 1999).

HIF-1 stimulates the transcription of a large number of genes by binding to sequences known as hypoxia-responsive elements (HREs) in the promoter regions of the target genes. Therefore, the strategy suggested to obtain hypoxia-specific transcription of a therapeutic gene would be to develop a promoter that is highly responsive to HIF-1 and that would therefore drive the expression of the therapeutic gene specifically in tumors. Expression of an enzyme not normally found in the human body (e.g., *E. coli*-derived cytosine deaminase) could, under the control of a hypoxia-responsive promoter, convert a nontoxic prodrug into a toxic drug in the tumor. Promoters using HREs from hypoxia-responsive genes have been developed (Greco and Dachs, 2001; Shibata *et al.*, 1998), and *in vivo* activity has been obtained in experimental tumor systems either by direct injection of adenoviral vectors containing the HRE promoters (Binley *et al.*, 2003) or by using tumor cells stably transfected with HRE-regulated prodrug-activating enzymes (Patterson *et al.*, 2002; Shibata *et al.*, 2002). Unfortunately, the latter systems, in which 100% of the tumor cells carry the hypoxia-responsive gene, are not realistic for the clinical situation. However, this issue may not be a critical limitation; various investigators have shown that, provided the active drug can diffuse from the cell in which it is generated to kill surrounding cells (the so-called "bystander effect," previously introduced in this chapter), efficient antitumor activity can be obtained with much lower percentages of transformed cells (Trinh *et al.*, 1995; Wilson *et al.*, 2002).

A further potential strategy for exploiting hypoxia in gene therapy is to deliver a one-electron reductase, such as cytochrome P450 reductase (P450R), as the prodrug-activating therapeutic gene, thus confining prodrug activation to hypoxic regions; this reductase can be combined with HIF-1 regulation of P450R expression to further enforce tumor selectivity (Patterson *et al.*, 2002). A similar approach relies on hypoxia-selective metabolism of AQ4N by CYP3A4 as an enzyme/prodrug system for GDEPT (McCarthy *et al.*, 2003).

A challenge with these approaches will be achieving efficient systemic delivery of vectors to HIF-1–expressing and/or hypoxic cells, which are generally found in regions distant from blood vessels. One possibility for delivery of the HRE-driven therapeutic protein to tumors would be to take advantage of the fact that macrophages are often recruited to tumors and that such macrophages exhibit elevated levels of HIF-1α in various human tumors (Burke *et al.*, 2002; Griffiths *et al.*, 2000).

7.3. Targeting HIF-1

The role of HIF-1 in angiogenesis, glucose utilization, and tumor cell survival (Semenza, 2002), its association with poor prognosis (Harris, 2002), and the fact that growth of mouse xenografts are inhibited by loss of HIF-1 activity (Semenza, 2002) all make it a potentially attractive tumor-specific target. It should be noted that, as the expression of HIF-1 is not restricted to hypoxic cells alone in many tumors but is also upregulated by oncogenic mutations in Ras, Src or Her2/Neu, targeting HIF-1 could also potentially target the better oxygenated cells in tumors (Giaccia *et al.*, 2003). Two general approaches could be used to exploit the high levels of HIF-1α in cancers.

First, inhibition of transactivation of HIF-1 target genes (such as the angiogenesis inducer VEGF) would be expected to have an antitumor effect. Proof of principle of this approach comes from studies of Kung *et al.* (2000), who showed that tumor cells infected with a polypeptide that disrupted the binding of HIF-1α to its transcriptional coactivators p300/CREB (thereby inhibiting hypoxia-induced transcription), markedly reduced the growth of these cells when transplanted into nude mice. It should be noted, however, that deletion or inhibition of HIF-1 does not always lead to suppression of tumor growth, and conflicting results have been reported depending on the cell type used, the HIF-1 subunit targeted, the tumor site, and the timing of the inhibition in relation to tumor growth (Melillo, 2006). Nonetheless, the positive data have led investigators to screen for small molecules that inhibit HIF-1 transcription, and such agents have been found, with the camptothecin analogue topotecan being one of the most active (Rapisarda *et al.*, 2002, 2004a). Antitumor activity attributable to HIF-1 inhibition has been reported for topotecan (Rapisarda *et al.*, 2004b) and for chetomin, a small molecule inhibitor of HIF-1 binding to the transcriptional coactivator p300 (Kung *et al.*, 2004). A clinical trial using topotecan to inhibit HIF-1 activity is underway (Melillo, 2006).

A second approach is to suppress HIF-1 protein levels, either by destabilizing them or inhibiting their production. The Hsp90 inhibitor geldanamycin has been shown to reduce HIF-1 protein levels by promoting its oxygen and von Hippel-Lindau (VHL)-independent degradation via the proteasome (Mabjeesh *et al.*, 2002; Sun *et al.*, 2001). However, it has yet to be demonstrated that this reduction occurs *in vivo* or that the antitumor activity of this compound is a direct result of reduced levels of HIF-1, as many other proteins are also affected. Targeting of HIF-1 by direct injection of an antisense construct to HIF-1α has been shown to eradicate a small transplanted thymic lymphoma and to enhance the efficacy of immunotherapy against larger tumors (Sun *et al.*, 2001). However, small molecule inhibitors of HIF-1 would be preferable, and two groups have reported success. Mabjeesh *et al.* (2003) reported that microtubule inhibitors, such as 2-methoxyestradiol (2ME2), vincristine, and Taxol, reduce HIF-1α levels *in vitro* apparently by inhibiting translation of HIF-1α messenger RNA (mRNA). These compounds can also reduce tumor growth and vascularity, though whether this is an effect of reduced levels of the HIF-1 or a direct effect on microtubules is not known. The second small molecule that has been reported to reduce HIF-1α levels and inhibit tumor growth is the soluble guanylyl cyclase stimulator YC-1 (Yeo *et al.*, 2003). Soluble guanylyl cyclase is the receptor for nitric oxide (NO), a molecule involved in many signaling pathways, including those regulating vascular tone and platelet function. However, the authors attribute the antitumor and anti-angiogenic effects of YC-1 to a reduction in HIF-1α protein levels (by an unknown posttranslational effect) rather than to an effect on NO signaling.

7.4. Exploiting tumor necrosis with obligate anaerobes

Brown and colleagues first suggested that the necrotic regions in human solid tumors could be used to target cancer therapy to tumors using a geneti-cally engineered nonpathogenic strain of the bacterial genus *Clostridium* (Fox *et al.*, 1996; Lemmon *et al.*, 1994, 1997). This genus comprises a large and heterogeneous group of Gram-positive, spore-forming bacteria that become vegetative and grow only in the absence (or at very low levels) of oxygen. Malmgren and Flanagan (1955) were the first to demonstrate this phenomenon by observing that tumor-bearing mice died of tetanus within 48 h of intravenous injection of *Clostridium tetani* spores, whereas non–tumor-bearing animals were unaffected. Möse and Möse (1959, 1964) later reported that a nonpathogenic clostridial strain, *Clostridium butyricum* M-55, localized and germinated in solid Ehrlich tumors, causing extensive lysis without any concomitant effect on normal tissues. Such observations were soon confirmed and extended by a number of investigators using tumors in mice, rats, hamsters, and rabbits (Engelbart and Gericke, 1964; Thiele *et al.*, 1964) and were followed by clinical studies with cancer patients (Carey *et al.*, 1967; Heppner and Mose, 1978; Heppner *et al.*, 1983). While the anaerobic bacteria did not significantly alter tumor control or eradica-tion, these clinical reports demonstrated that spores of nonpathogenic strains of clostridia could be given safely, that the spores germinate in the necrotic regions of tumors, and that lysis in these tumor regions can occur. This distinction is important over the similar approach using genetically modified, live-attenuated *Salmonella*, which, though producing excellent colonization of transplanted tumors in mice (Pawelek *et al.*, 1997), produced only mar-ginal colonization of human tumors in a phase I clinical trial (Toso *et al.*, 2002). The reasons for the difference between the rodent and human tumors in colonization by *Salmonella* are unknown. However, colonization by clostridia is different from that of *Salmonella* in that it is dependent on hypoxic necrotic regions, which are equally common in human and rodent tumors. In addition, as previously noted, excellent colonization of human tumors has been reported following intravenous injection of clostridial spores.

The *Clostridium* used in the clinical studies was a strain of *Clostridium sporogenes* renamed *Clostridium oncolyticum* to reflect the lysis that was produced in human tumors. This strain has been genetically modified to express the *E. coli* enzyme cytosine deaminase, which can convert the nontoxic 5-fluorocytosine to the toxic anticancer drug 5-fluorouracil. Animal experiments have demonstrated the efficacy of this approach (Liu *et al.*, 2002), and clinical studies are planned. In addition, other enzyme/prodrug systems for arming clostridia are in development, including CB 1954, which, when activated by *E. coli* nitroreductase, efficiently kills non-cycling cells (Bridgewater *et al.*, 1995) and is therefore expected to have greater activity against cells in hypoxic regions. (See recent review by Minton [2003] for further details of possible enzyme-prodrug combinations that can be used with clostridial targeting of tumors.) Although clostridial-dependent enzyme prodrug therapy (CDEPT) is similar to the strategy of antibody-dependent enzyme prodrug therapy (ADEPT), which is currently under clinical evaluation, it has a number of significant advantages, not the least of which is its favorable intratumor distribution. Because the prodrug-activating enzyme from clostridia will be at its highest concentration in areas adjacent to necrosis and far from blood vessels, it guarantees the highest active drug concentrations in the distant cells and minimizes the problem of leakage of activated drug back into the blood vessels, which has been reported to be a problem for ADEPT (Martin *et al.*, 1997).

REFERENCES

Anlezark, G. M., Melton, R. G., Sherwood, R. F., Coles, B., Friedlos, F., and Knox, R. J. (1992). The bioactivation of 5-(Aziridin-1-YL)-2,4-dinitrobenzamide (CB 1954)-I Purification and properties of a nitroreductase enzyme from *Escherichia Coli*—A potential enzyme for antibody-directed enzyme prodrug therapy (ADEPT). *Biochem. Pharmacol.* **44**, 2289–2295.

Batchelder, R. M., Wilson, W. R., Hay, M. P., and Denny, W. A. (1996). Oxygen dependence of the cytotoxicity of the enediyne anti-tumour antibiotic esperamicin A1. *Br. J. Cancer Suppl.* **27**, S52–S56.

Binley, K., Askham, Z., Martin, L., Spearman, H., Day, D., Kingsman, S., and Naylor, S. (2003). Hypoxia-mediated tumour targeting. *Gene Ther.* **10**, 540–549.

Bridgewater, J. A., Springer, C. J., Knox, R. J., Minton, N. P., Michael, N. P., and Collins, M. K. (1995). Expression of the bacterial nitroreductase enzyme in mammalian cells renders them selectively sensitive to killing by the prodrug CB1954. *Eur. J. Cancer* **31A**, 2362–2370.

Brizel, D. M., Scully, S. P., Harrelson, J. M., Layfield, L. J., Bean, J. M., Prosnitz, L. R., and Dewhirst, M. W. (1996). Tumor oxygenation predicts for the likelihood of distant metastases in human soft tissue sarcoma. *Cancer Res.* **56**, 941–943.

Brown, J. M. (1975). Selective radiosensitization of the hypoxic cells of mouse tumors with the nitroimidazoles metronidazole and Ro-7-0582. *Radiat. Res.* **64**, 633–647.

Brown, J. M. (1979). Evidence for acutely hypoxic cells in mouse tumours, and a possible mechanism of reoxygenation. *Br. J. Radiol.* **52**, 650–656.

Brown, J. M. (2002). Tumor microenvironment and the response to therapy. *Cancer Biol. Ther.* **1**, 453–458.

Brown, J. M., and Lemmon, M. J. (1990). Potentiation by the hypoxic cytotoxin SR 4233 of cell killing produced by fractionated irradiation of mouse tumors. *Cancer Res.* **50**, 7745–7749.

Brown, J. M., and Wilson, W. R. (2004). Exploiting tumour hypoxia in cancer treatment. *Nat. Rev. Cancer* **4**, 437–447.

Burke, B., Tang, N., Corke, K. P., Tazzyman, D., Ameri, K., Wells, M., and Lewis, C. E. (2002). Expression of HIF-1α by human macrophages: Implications for the use of macrophages in hypoxia-regulated cancer gene therapy. *J. Pathol.* **196**, 204–212.

Carey, R. W., Holland, J. F., Whang, H. Y., Neter, E., and Bryant, B. (1967). Clostridial oncolysis in man. *Europ. J. Cancer* **3**, 37–46.

Chaplin, D. J., Durand, R. E., and Olive, P. L. (1986). Acute hypoxia in tumors: Implications for modifiers of radiation effects. *Int. J. Radiat. Oncol. Biol. Phys.* **12**, 1279–1282.

Chaplin, D. J., Olive, P. L., and Durand, R. E. (1987). Intermittent blood flow in a murine tumor: Radiobiological effects. *Cancer Res.* **47**, 597–601.

Chung-Faye, G., Palmer, D., Anderson, D., Clark, J., Downes, M., Baddeley, J., Hussain, S., Murray, P. I., Searle, P., Seymour, L., Harris, P. A., Ferry, D., *et al.* (2001). Virus-directed, enzyme prodrug therapy with nitroimidazole reductase: A phase I and pharmacokinetic study of its prodrug, CB1954. *Clin. Cancer Res.* **7**, 2662–2668.

Churchill-Davidson, I., Foster, C. A., and Thomlinson, R. H. (1958). High-pressure oxygen and radiotherapy. *Med. World* **88**, 125–128.

Cobb, L. M., Connors, T. A., Elson, L. A., Khan, A. H., Mitchley, B. C. V., Ross, W. C. J., and Whisson, M. E. (1969). 2,4-Dinitro-5-ethyleneiminobenzamide (CB 1954): A potent and selective inhibitor of the growth of the Walker carcinoma 256. *Biochem. Pharmacol.* **18**, 1519–1527.

Comerford, K. M., Wallace, T. J., Karhausen, J., Louis, N. A., Montalto, M. C., and Colgan, S. P. (2002). Hypoxia-inducible factor-1–dependent regulation of the multidrug resistance (MDR1) gene. *Cancer Res.* **62**, 3387–3394.

Crabtree, H. G., and Cramer, W. (1933). The action of radium on cancer cells I. II.—Some factors determining the susceptibility of cancer cells to radium. *Proc. Roy. Soc. B.* **113**, 238–250.

Denekamp, J., and Harris, S. R. (1975). Tests of two electron-affinic radiosensitizers *in vivo* using regrowth of an experimental carcinoma. *Rad. Res.* **61**, 191–203.

Denekamp, J., and Joiner, M. C. (1982). The potential benefit from a perfect radiosensitizer and its dependence on reoxygenation. *Br. J. Radiol.* **55**, 657–663.

Dorie, M. J., and Brown, J. M. (1993). Tumor-specific, schedule-dependent interaction between tirapazamine (SR 4233) and cisplatin. *Cancer Res.* **53**, 4633–4636.

Durand, R. E. (1994). The influence of microenvironmental factors during cancer therapy. *In vivo* **8**, 691–702.

Engelbart, K., and Gericke, D. (1964). Oncolysis by clostridia V. Transplanted tumors of the hamster. *Cancer Res.* **24**, 239–243.

Fox, M. E., Lemmon, M. J., Mauchline, M. L., Davis, T. O., Giaccia, A. J., Minton, N. P., and Brown, J. M. (1996). Anaerobic bacteria as a delivery system for cancer gene therapy: Activation of 5-fluorocytosine by genetically engineered clostridia. *Gene Ther.* **3**, 173–178.

Giaccia, A., Siim, B. G., and Johnson, R. S. (2003). HIF-1 as a target for drug development. *Nat. Rev. Drug Discov.* **2**, 803–811.

Graeber, T. G., Osmanian, C., Jacks, T., Housman, D. E., Koch, C. J., Lowe, S. W., and Giaccia, A. J. (1996). Hypoxia-mediated selection of cells with diminished apoptotic potential in solid tumours. *Nature* **379**, 88–91.

Gray, L. H., Conger, A. D., Ebert, M., Hornsey, S., and Scott, O. C. (1953). Concentration of oxygen dissolved in tissues at the time of irradiation as a factor in radiotherapy. *Brit. J. Radiol.* **26**, 638–648.

Greco, O., and Dachs, G. U. (2001). Gene directed enzyme/prodrug therapy of cancer: Historical appraisal and future prospectives. *J. Cell Physiol.* **187**, 22–36.

Griffiths, L., Binley, K., Iqball, S., Kan, O., Maxwell, P., Ratcliffe, P., Lewis, C., Harris, A., Kingsman, S., and Naylor, S. (2000). The macrophage—a novel system to deliver gene therapy to pathological hypoxia. *Gene Ther.* **7**, 255–262.

Hall, E. J. (2000). "Radiobiology for the Radiologist." J.B. Lippincott Company, Philadelphia, PA.

Harris, A. L. (2002). Hypoxia—a key regulatory factor in tumour growth. *Nat. Rev. Cancer* **2**, 38–47.

Henk, J. M. (1981). Does hyperbaric oxygen have a future in radiation therapy? *Int. J. Radiat. Oncol. Biol. Phys.* **7**, 1125–1128.

Heppner, F., and Mose, J. R. (1978). The liquefaction (oncolysis) of malignant gliomas by a nonpathogenic clostridium. *Acta Neuro.* **12**, 123–125.

Heppner, F., Mose, J., Ascher, P. W., and Walter, G. (1983). Oncolysis of malignant gliomas of the brain. *13th Int. Cong. Chemother.* **226**, 38–45.

Hicks, K. O., Pruijn, F. B., Secomb, T. W., Hay, M. P., Hsu, R., Brown, J. M., Denny, W. A., Dewhirst, M. W., and Wilson, W. R. (2006). Use of three-dimensional tissue cultures to model extravascular transport and predict *in vivo* activity of hypoxia-targeted anticancer drugs. *J. Natl. Cancer Inst.* **98**, 1118–1128.

Hill, R. P. (1992). Experimental Radiotherapy. In "The Basic Science of Oncology" (I. F. Tannock and R. P. Hill, eds.), pp. 276–301. McGraw-Hill, Inc., Toronto, Canada.

Hill, S. A., Collingridge, D. R., Vojnovic, B., and Chaplin, D. J. (1998). Tumour radiosensitization by high-oxygen-content gases: Influence of the carbon dioxide content of the inspired gas on PO_2, microcirculatory function and radiosensitivity. *Int. J. Radiat. Oncol. Biol. Phys.* **40**, 943–951.

Hill, S. A., Pigott, K. H., Saunders, M. I., Powell, M. E., Arnold, S., Obeid, A., Ward, G., Leahy, M., Hoskin, P. J., and Chaplin, D. J. (1996). Microregional blood flow in murine and human tumours assessed using laser Doppler microprobes. *Brit. J. Cancer* **74**(Suppl 27), S260–S263.

Hodgkiss, R. J., Roberts, I. J., Watts, M. E., and Woodcock, M. (1987). Rapid-mixing studies of radiosensitivity with thiol-depleted mammalian cells. *Int. J. Radiat. Biol. Relat. Stud. Phys. Chem. Med.* **52**, 735–744.

Holthusen, H. (1921). Beitrage zur biologie der strahelnwirkung. Untershungen an askarideneiern. *Ptlugers Arch. f.d. ges. Physio.* **187**, 1–24.

Kallman, R. F. (1972). The phenomenon of reoxygenation and its implications for fractionated radiotherapy. *Radiology* **105**, 135–142.

Kimura, H., Braun, R. D., Ong, E. T., Hsu, R., Secomb, T. W., Papahadjopoulos, D., Hong, K., and Dewhirst, M. W. (1996). Fluctuations in red cell flux in tumor microvessels can lead to transient hypoxia and reoxygenation in tumor parenchyma. *Cancer Res.* **56**, 5522–5528.

Knox, R. J., Friedlos, F., Jarman, M., and Roberts, J. J. (1988b). A new cytotoxic, DNA interstrand crosslinking agent, 5-(Aziridin-1-YL)-4-hydroxylamino-2-nitrobenzamide, is formed from 5-(Aziridin-1-YL)-2,4-dinitrobenzamide (CB 1954) by a nitroreductase enzyme in Walker carcinoma cells. *Biochem. Pharm.* **37**, 4661–4669.

Knox, R. J., Boland, M. P., Friedlos, F., Coles, B., Southan, C., and Roberts, J. J. (1988a). The nitroreductase enzyme in Walker cells that activates 5-(Aziridin-1-YL)-2,4-dinitrobenzamide (CB 1954) to 5-(Aziridin-1-YL)-4-hydroxylamino-2-nitrobenzamide is a form of NAD(P)H dehydrogenase (quinone) (EC 1.6.99.2). *Biochem. Pharmacol.* **37**, 4671–4677.

Koch, C. J., Stobbe, C. C., and Bump, E. A. (1984). The effect on the Km for radiosensitization at 0° of thiol depletion by diethylmaleate pretreatment: Quantitative differences found using the radiation sensitizing agent misonidazole or oxygen. *Radiat. Res.* **98**, 141–153.

Kovacs, M. S., Hocking, D. J., Evans, J. W., Siim, B. G., Wouters, B. G., and Brown, J. M. (1999). Cisplatin anti-tumour potentiation by tirapazamine results from a hypoxia-dependent cellular sensitization to cisplatin. *Br. J. Cancer* **80**, 1245–1251.

Kung, A. L., Wang, S., Klco, J. M., Kaelin, W. G., and Livingston, D. M. (2000). Suppression of tumor growth through disruption of hypoxia-inducible transcription. *Nat. Med.* **6,** 1335–1340.

Kung, A. L., Zabludoff, S. D., France, D. S., Freedman, S. J., Tanner, E. A., Vieira, A., Cornell-Kennon, S., Lee, J., Wang, B., Wang, J., Memmert, K., Naegeli, H. U., et al. (2004). Small molecule blockade of transcriptional coactivation of the hypoxia-inducible factor pathway. *Cancer Cell* **6,** 33–43.

Lanzen, J., Braun, R. D., Klitzman, B., Brizel, D., Secomb, T. W., and Dewhirst, M. W. (2006). Direct demonstration of instabilities in oxygen concentrations within the extravascular compartment of an experimental tumor. *Cancer Res.* **66,** 2219–2223.

Le, Q. T., Denko, N. C., and Giaccia, A. J. (2004). Hypoxic gene expression and metastasis. *Cancer Met. Rev.* **23,** 293–310.

Lemmon, M. L., Van Zijl, P., Fox, M. E., Mauchline, M. L., Giaccia, A. J., Minton, N. P., and Brown, J. M. (1997). Anaerobic bacteria as a gene delivery system that is controlled by the tumor microenvironment. *Gene Ther.* **4,** 791–796.

Lemmon, M. J., Elwell, J. H., Brehm, J. K., Mauchline, M. L., N, M., Minton, N. P., Giaccia, A. J., and Brown, J. M. (1994). Anaerobic bacteria as a gene delivery system to tumors. *Proc. Am. Assoc. Cancer Res.* **35,** 374.

Liu, S. C., Minton, N. P., Giaccia, A. J., and Brown, J. M. (2002). Anticancer efficacy of systemically delivered anaerobic bacteria as gene therapy vectors targeting tumor hypoxia/necrosis. *Gene Therapy* **9,** 291–296.

Mabjeesh, N. J., Post, D. E., Willard, M. T., Kaur, B., Van Meir, E. G., Simons, J. W., and Zhong, H. (2002). Geldanamycin induces degradation of hypoxia-inducible factor 1alpha protein via the proteosome pathway in prostate cancer cells. *Cancer Res.* **62,** 2478–2482.

Mabjeesh, N. J., Escuin, D., LaVallee, T. M., Pribluda, V. S., Swartz, G. M., Johnson, M. S., Willard, M. T., Zhong, H., Simons, J. W., and Giannakakou, P. (2003). 2ME2 inhibits tumor growth and angiogenesis by disrupting microtubules and dysregulating HIF. *Cancer Cell* **3,** 363–375.

Malmgren, R. A., and Flanigan, C. C. (1955). Localization of the vegetative form of *Clostridium tetani* in mouse tumors following intravenous spore administration. *Cancer Res.* **15,** 473–478.

Martin, J., Stribbling, S. M., Poon, G. K., Begent, R. H., Napier, M., Sharma, S. K., and Springer, C. J. (1997). Antibody-directed enzyme prodrug therapy: Pharmacokinetics and plasma levels of prodrug and drug in a phase I clinical trial. *Cancer Chemother. Pharmacol.* **40,** 189–201.

McCarthy, H. O., Yakkundi, A., McErlane, V., Hughes, C. M., Keilty, G., Murray, M., Patterson, L. H., Hirst, D. G., McKeown, S. R., and Robson, T. (2003). Bioreductive GDEPT using cytochrome P450 3A4 in combination with AQ4N. *Cancer Gene Ther.* **10,** 40–48.

Melillo, G. (2006). Inhibiting hypoxia-inducible factor 1 for cancer therapy. *Mol. Cancer Res.* **4,** 601–605.

Minton, N. P. (2003). Clostridia in cancer therapy. *Nat. Rev. Microbiol.* **1,** 237–242.

Moeller, B. J., Cao, Y., Li, C. Y., and Dewhirst, M. W. (2004). Radiation activates HIF-1 to regulate vascular radiosensitivity in tumors: Role of reoxygenation, free radicals, and stress granules. *Cancer Cell* **5,** 429–441.

Möse, J. R., and Möse, G. (1959). Onkolyseversuche mit apathogenen anaeroben Sporenbildern am Ehrlich tumor des Maus. *Z. Krebsforsch* **63,** 63–74.

Möse, J. R., and Möse, G. (1964). Oncolysis by clostridia. I. Activity of *Clostridium butyricum* (M-55) and other nonpathogenic clostridia against the *Ehrlich carcinoma*. *Cancer Res.* **24,** 212–216.

Mottram, J. C. (1935). On the alteration in the sensitivity of cells towards radiation produced by cold and by anaerobiosis. *Brit. J. Radiol.* **8,** 32–39.

Moulder, J. E., and Rockwell, S. (1987). Tumor hypoxia: Its impact on cancer therapy. *Cancer Metastasis Rev.* **5,** 313–341.

Nordsmark, M., Overgaard, M., and Overgaard, J. (1996). Pretreatment oxygenation predicts radiation response in advanced squamous cell carcinoma of the head and neck. *Radiother. Oncol.* **41,** 31–40.

Nordsmark, M., Bentzen, S. M., Rudat, V., Brizel, D., Lartigau, E., Stadler, P., Becker, A., Adam, M., Molls, M., Dunst, J., Terris, D. J., Overgaard, J., *et al.* (2005). Prognostic value of tumor oxygenation in 397 head and neck tumors after primary radiation therapy. An international multi-center study. *Radiother. Oncol.* **77,** 18–24.

Patterson, L. H. (1993). Rationale for the use of aliphatic N-oxides of cytotoxic anthraquinones as prodrug DNA binding agents: A new class of bioreductive agent. *Cancer Metastasis Rev.* **12,** 119–134.

Patterson, L. H. (2002). Bioreductively activated antitumor N-oxides: The case of AQ4N, a unique approach to hypoxia-activated cancer chemotherapy. *Drug Metab. Rev.* **34,** 581–592.

Patterson, L. H., and McKeown, S. R. (2000). AQ4N: A new approach to hypoxia-activated cancer chemotherapy. *Br. J. Cancer* **83,** 1589–1593.

Patterson, L. H., and Murray, G. I. (2002). Tumour cytochrome P450 and drug activation. *Curr. Pharm. Des.* **8,** 1335–1347.

Patterson, L. H., McKeown, S. R., Ruparelia, K., Double, J. A., Bibby, M. C., Cole, S., and Stratford, I. J. (2000). Enhancement of chemotherapy and radiotherapy of murine tumours by AQ4N, a bioreductively activated anti-tumour agent. *Br. J. Cancer* **82,** 1984–1990.

Patterson, A. V., Williams, K. J., Cowen, R. L., Jaffar, M., Telfer, B. A., Saunders, M., Airley, R., Honess, D., Van Der Kogel, A. J., Wolf, C. R., and Stratford, I. J. (2002). Oxygen-sensitive enzyme-prodrug gene therapy for the eradication of radiation-resistant solid tumours. *Gene Ther.* **9,** 946–954.

Pawelek, J. M., Low, K. B., and Bermudes, D. (1997). Tumor-targeted *Salmonella* as a novel anticancer vector. *Cancer Res.* **57,** 4537–4544.

Petry, E. (1923). Zur kenntnis der degingungen der biologischen wirkung der Rontgenstrahlen III Mitteilung. Wirkung von oxydationsmitteln auf die empfindlichkeit. *Biochem. Z.* **135,** 353–383.

Rapisarda, A., Uranchimeg, B., Scudiero, D. A., Selby, M., Sausville, E. A., Shoemaker, R. H., and Melillo, G. (2002). Identification of small molecule inhibitors of hypoxia-inducible factor 1 transcriptional activation pathway. *Cancer Res.* **62,** 4316–4324.

Rapisarda, A., Uranchimeg, B., Sordet, O., Pommier, Y., Shoemaker, R. H., and Melillo, G. (2004a). Topoisomerase I-mediated inhibition of hypoxia-inducible factor 1: Mechanism and therapeutic implications. *Cancer Res.* **64,** 1475–1482.

Rapisarda, A., Zalek, J., Hollingshead, M., Braunschweig, T., Uranchimeg, B., Bonomi, C. A., Borgel, S. D., Carter, J. P., Hewitt, S. M., Shoemaker, R. H., and Melillo, G. (2004b). Schedule-dependent inhibition of hypoxia-inducible factor-1alpha protein accumulation, angiogenesis, and tumor growth by topotecan in U251-HRE glioblastoma xenografts. *Cancer Res.* **64,** 6845–6848.

Riballo, E., Kuhne, M., Rief, N., Doherty, A., Smith, G. C., Recio, M. J., Reis, C., Dahm, K., Fricke, A., Krempler, A., Parker, A. R., Jackson, S. P., *et al.* (2004). A pathway of double-strand break rejoining dependent upon ATM, Artemis, and proteins locating to γ-H2AX foci. *Mol. Cell* **16,** 715–724.

Rischin, D., Peters, L., Fisher, R., Macann, A., Denham, J., Poulsen, M., Jackson, M., Kenny, L., Penniment, M., Corry, J., Lamb, D., McClure, B., *et al.* (2005). Tirapazamine, cisplatin, and radiation versus fluorouracil, cisplatin, and radiation in patients with locally advanced head and neck cancer: A randomized phase II trial of the Trans-Tasman Radiation Oncology Group (TROG 98.02). *J. Clin. Oncol.* **23,** 79–87.

Romero-Ramirez, L., Cao, H., Nelson, D., Hammond, E., Lee, A. H., Yoshida, H., Mori, K., Glimcher, L. H., Denko, N. C., Giaccia, A. J., Le, Q. T., Koong, A. C., et al. (2004). XBP1 is essential for survival under hypoxic conditions and is required for tumor growth. Cancer Res. 64, 5943–5947.

Semenza, G. L. (2002). Involvement of hypoxia-inducible factor 1 in human cancer. Intern. Med. 41, 79–83.

Sheldon, P. W., Foster, J. L., and Fowler, J. F. (1974). Radiosensitization of C3H mouse mammary tumours by a 2-nitroimidazole drug. Br. J. Cancer 30, 560–565.

Shibata, T., Giaccia, A. J., and Brown, J. M. (2002). Hypoxia-inducible regulation of a prodrug-activating enzyme for tumor-specific gene therapy. Neoplasia 4, 40–48.

Shibata, T., Akiyama, N., Noda, M., Sasai, K., and Hiraoka, M. (1998). Enhancement of gene expression under hypoxic conditions using fragments of the human vascular endothelial growth factor and the erythropoietin genes. Int. J. Radiat. Oncol. Biol. Phys. 42, 913–916.

Suit, H. D., Sedlacek, R., Silver, G., Hsieh, C. C., Epp, E. R., Ngo, F. Q., Roberts, W. K., and Verhey, L. (1988). Therapeutic gain factors for fractionated radiation treatment of spontaneous murine tumors using fast neutrons, photons plus $O_2(1)$ or 3 ATA, or photons plus misonidazole. Radiat. Res. 116, 482–502.

Sun, X., Kanwar, J. R., Leung, E., Lehnert, K., Wang, D., and Krissansen, G. W. (2001). Gene transfer of antisense hypoxia inducible factor-1 alpha enhances the therapeutic efficacy of cancer immunotherapy. Gene Ther. 8, 638–645.

Talks, K. L., Turley, H., Gatter, K. C., Maxwell, P. H., Pugh, C. W., Ratcliffe, P. J., and Harris, A. L. (2000). The expression and distribution of the hypoxia-inducible factors HIF-1α and HIF-2α in normal human tissues, cancers, and tumor-associated macrophages. Am. J. Pathol. 157, 411–421.

Tannock, I. F. (1968). The relation between cell proliferation and the vascular system in a transplanted mouse mammary tumour. Brit. J. Cancer 22, 258–273.

Tannock, I. F. (1998). Conventional cancer therapy: Promise broken or promise delayed? Lancet 351(Suppl 2), SII9–SII16.

Teicher, B. A., Lazo, J. S., and Sartorelli, A. C. (1981). Classification of antineoplastic agents by their selective toxicities toward oxygenated and hypoxic tumor cells. Cancer Res. 41, 73–81.

Thiele, E. H., Arison, R. N., and Boxer, G. E. (1964). Oncolysis by clostridia. III. Effects of clostridia and chemotherapeutic agents on rodent tumors. Cancer Res. 24, 222–233.

Thomlinson, R. H., and Gray, L. H. (1955). The histological structure of some human lung cancers and the possible implications for radiotherapy. Brit. J. Cancer 9, 539–549.

Toso, J. F., Gill, V. J., Hwu, P., Marincola, F. M., Restifo, N. P., Schwartzentruber, D. J., Sherry, R. M., Topalian, S. L., Yang, J. C., Stock, F., Freezer, L. J., Morton, K. E., et al. (2002). Phase I study of the intravenous administration of attenuated Salmonella typhimurium to patients with metastatic melanoma. J. Clin. Oncol. 20, 142–152.

Trinh, Q. T., Austin, E. A., Murray, D. M., Knick, V. C., and Huber, B. E. (1995). Enzyme/prodrug gene therapy: Comparison of cytosine deaminase/5- fluorocytosine versus thymidine kinase/ganciclovir enzyme/prodrug systems in a human colorectal carcinoma cell line. Cancer Res. 55, 4808–4812.

Trotter, M. J., Chaplin, D. J., and Olive, P. L. (1989). Use of a carbocyanine dye as a marker of functional vasculature in murine tumours. Brit. J. Cancer 59, 706–709.

von Pawel, J., von Roemeling, R., Gatzemeier, U., Boyer, M., Elisson, L. O., Clark, P., Talbot, D., Rey, A., Butler, T. W., Hirsh, V., Olver, I., Bergman, B., et al. (2000). Tirapazamine plus cisplatin versus cisplatin in advanced non–small-cell lung cancer: A report of the international CATAPULT I study group. J. Clin. Oncol. 18, 1351–1359.

Wachsberger, P., Burd, R., and Dicker, A. P. (2003). Tumor response to ionizing radiation combined with antiangiogenesis or vascular targeting agents: Exploring mechanisms of interaction. Clin. Cancer Res. 9, 1957–1971.

Wartenberg, M., Ling, F. C., Muschen, M., Klein, F., Acker, H., Gassmann, M., Petrat, K., Putz, V., Hescheler, J., and Sauer, H. (2003). Regulation of the multidrug resistance transporter P-glycoprotein in multicellular tumor spheroids by hypoxia-inducible factor (HIF-1) and reactive oxygen species. *FASEB J.* **17,** 503–505.

Whillans, D. W., and Hunt, J. W. (1982). A rapid-mixing comparison on the mechanisms of radiosensitization by oxygen and misonidazole in CHO cells. *Radiat. Res.* **90,** 126–141.

Williams, K. J., Telfer, B. A., Xenaki, D., Sheridan, M. R., Desbaillets, I., Peters, H. J., Honess, D., Harris, A. L., Dachs, G. U., van der Kogel, A., and Stratford, I. J. (2005). Enhanced response to radiotherapy in tumours deficient in the function of hypoxia-inducible factor-1. *Radiother. Oncol.* **75,** 89–98.

Wilson, W. R., Pullen, S. M., Hogg, A., Helsby, N. A., Hicks, K. O., and Denny, W. A. (2002). Quantitation of bystander effects in nitroreductase suicide gene therapy using three-dimensional cell cultures. *Cancer Res.* **62,** 1425–1432.

Wilson, W. R., Edmunds, S. J., Valentine, S., Gu, Y., Myint, H., Hicks, K. O., Yang, S., Atwell, G. J., and Patterson, A. V. (2005). Mechanism of action and antitumour activity of PR-104, a dinitrobenzamide mustard pre-prodrug that is activated selectively under hypoxia. October 3. Poster at National Cancer Research Institute Conference, Birmingham, UK.

Wilson, W. R., Pullen, S. M., Degenkolbe, D. M., Ferry, D. M., Helsby, N. A., Hicks, K. O., Atwell, G. J., Yang, S., Denny, W. A., and Patterson, A. V. (2004). Water-soluble dinitrobenzamide mustard phosphate pre-prodrugs as hypoxic cytotoxins [abstract 496]. *Eur. J. Cancer Supp.* **2,** 151.

Wouters, B. G., and Brown, J. M. (1997). Cells at intermediate oxygen levels can be more important than the "hypoxic fraction" in determining tumor response to fractionated radiotherapy. *Radiat. Res.* **147,** 541–550.

Yeo, E. J., Chun, Y. S., Cho, Y. S., Kim, J., Lee, J. C., Kim, M. S., and Park, J. W. (2003). YC-1: A potential anticancer drug targeting hypoxia-inducible factor 1. *J. Natl. Cancer Inst.* **95,** 516–525.

Zeman, E. M., Brown, J. M., Lemmon, M. J., Hirst, V. K., and Lee, W. W. (1986). SR-4233: A new bioreductive agent with high selective toxicity for hypoxic mammalian cells. *Int. J. Radiat. Oncol. Biol. Phys.* **12,** 1239–1242.

Zhong, H., De Marzo, A. M., Laughner, E., Lim, M., Hilton, D. A., Zagzag, D., Buechler, P., Isaacs, W. B., Semenza, G. L., and Simons, J. W. (1999). Overexpression of hypoxia-inducible factor 1alpha in common human cancers and their metastases. *Cancer Res.* **59,** 5830–5835.

CHAPTER SIXTEEN

HIF GENE EXPRESSION IN CANCER THERAPY

Denise A. Chan, Adam J. Krieg, Sandra Turcotte, *and*
Amato J. Giaccia

Contents

1. Introduction	324
2. Experimental Procedures	326
2.1. Induction by hypoxia or chemical mimetics	326
2.2. Genetic mutations	327
2.3. HIF hydroxylation and other posttranslational modifications	328
2.4. HIF inhibitors	329
2.5. RNA interference	331
2.6. HIF knockout mice	332
2.7. Somatic cell HIF knockout	332
2.8. HIF activity (reporter assays)	333
2.9. HIF responsive elements	333
2.10. RNA analysis	333
2.11. Quantitative RT-PCR	334
2.12. Western blot analysis	334
2.13. Chromatin immunoprecipitation	334
2.14. HEEBO microarray analysis	335
3. Conclusions	337
Acknowledgments	337
References	338

Abstract

Tumor hypoxia is a feature common to almost all solid tumors due to malformed vasculature and inadequate perfusion. Tumor cells have evolved mechanisms that allow them to respond and adapt to a hypoxic microenvironment. The hypoxia-inducible transcription factor (HIF) family is comprised of oxygen-sensitive alpha (α) subunits that respond rapidly to decreased oxygen levels and oxygen-insensitive beta (β) subunits. HIF binds to specific recognition sequences in the genome and increases the transcription of genes involved in a variety of metabolic and enzymatic pathways that are necessary for cells to respond to an oxygen-poor environment. The critical role of this family of

Department of Radiation Oncology, Stanford University School of Medicine, Stanford, California

Methods in Enzymology, Volume 435
ISSN 0076-6879, DOI: 10.1016/S0076-6879(07)35016-7

transcriptional regulators in maintaining oxygen homeostasis is supported by multiple regulatory mechanisms that allow the cell to control the levels of HIF as well as its transcriptional activity. This review will focus on how the transcriptional activity of HIF is studied and how it can be exploited for cancer therapy.

1. INTRODUCTION

Hypoxia, or low oxygen tensions, is a critical characteristic of solid tumors. A negative prognostic factor, hypoxic tumors pose several significant barriers to the treatment of cancer. The key mediator of the cellular response to hypoxia is through the action of the HIF family of transcription factors (Giaccia *et al.*, 2003; Semenza, 2003). HIF-dependent gene transcription plays a key role in a variety of pathways, including oxygen homeostasis, glycolysis, tissue remodeling, migration, viability, proliferation, and angiogenesis. Thus, HIF and its downstream effectors may be attractive targets for novel therapeutics. This chapter focuses on HIF and its target genes as potential cancer therapeutics and the methods employed to determine whether a given gene is an HIF target.

Regulation of the HIF family of transcription factors occurs primarily at the level of posttranslational modification. The HIF transcription factor is composed of an oxygen-labile α subunit and an oxygen-insensitive β subunit (Fig. 16.1). Under normoxic conditions, HIF-1α is hydroxylated on either one of two conserved proline residues, proline 402 or proline 564 (Chan *et al.*, 2005; Ivan *et al.*, 2001; Jaakkola *et al.*, 2001; Masson *et al.*, 2001; Yu *et al.*, 2001). This hydroxylation reaction is catalyzed by a family of prolyl-4-hydroxylases (PHDs), which require both oxygen and 2-oxoglutarate as substrates as well as iron and ascorbate as cofactors (Bruick and McKnight, 2001; Epstein *et al.*, 2001). Hydroxylation of HIF-1α allows it to be recognized by the protein product of the von Hippel-Lindau (VHL) tumor suppressor gene. VHL, along with elongin B and elongin C, function as an E3 ubiquitin ligase, targeting HIF-1α to the proteasome for degradation. Conversely, when oxygen is limited under hypoxic conditions, the prolyl hydroxylases are inhibited, and HIF-1α escapes hydroxylation and recognition by VHL. Subsequently, HIF-1α is stabilized under hypoxic conditions and can dimerize with its heterodimeric binding partner, HIF-1β. Together, HIF-1α and HIF-1β bind to the enhancer regions of target genes, driving their transcription, which results in cellular adaptation to a low-oxygen environment.

It is estimated that approximately 1 to 2% of the entire genome is hypoxia-inducible (Denko *et al.*, 2003). Several groups have conducted global genomic screens to identify gene expression profile changes in

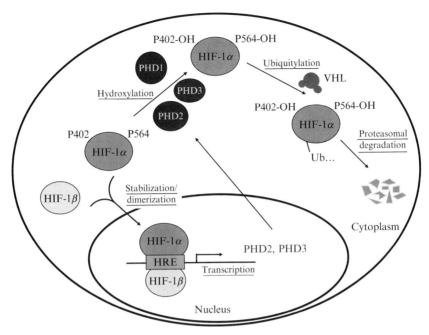

Figure 16.1 Regulation of hypoxia-inducible factors. Under normal oxygenated conditions, a family of prolyl hydroxylases (PHDs) hydroxylate proline 402 or proline 564 of hypoxia-inducible transcription factor alpha (HIF-1α). Hydroxylation of either residue results in recognition by the von Hippel-Lindau (VHL)/elongin B/elongin C complex, which functions as an E3 ubiquitin ligase, targeting HIF-1α to the proteasome for degradation. Conversely, under hypoxic conditions, HIF-1α is not hydroxylated and is able to dimerize with HIF-1β to drive transcription of target genes. Two of the target genes are PHD2 and PHD3, which feedback to regulate HIF-1α. (See color insert.)

response to hypoxia (Denko *et al.*, 2003; Koong *et al.*, 2000; Lal *et al.*, 2001; Scandurro *et al.*, 2001; Wykoff *et al.*, 2000). It has become increasingly clear that hypoxia and the HIF pathway, in particular, play essential roles in the development and malignant progression of cancer. HIF targets include those that are directly involved in chemotherapeutic resistance, genomic instability, proliferation, and metastasis (Ceradini *et al.*, 2004; Comerford *et al.*, 2002; Erler *et al.*, 2006; Koshiji *et al.*, 2005; Liao *et al.*, 2007; Staller *et al.*, 2003; Wartenberg *et al.*, 2003; Welford *et al.*, 2006). HIF has been shown to induce the expression of multidrug resistance (MDR1) gene, leading to the transport of chemotherapeutic drugs out of cell and drug resistance (Comerford *et al.*, 2002; Wartenberg *et al.*, 2003). Hypoxia increases genomic instability, leading to enhanced tumor aggressiveness (Bindra and Glazer, 2005, 2007; Huang *et al.*, 2007). Through competition with Myc, HIF-1 has been shown to repress MutSα, which specifically recognizes base mismatches, leading to genomic instability (Koshiji *et al.*,

2005). The migration inhibitor factor (MIF) is an HIF target, which is necessary and sufficient to delay replicative senescence, demonstrating a direct role of HIF in proliferation (Welford et al., 2006). In a transgenic model, HIF-1 has been shown to contribute to the metastatic potential of a tumor (Liao et al., 2007). Furthermore, through the action of lysyl oxidase (an HIF target), HIF contributes to the metastatic potential of a tumor by affecting invasion and migration (Erler et al., 2006). HIF and HIF gene expression, therefore, pose significant barriers to the treatment of solid tumors. However, HIF and HIF targets may also pose a unique opportunity to exploit in the specific targeting of solid tumors.

2. Experimental Procedures

2.1. Induction by hypoxia or chemical mimetics

HIF is a transcription factor primarily activated under hypoxic conditions through the stabilization of its α subunit. Transcriptional targets of HIF can also be induced by other agents that induce HIF stabilization, such as growth factors, PHD inhibitors, and proteasome inhibitors. Hypoxic conditions can be achieved through the use of specialized chambers, incubators, or chemical mimetics. Control cells are maintained at atmospheric oxygen (21% O_2) in a standard cell culture incubator. Stringent hypoxia (0.02% O_2) is attained through the use of an anaerobic chamber (Bactron Anaerobic/Environmental Chamber; Sheldon Manufacturing, Inc., Cornelius, OR). Cells can be maintained in mild hypoxia conditions (0.5–8% O_2) in a variable hypoxia chamber (InVivo400; Ruskinn, Inc., Cincinnati, OH). For long-term culturing (several weeks) in mildly hypoxic conditions (2–5% O_2), a standard incubator can be equipped with nitrogen gas and set to the desired hypoxic condition. HIF levels can also be induced by the addition of iron chelators or cobaltous ions to the cell culture media directly. For example, desferrioxamine (DFO, 100 μM; Sigma-Aldrich, St. Louis, MO), an iron chelator, inhibits the prolyl hydroxylation reaction since iron is a necessary cofactor, stabilizing HIF-α. Similarly, cobalt chloride ($CoCl_2$, 100 μM, Sigma) interferes with the prolyl hydoxylase by competing with iron for the binding site of the prolyl hydroxylase enzyme, again resulting in HIF-α stabilization. When inducing HIF and its target genes, the duration and stringency of hypoxia treatment are both factors that must be taken into consideration. Is the gene being examined only responsive to anoxia or can it be induced by milder hypoxia? Is the gene induction rapid (within hours) or does it require sustained exposure to hypoxia?

2.2. Genetic mutations

Experimentally additional factors, including genetic mutation, growth factors, transition metals, nitric oxide, reactive oxygen species, heat shock, and redox status of the cell, have all been shown to stabilize HIF-1α under normoxic conditions and have been reported to contribute to tumor aggressiveness (Table 16.1) (Gao *et al.*, 2002a,b; Katschinski *et al.*, 2002; Li *et al.*, 2004; Maxwell *et al.*, 1999; Mazure *et al.*, 1996; Minet *et al.*, 2001; Ohh *et al.*, 2000; Salceda and Caro, 1997; Skinner *et al.*, 2004; Srinivas *et al.*, 1999; Sugawara *et al.*, 2000). Genetic alterations have been shown to increase HIF-1α protein levels under normoxia either by loss-of-function mutations in tumor suppressors or gain-of-function mutations in oncogenes. Loss-of-function mutation in three tumor suppressors, VHL, tuberous sclerosis (TSC)2, and phosphate and tensin homolog (PTEN), have been linked to the stabilization of HIF-1α subunits and the development of cancer-prone syndromes (Brugarolas *et al.*, 2003; Liu *et al.*, 2003; Maxwell *et al.*, 1999;

Table 16.1 HIF protein stabilizers

Target	Name	References
Tumor suppressor	VHL	Maxwell *et al.*, 1999; Ohh *et al.*, 2000
	TSC	Brugarolas *et al.*, 2003; Liu *et al.*, 2003
	PTEN	Zundel *et al.*, 2000
Oncogene	Ras	Chen *et al.*, 2001; Mazure *et al.*, 1996; Mazure *et al.*, 1997; Rak *et al.*, 1995
	Akt	Hudson *et al.*, 2002; Mottet *et al.*, 2003
	v-Src	Jiang *et al.*, 1997
Metals	Cobalt	Chan *et al.*, 2002
	Arsenite	Gao *et al.*, 2004; Skinner *et al.*, 2004
	Vanadate	Gao *et al.*, 2002a
	Nickel	Davidson *et al.*, 2006; Li *et al.*, 2004
Stresses	Heat shock	Katschinski *et al.*, 2002
	Reactive oxygen species	Chandel *et al.*, 2000; Richard *et al.*, 2000; Sandau *et al.*, 2001
	Redox status	Salceda and Caro, 1997; Srinivas *et al.*, 1999
Point mutation (*in vitro*)	Proline 402	Chan *et al.*, 2005; Masson *et al.*, 2001
	Proline 564	Ivan *et al.*, 2001; Jaakkola *et al.*, 2001; Yu *et al.*, 2001

Ohh *et al.*, 2000; Zundel *et al.*, 2000). Mutations of these tumor suppressors are clinically manifested in the generation of highly vascular and highly aggressive cancers. At the intersection of growth factors, oncogenes, and tumor suppressors are two signaling pathways, PI3-K/Akt and Ras/Raf/MAPK (Bardos and Ashcroft, 2004). Dysregulation of these two signaling pathways results in inappropriate normoxic HIF-1 activity (Chan *et al.*, 2002; Chen *et al.*, 2001; Mazure *et al.*, 1996, 1997; Rak *et al.*, 1995; Richard *et al.*, 1999; Zundel *et al.*, 2000). The Ras/Raf/MAPK pathway appears to be sensitive to the severity of the hypoxic stress and dependent on the cell type. Oncogenic Ras, which is commonly found in human cancers, leads to increased levels of HIF-1α protein as well as HIF-1 activity. Thus, in addition to hypoxia, other pathophysiological stresses may be important and should be examined in inducing HIF and its downstream targets. Human cell lines containing these mutations have been instrumental in better understanding how HIF is regulated through the protein stabilization and protein synthesis. In particular, a number of renal cell lines deficient in VHL function exist. These parental cells, which harbor inactivated VHL, are frequently used in conjunction with cells that have wild-type VHL reintroduced, offering a tractable genetic system that can be used in addition to hypoxia to examine the induction of HIF or its targets.

2.3. HIF hydroxylation and other posttranslational modifications

Overexpression of a normoxically stabilized HIF-1α is another method to examine whether a gene of interest is an HIF target. HIF-1α is hydroxylated on two highly conserved proline residues, amino acids 402 or 564 (Chan *et al.*, 2005; Masson *et al.*, 2001). Mutation of proline 402 into an alanine or proline 564 into a glycine prevents the prolyl hydroxylases from modifying HIF-1α, resulting in stabilization of HIF-1α. A third highly conserved residue has also been shown to be modified by hydroxylation (Lando *et al.*, 2002b). Asparagine 803 is hydroxylated by factor inhibiting HIF (FIH) (Hewitson *et al.*, 2002; Lando *et al.*, 2002a; Mahon *et al.*, 2001; McNeill *et al.*, 2002). Hydroxylation of asparagine 803 prevents the necessary recruitment of coactivators, preventing HIF-1α from trans-activating (Lando *et al.*, 2002b). S-nitrosylation of cysteine 800 of HIF-1α, likewise, has been shown to be critical in the recruitment of the p300/CREB-binding protein (CBP) coactivators to the transactivation domains of HIF-1α, resulting in increased transcriptional activity (Yasinska and Sumbayev, 2003). Whereas proline mutation of either amino acid 402 or 564 results in stabilization, mutation of residue 803 results in a constitutively transactivation-competent HIF-1.

2.4. HIF inhibitors

As hypoxia and HIF have specifically been implicated in inducing genes that drive malignant growth, an active area of research has been the search for HIF inhibitors. Several categories of HIF inhibitors have emerged, including those that target HIF synthesis, stabilization, and transactivation (Fig. 16.2 and Table 16.2) (Belozerov and Van Meir, 2005; Giaccia *et al.*, 2003; Melillo, 2006; Semenza, 2003; Sutphin *et al.*, 2004). For example, geldanamycin and its chemical derivative, 17–allylamino–17–demethoxygeldanamycin (17-AAG), decrease the HIF-1 level by affecting its interaction with Hsp90 and increasing the proteasomal degradation of HIF-1α (Isaacs *et al.*, 2002). Using a novel screen to identify small molecule compounds that inhibit HIF-1 transcriptional activity, Rapisarda *et al.* (2002) identified topotecan, a topoisomerase I inhibitor. Subsequently, NSC 644221, a topoisomerase II inhibitor, was also found to inhibit HIF (Creighton-Gutteridge *et al.*, 2007). High-throughput screens to identify small molecules that disrupt HIF-1 interaction with p300 or DNA binding discovered chemotin and echinomycin, respectively (Kong *et al.*, 2005;

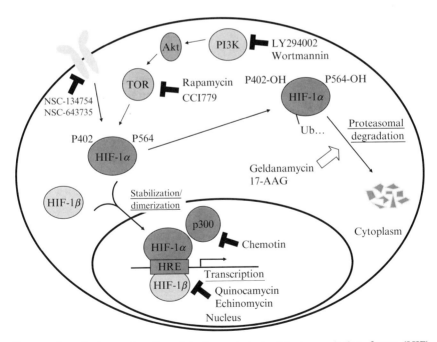

Figure 16.2 Pathways implicated in hypoxia-inducible transcription factor (HIF) inhibition. Several HIF inhibitors and their targets are shown in this schematic. (See color insert.)

Table 16.2 HIF Inhibitors

Target	Inhibitor name	References
Topoisomerase I	Topotecan	Rapisarda *et al.*, 2002
Topoisomerase II	NSC 644221	Creighton-Gutteridge *et al.*, 2007
Hsp90	Geldanamycin/ 17-AAG	Isaacs *et al.*, 2002
Microtubule stabilizer	Taxol	Mabjeesh *et al.*, 2003
	Vincristine	Mabjeesh *et al.*, 2003
Microtubule destabilizer	2ME	Mabjeesh *et al.*, 2003
Microtubule binding	Noscapine	Newcomb *et al.*, 2006
Thioredoxin reductase	PX-12, PX-478, Pleurotin	Welsh *et al.*, 2002, 2003, 2004
Soluble guanylyl cycle	YC-1	Chun *et al.*, 2001
Farnesyltransferase	SCH66336	Han *et al.*, 2005
ATPase	Oligomycin	Gong and Agani, 2005
Cyclooxygenase-2	NS398	Zhong *et al.*, 2004
c-Kit	Imatinib	Litz and Krystal, 2006
Proteasome	Bortezomib	Birle and Hedley, 2007
p300	Chemotin	Kung *et al.*, 2004
MEKK	PD98059	Berra *et al.*, 2000; Sodhi *et al.*, 2001
PI3K	LY294002, Wortmannin	Blancher *et al.*, 2001; Jiang *et al.*, 2001; Mazure *et al.*, 1997; Zhong *et al.*, 2000; Zundel *et al.*, 2000
TOR/protein translation	Rapamycin/ CCI779	Hudson *et al.*, 2002; Treins *et al.*, 2002; Zhong *et al.*, 2000
	103D5R	Tan *et al.*, 2005
Insulin-like growth factor	NSC 134754	Chau *et al.*, 2005
	NSC 643735	Chau *et al.*, 2005
Mitochondrial electron transport chain	Antimycin A1	Maeda *et al.*, 2006
Sp1	Nelfinavir	Pore *et al.*, 2006
Histone deacetylase	HDACi	Fath *et al.*, 2006; Kong *et al.*, 2006
HRE transcriptional activity	Quinocamycin	Rapisarda *et al.*, 2002
	Echinomycin	Kong *et al.*, 2005

Kung *et al.*, 2000, 2004). However, the clinical effectiveness of all of these compounds remains to be seen and is currently being followed.

2.5. RNA interference

To date, direct mutation of the conserved hydroxylation residues or other amino acids within HIF-1 have not been identified in clinical samples. Thus, disruption of upstream regulators, such as the PHDs, can provide yet another method for stabilizing HIF-1α under normoxic conditions. Of the three identified HIF prolyl hydroxylases, PHD2 has been reported to be the key oxygen sensor under normoxic conditions, as silencing PHD2 is sufficient for normoxic stabilization of HIF-1α (Berra *et al.*, 2003). Silencing of PHD2 can be achieved through the use of RNA interference (RNAi) by several different methods, such as antisense, short interfering oligos, or stable short hairpins. Small interfering RNAs (siRNAs) are 21- to 25-nucleotides in length, complementary to a target RNA sequence. When introduced into a cell, siRNAs hybridize to their target sequence and interact with the RNA-induced silencing complex (RISC), resulting in cleavage, degradation, and subsequent silencing of the target RNA.

Silencing RNA can be chemically synthesized, transcribed *in vitro*, expressed from a vector, or made by polymerase chain reaction (PCR). To identify siRNA target sequences, several design algorithms exist; however, the basic principle of a 21- to 25-nucleotide-long, double-stranded RNA remains constant. For transient assays, antisense or introduction of short interfering oligos is sufficient. Stable short hairpins (shRNAs) offer the advantage of sustained knockdown and, unlike short interfering oligos, are plasmid-based and therefore a renewable resource. Furthermore, several variations on shRNA vectors exist, including retroviral-based expression for cells that are difficult to transfect or inducible systems to temporally reduce gene expression.

RNA interference represents a powerful technology to study gene function. However, RNAi requires several key controls to ensure specificity of RNA silencing. Blast searches should be performed against the targeting sequence to remove the possibility of knocking down other genes. Scrambled targeting sequences as well as sequences targeting irrelevant genes should also be used to ensure knockdown specificity. Employing multiple sequences targeting different regions of the gene of interest is a standard control for specificity. If two independent constructs targeting different regions are both effective, then it is unlikely that they both have the same off-target effect that could account for it. Furthermore, rescue of the gene of interest by overexpression of a construct with silent mutations to the targeted region can also be used to validate specificity of knockdown.

RNA interference can also be used to silence HIF directly to determine whether a gene of interest is an HIF target. However, HIF represents a family of transcription factors (Semenza *et al.*, 1991; Wang and Semenza, 1995). Sharing a high level of protein homology and recognizing identical hypoxia response element (HRE) sequences, HIF-1 and HIF-2 differ primarily in the tissue expression pattern, with HIF-1 expressed ubiquitously and HIF-2 more restricted in its tissue expression pattern (Park *et al.*, 2003; Tian *et al.*, 1997; Wiesener *et al.*, 1998). A number of groups have attempted to identify HIF-1 versus HIF-2 targets, finding that the two have distinct, non-redundant targets. Furthermore, HIF-1α appears to play a more essential role in glycolysis than HIF-2α (Bracken *et al.*, 2006; Hu *et al.*, 2003; Wang *et al.*, 2005; Warnecke *et al.*, 2004). HIF-3α and its splice variant IPAS appear to be part of a negative feedback loop to regulate the expression of HIF (Gu *et al.*, 1998; Makino *et al.*, 2001, 2002). Depending on the tissue expression of HIF-1α and HIF-2α, knockdown of either gene is sometimes sufficient to deplete the HIF response and is useful if examining differences between HIF-1 and HIF-2. Alternatively, using siRNA to selectively degrade HIF-1β/aryl hydrocarbon receptor nuclear translocator (ARNT) represents a viable option to eliminate the HIF response.

2.6. HIF knockout mice

A more defined genetic model to examine the HIF response is the use of the HIF knockout mice. Several different knockouts for HIF-1α, HIF-2α, and HIF-1β were created with variable phenotypes depending on their genetic background (Carmeliet *et al.*, 1998; Iyer *et al.*, 1998; Maltepe *et al.*, 1997; Ryan *et al.*, 1998; Scortegagna *et al.*, 2003; Tian *et al.*, 1998). HIF-1α null mice are embryonically lethal at day E9.5 due to defects in vascular formation and central nervous system development (Iyer *et al.*, 1998; Ryan *et al.*, 1998). HIF-1β null mice are also embryonically lethal due to defects in placental differentiation and the neural tube (Maltepe *et al.*, 1997). Although HIF-2α null mice are viable in some genetic backgrounds, they have defects in catecholamine synthesis, vascular defects, lung surfactant production, and mitochondria homeostasis (Scortegagna *et al.*, 2003; Tian *et al.*, 1998). Heterozygous HIF-1α mice can be crossed with conditional cre mice to examine the role of HIF-1α in specific tissues. Whereas knockdown of a given target RNA is not complete, knockout mice allow for a completely null genetic background to examine whether HIF is involved in the regulation of a particular gene.

2.7. Somatic cell HIF knockout

Targeted homologous recombination of human somatic cells offers the additional advantage of examining the role of HIF in human cells in a particular tissue or cell line. Dang *et al.* (2006) used homologous recombination to delete HIF-1α in two colon cancer cell lines, HCT116 and RKO. Under hypoxic conditions, cellular proliferation and cell survival were impaired in both null cell lines. When the HCT116 HIF-1 null cells were implanted as xenograft tumors in nude mice, tumor growth was also impaired. In contrast, the RKO HIF-1 null cells displayed no difference in tumor growth, suggesting that the RKO tumor growth is HIF-independent. These somatic cell knock-outs of HIF-1 offer the investigator the ability to study the effects of either HIF-1 or HIF-2 loss of function. However, one needs to be wary of interpreting these results as the loss of HIF-1 or HIF-2 in the chronic condition can lead to cell adaptation and could be quite different than the acute loss of HIF-1 or HIF-2 where cells are not given the chance to adapt.

2.8. HIF activity (reporter assays)

HIF activity can be measured *in vitro* through the use of a reporter gene. In these assays, a luciferase gene, fluorescent gene, or β-galactosidase is placed downstream of tandem copies of a hypoxia responsive element and transfected into cells of interest. The next day, the cells are treated with an agent to stimulate HIF, such as hypoxia or DFO, and the cells are analyzed for reporter activity through quantitation of luciferase fluorescence or β-galactosidase activity. Each condition is performed in triplicate.

2.9. HIF responsive elements

Similarly, to investigate whether HIF is driving the transcription of a specific gene, a commonly used technique is to analyze the promoter region of the gene of interest. The promoter sequence can be directly scanned for canonical HREs. In parallel, the promoter region can be cloned upstream of the reporter gene and assayed for hypoxia responsiveness. Dissection of the promoter through deletion and mutation of the suspected HIF binding sites can determine whether the gene is dependent on HIF for transcription.

2.10. RNA analysis

Northern blot analysis is used to detect the presence, abundance, and turnover of a specific RNA transcript within cells. Total cellular RNA is isolated from tissue or cells. The RNA is denatured to prevent hydrogen bonding between base pairs, which allows the linear, unfolded RNA to be separated by gel electrophoresis according to size. The separate RNA is then transferred to a membrane of either nitrocellulose or nylon, which can then be hybridized by a specific DNA probe.

2.11. Quantitative RT-PCR

RNA from cells exposed to hypoxia is harvested in TRIzol (Invitrogen, Carlsbad, CA) according to manufacturer's instructions. One to two micrograms of total RNA is reverse transcribed (RT) with Moloney Murine Leukemia Virus MMLV reverse transcriptase (Invitrogen) with 5 μM-random primers (Invitrogen) according to the manufacturer's instructions. Approximately 0.5% of each RT reaction is added to quantitative RT (qRT)-PCR reactions containing the following in a total volume of 10 μl: 5 μl 2× SYBR Green master mix (ABI, Foster City, CA) and 200 nM forward and reverse primers specific for the genes of interest. Detection and data analysis is carried out with the ABI PRISM 7900 sequence detection system using 18S rRNA as an internal control. PCR primer sequences are obtained from the Primer Bank database (http://pga.mgh.harvard.edu/primerbank/) and synthesized in the Stanford Protein and Nucleic Acid Biotechnology Facility. Primers are tested against pooled complementary DNA (cDNA) samples and analyzed by agarose electrophoresis to verify formation of a single band after 40 rounds of PCR.

2.12. Western blot analysis

Western blot analysis is used to detect the presence and abundance of HIF-1α protein within a cell. Cells are lysed with urea buffer (9 M urea, 75 mM Tris-Cl (pH 7.5), 150 mM β-mercaptoethanol), and the cellular protein concentration is determined by Bradford protein assay. Equal amounts of whole-cell extract (generally, 50–100 μg) are separated based on charge and size through sodium dodecyl sulfate polyacrylamide gel electrophoresis (SDS-PAGE). Proteins are then transferred to either a nitrocellulose or polyvinylidene difluoride membrane by electrophoresis. Nonspecific protein binding is blocked by incubation of the membrane in 5% nonfat dried milk in Tris-buffered saline (TBS) for 1 h at room temperature or overnight at 4°. The membrane is then incubated with a primary antibody, which specifically recognizes HIF-1α. Following a brief washing period in TBS-Tween to eliminate excess primary antibody, the membrane is then incubated with an enzyme-conjugated secondary antibody that specifically recognizes the HIF-1α antibody. The enzyme conjugated to the secondary antibody allows for subsequent detection by either fluorescence or chemiluminescence.

2.13. Chromatin immunoprecipitation

Chromatin immunoprecipitation (ChIP) can be used to determine whether HIF-1 is directly bound to the endogenous promoter of a gene of interest. Cells or tissue are fixed with formaldehyde, resulting in crosslinking HIF-1 protein to the chromatin where it is bound. The cells are then lysed, and the chromatin DNA is sonicated to fragment the DNA into short pieces (100–1000 base pairs). The HIF protein crosslinked to DNA is then immunoprecipitated with an antibody that specifically recognizes HIF. The protein-DNA complexes are de-crosslinked, releasing the DNA fragments. The identity of these DNA fragments are then determined by PCR, using specific primers to the promoter that HIF is hypothesized to bind. A whole-genome approach using DNA microarrays can also be used to determine which promoters HIF is bound to in ChIP on Chips.

More specifically, ChIP assays are performed as described by Krieg *et al.* (2004) with the following modifications. For cells exposed to hypoxia, cells are fixed with 1% formaldehyde within the chamber to avoid reoxygenation. Fixed and lysed cells are sonicated with eight 10-second bursts using a Sonix Vibra-cell 130 sonicator set to 90% power and equipped with a 3-mm tip. Sonicates are measured for protein content using bicinchoninic acid (BCA) reagent with bovine serum albumin (BSA) as a standard (Pierce, Rockford, IL). After preclearing, approximately 1 mg of protein extract is incubated with 15 μg of anti–HIF-1α antibody (BD 610959; BD Biosciences, San Jose, CA), anti-HIF-2α antibody (NB 100–132; Novus Biologicals, Littleton, CO), or nonspecific mouse immunoglobulin G (IgG) overnight prior to addition of protein A Sepharose slurry (Sigma, St. Louis, MO). Five percent of the sample (approximately 10 μg of DNA) from each immunoprecipitation is reserved for input controls. Immunoprecipitated complexes are washed, eluted, and de-crosslinked, as previously described (Krieg *et al.*, 2004). DNA is purified with QIAquick PCR Purification columns according to the manufacturer's instructions (QIAgen Sciences, MA). One to 2.5% of each IP is assayed by PCR using primers specific for a region of interest. Amplified DNA is separated on a 2.2% agarose gel in 1$\times$ TAE, stained with ethidium bromide and imaged using a GelDoc (BioRad, Hercules, CA). For semi-quantitative PCR, Imagequant (GE Health Care, Piscataway, NJ) is used to measure the intensity of amplified bands, and sample quantity is interpolated from a series of titrated inputs and normalized to IgG signals. ChIP PCR primers are designed with Primer3 (http://frodo.wi.mit.edu/cgi-bin/primer3/primer3_www.cgi) using default settings altered in the following manner: Human Mispriming Library, Optimal Tm = 63° (range 60–65°), and GC clamp = 1.

2.14. HEEBO microarray analysis

We have recently adopted the use of human exonic evidence-based oligonucleotide (HEEBO) oligo microarrays for the use of two-color microarray expression analysis. These arrays are available through the Stanford Functional Genomics Facility (http://www.microarray.org/sfgf/) and contain 44,544 isothermal oligomers corresponding to over 30,000 unique human genes and several alternately transcribed messages. A number of unique control oligos are also printed on the arrays to enable better sample normalization. Although we have just recently adopted this technology, we have been happy with the quality and reproducibility of the data generated thus far.

Sample preparation and microarray hybridization is performed using modified protocols from Pat Brown's laboratory (http://cmgm.stanford.edu/pbrown/protocols/index.htm). Thirty micrograms of total RNA is mixed with RNA doping controls specific for the Cy dye to be used for detection (Stanford Functional Genomics Facility). RNA is annealed to 4 μg anchored oligo-dT primer (Invitrogen, Carlsbad, CA) and 4 μg random primers (Invitrogen) by mixing on ice, heating to 70° for 10 min, and chilling on ice for 10 min. Samples are mixed with RT cocktail (50 mM Tris [pH 8.3], 75 mM KCl, 3 mM MgCl2, 10 mM dithiothreitol [dTT], 200 nM 2′-deoxyadenosine 5′-triphosphate [dATP], 200 nM 2′-deoxycytidine 5′-triphosphate [dCTP], 200 nM 2′-deoxyguanosine 5′-triphosphate [dGTP], 100 nM 2′-deoxythymidine 5′-triphosphate [dTTP] [Invitrogen, Carlsbad, CA], 100 nM amino–allyl–dUTP [Ambion, Foster City, CA], and 200 units of Superscript II [Invitrogen]) and incubated at 42° for 1 h (final volume is 30 μl). Reactions are boosted with another 200 units of RT and incubated for an additional hour. The reaction is heat–inactivated at 95° for 5 min and cooled to room temperature, and the sample RNA is hydrolyzed by adjusting pH to 13 (add 13 μl 1 N NaOH and 1 μl 0.5 M ethylenediaminetetraacetic acid [EDTA]) and incubating at 67° for 5 min. Reaction is neutralized by addition of 50 μl 1 M HEPES (pH 7.0). The resulting cDNA is purified using Minelute Reaction Cleanup Columns (PN# 28206, QIAGEN, Valencia, CA) and eluted in 10 μl sodium phosphate (pH 8.5). Elution is repeated and the cDNA concentrated to 5 μl in a speed vac.

RNA is coupled to monofunctional NHS ester Cy dyes corresponding to the respective doping control used. (We generally use Cy5 for the experimental sample and Cy3 for the reference sample.) Cy dyes are resuspended in 5 μl dimethyl sulfoxide (DMSO), mixed with the cDNA/phosphate solution, and incubated in the dark for 90 min at room temperature. Samples are bound with Minelute columns and eluted in 10 μl of 10 mM Tris-Cl (pH 8.5). Elution is repeated, and Cy5-labeled sample is mixed with its corresponding the Cy3-labeled sample. Combined samples are mixed with 9.35 μl 20× SSC, 1.65 μl 10% sodium dodecyl sulfate (SDS), and 20 μg each of human Cot1 DNA (Invitrogen, Carlsbad, CA), polyA RNA (P9403; Sigma-Aldrich, St. Louis, MO), and yeast transfer RNA (tRNA) (Invitrogen). Total volume should be 55 μl. The sample is heated at 100° for 2 min and centrifuged at 14,000 rpm for 5 to 10 min. While samples are in the final stages of preparation, HEEBO arrays are placed into humidified hybridization chambers and covered with 22 × 60 M-series lifter slips (Erie Scientific, Portsmouth, NH). The sample is pipetted under the slip covers, and the chambers are closed, sealed, and incubated at 65° for 14 to 18 h.

The following day, arrays are removed from the chambers and rapidly disassembled in 500 ml Wash 1 buffer (2× SSC, 0.03% SDS) preheated to 60°. Arrays are immediately transferred to a slide rack immersed in a separate container of preheated Wash 1B (2× SSC, 0.03% SDS). Each array should be disassembled and placed in Wash 1B separately. All arrays are then agitated for 1 min. Up to five arrays can be processed per 500 ml of buffer. Arrays are then lifted individually from the slide rack with forceps, dipped into 500 ml Wash 1C (2× SSC, room temperature), transferred to a clean slide rack immersed in 500 ml Wash 2 (1× SSC), and agitated for 1 to 2 min at room temperature. The slide rack is transferred to 500 ml Wash 3 (0.2times; SSC) and carried to a centrifuge equipped with a swinging bucket rotor that accommodates microplates. The rack is transferred to the rotor and centrifuged for 5 to 10 min at 100 × g to dry. Arrays are immediately scanned on an Axon 4000B scanner. Images are gridded and analyzed using GenePix 6.0 (Molecular Devices, Sunnyvale, CA). Images, results, and settings are loaded onto the Stanford Microarray Database for normalization and analysis.

3. Conclusions

At the center of the cellular response to hypoxia is the HIF family of transcription factors. Thus, targeting HIF and its downstream effectors in cancer therapy is a burgeoning area of research. The ability to screen for small molecule therapeutics that disrupt HIF dimerization or interaction with coactivators, such as p300, would provide a targeted approach at inhibiting HIF transcriptional activity. However, the identification of small molecules that inhibit protein to protein interactions have been difficult to find. Perhaps a more feasible approach would be to inhibit specific HIF target genes through small molecule inhibitors, such as echinomycin, that block HIF binding. Future studies into how to better exploit HIF and HIF-mediated gene expression may lead to more directed therapeutics.

ACKNOWLEDGMENTS

We apologize to those we may have inadvertently failed to acknowledge. This work is supported by the National Cancer Institute (CA123823 to D. A. C. and CA67166, CA082566, and CA116685 to A. J. G.) and the Canadian Institute of Health Research (S. T.).

REFERENCES

Bardos, J. I., and Ashcroft, M. (2004). Hypoxia-inducible factor-1 and oncogenic signaling. *Bioessays* **26,** 262–269.

Belozerov, V. E., and Van Meir, E. G. (2005). Hypoxia inducible factor-1: A novel target for cancer therapy. *Anticancer Drugs* **16,** 901–909.

Berra, E., Pages, G., and Pouyssegur, J. (2000). MAP kinases and hypoxia in the control of VEGF expression. *Cancer Metastasis Rev.* **19,** 139–145.

Berra, E., Benizri, E., Ginouves, A., Volmat, V., Roux, D., and Pouyssegur, J. (2003). HIF prolyl-hydroxylase 2 is the key oxygen sensor setting low steady-state levels of HIF-1α in normoxia. *EMBO J.* **22,** 4082–4090.

Bindra, R. S., and Glazer, P. M. (2005). Genetic instability and the tumor microenvironment: Towards the concept of microenvironment-induced mutagenesis. *Mutat. Res.* **569,** 75–85.

Bindra, R. S., and Glazer, P. M. (2007). Co-repression of mismatch repair gene expression by hypoxia in cancer cells: Role of the Myc/Max network. *Cancer Lett.* **252,** 93–103.

Birle, D. C., and Hedley, D. W. (2007). Suppression of the hypoxia-inducible factor-1 response in cervical carcinoma xenografts by proteasome inhibitors. *Cancer Res.* **67,** 1735–1743.

Blancher, C., Moore, J. W., Robertson, N., and Harris, A. L. (2001). Effects of ras and von Hippel-Lindau (VHL) gene mutations on hypoxia-inducible factor (HIF)-1α, HIF-2α, and vascular endothelial growth factor expression and their regulation by the phosphatidylinositol 3′-kinase/Akt signaling pathway. *Cancer Res.* **61,** 7349–7355.

Bracken, C. P., Fedele, A. O., Linke, S., Balrak, W., Lisy, K., Whitelaw, M. L., and Peet, D. J. (2006). Cell-specific regulation of hypoxia-inducible factor (HIF)-1α and HIF-2α stabilization and transactivation in a graded oxygen environment. *J. Biol. Chem.* **281,** 22575–22585.

Brugarolas, J. B., Vazquez, F., Reddy, A., Sellers, W. R., and Kaelin, W. G., Jr. (2003). TSC2 regulates VEGF through mTOR-dependent and -independent pathways. *Cancer Cell* **4,** 147–158.

Bruick, R. K., and McKnight, S. L. (2001). A conserved family of prolyl-4-hydroxylases that modify HIF. *Science* **294,** 1337–1340.

Carmeliet, P., Dor, Y., Herbert, J. M., Fukumura, D., Brusselmans, K., Dewerchin, M., Neeman, M., Bono, F., Abramovitch, R., Maxwell, P., Koch, C. J., Ratcliffe, P., *et al.* (1998). Role of HIF-1α in hypoxia-mediated apoptosis, cell proliferation and tumour angiogenesis. *Nature* **394,** 485–490.

Ceradini, D. J., Kulkarni, A. R., Callaghan, M. J., Tepper, O. M., Bastidas, N., Kleinman, M. E., Capla, J. M., Galiano, R. D., Levine, J. P., and Gurtner, G. C. (2004). Progenitor cell trafficking is regulated by hypoxic gradients through HIF-1 induction of SDF-1. *Nat. Med.* **10,** 858–864.

Chan, D. A., Sutphin, P. D., Denko, N. C., and Giaccia, A. J. (2002). Role of prolyl hydroxylation in oncogenically stabilized hypoxia-inducible factor-1alpha. *J. Biol. Chem.* **277,** 40112–40117.

Chan, D. A., Sutphin, P. D., Yen, S. E., and Giaccia, A. J. (2005). Coordinate regulation of the oxygen-dependent degradation domains of hypoxia-inducible factor 1 alpha. *Mol. Cell Biol.* **25,** 6415–6426.

Chandel, N. S., McClintock, D. S., Feliciano, C. E., Wood, T. M., Melendez, J. A., Rodriguez, A. M., and Schumacker, P. T. (2000). Reactive oxygen species generated at mitochondrial complex III stabilize hypoxia-inducible factor-1alpha during hypoxia: A mechanism of O_2 sensing. *J. Biol. Chem.* **275,** 25130–25138.

Chau, N. M., Rogers, P., Aherne, W., Carroll, V., Collins, I., McDonald, E., Workman, P., and Ashcroft, M. (2005). Identification of novel small molecule inhibitors of hypoxia-inducible factor-1 that differentially block hypoxia-inducible factor-1 activity and hypoxia-inducible factor-1alpha induction in response to hypoxic stress and growth factors. *Cancer Res.* **65**, 4918–4928.

Chen, C., Pore, N., Behrooz, A., Ismail-Beigi, F., and Maity, A. (2001). Regulation of glut1 mRNA by hypoxia-inducible factor-1. Interaction between H-ras and hypoxia. *J. Biol. Chem.* **276**, 9519–9525.

Chun, Y. S., Yeo, E. J., Choi, E., Teng, C. M., Bae, J. M., Kim, M. S., and Park, J. W. (2001). Inhibitory effect of YC-1 on the hypoxic induction of erythropoietin and vascular endothelial growth factor in Hep3B cells. *Biochem. Pharmacol.* **61**, 947–954.

Comerford, K. M., Wallace, T. J., Karhausen, J., Louis, N. A., Montalto, M. C., and Colgan, S. P. (2002). Hypoxia-inducible factor-1-dependent regulation of the multidrug resistance (MDR1) gene. *Cancer Res.* **62**, 3387–3394.

Creighton-Gutteridge, M., Cardellina, J. H., 2nd, Stephen, A. G., Rapisarda, A., Uranchimeg, B., Hite, K., Denny, W. A., Shoemaker, R. H., and Melillo, G. (2007). Cell type-specific, topoisomerase II-dependent inhibition of hypoxia-inducible factor-1alpha protein accumulation by NSC 644221. *Clin. Cancer Res.* **13**, 1010–1018.

Dang, D. T., Chen, F., Gardner, L. B., Cummins, J. M., Rago, C., Bunz, F., Kantsevoy, S. V., and Dang, L. H. (2006). Hypoxia-inducible factor-1alpha promotes nonhypoxia-mediated proliferation in colon cancer cells and xenografts. *Cancer Res.* **66**, 1684–1936.

Davidson, T. L., Chen, H., Di Toro, D. M., D'Angelo, G., and Costa, M. (2006). Soluble nickel inhibits HIF-prolyl-hydroxylases creating persistent hypoxic signaling in A549 cells. *Mol. Carcinog.* **45**, 479–489.

Denko, N. C., Fontana, L. A., Hudson, K. M., Sutphin, P. D., Raychaudhuri, S., Altman, R., and Giaccia, A. J. (2003). Investigating hypoxic tumor physiology through gene expression patterns. *Oncogene* **22**, 5907–5914.

Epstein, A. C., Gleadle, J. M., McNeill, L. A., Hewitson, K. S., O'Rourke, J., Mole, D. R., Mukherji, M., Metzen, E., Wilson, M. I., Dhanda, A., Tian, Y. M., Masson, N., *et al.* (2001). *C. elegans* EGL-9 and mammalian homologs define a family of dioxygenases that regulate HIF by prolyl hydroxylation. *Cell* **107**, 43–54.

Erler, J. T., Bennewith, K. L., Nicolau, M., Dornhofer, N., Kong, C., Le, Q. T., Chi, J. T., Jeffrey, S. S., and Giaccia, A. J. (2006). Lysyl oxidase is essential for hypoxia-induced metastasis. *Nature* **440**, 1222–1226.

Fath, D. M., Kong, X., Liang, D., Lin, Z., Chou, A., Jiang, Y., Fang, J., Caro, J., and Sang, N. (2006). Histone deacetylase inhibitors repress the transactivation potential of hypoxia-inducible factors independently of direct acetylation of HIF-α. *J. Biol. Chem.* **281**, 13612–13619.

Gao, N., Shen, L., Zhang, Z., Leonard, S. S., He, H., Zhang, X. G., Shi, X., and Jiang, B. H. (2004). Arsenite induces HIF-1α and VEGF through PI3K, Akt and reactive oxygen species in DU145 human prostate carcinoma cells. *Mol. Cell Biochem.* **255**, 33–45.

Gao, N., Ding, M., Zheng, J. Z., Zhang, Z., Leonard, S. S., Liu, K. J., Shi, X., and Jiang, B. H. (2002a). Vanadate-induced expression of hypoxia-inducible factor 1 alpha and vascular endothelial growth factor through phosphatidylinositol 3-kinase/Akt pathway and reactive oxygen species. *J. Biol. Chem.* **277**, 31963–31971.

Gao, N., Jiang, B. H., Leonard, S. S., Corum, L., Zhang, Z., Roberts, J. R., Antonini, J., Zheng, J. Z., Flynn, D. C., Castranova, V., and Shi, X. (2002b). p38 signaling-mediated hypoxia-inducible factor 1alpha and vascular endothelial growth factor induction by Cr (VI) in DU145 human prostate carcinoma cells. *J. Biol. Chem.* **277**, 45041–45048.

Giaccia, A., Siim, B. G., and Johnson, R. S. (2003). HIF-1 as a target for drug development. *Nat. Rev. Drug Discov.* **2**, 803–811.

Gong, Y., and Agani, F. H. (2005). Oligomycin inhibits HIF-1α expression in hypoxic tumor cells. *Am. J. Physiol. Cell Physiol.* **288**, C1023–C1029.

Gu, Y. Z., Moran, S. M., Hogenesch, J. B., Wartman, L., and Bradfield, C. A. (1998). Molecular characterization and chromosomal localization of a third alpha-class hypoxia inducible factor subunit, HIF-3α. *Gene Expr.* **7,** 205–213.

Han, J. Y., Oh, S. H., Morgillo, F., Myers, J. N., Kim, E., Hong, W. K., and Lee, H. Y. (2005). Hypoxia-inducible factor 1alpha and antiangiogenic activity of farnesyltransferase inhibitor SCH66336 in human aerodigestive tract cancer. *J. Natl. Cancer Inst.* **97,** 1272–1286.

Hewitson, K. S., McNeill, L. A., Riordan, M. V., Tian, Y. M., Bullock, A. N., Welford, R. W., Elkins, J. M., Oldham, N. J., Bhattacharya, S., Gleadle, J. M., Ratcliffe, P. J., Pugh, C. W., *et al.* (2002). Hypoxia-inducible factor (HIF) asparagine hydroxylase is identical to factor inhibiting HIF (FIH) and is related to the cupin structural family. *J. Biol. Chem.* **277,** 26351–26355.

Hu, C. J., Wang, L. Y., Chodosh, L. A., Keith, B., and Simon, M. C. (2003). Differential roles of hypoxia-inducible factor 1alpha (HIF-1α) and HIF-2α in hypoxic gene regulation. *Mol. Cell Biol.* **23,** 9361–9374.

Huang, L. E., Bindra, R. S., Glazer, P. M., and Harris, A. L. (2007). Hypoxia-induced genetic instability–a calculated mechanism underlying tumor progression. *J. Mol. Med.* **85,** 139–148.

Hudson, C. C., Liu, M., Chiang, G. G., Otterness, D. M., Loomis, D. C., Kaper, F., Giaccia, A. J., and Abraham, R. T. (2002). Regulation of hypoxia-inducible factor 1alpha expression and function by the mammalian target of rapamycin. *Mol. Cell Biol.* **22,** 7004–7014.

Isaacs, J. S., Jung, Y. J., Mimnaugh, E. G., Martinez, A., Cuttitta, F., and Neckers, L. M. (2002). Hsp90 regulates a von Hippel Lindau-independent hypoxia-inducible factor-1 alpha-degradative pathway. *J. Biol. Chem.* **277,** 29936–29944.

Ivan, M., Kondo, K., Yang, H., Kim, W., Valiando, J., Ohh, M., Salic, A., Asara, J. M., Lane, W. S., and Kaelin, W. G., Jr. (2001). HIFα targeted for VHL-mediated destruction by proline hydroxylation: Implications for O_2 sensing. *Science* **292,** 464–468.

Iyer, N. V., Kotch, L. E., Agani, F., Leung, S. W., Laughner, E., Wenger, R. H., Gassmann, M., Gearhart, J. D., Lawler, A. M., Yu, A. Y., and Semenza, G. L. (1998). Cellular and developmental control of O_2 homeostasis by hypoxia-inducible factor 1 alpha. *Genes Dev.* **12,** 149–162.

Jaakkola, P., Mole, D. R., Tian, Y. M., Wilson, M. I., Gielbert, J., Gaskell, S. J., Kriegsheim, A., Hebestreit, H. F., Mukherji, M., Schofield, C. J., Maxwell, P. H., Pugh, C. W., *et al.* (2001). Targeting of HIF-α to the von Hippel-Lindau ubiquitylation complex by O_2-regulated prolyl hydroxylation. *Science* **292,** 468–472.

Jiang, B. H., Agani, F., Passaniti, A., and Semenza, G. L. (1997). V-SRC induces expression of hypoxia-inducible factor 1 (HIF-1) and transcription of genes encoding vascular endothelial growth factor and enolase 1: Involvement of HIF-1 in tumor progression. *Cancer Res.* **57,** 5328–5335.

Jiang, B. H., Jiang, G., Zheng, J. Z., Lu, Z., Hunter, T., and Vogt, P. K. (2001). Phosphatidylinositol 3-kinase signaling controls levels of hypoxia-inducible factor 1. *Cell Growth Differ.* **12,** 363–369.

Katschinski, D. M., Le, L., Heinrich, D., Wagner, K. F., Hofer, T., Schindler, S. G., and Wenger, R. H. (2002). Heat induction of the unphosphorylated form of hypoxia-inducible factor-1alpha is dependent on heat shock protein-90 activity. *J. Biol. Chem.* **277,** 9262–9267.

Kong, D., Park, E. J., Stephen, A. G., Calvani, M., Cardellina, J. H., Monks, A., Fisher, R. J., Shoemaker, R. H., and Melillo, G. (2005). Echinomycin, a small-molecule inhibitor of hypoxia-inducible factor-1 DNA-binding activity. *Cancer Res.* **65,** 9047–9055.

Kong, X., Lin, Z., Liang, D., Fath, D., Sang, N., and Caro, J. (2006). Histone deacetylase inhibitors induce VHL and ubiquitin-independent proteasomal degradation of hypoxia-inducible factor 1alpha. *Mol. Cell Biol.* **26,** 2019–2028.

Koong, A. C., Denko, N. C., Hudson, K. M., Schindler, C., Swiersz, L., Koch, C., Evans, S., Ibrahim, H., Le, Q. T., Terris, D. J., and Giaccia, A. J. (2000). Candidate genes for the hypoxic tumor phenotype. *Cancer Res.* **60,** 883–887.

Koshiji, M., To, K. K., Hammer, S., Kumamoto, K., Harris, A. L., Modrich, P., and Huang, L. E. (2005). HIF-1α induces genetic instability by transcriptionally down-regulating MutSα expression. *Mol. Cell* **17,** 793–803.

Krieg, A. J., Krieg, S. A., Ahn, B. S., and Shapiro, D. J. (2004). Interplay between estrogen response element sequence and ligands controls *in vivo* binding of estrogen receptor to regulated genes. *J. Biol. Chem.* **279,** 5025–5034.

Kung, A. L., Wang, S., Klco, J. M., Kaelin, W. G., and Livingston, D. M. (2000). Suppression of tumor growth through disruption of hypoxia-inducible transcription. *Nat. Med.* **6,** 1335–1340.

Kung, A. L., Zabludoff, S. D., France, D. S., Freedman, S. J., Tanner, E. A., Vieira, A., Cornell-Kennon, S., Lee, J., Wang, B., Wang, J., Memmert, K., Naegeli, H. U., *et al.* (2004). Small molecule blockade of transcriptional coactivation of the hypoxia-inducible factor pathway. *Cancer Cell* **6,** 33–43.

Lal, A., Peters, H., St Croix, B., Haroon, Z. A., Dewhirst, M. W., Strausberg, R. L., Kaanders, J. H., van der Kogel, A. J., and Riggins, G. J. (2001). Transcriptional response to hypoxia in human tumors. *J. Natl. Cancer Inst.* **93,** 1337–1343.

Lando, D., Peet, D. J., Gorman, J. J., Whelan, D. A., Whitelaw, M. L., and Bruick, R. K. (2002a). FIH-1 is an asparaginyl hydroxylase enzyme that regulates the transcriptional activity of hypoxia-inducible factor. *Genes Dev.* **16,** 1466–1471.

Lando, D., Peet, D. J., Whelan, D. A., Gorman, J. J., and Whitelaw, M. L. (2002b). Asparagine hydroxylation of the HIF transactivation domain: a hypoxic switch. *Science* **295,** 858–861.

Li, J., Davidson, G., Huang, Y., Jiang, B. H., Shi, X., Costa, M., and Huang, C. (2004). Nickel compounds act through phosphatidylinositol-3-kinase/Akt-dependent, p70 (S6k)-independent pathway to induce hypoxia inducible factor transactivation and Cap43 expression in mouse epidermal Cl41 cells. *Cancer Res.* **64,** 94–101.

Liao, D., Corle, C., Seagroves, T. N., and Johnson, R. S. (2007). Hypoxia-inducible factor-1alpha is a key regulator of metastasis in a transgenic model of cancer initiation and progression. *Cancer Res.* **67,** 563–572.

Litz, J., and Krystal, G. W. (2006). Imatinib inhibits c-Kit-induced hypoxia-inducible factor-1alpha activity and vascular endothelial growth factor expression in small cell lung cancer cells. *Mol. Cancer Ther.* **5,** 1415–1422.

Liu, M. Y., Poellinger, L., and Walker, C. L. (2003). Up-regulation of hypoxia-inducible factor 2alpha in renal cell carcinoma associated with loss of Tsc-2 tumor suppressor gene. *Cancer Res.* **63,** 2675–2680.

Mabjeesh, N. J., Escuin, D., LaVallee, T. M., Pribluda, V. S., Swartz, G. M., Johnson, M. S., Willard, M. T., Zhong, H., Simons, J. W., and Giannakakou, P. (2003). 2ME2 inhibits tumor growth and angiogenesis by disrupting microtubules and dysregulating HIF. *Cancer Cell* **3,** 363–375.

Maeda, M., Hasebe, Y., Egawa, K., Shibanuma, M., and Nose, K. (2006). Inhibition of angiogenesis and HIF-1α activity by antimycin A1. *Biol. Pharm. Bull.* **29,** 1344–1348.

Mahon, P. C., Hirota, K., and Semenza, G. L. (2001). FIH-1: A novel protein that interacts with HIF-1α and VHL to mediate repression of HIF-1 transcriptional activity. *Genes Dev.* **15,** 2675–2686.

Makino, Y., Kanopka, A., Wilson, W. J., Tanaka, H., and Poellinger, L. (2002). Inhibitory PAS domain protein (IPAS) is a hypoxia-inducible splicing variant of the hypoxia-inducible factor-3alpha locus. *J. Biol. Chem.* **277,** 32405–32408.

Makino, Y., Cao, R., Svensson, K., Bertilsson, G., Asman, M., Tanaka, H., Cao, Y., Berkenstam, A., and Poellinger, L. (2001). Inhibitory PAS domain protein is a negative regulator of hypoxia-inducible gene expression. *Nature* **414,** 550–554.

Maltepe, E., Schmidt, J. V., Baunoch, D., Bradfield, C. A., and Simon, M. C. (1997). Abnormal angiogenesis and responses to glucose and oxygen deprivation in mice lacking the protein ARNT. *Nature* **386,** 403–407.

Masson, N., Willam, C., Maxwell, P. H., Pugh, C. W., and Ratcliffe, P. J. (2001). Independent function of two destruction domains in hypoxia-inducible factor-alpha chains activated by prolyl hydroxylation. *EMBO J.* **20,** 5197–5206.

Maxwell, P. H., Wiesener, M. S., Chang, G. W., Clifford, S. C., Vaux, E. C., Cockman, M. E., Wykoff, C. C., Pugh, C. W., Maher, E. R., and Ratcliffe, P. J. (1999). The tumour suppressor protein VHL targets hypoxia-inducible factors for oxygen-dependent proteolysis. *Nature* **399,** 271–275.

Mazure, N. M., Chen, E. Y., Laderoute, K. R., and Giaccia, A. J. (1997). Induction of vascular endothelial growth factor by hypoxia is modulated by a phosphatidylinositol 3-kinase/Akt signaling pathway in Ha-ras-transformed cells through a hypoxia inducible factor-1 transcriptional element. *Blood* **90,** 3322–3331.

Mazure, N. M., Chen, E. Y., Yeh, P., Laderoute, K. R., and Giaccia, A. J. (1996). Oncogenic transformation and hypoxia synergistically act to modulate vascular endothelial growth factor expression. *Cancer Res.* **56,** 3436–3440.

McNeill, L. A., Hewitson, K. S., Claridge, T. D., Seibel, J. F., Horsfall, L. E., and Schofield, C. J. (2002). Hypoxia-inducible factor asparaginyl hydroxylase (FIH-1) catalyses hydroxylation at the β-carbon of asparagine-803. *Biochem. J.* **367,** 571–575.

Melillo, G. (2006). Inhibiting hypoxia-inducible factor 1 for cancer therapy. *Mol. Cancer Res.* **4,** 601–605.

Minet, E., Michel, G., Mottet, D., Raes, M., and Michiels, C. (2001). Transduction pathways involved in hypoxia-inducible factor-1 phosphorylation and activation. *Free Radic. Biol. Med.* **31,** 847–855.

Mottet, D., Dumont, V., Deccache, Y., Demazy, C., Ninane, N., Raes, M., and Michiels, C. (2003). Regulation of hypoxia-inducible factor-1alpha protein level during hypoxic conditions by the phosphatidylinositol 3-kinase/Akt/glycogen synthase kinase 3β pathway in HepG2 cells. *J. Biol. Chem.* **278,** 31277–31285.

Newcomb, E. W., Lukyanov, Y., Schnee, T., Ali, M. A., Lan, L., and Zagzag, D. (2006). Noscapine inhibits hypoxia-mediated HIF-1alpha expression and angiogenesis *in vitro*: A novel function for an old drug. *Int. J. Oncol.* **28,** 1121–1130.

Ohh, M., Park, C. W., Ivan, M., Hoffman, M. A., Kim, T. Y., Huang, L. E., Pavletich, N., Chau, V., and Kaelin, W. G. (2000). Ubiquitination of hypoxia-inducible factor requires direct binding to the beta-domain of the von Hippel-Lindau protein. *Nat. Cell Biol.* **2,** 423–427.

Park, S. K., Dadak, A. M., Haase, V. H., Fontana, L., Giaccia, A. J., and Johnson, R. S. (2003). Hypoxia-induced gene expression occurs solely through the action of hypoxia-inducible factor 1alpha (HIF-1α): Role of cytoplasmic trapping of HIF-2α. *Mol. Cell Biol.* **23,** 4959–4971.

Pore, N., Gupta, A. K., Cerniglia, G. J., Jiang, Z., Bernhard, E. J., Evans, S. M., Koch, C. J., Hahn, S. M., and Maity, A. (2006). Nelfinavir down-regulates hypoxia-inducible factor 1alpha and VEGF expression and increases tumor oxygenation: Implications for radiotherapy. *Cancer Res.* **66,** 9252–9259.

Rak, J., Mitsuhashi, Y., Bayko, L., Filmus, J., Shirasawa, S., Sasazuki, T., and Kerbel, R. S. (1995). Mutant ras oncogenes upregulate VEGF/VPF expression: Implications for induction and inhibition of tumor angiogenesis. *Cancer Res.* **55**, 4575–4580.

Rapisarda, A., Uranchimeg, B., Scudiero, D. A., Selby, M., Sausville, E. A., Shoemaker, R. H., and Melillo, G. (2002). Identification of small molecule inhibitors of hypoxia-inducible factor 1 transcriptional activation pathway. *Cancer Res.* **62**, 4316–4324.

Richard, D. E., Berra, E., and Pouyssegur, J. (2000). Nonhypoxic pathway mediates the induction of hypoxia-inducible factor 1alpha in vascular smooth muscle cells. *J. Biol. Chem.* **275**, 26765–26771.

Richard, D. E., Berra, E., Gothie, E., Roux, D., and Pouyssegur, J. (1999). p42/p44 mitogen-activated protein kinases phosphorylate hypoxia-inducible factor 1alpha (HIF-1α) and enhance the transcriptional activity of HIF-1. *J. Biol. Chem.* **274**, 32631–32637.

Ryan, H. E., Lo, J., and Johnson, R. S. (1998). HIF-1α is required for solid tumor formation and embryonic vascularization. *EMBO J.* **17**, 3005–3015.

Salceda, S., and Caro, J. (1997). Hypoxia-inducible factor 1alpha (HIF-1α) protein is rapidly degraded by the ubiquitin-proteasome system under normoxic conditions. Its stabilization by hypoxia depends on redox-induced changes. *J. Biol. Chem.* **272**, 22642–22647.

Sandau, K. B., Fandrey, J., and Brune, B. (2001). Accumulation of HIF-1α under the influence of nitric oxide. *Blood* **97**, 1009–1015.

Scandurro, A. B., Weldon, C. W., Figueroa, Y. G., Alam, J., and Beckman, B. S. (2001). Gene microarray analysis reveals a novel hypoxia signal transduction pathway in human hepatocellular carcinoma cells. *Int. J. Oncol.* **19**, 129–135.

Scortegagna, M., Ding, K., Oktay, Y., Gaur, A., Thurmond, F., Yan, L. J., Marck, B. T., Matsumoto, A. M., Shelton, J. M., Richardson, J. A., Bennett, M. J., and Garcia, J. A. (2003). Multiple organ pathology, metabolic abnormalities and impaired homeostasis of reactive oxygen species in Epas1$^{-/-}$ mice. *Nat. Genet.* **35**, 331–340.

Semenza, G. L. (2003). Targeting HIF-1 for cancer therapy. *Nat. Rev. Cancer* **3**, 721–732.

Semenza, G. L., Nejfelt, M. K., Chi, S. M., and Antonarakis, S. E. (1991). Hypoxia-inducible nuclear factors bind to an enhancer element located 3′ to the human erythropoietin gene. *Proc. Natl. Acad. Sci. USA* **88**, 5680–5684.

Skinner, H. D., Zhong, X. S., Gao, N., Shi, X., and Jiang, B. H. (2004). Arsenite induces p70S6K1 activation and HIF-1α expression in prostate cancer cells. *Mol. Cell Biochem.* **255**, 19–23.

Sodhi, A., Montaner, S., Miyazaki, H., and Gutkind, J. S. (2001). MAPK and Akt act cooperatively but independently on hypoxia inducible factor-1alpha in rasV12 upregulation of VEGF. *Biochem. Biophys. Res. Commun.* **287**, 292–300.

Srinivas, V., Zhang, L. P., Zhu, X. H., and Caro, J. (1999). Characterization of an oxygen/redox-dependent degradation domain of hypoxia-inducible factor alpha (HIF-α) proteins. *Biochem. Biophys. Res. Commun.* **260**, 557–561.

Staller, P., Sulitkova, J., Lisztwan, J., Moch, H., Oakeley, E. J., and Krek, W. (2003). Chemokine receptor CXCR4 downregulated by von Hippel-Lindau tumour suppressor pVHL. *Nature* **425**, 307–311.

Sugawara, J., Tazuke, S. I., Suen, L. F., Powell, D. R., Kaper, F., Giaccia, A. J., and Giudice, L. C. (2000). Regulation of insulin-like growth factor-binding protein 1 by hypoxia and 3′, 5′-cyclic adenosine monophosphate is additive in HepG2 cells. *J. Clin. Endocrinol. Metab.* **85**, 3821–3827.

Sutphin, P. D., Chan, D. A., and Giaccia, A. J. (2004). Dead cells don't form tumors: HIF-dependent cytotoxins. *Cell Cycle* **3**, 160–163.

Tan, C., de Noronha, R. G., Roecker, A. J., Pyrzynska, B., Khwaja, F., Zhang, Z., Zhang, H., Teng, Q., Nicholson, A. C., Giannakakou, P., Zhou, W., Olson, J. J., *et al.* (2005). Identification of a novel small-molecule inhibitor of the hypoxia-inducible factor 1 pathway. *Cancer Res.* **65,** 605–612.

Tian, H., McKnight, S. L., and Russell, D. W. (1997). Endothelial PAS domain protein 1 (EPAS1), a transcription factor selectively expressed in endothelial cells. *Genes Dev.* **11,** 72–82.

Tian, H., Hammer, R. E., Matsumoto, A. M., Russell, D. W., and McKnight, S. L. (1998). The hypoxia-responsive transcription factor EPAS1 is essential for catecholamine homeostasis and protection against heart failure during embryonic development. *Genes Dev.* **12,** 3320–3324.

Treins, C., Giorgetti-Peraldi, S., Murdaca, J., Semenza, G. L., and Van Obberghen, E. (2002). Insulin stimulates hypoxia-inducible factor 1 through a phosphatidylinositol 3-kinase/target of rapamycin-dependent signaling pathway. *J. Biol. Chem.* **277,** 27975–27981.

Wang, G. L., and Semenza, G. L. (1995). Purification and characterization of hypoxia-inducible factor 1. *J. Biol. Chem.* **270,** 1230–1237.

Wang, V., Davis, D. A., Haque, M., Huang, L. E., and Yarchoan, R. (2005). Differential gene up-regulation by hypoxia-inducible factor-1alpha and hypoxia-inducible factor-2alpha in HEK293T cells. *Cancer Res.* **65,** 3299–3306.

Warnecke, C., Zaborowska, Z., Kurreck, J., Erdmann, V. A., Frei, U., Wiesener, M., and Eckardt, K. U. (2004). Differentiating the functional role of hypoxia-inducible factor (HIF)-1alpha and HIF-2α (EPAS-1) by the use of RNA interference: Erythropoietin is a HIF-2α target gene in Hep3B and Kelly cells. *FASEB J.* **18,** 1462–1464.

Wartenberg, M., Ling, F. C., Muschen, M., Klein, F., Acker, H., Gassmann, M., Petrat, K., Putz, V., Hescheler, J., and Sauer, H. (2003). Regulation of the multidrug resistance transporter P-glycoprotein in multicellular tumor spheroids by hypoxia-inducible factor (HIF-1) and reactive oxygen species. *FASEB J.* **17,** 503–505.

Welford, S. M., Bedogni, B., Gradin, K., Poellinger, L., Broome Powell, M., and Giaccia, A. J. (2006). HIF1α delays premature senescence through the activation of MIF. *Genes Dev.* **20,** 3366–3371.

Welsh, S., Williams, R., Kirkpatrick, L., Paine-Murrieta, G., and Powis, G. (2004). Antitumor activity and pharmacodynamic properties of PX-478, an inhibitor of hypoxia-inducible factor-1alpha. *Mol. Cancer Ther.* **3,** 233–244.

Welsh, S. J., Bellamy, W. T., Briehl, M. M., and Powis, G. (2002). The redox protein thioredoxin-1 (Trx-1) increases hypoxia-inducible factor 1alpha protein expression: Trx-1 overexpression results in increased vascular endothelial growth factor production and enhanced tumor angiogenesis. *Cancer Res.* **62,** 5089–5095.

Welsh, S. J., Williams, R. R., Birmingham, A., Newman, D. J., Kirkpatrick, D. L., and Powis, G. (2003). The thioredoxin redox inhibitors 1-methylpropyl 2-imidazolyl disulfide and pleurotin inhibit hypoxia-induced factor 1alpha and vascular endothelial growth factor formation. *Mol. Cancer Ther.* **2,** 235–243.

Wiesener, M. S., Turley, H., Allen, W. E., Willam, C., Eckardt, K. U., Talks, K. L., Wood, S. M., Gatter, K. C., Harris, A. L., Pugh, C. W., Ratcliffe, P. J., and Maxwell, P. H. (1998). Induction of endothelial PAS domain protein-1 by hypoxia: Characterization and comparison with hypoxia-inducible factor-1alpha. *Blood* **92,** 2260–2268.

Wykoff, C. C., Pugh, C. W., Maxwell, P. H., Harris, A. L., and Ratcliffe, P. J. (2000). Identification of novel hypoxia dependent and independent target genes of the von Hippel-Lindau (VHL) tumour suppressor by mRNA differential expression profiling. *Oncogene* **19,** 6297–6305.

Yasinska, I. M., and Sumbayev, V. V. (2003). S-nitrosation of Cys-800 of HIF-1α protein activates its interaction with p300 and stimulates its transcriptional activity. *FEBS Lett.* **549,** 105–109.

Yu, F., White, S. B., Zhao, Q., and Lee, F. S. (2001). HIF-1α binding to VHL is regulated by stimulus-sensitive proline hydroxylation. *Proc. Natl. Acad. Sci. USA* **98,** 9630–9635.

Zhong, H., Willard, M., and Simons, J. (2004). NS398 reduces hypoxia-inducible factor (HIF)-1alpha and HIF-1 activity: Multiple-level effects involving cyclooxygenase-2 dependent and independent mechanisms. *Int. J. Cancer* **112,** 585–595.

Zhong, H., Chiles, K., Feldser, D., Laughner, E., Hanrahan, C., Georgescu, M. M., Simons, J. W., and Semenza, G. L. (2000). Modulation of hypoxia-inducible factor 1alpha expression by the epidermal growth factor/phosphatidylinositol 3-kinase/PTEN/AKT/FRAP pathway in human prostate cancer cells: Implications for tumor angiogenesis and therapeutics. *Cancer Res.* **60,** 1541–1545.

Zundel, W., Schindler, C., Haas-Kogan, D., Koong, A., Kaper, F., Chen, E., Gottschalk, A. R., Ryan, H. E., Johnson, R. S., Jefferson, A. B., Stokoe, D., and Giaccia, A. J. (2000). Loss of PTEN facilitates HIF-1-mediated gene expression. *Genes Dev.* **14,** 391–396.

CHAPTER SEVENTEEN

ANALYSIS OF HYPOXIA-INDUCIBLE FACTOR 1α EXPRESSION AND ITS EFFECTS ON INVASION AND METASTASIS

Balaji Krishnamachary *and* Gregg L. Semenza

Contents

1. Introduction	347
2. Protocol 1: HIF-1α Immunohistochemistry	349
3. Protocol 2: Invasion Assay	350
4. Protocol 3: Transepithelial Resistance Measurement of Cell–Cell Adhesion	351
5. Protocol 4: Analysis of mRNA Expression by qRT-PCR	351
References	352

Abstract

Hypoxia-inducible factor 1 (HIF-1) plays an important role in human cancer cell invasion and metastasis. As a result, overexpression of the HIF-1α subunit in biopsy specimens is associated with increased patient mortality in several common cancers, including breast adenocarcinoma and oropharyngeal squamous cell carcinoma. Here, we describe methods for immunohistochemical detection of HIF-1α in tumor biopsy sections and *ex vivo* assays for analyzing the effects of hypoxia and HIF-1 on cancer cell invasiveness and cell–cell adhesion.

1. INTRODUCTION

Genetic alterations promote tumor cell proliferation and survival by inducing pathophysiological alterations within tumor cells (e.g., dysregulation of apoptosis, cell cycle, and growth factor signaling pathways) and

Vascular Biology Program, Institute for Cell Engineering, Departments of Pediatrics, Medicine, Oncology, and Radiation Oncology, and McKusick-Nathans Institute of Genetic Medicine, The Johns Hopkins University School of Medicine, Baltimore, Maryland

Methods in Enzymology, Volume 435
ISSN 0076-6879, DOI: 10.1016/S0076-6879(07)35017-9

stromal cells (e.g., stimulation of angiogenesis) (Hanahan and Weinberg, 2000). The resulting pathological increase in cell number defines a tumor. In contrast, cancer is defined by the ability of cells to penetrate the extracellular matrix (ECM) of basement membrane and underlying stroma and to invade into surrounding tissue (Liotta and Kohn, 2000).

A consequence of increased cell number within a tumor is a corresponding increase in O_2 consumption. Tumor progression and patient mortality are correlated with intratumoral hypoxia (Hockel and Vaupel, 2001). HIF-1 is a transcriptional activator composed of O_2-regulated HIF-1α and constitutively-expressed HIF-1β subunits (Wang et al., 1995) that functions as a master regulator of O_2 homeostasis (Iyer et al., 1998). Immunohistochemical analysis (see "Protocol 1" section) has demonstrated that HIF-1α is overexpressed in primary and metastatic human cancers (Talks et al., 2000; Zhong et al., 1999) and that the level of expression is correlated with patient mortality in many cancers, including breast adenocarcinoma and oropharyngeal squamous cell carcinoma, among others (Aebersold et al., 2001; Bachtiary et al., 2003; Birner et al., 2000, 2001a,b; Bos et al., 2001; Korkolopoulou et al., 2004; Kronblad et al., 2006; Schindl et al., 2002; Theodoropoulos et al., 2004; Vleugel et al., 2005). HIF-1α overexpression is observed in human brain and colon cancer biopsies at the invading tumor margin (Zagzag et al., 2000; Zhong et al., 1999). Colon cancer cells subjected to hypoxia or HIF-1α overexpression demonstrate increased invasion through matrigel, an experimental basement membrane (see "Protocol 2" section). In contrast, inhibition of HIF-1α expression by transfection of cells with small interfering RNA (siRNA) blocks hypoxia-induced invasion (Krishnamachary et al., 2003).

Epithelial cell–cell adhesion in humans and other mammals is mediated by intercellular junctional complexes consisting of tight junctions, adherens junctions, and desmosomes. E-cadherin, which is the principal component of adherens junctions and desmosomes in epithelial cells, mediates adhesion by homophilic interactions between cells (Cavallaro and Christofori, 2004). A defining step in the pathogenesis of carcinomas is the epithelial-mesenchymal transition, during which E-cadherin–mediated cell–cell adhesion is lost and cells acquire invasive and metastatic properties (Behrens et al., 1989; Thiery, 2002). In addition to its direct effects on cell–cell adhesion, E-cadherin loss of function in cancer cells also activates signal transduction pathways that promote proliferation, invasion, and metastasis (Cavallaro and Christofori, 2004).

Renal clear-cell carcinoma (RCC) is characterized by loss of function of the von Hippel-Lindau tumor suppressor (VHL), which negatively regulates HIF-1. Loss of E-cadherin expression and decreased cell–cell adhesion (see "Protocol 3" section) in VHL-null RCC4 cells was corrected by enforced expression of either VHL, a dominant-negative form of HIF-1α or a short hairpin RNA directed against HIF-1α (Krishnamachary et al., 2006). In human RCC biopsies, expression of E-cadherin and HIF-1α, as determined by immunohistochemistry (see "Protocol 1" section), was mutually exclusive (Krishnamachary *et al.*, 2006). VHL inactivation in precancerous lesions in kidneys from patients with VHL syndrome correlated with marked downregulation of E-cadherin (Esteban *et al.*, 2006). The expression of messenger RNAs (mRNAs) (see "Protocol 4" section) encoding SNAIL, transcription factor (TCF)3 (also known as E12/E47), ZFHX1A (also known as δEF1 or ZEB1), and ZFHX1B (also known as SIP1 or ZEB2), which repress *E-cadherin* gene transcription, was increased in VHL-null RCC4 cells in an HIF-1–dependent manner (Evans *et al.*, 2007; Krishnamachary *et al.*, 2006). Thus, HIF-1 contributes to the epithelial-mesenchymal transition in VHL-null RCC by indirect repression of *E-cadherin*.

2. PROTOCOL 1: HIF-1α IMMUNOHISTOCHEMISTRY

Tissue section preparation, processing, duration of fixation, type of fixative, storage of the tissue, antigen of interest, and detection method are critical variables that influence the quality of immunohistochemical analysis. HIF-1α is best detected in tissue sections using catalyzed signal amplification (CSA) techniques. Aldehyde-based fixatives like formalin create formation of methylene bridges between reactive sites, such as primary amines, amides groups, thiols, alcoholic hydroxyl groups, and cyclic aromatic rings, in tissue proteins. The degree of masking of the antigenic sites depends upon the duration of fixation, temperature, concentration of fixative, and the availability of other nearby proteins for crosslinking. In order to unmask these antigenic sites, a range of antigen retrieval techniques are available. Heat-mediated unmasking is a preferred method, as it removes the weaker Schiff bases but does not affect methylene bridges so that the resulting protein conformation is intermediate between fixed and unfixed (Shi *et al.*, 1991). The immunohistochemistry protocol for HIF-1α that follows is a slightly modified and more detailed version of published methods (Zagzag *et al.*, 2000; Zhong *et al.*, 1999).

1. Paraffin-embedded tissue sections should be stored at −20° or below. The sections are first incubated at 65° to melt the paraffin, followed by passage of the slides through Xylene (5 min × 2), then graded alcohol (100, 95, and 70% for 30 sec each) to hydrate the tissue.
2. The slides are rinsed with 0.1% Tween 20 in phosphate-buffered saline, pH 7.4 (PBST; catalog number P-3563; Sigma-Aldrich, St. Louis, MO) and then placed in a Coplin or screw-capped Pap jar containing pre-warmed citrate buffer (antigen unmasking solution; catalog number H-3300 [Vector, Burlingame, CA] or X1699 [Dako, Glostrup, Denmark]) at 95°. The slides are heated in a pressure cooker/steamer for 50 min.

3. Upon cooling, start the CSA kit (Dako, catalogue number K1500; specificity for rats and humans) protocol by quenching endogenous peroxidase with 3% hydrogen peroxide in water. This step is followed by biotin blocking for 10 min using vial 1 and 2 supplied in the kit. Apply HIF-1α primary monoclonal antibody (H1α67; catalog number NB-100-102; Novus Biologicals, Littleton, CO) at 1:10,000 in diluent (Dako, catalog number S0809) and incubate the section for 30 min at room temperature or overnight at 4°.

4. In order to enhance the signal, a link antibody (vial 4) is added to the tissue section for 15 min following incubation with primary antibody. Prepare the primary streptavidin-biotin complex according to kit instructions and apply to slide. This step is followed by incubation with biotinylated tyramide diluted 1:10 in TNB blocking solution (catalog number NEL 747B; NEN Life Sciences, Boston, MA) and then streptavidin-peroxidase secondary antibody, respectively, for 15 min each. Each step is followed by rinsing with PBST twice. Diaminobenzidine (Vector, catalog number SK 4100) is applied for 5 min, and the sections are rinsed twice with 0.1% Tween 20 in Tris-buffered saline, pH 8 (Sigma-Aldrich, catalog number T-9039).

5. The sections are counterstained with either hematoxylin QS (Vector, catalog number H-3404) or methyl green (Vector, catalog number H-3402) for 30 sec or 1 min, respectively. Following rinsing with PBST, the sections are dehydrated by passage through graded alcohol series (from low to high) and then through xylene series before mounting.

3. PROTOCOL 2: INVASION ASSAY

To ascertain the effects of hypoxia or HIF-1α overexpression on the ability of cancer cells to invade basement membrane, a matrigel invasion assay (Krishnamachary et al., 2003) is performed as follows.

1. HCT-116 or other human cancer cells are transfected with expression vector–encoding HIF-1α or siRNA against HIF-1α or empty vector. At the same time, 12-mm polycarbonate filters with 12-μm pore size in a modified Boyden chamber (BD Biosciences, San Jose, CA) are coated with a 1:20 dilution of Matrigel (Sigma-Aldrich, St. Louis, MO) and air-dried in a laminar flow hood overnight.

2. On day 2, cells are trypsinized, and 5×10^4 transfected HCT-116 cells in a 200-μl volume are seeded directly into the rehydrated inner chamber. About 600 μl of media is added to the outer chamber and the plates incubated at 37° in a 5% CO_2/95% air incubator (20% O_2 condition) for 24 hours. Assays for each experimental condition are performed in triplicate.

3. For hypoxic treatment, one set of the plates is placed in a modular incubator chamber (Billups-Rothenberg Inc., Del Mar, CA) and flushed for 3 min with a gas mixture consisting of 1% O_2, 5% CO_2, and 94% N_2, sealed, and incubated at 37° for 24 hours.

4. At the end of the incubation, media in the lower chamber is saved for any floating cells, cells sticking on the lower side of the filter are gently scraped with a rubber spatula, and the media is collected. Cells from the lower chamber are trypsinized, pooled with cells from the media and filter, and pelleted. The cell pellet is resuspended in 50 μl of media, and the cell number is determined using a hemocytometer.

4. PROTOCOL 3: TRANSEPITHELIAL RESISTANCE MEASUREMENT OF CELL–CELL ADHESION

To investigate the functional consequences associated with the presence or absence of E-cadherin expression in human renal clear-cell carcinoma RCC4 cells, we analyzed transepithelial resistance as a measure of cell–cell adhesion in parental RCC4 cells and in RCC4 subclones expressing VHL, a dominant-negative form of HIF-1α, or short hairpin RNA targeting HIF-1α or HIF-2α over 6 days in culture. VHL loss of function was associated with a significant reduction in transepithelial resistance at all time points, which was reversed by VHL expression or HIF-1 loss of function, demonstrating that HIF-1 activation is necessary and sufficient for the epithelial-mesenchymal transformation of these cells (Krishnamachary *et al.*, 2006).

1. 1×10^4 cells (RCC4 parent line and derived subclones) are resuspended in 2.5 ml of media and added to the inner chamber of a modified Boyden chamber (BD Biosciences, San Jose, CA). 3.5 ml of complete media is added to the outer chamber.

2. Resistance is measured using an epithelial voltohmmeter (EVOM, World Precision Instruments, Sarasota, FL). A blank chamber with media is maintained as a control for background correction. Electrical measurements are taken for each chamber at three different regions daily for each condition for over a period of 6 days. Resistance (R) per unit area is calculated as $(R_{sample} - R_{blank}) \times \pi d^2/4$, where d = diameter (in cm) of the chamber.

5. PROTOCOL 4: ANALYSIS OF MRNA EXPRESSION BY qRT-PCR

1. Tissue samples are immersed in RNAlater (QIAGEN, Valencia, CA) and stored at −20°.

2. The samples (tissue or cultured cells) are homogenized in Trizol (Invitrogen, Carlsbad, CA), and total RNA is extracted, precipitated

by addition of isopropanol, and purified using RNeasy Mini columns (QIAGEN) with on-column DNase I digestion.

3. First-strand complementary DNA (cDNA) synthesis is performed with 5 μg of total RNA in 100 μl reactions using the iScript cDNA Synthesis Kit (BioRad Laboratories, Hercules, CA) with oligo-dT and random hexamer primers.

4. Real-time PCR is performed in an iCycler (BioRad) using iQ SYBR Green Supermix (BioRad). Primers were designed using Beacon Designer software (BioRad). For each set of primers, gradient PCR is performed for determination of the optimal annealing temperature. cDNA was diluted 1:10, and 2 μl was added to each PCR.

5. Stringent criteria are imposed to assure linearity of all qualitative reverse transcriptase polymerase chain reaction (qRT-PCR) assays.

   a. Serial dilutions of the cDNA samples are analyzed to determine the efficiency and dynamic range of the PCR. An assay requirement is that the standard deviation for the cycle threshold (C_T) among 3 to 5 replicate samples is less than 0.3.

   b. C_T is plotted versus log (ng input RNA), and the best-fit line is constructed. An assay requirement is that the correlation coefficient of the line is more than 0.99.

   c. The slope (m) of the line is used to determine PCR efficiency (E) based on the formula $E = (10^{-1/m}) - 1$. In order for samples to be compared, the efficiencies are required to vary by no more than 5%.

6. The expression (R) of each target gene mRNA relative to 18S ribosomal RNA (rRNA) is calculated by the cycle threshold method based on the formula $R = 2^{-\Delta(\Delta C_T)}$, where $\Delta C_T = C_{T,target} - C_{T,18S}$ and $\Delta(\Delta C_T) = \Delta C_{T,test\ condition} - \Delta C_{T,control}$.

REFERENCES

Aebersold, D. M., Burri, P., Beer, K. T., Laissue, J., Djonov, V., Greiner, R. H., and Semenza, G.L (2001). Expression of hypoxia-inducible factor 1α: A novel predictive and prognostic parameter in the radiotherapy of oropharyngeal cancer. *Cancer Res.* **61,** 2911–2916.

Bachtiary, B., Schindl, M., Potter, R., Dreier, B., Knocke, T. H., Hainfellner, J. A., Horvat, R., and Birner, P. (2003). Overexpression of hypoxia-inducible factor 1alpha indicates diminished response to radiotherapy and unfavorable prognosis in patients receiving radical radiotherapy for cervical cancer. *Clin. Cancer Res.* **9,** 2234–2240.

Behrens, J., Mareel, M. M., Van Roy, F. M., and Birchmeier, W. (1989). Dissecting tumor cell invasion: Epithelial cells acquire invasive properties after the loss of uvomorulin–mediated cell–cell adhesion. *J. Cell Biol.* **108,** 2435–2447.

Birner, P., Schindl, M., Obermair, A., Plank, C., Breitenecker, G., and Oberhuber, G. (2000). Overexpression of hypoxia-inducible factor 1α is a marker for an unfavorable prognosis in early-stage invasive cervical cancer. *Cancer Res.* **60,** 4693–4696.

Birner, P., Schindl, M., Obermair, A., Breitenecker, G., and Oberhuber, G. (2001b). Expression of hypoxia-inducible factor 1α in epithelial ovarian tumors: Its impact on prognosis and on response to chemotherapy. *Clin. Cancer Res.* **7,** 1661–1668.

Birner, P., Gatterbauer, B., Oberhuber, G., Schindl, M., Rossler, K., Prodinger, A., Budka, H., and Hainfellner, J. A. (2001a). Expression of hypoxia-inducible factor-1α in oligodendrogliomas: Its impact on prognosis and on neoangiogenesis. *Cancer* **92,** 165–171.

Bos, R., Zhong, H., Hanrahan, C. F., Mommers, E. C., Semenza, G. L., Pinedo, H. M., Abeloff, M. D., Simons, J. W., van Diest, P. J., and van der Wall, E. (2001). Levels of hypoxia-inducible factor 1α during breast carcinogenesis. *J. Natl. Cancer Inst.* **93,** 309–314.

Cavallaro, U., and Christofori, G. (2004). Cell adhesion and signalling by cadherins and Ig-CAMs in cancer. *Nat. Rev. Cancer* **4,** 118–132.

Esteban, M. A., Tran, M. G., Harten, S. K., Hill, P., Castellanos, M. C., Chandra, A., Raval, R., O'Brien, T. S., and Maxwell, P. H. (2006). Regulation of E-cadherin expression by VHL and hypoxia-inducible factor. *Cancer Res.* **66,** 3567–3575.

Evans, A. J., Russell, R. C., Roche, O., Burry, T. N., Fish, J. E., Chow, V. W., Kim, W. Y., Saravanan, A., Maynard, M. A., Gervais, M. L., Sufan, R. I., Roberts, A. M., *et al.* (2007). VHL promotes E2 box-dependent E-cadherin transcription by HIF-mediated regulation of SIP1 and Snail. *Mol. Cell Biol.* **27,** 157–169.

Hanahan, D., and Weinberg, R. A. (2000). The hallmarks of cancer. *Cell* **100,** 57–70.

Hockel, M., and Vaupel, P. (2001). Tumor hypoxia: Definitions and current clinical, biologic, and molecular aspects. *J. Natl. Cancer Inst.* **93,** 266–276.

Iyer, N. V., Kotch, L. E., Agani, F., Leung, S. W., Laughner, E., Wenger, R. H., Gassmann, M., Gearhart, J. D., Lawler, A. M., Yu, A. Y., and Semenza, G. L. (1998). Cellular and developmental control of O_2 homeostasis by hypoxia-inducible factor 1α. *Genes Dev.* **12,** 149–162.

Korkolopoulou, P., Patsouris, E., Konstantinidou, A. E., Pavlopoulos, P. M., Kavantzas, N., Boviatsis, E., Thymara, I., Perdiki, M., Thomas-Tsagli, E., Angelidakis, D., Rologis, D., and Sakkas, D. (2004). Hypoxia-inducible factor 1α/vascular endothelial growth factor axis in astrocytomas: Associations with microvessel morphometry, proliferation and prognosis. *Neuropathol. Appl. Neurobiol.* **30,** 267–278.

Krishnamachary, B., Zagzag, D., Nagasawa, H., Rainey, K., Okuyama, H., Baek, J. H., and Semenza, G. L. (2006). Hypoxia-inducible factor-1-dependent repression of E-cadherin in von Hippel-Lindau tumor suppressor-null renal cell carcinoma mediated by TCF3, ZFHX1A, and ZFHX1B. *Cancer Res.* **66,** 2725–2731.

Krishnamachary, B., Berg-Dixon, S., Kelly, B., Agani, F., Feldser, D., Ferreira, G., Iyer, N., LaRusch, J., Pak, B., Taghavi, P., and Semenza, G. L. (2003). Regulation of colon carcinoma cell invasion by hypoxia-inducible factor 1. *Cancer Res.* **63,** 1138–1143.

Kronblad, A., Jirstrom, K., Ryden, L., Nordenskjold, B., and Landberg, G. (2006). Hypoxia inducible factor 1α is a prognostic marker in premenopausal patients with intermediate to highly differentiated breast cancer but not a predictive marker for tamoxifen response. *Int. J. Cancer* **118,** 2609–2916.

Liotta, L. A., and Kohn, E. C. (2000). Invasion and metastasis. *In* "Cancer Medicine, edn. 5" (R. C. Bast, Jr., D. W. Kufe, and R. E. Pollock *et al.*, eds.) pp. 121–131. B.C. Decker, Ontario, Canada.

Schindl, M., Schoppmann, S. F., Samonigg, H., Hausmaninger, H., Kwasny, W., Gnant, M., Jakesz, R., Kubista, E., Birner, P., and Oberhuber, G.; Austrian Breast Colorectal Cancer Study Group (2002). Overexpression of hypoxia-inducible factor 1α is associated with an unfavorable prognosis in lymph node-positive breast cancer. *Clin. Cancer Res.* **8,** 1831–1837.

Shi, S. R., Key, M. E., and Kalra, K. L. (1991). Antigen retrieval in formalin-fixed, paraffin-embedded tissues: An enhancement method for immunohistochemical staining based on microwave oven heating of tissue sections. *J. Histochem. Cytochem.* **39,** 741–748.

Talks, K. L., Turley, H., Gatter, K. C., Maxwell, P. H., Pugh, C. W., Ratcliffe, P. J., and Harris, A. L. (2000). The expression and distribution of the hypoxia-inducible factors HIF-1α and HIF-2α in normal human tissues, cancers, and tumor-associated macrophages. *Am. J. Pathol.* **157,** 411–421.

Theodoropoulos, V. E., Lazaris, A. Ch., Sofras, F., Gerzelis, I., Tsoukala, V., Ghikonti, I., Manikas, K., and Kastriotis, I. (2004). Hypoxia-inducible factor 1α expression correlates with angiogenesis and unfavorable prognosis in bladder cancer. *Eur. Urol.* **46,** 200–208.

Thiery, J. P. (2002). Epithelial-mesenchymal transitions in tumor progression. *Nat. Rev. Cancer* **2,** 442–454.

Vleugel, M. M., Greijer, A. E., Shvarts, A., van der Groep, P., van Berkel, M., Aarbodem, Y., van Tinteren, H., Harris, A. L., van Diest, P. J., and van der Wall, E. (2005). Differential prognostic impact of hypoxia-induced and diffuse HIF-1α expression in invasive breast cancer. *J. Clin. Pathol.* **58,** 172–177.

Wang, G. L., Jiang, B.-H., Rue, E. A., and Semenza, G. L. (1995). Hypoxia-inducible factor 1 is a basic helix-loop-helix PAS heterodimer regulated by cellular O_2 tension. *Proc. Natl. Acad. Sci. USA* **92,** 5510–5514.

Zagzag, D., Zhong, H., Scalzitti, J. M., Laughner, E., Simons, J.W, and Semenza, G. L. (2000). Expression of hypoxia-inducible factor 1α in brain tumors: Association with angiogenesis, invasion, and progression. *Cancer* **88,** 2606–2618.

Zhong, H., De Marzo, A. M., Laughner, E., Lim, M., Hilton, D. A., Zagzag, D., Buechler, P., Isaacs, W. B., Semenza, G. L., and Simons, J. W. (1999). Overexpression of hypoxia-inducible factor 1α in common human cancers and their metastases. *Cancer Res.* **59,** 5830–5835.

CHAPTER EIGHTEEN

Macrophage Migration Inhibitory Factor Manipulation and Evaluation in Tumoral Hypoxic Adaptation

Millicent Winner,[†] Lin Leng,[*] Wayne Zundel,[†] *and* Robert A. Mitchell[†]

Contents

1. Introduction	356
2. Modulation of MIF Levels by Targeted shRNAs and Assessment of Knockdown Efficiency	357
2.1. MIF-specific shRNA transfection	357
2.2. Assessment of MIF knockdown and associated phenotypes by RT-PCR	359
2.3. Assessment of MIF knockdown and associated loss of HIF-α stability by Western blotting	360
2.4. Enzymatic analyses	361
3. Analysis of MIF-Dependent CSN5 and COP9 Signalosome Function	362
3.1. CSN5 co-immunoprecipitations	363
3.2. CSN-dependent deneddylation	363
3.3. Determination of CSN-associated versus -disassociated CSN5	364
4. Determination of Tumor-Associated MIF Expression and MIF Polymorphic Disparity	364
4.1. Immunohistochemistry of MIF and CSN5 tumor expression levels and correlation to hypoxic adaptation	365
4.2. MIF plasma analysis and genomic DNA extraction	365
4.3. Genotyping of MIF −173 G/C and MIF$^{5-8}$ CATT repeats from human samples	366
5. Conclusions	367
References	367

[*] Yale University, New Haven, Connecticut
[†] Molecular Targets Program, JG Brown Cancer Center, University of Louisville, Louisville, Kentucky

Methods in Enzymology, Volume 435
ISSN 0076-6879, DOI: 10.1016/S0076-6879(07)35018-0

Abstract

Increasingly clear is an important regulatory role for hypoxia-inducible factor 1alpha (HIF-1α) in the expression of the cytokine/growth factor macrophage migration inhibitory factor (MIF). The functional significance of hypoxia-induced MIF expression is revealed by findings demonstrating that HIF-1α–dependent MIF expression is necessary for hypoxia-induced evasion from cell senescence and that MIF is necessary for HIF-1α stabilization induced by hypoxia and prolyl hydroxylase (PHD) inhibitors. Both of these activities attributed to MIF likely involve the modulation of protein degratory pathways mediated by cullin-dependent E3 ubiquitin ligase complexes and their regulation by the COP9 signalosome (CSN). As the importance of MIF in hypoxic adaptation of human tumors is now becoming fully realized, we review protocols designed to evaluate MIF expression, activity, and functional consequences in hypoxic environments.

1. Introduction

Since the cloning of the factor responsible for a transcriptional activity associated with hypoxic adaptation (Wang *et al.*, 1995), the molecular events involved in the degradation of HIF-α have been extensively characterized. These studies led to the eventual identification of a family of PHDs that serve to act as dioxygen sensors (Kim and Kaelin, 2003). Under normoxic conditions, HIF-1α undergoes trans-4-hydroxylation at Pro-564 (CODD, or C-terminal ODD) and Pro-402 (NODD, or N-terminal ODD), which form part of highly conserved LXXLAP motifs in oxygen-dependent degradation domains (ODDs) (Chan *et al.*, 2005; Kim and Kaelin, 2003). Hydroxylation allows recognition of HIF-1α by the von Hippel-Lindau tumor suppressor protein (pVHL), which serves as the recognition component of the E3 ubiquitin ligase complex consisting of VHL/elongin C/elongin B (VCB), cullin 2, and the RING-H2 finger protein Rbx-1 (Hon *et al.*, 2002). Importantly, structural analysis of the HIF-CODD and pVHL reveals that all five pVHL residues lining the 4-hydroxyproline–binding pocket are affected by missense mutations in VHL disease (Kim and Kaelin, 2004). This characteristic suggests that failure to capture HIF-1α and/or other hydroxylated targets is important to the tumor-promoting mechanism associated with VHL disease. Subsequent ubiquitylation of HIF-α by the Cdc34/Ubc5 E2 ubiquitin-conjugating complex targets HIF-α for transport to the proteasome and degradation.

COP9 signalosome subunit 5 (CSN5) is an essential component of CSN, which is composed of eight subunits designated CSN1 to CSN8 (Wolf *et al.*, 2003). Until recently, the function of the CSN was obscure, though it appeared to control proteins that had high turnover rates.

The conjugation of the small ubiquitin-like protein Nedd8 to cullins is thought to be required for E2 recruitment and targeted ubiquitylation. CSN5 contains a JAB-1/MPN domain metalloenzyme motif (JAMM), which forms the catalytic region of the isopeptidase. In CSN5, the JAMM domain is responsible for the cleavage of Nedd8 from cullins, resulting in an inability of cullin complexes to catalyze ubiquitin ligation to target proteins, such as HIF-1α, for 26S-dependent degradation. This responsibility directly connects the CSN in dynamically preventing ubiquitylation of certain proteins and subsequent 26S proteasome–dependent degradation.

One of the first cytokine activities ever described, MIF is found overexpressed in a wide variety of human tumors, and MIF intratumoral expression strongly correlates with angiogenic growth factor expression, tumor vessel density, and risk of recurrence after resection (Chesney et al., 1999; Hira et al., 2005; Ren et al., 2005; Shun et al., 2005; White et al., 2003). Given the importance of hypoxic adaptation in malignant disease severity and progression (Melillo, 2006), in-depth studies of intrinsic regulators of hypoxia-dependent HIF-1α expression are critical. Here, we discuss detailed methods for analyzing hypoxia-mediated MIF expression, MIF-dependent HIF-1α stability, and regulation of cullin-dependent E3 ubiquitin ligase complex function by MIF and CSN5 (Winner et al., 2007).

2. Modulation of MIF Levels by Targeted shRNAs and Assessment of Knockdown Efficiency

RNA interference has proven to be a powerful tool for studying gene product function in cancer-related pathways (Fuchs and Borkhardt, 2007). Using the siDesign center from Dharmacon (www.dharmacon.com), short hairpin (shRNA) target sequences can be identified and double-stranded small interfering (siRNA) oligos tested for MIF assessment of knockdown efficiency. Detailed protocols for MIF-specific shRNA-mediated knockdown and assessment of expression are described next.

2.1. MIF-specific shRNA transfection

For shRNA transfections, cells are seeded in growth media containing 10% heat-inactivated fetal bovine serum, 1% gentamicin, and 1% L-glutamine (Invitrogen, Carlsbad, CA) at ~3 × 10^5 cells (20–30% confluency) on 10-cm tissue culture dishes in 6 ml of medium. It is important to initiate shRNA transfections beginning with low cell density to ensure efficient transfection efficiencies. In our experience, higher cell densities generally result in lower shRNA transfection efficiency. After seeding, cells should be allowed to adhere overnight in a humidified incubator of 5% CO_2 at 37°.

Of the three MIF-specific siRNAs we initially tested, two were found to very efficiently inhibit MIF expression, and one is described in detail here. The human MIF-targeted base sequence is 5′-CCTTCTGGTGGGGAGAAAT-3′, and the derivative scrambled oligo (NS) sequence is 5′-CCTTCTGGT GGGGAGAAAT-3′. Annealed siRNA oligos (Dharmacon, Lafayette, CO) should be resuspended in siRNA buffer (20 mM KCl, 6 mM HEPES pH 7.5, 0.2 mM $MgCl_2$) made with RNase-free water to give a final 20-mM stock. RNA working precautions should be exercised when working with shRNAs. For transfections, dilute Oligofectamine (Invitrogen) in OPTIMEM media (Invitrogen) at a final ratio of 1:2.75. Mix by gentle pipetting, then incubate this mixture for 10 min. In another tube, dilute each siRNA oligo in 182.5 μl of OPTIMEM for each milliliter of medium to a final concentration of 50 nM. After incubating for 10 min, add 15 μl of the diluted Oligofectamine for each milliliter of medium to the tube containing the diluted siRNA oligo and mix by gentle vortexing. After incubating for an additional 20 min, remove the equivalent amount of media from cells that will be added, and add 200 μl per each milliliter of medium to the cells in a drop-wise fashion. Incubate the transfected cells for 48 to 72 h at 5% CO_2 at 37°.

In order to rule out off-target effects of MIF shRNAs, studies should be performed to add-back soluble MIF to evaluate rescue of HIF-α expression. Repeated attempts by our laboratory to rescue the HIF destabilization phenotype induced by MIF knockdown by prokaryotically expressed recombinant MIF were unsuccessful. However, conditioned supernatants from cells expressing MIF can be used to rescue HIF-destabilizing effects associated with loss of MIF. The specificity of this rescue must be tested by adding neutralizing monoclonal antibodies or small molecule MIF inhibitors. For human MIF reconstitution, conditioned supernatants from non-confluent cultures are preincubated for 1 h with a control monoclonal antibody (mAb, 50 μg/ml; R&D Systems, Minneapolis, MN), anti–MIF mAb (50 μg/ml; R&D Systems), dimethyl sulfoxide (DMSO), or the small molecule antagonist (ISO-1, 100 μM; EMD Biosciences, San Diego, CA). After preincubation with control or MIF–inhibiting substances, conditioned supernatants are added at a 1:1 ratio with fresh media and placed onto MIF shRNA-transfected cells.

Short hairpin RNA–mediated depletion of MIF or MIF$^{-/-}$ murine embryonic fibroblasts displays markedly defective HIF-1α stabilization when challenged with hypoxia (1% O_2), anoxia (< 0.2% O_2), and inhibitors of prolyl hydroxylases (cobalt chloride, $CoCl_2$). In the case of shRNA-transfected cells 48 to 72 h post-transfection, media is replaced with fresh media, and different groups of plates are exposed to ambient pO_2 (normoxia), hypoxia (1% O_2), anoxia (< 0.2% O_2), and/or $CoCl_2$ (150 μM final concentration) for periods ranging between 4 and 16 h. Hypoxic or anoxic conditions are created by placing the cells in a Sheldon Bactron Anaerobic/Environment chamber.

2.2. Assessment of MIF knockdown and associated phenotypes by RT-PCR

Initial studies to evaluate knockdown efficiency for MIF should include a stringent evaluation of MIF messenger RNA (mRNA) levels. Quantitation polymerase chain reaction (q-PCR) is routinely used to evaluate not only knockdown efficiencies in cells transfected with shRNAs but also as a means of measuring HIF-1α–dependent MIF and vascular endothelial growth factor (VEGF) induction. For total RNA isolation, we use the RNeasy Mini Kit (Qiagen, Valencia, CA). Cell culture medium is removed 48 to 72 h post-shRNA transfection, and 600 μl of Buffer RLT containing 10 μl of beta (β)-mercaptoethanol is added to each plate. Plates are rotated for 10 min, and cell lysates are collected with a rubber policeman and transferred to a microcentrifuge tube. Samples are homogenized by passing the lysate through a 23-gauge needle (Becton Dickinson, Franklin Lakes, NJ) four to five times. Six hundred microliters of 70% ethanol is added and mixed by inversion. Seven hundred microliters of the lysate is then added to an RNeasy mini-column and placed in a 2-ml collection tube. After centrifuging for 15 s at a minimum of 10,000 rpm, the flow-through is discarded, and the rest of the lysate is added to the column. Repeat the centrifugation. Add 700 μl of Buffer RW1 to the column, repeat the centrifugation, and discard the flow-through and collection tube. To wash the column, add Buffer RPE onto the column (placed on a new collection tube) and centrifuge for 15 s at a minimum of 10,000 rpm. Add another 500 μl of Buffer RPE to the column and centrifuge for 2 min at a minimum of 10,000 rpm. Add 40 μl of RNase-free water to the column placed on a new 1.5-ml collection tube and centrifuge for 1 min at a minimum of 10,000 rpm. Determine RNA concentration by adding 5 μl of RNA to 995 μl of water in quartz cuvettes and measuring the absorbance at 260 nm and 280 nm with a Varian Cary 50 Bio ultraviolet (UV) spectrophotometer. Determine the volume needed for 1 μg of RNA, and bring the total volume up to 12.75 μl with RNase-free water.

For complementary DNA (cDNA) synthesis, make a master mix sufficient for all samples using the Omniscript RT kit (QIAGEN) containing 2 μl of RT Buffer, 2 μl of Deoxyribonucleotide triphosphates (dNTPs), 2 μl oligo (dT) (Sigma, St. Louis, MO), 0.25 μl RNase inhibitor (Promega, Madison, WI), and 1 μl of reverse transcriptase for each reaction. After pipetting up and down, centrifuge briefly to collect liquid at the bottom of the tubes. Add 7.25 μl of this master mix to sterile, RNase/DNase-free microcentrifuge tubes followed by the addition of 12.75 μl RNA into the appropriate tubes. Mix while incubating at 37° for 1 h in an Eppendorf thermomixer.

Amplification is carried out by making a master mix of 5 μl of 5 × Takara PCR mix (Takara Bio Inc, Otsu, Shiga, Japan), 0.3 μM final concentration of forward and reverse primers (Invitrogen; discussed later), SYBr Green

(Molecular Probes) diluted to a ratio of 1:25,000, and 15 μl of water, to bring the volume up to 23.5 μl for each reaction. Aliquot 23.5 μl of the mixture into 25 μl SmartCycler tubes (Cepheid, Sunnyvale, CA) and add 1.5 μl of the template DNA to the appropriate tubes. The specific primer sequences used are:

MIF: Forward 5'-AGAACCGCTCCTACAGCAAG-3'
Reverse 5'-TAGGCGAAGGTGGAGTTGTT-3'
VEGF: Forward 5' CAACATCACCATGCAGATTATGC 3'
Reverse 5'-GCTTTCGTTTTTGCCCCTTTC-3'
β-actin: Forward 5'-CAAGGCCAACCGCGAGAAGA-3'
Reverse 5'-GGATAGCACAGCCTGGATAG-3'
HIF-1α: Forward 5'-CGTTCCTTCGATCAGTTGTC-3'
Reverse 5'-TCAGTGGTGGCAGTGGTAGT-3'

For real time analyses, we use a DNA Engine Opticon (BioRad, Hercules, CA) to perform the PCR amplification. Relative expression levels of mRNAs are determined using the delta C_T method. The $\Delta\Delta C_T$ is calculated as the difference between the normalized C_T values (ΔC_T) of the treatment and the control samples: $\Delta\Delta C_T = \Delta C_{T\ treatment} - \Delta C_{T\ control}$. $\Delta\Delta C_T$ is then converted to fold change by the following formula: fold change $= 2^{-\Delta\Delta C_T}$.

2.3. Assessment of MIF knockdown and associated loss of HIF-α stability by Western blotting

Because the requirement for MIF in HIF-α stabilization is at the level of protein stability (Winner *et al.*, 2007), Western blotting techniques are routinely used to monitor not only MIF knockdown efficiency but also hypoxia, anoxia, or PHD inhibitor-induced HIF-α stability and associated transcriptional activities. For assessment of MIF expression in knockdown cells or in cells pulsed with hypoxia, whole cell lysates are used. After aspirating the media away from the treated and untreated cells, 250 μl of 1$\times$ radioimmunoprecipitation assay (RIPA) lysis buffer (Upstate, Charlottesville, VA) is added to 10-cm dishes. All lysis buffers contain 1 mM NaVO$_4$, 2 mM NaF, and a protease inhibitor cocktail (Roche Biochemical, Indianapolis, IN). Plates are placed on a rocker for 5 min, and then cell lysates are scraped and transferred to microcentrifuge tubes. Samples should be further homogenized by passing the samples through a 23-gauge needle 9 to 10 times. Centrifuge the samples for 3 min at 6000 rpm to pellet the insoluble material, and then pipette the supernatant into fresh microcentrifuge tubes. Add Laemmli sample buffer (BioRad) in a 1:1 ratio to the lysate. Mix, centrifuge briefly, and boil at 100° for 10 min.

For nuclear protein determination (HIF-1α and HIF-2α), cytoplasmic and nuclear extracts are prepared using components from the NE-PER kit

(Pierce, Rockford, IL). Media should be completely aspirated off of the cells. Two hundred microliters of CER buffer is then added. Cells are scraped off of plates with a rubber policeman and transferred into microcentrifuge tubes. After incubating on ice for 10 min, add 11 μl of CERII, vortex for 5 s, incubate on ice 1 min, and vortex another 5 s. Centrifuge immediately in a microcentrifuge for 5 min at 13,000 rpm at 4°. Collect the cytoplasmic fraction into fresh microcentrifuge tubes and store at −80°. Resuspend the nuclear pellet in 100 μl of NER by pipetting up and down. Vortex for 15 s, incubate on ice for 10 min, and repeat four times. Centrifuge at 13,000 rpm for 10 min at 4°. The supernatant is the nuclear lysate and should be aliquoted into another set of microcentrifuge tubes. These can be stored at −80°. Samples to be examined by Western blotting should be normalized for protein concentration by using the DC protein assay kit (BioRad). Briefly, prepare a standard curve ranging from 2 mg/ml BSA and diluting down to 0.1 mg/ml BSA in appropriate lysis buffer. Prepare reagent A′ by adding 20 μl of reagent S for every 1 ml of reagent A. In a 96-well plate, add 5 μl of sample or standard, 25 μl of reagent A′, and 200 μl of reagent B. Mix and incubate the plate at room temperature for 15 min. Measure the absorbance with a Biotek plate reader at 750 nm.

Load precast 10% or 10 to 20% SDS-Polyacrylamide gels (BioRad), depending on the size of the protein to be detected. Run the gel at 120V on a BioRad Mini-PROTEAN 3 cell and transfer at 350 mA to a PVDF membrane (Millipore, Charlottesville, VA). Block membranes for 1 h in blocking solution (5% nonfat dried milk, 0.2% Tween-20, 1% Goat Serum in TBS). Add primary antibody to the blocking solution and probe for 1h. Wash three times for 5 min each with TBS + Tween-20 (TBS-T, 0.2% Tween-20), add secondary antibody at a 1:8000 dilution to blocking solution, and incubate 1 h. Wash the membrane four times for 5 min and visualize with ECL reagent (Pierce).

Antibodies used for immunoblotting include polyclonal and monoclonal antibodies directed against MIF (Santa Cruz Biotechnology, Santa Cruz, CA and R&D Systems, Minneapolis, MN, respectively), HIF-1α (BD Transduction Laboratories, San Jose, CA for human and Novus Biologicals, Littleton, CO for mouse HIF-1α), CSN5 (Bethyl Laboratories, Montgomery, TX), LDH-A (Abcam, Cambridge, MA), inducible nitric oxide synthase (iNOS; Santa Cruz, CA), β-actin (Sigma), Nedd8 (Boston Biochem, Cambridge, MA), VHL (BD Pharmingen), HIF-2α (Stratagene, La Jolla, CA), and α-tubulin (Sigma).

2.4. Enzymatic analyses

Three-dimensional X-ray crystallographic studies have revealed that human MIF exists as a homotrimer and is structurally related to the bacterial enzymes 4-oxalocrotonate tautomerase and 5-carboxymethyl-2-hydroxymuconate isomerase (Sugimoto *et al.*, 1996; Sun *et al.*, 2005). MIF possesses the unusual ability to catalyze the tautomerization of the non-physiological substrates D-dopachrome and L-dopachrome methyl ester into their corresponding indole derivatives (Rosengren *et al.*, 1996). Measuring MIF enzymatic activity in cell and tissue lysates represents a semiquantitative and facile way to evaluate MIF cell or tissue concentrations of enzymatically active MIF. Cell or tissue samples are Dounce-homogenized in a phosphate-buffered saline (PBS)/ protease inhibitor solution. After protein determination by standard Bradford assay (BioRad), 10, 25, 50, and 100 μg of total protein is diluted into a final volume of 700 μl PBS in a 24-well plate. In parallel, 0.1, 0.2, 0.3, 0.4, and 0.5 μM recombinant MIF (rMIF) should be diluted into 700 μl PBS in the same 24-well plate to be used as a standard curve. Just prior to setting up the assay, combine 4 mM L-3,4-dihydroxyphenylalanine methyl ester (Sigma) with 8 mM sodium periodate (Sigma) in a 3:2 ratio in order to obtain L-dopachrome methyl ester, which becomes a reddish color. This solution is stable for no more than 30 min. Add 300 μl of L-dopachrome methyl ester to the cuvette, mix by pipette, and incubate 5 min at room temperature. Absorbances are read in a BioTek Power Wave multi-plate reader at OD_{475} nm. From the rMIF standard curve, OD_{475} test measurements that fall within the linear range of the standard curve are extrapolated against rMIF enzyme concentrations.

3. ANALYSIS OF MIF-DEPENDENT CSN5 AND COP9 SIGNALOSOME FUNCTION

Macrophage migration inhibitory factor–dependent stabilization of HIF-1α induced by hypoxia or PHD inhibitors involves the functional modulation of CSN5 (Winner *et al.*, 2007). CSN5-mediated HIF stabilization is proposed to be independent of its deneddylating activity and is likely independent of the CSN holocomplex (Bemis *et al.*, 2004). In contrast, MIF was originally demonstrated to bind to and inhibit CSN5-dependent turnover of p27[Kip1] (Kleemann *et al.*, 2000) in a process that is likely CSN-dependent. While these two findings may at first appear paradoxical, combined they may help to explain how MIF acts to modulate CSN function. Studies have shown that, while the majority of cellular CSN5 exists in large, CSN-associated complexes, CSN5 is also found in smaller complexes or in monomeric form outside of the CSN (Kwok *et al.*, 1998; Tomoda *et al.*, 2004). It has been postulated that, depending on expression levels of CSN5, the levels of free or small complex CSN5 can act independently of the CSN (Richardson and Zundel, 2005). In the case of CSN5-dependent HIF-1α stabilization, the net effect of CSN5 overexpression would increase small complex or monomeric CSN5 that would be accessible to HIF-1α binding and subsequent stabilization (Bemis *et al.*, 2004). Because MIF inhibits CSN5-dependent, CSN-mediated actions while promoting CSN-independent CSN5-mediated functions (Kleemann *et al.*, 2000; Winner *et al.*, 2007), it is possible that MIF modulates CSN5 functions by promoting the levels or activity of free or small complex CSN5. Protocols for the investigation of MIF-dependent CSN5/JAB1 function are described next.

3.1. CSN5 co-immunoprecipitations

To investigate CSN5/HIF-1α and CSN5/MIF association in normoxic or hypoxic cells, primary immunoprecipitation of CSN5 is used. Briefly, 1 × IP lysis buffer (20 mM Tris-HCL pH 8.0, 137 mM NaCl, 1 mM ethylene glycol tetraacetic acid (EGTA), 1% Triton X-100, 10% glycerol, and 1.5 mM MgCl$_2$) is added to 10-cm tissue culture dishes and rocked for 5 min. After scraping cells into lysis buffer and transferring to microcentrifuge tubes, lysates are homogenized with a 23-gauge needle and spun at 6000 rpm for 3 min. Supernatant is transferred to a new tube, and 1 μg of the polyclonal CSN5 antibody (Bethyl Laboratories, Montgomery, TX) is added to lysates. In parallel, lysis buffer alone should be incubated with 1 μg of CSN5 antibody to serve as an antibody background control during Western blotting. Rotate samples for 3 h to overnight at 4° and then add 30 μl of Protein A/G agarose beads (Santa Cruz) for 1 h at 4°. Pellet the beads by centrifuging for 30 s at 6000 rpm. Carefully aspirate supernatant and wash the pellet by adding 500 μl of 1× IP lysis buffer, mixing, and centrifuging for 30 s at 6000 rpm. Repeat the washing five times. To a dry pellet, add 35 μl of 2× Laemmli sample buffer and dissociate the proteins from the beads by boiling at 100° for 10 min. Spin the samples briefly to pellet the beads and load the entire volume. Immunoblotting for either HIF-1α or MIF should be performed as previously described.

3.2. CSN-dependent deneddylation

The CSN serves to modulate the activities of Skp1/cullin/F-box protein (SCF)-containing ubiquitin ligase complexes by removing ubiquitin-like Nedd8 from SCF scaffolding proteins, cullins (Lyapina *et al.*, 2001). Disruption of the CSN results in enhanced cullin neddylation and the loss of SCF ubiquitin ligase activity due to increased turnover of F-box–containing proteins (Cope and Deshaies, 2006). Neddylated cullins are distinguishable by their slower mobility in sodium dodecyl sulfate polyacrylamide gel electrophoresis (SDS-PAGE) immunoblots. Cell lysates from cells transfected with scrambled or MIF shRNA oligos are run for an extended time on 10% pre-cast polyacrylamide gels to adequately resolve the neddylated and deneddylated cullins. Immunoblotting against Cul2 (Zymed Laboratories, San Francisco, CA) will reveal distinct upper and lower bands representing neddylated and deneddylated Cul2, respectively. Cul1 and Cul4A neddylation patterns can be assessed in a similar fashion.

3.3. Determination of CSN-associated versus -disassociated CSN5

As previously discussed, a potential explanation for the differently observed activities of MIF on CSN5 may lie in the relative CSN-associated versus -disassociated CSN5 levels or activities. To determine potential MIF influences on the levels of small complex or "free" CSN5, gel filtration analysis of MIF-containing and depleted cell lysates is performed. A protocol for gel filtration of Arabidopsis Jab1/CSN5 homologs (Kwok *et al.*, 1998) has been adapted for mammalian cell studies. Briefly, two 10-cm dishes of cells transfected with scrambled or MIF shRNA are scraped into a buffer containing 25 mM Tris (pH 8.0), 10 mM NaCl, 10 mM MgCl$_2$, 5 mM EDTA, 10 mM β-mercaptoethanol, and protease inhibitors. On ice, lysates should be homogenized with 15 strokes of a Dounce homogenizer followed by centrifugation at 3000 rpm for 5 min. Soluble protein extract (250 μg) is fractionated by a Superose 6 10/300 gel filtration column fitted onto a BioRad BioLogic DuoFlow FPLC. Fractions (0.5 ml) are collected, TCA-precipitated, and analyzed by Western blotting for CSN5 and CSN8 (Bethyl Laboratories).

4. DETERMINATION OF TUMOR-ASSOCIATED MIF EXPRESSION AND MIF POLYMORPHIC DISPARITY

Hypoxia is a strong physiologic regulator of MIF transcription in normal and malignant cells, and HIF-1α has now been shown to be necessary for hypoxia-induced MIF expression (Baugh *et al.*, 2006; Welford *et al.*, 2006; Winner *et al.*, 2007). Numerous studies have additionally demonstrated that MIF is found overexpressed in many different human tumor types (Mitchell, 2004). Combined with our findings that plasma MIF levels are elevated in pancreatic cancer patients (Winner *et al.*, 2007), it is conceivable that MIF may represent a novel biomarker for intratumoral hypoxia. Additionally, because steady state levels of MIF dictate the stabilization of hypoxia-induced HIF-1α in human cancer cells, it is likely that relative MIF and HIF levels closely correlate with each other. Studies designed to evaluate MIF and HIF expression levels from tumor biopsy sections are described next.

A polymorphism in the promoter of the *Mif* gene regulates MIF expression and both immune- and inflammatory-associated disease severity

(Barton *et al.*, 2003; Baugh *et al.*, 2002; Hizawa *et al.*, 2004). It was found that a tetranucleotide CATT repeat polymorphism exists in the human MIF promoter region at position −794 (Baugh *et al.*, 2002). Identified individuals who are homozygous or heterozygous for these repeats are designated individual alleles as *Mif* 5-CATT, 6-CATT, 7-CATT, or 8-CATT. Functionally, the lowest number of CATT repeats correlates with low MIF expression while increased numbers of repeats correlates with increased expression. Increased repeat polymorphic individuals have been described as having an increased risk or severity of rheumatoid arthritis (Baugh *et al.*, 2002), inflammatory polyarthritis (Barton *et al.*, 2003), juvenile idiopathic arthritis (Hizawa *et al.*, 2004), or systemic lupus erythematosus (Sanchez *et al.*, 2006). Given the importance of MIF in modulating HIF-1α–dependent responses and tumoral hypoxic adaptation, procedures designed to determine MIF polymorphism disparity in cancer patients versus HIF expression and hypoxic adaptation are included in the following section.

4.1. Immunohistochemistry of MIF and CSN5 tumor expression levels and correlation to hypoxic adaptation

Serial human tumor sections can be analyzed for MIF, CSN5, HIF-1α, GLUT-1 (to evaluate HIF–dependent expression), and blood vessel content by antifactor VIII using established techniques and antibodies. Antibodies reactive against dewaxed paraffin sections are as follows: MIF (R&D Systems), CSN5 (Zymed), HIF-1α, HIF-2α, Factor VIII, and GLUT-1 (Pharmingen). The antigen retrieval process after dewaxing follows a method adapted from Boddy *et al.* (2005). Dewaxed slides undergo pressurized steam treatment at 15 psi for 3 min. Peroxidases from slide samples are then blocked by incubating slides in 0.3% hydrogen peroxidase in 0.1% sodium azide for 10 min. After blocking and staining with the appropriate antibodies (1:50 dilutions for all), appropriate secondary antibodies are added, followed by DAB development. After counterstaining with hematoxylin, three to seven fields from each slide are examined at ×200 magnification and scored on a scale of 0 to 3+. Cytoplasmic and nuclear HIF-α expression should be noted and scored independently.

4.2. MIF plasma analysis and genomic DNA extraction

For plasma MIF assessment, two independent analyses can be performed. The first utilizes patient plasma samples to determine circulating MIF levels. Plasma is tested in triplicate against normal donor controls in a commercially available human MIF enzyme-linked immunosorbent assay (ELISA) following the manufacturer's instructions (R&D Systems). An additional aliquot from the patient's plasma is used for the promoter polymorphism studies.

Genomic DNA can be extracted from as little as 10 to 20 μl of patient plasma collected in sodium citrate using an Easy-DNA Kit (Invitrogen) and following instructions for Protocol #1—Small Blood Samples and Hair Follicles. Care should be taken not to collect plasma in EDTA as it will interfere with the analysis. To extract genomic DNA from paraffin, slides are first deparaffinized in xylene and ethanol and allowed to air dry overnight. Tissue is scraped from slides into RNase/DNase-free microcentrifuge tubes containing Proteinase K buffer (10 mM Tris, pH 7.4, 2 mM EDTA, and 0.5% Tween-20). Heat the tubes to 99° for 1 h and cool to room temperature. Add 10 μl of a 2-mg/ml stock of Proteinase K (QIAGEN) and digest for 48 h at 55°.

4.3. Genotyping of MIF −173 G/C and MIF$^{5-8}$ CATT repeats from human samples

In addition to the CATT repeat MIF promoter polymorphism, a G → C polymorphism at −173 was independently discovered and frequently links with the 7 CATT repeat polymorphism (Barton *et al.*, 2003; Sanchez *et al.*, 2006). The MIF promoter −173 polymorphism analysis was achieved using an Assay-on-Demand Allelic Discrimination Assay Kit (Applied Biosystems, Foster City, CA). The PCR contains 10 to 50 ng of genomic DNA, 2.5 μl of TaqMan master mix (Applied Biosystems), 0.25 μl 20× Assay-on-Demand SNP Genotyping Assay Mix, and H$_2$O to a total of 5 μl. PCR is performed using 384-well plates on an ABI thermal cycler (Applied Biosystems). The reaction conditions used are 95° for 10 min followed by 50 cycles of 92° for 15 sec and 60° for 1 min.

Genotyping of MIF$^{5-8}$ CATT repeat alleles is performed using a fluorescence-labeled PCR primer and capillary electrophoresis. Genomic DNA (50 ng in 0.5 μl) is amplified by PCR in a total reaction volume of 25 μl containing 5 pmols of both forward and reverse primers: forward, 5′-TGCAGGAACCAATACCCATAGG-3′; reverse, 5′-AATGGTAAACTC GGGGAC.

The reverse primer is pre-labeled with 6-carboxy-fluoricine (FAM) fluorescent dye. The PCR reaction contains 22.5 μl of PCR Supermix (Invitrogen), 0.5 μl each of primer, and 1 μl of H$_2$O. The PCR is performed in 96-well plates on a PTC-100 Peltier Thermal Cycler (BioRad). Forty PCR cycles are carried out, each with denaturation for 30 s at 95°, primer annealing at 54° for 30 s, and extension for 60 s at 72°, and completed with a final extension at 72° for 10 min. The PCR product (anticipated size of 340–352 base pairs) is diluted 1:10 with H$_2$O and subjected to capillary electrophoresis followed by data analysis retrieval using Genemapper (Applied Biosystems).

5. CONCLUSIONS

The protocols provided should allow investigators to evaluate MIF functional modulation of HIF-1α stabilization and associated tumoral hypoxic adaptation. Moreover, the combined analyses of CSN5, HIF-1α, HIF-1α target genes, and extra- and intracellular MIF using these methods will provide important insight into the coordinate regulation of tumoral hypoxic responses and adaptation.

REFERENCES

Barton, A., Lamb, R., Symmons, D., Silman, A., Thomson, W., Worthington, J., and Donn, R. (2003). Macrophage migration inhibitory factor (MIF) gene polymorphism is associated with susceptibility to but not severity of inflammatory polyarthritis. *Genes Immun.* **4**, 487–491.

Baugh, J. A., Gantier, M., Li, L., Byrne, A., Buckley, A., and Donnelly, S. C. (2006). Dual regulation of macrophage migration inhibitory factor (MIF) expression in hypoxia by CREB and HIF-1. *Biochem. Biophys. Res. Commun.* **347**, 895–903.

Baugh, J. A., Chitnis, S., Donnelly, S. C., Monteiro, J., Lin, X., Plant, B. J., Wolfe, F., Gregersen, P. K., and Bucala, R. (2002). A functional promoter polymorphism in the macrophage migration inhibitory factor (MIF) gene associated with disease severity in rheumatoid arthritis. *Genes Immun.* **3**, 170–176.

Bemis, L., Chan, D. A., Finkielstein, C. V., Qi, L., Sutphin, P. D., Chen, X., Stenmark, K., Giaccia, A. J., and Zundel, W. (2004). Distinct aerobic and hypoxic mechanisms of HIF-α regulation by CSN5. *Genes Dev.* **18**, 739–744.

Boddy, J. L., Fox, S. B., Han, C., Campo, L., Turley, H., Kanga, S., Malone, P. R., and Harris, A. L. (2005). The androgen receptor is significantly associated with vascular endothelial growth factor and hypoxia sensing via hypoxia-inducible factors HIF-1α, HIF-2α, and the prolyl hydroxylases in human prostate cancer. *Clin. Cancer Res.* **11**, 7658–7663.

Chan, D. A., Sutphin, P. D., Yen, S. E., and Giaccia, A. J. (2005). Coordinate regulation of the oxygen-dependent degradation domains of hypoxia-inducible factor 1 alpha. *Mol. Cell Biol.* **25**, 6415–6426.

Chesney, J., Metz, C., Bacher, M., Peng, T., Meinhardt, A., and Bucala, R. (1999). An essential role for macrophage migration inhibitory factor (MIF) in angiogenesis and the growth of a murine lymphoma. *Mol. Med.* **5**, 181–191.

Cope, G. A., and Deshaies, R. J. (2006). Targeted silencing of Jab1/Csn5 in human cells downregulates SCF activity through reduction of F-box protein levels. *BMC Biochem.* **7**, 1.

Fuchs, U., and Borkhardt, A. (2007). The application of siRNA technology to cancer biology discovery. *Adv. Cancer Res.* **96**, 75–102.

Hira, E., Ono, T., Dhar, D. K., El Assal, O. N., Hishikawa, Y., Yamanoi, A., and Nagasue, N. (2005). Overexpression of macrophage migration inhibitory factor induces angiogenesis and deteriorates prognosis after radical resection for hepatocellular carcinoma. *Cancer* **103**, 588–598.

Hizawa, N., Yamaguchi, E., Takahashi, D., Nishihira, J., and Nishimura, M. (2004). Functional polymorphisms in the promoter region of macrophage migration inhibitory factor and atopy. *Am. J. Respir. Crit. Care Med.* **169**, 1014–1018.

Hon, W. C., Wilson, M. I., Harlos, K., Claridge, T. D., Schofield, C. J., Pugh, C. W., Maxwell, P. H., Ratcliffe, P. J., Stuart, D. I., and Jones, E. Y. (2002). Structural basis for the recognition of hydroxyproline in HIF-1α by pVHL. *Nature* **417,** 975–978.

Kim, W., and Kaelin, W. G., Jr. (2003). The von Hippel-Lindau tumor suppressor protein: New insights into oxygen sensing and cancer. *Curr. Opin. Genet. Dev.* **13,** 55–60.

Kim, W. Y., and Kaelin, W. G. (2004). Role of VHL gene mutation in human cancer. *J. Clin. Oncol.* **22,** 4991–5004.

Kleemann, R., Hausser, A., Geiger, G., Mischke, R., Burger-Kentischer, A., Flieger, O., Johannes, F. J., Roger, T., Calandra, T., Kapurniotu, A., Grell, M., Finkelmeier, D., Brunner, H., and Bernhagen, J. (2000). Intracellular action of the cytokine MIF to modulate AP-1 activity and the cell cycle through Jab1. *Nature* **408,** 211–216.

Kwok, S. F., Solano, R., Tsuge, T., Chamovitz, D. A., Ecker, J. R., Matsui, M., and Deng, X. W. (1998). Arabidopsis homologs of a c-Jun coactivator are present both in monomeric form and in the COP9 complex, and their abundance is differentially affected by the pleiotropic cop/det/fus mutations. *Plant Cell* **10,** 1779–1790.

Lyapina, S., Cope, G., Shevchenko, A., Serino, G., Tsuge, T., Zhou, C., Wolf, D. A., Wei, N., Shevchenko, A., and Deshaies, R. J. (2001). Promotion of NEDD-CUL1 conjugate cleavage by COP9 signalosome. *Science* **292,** 1382–1385.

Melillo, G. (2006). Inhibiting hypoxia-inducible factor 1 for cancer therapy. *Mol. Cancer Res.* **4,** 601–605.

Mitchell, R. A. (2004). Mechanisms and effectors of MIF-dependent promotion of tumour-igenesis. *Cell Signal.* **16,** 13–19.

Ren, Y., Law, S., Huang, X., Lee, P. Y., Bacher, M., Srivastava, G., and Wong, J. (2005). Macrophage migration inhibitory factor stimulates angiogenic factor expression and correlates with differentiation and lymph node status in patients with esophageal squamous cell carcinoma. *Ann. Surg.* **242,** 55–63.

Richardson, K. S., and Zundel, W. (2005). The emerging role of the COP9 signalosome in cancer. *Mol. Cancer Res.* **3,** 645–653.

Rosengren, E., Bucala, R., Aman, P., Jacobsson, L., Odh, G., Metz, C. N., and Rorsman, H. (1996). The immunoregulatory mediator macrophage migration inhibitory factor (MIF) catalyzes a tautomerization reaction. *Mol. Med.* **2,** 143–149.

Sanchez, E., Gomez, L. M., Lopez-Nevot, M. A., Gonzalez-Gay, M. A., Sabio, J. M., Ortego-Centeno, N., de Ramon, E., Anaya, J. M., Gonzalez-Escribano, M. F., Koeleman, B. P., and Martin, J. (2006). Evidence of association of macrophage migration inhibitory factor gene polymorphisms with systemic lupus erythematosus. *Genes Immun.* **7,** 433–436.

Shun, C. T., Lin, J. T., Huang, S. P., Lin, M. T., and Wu, M. S. (2005). Expression of macrophage migration inhibitory factor is associated with enhanced angiogenesis and advanced stage in gastric carcinomas. *World J. Gastroenterol.* **11,** 3767–3771.

Sugimoto, H., Suzuki, M., Nakagawa, A., Tanaka, I., and Nishihira, J. (1996). Crystal structure of macrophage migration inhibitory factor from human lymphocyte at 2.1 Å resolution. *FEBS Lett.* **389,** 145–148.

Sun, B., Nishihira, J., Yoshiki, T., Kondo, M., Sato, Y., Sasaki, F., and Todo, S. (2005). Macrophage migration inhibitory factor promotes tumor invasion and metastasis via the Rho-dependent pathway. *Clin. Cancer Res.* **11,** 1050–1058.

Tomoda, K., Yoneda-Kato, N., Fukumoto, A., Yamanaka, S., and Kato, J. Y. (2004). Multiple functions of Jab1 are required for early embryonic development and growth potential in mice. *J. Biol. Chem.* **279,** 43013–43018.

Wang, G. L., Jiang, B. H., Rue, E. A., and Semenza, G. L. (1995). Hypoxia-inducible factor 1 is a basic helix-loop-helix PAS heterodimer regulated by cellular O_2 tension. *Proc. Natl. Acad. Sci. USA* **92,** 5510–5514.

Welford, S. M., Bedogni, B., Gradin, K., Poellinger, L., Broome, P. M., and Giaccia, A. J. (2006). HIF-1α delays premature senescence through the activation of MIF. *Genes Dev.* **20,** 3366–3371.

White, E. S., Flaherty, K. R., Carskadon, S., Brant, A., Iannettoni, M. D., Yee, J., Orringer, M. B., and Arenberg, D. A. (2003). Macrophage migration inhibitory factor and CXC chemokine expression in non-small cell lung cancer: Role in angiogenesis and prognosis. *Clin. Cancer Res.* **9,** 853–860.

Winner, M., Koong, A. C., Rendon, B. E., Zundel, W., and Mitchell, R. A. (2007). Amplification of tumor hypoxic responses by macrophage migration inhibitory factor-dependent hypoxia-inducible factor stabilization. *Cancer Res.* **67,** 186–193.

Wolf, D. A., Zhou, C., and Wee, S. (2003). The COP9 signalosome: An assembly and maintenance platform for cullin ubiquitin ligases? *Nat. Cell Biol.* **5,** 1029–1033.

THE VON HIPPEL-LINDAU TUMOR SUPPRESSOR PROTEIN: AN UPDATE

William G. Kaelin, Jr.

Contents

1. Introduction	371
2. Regulation of Epithelial Differentiation by pVHL	373
2.1. E-cadherin	373
2.2. The primary cilium	373
3. Crosstalk between c-Met and VHL	374
4. Regulation of Neuronal Apoptosis by pVHL	375
5. Possible Links Between p53 and pVHL	376
6. Regulation of pVHL by Phosphorylation	376
7. Polyubiquitylation of pVHL	377
8. Mouse Models for Studying pVHL Function	377
References	378

Abstract

Inactivation of the von Hippel-Lindau (VHL) tumor suppressor has been linked to a variety of tumors, including clear cell renal carcinoma, retinal and cerebellar hemangioblastoma, and pheochromocytoma. The best documented function of VHL protein (pVHL) relates to its ability to target the hypoxia-inducible transcription factor (HIF) for polyubiquitylation and proteasomal degradation. This chapter focuses on studies published over the past 2 years related to pVHL.

These studies include those describing genetically engineered mice that were used to interrogate the relationship between pVHL and HIF *in vivo* and cell culture studies that underscore the importance of pVHL in epithelial differentiation and maintenance of the primary cilium. In addition, recent work suggests that pVHL regulates neuronal apoptosis in an HIF-independent manner, and this activity is linked to the risk of developing pheochromocytoma.

Howard Hughes Medical Institute, Harvard Medical School, Boston, Massachusetts

Methods in Enzymology, Volume 435
ISSN 0076-6879, DOI: 10.1016/S0076-6879(07)35019-2

1. INTRODUCTION

von Hippel-Lindau disease is a multisystem disorder caused by inherited (or, more rarely, *de novo*) mutations that inactivate the *VHL* tumor suppressor gene, which is located on chromosome 3p25. The classical manifestations of VHL disease include blood vessel tumors (hemangioblastomas) of the retina, cerebellum, or spinal cord, visceral cysts (especially of the kidney and pancreas), and a variety of solid tumors, including clear cell renal carcinomas, pheochromocytomas, pancreatic islet cell tumors, endolymphatic sac tumors, and papillary tumors of the broad ligament (women) or epididymis (men). The development of pathology, including preneoplastic changes, in VHL disease is linked to inactivation of the remaining wild-type *VHL* allele and, consequently, the loss of wild-type *VHL* gene product, pVHL. (In actuality, two VHL gene products are produced in cells as a result of alternative translation initiation from an in-frame ATG at codon 54; for simplicity, "pVHL" will be used to refer to these two isoforms generically throughout this review) (Blankenship *et al.*, 1999; Iliopoulos *et al.*, 1998; Schoenfeld *et al.*, 1998). As would be expected based on the Knudson 2-Hit Model, biallelic inactivation of *VHL* is also common in sporadic clear cell renal carcinoma and hemangioblastoma (Kim and Kaelin, 2004). It is now also clear that some cases of familial pheochromocytoma, without the other stigmata of VHL disease, are also caused by germline *VHL* mutations (Type 2C VHL disease; further discussed later) and that some individuals with hereditary polycythemia carry two hypomorphic VHL alleles in their germline (Ang *et al.*, 2002; Bento *et al.*, 2005; Gordeuk *et al.*, 2004; Perrotta *et al.*, 2006; Smith *et al.*, 2006).

The best documented function of pVHL relates to control of the heterodimeric transcription factor HIF, which consists of an unstable alpha (α) subunit and a stable beta (β) subunit. In the presence of oxygen, the α subunit becomes hydroxylated on one (or both) of two conserved prolyl residues, which generates a binding site for pVHL (Ivan *et al.*, 2001; Jaakkola *et al.*, 2001; Yu *et al.*, 2001). pVHL is part of an E3 ubiquitin ligase complex that then targets HIF-α for proteasomal degradation. In cells lacking pVHL or exposed to hypoxia, HIF-α accumulates, binds to HIF-β, and transcriptionally activates over 100 genes, including genes linked to erythropoiesis, angiogenesis, and cell metabolism. *VHL* alleles linked to hemangioblastoma and clear cell renal carcinoma are invariably defective with respect to HIF regulation. In xenograft assays, overexpression HIF-2α (but not HIF-1α) can override the ability of pVHL to suppress renal carcinoma growth (Kondo *et al.*, 2002, 2003; Maranchie *et al.*, 2002; Raval *et al.*, 2005), while elimination of HIF-2α is sufficient to impede $VHL^{-/-}$ renal carcinoma growth (Kondo *et al.*, 2003; Zimmer *et al.*, 2004). Collectively, these results suggest that deregulation of HIF-α, especially HIF-2α, plays an important role in the development of pVHL-defective hemangioblastomas and clear cell renal carcinomas. This chapter will focus on relatively recent insights into the functions of pVHL and how the loss of those functions contributes to tumorigenesis.

2. REGULATION OF EPITHELIAL DIFFERENTIATION BY pVHL

2.1. E-cadherin

Earlier studies showed that loss of pVHL leads to an epithelial-to-mesenchymal transition in renal carcinoma cells and loss of differentiation (Davidowitz *et al.*, 2001; Lieubeau-Teillet *et al.*, 1998). Several recent studies showed that one of the hallmarks of pVHL-defective renal epithelial cells is loss of intercellular structures called adherens junctions and tight junctions that play an important role in epithelial cell homeostasis (Calzada *et al.*, 2006; Esteban *et al.*, 2006b; Evans *et al.*, 2007; Krishnamachary *et al.*, 2006). Adherens junctions contain transmembrane proteins called Cadherins, which physically interact with the actin cytoskeleton and which also play signaling roles in concert with members of the Catenin family (discussed later). Downregulation of E-Cadherin in pVHL-defective cells seems to be primarily mediated by HIF, which transcriptionally activates genes, such as Snail, ZFHX1A, ZFHX1B, and/or TCF-3 (Esteban *et al.*, 2006b; Evans *et al.*, 2007; Krishnamachary *et al.*, 2006). The products of these genes, in turn, repress E-Cadherin transcription. Loss of E-Cadherin is associated with increased cell migration and invasion and diminished epithelial barrier function (Evans *et al.*, 2007; Krishnamachary *et al.*, 2006). Regulation of intercellular junctions by pVHL appears to also involve HIF-independent pathways in addition to HIF-dependent changes in E-Cadherin levels (Calzada *et al.*, 2006).

2.2. The primary cilium

The kidneys of VHL patients frequently contain hundreds of renal cysts. Many other causes of polycystic renal disease are associated with abnormalities of a specialized structure called the primary cilium, which is found on renal epithelial cells as well as other cell types (Bisgrove and Yost, 2006; Singla and Reiter, 2006). These microtubule-based structures are thought to be mechanosensors that, in the case of renal epithelial cells, deliver calcium-dependent signals in response to changes in urine flow. These signals, in turn, are believed to affect both cytoarchitecture and cell proliferation. Several groups reported that loss of pVHL leads to loss of the primary cilium (Esteban *et al.*, 2006a; Lutz and Burk, 2006; Schermer *et al.*, 2006). One study found that loss of the cilium was HIF-dependent (Esteban *et al.*, 2006a), while another suggested an HIF-independent mechanism (Lutz and Burk, 2006). With regard to the latter, pVHL can associate with microtubules (Hergovich *et al.*, 2003), which, in one study, translated into microtubule stabilization (Hergovich *et al.*, 2003) and, in another, directional growth of microtubules toward the cell periphery (Schermer *et al.*, 2006). Therefore pVHL might affect ciliogenesis through changes in microtubule dynamics.

3. Crosstalk between c-Met and VHL

Clear cell renal carcinoma and papillary renal carcinoma are the two most common forms of kidney cancer (Linehan and Zbar, 2004). Most hereditary and many sporadic clear cell renal carcinomas are linked to inactivation of the *VHL* gene (Kim and Kaelin, 2004). Some hereditary and occasional sporadic papillary renal cancers are caused by activating mutations of Mesenchymal epithelial transition factor (c-Met) (Linehan and Zbar, 2004). Despite the histological differences between these two tumor types, mounting evidence exists for crosstalk between c-Met and VHL.

In one early study, $VHL^{-/-}$ renal carcinoma cells were found to be hypersensitive to the c-Met ligand, hepatocyte growth factor (HGF) (Koochekpour *et al.*, 1999). This observation has now been confirmed by others (Peruzzi *et al.*, 2006). Several recent reports found evidence for enhanced c-Met activity in $VHL^{-/-}$ renal carcinoma cells, although the mechanisms put forth in these studies differ somewhat. Pennacchietti *et al.* (2003) reported that *c-Met* was, itself, induced by hypoxia and suggested that this was at least partially due to HIF. In contrast, Nakaigawa *et al.* (2006) found evidence that c-Met activation, as determined by c-Met phosphorylation, rather than c-Met protein levels were induced by pVHL loss. Peruzzi *et al.* (2006) suggested that enhanced c-Met signaling in pVHL-defective cells was actually due to abnormalities downstream of the receptor involving cytoplasmic β-catenin, perhaps as a consequence of altered adherens junctions as previously described. They, along with another group (Lutz and Burk, 2006), documented increased cytoplasmic β-catenin in $VHL^{-/-}$ cells. This pool of β-catenin, once translocated to the nucleus, can transcriptionally activate Tcf/LEF-sensitive genes in $VHL^{-/-}$ cells and enhance HGF signaling (Peruzzi *et al.*, 2006).

Mutations of the fumarate hydratase gene are another cause of hereditary papillary renal carcinoma (Linehan and Zbar, 2004). Interestingly, the accumulation of fumarate in cells lacking wild-type fumarate hydratase blocks the activity of the HIF prolyl hydroxylases (Isaacs *et al.*, 2005), which are the enzymes that mark HIF for destruction by pVHL. Accordingly, HIF-α protein levels are increased in *Fumarate Hydratase*$^{-/-}$ papillary renal carcinomas (Isaacs *et al.*, 2005; Kim *et al.*, 2006a). Collectively, these results suggest that HIF is important in the pathogenesis in at least a subset of papillary renal carcinomas, and c-Met activation might play a contributory role in some clear cell renal carcinomas.

4. Regulation of Neuronal Apoptosis by pVHL

Some germline *VHL* mutations predispose to the development of pheochromocytomas, which are intraadrenal paragangliomas without the other stigmata of VHL disease. Such families are said to have type 2C VHL disease. The products of type 2C *VHL* alleles, when tested, appear to be wild-type with respect to the regulation of HIF, suggesting that loss of an HIF-independent pVHL function is relevant to the pathogenesis of these tumors (Clifford *et al.*, 2001; Hoffman *et al.*, 2001). A significant fraction of sporadic pheochromocytomas is also due to previously unsuspected germline *VHL* mutations (Nakamura and Kaelin, 2006). Germline *NF-1, c-Ret, SDH B,* and *SDH D* mutations have also been linked to the development of paragangliomas (Nakamura and Kaelin, 2006). Contrary to what might be expected based on the Knudson 2-Hit Model, somatic mutations of *VHL, NF-1, c-Ret, SDH B,* or *SDH D* are rare in sporadic paragangliomas, however, unless they are first mutated in the germline (Nakamura and Kaelin, 2006).

Paragangliomas are tumors of the sympathetic nervous system. Primitive sympathetic neuronal progenitor cells compete for growth factors, such as nerve growth factor (NGF), during embryological development, with the losers undergoing c-Jun–dependent apoptosis. pVHL was previously reported to inhibit atypical protein kinase C (aPKC) activity (Okuda *et al.*, 1999, 2001; Pal *et al.*, 1997, 1998), possibly by targeting aPKC for polyubiquitination and proteasomal degradation (Okuda *et al.*, 2001), and aPKC has been implicated in neuroblast survival (Lee *et al.*, 2006). Lee *et al.* (2005) showed that type 2C pVHL mutants fail to inhibit aPKC and that, as expected from earlier studies (Kieser *et al.*, 1996), this leads to an increase in JunB levels. The increase in JunB, in turn, inhibits c-Jun from inducing apoptosis upon NGF withdrawal (Lee *et al.*, 2005). Interestingly, it was shown earlier that loss of the Neurofibromatosis type 1 (NF1) gene product, Neurofibromin, likewise promotes survival in neurons subjected to NGF depletion (Vogel, *et al.*, 1995), and Lee *et al.* (2005) confirmed the earlier speculation (Dechant, 2002) that activated c-Ret might promote survival in this setting as well. Earlier work showed that EglN3 (also called prolyl-hydroxylase domain [PHD]3), which is a paralog of the primary HIF PHD

EglN1 (also called [PHD2]) (Berra *et al.*, 2003), also plays a role in neuronal apoptosis (Lipscomb *et al.*, 1999, 2001; Straub *et al.*, 2003). Lee *et al.* (2005) showed that EglN3, but not EglN1, is necessary and sufficient for neuronal apoptosis after NGF withdrawal and acts downstream of c-Jun. Moreover, they showed that the accumulation of succinate consequent to succinate dehydrogenase (SDH) inactivation blocks EglN3-induced apoptosis (Lee *et al.*, 2005). Hence, it appears that all the familial paraganglioma genes define a pathway that is critical for culling neuroblasts during embryological development. This characteristic might explain why mutations of these same genes are rare in sporadic tumors, which are presumably initiated after this developmental window has passed.

5. POSSIBLE LINKS BETWEEN p53 AND pVHL

Roe *et al.* (2006) reported that p53 and pVHL can bind to one another in cells, leading to p53 stabilization and enhanced p53-dependent transcriptional activation, especially in response to genotoxic stress. In an earlier study, pVHL was reported to enhance p53 translation through its interaction with the messenger RNA (mRNA)-binding protein Hur (Galban *et al.*, 2003). However, both of these studies require independent corroboration, and Stickle *et al.* (2005) failed to detect differences in p53 accumulation in isogenic cell line pairs that did or did not produce wild-type pVHL.

6. REGULATION OF pVHL BY PHOSPHORYLATION

Two VHL gene products are detected in cells as a result of alternative translation initiation from an in-frame methionine at codon 54 (Blankenship *et al.*, 1999; Iliopoulos *et al.*, 1998; Schoenfeld *et al.*, 1998). Lolkema *et al.* (2005) reported that the long form, but not the short form, is phosphorylated by casein kinase-2, probably at Ser residues 33, 38, and 43. A pVHL variant lacking these sites retained the ability to ubiquitinate HIF, but displayed abnormalities with respect to fibronectin matrix assembly (Lolkema *et al.*, 2005). Interestingly, this variant was also compromised with respect to its ability to inhibit $VHL^{-/-}$ renal carcinoma growth in nude mice xenograft assays (Lolkema *et al.*, 2005).

Hergovich *et al.* (2006) showed that pVHL is phosphorylated on Ser 68 by glycogen synthase kinase 3 (GSK3) after a priming phosphorylation at Ser 72 by an unidentified kinase. These phosphorylation events do not disrupt the interaction of pVHL with microtubules, but do prevent pVHL from stabilizing them. Moreover, a pVHL variant with phosphomimetic substitutions at Ser 68 and Ser 72 was impaired with respect to HIF polyubiquitylation. These results suggest that GSK3 regulates at least two pVHL functions.

7. POLYUBIQUITYLATION OF PVHL

Free pVHL, not bound to the elongins, is relatively unstable (Feldman *et al.*, 1999; Kamura *et al.*, 2002; Schoenfeld *et al.*, 2000). Jung *et al.* (2006) showed that E2-EPF UCP associates with free pVHL and targets it for polyubiquitylation and proteasomal degradation. E2-EPF ubiquitin carrier protein (UCP) levels, which are frequently elevated in cancer, were found to inversely correlate with pVHL levels in cell lines and tissues (Jung *et al.*, 2006). Overexpression of E2-EPF in cells containing pVHL promoted the accumulation of HIF and enhanced tumor growth, invasion, and metastasis.

8. MOUSE MODELS FOR STUDYING PVHL FUNCTION

VHL mice develop hemangioma-like lesions in their livers (Haase *et al.*, 2001). These lesions appear to arise as a result of stochastic inactivation of the remaining wild-type VHL allele and are associated with increased accumulation of HIF and the products of HIF target genes. Similar lesions arise, albeit with shorter latency and higher penetrance, in mice that are homozygous for a conditional ("floxed") VHL allele after hepatic expression of Cre recombinase (Haase *et al.*, 2001; Ma *et al.*, 2003).

Rankin *et al.* (2005) showed that these hepatic lesions do not occur in mice lacking Arnt1, which encodes a heterodimeric partner for HIF-1α and HIF-2α, but do occur in mice specifically lacking HIF-1α. In a complementary set of experiments, Kim *et al.* (2006b) showed that liver pathology observed after pVHL loss in the mouse could be largely recapitulated by the expression of a stabilized version of HIF-2α. Collectively, these results add to the growing body of evidence that HIF-2α plays an important role in the pVHL-defective tumor formation. Notably, Kim *et al.* (2006b) also found that some transcriptional and pathological changes observed after pVHL loss were due to both HIF-1α and HIF-2α, which is in keeping with reports that HIF-1α and HIF-2α control an overlapping, but not identical, set of genes (Carroll and Ashcroft, 2006; Raval *et al.*, 2005; Sowter *et al.*, 2003). Recently, Covello *et al.* (2005) showed that replacement of HIF-1α with HIF-2α by homologous recombination led to increased tumor growth by embryonic stem (ES) cells and demonstrated that HIF-2α, but not HIF-1α, transcriptionally activated the stem cell factor Oct4 (Covello *et al.*, 2006).

One potential difference between HIF-1α and HIF-2α relates to the factor–inhibiting HIF (FIH) 1, which hydroxylates a conserved asparagine residue within the HIF-α C-terminal transactivation domain (C-TAD) and thereby inhibits HIF-α–dependent transcription (Hewitson *et al.*, 2002; Lando *et al.*, 2002; Mahon *et al.*, 2001). HIF-2α appears to be less sensitive than HIF-1α to the inhibitory effects of FIH1 *in vitro* (Bracken *et al.*, 2006) and *in cellulo* (Yan *et al.*, 2007). Yan *et al.* (2007) showed that both the N-terminal transactivation domain and C-TAD contributed to the tumor-promoting effects of HIF-2α in nude mice xenograft studies.

Drugs that inhibit the HIF-responsive angiogenic factor vascular endo-thelial growth factor (VEGF), or its receptor KDR, have proven to be efficacious (but not curative) in the treatment of clear cell renal carcinoma in humans (Kaelin, 2007). In fact, two KDR inhibitors (sorafenib and sunitinib) have now been approved for the treatment of this disease (Kaelin, 2007). Surprisingly, Blouw *et al.* (2007) showed that genetic abla-tion of VEGF inhibited angiogenesis, but not tumor growth, by $VHL^{-/-}$ astrocytes transformed with SV40 T antigen and Ras. It is not clear, however, that these results can be readily extrapolated to human renal cancer because of the different cell types involved and because of potential confounding effects of T and Ras.

von Hippel-Lindau protein can physically associate with fibronectin (Clifford *et al.*, 2001; Hoffman *et al.*, 2001; Ohh *et al.*, 1998), and $VHL^{-/-}$ embryos and cells display defects in fibronectin matrix assembly, (He *et al.*, 2004; Lieubeau-Teillet *et al.*, 1998; Ohh *et al.*, 1998) though the precise mechanism(s) remain obscure. Tang *et al.* (2006) reported that inactivation of *VHL* in murine endothelial cells led to defective vasculogenesis that was associated with impaired fibronectin matrix deposition *in vivo*. Importantly, $VHL^{-/-}$ endothelial cells also produced less fibronectin *in vitro*, and exoge-nous fibronectin corrected their impaired migration and adhesion, implying that loss of fibronectin contributes to, and does not merely correlate with, these phenotypes (Tang *et al.*, 2006).

In an attempt to create a mouse model that more closely resembles human VHL disease, Rankin *et al.* (2006) inactivated VHL in proximal renal tubular epithelial cells and in hepatocytes using Cre recombinase under the control of phosphoenolpyruvate carboxykinase (PEPCK) pro-moter. These animals developed polycythemia and renal cysts. Importantly, these lesions were prevented by simultaneous inactivation of Arnt1.

REFERENCES

Ang, S. O., Chen, H., Hirota, K., Gordeuk, V. R., Jelinek, J., Guan, Y., Liu, E., Sergueeva, A. I., Miasnikova, G. Y., Mole, D., Maxwell, P. H., Stockton, D. W., *et al.* (2002). Disruption of oxygen homeostasis underlies congenital Chuvash polycythemia. *Nat. Genet.* **32**, 614–621.

Bento, M. C., Chang, K. T., Guan, Y., Liu, E., Caldas, G., Gatti, R. A., and Prchal, J. T. (2005). Congenital polycythemia with homozygous and heterozygous mutations of von Hippel-Lindau gene: Five new Caucasian patients. *Haematologica* **90**, 128–129.

Berra, E., Benizri, E., Ginouves, A., Volmat, V., Roux, D., and Pouyssegur, J. (2003). HIF prolyl-hydroxylase 2 is the key oxygen sensor setting low steady-state levels of HIF-1alpha in normoxia. *EMBO J.* **22**, 4082–4090.

Bisgrove, B. W., and Yost, H. J. (2006). The roles of cilia in developmental disorders and disease. *Development* **133**, 4131–4143.

Blankenship, C., Naglich, J., Whaley, J., Seizinger, B., and Kley, N. (1999). Alternate choice of initiation codon produces a biologically active product of the von Hippel Lindau gene with tumor suppressor activity. *Oncogene* **18**, 1529–1535.

Blouw, B., Haase, V. H., Song, H., Bergers, G., and Johnson, R. S. (2007). Loss of vascular endothelial growth factor expression reduces vascularization, but not growth, of tumors lacking the Von Hippel-Lindau tumor suppressor gene. *Oncogene* **26**, 4531–4540.

Bracken, C. P., Fedele, A. O., Linke, S., Balrak, W., Lisy, K., Whitelaw, M. L., and Peet, D. J. (2006). Cell-specific regulation of hypoxia-inducible factor (HIF)-1alpha and HIF-2alpha stabilization and transactivation in a graded oxygen environment. *J. Biol. Chem.* **281**, 22575–22585.

Calzada, M. J., Esteban, M. A., Feijoo-Cuaresma, M., Castellanos, M. C., Naranjo-Suarez, S., Temes, E., Mendez, F., Yanez-Mo, M., Ohh, M., and Landazuri, M. O. (2006). von Hippel-Lindau tumor suppressor protein regulates the assembly of intercellular junctions in renal cancer cells through hypoxia-inducible factor-independent mechanisms. *Cancer Res.* **66**, 1553–1560.

Carroll, V. A., and Ashcroft, M. (2006). Role of hypoxia-inducible factor (HIF)-1alpha versus HIF-2alpha in the regulation of HIF target genes in response to hypoxia, insulin-like growth factor-I, or loss of von Hippel-Lindau function: Implications for targeting the HIF pathway. *Cancer Res.* **66**, 6264–6270.

Clifford, S., Cockman, M., Smallwood, A., Mole, D., Woodward, E., Maxwell, P., Ratcliffe, P., and Maher, E. (2001). Contrasting effects on HIF-1alpha regulation by disease-causing pVHL mutations correlate with patterns of tumourigenesis in von Hippel-Lindau disease. *Hum. Mol. Genet.* **10**, 1029–1038.

Covello, K. L., Kehler, J., Yu, H., Gordan, J. D., Arsham, A. M., Hu, C. J., Labosky, P. A., Simon, M. C., and Keith, B. (2006). HIF-2alpha regulates Oct-4: Effects of hypoxia on stem cell function, embryonic development, and tumor growth. *Genes Dev.* **20**, 557–570.

Covello, K. L., Simon, M. C., and Keith, B. (2005). Targeted replacement of hypoxia-inducible factor-1alpha by a hypoxia-inducible factor-2alpha knock-in allele promotes tumor growth. *Cancer Res.* **65**, 2277–2286.

Davidowitz, E., Schoenfeld, A., and Burk, R. (2001). VHL induces renal cell differentiation and growth arrest through integration of cell-cell and cell-extracellular matrix signaling. *Mol. Cell Biol.* **21**, 865–874.

Dechant, G. (2002). Chat in the trophic web: NGF activates Ret by inter-RTK signaling. *Neuron* **33**, 156–158.

Esteban, M. A., Harten, S. K., Tran, M. G., and Maxwell, P. H. (2006a). Formation of primary cilia in the renal epithelium is regulated by the von Hippel-Lindau tumor suppressor protein. *J. Am. Soc. Nephrol.* **17**, 1801–1806.

Esteban, M. A., Tran, M. G., Harten, S. K., Hill, P., Castellanos, M. C., Chandra, A., Raval, R., O'Brien, T. S., and Maxwell, P. H. (2006b). Regulation of E-cadherin expression by VHL and hypoxia-inducible factor. *Cancer Res.* **66**, 3567–3575.

Evans, A. J., Russell, R. C., Roche, O., Burry, T. N., Fish, J. E., Chow, V. W., Kim, W. Y., Saravanan, A., Maynard, M. A., Gervais, M. L., Sufan, R. I., Roberts, A. M., et al. (2007). VHL promotes E2 box-dependent E-cadherin transcription by HIF-mediated regulation of SIP1 and snail. *Mol. Cell Biol.* **27**, 157–169.

Feldman, D., Thulasiraman, V., Ferreyra, R., and Frydman, J. (1999). Formation of the VHL-elongin BC tumor suppressor complex is mediated by the chaperonin TRiC. *Mol. Cell* **4**, 1051–1061.

Galban, S., Martindale, J. L., Mazan-Mamczarz, K., Lopez de Silanes, I., Fan, J., Wang, W., Decker, J., and Gorospe, M. (2003). Influence of the RNA-binding protein HuR in pVHL-regulated p53 expression in renal carcinoma cells. *Mol. Cell Biol.* **23**, 7083–7095.

Gordeuk, V. R., Sergueeva, A. I., Miasnikova, G. Y., Okhotin, D., Voloshin, Y., Choyke, P. L., Butman, J. A., Jedlickova, K., Prchal, J. T., and Polyakova, L. A. (2004). Congenital disorder of oxygen sensing: Association of the homozygous chuvash polycythemia VHL mutation with thrombosis and vascular abnormalities but not tumors. *Blood* **103**, 3924–3932.

Haase, V., Glickman, J., Socolovsky, M., and Jaenisch, R. (2001). Vascular tumors in livers with targeted inactivation of the von Hippel-Lindau tumor suppressor. *Proc. Natl. Acad. Sci. USA* **98**, 1583–1588.

He, Z., Liu, S., Guo, M., Mao, J., and Hughson, M. D. (2004). Expression of fibronectin and HIF-1alpha in renal cell carcinomas: Relationship to von Hippel-Lindau gene inactivation. *Cancer Genet. Cytogenet.* **152**, 89–94.

Hergovich, A., Lisztwan, J., Barry, R., Ballschmieter, P., and Krek, W. (2003). Regulation of microtubule stability by the von Hippel-Lindau tumour suppressor protein pVHL. *Nat. Cell Biol.* **5**, 64–70.

Hergovich, A., Lisztwan, J., Thoma, C. R., Wirbelauer, C., Barry, R. E., and Krek, W. (2006). Priming-dependent phosphorylation and regulation of the tumor suppressor pVHL by glycogen synthase kinase 3. *Mol. Cell Biol.* **26**, 5784–5796.

Hewitson, K. S., McNeill, L. A., Riordan, M. V., Tian, Y. M., Bullock, A. N., Welford, R. W., Elkins, J. M., Oldham, N. J., Bhattacharya, S., Gleadle, J. M., Ratcliffe, P. J., Pugh, C. W., *et al.* (2002). Hypoxia-inducible factor (HIF) asparagine hydroxylase is identical to factor inhibiting HIF (FIH) and is related to the cupin structural family. *J. Biol. Chem.* **277**, 26351–26355.

Hoffman, M., Ohh, M., Yang, H., Klco, J., Ivan, M., and Kaelin, W. J. (2001). von Hippel-Lindau protein mutants linked to type 2C VHL disease preserve the ability to down-regulate HIF. *Hum. Mol. Genet.* **10**, 1019–1027.

Iliopoulos, O., Ohh, M., and Kaelin, W. (1998). pVHL19 is a biologically active product of the von Hippel-Lindau gene arising from internal translation initiation. *Proc. Natl. Acad. Sci. USA* **95**, 11661–11666.

Isaacs, J. S., Jung, Y. J., Mole, D. R., Lee, S., Torres-Cabala, C., Chung, Y. L., Merino, M., Trepel, J., Zbar, B., Toro, J., Ratcliffe, P. J., Linehan, W. M., *et al.* (2005). HIF overexpression correlates with biallelic loss of fumarate hydratase in renal cancer: Novel role of fumarate in regulation of HIF stability. *Cancer Cell* **8**, 143–153.

Ivan, M., Kondo, K., Yang, H., Kim, W., Valiando, J., Ohh, M., Salic, A., Asara, J., Lane, W., and Kaelin, W. J. (2001). HIFalpha targeted for VHL-mediated destruction by proline hydroxylation: implications for O_2 sensing. *Science* **292**, 464–468.

Jaakkola, P., Mole, D., Tian, Y., Wilson, M., Gielbert, J., Gaskell, S., Kriegsheim, A., Hebestreit, H., Mukherji, M., Schofield, C., Maxwell, P., Pugh, C., *et al.* (2001). Targeting of HIF-alpha to the von Hippel-Lindau ubiquitylation complex by O_2-regulated prolyl hydroxylation. *Science* **292**, 468–472.

Jung, C. R., Hwang, K. S., Yoo, J., Cho, W. K., Kim, J. M., Kim, W. H., and Im, D. S. (2006). E2-EPF UCP targets pVHL for degradation and associates with tumor growth and metastasis. *Nat. Med.* **12**, 809–816.

Kaelin, W. G., Jr. (2007). The von Hippel-Lindau tumor suppressor protein and clear cell renal carcinoma. *Clin. Cancer Res.* **13**, 680s–684s.

Kamura, T., Brower, C. S., Conaway, R. C., and Conaway, J. W. (2002). A molecular basis for stabilization of the von Hippel-Lindau (VHL) tumor suppressor protein by components of the VHL ubiquitin ligase. *J. Biol. Chem.* **277**, 30388–30393.

Kieser, A., Seitz, T., Adler, H. S., Coffer, P., Kremmer, E., Crespo, P., Gutkind, J. S., Henderson, D. W., Mushinski, J. F., Kolch, W., and Mischak, H. (1996). Protein kinase C-zeta reverts v-raf transformation of NIH-3T3 cells. *Genes. Dev.* **10**, 1455–1466.

Kim, C. M., Vocke, C., Torres-Cabala, C., Yang, Y., Schmidt, L., Walther, M., and Linehan, W. M. (2006a). Expression of hypoxia inducible factor-1alpha and 2alpha in genetically distinct early renal cortical tumors. *J. Urol.* **175**, 1908–1914.

Kim, W. Y., and Kaelin, W. G. (2004). Role of VHL Gene Mutation in Human Cancer. *J. Clin. Onc.* **22**, 4991–5004.

Kim, W. Y., Safran, M., Buckley, M. R., Ebert, B. L., Glickman, J., Bosenberg, M., Regan, M., and Kaelin, W. G., Jr. (2006b). Failure to prolyl hydroxylate hypoxia-inducible factor alpha phenocopies VHL inactivation *in vivo. EMBO J.* **25**, 4650–4662.

Kondo, K., Kim, W. Y., Lechpammer, M., and Kaelin, W. G., Jr. (2003). Inhibition of HIF-2α is sufficient to suppress pVHL-defective tumor growth. *PLoS Biol.* **1**, E83.

Kondo, K., Klco, J., Nakamura, E., Lechpammer, M., and Kaelin, W. G. (2002). Inhibition of HIF is necessary for tumor suppression by the von Hippel-Lindau protein. *Cancer Cell.* **1**, 237–246.

Koochekpour, S., Jeffers, M., Wang, P., Gong, C., Taylor, G., Roessler, L., Stearman, R., Vasselli, J., Stetler-Stevenson, W., Kaelin, W. J., Linehan, W., Klausner, R., *et al.* (1999). The von Hippel-Lindau tumor suppressor gene inhibits hepatocyte growth factor/scatter factor-induced invasion and branching morphogenesis in renal carcinoma cells. *Mol. Cell Biol.* 1999 **19**, 5902–5912.

Krishnamachary, B., Zagzag, D., Nagasawa, H., Rainey, K., Okuyama, H., Baek, J. H., and Semenza, G. L. (2006). Hypoxia-inducible factor-1-dependent repression of E-cadherin in von Hippel-Lindau tumor suppressor-null renal cell carcinoma mediated by TCF3, ZFHX1A, and ZFHX1B. *Cancer Res.* **66**, 2725–2731.

Lando, D., Peet, D., Gorman, J., Whelan, D., Whitelaw, M., and Bruick, R. (2002). FIH-1 is a an asparaginyl hydroxylase that regulates the transcriptional activity of hypoxia inducible factor. *Genes Dev.* **16**, 1466–1471.

Lee, C. Y., Robinson, K. J., and Doe, C. Q. (2006). Lgl, Pins and aPKC regulate neuroblast self-renewal versus differentiation. *Nature* **439**, 594–598.

Lee, S., Nakamura, E., Yang, H., Wei, W., Linggi, M. S., Sajan, M. P., Farese, R. V., Freeman, R. S., Carter, B. D., Kaelin, W. G., Jr., and Schlisio, S. (2005). Neuronal apoptosis linked to EglN3 prolyl hydroxylase and familial pheochromocytoma genes: Developmental culling and cancer. *Cancer Cell* **8**, 155–167.

Lieubeau-Teillet, B., Rak, J., Jothy, S., Iliopoulos, O., Kaelin, W., and Kerbel, R. (1998). von Hippel-Lindau gene-mediated growth suppression and induction of differentiation in renal cell carcinoma cells grown as multicellular tumor spheroids. *Cancer Res.* **58**, 4957–4962.

Linehan, W. M., and Zbar, B. (2004). Focus on kidney cancer. *Cancer Cell* **6**, 223–228.

Lipscomb, E., Sarmiere, P., Crowder, R., and Freeman, R. (1999). Expression of the SM-20 gene promotes death in nerve growth factor-dependent sympathetic neurons. *J. Neurochem.* **73**, 429–432.

Lipscomb, E., Sarmiere, P., and Freeman, R. (2001). SM-20 is a novel mitochondrial protein that causes caspase-dependent cell death in nerve growth factor-dependent neurons. *J. Biol. Chem.* **276**, 11775–11782.

Lolkema, M. P., Gervais, M. L., Snijckers, C. M., Hill, R. P., Giles, R. H., Voest, E. E., and Ohh, M. (2005). Tumor suppression by the von Hippel-Lindau protein requires phosphorylation of the acidic domain. *J. Biol. Chem.* **280**, 22205–22211.

Lutz, M. S., and Burk, R. D. (2006). Primary cilium formation requires von Hippel-Lindau gene function in renal-derived cells. *Cancer Res.* **66**, 6903–6907.

Ma, W., Tessarollo, L., Hong, S. B., Baba, M., Southon, E., Back, T. C., Spence, S., Lobe, C. G., Sharma, N., Maher, G. W., Pack, S., Vortmeyer, A. O., et al. (2003). Hepatic vascular tumors, angiectasis in multiple organs, and impaired spermatogenesis in mice with conditional inactivation of the VHL gene. Cancer Res. **63**, 5320–5328.

Mahon, P., Hirota, K., and Semenza, G. (2001). FIH-1: A novel protein that interacts with HIF-1alpha and VHL to mediate repression of HIF-1 transcriptional activity. Genes Dev. **15**, 2675–2686.

Maranchie, J. K., Vasselli, J. R., Riss, J., Bonifacino, J. S., Linehan, W. M., and Klausner, R. D. (2002). The contribution of VHL substrate binding and HIF1-alpha to the phenotype of VHL loss in renal cell carcinoma. Cancer Cell. **1**, 247–255.

Nakaigawa, N., Yao, M., Baba, M., Kato, S., Kishida, T., Hattori, K., Nagashima, Y., and Kubota, Y. (2006). Inactivation of von Hippel-Lindau gene induces constitutive phosphorylation of MET protein in clear cell renal carcinoma. Cancer Res. **66**, 3699–3705.

Nakamura, E., and Kaelin, W. G., Jr. (2006). Recent insights into the molecular pathogenesis of pheochromocytoma and paraganglioma. Endocr. Pathol. **17**, 97–106.

Ohh, M., Yauch, R. L., Lonergan, K. M., Whaley, J. M., Stemmer-Rachamimov, A. O., Louis, D. N., Gavin, B. J., Kley, N., Kaelin, W. G., Iliopoulos, O., and Kaelin, W. G. (1998). The von Hippel-Lindau tumor suppressor protein is required for proper assembly of an extracellular fibronectin matrix. Mol. Cell **1**, 959–968.

Okuda, H., Hirai, S., Takaki, Y., Kamada, M., Baba, M., Sakai, N., Kishida, T., Kaneko, S., Yao, M., Ohno, S., and Shuin, T. (1999). Direct interaction of the beta-domain of VHL tumor suppressor protein with the regulatory domain of atypical PKC isotypes. Biochem. Biophys. Res. Commun. **263**, 491–497.

Okuda, H., Saitoh, K., Hirai, S., Iwai, K., Takaki, Y., Baba, M., Minato, N., Ohno, S., and Shuin, T. (2001). The von Hippel-Lindau tumor suppressor protein mediates ubiquitination of activated atypical protein kinase C. J. Biol. Chem. **276**, 43611–43617.

Pal, S., Claffey, K., Cohen, H., and Mukhopadhyay, D. (1998). Activation of Sp1-mediated vascular permeability factor/vascular endothelial growth factor transcription requires specific interaction with protein kinase Cζ. J. Bio. Chem. **273**, 26277–26280.

Pal, S., Claffey, K., Dvorak, H., and Mukhopadhyay, D. (1997). The von Hippel-Lindau gene product inhibits vascular permeability factor/vascular endothelial growth factor expression in renal cell carcinoma by blocking protein kinase C pathways. J. Biol. Chem. **272**, 27509–27512.

Pennacchietti, S., Michieli, P., Galluzzo, M., Mazzone, M., Giordano, S., and Comoglio, P. M. (2003). Hypoxia promotes invasive growth by transcriptional activation of the met protooncogene. Cancer Cell **3**, 347–361.

Perrotta, S., Nobili, B., Ferraro, M., Migliaccio, C., Borriello, A., Cucciolla, V., Martinelli, V., Rossi, F., Punzo, F., Cirillo, P., Parisi, G., Zappia, V., et al. (2006). Von Hippel-Lindau-dependent polycythemia is endemic on the island of Ischia: Identification of a novel cluster. Blood **107**, 514–519.

Peruzzi, B., Athauda, G., and Bottaro, D. P. (2006). The von Hippel-Lindau tumor suppressor gene product represses oncogenic beta-catenin signaling in renal carcinoma cells. Proc. Natl. Acad. Sci. USA **103**, 14531–14536.

Rankin, E. B., Higgins, D. F., Walisser, J. A., Johnson, R. S., Bradfield, C. A., and Haase, V. H. (2005). Inactivation of the arylhydrocarbon receptor nuclear translocator (Arnt) suppresses von Hippel-Lindau disease-associated vascular tumors in mice. Mol. Cell Biol. **25**, 3163–3172.

Rankin, E. B., Tomaszewski, J. E., and Haase, V. H. (2006). Renal cyst development in mice with conditional inactivation of the von Hippel-Lindau tumor suppressor. Cancer Res. **66**, 2576–2583.

Raval, R. R., Lau, K. W., Tran, M. G., Sowter, H. M., Mandriota, S. J., Li, J. L., Pugh, C. W., Maxwell, P. H., Harris, A. L., and Ratcliffe, P. J. (2005). Contrasting properties of hypoxia-inducible factor 1 (HIF-1) and HIF-2 in von Hippel-Lindau-associated renal cell carcinoma. *Mol. Cell. Biol.* **25,** 5675–5686.

Roe, J. S., Kim, H., Lee, S. M., Kim, S. T., Cho, E. J., and Youn, H. D. (2006). p53 stabilization and transactivation by a von Hippel-Lindau protein. *Mol. Cell* **22,** 395–405.

Schermer, B., Ghenoiu, C., Bartram, M., Muller, R. U., Kotsis, F., Hohne, M., Kuhn, W., Rapka, M., Nitschke, R., Zentgraf, H., Fliegauf, M., Omran, H., *et al.* (2006). The von Hippel-Lindau tumor suppressor protein controls ciliogenesis by orienting microtubule growth. *J. Cell Biol.* **175,** 547–554.

Schoenfeld, A., Davidowitz, E., and Burk, R. (1998). A second major native von Hippel-Lindau gene product, initiated from an internal translation start site, functions as a tumor suppressor. *Proc. Natl. Acad. Sci. USA* **195,** 8817–8822.

Schoenfeld, A. R., Davidowitz, E. J., and Burk, R. D. (2000). Elongin BC complex prevents degradation of von Hippel-Lindau tumor suppressor gene products. *Proc. Natl. Acad. Sci. USA* **97,** 8507–8512.

Singla, V., and Reiter, J. F. (2006). The primary cilium as the cell's antenna: Signaling at a sensory organelle. *Science* **313,** 629–633.

Smith, T. G., Brooks, J. T., Balanos, G. M., Lappin, T. R., Layton, D. M., Leedham, D. L., Liu, C., Maxwell, P. H., McMullin, M. F., McNamara, C. J., Percy, M. J., Pugh, C. W., *et al.* (2006). Mutation of von Hippel-Lindau tumour suppressor and human cardiopulmonary physiology. *PLoS Med.* **3,** e290.

Sowter, H. M., Raval, R., Moore, J., Ratcliffe, P. J., and Harris, A. L. (2003). Predominant role of hypoxia-inducible transcription factor (HIF)-1alpha versus HIF-2alpha in regulation of the transcriptional response to hypoxia. *Cancer Res.* **63,** 6130–6134.

Stickle, N. H., Cheng, L. S., Watson, I. R., Alon, N., Malkin, D., Irwin, M. S., and Ohh, M. (2005). Expression of p53 in renal carcinoma cells is independent of pVHL. *Mutat. Res.* **578,** 23–32.

Straub, J. A., Lipscomb, E. A., Yoshida, E. S., and Freeman, R. S. (2003). Induction of SM-20 in PC12 cells leads to increased cytochrome c levels, accumulation of cytochrome c in the cytosol, and caspase-dependent cell death. *J. Neurochem.* **85,** 318–328.

Tang, N., Mack, F., Haase, V. H., Simon, M. C., and Johnson, R. S. (2006). pVHL Function Is Essential for Endothelial Extracellular Matrix Deposition. *Mol. Cell Biol.* **26,** 2519–2530.

Vogel, K. S., Brannan, C. I., Jenkins, N. A., Copeland, N. G., and Parada, L. F. (1995). Loss of neurofibromin results in neurotrophin-independent survival of embryonic sensory and sympathetic neurons. *Cell* **82,** 733–742.

Yan, Q., Bartz, S., Mao, M., Li, L., and Kaelin, W. G., Jr. (2007). The HIF-2alpha N-terminal and C-terminal transactivation domains cooperate to promote renal tumorigenesis *in vivo. Mol. Cell. Biol.* **27,** 2092–2102.

Yu, F., White, S., Zhao, Q., and Lee, F. (2001). HIF-1alpha binding to VHL is regulated by stimulus-sensitive proline hydroxylation. *Proc. Natl. Acad. Sci. (USA)* **98,** 9630–5.

Zimmer, M., Doucette, D., Siddiqui, N., and Iliopoulos, O. (2004). Inhibition of hypoxia-inducible factor is sufficient for growth suppression of VHL-/- tumors. *Mol. Cancer. Res.* **2,** 89–95.

CHAPTER TWENTY

HYPOXIA-INDUCIBLE FACTOR 1 INHIBITORS

Giovanni Melillo

Contents

1. Introduction	386
2. Cell-Based High Throughput Screens	387
2.1. Cell-based HTS protocol	389
2.2. Validation of active "hits" from HTS	390
2.3. Small molecule inhibitors of HIF-1 identified in cell-based assays	392
3. Cell-Free Assays	393
3.1. Inhibition of HIF-1 DNA binding	393
3.2. Inhibition of protein–protein interaction	395
3.3. Inhibition of HIF-1 transcriptional activity	397
4. Bioassay-Directed Isolation of Natural Product HIF-1 Inhibitors	398
5. Conclusions	399
Acknowledgments	399
References	400

Abstract

The tremendous progress in our understanding of the molecular mechanisms underlying the presence and consequences of hypoxia in human cancers has been accompanied by renewed enthusiasm for the development of therapeutic strategies targeting hypoxic cells signaling pathways. Hypoxia-inducible factor 1 (HIF-1), a key transcriptional activator that mediates hypoxic responses, has been the focus of intense investigation and efforts to identify small molecule inhibitors or novel strategies for HIF-1 inhibition have multiplied over the last few years. Despite challenges associated with targeting transcription factors, which hamper these efforts, several strategies have been pursued. In this chapter, protocols related to screening assays, both cell-based and cell-free, are described and discussed in the context of their application for the identification of HIF-1 inhibitors. While cell-based assays offer the opportunity to reveal unidentified components of the hypoxic cell signaling pathway, cell-free targeted approaches may lead to the identification of more selective HIF-1

Developmental Therapeutics Program, SAIC Frederick, Inc., National Cancer Institute at Frederick, Frederick, Maryland

Methods in Enzymology, Volume 435
ISSN 0076-6879, DOI: 10.1016/S0076-6879(07)35020-9

inhibitors. Validation of "hits" and characterization of their mechanism of action are essential for a rational development of putative HIF-1 inhibitors in preclinical models and early clinical trials.

1. INTRODUCTION

Transcription factors (TFs) are attractive targets for the development of cancer therapeutics, for they mediate transcriptional programs of multiple aberrant signaling pathways that are ultimately responsible for maintaining the malignant phenotype (Darnell, 2002). However, most of the functions mediated by transcription factors involve protein–protein interaction and binding to the DNA, conventionally considered difficult if not impractical targets for the discovery of small molecule inhibitors.

Hypoxia-inducible factor 1 is a basic helix–loop–helix TF that is critical for the response of mammalian cells to oxygen deprivation (Wang et al., 1995). Since the HIF–alpha (α) subunit is tightly regulated by oxygen levels, HIF-1 is primarily activated in pathophysiological conditions in which oxygen is limited and adaptive responses are required for cell survival (Melillo, 2004). In this regard, over the last decade or so, evidence has accumulated indicating that HIF-1 plays an important role in cancer pro- gression by affecting the biological behavior of cancer cells and their response to therapy (Giaccia et al., 2003; Melillo, 2006; Semenza, 2003).

Several approaches to inhibit HIF-1 have been proposed and explored. Cell-based high throughput screens (HTS) exploit the ability of HIF-1 to activate transcription of reporter genes either when cells are placed under hypoxic conditions (in general, 0.1–1% O_2) or when they are exposed to HIF-1α activators, such as desferrioxamine (DFO) or cobalt chloride (CoCl) (Wang et al., 1993a). Cell-free assays have also been developed to target protein–protein interaction, protein DNA binding, or transcriptional activation. Advantages of cell-based screens are that they present the target in the cellular context and have the potential to reveal unidentified signaling molecules relevant to the target under investigation, but "hits" identified in the primary screen may require extensive mechanistic investigation (Shoemaker et al., 2002). In contrast, cell-free assays are amenable to very HTS; however, "hits" need to be validated in intact cells. The majority of HIF-1 inhibitors described in the literature have been identified by either cell-based screens or empirical discoveries (Melillo, 2006). Examples of inhibitors targeting DNA binding or transcriptional activity have also been reported, and strategies employed for their identification will be discussed.

2. Cell-Based High Throughput Screens

Hypoxia-inducible factor 1–targeted cell-based HTS rely on the ability of HIF-1 to bind a recognition sequence in the DNA (so-called hypoxia responsive element [HRE]) and activate transcription in a sequence-specific fashion (Jiang *et al.*, 1996). Earlier experiments, performed by transiently transfecting cancer cell lines with constructs containing reporter genes under control of multiple copies of HRE, had demonstrated that virtually all mammalian cells respond to oxygen deprivation with induction of HIF-1 transcriptional activity (Wang *et al.*, 1993b). Although this is generally true, culture conditions and genetic background of the cells may greatly influence the inducibility of HIF-1α and HIF-1 transcriptional activity. For instance, von Hippel-Lindau (VHL) status and confluence of cells will significantly affect the inducibility of HIF-1α and reporter genes.

Hypoxia-inducible factor–dependent transcription can be demonstrated at various oxygen concentrations, with a gradient effect starting around 5% O_2 and becoming maximal at 0.5 to 1% O_2. Different reporter genes, HRE sequences, and vectors have been used, all of which appear to efficiently detect HIF-1 activation. Cell-based screens have been performed by culturing cells under hypoxic conditions (0.1–1% O_2) or by treating cells with HIF-1 inducers (DFO, CoCl). DFO induces HIF-1α by inhibiting the activity of prolyl hydroxylases, enzymes involved in the proteasomal degradation of HIF-α. These agents, conventionally defined hypoxia mimetic, including DFO and CoCl, should not be considered equivalent to hypoxia, but rather HIF-1α activators, and results should be interpreted accordingly.

Luciferase is the most common reporter gene that has been used for screening purposes, for it can be easily measured in high throughput assays as well as in animal models by noninvasive imaging of bioluminescence. Stable transfection of HRE-luciferase reporter plasmids in mammalian cancer cell lines has been extensively employed to generate useful reagents for screening purposes. We have successfully generated several human cancer cell lines stably transfected with luciferase reporter plasmids containing three copies of a canonical HRE from the inducible nitric oxide synthase gene (Fig. 20.1A) (Rapisarda *et al.*, 2002). Although inducibility of the reporter gene in stable cell lines may not necessarily parallel the one that can be achieved in transient transfection experiments, it is generally robust enough to be used in HTS. As an example, the U251-HRE cell line that has been generated in our laboratory consistently shows an 8- to 12-fold induction of luciferase expression when cells are cultured under hypoxic conditions (1% O_2) or in the presence of DFO (100 μM) relative to untreated normoxic cells (see Fig. 20.1B) (Rapisarda *et al.*, 2002).

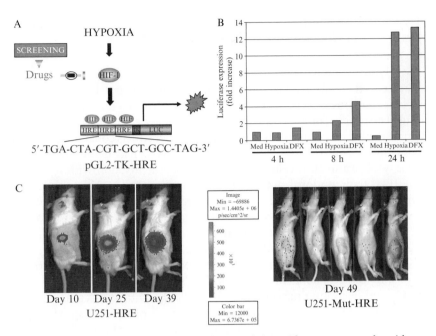

Figure 20.1 Targeting HIF-1 *in vitro* and *in vivo*. (A) Luciferase reporter plasmid containing 3 copies of a canonical HRE from the iNOS-HRE promoter. (B) Luciferase expression in U251-HRE cells cultured under normoxic conditions or in the presence of desferrioxamine (100 μM). (C) U251-HRE or U251-Mut-HRE where implanted subcutaneously in SCID or nude mice. Imaging was performed with a Xenogen® instrument as described in Rapisarda *et al.*, 2004).

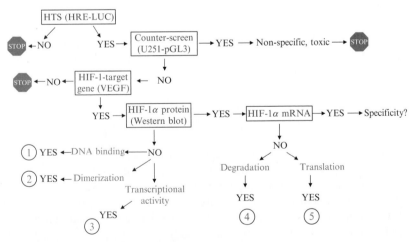

Figure 20.2 Simplified algorithm for validation of "hits" identified in a HTS targeting HIF-1.

Cell lines engineered to express luciferase in an HRE-dependent fashion can also be used to image bioluminescence in xenograft or orthotopic mouse models by using commercially available imaging techniques (Xenogen®) (Rapisarda *et al.*, 2004b). *In vivo* noninvasive imaging of HIF-1–inducible gene expression is particularly useful for the validation of putative HIF-1 inhibitors in a relevant biological model and can be assessed concomitantly with antitumor activity (see Fig. 20.1C) (Rapisarda *et al.*, 2004b).

2.1. Cell-based HTS protocol

The following is the protocol used at the National Cancer Institute at Frederick for HIF-1 HTS using U251-HRE cells.

- Generation of U251-HRE stable cell lines: U251 human glioma cells were co-transfected with pGL2-TK-HRE containing three copies of the inducible nitric oxide synthase (iNOS)-HRE in front of TK-luciferase and an expression vector carrying the neomycin resistance gene for selection in mammalian cells (ratio 1:100). Twenty-four hours following transfection, reagents were removed and cells were allowed to recover for 24 h before addition of selection medium containing the antibiotic G418 at 500 μM (Invitrogen-Life Technologies, Inc., Carlsbad, CA). U251-HRE are routinely maintained in RPMI 1640 medium (Whittaker Bioproducts, Walkersville, MD) supplemented with 5% heat-inactivated fetal bovine serum (Hyclone Laboratories, Logan, UT), penicillin (50 IU/ml), streptomycin (50 μg/ml), 2 mM glutamine (all from Invitrogen-Life Technologies, Inc., Carlsbad, CA), and 200 μg/ml G418. Cells are maintained at 37° in a humidified incubator containing 21% O_2, 5% CO_2 in air (referred to as normoxic conditions). Hypoxia treatment is performed by placing cells in a modular incubator chamber (Billups-Rotheberg Inc., Del Mar, CA), flushing them with a mixture of 1% O_2, 5% CO_2, and 94% nitrogen for 20 min, and then placing them at 37°. Plasmids with mutated HRE sequences can be used to confirm that the expression of luciferase is indeed HIF-1–dependent.
- U251-pGL3 control cell lines: U251 cells were transfected as previously stated with pGL3-control (Promega, Madison, WI), which contains the firefly luciferase coding sequence under control of the SV40 promoter and enhancer sequences. A secondary screen using control cell lines essential to validate the specificity of HIF-1 inhibitors can be run in parallel to a primary screen or to validate "hits" and generate a dose-response curve.
- U251-HRE cells are inoculated into 384-well white, flat-bottom plates (Costar, Catalog # 3704) at 3000 cells/well with a Beckman Biomek

2000 Laboratory Automation Workstation in a volume of 25 μl and incubated for 24 h at 37°, 5 % CO_2, and ambient O_2.

- Experimental agents at the appropriate concentrations are added in a volume of 25 μl using the Biomek 2000 and, after a 20-h incubation in a hypoxia chamber (Billups Rothenberg, MIC 101) at 37°, 5 % CO_2, and 1 % O_2, the plates are removed and incubated at room temperature, 5 % CO_2, and ambient O_2 for 1.5 h.
- 40 μl of Bright Glo luciferase reagent (Promega, Catalog # E26500) is added with the Biomek 2000, and, after 3 min, luminescence is measured using a Tecan Ultra Multifunction Plate Reader in luminescence mode.
- Appropriate control cells are treated identically except that they are incubated at 37°, 5 % CO_2, and ambient O_2.
- Compound toxicity is assayed in parallel utilizing the sulphorhodamine B or Alamar blue assays.

2.2. Validation of active "hits" from HTS

The main advantage of cell-based HIF-1–targeted HTS is the potential to identify small molecules that inhibit unidentified components of the hypoxic cell signaling pathway. An inevitable consequence is that a systematic effort to elucidate the mechanism of action of "hits" is required, with substantial investment of time and resources. A number of validation assays has been utilized to elucidate the mechanism of action of small molecule inhibitors identified in HTS (Figure 2). The following tests are routinely performed in our laboratory and may constitute a useful algorithm for validation purposes.

- Since HIF-1–targeted HTS rely on synthetic promoters driving the expression of a reporter gene (e.g., luciferase), control cell lines and/or plasmids in which luciferase is constitutively expressed should be used to rule out nonspecific effects. If multiple doses of the compounds are used, the ratio of the EC50 observed in control versus HRE cells (e.g., U251-pGL3/U251-HRE) can be used as a fairly reliable indicator of specificity. These studies will also provide information on potential toxicity of the compounds, which will be reflected in a decreased expression of luciferase.
- The first level of validation of putative HIF-1 inhibitors should be the evaluation of their effect on the hypoxic induction of messenger RNA (mRNA) expression of endogenous HIF-1 target genes (Rapisarda et al., 2002). Prototype genes that are generally expressed in an HIF-1–dependent fashion in the majority of mammalian cells include vascular endothelial growth factor (VEGF), glucose transporter 1 (Glut-1), and glycolytic enzymes. It should be emphasized, however, that the involvement of HIF-1 in the induction of some of these genes may vary among

cell lines tested, so that a direct correlation with HIF-1 inhibition can only be inferred if the experimental system has been rigorously validated.

- Further studies to elucidate the mechanism of action of HIF inhibition may include Western blot analysis to measure the levels of HIF-1α protein in cells cultured under hypoxia or in the presence of HIF-α activators (DFO, CoCl). Examples of small molecules that block induction of HIF-1α by either hypoxia or chemical activators (but not both) have been described (Chau et al., 2005) that point to selective mechanisms of action that should be further investigated.

- Decreased levels of HIF-1α protein may reflect an impairment of protein translation or an increase of its degradation. To address the involvement of VHL-dependent proteasomal degradation, experiments can be conducted in the presence of proteasome inhibitors, such as PS-341 or MG-132, or in cells with functional impairment of the VHL protein, in which HIF-α is constitutively expressed under normoxic conditions. Interestingly, evidence has been provided that, in the presence of proteasome inhibitors, HIF-1 is paradoxically transcriptionally inactive (Kaluz et al., 2006). To further investigate involvement of protein translation or degradation, more sophisticated experiments may be required, including pulse-chase experiments and assessment of de novo HIF-1α protein synthesis. Levels of HIF-1α mRNA should be evaluated to rule out a direct effect on mRNA expression, which, if present, requires further experiments to address specificity (Rapisarda et al., 2004a).

- If the compound does not inhibit hypoxic induction of HIF-1α protein accumulation, potential mechanisms of action to consider include inhibition of DNA binding, dimerization, and/or transcriptional activity. Inhibition of HIF-1 DNA binding activity can be assessed by using enzyme-linked immunosorbent assay (ELISA) DNA binding assays (described in the next paragraph), electromobility gel shift assays (EMSA), which can be performed using nuclear extracts or expressed protein, and chromatin immunoprecipitation (ChIP) experiments, which provide evidence of activity on endogenous promoters (Kong et al., 2005).

- Inhibition of protein–protein interaction may be detected in cell-free assays using expressed proteins (described in next section) or immuno-precipitation experiments of nuclear proteins from hypoxia-treated cells (Park et al., 2006).

- Finally, mammalian two-hybrid systems, based on the commercially available GAL4 binding reporter system, have been used to test the interaction between the CH1 domain of p300 and the C-TAD of HIF-1α and to assess the potential effect of small molecules on transcriptional activity. Under these conditions, luciferase is expressed only in the presence of a physical interaction between these two domains of p300 and HIF-1α, which may be blocked by a small molecule inhibitor. Additionally, cell-free assays using the CH1 domain of p300 and an HIF-1α C-TAD peptide have also been used to screen large chemical libraries or to validate inhibitors identified in cell-based assays (Kung *et al.*, 2004).

2.3. Small molecule inhibitors of HIF-1 identified in cell-based assays

The great majority of HIF-1 inhibitors described so far have been identified either in cell-based screens or by empirical discoveries during evaluation of HIF-1 activity in cultured cancer cell lines (Melillo, 2006).

One of the first cell-based HTS-targeting HIF-1 has been developed and implemented at the Developmental Therapeutics Program of the National Cancer Institute (NCI) at Frederick. By using U251-HRE glioma cells engineered to express luciferase under control of three copies of a canonical hypoxia responsive element, several chemical libraries were screened to identify HIF-1 inhibitors (Rapisarda *et al.*, 2002). A parallel screen using U251-pGL3 cells engineered to express luciferase under control of a constitutive oxygen-independent promoter was also conducted to assess specificity. A pilot screen of the NCI "Diversity Set," comprising approximately 2000 compounds, led to the identification of topotecan, a camptothecin analog and topoisomerase I inhibitor, as a potent inhibitor of HIF-1α protein translation (Rapisarda *et al.*, 2004a). Screening of larger libraries have confirmed that camptothecin analogs and topoisomerase I inhibitors are a consistent and potent class of HIF-1α inhibitors. Topotecan is an example of HIF-1α inhibitor that blocks its translation. A similar mechanism of action has been proposed for 2ME2, an inhibitor of microtubule polymerization currently in clinical development as an antiangiogenic agent (Mabjeesh *et al.*, 2003). Inhibition of the following targets has instead been associated with increased HIF-1α degradation: Hsp90, a molecular chaperone whose inhibition by geldanamycin, or its analogs 17-AAG and 17-DMAG, leads to HIF-α degradation by a VHL-independent proteasome-dependent pathway (Isaacs *et al.*, 2002; Mabjeesh *et al.*, 2002) histone deacetylase (HDAC) inhibition, by virtue of direct acetylation of HIF-1α or increased acetylation of Hsp90 and indirect degradation of HIF-1α (Kong *et al.*, 2006; Qian *et al.*, 2006). Other small molecule inhibitors of HIF-1α have been associated with inhibition of HIF-1α protein accumulation, though the exact effect on HIF-1α translation and/or degradation has not been conclusively shown. These inhibitors include YC-1, a guanylate cyclase activator (Yeo *et al.*, 2003), PX-478, which showed potent antitumor activity in multiple xenograft models (Welsh *et al.*, 2004), and thioredoxin 1 inhibitors, which can inhibit either HIF-1α protein accumulation or transcriptional activity (Jones *et al.*, 2006; Welsh *et al.*, 2003).

Several inhibitors of signaling pathways frequently dysregulated in human cancers have also been associated with inhibition of HIF-1 activity, including inhibitors of the mTOR pathway, AKT, EGFR/Her2Neu, and BCR-ABL/c-Kit (Melillo, 2006). These agents may inhibit signaling pathways leading to normoxic induction of HIF-1α by growth factor, which is thought to primarily increase HIF-1α protein translation. Most of these agents are either approved for treatment of human cancers or are in clinical or advanced preclinical development. It will be important to establish whether inhibition of HIF contributes to the therapeutic activity of these agents or may be used as a potential biomarker of biological activity in order to identify responsive patients.

3. CELL-FREE ASSAYS

A number of cell-free assays have been designed and utilized to identify inhibitors of HIF-1 activity. Unlike cell-based HTS, these assays target domains of HIF-1α that have been associated with specific functions. Although these assays are more targeted and do not require elaborate mechanistic investigation relative to cell-based HTS, their degree of success has been, perhaps not surprisingly, very dismal. If, on the one hand, this poor output bears the challenge of targeting protein–protein interaction or protein–DNA binding, on the other hand, it also reflects the lack of detailed structural information regarding domains of HIF-1 involved in these functions (Yang *et al.*, 2005). Recent progress in the elucidation of the structure of the Per/ARNT/Sim (PAS)-B domain of HIF-2α may represent an encouraging example for a more rational strategic approach to the identification of compounds that may specifically interact with this domain (Erbel *et al.*, 2003).

3.1. Inhibition of HIF-1 DNA binding

Following accumulation and translocation to the nucleus, HIF-α dimerizes with HIF-1beta (*β*)/ARNT and binds DNA in regions containing the core sequence 5′-RCGTG-3′ (HRE), which is present in the regulatory regions of HIF-1 target genes (Jiang *et al.*, 1996). The possibility to modulate DNA binding of transcription factors is intellectually attractive and has been explored by two different approaches: design of sequence-specific polyamides and screening using DNA-binding assays. Both approaches are elaborated in the following sections. Several methods are routinely used to detect binding of transcription factors to specific DNA sequences, including EMSA and ChIP. However, these assays are not amenable to HTS and may be used for validation and further mechanistic investigation of "hits" identified in HTS. Next, an ELISA assay based on binding to a canonical HRE of expressed HIF-α and HIF-β proteins is described (Kong *et al.*, 2005).

3.1.1. ELISA HIF-1 DNA binding assay

- Preparation of recombinant proteins: the basic helix-loop-helix (bHLH) PAS domain of the human HIF-1α complementary DNA (cDNA; 1167 bp) is polymerase chain reaction (PCR)-amplified and cloned into pTriEx4-His tag expression vector.
- Protein is expressed in SF9 insect cells with recombinant baculoviruses, which are prepared by co-transfecting SF9 cells with pTriEx4-His tag expression vector and BacVector®-3000 Triple Cut Virus DNA using Eufectin™ transfection reagent (Novagen, Madison, WI).
- Expressed protein is purified by affinity chromatography on nickel resin columns.
- The bHLH-PAS domain of the human HIF-1β cDNA (1425 bp) is PCR-amplified and cloned into pFLAG1 expression vector (Sigma).
- Protein is expressed in *Escherichia coli* DH5a cells and purified by M2 affinity chromatography.
- About 30 pmol of 5′ end biotin–labeled double-stranded DNA oligonucleotide encompassing a canonical HIF-1–binding site is added onto a streptavidin-immobilized ELISA plate (Pierce, Rockford, IL).
- Sequences of the sense strand of the double-stranded oligonucleotides used are as follows: wild-type HRE, 5′-GTGCTACGTGCTGCCTAG-3′; mutant HRE, 5′-GTGCTAAAAGCTGCCTAG-3′.
- Recombinant HIF-1α–bHLH-PAS (10 pmol) and HIF-1β–bHLH-PAS (6 pmol) are mixed in 50 μl of 1× buffer (25 mM Tris-HCl [pH 7.6], 100 mM KCl, 0.2 mM ethylenediaminetetraacetic acid [EDTA], 20% glycerol, 5 mM dithiothreitol).
- After incubation for 5 min at room temperature, protein mix is added to the plate and incubated at room temperature for 1 h.
- The protein complex is detected by anti-His tag monoclonal antibody (mAb) (Novagen, Madison, WI) at a dilution of 1:2000 and anti-mouse horseradish peroxidase (HRP)–conjugated Ab (Sigma) at a dilution of 1:10000.
- After incubating all wells with 100 μl of TMB substrate solution (Sigma) for 10 min at room temperature, 100 μl of 0.5 M H_2SO_4 is added and absorbance is read at 450 nm with a reference wavelength of 655 nm.
- Oligonucleotide competition experiments are performed with 50 pmol of non-biotinylated oligonucleotides.

3.1.2. Small molecule inhibitors of HIF-1 DNA binding

Two main examples of compounds that inhibit HIF-1 by blocking its binding to the HRE have been published. Polyamides, oligomers that can be designed to specifically modulate the binding of TFs to DNA, have been pioneered by Peter Dervan's group. Evidence has been provided that specific polyamides can be designed to efficiently block binding of HIF-1 to the cognate sequence present in the VEGF promoter, thus blocking VEGF expression (Olenyuk et al., 2004). A potential limitation of poly-amides is their delivery in the nucleus, where their activity takes place. However, significant progress in the delivery of polyamide has been recently reported, which should now facilitate the preclinical testing of this approach (Nickols et al., 2007). The second example of agents that interfere with HIF-1 DNA binding is echinomycin, a small molecule identified using an ELISA DNA binding assay (previously described) (Kong et al., 2005) and previously known to bind DNA in a sequence-specific fashion (Van Dyke and Dervan, 1984). Inhibition of DNA-binding activity by echinomycin was shown in ELISA as well as in EMSA and ChIP (Kong et al., 2005). Echinomycin inhibited HIF-1, but not AP-1– or NF-κB–binding to the corresponding recognition sequences, and blocked HIF-1–binding to the HRE present in the VEGF promoter. Both HIF-1–targeted polyamides and echinomycin only inhibited a relatively small number of genes when global transcriptional profile was assessed by gene array experiments and coherently inhibited HIF-1–dependent targeted genes (Kong et al., 2005).

3.2. Inhibition of protein–protein interaction

Several domains of HIF-α, including the HLH and PAS domains, mediate dimerization with ARNT, which is required for HIF-1 transcriptional activity (Erbel et al., 2003; Jiang et al., 1996). The lack of information regarding the structural conformation of these domains significantly ham-pers efforts aimed at identifying small molecule inhibitors of HIF dimeriza-tion. Recently, structural information regarding HIF-2α PAS-B domain was published and may provide the foundation for a more rational approach to identifying small molecules that inhibit PAS-mediated dimerization (Erbel et al., 2003). Efforts to better characterize domains involved in protein dimerization, including the PAS-A and the HLH domains, might provide further therapeutic opportunities.

Screening approaches using cell-free assays based on recombinant expression of PAS-A or PAS-B domains of HIF-1α and ARNT have also been described, and protocols used are detailed in the following section (Park et al., 2006).

3.2.1. HIF-1α/ARNT PAS-A protein–protein interaction ELISA

- Expression of recombinant proteins: the bHLH (amino acids 1–86) and PAS-A (amino acids 86–165) domains of the human HIF-1α cDNA are PCR-amplified and cloned into pET28b (+)-His tag expression vector (Novagen, Madison, WI).
- For expression in mammalian cells, the PAS-A and bHLH-PAS (amino acids 1–380) domains were cloned into pTriEx-4 vector (Novagen).

- The PAS-A (amino acids 159–240) and bHLH-PAS (amino acids 1–475) domains of the human HIF-1β cDNA were PCR-amplified and cloned into pFlag1 expression vector (Sigma, St. Louis, MO).
- All plasmids are verified by DNA sequencing.
- Recombinant HIF-1α proteins are expressed in *E. coli* BL21 (DE3) cells with isopropyl-β-D-thiogalactopyranoside (IPTG).
- HIF-1α recombinant proteins are extracted using a Bugbuster protein extraction reagent (Novagen) following the manufacturer's protocol.
- HIF-1α recombinant proteins are purified by fast protein liquid chromatography (FPLC) with HisTrp^{TM}-HPE (Amersham Biosciences, Uppsala, Sweden) and by high performance liquid chromatography (HPLC) with C18 column (4.6 × 250 mm, Vydac^{TM}) (Grace Vydac, Columbia, MD).
- Recombinant HIF-1β proteins are expressed in *E. coli* DH5α cells in the presence of IPTG.
- The soluble fractions of HIF-1β proteins are prepared by CelLytic^{TM} B-II (Sigma) reagent and purified by anti–FLAG-M2 agarose affinity chromatography (Sigma).
- The molecular weight of expressed HIF-1α–bHLH, HIF-1α–PAS-A, HIF-1β–PAS-A, and HIF-1β–bHLH-PAS protein is 10, 12, 11, and 56 kDa, respectively.
- Anti-FLAG M2 monoclonal antibody (0.4 μg per well) is immobilized on a high-binding ELISA plate (Corning Inc., Corning, NY) in the presence of 50 mM sodium carbonate buffer, pH 9.7 at 4°.
- After blocking with 3% BSA in phosphate-buffered saline (PBS) solution for 1.5 h, the plate is washed with PBS.
- The soluble fraction of expressed HIF-1β–PAS-A or HIF-1β–bHLH-PAS is added to the plate for 1 h at room temperature.
- Plates are then extensively washed with washing buffer (1× buffer, 200 mM HEPES, pH 7.0, 500 mM KCl, 30 mM MgCl$_2$, 2 mM dithiothreitol [DTT], and 10 mM EDTA) to remove binding of nonspecific proteins.
- HIF-1α–PAS-A or HIF-1α–bHLH is added to the well, and the complex with HIF-1β protein is allowed to occur in the presence or absence of test compounds.
- Unbound HIF-1α protein is washed out with 0.5× washing buffer.
- HIF-1α–His/HIF-1α complex is detected using anti–His-HRP conjugate solution (QIAGEN, Inc., Valencia, CA) in the presence of TMB substrate (Sigma).
- Validation assays to exclude compounds that interfere with the detection system are developed by immobilizing HIF-1α–PAS-A protein in the absence of HIF-1β–PAS-A on a 96-well plate at 4° for 16 h.
- After blocking with 3% BSA/PBS, compounds are added in 0.5× buffer for 1 h at room temperature.

- Anti–His-HRP conjugate/TMB reagent was added to detect HIF-1α–PAS-A–His protein in the absence or presence of test compounds.

3.2.2. Inhibitors of HIF-1α/ARNT dimerization

In the absence of structural information regarding the domains involved in protein dimerization, the likelihood of success of screening efforts is substantially decreased. Our group at NCI has used a PAS-A–based assay to validate "hits" identified in a cell-based HTS. Although transient transfection experiments did show that the PAS-A domain may be targeted for inhibition and that its inhibition has functional consequences on HIF-1 activity, only one compound (NSC 50352) was found to be active in the cell-free assay (Park *et al.*, 2006) but not in cell-based assays that were used for validation purposes. Although it is possible that screening of large chemical libraries using assays based on protein–protein interaction might lead to the identification of active compounds, more targeted approaches relying on a better understanding of the structural conformation of PAS domains may be a more successful strategy.

3.3. Inhibition of HIF-1 transcriptional activity

Hypoxia-inducible factor 1 transcriptional activity is mediated by two transactivation domains (N-TAD and C-TAD). Hydroxylation of an asparagine residue at position 803 of the C-TAD, an oxygen-dependent reaction mediated by the enzyme factor-inhibiting HIF-1 (FIH-1), prevents the recruitment of coactivators p300/CREB-binding protein (CBP) under normoxic conditions. Under hypoxic conditions, FIH-1 no longer hydroxylates Asn803, which allows the binding of HIF-1α to p300/CBP and maximal transcriptional activity.

An HTS for the identification of small molecules capable of disrupting the interaction of HIF-1α with p300 has been described (Kung *et al.*, 2004). The assay was based on the interaction between a biotinylated, 41 amino acid polypeptide corresponding to the minimal p300/CBP–binding domain of HIF-1α (TADC; residues 786–826) and the minimal HIF-1α–binding domain of p300 (CH1; residues 302–423) fused to glutathione-S-transferase (GST-CH1). The HIF-1α peptide was immobilized on streptavidin-coated multiwell plates, and binding of GST-CH1 to immobilized TADC, in the presence or absence of compounds, was probed by europium-conjugated anti-GST antibody and detected by time-resolved fluorescence (Kung *et al.*, 2004).

Unfortunately, because of the unstructured conformation of C-TAD before its interaction with p300 (Dames *et al.*, 2002), these efforts have been largely disappointing. Chetomin, a dithiodiketopiperazine metabolite of the fungus *Chaetomium* species, was shown to inhibit HIF-1 transcriptional activity and to mediate antitumor activity in xenograft models

(Kung *et al.*, 2004). These results provided proof-of-principle that HIF-1 transcriptional activity can be inhibited and that this inhibition is sufficient to mediate antitumor effects.

4. Bioassay-Directed Isolation of Natural Product HIF-1 Inhibitors

Natural products are an invaluable source of active compounds, including anticancer agents, many of which are currently used in the clinic. An increasing number of active natural products are being reported in the literature. For many of these compounds, the exact mechanism of action is not completely clear; however, they represent a potential source of information regarding novel targets and mechanisms of action that may be associated with HIF inhibition.

Bioactivity-driven research with natural products is frequently criticized as too cumbersome and slow to keep pace with a high-throughput screening environment; however, technological and methodological advances over the past decade have made it possible to pursue natural product leads from HTS in a very efficient manner. At NCI, the phases of work in natural products are comprised of:

- Extraction: both aqueous and organic extracts are prepared for each specimen collected (whether microorganism, plant, or marine invertebrate or alga). These extracts are not pretreated or pre-fractionated prior to screening.
- Biological screening: all extracts are plated initially into 96-well master plates, from which 96-, 384-, or even 1536-well daughter plates can be prepared for various assays.
- Dereplication: confirmed bioactive extracts are then subjected to preparative HPLC using a multiplicity of detectors—evaporative light-scattering detection (ELSD), ultraviolet (UV)/vis, $+/-$ electrospray ionization mass spectrometry (ESIMS), and bioassay. The structural information collected for each active fraction (UV, MW, relative polarity) coupled with taxonomic information on the source organism can then be used to search available databases (e.g., Beilstein, Berdy, Chapman & Hall, MarinLit) to identify known compounds responsible for the observed bioactivity. In a majority of cases, a putative identification can be made at this point and could be confirmed by nuclear magnetic resonance (NMR) analysis of the residual material in the active fraction(s).
- Isolation and identification: should the bioactivity not be attributable to a previously known natural product during the dereplication analysis, then isolation, purification, and structure elucidation of the active constituent(s) must be undertaken. Using the chromatographic results

of the dereplication analysis as a guide, one can usually obtain pure bioactive compounds in two to three separation steps, usually a combination of gel permeation and vacuum liquid chromatographies and HPLC. Two dimensional (2D) NMR and MS analysis generally suffice for the identification of the isolated novel compounds. Once the bioactivity of the new compound(s) is confirmed, scaleup chromatography can provide the material necessary for secondary biological characterization. An example of this process at work to identify HIF-1 inhibitory natural products can be found in the protocol summarized by Klausmeyer *et al.* (2007).

5. Conclusions

Identification of HIF-1 inhibitors has been attempted using several different experimental approaches. Cell-based HTS have provided the majority of hits and the most useful information regarding novel pathways that are involved in HIF-1 regulation. The majority of inhibitors identified in cell-based assays are nonselective, and inhibition of their primary target is also associated with HIF-1–independent effects.

More targeted approaches have also been used, including inhibition of PAS domains and TAD; they have provided useful information regarding the role of these domains as drug targets. Unfortunately, none of the compounds identified in these more targeted approaches has progressed through the preclinical models to be found of interest for further clinical development. A better understanding of the molecular structure of domains that mediate critical functions of HIF-1 will undoubtedly lead to the identification of active compound that may be of interest for preclinical and clinical testing.

It can be anticipated that results of cell-based screens as well as empirical discoveries will increase the number of HIF inhibitors over the next few years and will uncover novel targets involved in the regulation of HIF activity. The rational development of these active leads in relevant preclinical models and tailored early clinical trials is essential for maximizing the results of this exciting area of drug development.

ACKNOWLEDGMENTS

I would like to thank John H. Cardellina for his contribution to the section on natural products, Robert H. Shoemaker for his support and helpful discussions, Nick Scudiero for performing the screen, and all of the members of the Screening Technologies Branch, DTP, NCI, who have contributed to this project. I also would like to thank all of the members of my laboratory who have contributed over the years to this work, in particular Annamaria

Rapisarda, Dehe Kong, Eun Jung Park, Badarch Uranchimeg, Maura Calvani, Mark Creighton-Gutteridge, and Daniela Trisciuoglio.
This project has been funded in whole or in part with federal funds from the National Cancer Institute, National Institutes of Health, under contract N01-CO-12400. The content of this publication does not necessarily reflect the views or policies of the Department of Health and Human Services, nor does mention of trade names, commercial products, or organizations imply endorsement by the US Government. This research was supported (in part) by the Developmental Therapeutics Program in the Division of Cancer Treatment and Diagnosis of the National Cancer Institute.

REFERENCES

Chau, N. M., Rogers, P., Aherne, W., Carroll, V., Collins, I., McDonald, E., Workman, P., and Ashcroft, M. (2005). Identification of novel small molecule inhibitors of hypoxia-inducible factor-1 that differentially block hypoxia-inducible factor-1 activity and hypoxia-inducible factor-1alpha induction in response to hypoxic stress and growth factors. *Cancer Res.* **65**, 4918–4928.

Dames, S. A., Martinez-Yamout, M., De Guzman, R. N., Dyson, H. J., and Wright, P. E. (2002). Structural basis for HIF-1α/CBP recognition in the cellular hypoxic response. *Proc. Natl Acad. Sci. USA* **99**, 5271–5276.

Darnell, J. E., Jr. (2002). Transcription factors as targets for cancer therapy. *Nat. Rev. Cancer* **2**, 740–749.

Erbel, P. J., Card, P. B., Karakuzu, O., Bruick, R. K., and Gardner, K. H. (2003). Structural basis for PAS domain heterodimerization in the basic helix-loop-helix PAS transcription factor hypoxia-inducible factor. *Proc. Natl. Acad. Sci. USA* **100**, 15504–15509.

Giaccia, A., Siim, B. G., and Johnson, R. S. (2003). HIF-1 as a target for drug development. *Nat. Rev. Drug Discov.* **2**, 803–811.

Isaacs, J. S., Jung, Y. J., Mimnaugh, E. G., Martinez, A., Cuttitta, F., and Neckers, L. M. (2002). Hsp90 regulates a von Hippel-Lindau–independent hypoxia-inducible factor-1 alpha-degradative pathway. *J. Biol. Chem.* **277**, 29936–29944.

Jiang, B. H., Rue, E., Wang, G. L., Roe, R., and Semenza, G. L. (1996). Dimerization, DNA binding, and transactivation properties of hypoxia-inducible factor 1. *J. Biol. Chem.* **271**, 17771–17778.

Jones, D. T., Pugh, C. W., Wigfield, S., Stevens, M. F. G., and Harris, A. L. (2006). Novel thioredoxin inhibitors paradoxically increase hypoxia-inducible factor-α expression but decrease functional transcriptional activity, DNA binding, and degradation. *Clin. Cancer Res.* **12**, 5384–5394.

Kaluz, S., Kaluzova, M., and Stanbridge, E. J. (2006). Proteasomal inhibition attenuates transcriptional activity of hypoxia-inducible factor 1 (HIF-1) via specific effect on the HIF-1α C-terminal activation domain. *Mol. Cell Biol.* **26**, 5895–5907.

Klausmeyer, P., McCloud, T. G., Melillo, G., Scudiero, D. A., Cardellina, J. H., and Shoemaker, R. H. (2007). Identification of a new natural camptothecin analogue in targeted screening for HIF-1α inhibitors. *Planta. Med.* **73**, 49–52.

Kong, D., Park, E. J., Stephen, A. G., Calvani, M., Cardellina, J. H., Monks, A., Fisher, R. J., Shoemaker, R. H., and Melillo, G. (2005). Echinomycin, a small-molecule inhibitor of hypoxia-inducible factor-1 DNA-binding activity. *Cancer Res.* **65**, 9047–9055.

Kong, X., Lin, Z., Liang, D., Fath, D., Sang, N., and Caro, J. (2006). Histone deacetylase inhibitors induce VHL and ubiquitin-independent proteasomal degradation of hypoxia-inducible factor 1alpha. *Mol. Cell Biol.* **26**, 2019–2028.

Kung, A. L., Zabludoff, S. D., France, D. S., Freedman, S. J., Tanner, E. A., Vieira, A., Cornell-Kennon, S., Lee, J., Wang, B., Wang, J., Memmert, K., Naegeli, H. U., et al. (2004). Small molecule blockade of transcriptional coactivation of the hypoxia-inducible factor pathway. Cancer Cell **6**, 33–43.

Mabjeesh, N. J., Post, D. E., Willard, M. T., Kaur, B., Van Meir, E. G., Simons, J. W., and Zhong, H. (2002). Geldanamycin induces degradation of hypoxia-inducible factor 1alpha protein via the proteosome pathway in prostate cancer cells. Cancer Res. **62**, 2478–2482.

Mabjeesh, N. J., Escuin, D., LaVallee, T. M., Pribluda, V. S., Swartz, G. M., Johnson, M. S., Willard, M. T., Zhong, H., Simons, J. W., and Giannakakou, P. (2003). 2ME2 inhibits tumor growth and angiogenesis by disrupting microtubules and dysregulating HIF. Cancer Cell **3**, 363–375.

Melillo, G. (2004). HIF-1: A target for cancer, ischemia and inflammation—too good to be true? Cell Cycle **3**, 154–155.

Melillo, G. (2006). Inhibiting hypoxia-inducible factor 1 for cancer therapy. Mol. Cancer Res. **4**, 601–605.

Nickols, N. G., Jacobs, C. S., Farkas, M. E., and Dervan, P. B. (2007). Improved nuclear localization of DNA-binding polyamides. Nucleic Acids Res. **35**, 363–370.

Olenyuk, B. Z., Zhang, G. J., Klco, J. M., Nickols, N. G., Kaelin, W. G., Jr., and Dervan, P. B. (2004). Inhibition of vascular endothelial growth factor with a sequence-specific hypoxia response element antagonist. Proc. Natl. Acad. Sci. USA **101**, 16768–16773.

Park, E. J., Kong, D., Fisher, R., Cardellina, J., Shoemaker, R. H., and Melillo, G. (2006). Targeting the PAS-A domain of HIF-1α for development of small molecule inhibitors of HIF-1. Cell Cycle **5**, 1847–1853.

Qian, D. Z., Kachhap, S. K., Collis, S. J., Verheul, H. M. W., Carducci, M. A., Atadja, P., and Pili, R. (2006). Class II histone deacetylases are associated with VHL-independent regulation of hypoxia-inducible factor 1α. Cancer Res. **66**, 8814–8821.

Rapisarda, A., Uranchimeg, B., Scudiero, D. A., Selby, M., Sausville, E. A., Shoemaker, R. H., and Melillo, G. (2002). Identification of small molecule inhibitors of hypoxia-inducible factor 1 transcriptional activation pathway. Cancer Res. **62**, 4316–4324.

Rapisarda, A., Uranchimeg, B., Sordet, O., Pommier, Y., Shoemaker, R. H., and Melillo, G. (2004a). Topoisomerase I-mediated inhibition of hypoxia-inducible factor 1: Mechanism and therapeutic implications. Cancer Res. **64**, 1475–1482.

Rapisarda, A., Zalek, J., Hollingshead, M., Braunschweig, T., Uranchimeg, B., Bonomi, C. A., Borgel, S. D., Carter, J. P., Hewitt, S. M., Shoemaker, R. H., and Melillo, G. (2004b). Schedule-dependent inhibition of hypoxia-inducible factor-1alpha protein accumulation, angiogenesis, and tumor growth by topotecan in U251-HRE glioblastoma xenografts. Cancer Res. **64**, 6845–6848.

Semenza, G. L. (2003). Targeting HIF-1 for cancer therapy. Nat. Rev. Cancer **3**, 721–732.

Shoemaker, R. H., Scudiero, D. A., Melillo, G., Currens, M. J., Monks, A. P., Rabow, A. A., Covell, D. G., and Sausville, E. A. (2002). Application of high-throughput, molecular-targeted screening to anticancer drug discovery. Curr. Top. Med. Chem. **2**, 229–246.

Van Dyke, M. M., and Dervan, P. B. (1984). Echinomycin binding sites on DNA. Science **225**, 1122–1127.

Wang, G. L., and Semenza, G. L. (1993a). Desferrioxamine induces erythropoietin gene expression and hypoxia-inducible factor 1 DNA-binding activity: Implications for models of hypoxia signal transduction. Blood **82**, 3610–3615.

Wang, G. L., and Semenza, G. L. (1993b). General involvement of hypoxia-inducible factor 1 in transcriptional response to hypoxia. Proc. Natl Acad. Sci. USA **90**, 4304–4308.

Wang, G. L., Jiang, B. H., Rue, E. A., and Semenza, G. L. (1995). Hypoxia-inducible factor 1 is a basic helix-loop-helix PAS heterodimer regulated by cellular O_2 tension. *Proc. Natl. Acad. Sci. USA* **92,** 5510–5514.

Welsh, S., Williams, R., Kirkpatrick, L., Paine-Murrieta, G., and Powis, G. (2004). Antitumor activity and pharmacodynamic properties of PX-478, an inhibitor of hypoxia-inducible factor-1alpha. *Mol. Cancer Ther.* **3,** 233–244.

Welsh, S. J., Williams, R. R., Birmingham, A., Newman, D. J., Kirkpatrick, D. L., and Powis, G. (2003). The thioredoxin redox inhibitors 1-methylpropyl 2-imidazolyl disulfide and pleurotin inhibit hypoxia-induced factor 1alpha and vascular endothelial growth factor formation. *Mol. Cancer Ther.* **2,** 235–243.

Yang, J., Zhang, L., Erbel, P. J., Gardner, K. H., Ding, K., Garcia, J. A., and Bruick, R. K. (2005). Functions of the Per/ARNT/Sim domains of the hypoxia-inducible factor. *J. Biol. Chem.* **280,** 36047–36054.

Yeo, E. J., Chun, Y. S., Cho, Y. S., Kim, J., Lee, J. C., Kim, M. S., and Park, J. W. (2003). YC-1: A potential anticancer drug targeting hypoxia-inducible factor 1. *J. Natl. Cancer Inst.* **95,** 516–525.

HYPOXIA AND INFLAMMATORY MEDIATORS

REGULATION OF HYPOXIA-INDUCIBLE FACTORS DURING INFLAMMATION

Stilla Frede, Utta Berchner-Pfannschmidt, *and* Joachim Fandrey

Contents

1. Introduction	406
2. Regulation of HIF at the Transcriptional Level	408
3. Regulation of HIF at the Translational Level	410
4. Regulation of HIF-1α at the Posttranslational Level	411
5. Regulation of HIF-1 Activity	413
6. Perspectives	414
7. Conclusions	414
Acknowledgments	415
References	415

Abstract

The microenvironment of inflamed and injured tissue is characterized by low levels of oxygen and glucose and high levels of inflammatory cytokines, reactive oxygen, and nitrogen species and metabolites. The transcription factor complex hypoxia-inducible factor (HIF)-1 is regulated by hypoxia as well as by a broad variety of inflammatory mediators. In cells of the innate and adaptive immune system, HIF-1 is upregulated by bacterial and viral compounds, even under normoxic conditions. This upregulation prepares these cells to migrate to and to function in hypoxic and inflamed tissues. Once extravasated from the vasculature, the activity of cells is further enhanced by stimulation of HIF-1 by proinflammatory cytokines like interleukin (IL)-1beta (β) and tumor necrosis factor (TNF) alpha (α), and locally expressed tissue factors. Crosstalk between hypoxic induction of HIF-1 and other signaling pathways activated by inflammation ensures a cell type–specific and stimulus-adequate cellular response. Prolonged activation of HIF-1 under conditions of inflammation, however, may contribute to the survival of damaged tissue and cells, thus promoting the development of tumors.

Institut für Physiologie, Universität Duisburg-Essen, Essen, Germany

Methods in Enzymology, Volume 435
ISSN 0076-6879, DOI: 10.1016/S0076-6879(07)35021-0

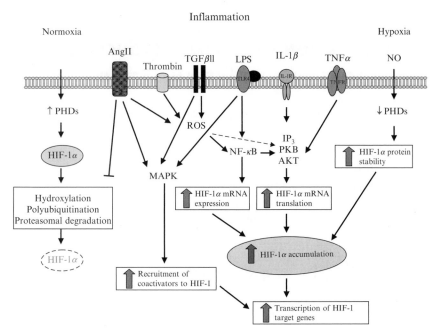

Figure 21.1 Schematic drawing of the influence of inflammatory mediators on the hypoxia-inducible factor (HIF) system.

 1. Introduction

Microenvironmental conditions found in inflamed and injured tissues are characterized by high concentrations of inflammatory mediators as well as low levels of oxygen and glucose (Cramer *et al.*, 2003; Karhausen *et al.*, 2005). To successfully fight infections, mechanisms have evolved that enable cells of the immune response to adapt to low oxygen and glucose conditions and even function under these conditions. The majority of immune cells found at the sites of inflammation are not resident at these sites, but rather recruited to them guided by gradients of oxygen and nutrients (Murdoch *et al.*, 2005; Sitkovsky and Lukashev, 2005). The fact that neutrophils and macrophages depend on generation of energy from glycolysis was known from the beginning of the last century when it was shown that these cells are highly adapted to function in areas of low oxygen tension (Bakker, 1927; Fleischmann and Kubowitz, 1927). More specifically, hypoxic conditions influence a broad range of myeloid and lymphoid cell properties, like survival, phagocytosis, migration, expression, and presentation of cell surface markers for adhesion and immune cell communication (Burke *et al.*, 2002; Caldwell *et al.*, 2001; Demasi *et al.*, 2003;

Kojima *et al.*, 2002; Kong *et al.*, 2004; Louis *et al.*, 2005; Lukashev *et al.*, 2001, 2006; Murdoch *et al.*, 2004; Neumann *et al.*, 2005; Sitkovsky *et al.*, 2004; Zünd *et al.*, 1997). Elevated levels of HIF-1α were found in activated macrophages in inflamed joints of patients suffering from rheumatoid arthritis (Hollander *et al.*, 2001). Induction of HIF-1α in primary inflammatory cells of healing wounds was at least partly caused by inflammatory cytokines (Albina *et al.*, 2001). Whereas in all these situations, hypoxia may also exist to activate HIF-1, hypoxia-independent induction of HIF-1α by inflammatory mediators will enhance the activity of the transcription factor. Similar observations have been made in tumor cell lines (Frede *et al.*, 2005; Hellwig-Bürgel *et al.*, 1999; Metzen *et al.*, 2003), epithelial cells (Sandau *et al.*, 2001), and vascular smooth muscle cells (Richard *et al.*, 2000).

Since HIFs are highly involved in the adaptation of cells to hypoxic conditions, a crucial role for HIFs in the initiation, regulation, and coordination of cell responses during inflammation may be envisioned.

First hints for interference between hypoxic- and inflammatory-signaling pathways came from clinical observations in patients suffering from chronic renal failure or other syndromes of chronic inflammation (Baer *et al.*, 1987; Means and Krantz, 1992; Ward *et al.*, 1971). These patients developed an anemia characterized by reduced iron transport capacity and shortened life time of erythrocytes. Despite their anemia, very low serum levels of erythropoietin were measured (Frede *et al.*, 1997; Jelkmann, 1998). As soon as HIF-1 was characterized as the key transcription factor for erythropoietin gene expression, experiments were performed to show the influence of inflammatory mediators on HIF activity (Hellwig-Bürgel *et al.*, 1999). The simple hypothesis that reduction of HIF activity under conditions of inflammation should be the reason for the blunted expression of erythropoietin in the anemia of chronic diseases could, however, not be proven (Hellwig-Bürgel *et al.*, 1999). In contrast, depending on the model used, an increase in HIF-1 activation was found (Frede *et al.*, 2005; Hellwig-Bürgel *et al.*, 1999). A clear contradiction became obvious: despite hypoxic conditions doubtlessly found in inflamed tissues, a variety of HIF target genes, including erythropoietin, were expressed to an inappropriately low degree; on the other side, HIF activity was increased by inflammatory mediators leading to increased expression of other HIF target genes despite non-hypoxic conditions. As undesirable effects of this HIF activation, an increase in vascular leakage, tumor vascularization, or metastasis were observed (Pollard, 2004). Therefore, discrimination between positive and negative effects of HIF activation during inflammation became of interest for many researchers.

As reviewed in the previous chapters, HIF-1α messenger RNA (mRNA) is constitutively expressed in most cells and tissues under normoxic conditions, and the increase of HIF-1 abundance under hypoxia is predominantly caused by reduced degradation due to a lack of posttranslational hydroxylation (Fandrey *et al.*, 2006). This results firstly in accumulation and nuclear translocation of the α-subunit, dimerization with the HIF-1β–subunit, and secondly in increased binding of transcriptional coactivators, which consecutively leads to increased expression of HIF target genes (Arany *et al.*, 1996; Sang *et al.*, 2003).

Recent studies have shown the influence of inflammatory mediators on HIF-1 activity on survival, differentiation, and functionality of immune cells. Because cells of the myeloid lineage in particular are critically involved in the early immune response, investigations concerning the role of HIF-1 activation focused on this cell type. By generating a myeloid-specific HIF-1α knockout (KO) in mice, Cramer and Johnson (2003) have recently identified HIF-1α as an indispensable transcription factor that coordinates gene expression in myeloid cells during the process of extravasation. In addition, bacterial killing as well as the release of TNFα was significantly reduced in the HIF-1α KO myeloid cells (Peyssonnaux *et al.*, 2005).

In contrast to the hypoxic activation by reduced posttranslational hydroxylation, the induction of HIF-1α by non-hypoxic, proinflammatory stimuli seems to be much more diverse. It includes modification at the transcriptional, the translational, and the posttranslational levels. Examples for each regulatory type are reviewed in the following sections.

2. Regulation of HIF at the Transcriptional Level

Shortly after the discovery of HIF-1α, it was commonly accepted that HIF-1α is not regulated at the transcriptional level, since early experiments performed with hepatoma cells acutely exposed to hypoxic conditions gave no hint for a regulation of HIF-1α mRNA levels by hypoxia (Wang and Semenza, 1993). Effects of HIF-1α mRNA as a result of increased transcription were first observed *in vivo* when experiments were performed with animals exposed to prolonged or intermittent hypoxia (Semenza, 2000). Investigations in the following years demonstrated that, in contrast to the original concept of an exclusively posttranslational regulation of HIF-1α protein, a variety of substances were able to induce HIF-1α mRNA depending on the different models used. With respect to the role of HIF-1 in inflammation, the induction of HIF-1α mRNA by bacterial lipopolysaccharides (LPS) under normoxic conditions may be of particular importance since it prepares monocytes for survival and function in a hypoxic microenvironment before they extravasate from the vasculature into the tissue (Blouin *et al.*, 2004; Frede *et al.*, 2006). LPS are components of the cell wall of gram-negative bacteria. LPS bind predominantly to the cell surface receptors CD14 and Toll-like receptor (TLR) 4 (Moynagh, 2003).

Binding of LPS to this receptor induces different intracellular signaling cascades, leading, among other effects, to the activation of the transcription factors NF-κB and AP-1, which are both known to be critical for the initiation and regulation of an appropriate immune response (Karin, 2006; Karin and Greten, 2005). A consecutively performed analysis of the HIF-1α promoter revealed binding sites for both transcription factors in the early promoter of the HIF-1α gene. Inhibition of NF-κB abolished the LPS-induced increase in HIF-1α mRNA, but also reduced constitutive HIF-1α expression (Frede et al., 2006). Induction of HIF-1α by LPS directly occurs at the transcriptional level since the LPS-induced HIF-1α accumulation under normoxic conditions was completely blocked in experiments in which transcriptional inhibitors were applied (Frede et al., 2006). Using reporter gene assays, Michiels et al. (Minet et al., 1999a) have shown that NF-κB is necessary for the constitutive expression of HIF-1α mRNA (Minet et al., 1999b). Furthermore, a close connection between HIF-1 and NF-κB was demonstrated by Walmsley et al. (2005) for neutrophils. They identified the NF-κB p65 subunit and the IκBα regulator IKKα as HIF-1 target genes, implicating that NF-κB is an important downstream regulator of the hypoxic response found in neutrophils. These observations further support the notion that HIF-1 and NF-κB are parts of a positive-enhancer loop activated under conditions of hypoxia and inflammation. In addition, it was found by Oda et al. (2006) that HIF-1α mRNA is upregulated during the differentiation of monocytes to macrophages. Responsible for this increase in HIF-1α gene expression was the activation of protein kinase C (PKC)- and mitogen-activated protein kinase (MAPK)-signaling pathways. Studies previously performed (Berra et al., 2000; Minet et al., 2000; Richard et al., 1999) had already demonstrated a distinct role for the MAPK extracellular signal-regulated kinases 1 and 2 (ERK1/2) in HIF-1 activation. Although none of these authors demonstrated a direct influence of ERK1/2 activation on the expression of HIF-1α mRNA, it is likely that, depending on the cell type used, ERK1/2 is critically involved in upregulation of HIF-1α mRNA as well.

In a cell culture model using human monocytic cells differentiated in vitro to a macrophage-like phenotype, it was shown that HIF-1α mRNA levels in macrophages are about twofold higher than in undifferentiated monocytes, underlining that HIF-1α mRNA expression is influenced by the differentiation status of these cells (Frede et al., 2006). In addition to LPS, thrombin and angiotensin II were reported to induce HIF-1α at the transcriptional level in resident macrophages of the lung (Blouin et al., 2004; Page et al., 2002). In these cells, hypoxia alone even failed to increase HIF-1α mRNA levels. Taken together, upregulation of HIF-1α mRNA by inflammatory mediators like LPS is useful to overcome the constitutive degradation of the HIF-1α protein under normoxic conditions. This characteristic may improve the survival and activity of cells of the innate immune system in the inflamed and hypoxic microenvironment to which they have to extravasate from the vasculature.

3. REGULATION OF HIF AT THE TRANSLATIONAL LEVEL

Apart from upregulation of HIF-1α mRNA in an inflammatory setting by LPS or thrombin under normoxic conditions, inflammatory mediators are capable of enhancing the hypoxia-induced HIF activity by increasing the translation of HIF-1α mRNA. This phenomenon was observed in different cell types, including renal proximal tubular cells (Zhou *et al.*, 2003b), vascular smooth muscle cells (BelAiba *et al.*, 2004), and tumor cells of different tissue origin (Bert *et al.*, 2006; Frede *et al.*, 2005; Stiehl *et al.*, 2002). Most of these cell types themselves express proinflammatory cytokines, like IL-1β and TNFα. They thus create an environment where inflammatory mediators will be able to induce HIF-1 in a paracrine or autocrine manner in addition to a general hypoxic stimulus. An increasing number of reports regarding the effect of cytokines, like IL-1β or TNFα, on HIF-1α translation were recently published. In ovarian carcinoma cells, IL-1β clearly increased the translation of HIF-1α mRNA rather than reduced the degradation of HIF-1α protein, since no changes in HIF-1α hydroxylation were observed and blocking the translation abolished the IL-1β–induced increase in HIF-1α protein accumulation and HIF-1 activity (Frede *et al.*, 2005). Similar results were obtained when human embryonic kidney (HEK) cells were treated with TNFα (Zhou *et al.*, 2003a). Studies using a bicistronic reporter vector containing the internal ribosomal entry site (IRES) located within the first exon of the HIF-1α mRNA cloned between the firefly and the renilla luciferase demonstrated that TNFα enhanced the cap-independent translation of HIF-1α mRNA (Bert *et al.*, 2006). Further investigations are necessary to elucidate whether cap-dependent translation is also increased by inflammatory cytokines. First hints for this mode of action to increase HIF-1α synthesis were provided by investigations on intracellular signaling by activation of the PI3 kinase (PI3K)–AKT–mammalian target of rapamycin (mTOR) pathway. Signaling through this pathway normally culminates in the activation of p70S6 kinase, which is involved in the activation of cap-dependent translation. Görlach *et al.* (2001) demonstrated that the PI3K-AKT-mTOR pathway was activated when vascular smooth muscle cells of the lung were stimulated with thrombin. Subsequently, the involvement of this pathway in IL-1β- and insulin-induced normoxic induction of HIF-1α was proposed by Stiehl *et al.* (2002) for hepatoma cells.

With respect to cells of the adaptive immune system, it was shown that viral compounds were able to induce HIF-1α in T-lymphocytes under hypoxic conditions. Activation of the T-cell receptor by stimulation with phytohemagglutinin (PHA) induced HIF-1α likewise by increasing the translation of HIF-1α protein via PI3K-AKT-mTOR pathway (Nakamura *et al.*, 2005). In other cell types, including lymphoid cells and tumor cell lines, infection with Epstein-Barr virus (EBV) induced the upregulation of p42/44 MAPK and increased levels of hydrogen peroxide, which then induced an increase in HIF-1α protein synthesis. EBV latent membrane protein 1 (LMP-1) was identified as the viral compound responsible for this effect (Wakisaka *et al.*, 2004). In contrast to cells of the innate immune response, NF-κB and PI3K pathways seemed to play no important role in EBV-induced HIF-1α synthesis. The question of whether this increase in HIF-1α synthesis is cap-dependent or IRES-mediated remains unresolved. All these reports demonstrate that the intracellular signaling pathways involved in HIF-1α synthesis are highly dependent on cell type and stimulus.

Upregulation of HIF-1α protein synthesis, therefore, contributes to the increase in HIF-dependent target gene expression under conditions of inflammation and hypoxia. First, under normoxic conditions, increased HIF-1α synthesis enables the cells to overcome the normal degradation of the HIF-1α protein, which is at an equilibrium between synthesis and the activity of the HIF-1α-degrading machinery. Second, under hypoxic conditions, increased translation of the HIF-1α mRNA seems to be a useful tool for the cells to better adapt to short-term changes in the microenvironment.

4. REGULATION OF HIF-1α AT THE POSTTRANSLATIONAL LEVEL

Since it is generally accepted that the activity of HIF-1 is regulated by the abundance of the HIF-1α subunit, much attention has been paid to the question of whether inflammatory mediators interfere with the proline hydroxylases (PHDs) responsible for HIF-1α hydroxylation or with the von Hippel-Lindau protein (pVHL) targeting HIF-1α for ubiquitination and consecutive proteasomal degradation. A common experiment to answer this question is to analyze the stability of HIF-1α protein under conditions of inflammation and hypoxia. Very recently, it was described that differentiation of monocytes to macrophages as it occurs during the inflammatory response changes the labile iron pool of these cells, resulting in normoxic stabilization of HIF-1α due to reduced activity of the PHDs (Knowles *et al.*, 2006). Jung *et al.* (2003) directly showed increased stability of HIF-1α protein after stimulation of alveolar epithelial cells with IL-1β. They postulated an inhibition of pVHL activity by prostaglandin E_2, the major physiologic product of cyclooxygenase-2 that was induced by IL-1β. In a model of experimental colitis, an increase was observed in HIF-1α stability after ischemia (Koury *et al.*, 2004). In line with these findings, Karhausen *et al.* (2004) reported that constitutive expression of HIF-1 in colon epithelial cells induced by a conditional epithelial specific KO of pVHL was protective in ischemia reperfusion. They found a significant protection from ischemia reperfusion–induced breakdown of the colon epithelial barrier by upregulation of barrier protective genes in an HIF-1–dependent manner. On the other hand, this initially protective effect of HIF-1 upregulation under conditions of colitis may contribute later to the development and progression of colon carcinoma.

Another factor found in inflammatory microenvironment is transforming growth factor β (TGFβ). Induction of HIF-1 accumulation by TGFβ was previously described (Görlach *et al.*, 2001). Recent studies revealed that selective inhibition of PHD2 and, thus, decreased hydroxylation and degradation of HIF-1α were responsible for the increase in HIF-1α protein and HIF-1 activity after TGFβ1 exposure (McMahon *et al.*, 2006).

An inflammatory mediator found under almost all conditions of inflammation, ischemia reperfusion, or hypoxia is nitric oxide (NO). Intracellular signaling pathways induced by NO include activation of NF-κB, c-jun, and p38 MAPK culminating in the activation of early responses to inflammatory stimuli. With respect to HIF-1α abundance and activity, NO appears to play a dual role. Under normoxia, NO induces HIF-1α stabilization and transcriptional activation of HIF-1 target gene expression (Kimura *et al.*, 2000; Sandau *et al.*, 2001; Zhou *et al.*, 2003b). Decreased degradation of HIF-1α by NO-dependent inhibition of PHD activity was found to be responsible for this effect (Metzen, *et al.*, 2003; Zhou *et al.*, 2003). However, under hypoxic conditions, NO revealed an opposite role in regulating HIF-1α because several NO donors were found to decrease HIF-1α stabilization and HIF-1 activity (Huang *et al.*, 1999; Liu *et al.*, 1998; Sogawa *et al.*, 1998). The inhibitory effect of NO on hypoxia-induced HIF-1α was explained by increased NO-derived species and/or reactive oxygen species that contribute to destabilization of HIF-1α by reactivation of PHD activity (Callapina *et al.*, 2005; Kohl *et al.*, 2006; Wellman *et al.*, 2003). In addition, it was reported that mitochondria were involved in NO regulation of HIF expression under hypoxia (Agani *et al.*, 2002; Hagen *et al.*, 2003; Mateo *et al.*, 2003; Quintero *et al.*, 2006). NO is a known inhibitor of cytochromes in the respiration chain and, consequently, of O_2 consumption. It has thus been proposed that reduction of HIF-1α by NO results from redistribution of O_2 from the mitochondrial cytochromes to the PHDs to increase hydroxylation and HIF-1α degradation under hypoxia (Hagen *et al.*, 2003). Very recently, a biphasic effect of NO on HIF-1α was observed by early induction through inhibition of PHD activity and later reduction by HIF-1 dependently increased PHD2 protein levels (Berchner-Pfannschmidt *et al.*, 2007). Thus, induction of PHD2 contributes to NO-induced degradation of HIF-1α under hypoxic conditions. In summary, the opposing effects of NO on HIF-1α stabilization may be explained by NO as putative activator of the HIF-1α–PHD2 autoregulatory loop (Berchner-Pfannschmidt et al., 2007) and would synergize with higher O_2 availability due to inhibition of the respiratory chain by NO (Hagen et al., 2003).

Another short-life mediator of inflammation is hydrogen peroxide (H_2O_2) generated by neutrophils upon stimulation by bacteria or bacterial compounds. The role of H_2O_2 in hypoxic stabilization of HIF-1α and HIF-1 activation is incompletely understood and therefore vividly discussed (Acker et al., 2006). It was clearly demonstrated that physiological concentrations of H_2O_2 were able to increase HIF-1α protein accumulation by increasing HIF-1α stability (BelAiba et al., 2004; Frede et al., unpublished data). Since cells deficient in jun-D and devoid of major antioxidant-protecting enzymes showed significantly higher HIF-1α levels, it was concluded that oxidation of the ferrous iron in the catalytically active center of PHDs may reduce their capacity to hydroxylate HIF-α (Gerald et al., 2004).

5. REGULATION OF HIF-1 ACTIVITY

Independent of the accumulation and stability of HIF-1α protein, the activity of the transcription factor complex HIF-1 is influenced by interfering with members of other signaling pathways activated by inflammation. The factor-inhibiting HIF (FIH-1) was identified as an asparagine hydroxylase controlling the recruitment of the transcriptional coactivators p300/CREB-binding protein (CBP) to HIF-1 and thus regulating HIF-1 transcriptional activity (Lando et al., 2002; Mahon et al., 2001). Under conditions of normoxia, FIH-1 hydroxylates the asparagine residue 802 within the transactivation domain of HIF-1α and thus prevents the binding of p300/CBP. FIH-1 would be an excellent target for such signaling cascades. Nevertheless, until now, studies trying to detect a direct influence of inflammatory mediators on FIH-1 abundance or activity have failed (Stolze et al., 2004).

NF-κB has been studied extensively for its role in innate immunity, stress response, cell survival, and development and is thought to be one of the central regulators of the inflammatory response. Therefore, interference of HIF-1 and NF-κB signaling is of special interest (Cummins and Taylor, 2005; Michiels et al., 2002). A coincidence of HIF-1α and NF-κB activation increased the release of proinflammatory cytokines from desferrioxamine (DFO)-stimulated macrophages (Jeong et al., 2003). Apart from its effects on HIF-1α gene expression and protein synthesis previously described, NF-κB exerts effects by modulating HIF-1 activity. The NF-κB essential modulator (NEMO) was shown to regulate the activity of HIF-2α rather than the activity of HIF-1α by enhancing the recruitment of p300/CBP to the transactivation domain of HIF-2α. This example displays a possible cell type–specific regulation of the different HIF isoforms (Bracken *et al.*, 2005).

The protein chaperones heat shock protein 70 (Hsp70) and 90 (Hsp90) were earlier discussed to modulate HIF-1α activity. While the role of Hsp90 most likely facilitates nuclear translocation of HIF-1α (Minet *et al.*, 1999b), the interaction of HIF-1 with Hsp70 was described to be responsible for the increased resistance of macrophages exposed to chronic hypoxia to hypoxia-induced apoptosis as well as to the increase in inflammatory mediator release (Yun *et al.*, 1997). Regulation of HIF-1 on multiple levels enables the cell to adapt immediately to changes in the environment and to integrate different signaling cascades for an appropriate cellular function and response to extracellular stimuli.

6. Perspectives

An increasing number of reports demonstrated that hypoxia not only acts on the HIF system but also independently of HIF on a variety of inflammatory pathways. Such a crosstalk between parts of the HIF system and NF-κB–activating signaling cascades was very recently described by Cummins *et al.* (2006). In their study, they demonstrated that the activity of the IκB kinase-β (IKKβ) was downregulated by PHD1-dependent hydroxylation, thus leading to an increase in NF-κB activity under hypoxic conditions. The IKKβ contains a hydroxylation motif (LXXLAP) similar to that found in the oxygen-dependent degradation domain (ODD) of HIF-1α. Searching for new substrates of the PHDs and their role in inflammation opens a new and interesting field with respect to HIF-1α–independent effects under conditions of hypoxia and inflammation.

7. Conclusions

HIF-1 is a transcription factor complex regulated by hypoxia and conditions found in inflamed and injured tissues characterized by high concentrations of inflammatory mediators and metabolites and low levels of glucose. HIF-1 has been shown to be upregulated at the transcriptional, the translational, and the posttranslational levels (summarized in Fig. 21.1). While upregulation at the transcriptional level most likely is involved in preparing cells for function in a hypoxic microenvironment, regulation of HIF-1α at the translational and posttranslational level ensures an immediate and adequate response to changes in the environment. Apart from the beneficial effects of HIF activation in inflammation, the bad side of this inflammation-induced (and often –prolonged) HIF-1 activation may be the initiation of tumor growth, progression, and proliferation.

ACKNOWLEDGMENTS

Our work is supported by grants from the Deutsche Forschungsgemeinschaft (DFG FA225/18/19/20 and GRK 1431).

REFERENCES

Acker, T., Fandrey, J., and Acker, H. (2006). The good, the bad, and the ugly in oxygen-sensing: ROS, cytochromes and prolyl-hydroxylases. *Cardiovasc. Res.* **71**, 195–207.

Albina, J. E., Mastrofrancesco, B., Vessella, J. A., Louis, C. A., Henry, W. L., Jr., and Reichner, J. S. (2001). HIF-1 expression in healing wounds: HIF-1α induction in primary inflammatory cells by TNFα. *Am. J. Physiol. Cell Physiol.* **281**, C1971–C1977.

Arany, Z., Huang, L. E., Eckner, R., Bhattacharya, S., Jiang, C., Goldberg, M. A., Bunn, H. F., and Livingston, D. M. (1996). An essential role for p300/CBP in the cellular response to hypoxia. *Proc. Natl. Acad. Sci. USA* **93**, 12969–12973.

Baer, A. N., Dessypris, E. N., Goldwasser, E., and Krantz, S. B. (1987). Blunted erythro-poietin response to anaemia in rheumatoid arthritis. *Br. J. Haematol.* **66**, 559–564.

Bakker, A. (1927). Einige Übereinstimmungen im Stoffwechsel der Carcinomzellen und der Exsudatleukocyten. *Klinische Wochenschrift* **6**, 252.

BelAiba, R. S., Djordjevic, T., Bonello, S., Flugel, D., Hess, J., Kietzmann, T., and Görlach, A. (2004). Redox-sensitive regulation of the HIF pathway under non-hypoxic conditions in pulmonary artery smooth muscle cells. *Biol. Chem.* **385**, 249–257.

Berchner-Pfannschmidt, U., Yamac, H., Trinidad, B., and Fandrey, J. (2007). Nitric oxide modulates oxygen sensing by HIF-1 dependent induction of prolyl hydroxylase 2. *J. Biol. Chem.* **282**, 1788–1796.

Berra, E., Milanini, J., Richard, D. E., Le Gall, M., Vinals, F., Gothie, E., Roux, D., Pages, G., and Pouyssegur, J. (2000). Signaling angiogenesis via p42/p44 MAP kinase and hypoxia. *Biochem. Pharmacol.* **60**, 1171–1178.

Bert, A. G., Grepin, R., Vadas, M. A., and Goodall, G. J. (2006). Assessing IRES activity in the HIF-1α and other cellular 5′UTRs. *RNA* **12**, 1074–1083.

Blouin, C. C., Page, E. L., Soucy, G. M., and Richard, D. E. (2004). Hypoxic gene activation by lipopolysaccharide in macrophages: Implication of hypoxia-inducible factor 1alpha. *Blood* **103**, 1124–1130.

Bracken, C. P., Whitelaw, M. L., and Peet, D. J. (2005). Activity of hypoxia-inducible factor 2alpha is regulated by association with the NF-κB essential modulator. *J. Biol. Chem.* **280**, 14240–14251.

Burke, B., Tang, N., Corke, K. P., Tazzyman, D., Ameri, K., Wells, M., and Lewis, C. E. (2002). Expression of HIF-1α by human macrophages: Implications for the use of macrophages in hypoxia-regulated cancer gene therapy. *J. Pathol.* **196**, 204–212.

Caldwell, C. C., Kojima, H., Lukashev, D., Armstrong, J., Farber, M., Apasov, S. G., and Sitkovsky, M. V. (2001). Differential effects of physiologically relevant hypoxic condi-tions on T-lymphocyte development and effector functions. *J. Immunol.* **167**, 6140–6149.

Callapina, M., Zhou, J., Schmid, T., Kohl, R., and Brune, B. (2005). NO restores HIF-1α hydroxylation during hypoxia: Role of reactive oxygen species. *Free Radic. Biol. Med.* **39**, 925–936.

Cramer, T., and Johnson, R. S. (2003). A novel role for the hypoxia inducible transcription factor HIF-1α: Critical regulation of inflammatory cell function. *Cell Cycle.* **2,** 192–3.

Cramer, T., Yamanishi, Y., Clausen, B. E., Forster, I., Pawlinski, R., Mackman, N., Haase, V. H., Jaenisch, R., Corr, M., Nizet, V., Firestein, G. S., Gerber, H. P., *et al.* (2003). HIF-1α is essential for myeloid cell-mediated inflammation. *Cell* **112,** 645–657.

Cummins, E. P., and Taylor, C. T. (2005). Hypoxia-responsive transcription factors. *Pflugers Arch.* **450,** 363–371.

Cummins, E. P., Berra, E., Comerford, K. M., Ginouves, A., Fitzgerald, K. T., Seeballuck, F., Godson, C., Nielsen, J. E., Moynagh, P., Pouyssegur, J., and Taylor, C. T. (2006). Prolyl hydroxylase-1 negatively regulates IκB kinase-β, giving insight into hypoxia-induced NF-κB activity. *Proc. Natl. Acad. Sci. USA* **103,** 18154–18159.

Demasi, M., Cleland, L. G., Cook-Johnson, R. J., Caughey, G. E., and James, M. J. (2003). Effects of hypoxia on monocyte inflammatory mediator production: Dissociation between changes in cyclooxygenase-2 expression and eicosanoid synthesis. *J. Biol. Chem.* **278,** 38607–38616.

Fandrey, J., Gorr, T. A., and Gassmann, M. (2006). Regulating cellular oxygen sensing by hydroxylation. *Cardiovasc. Res.* **71,** 642–651.

Fleischmann, W., and Kubowitz, F. (1927). Über den Stoffwechsel der Leukocyten. *Biochem. Z.* **181,** 395.

Frede, S., Stockmann, C., Freitag, P., and Fandrey, J. (2006). Bacterial lipopolysaccharide induces HIF-1 activation in human monocytes via p44/42 MAPK and NF-κB. *Biochem. J.* **396,** 517–527.

Frede, S., Fandrey, J., Pagel, H., Hellwig, T., and Jelkmann, W. (1997). Erythropoietin gene expression is suppressed after lipopolysaccharide or interleukin-1 beta injections in rats. *Am. J. Physiol.* **273,** R1067–R1071.

Frede, S., Freitag, P., Otto, T., Heilmaier, C., and Fandrey, J. (2005). The proinflammatory cytokine interleukin 1beta and hypoxia cooperatively induce the expression of adreno-medullin in ovarian carcinoma cells through hypoxia inducible factor 1 activation. *Cancer Res.* **65,** 4690–4697.

Gerald, D., Berra, E., Frapart, Y. M., Chan, D. A., Giaccia, A. J., Mansuy, D., Pouyssegur, J., Yaniv, M., and Mechta-Grigoriou, F. (2004). JunD reduces tumor angiogenesis by protecting cells from oxidative stress. *Cell* **118,** 781–794.

Görlach, A., Diebold, I., Schini-Kerth, V. B., Berchner-Pfannschmidt, U., Roth, U., Brandes, R. P., Kietzmann, T., and Busse, R. (2001). Thrombin activates the hypoxia-inducible factor-1 signaling pathway in vascular smooth muscle cells: Role of the p22 (phox)-containing NADPH oxidase. *Circ. Res.* **89,** 47–54.

Hellwig-Bürgel, T., Rutkowski, K., Metzen, E., Fandrey, J., and Jelkmann, W. (1999). Interleukin-1beta and tumor necrosis factor-alpha stimulate DNA binding of hypoxia-inducible factor-1. *Blood* **94,** 1561–1567.

Hollander, A. P., Corke, K. P., Freemont, A. J., and Lewis, C. E. (2001). Expression of hypoxia-inducible factor 1alpha by macrophages in the rheumatoid synovium: Implications for targeting of therapeutic genes to the inflamed joint. *Arthritis Rheum.* **44,** 1540–1544.

Jelkmann, W. (1998). Proinflammatory cytokines lowering erythropoietin production. *J. Interferon Cytokine Res.* **18,** 555–559.

Jeong, H. J., Chung, H. S., Lee, B. R., Kim, S. J., Yoo, S. J., Hong, S. H., and Kim, H. M. (2003). Expression of proinflammatory cytokines via HIF-1α and NF-κB activation on desferrioxamine-stimulated HMC-1 cells. *Biochem. Biophys. Res. Commun.* **306,** 805–811.

Jung, Y. J., Isaacs, J. S., Lee, S., Trepel, J., and Neckers, L. (2003). IL-β-mediated upregulation of HIF-1α via an NF-κB/COX-2 pathway identifies HIF-1 as a critical link between inflammation and oncogenesis. *FASEB J.* **17,** 2115–2117.

Karhausen, J., Haase, V. H., and Colgan, S. P. (2005). Inflammatory hypoxia: Role of hypoxia-inducible factor. *Cell Cycle* **4,** 256–258.

Karhausen, J., Furuta, G. T., Tomaszewski, J. E., Johnson, R. S., Colgan, S. P., and Haase, V. H. (2004). Epithelial hypoxia-inducible factor-1 is protective in murine experimental colitis. *J. Clin. Invest.* **114,** 1098–1106.

Karin, M. (2006). Nuclear factor-kappaB in cancer development and progression. *Nature* **441,** 431–436.

Karin, M., and Greten, F. R. (2005). NF-κB: Linking inflammation and immunity to cancer development and progression. *Nat. Rev. Immunol.* **5,** 749–759.

Knowles, H. J., Mole, D. R., Ratcliffe, P. J., and Harris, A. L. (2006). Normoxic stabilization of hypoxia-inducible factor-1alpha by modulation of the labile iron pool in differentiating U937 macrophages: effect of natural resistance-associated macrophage protein 1. *Cancer Res.* **66,** 2600–2607.

Kojima, H., Gu, H., Nomura, S., Caldwell, C. C., Kobata, T., Carmeliet, P., Semenza, G. L., and Sitkovsky, M. V. (2002). Abnormal B lymphocyte development and autoimmunity in hypoxia-inducible factor 1alpha-deficient chimeric mice. *Proc. Natl. Acad. Sci. USA* **99,** 2170–2174.

Kong, T., Eltzschig, H. K., Karhausen, J., Colgan, S. P., and Shelley, C. S. (2004). Leukocyte adhesion during hypoxia is mediated by HIF-1–dependent induction of β2 integrin gene expression. *Proc. Natl. Acad. Sci. USA* **101,** 10440–10445.

Lando, D., Peet, D. J., Gorman, J. J., Whelan, D. A., Whitelaw, M. L., and Bruick, R. K. (2002). FIH-1 is an asparaginyl hydroxylase enzyme that regulates the transcriptional activity of hypoxia-inducible factor. *Genes Dev.* **16,** 1466–1471.

Louis, N. A., Hamilton, K. E., Kong, T., and Colgan, S. P. (2005). HIF-dependent induction of apical CD55 coordinates epithelial clearance of neutrophils. *FASEB J.* **19,** 950–959.

Lukashev, D., Caldwell, C., Ohta, A., Chen, P., and Sitkovsky, M. (2001). Differential regulation of two alternatively spliced isoforms of hypoxia-inducible factor-1 alpha in activated T lymphocytes. *J. Biol. Chem.* **276,** 48754–48763.

Lukashev, D., Klebanov, B., Kojima, H., Grinberg, A., Ohta, A., Berenfeld, L., Wenger, R. H., Ohta, A., and Sitkovsky, M. (2006). Cutting edge: Hypoxia-inducible factor 1alpha and its activation-inducible short isoform I.1 negatively regulate functions of CD4+ and CD8+ T lymphocytes. *J. Immunol.* **177,** 4962–4965.

Mahon, P. C., Hirota, K., and Semenza, G. L. (2001). FIH-1: A novel protein that interacts with HIF-1α and VHL to mediate repression of HIF-1 transcriptional activity. *Genes Dev.* **15,** 2675–2686.

McMahon, S., Charbonneau, M., Grandmont, S., Richard, D. E., and Dubois, C. M. (2006). Transforming growth factor beta1 induces hypoxia-inducible factor-1 stabilization through selective inhibition of PHD2 expression. *J. Biol. Chem.* **281,** 24171–24181.

Means, R. T., Jr., and Krantz, S. B. (1992). Progress in understanding the pathogenesis of the anemia of chronic disease. *Blood* **80,** 1639–1647.

Metzen, E., Zhou, J., Jelkmann, W., Fandrey, J., and Brune, B. (2003). Nitric oxide impairs normoxic degradation of HIF-1α by inhibition of prolyl hydroxylases. *Mol. Biol. Cell.* **14,** 3470–3481.

Michiels, C., Minet, E., Mottet, D., and Raes, M. (2002). Regulation of gene expression by oxygen: NF-κB and HIF-1, two extremes. *Free Radic. Biol. Med.* **33,** 1231–1242.

Minet, E., Ernest, I., Michel, G., Roland, I., Remacle, J., Raes, M., and Michiels, C. (1999a). HIF-1α gene transcription is dependent on a core promoter sequence encompassing activating and inhibiting sequences located upstream from the transcription initiation site and *cis* elements located within the 5′UTR. *Biochem. Biophys. Res. Commun.* **261,** 534–540.

Minet, E., Mottet, D., Michel, G., Roland, I., Raes, M., Remacle, J., and Michiels, C. (1999b). Hypoxia-induced activation of HIF-1: Role of HIF-1α–Hsp90 interaction. *FEBS Lett.* **460,** 251–256.

Minet, E., Arnould, T., Michel, G., Roland, I., Mottet, D., Raes, M., Remacle, J., and Michiels, C. (2000). ERK activation upon hypoxia: Involvement in HIF-1 activation. *FEBS Lett.* **468,** 53–58.

Moynagh, P. N. (2003). Toll-like receptor signalling pathways as key targets for mediating the anti-inflammatory and immunosuppressive effects of glucocorticoids. *J. Endocrinol.* **179,** 139–144.

Murdoch, C., Giannoudis, A., and Lewis, C. E. (2004). Mechanisms regulating the recruitment of macrophages into hypoxic areas of tumors and other ischemic tissues. *Blood* **104,** 2224–2234.

Murdoch, C., Muthana, M., and Lewis, C. E. (2005). Hypoxia regulates macrophage functions in inflammation. *J. Immunol.* **175,** 6257–6263.

Nakamura, H., Makino, Y., Okamoto, K., Poellinger, L., Ohnuma, K., Morimoto, C., and Tanaka, H. (2005). TCR engagement increases hypoxia-inducible factor-1 alpha protein synthesis via rapamycin-sensitive pathway under hypoxic conditions in human peripheral T cells. *J. Immunol.* **174,** 7592–7599.

Neumann, A. K., Yang, J., Biju, M. P., Joseph, S. K., Johnson, R. S., Haase, V. H., Freedman, B. D., and Turka, L. A. (2005). Hypoxia inducible factor 1 alpha regulates T cell receptor signal transduction. *Proc. Natl. Acad. Sci. USA* **102,** 17071–17076.

Oda, T., Hirota, K., Nishi, K., Takabuchi, S., Oda, S., Yamada, H., Arai, T., Fukuda, K., Kita, T., Adachi, T., Semenza, G. L., Nohara, R., *et al.* (2006). Activation of hypoxia-inducible factor 1 during macrophage differentiation. *Am. J. Physiol. Cell Physiol.* **291,** C104–C113.

Pagé, E. L., Robitaille, G. A., Pouysségur, J., and Richard, D. E. (2002). Induction of hypoxia-inducible factor-1alpha by transcriptional and translational mechanisms. *J. Biol. Chem.* **277,** 48403–48409.

Peyssonnaux, C., Datta, V., Cramer, T., Doedens, A., Theodorakis, E. A., Gallo, R. L., Hurtado-Ziola, N., Nizet, V., and Johnson, R. S. (2005). HIF-1α expression regulates the bactericidal capacity of phagocytes. *J. Clin. Invest.* **115,** 1806–1815.

Pollard, J. W. (2004). Tumour-educated macrophages promote tumour progression and metastasis. *Nat. Rev. Cancer.* **4,** 71–78.

Richard, D. E., Berra, E., and Pouyssegur, J. (2000). Nonhypoxic pathway mediates the induction of hypoxia-inducible factor 1alpha in vascular smooth muscle cells. *J. Biol. Chem.* **275,** 26765–26771.

Richard, D. E., Berra, E., Gothie, E., Roux, D., and Pouyssegur, J. (1999). p42/p44 mitogen-activated protein kinases phosphorylate hypoxia-inducible factor 1alpha (HIF-1α). and enhance the transcriptional activity of HIF-1. *J. Biol. Chem.* **274,** 32631–32637.

Sandau, K. B., Fandrey, J., and Brune, B. (2001). Accumulation of HIF-1α under the influence of nitric oxide. *Blood* **97,** 1009–1015.

Sang, N., Stiehl, D. P., Bohensky, J., Leshchinsky, I., Srinivas, V., and Caro, J. (2003). MAPK signaling up-regulates the activity of hypoxia-inducible factors by its effects on p300. *J. Biol. Chem.* **278,** 14013–14019.

Semenza, G. L. (2000). Surviving ischemia: Adaptive responses mediated by hypoxia-inducible factor 1. *J. Clin. Invest.* **106,** 809–812.

Sitkovsky, M., and Lukashev, D. (2005). Regulation of immune cells by local-tissue oxygen tension: HIF1α and adenosine receptors. *Nat. Rev. Immunol.* **5,** 712–721.

Sitkovsky, M. V., Lukashev, D., Apasov, S., Kojima, H., Koshiba, M., Caldwell, C., Ohta, A., and Thiel, M. (2004). Physiological control of immune response and inflammatory tissue damage by hypoxia-inducible factors and adenosine A2A receptors. *Annu. Rev. Immunol.* **22,** 657–682.

Stiehl, D. P., Jelkmann, W., Wenger, R. H., and Hellwig-Burgel, T. (2002). Normoxic induction of the hypoxia-inducible factor 1alpha by insulin and interleukin-1beta involves the phosphatidylinositol 3-kinase pathway. *FEBS Lett.* **512,** 157–162.

Stolze, I. P., Tian, Y. M., Appelhoff, R. J., Turley, H., Wykoff, C. C., Gleadle, J. M., and Ratcliffe, P. J. (2004). Genetic analysis of the role of the asparaginyl hydroxylase factor inhibiting hypoxia-inducible factor (HIF) in regulating HIF transcriptional target genes. *J. Biol. Chem.* **279,** 42719–42725.

Walmsley, S. R., Print, C., Farahi, N., Peyssonnaux, C., Johnson, R. S., Cramer, T., Sobolewski, A., Condliffe, A. M., Cowburn, A. S., Johnson, N., and Chilvers, E. R. (2005). Hypoxia-induced neutrophil survival is mediated by HIF-1α–dependent NF-κB activity. *J. Exp. Med.* **201,** 105–115.

Wang, G. L., and Semenza, G. L. (1993). General involvement of hypoxia-inducible factor 1 in transcriptional response to hypoxia. *Proc. Natl. Acad. Sci. USA* **90,** 4304–4308.

Ward, H. P., Kurnick, J. E., and Pisarczyk, M. J. (1971). Serum level of erythropoietin in anemias associated with chronic infection, malignancy, and primary hematopoietic disease. *J. Clin. Invest.* **50,** 332–335.

Yun, J. K., McCormick, T. S., Villabona, C., Judware, R. R., Espinosa, M. B., and Lapetina, E. G. (1997). Inflammatory mediators are perpetuated in macrophages resistant to apoptosis induced by hypoxia. *Proc. Natl. Acad. Sci. USA* **94,** 13903–13908.

Zhou, J., Schmid, T., and Brune, B. (2003a). Tumor necrosis factor-alpha causes accumulation of a ubiquitinated form of hypoxia inducible factor-1alpha through a nuclear factor-kappaB-dependent pathway. *Mol. Biol. Cell.* **14,** 2216–2225.

Zhou, J., Fandrey, J., Schumann, J., Tiegs, G., and Brune, B. (2003b). NO and TNF-α released from activated macrophages stabilize HIF-1α in resting tubular LLC-PK1 cells. *Am. J. Physiol. Cell Physiol.* **284,** C439–C446.

Zünd, G., Uezono, S., Stahl, G. L., Dzus, A. L., McGowan, F. X., Hickey, P. R., and Colgan, S. P. (1997). Hypoxia enhances induction of endothelial ICAM-1: Role for metabolic acidosis and proteasomes. *Am. J. Physiol.* **273,** C1571–C1580.

CHAPTER TWENTY-TWO

SUPEROXIDE AND DERIVED REACTIVE OXYGEN SPECIES IN THE REGULATION OF HYPOXIA-INDUCIBLE FACTORS

Agnes Görlach* *and* Thomas Kietzmann[†]

Contents

1. Introduction		422
2. Reactive Oxygen Species Act as Signaling Molecules		423
3. HIFs are Sensitive to Oxygen		424
4. Reactive Oxygen Species Modulate HIF		425
5. How are HIFs Regulated by Reactive Oxygen Species?		427
	5.1. Regulation of HIF-α synthesis by reactive oxygen species	427
	5.2. Direct regulation of HIF-α by reactive oxygen species	428
	5.3. Regulation of HIF by reactive oxygen species via interference with a regulatory signaling pathway	428
6. Summary		431
7. Methods		431
8. The Cytochrome C Reduction Assay for Detection of Extracellular Reactive Oxygen Species		431
9. Chemiluminescence Assay for Detection of Extracellular Reactive Oxygen Species		432
10. Measuring Intracellular Production of Reactive Oxygen Species using Fluorescent Dyes		433
	10.1. DCF fluorescence	434
	10.2. Hydroethidine fluorescence	434
	10.3. Dihydrorhodamine fluorescence	435
	10.4. ROS measurements in tissues	436
11. Detection of Reactive Oxygen Species by Electron Paramagnetic Resonance		436
Acknowledgments		438
References		438

* Experimental Pediatric Cardiology, German Heart Center Munich, Munich, Germany
[†] Faculty of Chemistry, Department of Biochemistry, University of Kaiserslautern, Kaiserslautern, Germany

Methods in Enzymology, Volume 435
ISSN 0076-6879, DOI: 10.1016/S0076-6879(07)35022-2

Abstract

Superoxide and its derived reactive oxygen species (ROS) have been considered for a long time to be generated as toxic byproducts of metabolic events. More recently, it has been acknowledged that ROS generated in low amounts are also able to act as signaling molecules in a variety of responses. One of the major pathways regulated by the ambient concentration of oxygen relies on the activity of hypoxia-inducible transcription factors (HIF).

Originally described to be only induced and activated under hypoxia, accumulating evidence suggests that HIFs play a more general role in the response to a variety of cellular activators and stressors, many of which use ROS as signal transducers. Indeed, ROS have been found to modulate the levels of HIF not only under hypoxia, but also in response to many factors and under different stress conditions. However, the underlying regulatory mechanisms by which superoxide and derived ROS control HIF are only slowly beginning to be elucidated. We summarize here current knowledge about the mechanisms by which ROS can regulate HIF and give additional information about useful methods to determine ROS under various conditions.

1. INTRODUCTION

An adequate supply of oxygen is mandatory for the function of diverse processes within all aerobic organisms. Thereby, O_2 can often be transformed into highly reactive derivatives, ROS. In the eucaryotic cell, ROS can be generated through multiple sources, including the electron transport chain in mitochondria, ionizing radiation, and enzymes producing superoxide anion radicals. Superoxide anion radical formation is often the initial step in ROS generation. Given that superoxide anion radicals and ROS are cytotoxic, cells have developed antioxidant mechanisms, which include enzymes that dismutate O_2^- into H_2O_2 (superoxide dismutases) or degrade H_2O_2 (catalase, glutathione peroxidases, and peroxiredoxins). When cellular production of ROS overwhelms its antioxidant capacity, a state of oxidative stress is reached, leading to serious cellular injuries contributing to the pathogenesis of a number of diseases. Nevertheless, when generated in lower concentrations, ROS can act as second messengers in signal transduction and gene regulation in a variety of cell types and under several biological conditions.

Due to the importance of oxygen for cell metabolism, elaborate mechanisms have been evolved to allow adaptation when oxygen availability drops. These mechanisms include responses ensuring energy and oxygen supply to compensate the drop in O_2 tension. Although the identity of a cellular oxygen sensor remains elusive, a family of transcription factors,

HIFs, has been reported to primarily mediate the cellular response to hypoxia.

Interestingly, a number of factors, including growth, coagulation, hormones, and inflammatory cytokines, as well as cellular stress factors (e.g., physical and chemical stress), have been shown to activate HIF transcription factors by using ROS as signaling molecules. In the following sections, we will therefore summarize the role of ROS in the regulation of HIF.

2. REACTIVE OXYGEN SPECIES ACT AS SIGNALING MOLECULES

Transfer of one electron to O_2 results in the production of superoxide anion radicals ($O_2^-\bullet$), which often are the precursors for formation of other reactive species, such as hydrogen peroxide (H_2O_2), hydroxyl radicals (OH•), peroxynitrite ($ONOO^-$), hypochlorous acid (HOCl), and singlet oxygen ($^1O_2$). In mammalian cells, ROS are formed in response to toxic reagents or as (by-) products of enzyme reactions. One interesting enzyme that actively generates $O_2^-\bullet$ is a multi-protein complex known as nicotinamide adenine dinucleotide phosphate (NADPH)-oxidase. Although initially presumed to be an enzyme existing only in neutrophils, NADPH oxidases have also been identified in non-phagocytic cells, including vascular cells and tumor cells (Bokoch and Knaus, 2003; Görlach et al., 2000, 1993; Griendling, 2004). Meanwhile, several families of NADPH oxidases (NOX family, with members NOX1 to NOX5) and an evolutionary, more distinct group named DUOX (dual oxidase) have been described (Babior, 2004; BelAiba et al., 2007; Bokoch and Knaus, 2003; Görlach et al., 2002; Griendling, 2004; Lambeth, 2004).

These non-phagocyte NADPH oxidases generate much lower amounts of ROS than the neutrophil enzyme, though they are all considered to catalyze the same two-step reaction necessary for superoxide anion radical production. In the first step, electrons are transferred from NADPH to flavin adenine dinucleotide (FAD), followed by a second step in which they are transferred onto a heme group leading to the reduction of molecular oxygen to superoxide anion radicals (Groemping and Rittinger, 2004).

In addition, $O_2^-\bullet$ or H_2O_2 can be produced by transfer of one or two electrons to molecular oxygen, respectively, as reported for oxidases and other enzymes, such as those of the arachidonic acid pathway, the cytochrome P450 family, glucose oxidase, amino acid oxidases, xanthine oxidase, NADH/NADPH oxidases, or NO synthases (Cai and Harrison, 2000; Finkel, 2001).

Importantly, a number of stimuli, including growth factors, cytokines, hormones, vasoactive factors, and coagulation factors, among them tumor necrosis factor-1alpha (TNF-1α), interleukin-1beta (IL-1β), and thrombin, have been shown to activate NADPH oxidases (Bokoch and Knaus, 2003; Görlach *et al.*, 2002; Herkert *et al.*, 2004). Thus, NADPH oxidases have been considered to act as sources of low amounts of $O_2^-\bullet$ and other ROS, which can activate specific signaling pathways.

Physiologically, an overshoot in ROS production is counterbalanced by the endogenous antioxidant defense systems, which are comprised of superoxide dismutases (SOD-1, -2, and -3), glutathione peroxidases (GPX), catalase, peroxiredoxins, thioredoxin, and exogenously taken-up micronutrients and vitamins (Brigelius-Flohe *et al.*, 2003; Freeman and Crapo, 1982). However, when ROS production exceeds the antioxidant capacity, the resulting oxidative stress can cause damage of cellular components, resulting in uncontrolled proliferation or apoptosis (Freeman and Crapo, 1982).

Whereas $O_2^-\bullet$ is less likely to act as second messenger since it is not freely diffusible, its freely diffusible dismutation product H_2O_2 is more suitable to fulfill such a function. Usually, excess H_2O_2 is degraded by GPX or by catalase, but it may, in the presence of Fe(II), undergo a Fenton reaction where hydroxyl anions and highly reactive hydroxyl radicals are produced (Walling, 1975), which in turn can affect signaling proteins, such as transcription factors like HIF-1.

3. HIFs ARE SENSITIVE TO OXYGEN

The first identified HIF transcription factor was HIF-1 (Beck *et al.*, 1993; Semenza and Wang, 1992; Wang and Semenza, 1995; Wang *et al.*, 1995a,b,c); however, now this group represents a small family with the additional members HIF-2 and HIF-3. All of these factors are heterodimers consisting of an α-subunit and a β-subunit, which both are basic helix-loop-helix and Per-ARNT-Sim (bHLH-PAS) domain–consisting proteins. While the α-subunits were found to be new proteins named HIF-1α, HIF-2α and HIF-3α, the β-subunits are represented by arylhydrocarbon–receptor nuclear translocators, ARNT (Hoffman *et al.*, 1991), from which three different isoforms (ARNT, ARNT2, and ARNT3) have been described (reviewed by Wenger [2002]).

Whereas the ARNT subunit is constitutively expressed, HIF-1α is regulated by oxygen, thus mediating the sensitivity towards environmental O_2-levels. Indeed, the HIF-1 dimer appears to be a major regulator of more than 100 physiologically important genes in response to hypoxia (Semenza, 2006).

Although HIF-2α is structurally similar to HIF-1α, newer studies revealed that HIF-2α and HIF-1a functions are not redundant (Hu *et al.*, 2003;

Sowter *et al.*, 2003), showing that the HIF-α proteins each have a defined role in the regulation of gene expression. By contrast, HIF-3α appears not to be as efficient in mediating the hypoxic response, or even to act as a negative regulator (Hara *et al.*, 2001; Kietzmann *et al.*, 2001).

Although regulation at the transcriptional and translational level was shown (Bonello *et al.*, 2007; Görlach *et al.*, 2000b; Gross *et al.*, 2003; Heidbreder *et al.*, 2003; Kietzmann *et al.*, 2001; Pascual *et al.*, 2001; Wang *et al.*, 1995c), the predominant mode of HIF-1α regulation appears to be posttranslational. This characteristic is brought about by the HIF-1α oxygen-dependent degradation domain (ODD) (Huang *et al.*, 1998) and two transactivation domains (TADs) referred to as amino-terminal TAD (TADN) and carboxy-terminal TAD (TADC) (Jiang *et al.*, 1997; Pugh *et al.*, 1997). The ODD and the TADN partially overlap and are mainly responsible for HIF-1α protein stability.

Thereby, normoxia initiates proline hydroxylase (PHD)-dependent hydroxylation of HIF-1α on at least two proline residues within the ODD (Jaakkola *et al.*, 2001; Masson *et al.*, 2001). This action allows binding of the von Hippel-Lindau tumor suppressor protein (pVHL) (Maxwell *et al.*, 1999; Min *et al.*, 2002; Ohh *et al.*, 2000; Tanimoto *et al.*, 2000), which initiates ubiquitinylation and proteasomal degradation (Ivan *et al.*, 2001; Kaelin and Maher, 1998; Kallio *et al.*, 1999; Kamura *et al.*, 1998; Salceda and Caro, 1997; Tanimoto *et al.*, 2000). To date, four PHD domain-containing proteins (PHD1, -2, -3, -4), also known as HIF-prolyl-4-hydroxylases (HIF-P4H), have been identified, though the experimental support for involvement of the latter hydroxylase (PHD4) in the regulation of HIF-1α is limited (Oehme *et al.*, 2002). The action of the hydroxylases is dependent on the presence of oxygen, Fe^{2+}, 2-oxoglutarate, and ascorbate. By contrast, another hydroxylase, named factor-inhibiting HIF (FIH), hydroxylates an asparaginyl residue in the TADC, which prevents the recruitment of the coactivator CREB binding protein (CBP)/p300 (Hewitson *et al.*, 2002; Lando *et al.*, 2002b; Mahon *et al.*, 2001). In addition, redox factor-1 (Ref-1) and other coactivators, such as steroid receptor coactivator 1 (SRC-1) and transcription intermediary factor 2 (TIE-2), interacted with TADC in a redox-dependent manner (Carrero *et al.*, 2000; Ema *et al.*, 1999; Huang *et al.*, 1996), thus suggesting a role for ROS in modulating HIF-1 activity.

4. REACTIVE OXYGEN SPECIES MODULATE HIF

In addition to hypoxia, HIF-1α is also responsive to a variety of non-hypoxic stimuli, such as insulin (Kietzmann *et al.*, 2003b; Treins *et al.*, 2002; Zelzer *et al.*, 1998), platelet-derived growth factor (PDGF), transforming growth factor (TGF)-β, insulin-like growth factor (IGF)-1 (Fukuda *et al.*, 2002; Görlach *et al.*, 2001; Richard *et al.*, 2000), epidermal growth factor (EGF) (Liu *et al.*, 2006), thrombin (Görlach *et al.*, 2001), angiotensin-II (Richard *et al.*, 2000), cytokines (Stiehl *et al.*, 2002), carbachol (Hirota *et al.*, 2004), chromium arsenite (Duyndam *et al.*, 2001; Gao *et al.*, 2002b), mechanical stress (Kim *et al.*, 2002), and "hypoxic mimetic" $CoCl_2$ (Chandel *et al.*, 2000). Interestingly, many of these factors use ROS as part of their signaling cascades, and some evidence indicates that ROS act as messengers of these non-hypoxic stimuli to regulate HIF-1α. (BelAiba *et al.*, 2004; Knowles *et al.*, 2003). To this end, treatment of cells with antioxidants, such as ascorbate, vitamin E, and pyrroli-dinedithiocarbamate (PDTC), completely prevented HIF-1α protein expression and nuclear translocation in response to thrombin, PDGF, TGF-β, activated platelets (Görlach *et al.*, 2001), $CoCl_2$ (Chandel *et al.*, 2000), and *Helicobacter pylori* infection (Park *et al.*, 2003). Importantly, this reaction was confirmed *in vivo* since pigs receiving vitamin C and E along with their high cholesterol diet did not show elevated ROS, HIF-1α, or vascular endothelial growth factor (VEGF) levels (Zhu *et al.*, 2004).

In addition, the endogenous glutathione system appears to have also a modulatory role in non-hypoxic HIF-1α regulation. According to this system, N-acetyl cysteine (NAC), a precursor of glutathione, which is the substrate of GPX, diminished HIF-1α protein accumulation in response to thrombin (Görlach *et al.*, 2001), $CoCl_2$ (Kim *et al.*, 2002b), oxidized low-density lipo-protein (oxLDL) (Shatrov *et al.*, 2003), arsenite (Duyndam *et al.*, 2001), and hepatocyte growth factor (HGF) (Tacchini *et al.*, 2001) in different cells. In line, when GPX was overexpressed, it inhibited HIF-1α accumulation in response to thrombin and $CoCl_2$ (BelAiba *et al.*, 2004), whereas, under unstimulated conditions, overexpression of GPX (BelAiba *et al.*, 2004) or treatment with NAC enhanced HIF-1α levels (Haddad *et al.*, 2000). Similarly, cystamine increased HIF-1α levels in the duodenum of rats (Khomenko *et al.*, 2004).

Degradation of H_2O_2 by overexpression of catalase had no effect on basal HIF-1α levels (BelAiba *et al.*, 2004; Srinivas *et al.*, 2001), whereas it prevented the thrombin- and $CoCl_2$-dependent induction of HIF-1α (BelAiba *et al.*, 2004). Similarly, exogenous addition of catalase-decreased angiotensin-II (Richard *et al.*, 2000), chromium (Gao *et al.*, 2002b), and EGF induced HIF-1α levels (Liu *et al.*, 2006), but had no effect on arsenite-induced HIF-1α levels (Duyndam *et al.*, 2001).

By contrast, overexpression of Cu/Zn superoxide dismutase (Cu/ZnSOD) promoted formation of H_2O_2 and increased HIF-1α levels (BelAiba *et al.*, 2004). In addition, cell-permeable SOD mimetics enhanced HIF-1α levels under normoxia in rat renal medullary interstitial cells and human cerebral vascular smooth muscle cells (Wellman *et al.*, 2004). Further, increased HIF-1α protein levels were found in H_2O_2-treated aortic and pulmonary artery smooth muscle cells (Bonello *et al.*, 2007; Görlach *et al.*, 2001) and Hep3B cells (Chandel *et al.*, 2000).

Remarkably, H_2O_2 did not induce HIF-1α in urinary bladder carcinoma ECV304 cells (Görlach *et al.*, 2001), and extracellular SOD decreased $CoCl_2$-induced HIF-1α levels in Hep3B cells (Zelko and Folz, 2005). In addition, xanthine/xanthine oxidase significantly inhibited HIF-1α stability and transcriptional activity in control cells and cells exposed to $CoCl_2$ (Wellman *et al.*, 2004), suggesting that the sensitivity of the HIF system toward ROS may also depend on the cell type.

Various studies identified NADPH oxidases as important sources of ROS in the regulation of HIF-1α (Bonello *et al.*, 2007; Görlach *et al.*, 2001; Hirota *et al.*, 2004; Kim *et al.*, 2002a; Richard *et al.*, 2000). The finding that NADPH oxidases regulate HIF-1α was also substantiated by the *in vivo* observation that HIF-α levels were elevated in carotid lesions of mice overexpressing p22phox in smooth muscle cells (Khatri *et al.*, 2004). It appears that NOX4 may play an important role, since depletion of NOX4 by small interfering RNA (siRNA) diminished thrombin–induced HIF-1α levels (Bonello *et al.*, 2007). In line, depletion of NOX4 or NOX1 also decreased HIF-2α levels in VHL-deficient 786-O renal carcinoma cells, suggesting that ROS may act via a VHL-dependent pathway (Block *et al.*, 2007; Maranchie and Zhan, 2005).

Since low concentrations of ROS (10–50 μM H_2O_2) appear to have the ability to upregulate HIF-1α while high concentrations of H_2O_2 prevented HIF-1α accumulation (Bonello *et al.*, 2007; Görlach *et al.*, 2001), a threshold concentration of ROS may exist. Thus, only slight changes in the redox state may be required to activate the HIF pathway.

5. How are HIFs Regulated by Reactive Oxygen Species?

Reactive oxygen species may influence HIFs at different levels: they may influence mechanisms regulating HIF-α synthesis or HIF-α stability; they may act directly on HIF-α, thereby modifying stability or activity; or they may act via interference with a regulatory signaling pathway farther upstream.

5.1. Regulation of HIF-α synthesis by reactive oxygen species

So far, the knowledge about transcriptional regulation of HIF-α is limited. However, a recent study showed that thrombin, NOX4, and H_2O_2 regulate HIF-1α transcription. This regulation was mediated by ROS-sensitive activation of NF-κB, which bound itself to a specific consensus site in the HIF-1α promoter (Bonello *et al.*, 2007). Enhanced levels of HIF-1α messenger RNA (mRNA) have also been observed in response to IL-1, HGF, angiotensin-II, or lipopolysaccharides (Blouin *et al.*, 2004; Frede *et al.*, 2006; Page *et al.*, 2002; Tacchini *et al.*, 2004; Thornton *et al.*, 2000).

Since these factors are also known to enhance ROS levels, transcriptional regulation of HIF-1α by ROS-sensitive activation of NF-κB may represent an important mechanism on how agonists can induce HIF-1. Thus, those data provide a direct link between these two important redox-sensitive transcription factors.

In addition to this direct link between NF-κB and HIF-1α in response to ROS, it was proposed that the phosphatidylinositol 3-kinase (PI3K) pathway in conjunction with NF-κB may be involved in the translational regulation of HIF-1α in response to TNF-α (Zhou *et al.*, 2004). Likewise, ROS-dependent activation of the PI3K was also found to be involved in the increase in HIF-1α translation by angiotensin-II (Page *et al.*, 2002).

In contrast, Epstein-Barr virus (EBV) and its major oncoprotein, LMP1, induced HIF-1α in a ROS-dependent but NF-κB–independent way in epithelial cells (Wakisaka and Pagano, 2003), suggesting that the link between NF-κB and HIF-1α may be cell-type and stimulus specific.

5.2. Direct regulation of HIF-α by reactive oxygen species

First evidence that HIF-1 is redox-sensitive came from studies in which treatment of purified HIF-1 with H_2O_2, diamide, and N-ethyl-maleimide led to a loss of DNA-binding activity. This loss was counteracted by prior addition of dithiothreitol, suggesting that HIF-1 DNA binding requires reducing conditions (Wang *et al.*, 1995c). Furthermore, Ref-1 and its regulator Trx were shown to enhance HIF-1α protein levels (Huang *et al.*, 1996; Welsh *et al.*, 2002). In contrast, a redox-inactive Trx (C32S/C35S) decreased HIF-1α protein levels (Welsh *et al.*, 2002). Subsequently, binding of Ref-1 to TADN and TADC could be detected (Carrero *et al.*, 2000; Ema *et al.*, 1999), and it appears that the oxidation/reduction states of cysteine 800 in HIF-1α and cysteine 848 in HIF-2α are critical (Carrero *et al.*, 2000; Ema *et al.*, 1999; Huang *et al.*, 1996), since mutation of cysteine 800 prevented the decrease in HIF-1α TADC activity in response to hydroxyl radicals (OH•) (Liu *et al.*, 2004).

5.3. Regulation of HIF by reactive oxygen species via interference with a regulatory signaling pathway

5.3.1. Reactive oxygen species and prolyl hydroxylases

The HIF-1α degradation process is initiated by hydroxylation of proline 402 and proline 564. In addition, a second hydroxylation occurs at asparagine 803. While the hydroxylation at the proline residues can be carried out by

PHD1, -2, and -3, the asparaginyl hydroxylation is mediated by a protein known as FIH-1 (Hewitson *et al.*, 2002; Lando *et al.*, 2002a; Lee *et al.*, 2002b; Mahon *et al.*, 2001; McNeill *et al.*, 2002).

All these enzymes belong to a family of dioxygenases which require O_2, 2-oxoglutarate, ascorbate, and Fe(II) (Ivan *et al.*, 2002; Jaakkola *et al.*, 2001; Yu *et al.*, 2001). Thus, changed levels of these cofactors would affect hydroxylase activity as seen with ascorbate, which mediated HIF-1α degradation (Görlach *et al.*, 2001) likely due to increased PHD activity (Knowles *et al.*, 2003). Since these enzymes act only when iron is in the Fe(II) state, which is lost during hydroxylation of the substrate, these enzymes require a radical cycling system for regeneration of Fe(II) (Epstein *et al.*, 2001). Although the mechanism is not entirely known yet, it is proposed that ascorbate reduces Fe(III) to Fe(II) within the enzyme active site, thus rendering the enzyme active. Additionally, ascorbate might enhance the provision of Fe(II) from an intracellular pool, such as ferritin, by conversion of Fe(III) into Fe(II). Moreover, addition of iron to cultured human prostate adenocarcinoma PC3 cells stimulated HIF-1α degradation (Knowles *et al.*, 2003), possibly mediated by OH• generated in a Fenton reaction. Indeed, hydroxyl radicals colocalized with HIF-1α and VHL at the endoplasmic reticulum (ER) (Liu *et al.*, 2004; Schoenfeld *et al.*, 2001), and Ref-1 mediated reduction of Cys800 in the TADC (Liu *et al.*, 2004). These radicals could increase PHD activity by reconversion of ferric iron into ferrous iron or by facilitating conversion of dehydroascorbate into ascorbate. In line, in *junD*-deficient cells that exhibit chronic oxidative stress, it was shown that ROS interfered with Fe(II) availability in the HIF PHD catalytic site, possibly by a Fenton-type reaction, thus inhibiting activity under normoxia (Gerald *et al.*, 2004).

Therefore, ROS may interfere with the complex regulation of PHDs responding with higher or lower activity. This may in part explain some of the contradictory results on the regulation of HIF-1α exerted by changed redox conditions. Interestingly, the redox cycler DMNQ was shown to prevent accumulation of HIF-1α induced by TNF-α (Sandau *et al.*, 2001), whereas DMNQ alone may lead to HIF-1α accumulation via reduction of VHL binding, suggesting that DMNQ was interfering with PHD activity (Kohl *et al.*, 2006). Thus, it turns out that the HIF system is very sensitive towards subtle changes in the redox state of the cells.

5.3.2. Reactive oxygen species, phosphatases, and kinases

It is now well established that ROS also influence protein phosphatases and protein kinases. Thereby, phosphatases may become inactive due to transient oxidation of thiols and formation of intramolecular disulfide bridges or sulfonyl-amide bonds. Conversely, protein kinases become active upon oxidation by direct modification of sulfhydryl groups or by indirect concomitant inhibition of their phosphatases (Chiarugi, 2005; Tonks, 2005). Those phosphorylation reactions that can be modified by ROS appear to directly modify HIF-1, since treatment of nuclear extracts from hypoxic Hep3B cells with phosphatase disrupts the HIF-1/DNA complex (Wang et al., 1995a).

Interestingly, phosphatases like mitogen-activated protein kinase (MAP) phosphatase (MKP)-1, MKP-3, and protein serine/threonine phosphatase type 2A (PP2A) appear to be regulated by ROS, mainly derived from NADPH oxidases activated in response to hormones and growth factors (Furst et al., 2005). Furthermore, downregulation of MKP-1 enhanced HIF-1α phosphorylation (Liu et al., 2003), whereas overexpression of MKP-3 reduced HIF-1α phosphorylation (Marchetti et al., 2004), suggesting a complex regulatory network of HIF-1α by different phosphatases and their redox state.

Reciprocally, several kinases, including PI3K, protein kinase B (PKB/Akt), and MAP kinases, influence HIF-α activity in a redox-sensitive manner (Lee et al., 2002a; Mazure et al., 1997; Richard et al., 1999, 2000; Sodhi et al., 2001). Subsequently, antioxidants or inhibitors of NADPH oxidases prevented signaling via PI3K/PKB to HIF-1α and expression of some HIF target genes (Görlach et al., 2001). Similarly, catalase diminished activation of PI3K/PKB by vanadate or arsenite in prostate cancer cells (Gao et al., 2002a, 2004), suggesting a role for H_2O_2, which solely can activate PKB in different cells (Djordjevic et al., 2005).

In addition to PI3K/PKB, MAP kinases also have been shown to be involved in the regulation of HIF-1α (Richard et al., 1999). In line, extracellular signal-regulated kinases 1 and 2 (ERK1/2) also contributed to HIF-1α induction by a number of non-hypoxic stimuli known to induce ROS production, like angiotensin-II (Richard et al., 2000), IL-1β (Qian et al., 2004), prostaglandin E2 (PGE2) (Fukuda et al., 2003), or shock waves (Wang et al., 2004). However, it appears that the ROS sensitivity of ERK1/2 and the subsequent induction of HIF-1α are stimulus- and cell-type–dependent (Görlach et al., 2001; Wang et al., 2004).

Furthermore, the p38 MAP kinases are activated via ROS by many growth factors, hormones, and extracellular stressors, and overexpression of the p38 upstream kinases MAP kinase kinase 3 (MKK3) and MKK6 enhanced HIF-1α levels and activity (Kietzmann et al., 2003a). Involvement of ROS-dependent p38 MAP kinase activation was shown for the induction of HIF-1α by thrombin (Görlach et al., 2001) and chromium (VI) (Gao et al., 2002b). Although the exact sites for phosphorylation of HIF-1α by ERK1/2 or p38 MAP kinase have not been identified, these findings demonstrate an important role of these kinases in the redox-sensitive regulation of HIF-α.

6. Summary

Together, a vast body of results has shown that ROS derived from superoxide anion radicals has a profound effect on the HIF system, which appears to react sensitively toward changes in the cellular redox state. Thereby, a threshold concentration of ROS may drive the response in dependence of the cell type and/or stimulus. Thus, only slight changes in the redox state may be required to activate the HIF pathway.

7. Methods

Since changes in ROS levels appear to be critical for HIF induction, the determination of ROS levels under various conditions is essential for understanding the involvement of ROS in the regulation of HIF. ROS can interfere with a variety of substances, thus requiring controlled conditions when measuring ROS levels in cells. In the following sections, we summarize some techniques that appear to be suitable for ROS measurements alone or in combination.

8. The Cytochrome C Reduction Assay for Detection of Extracellular Reactive Oxygen Species

Cytochrome C is a heme protein and part of the mitochondrial electron transport chain. The ferricytochrome C can be directly reduced by superoxide to ferrocytochrome C:

$$Fe^{3+} - Cyt.c + O^{2-\bullet} \rightarrow Fe^{2+} - Cyt.c + O^2 \qquad (22.1)$$

This reduction leads to an increase in the absorbance spectrum of cytochrome C at 550 nm, which can be spectrophotometrically monitored. Since the reduction of cytochrome C is proportional to the amount of the reducing agent, it is possible to quantify the ROS production (Johnston et al., 1978; Murrant and Reid, 2001).

In order to determine ROS levels by this assay, cultivated cells are washed with Hanks' Balanced Salt Solution (HBSS; Gibco, Invitrogen, Carlsbad, CA) to remove media and sera. As cytochrome C is easily soluble in water, it is not necessary to use additional organic solvents, such as dimethyl sulfoxide (DMSO) or dimethylformamide (DMF), which could interfere with cellular function. On the other hand, the highly polar nature and high-molecular weight of cytochrome C prohibits cell membrane crossing (Murrant and Reid, 2001); therefore, this assay detects only extracellular ROS.

Cytochrome C (BIOMOL International, Plymouth Meeting, PA) is dissolved in HBSS to give an 800 -μM stock solution. Cytochrome C is added to the cells in a final concentration of 80 μM, then cells are incubated at 37° for 5 min, and the absorbance is measured in a microplate reader at a wavelength of 550 nm. Background readings at 540 and 560 nm are subtracted.

In addition to superoxide, cytochrome C can also be reduced by other radicals, such as OH^- and NO (Murrant and Reid, 2001). Thus, in order to specifically determine the amount of O_2^- generated, parallel measurements need to be conducted in the presence of superoxide dismutase (Cu/ZnSOD) from bovine erythrocytes (Calbiochem-Bachem, Bubendorf, Switzerland) at a final concentration of 50 $\mu g/ml$.

The SOD-inhibitable cytochrome C reduction is calculated by subtracting the absorbance measured in the solution supplemented with SOD from the values of the solution without SOD using an extinction coefficient $\varepsilon = 21$ mmol/l*cm^{-1} when measuring at 25°. With the Lambert-Beer's law, E(2) $E = \varepsilon \bullet c \bullet d$ (E = extinction, ε = extinction coefficient, c = sample concentration, d = layer thickness), the concentration of produced superoxide can be calculated: E(3) $c_{O_2^-} = E_{-SOD} - E_{+SOD}/\varepsilon \cdot d$ (E_{-SOD} = extinction without SOD, E_{+SOD} = extinction with SOD, $c_{O_2^-}$ = superoxide concentration). For absolute quantification, the total protein concentration of the cells is determined.

9. CHEMILUMINESCENCE ASSAY FOR DETECTION OF EXTRACELLULAR REACTIVE OXYGEN SPECIES

Chemiluminescence methods for superoxide detection have been frequently employed because of the alleged specificity of the reaction of the chemiluminescent probe with superoxide, minimal cellular toxicity, and relatively high sensitivity. Frequently used probes are luminol, lucigenin, or the chemiluminescence *Cypridina* luciferin derivative, 2-methyl-6-(4-methoxyphenyl)-3,7-dihydroimidazo[1,2-a]pyrazin-3-one, hydrochloride (MCLA, Invitrogen, Carlsbad, CA).

MCLA is used to monitor superoxide production in the extracellular milieu. Superoxide-stimulated chemiluminescence occurs after direct oxidation of the MCLA luminophor by extracellular O_2^-, followed by electrons returning to the electronic ground state from an excited state and emission of the light, which can be recorded. The advantage of MCLA compared to luminol is that it is not cell-permeable and therefore specific for detection of extracellular O_2^-. Thus, different than luminol, it is also useful for detecting superoxide released from the live cells. MCLA is also superior compared to lucigenin, since lucigenin can reportedly sensitize superoxide production, leading to false-positive results. Finally, the pH optimum for MCLA luminescence generation is closer to the physiological neutral range than the pH optima of luminol and lucigenin.

To execute the measurements with MCLA, cells are detached by trypsinization, and 0.5×10^6 cells are resuspended in HBSS containing 10 μM MCLA. Cells are then incubated for 5 min at 37° in the dark, and the luminescence is detected in a luminometer (Berthold Technologies, Oak Ridge, TN) at 37°. MCLA alone in HBSS buffer is measured as negative control for background levels of MCLA chemiluminescence. The findings are expressed as the counts of MCLA luminescence above the background level. Although it is difficult to convert MCLA chemiluminescence into absolute values of O_2^- (Shimmura *et al.*, 1992), it can be expressed as the amount of O_2^- as the equivalent concentration of xanthine oxidase (XO) that reacts with xanthine to generate O_2^- (Kimura *et al.*, 2001). The MCLA chemiluminescence induced by xanthine/XO depends linearly on the XO concentration.

However, some limitations also exist for this method. First, due to the high auto-oxidation potential of MCLA, strong background signals are likely. The obtained measurement values must always be corrected against the values obtained with the blank (buffer and MCLA). Second, although MCLA chemiluminescence is highly specific for extracellular O_2^-, it can also detect singlet oxygen $^1O_2$ (Kimura *et al.*, 2001). To avoid this, superoxide dismutase can be added to distinguish between chemiluminescence of $O_2^- \bullet$ and $^1O_2$ because SOD does not scavenge $^1O_2$.

10. Measuring Intracellular Production of Reactive Oxygen Species using Fluorescent Dyes

Intracellularly formed ROS can be determined using fluorescent dyes. These dyes, including dihydroethidine (DHE), dihydroflorescein (H_2DCF), and dihydrorhodomine 123 (DHR), are initially nonfluorescent; however, they become fluorescent upon reaction with ROS to give hydroethidium (HE), dichlorofluorescin (DCF), and rhodamine 123, respectively. Subsequently, fluorescence intensity can be easily determined using a fluorometer or fluorescence reader.

10.1. DCF fluorescence

Dihydroflorescein is oxidized by various ROS, mainly peroxides, like H_2O_2 and peroxynitrite; however, hydroxyl anions (OH•) and hypochlorous acid (HOCl) also are known to react with this dye, forming the fluorescent DCF (Tarpey et al., 2004). The H_2O_2-mediated reaction of H_2DCF to DCF is rather slow, and ferrous ions seem to be elementary during this conversion. Indeed, intracellular iron trafficking appears to critically influence the H_2O_2-dependent DCF fluorescence (Rothe and Valet, 1990). In addition, heme-containing substances (e.g., hematin, peroxidases, or cytochrome C) can enhance DCF formation independently from H_2O_2 (LeBel et al., 1992; Rothe and Valet, 1990). Since H_2DCF and even more DCF can be easily retransported out of the cell—an action that could interfere with the measurements, thus leading to misinterpretations—chloromethyl (CM)-H_2DCFDA is often used. CM-H_2DCFDA is taken up by the cell, de-esterified by intracellular esterases, and is, due to its reactive thiol-chloromethyl groups, covalently added to intracellular glutathione and other thiols.

To determine ROS by DCF fluorescence, cultivated cells are washed with HBSS and loaded with CM-H_2DCFDA (final concentration of 8.64 μM; Molecular Probes, Invitrogen, Carlsbad, CA) for 5 to 10 min, depending on the cell type, at 37° in the dark. After washing with HBSS, the specific fluorescence (excitation 480 nm/emission 540 nm) is recorded with a fluorescence multi-well plate reader (Tecan, Zurich, Switzerland). The increase in fluorescence correlates with ROS production. Since plastic cell culture material possesses autofluorescent properties, this material should be subtracted or special black plastic material should be used. In addition, auto-oxidation of DCF in the medium should be determined in the absence of cells. After measuring, the fluorescence values should be normalized to the number of cells present or to the total amount of protein.

Since superoxide anion radicals can quickly react with NO• to form peroxynitrite, samples containing cells that are considered to also generate NO, such as endothelial cells, need to be preincubated with inhibitors of NO synthases (NG-nitro-L-arginine [L-NNA], 10 μM, 15 min preincubation) before determining DCF fluorescence (Tarpey et al., 2004).

10.2. Hydroethidine fluorescence

The cell-permeable dihydroethidine (DHE, Sigma-Aldrich, St. Louis, MO) can undergo oxidation to form the DNA-binding fluorophores ethidium (E^+) or 2-hydroxyethidium (2-OH-E^+). The latter reaction is relatively specific for superoxide, with minimal oxidation induced by H_2O_2 or HOCl (Tarpey et al., 2004; Zhao et al., 2003). In order to discriminate ethidium from 2-OH-E^+, several high performance liquid chromatography

(HPLC)-based protocols have been established (though increased specificity for superoxide can also be achieved using fluorescence-based techniques) (Zhao et al., 2005, 2003).

With an excitation wavelength of 490 nm, ethidium shows a fluorescence emission peak at 590 nm, whereas 2-OH-E$^+$ has a maximum of emission at 560 to 570 nm. Therefore, the fluorescence between 560 and 570 nm at an excitation wavelength of 490 nm should be measured in order to minimize spectral interference of these two compounds. In addition, a mitochondrial-specific derivative of E$^+$ has been designed. This compound shows a specific absorption peak at 396 nm for its 2-OH-E$^+$ product (Robinson et al., 2006). This characteristic would make it possible to distinguish easily between E$^+$ and 2-OH-E$^+$ using excitation wavelengths of 396 nm and measuring emission at 510 nm. This compound is therefore very valuable in order to judge mitochondrial superoxide production specifically.

In contrast to other probes for superoxide, little evidence exists that shows that DHE can lead to the formation of superoxide due to redox cycling (Benov et al., 1998). However, it has been reported that cytochrome C is capable of oxidizing DHE to E$^+$, which has to be considered especially in situations where mitochondria are the major source for superoxide or in situations where the stimulus leads to apoptosis resulting in an efflux of cytochrome C. In addition, the ability of E$^+$ to accumulate in mitochondria has to be considered, especially if high concentrations of DHE are used. If the amount of E$^+$ exceeds the binding capacity of mitochondrial nucleic acids, E$^+$ can accumulate in the mitochondrial matrix and can be released to the cytosol due to a breakdown of the mitochondrial membrane potential. Therefore, E$^+$ could now bind to nuclear DNA with marked increase of fluorescence (Budd et al., 1997).

10.3. Dihydrorhodamine fluorescence

The reduction product of the mitochondria-selective dye rhodamine is the uncharged nonfluorescent dihydrorhodamine. This light-sensitive molecule can passively diffuse across the plasma membrane, where it is oxidized to the cationic rhodamine 123 in the presence of H_2O_2 and peroxidases, cytochrome C, and ferrous iron (Henderson and Chappell, 1993). In addition to H_2O_2, dihydrorhodamine also reacts with HOCl and peroxynitrite (Kooy et al., 1994). Since the oxidation of dihydrorhodamine requires peroxidases, changes in peroxidase activity will directly influence the measurement; furthermore, auto-oxidation and changes in ferrous iron levels can affect the measurements. In addition, collapse of the mitochondrial potential can lead to efflux of the fluorescent dye.

Today two different forms of dihydrorhodamine are used. The most widely used is dihydrorhodamine 123 (Invitrogen, Carlsbad, CA), which is excited at 488 nm and whose emission is detected at 515 nm. The other one used is dihydrorhodamine 6G (Invitrogen), which has a longer wavelength spectra (emission at 575 nm), making this probe useful for multicolor applications or in some autofluorescent cells or tissue (Wersto *et al.*, 1996).

10.4. ROS measurements in tissues

Dihydroethidine or CM-H$_2$DCFDA could also be used to determine ROS levels *in situ*. To this end, tissue samples are rapidly excised and snap-frozen in liquid nitrogen. The samples are cut in 10- to 20 -μm slices using a cryotome. Sectioned samples are stained using DHE (50 μM) or CM-H$_2$DCFDA (8.6 μM) for 10 min in the dark. After washing with HBSS, the fluorescence can be detected using fluorescence microscopy. As a control, unstained tissue should be used. In addition, preincubation of the sample with antioxidants like (PEG)-SOD and catalase should be performed to improve specificity of the dyes. In addition to the previously mentioned problems, the histology of the tissue may be difficult to judge, especially if more complex tissues are stained. Therefore, staining of the nuclei with Hoechst dyes can be considered. Care has to be taken that these dyes will not interfere with regard to their spectroscopic properties.

11. DETECTION OF REACTIVE OXYGEN SPECIES BY ELECTRON PARAMAGNETIC RESONANCE

Electron paramagnetic resonance (EPR) is a spectroscopic technique that detects paramagnetic species that have unpaired electrons, including free radicals, many transition metal ions, and defects in materials. It is also often called electron spin resonance (ESR).

Electron paramagnetic resonance is a magnetic resonance technique used to detect the transitions of unpaired electrons in an applied magnetic field. Like a proton, the electron has "spin," which gives it a magnetic property known as a magnetic moment. When we supply an external magnetic field, the paramagnetic electron can orient in a direction parallel or antiparallel to the direction of the magnetic field. This orientation creates two distinct energy levels for the unpaired electrons and allows us to measure them as they are driven between.

Most of the free radicals are very reactive species with very short half-life times, whose radical character has to be stabilized for EPR measurement. Oxygen-derived free radicals can be detected by a spin-trapping technique with EPR spectroscopy. With this technique, very short-lived oxygen-derived free radicals react with spin traps to yield longer-lived radicals.

Nitrones have emerged as the most popular spin traps. The cyclic 5,5-dimethyl-1-pyrroline N-oxide (DMPO) and its derivate 5-diethoxyphosphoryl-5-methyl-1-pyrroline N-oxide (DEPMPO) have received the most attention since they yield distinct and characteristic spin adducts with $O_2^-\bullet$ and HO• radicals. However, the use of DMPO or DEPMPO in a biological milieu is not without its limitations. Reaction of the spin traps with superoxide is rather slow. Therefore, high concentrations of spin traps are demanded, which can cause toxic effects in biological systems. Moreover several cellular components and even superoxide itself are able to reduce the formed spin adducts into diamagnetic, EPR-silent species (Dikalov et al., 2006; Frejaville et al., 1995).

To overcome these disadvantages, a new technique for detection of $O_2^-\bullet$ in vitro and in vivo has emerged. A strength of these new spin probes is that cyclic hydroxylamines, such as 1-hydroxy-3-methoxycarbonyl-2,2,5,5-tetramethylpyrrolidine (CMH), are cell permeable and are very effective scavengers of $O_2^-\bullet$ radicals. The long half life of the nitroxide radical produced with ROS is a distinct advantage over nitrone spin traps, which form unstable $O_2^-\bullet$ radical adducts in biological samples; nitroxides, on the other hand, are stable in the presence of reducing agents. Therefore, cyclic hydroxylamines provide quantitative measurements of $O_2^-\bullet$ radicals with higher sensitivity than the nitrone spin traps. A weakness of the cyclic hydroxylamines is that the product formed upon reaction with different ROS yields the same nitroxide, regardless of the ROS trapped, and the resultant EPR spectra are therefore identical for several different ROS. In order to overcome this limitation, it is necessary to perform additional studies using antioxidants and inhibitors specific for the various ROS. For example, $O_2^-\bullet$ radicals can be determined as a SOD-inhibitable CM-nitroxide formation. Since CMH detects both extra- and intracellular ROS, extracellular O_2^- can be quantified by addition of Cu,Zn-SOD, while supplementation of cells with cell-permeable PEG-SOD will inhibit detection of both extra- and intracellular O_2^-.

To determine $O_2^-\bullet$ production by EPR spectroscopy, the spin probe CMH (Alexis Corp., San Diego, CA) is used. Oxidation of CMH by ROS forms stable 3-carboxymethoxyl nitroxide radicals (CM•). Cells (800,000 cells/ml) are suspended in Krebs-Hepes Buffer containing 25 μM deferoxamine and 5 μM DETC (sodium diethyldithiocarbamate) as chelating agents and SOD (50 U/ml). To 45 μl cell suspension, 2.5 μl CMH stock solution (4 mM) is added to obtain a final concentration of 200 μM CMH. The samples are transferred to 50 -μl glass capillaries (Brand, Wertheim, Germany) and sealed with Hematocrit Sealing Compound (Brand, Wertheim, Germany). The EPR spectra were recorded using an e-scan spectrometer (Bruker, Billerica, MA) with the instrumental settings as follows: field sweep, 50 G; microwave frequency, 9.45 GHz; microwave power, 22 mW; modulation amplitude, 2 G; conversion time, 10.24 ms; time constant, 40.96 ms; receiver gain, 1×10^3.

ACKNOWLEDGMENTS

This work was supported by DFG GO709/4–4, the 6th European framework program (EUROXY), and Fondation Leducq to A. G. and by DFG SFB 402 A1, GRK 335, Deutsche Krebshilfe (106429), Fonds der Chemischen Industrie, and Fondation Leducq to T. K.

REFERENCES

Agani, F. H., Puchowicz, M., Chavez, J. C., Pichiule, P., and LaManna, J. (2002). Role of nitric oxide in the regulation of HIF-1alpha expression during hypoxia. *Am. J. Physiol. Cell. Physiol* **283,** C178–C186.

Babior, B. M. (2004). NADPH oxidase. *Curr. Opin. Immunol.* **16,** 42–47.

Beck, I., Weinmann, R., and Caro, J. (1993). Characterization of hypoxia-responsive enhancer in the human erythropoietin gene shows presence of hypoxia-inducible 120-Kd nuclear DNA-binding protein in erythropoietin-producing and nonproducing cells. *Blood* **82,** 704–711.

BelAiba, R. S., Djordjevic, T., Bonello, S., Flugel, D., Hess, J., Kietzmann, T., and Görlach, A. (2004). Redox-sensitive regulation of the HIF pathway under non-hypoxic conditions in pulmonary artery smooth muscle cells. *Biol. Chem.* **385,** 249–257.

BelAiba, R. S., Djordjevic, T., Petry, A., Diemer, K., Bonello, S., Hess, J., Pogrebniak, A., Bickel, C., and Görlach, A. (2007). NOX5 variants are functionally active in endothelial cells. *Free Radic. Biol. Med.* **42,** 446–459.

Benov, L., Sztejnberg, L., and Fridovich, I. (1998). Critical evaluation of the use of hydroethidine as a measure of superoxide anion radical. *Free Radic. Biol. Med.* **25,** 826–831.

Block, K., Gorin, Y., Hoover, P., Williams, P., Chelmicki, T., Clark, R. A., Yoneda, T., and Abboud, H. E. (2007). NAD(P)H oxidases regulate HIF-2α protein expression. *J. Biol. Chem.* **282,** 8019–8026.

Blouin, C. C., Page, E. L., Soucy, G. M., and Richard, D. E. (2004). Hypoxic gene activation by lipopolysaccharide in macrophages: Implication of hypoxia-inducible factor 1alpha. *Blood* **103,** 1124–1130.

Bokoch, G. M., and Knaus, U. G. (2003). NADPH oxidases: Not just for leukocytes anymore! *Trends Biochem. Sci.* **28,** 502–508.

Bonello, S., Zähringer, C., BelAiba, R. S., Djordjevic, T., Hess, J., Michiels, C., Kietzmann, T., and Görlach, A. (2007). Reactive oxygen species activate the HIF-1α promoter via a functional NF-κB site. *Arterioscler. Thromb. Vasc. Biol.* **27,** 755–761.

Brigelius-Flohe, R., Banning, A., and Schnurr, K. (2003). Selenium-dependent enzymes in endothelial cell function. *Antioxid. Redox. Signal.* **5,** 205–215.

Budd, S. L., Castilho, R. F., and Nicholls, D. G. (1997). Mitochondrial membrane potential and hydroethidine-monitored superoxide generation in cultured cerebellar granule cells. *FEBS Lett.* **415,** 21–24.

Cai, H., and Harrison, D. G. (2000). Endothelial dysfunction in cardiovascular diseases: The role of oxidant stress. *Circ. Res.* **87,** 840–844.

Carrero, P., Okamoto, K., Coumailleau, P., O'Brien, S., Tanaka, H., and Poellinger, L. (2000). Redox-regulated recruitment of the transcriptional coactivators CREB-binding protein and SRC-1 to hypoxia-inducible factor 1alpha. *Mol. Cell Biol.* **20,** 402–415.

Chandel, N. S., McClintock, D. S., Feliciano, C. E., Wood, T. M., Melendez, J. A., Rodriguez, A. M., and Schumacker, P. T. (2000). Reactive oxygen species generated at mitochondrial complex III stabilize hypoxia-inducible factor 1alpha during hypoxia: A mechanism of O_2 sensing. *J. Biol. Chem.* **275**, 25130–25138.

Chiarugi, P. (2005). PTPs versus PTKs: The redox side of the coin. *Free Radic.Res.* **39**, 353–364.

Dikalov, S. I., Li, W., Mehranpour, P., Wang, S. S., and Zafari, A. M. (2006). Production of extracellular superoxide by human lymphoblast cell lines: Comparison of electron spin resonance techniques and cytochrome C reduction assay. *Biochem. Pharmacol.* **73**, 972–980.

Djordjevic, T., Pogrebniak, A., BelAiba, R. S., Bonello, S., Wotzlaw, C., Acker, H., Hess, J., and Görlach, A. (2005). The expression of the NADPH oxidase subunit p22phox is regulated by a redox-sensitive pathway in endothelial cells. *Free Radic. Biol. Med.* **38**, 616–630.

Duyndam, M. C., Hulscher, T. M., Fontijn, D., Pinedo, H. M., and Boven, E. (2001). Induction of vascular endothelial growth factor expression and hypoxia-inducible factor 1alpha protein by the oxidative stressor arsenite. *J. Biol. Chem.* **276**, 48066–48076.

Ema, M., Hirota, K., Mimura, J., Abe, H., Yodoi, J., Sogawa, K., Poellinger, L., and Fujii-Kuriyama, Y. (1999). Molecular mechanisms of transcription activation by HLF and HIF-1α in response to hypoxia: Their stabilization and redox signal-induced interaction with CBP/p300. *EMBO J.* **18**, 1905–1914.

Epstein, A. C., Gleadle, J. M., McNeill, L. A., Hewitson, K. S., O'Rourke, J., Mole, D. R., Mukherji, M., Metzen, E., Wilson, M. I., Dhanda, A., Tian, Y. M., Masson, N., *et al.* (2001). *C. elegans* EGL-9 and mammalian homologs define a family of dioxygenases that regulate HIF by prolyl hydroxylation. *Cell* **107**, 43–54.

Finkel, T. (2001). Reactive oxygen species and signal transduction. *IUBMB Life* **52**, 3–6.

Frede, S., Stockmann, C., Freitag, P., and Fandrey, J. (2006). Bacterial lipopolysaccharide induces HIF-1 activation in human monocytes via p44/42 MAPK and NF-κB. *Biochem. J.* **396**, 517–527.

Freeman, B. A., and Crapo, J. D. (1982). Free-radicals and tissue-injury. *Lab. Invest.* **47**, 412–426.

Frejaville, C., Karoui, H., Tuccio, B., Le Moigne, F., Culcasi, M., Pietri, S., Lauricella, R., and Tordo, P. (1995). 5-(Diethoxyphosphoryl)-5-methyl-1-pyrroline N-oxide: A new efficient phosphorylated nitrone for the *in vitro* and *in vivo* spin trapping of oxygen-centered radicals. *J. Med. Chem.* **38**, 258–265.

Fukuda, R., Kelly, B., and Semenza, G. L. (2003). Vascular endothelial growth factor gene expression in colon cancer cells exposed to prostaglandin E2 is mediated by hypoxia-inducible factor 1. *Cancer Res.* **63**, 2330–2334.

Fukuda, R., Hirota, K., Fan, F., Jung, Y. D., Ellis, L. M., and Semenza, G. L. (2002). Insulin-like growth factor 1 induces hypoxia-inducible factor 1-mediated vascular endothelial growth factor expression, which is dependent on MAP kinase and phosphatidyl-inositol 3-kinase signaling in colon cancer cells. *J. Biol. Chem.* **277**, 38205–38211.

Furst, R., Brueckl, C., Kuebler, W. M., Zahler, S., Krotz, F., Görlach, A., Vollmar, A. M., and Kiemer, A. K. (2005). Atrial natriuretic peptide induces mitogen-activated protein kinase phosphatase-1 in human endothelial cells via Rac1 and NAD(P)H oxidase/Nox2-activation. *Circ. Res.* **96**, 43–53.

Gao, N., Shen, L., Zhang, Z., Leonard, S. S., He, H., Zhang, X. G., Shi, X., and Jiang, B. H. (2004). Arsenite induces HIF-1α and VEGF through PI3K, Akt and reactive oxygen species in DU145 human prostate carcinoma cells. *Mol. Cell Biochem.* **255**, 33–45.

Gao, N., Ding, M., Zheng, J. Z., Zhang, Z., Leonard, S. S., Liu, K. J., Shi, X., and Jiang, B. H. (2002a). Vanadate-induced expression of hypoxia-inducible factor 1 alpha and vascular endothelial growth factor through phosphatidylinositol 3-kinase/Akt pathway and reactive oxygen species. *J. Biol. Chem.* **277**, 31963–31971.

Gao, N., Jiang, B. H., Leonard, S. S., Corum, L., Zhang, Z., Roberts, J. R., Antonini, J., Zheng, J. Z., Flynn, D. C., Castranova, V., and Shi, X. (2002b). p38 signaling-mediated hypoxia-inducible factor 1alpha and vascular endothelial growth factor induction by Cr(VI) in DU145 human prostate carcinoma cells. *J. Biol. Chem.* **277,** 45041–45048.

Gerald, D., Berra, E., Frapart, Y. M., Chan, D. A., Giaccia, A. J., Mansuy, D., Pouyssegur, J., Yaniv, M., and Mechta-Grigoriou, F. (2004). JunD reduces tumor angiogenesis by protecting cells from oxidative stress. *Cell* **118,** 781–794.

Görlach, A., Kietzmann, T., and Hess, J. (2002). Redox signaling through NADPH oxidases: Involvement in vascular proliferation and coagulation. *Ann. N.Y. Acad. Sci.* **973,** 505–507.

Görlach, A., Brandes, R. P., Nguyen, K., Amidi, M., Dehghani, F., and Busse, R. (2000a). A gp91phox containing NADPH oxidase selectively expressed in endothelial cells is a major source of oxygen radical generation in the arterial wall. *Circ. Res.* **87,** 26–32.

Görlach, A., Camenisch, G., Kvietikova, I., Vogt, L., Wenger, R. H., and Gassmann, M. (2000b). Efficient translation of mouse hypoxia-inducible factor-1alpha under normoxic and hypoxic conditions. *Biochim. Biophys. Acta* **1493,** 125–134.

Görlach, A., Holtermann, G., Jelkmann, W., Hancock, J. T., Jones, S. A., Jones, O. T., and Acker, H. (1993). Photometric characteristics of haem proteins in erythropoietin-producing hepatoma cells (HepG2). *Biochem. J.* **290,** 771–776.

Görlach, A., Diebold, I., Schini-Kerth, V. B., Berchner-Pfannschmidt, U., Roth, U., Brandes, R. P., Kietzmann, T., and Busse, R. (2001). Thrombin activates the hypoxia-inducible factor-1 signaling pathway in vascular smooth muscle cells: Role of the p22(phox)-containing NADPH oxidase. *Circ. Res.* **89,** 47–54.

Griendling, K. K. (2004). Novel NAD(P)H oxidases in the cardiovascular system. *Heart* **90,** 491–493.

Groemping, Y., and Rittinger, K. (2004). Activation and assembly of the NADPH oxidase: A structural perspective. *Biochem. J.* **386,** 401–416.

Gross, J., Rheinlander, C., Fuchs, J., Mazurek, B., Machulik, A., Andreeva, N., and Kietzmann, T. (2003). Expression of hypoxia-inducible factor 1 in the cochlea of newborn rats. *Heart Res.* **183,** 73–83.

Haddad, J. J., Olver, R. E., and Land, S. C. (2000). Antioxidant/pro-oxidant equilibrium regulates HIF-1α and NF-κB redox sensitivity. Evidence for inhibition by glutathione oxidation in alveolar epithelial cells. *J. Biol. Chem.* **275,** 21130–21139.

Hagen, T., Taylor, C. T., Lam, F., and Moncada, S. (2003). Redistribution of intracellular oxygen in hypoxia by nitric oxide: Effect on HIF1α. *Science.* **302,** 1975–1978.

Hara, S., Hamada, J., Kobayashi, C., Kondo, Y., and Imura, N. (2001). Expression and characterization of hypoxia-inducible factor (HIF)-3alpha in human kidney: Suppression of HIF-mediated gene expression by HIF-3α. *Biochem. Biophys. Res. Commun.* **287,** 808–813.

Heidbreder, M., Frohlich, F., Johren, O., Dendorfer, A., Qadri, F., and Dominiak, P. (2003). Hypoxia rapidly activates HIF-3α mRNA expression. *FASEB J.* **17,** 1541–1543.

Henderson, L. M., and Chappell, J. B. (1993). Dihydrorhodamine 123: A fluorescent probe for superoxide generation? *Eur. J. Biochem.* **217,** 973–980.

Herkert, O., Djordjevic, T., BelAiba, R. S., and Görlach, A. (2004). Insights into the redox control of blood coagulation: Role of vascular NADPH oxidase-derived reactive oxygen species in the thrombogenic cycle. *Antioxid. Redox. Signal.* **6,** 765–776.

Hewitson, K. S., McNeill, L. A., Riordan, M. V., Tian, Y. M., Bullock, A. N., Welford, R. W., Elkins, J. M., Oldham, N. J., Bhattacharya, S., Gleadle, J. M., Ratcliffe, P. J., Pugh, C. W., *et al.* (2002). Hypoxia-inducible factor (HIF) asparagine hydroxylase is identical to factor inhibiting HIF (FIH) and is related to the cupin structural family. *J. Biol. Chem.* **277,** 26351–26355.

Hirota, K., Fukuda, R., Takabuchi, S., Kizaka-Kondoh, S., Adachi, T., Fukuda, K., and Semenza, G. L. (2004). Induction of hypoxia-inducible factor 1 activity by muscarinic acetylcholine receptor signaling. *J. Biol. Chem.* **279**, 41521–41528.

Hoffman, E. C., Reyes, H., Chu, F. F., Sander, F., Conley, L. H., Brooks, B. A., and Hankinson, O. (1991). Cloning of a factor required for activity of the Ah (dioxin) receptor. *Science* **252**, 954–958.

Hu, C. J., Wang, L. Y., Chodosh, L. A., Keith, B., and Simon, M. C. (2003). Differential roles of hypoxia-inducible factor 1alpha (HIF-1α) and HIF-2α in hypoxic gene regulation. *Mol. Cell Biol.* **23**, 9361–9374.

Huang, L. E., Arany, Z., Livingston, D. M., and Bunn, H. F. (1996). Activation of hypoxia-inducible transcription factor depends primarily upon redox-sensitive stabilization of its alpha subunit. *J. Biol. Chem.* **271**, 32253–32259.

Huang, L. E., Gu, J., Schau, M., and Bunn, H. F. (1998). Regulation of hypoxia-inducible factor 1alpha is mediated by an O_2-dependent degradation domain via the ubiquitin-proteasome pathway. *Proc. Natl. Acad. Sci. USA* **95**, 7987–7992.

Huang, L. E., Willmore, W. G., Gu, J., Goldberg, M. A., and Bunn, H. F. (1999). Inhibition of hypoxia-inducible factor 1 activation by carbon monoxide and nitric oxide. Implications for oxygen sensing and signaling. *J. Biol. Chem.* **274**, 9038–9044.

Ivan, M., Kondo, K., Yang, H., Kim, W., Valiando, J., Ohh, M., Salic, A., Asara, J. M., Lane, W. S., and Kaelin, W. G., Jr. (2001). HIF-α targeted for VHL-mediated destruction by proline hydroxylation: Implications for O_2 sensing. *Science* **292**, 464–468.

Ivan, M., Haberberger, T., Gervasi, D. C., Michelson, K. S., Gunzler, V., Kondo, K., Yang, H., Sorokina, I., Conaway, R. C., Conaway, J. W., and Kaelin, W. G., Jr. (2002). Biochemical purification and pharmacological inhibition of a mammalian prolyl hydroxylase acting on hypoxia-inducible factor. *Proc. Natl. Acad. Sci. USA* **99**, 13459–13464.

Jaakkola, P., Mole, D. R., Tian, Y. M., Wilson, M. I., Gielbert, J., Gaskell, S. J., von Kriegsheim, A., Hebestreit, H. F., Mukherji, M., Schofield, C. J., Maxwell, P. H., Pugh, C. W., *et al.* (2001). Targeting of HIF-α to the von Hippel-Lindau ubiquitylation complex by O_2-regulated prolyl hydroxylation. *Science* **292**, 468–472.

Jiang, B. H., Zheng, J. Z., Leung, S. W., Roe, R., and Semenza, G. L. (1997). Transactivation and inhibitory domains of hypoxia-inducible factor 1alpha. Modulation of transcriptional activity by oxygen tension. *J. Biol. Chem.* **272**, 19253–19260.

Johnston, R. B., Jr., Godzik, C. A., and Cohn, Z. A. (1978). Increased superoxide anion production by immunologically activated and chemically elicited macrophages. *J. Exp. Med.* **148**, 115–127.

Kaelin, W. G., Jr., and Maher, E. R. (1998). The VHL tumour-suppressor gene paradigm. *Trends Genet.* **14**, 423–426.

Kallio, P. J., Wilson, W. J., O'Brien, S., Makino, Y., and Poellinger, L. (1999). Regulation of the hypoxia-inducible transcription factor 1alpha by the ubiquitin-proteasome pathway. *J. Biol. Chem.* **274**, 6519–6525.

Kamura, T., Sato, S., Haque, D., Liu, L., Kaelin-WG, J., Conaway, R. C., and Conaway, J. W. (1998). The Elongin BC complex interacts with the conserved SOCS-box motif present in members of the SOCS, ras, WD-40 repeat, and ankyrin repeat families. *Genes Dev.* **12**, 3872–3881.

Khatri, J. J., Johnson, C., Magid, R., Lessner, S. M., Laude, K. M., Dikalov, S. I., Harrison, D. G., Sung, H. J., Rong, Y., and Galis, Z. S. (2004). Vascular oxidant stress enhances progression and angiogenesis of experimental atheroma. *Circulation* **109**, 520–525.

Khomenko, T., Deng, X., Sandor, Z., Tarnawski, A. S., and Szabo, S. (2004). Cysteamine alters redox state, HIF-1α transcriptional interactions and reduces duodenal mucosal oxygenation: Novel insight into the mechanisms of duodenal ulceration. *Biochem. Biophys. Res. Commun.* **317**, 121–127.

Kietzmann, T., Jungermann, K., and Görlach, A. (2003a). Regulation of the hypoxia-dependent plasminogen activator inhibitor 1 expression by MAP kinases in HepG2 cells. *Thromb. Haemost.* **89,** 666–674.

Kietzmann, T., Samoylenko, A., Roth, U., and Jungermann, K. (2003b). Hypoxia-inducible factor-1 and hypoxia response elements mediate the induction of plasminogen activator inhibitor-1 gene expression by insulin in primary rat hepatocytes. *Blood* **101,** 907–914.

Kietzmann, T., Cornesse, Y., Brechtel, K., Modaressi, S., and Jungermann, K. (2001). Perivenous expression of the mRNA of the three hypoxia-inducible factor alpha-subunits, HIF-1α, HIF-2α and HIF-3α, in rat liver. *Biochem. J.* **354,** 531–537.

Kim, C. H., Cho, Y. S., Chun, Y. S., Park, J. W., and Kim, M. S. (2002a). Early expression of myocardial HIF-1α in response to mechanical stresses: Regulation by stretch-activated channels and the phosphatidylinositol 3-kinase signaling pathway. *Circ. Res.* **90,** E25–E33.

Kim, H. H., Lee, S. E., Chung, W. J., Choi, Y., Kwack, K., Kim, S. W., Kim, M. S., Park, H., and Lee, Z. H. (2002b). Stabilization of hypoxia-inducible factor-1alpha is involved in the hypoxic stimuli-induced expression of vascular endothelial growth factor in osteoblastic cells. *Cytokine* **17,** 14–27.

Kimura, C., Oike, M., Koyama, T., and Ito, Y. (2001). Impairment of endothelial nitric oxide production by acute glucose overload. *Am. J. Physiol. Endocrinol. Metab.* **280,** E171–E178.

Kimura, H., Weisz, A., Kurashima, Y., Hashimoto, K., Ogura, T., D'Acquisto, F., Addeo, R., Makuuchi, M., and Esumi, H. (2000). Hypoxia response element of the human vascular endothelial growth factor gene mediates transcriptional regulation by nitric oxide: Control of hypoxia-inducible factor-1 activity by nitric oxide. *Blood* **95,** 189–197.

Knowles, H. J., Raval, R. R., Harris, A. L., and Ratcliffe, P. J. (2003). Effect of ascorbate on the activity of hypoxia-inducible factor in cancer cells. *Cancer Res.* **63,** 1764–1768.

Kohl, R., Zhou, J., and Brune, B. (2006). Reactive oxygen species attenuate nitric-oxide-mediated hypoxia-inducible factor-1alpha stabilization. *Free Radic. Biol. Med.* **40,** 1430–1442.

Koury, J., Deitch, E. A., Homma, H., Abungu, B., Gangurde, P., Condon, M. R., Lu, Q., Xu, D. Z., and Feinman, R. (2004). Persistent HIF-1alpha activation in gut ischemia/reperfusion injury: Potential role of bacteria and lipopolysaccharide. *Shock* **22,** 270–277.

Kooy, N. W., Royall, J. A., Ischiropoulos, H., and Beckman, J. S. (1994). Peroxynitrite-mediated oxidation of dihydrorhodamine 123. *Free Radic. Biol. Med.* **16,** 149–156.

Lambeth, J. D. (2004). NOX enzymes and the biology of reactive oxygen. *Nat. Rev. Immunol.* **4,** 181–189.

Lando, D., Peet, D. J., Whelan, D. A., Gorman, J. J., and Whitelaw, M. L. (2002a). Asparagine hydroxylation of the HIF transactivation domain: A hypoxic switch. *Science* **295,** 858–861.

Lando, D., Peet, D. J., Gorman, J. J., Whelan, D. A., Whitelaw, M. L., and Bruick, R. K. (2002b). FIH-1 is an asparaginyl hydroxylase enzyme that regulates the transcriptional activity of hypoxia-inducible factor. *Genes Dev.* **16,** 1466–1471.

LeBel, C. P., Ischiropoulos, H., and Bondy, S. C. (1992). Evaluation of the probe 2′,7′-dichlorofluorescin as an indicator of reactive oxygen species formation and oxidative stress. *Chem. Res. Toxicol.* **5,** 227–231.

Lee, E., Yim, S., Lee, S. K., and Park, H. (2002a). Two transactivation domains of hypoxia-inducible factor-1alpha regulated by the MEK-1/p42/p44 MAPK pathway. *Mol. Cells* **14,** 9–15.

Lee, C., Kim, S. J., Jeong, D. G., Lee, S. M., and Ryu, S. E. (2002b). Structure of human FIH-1 reveals a unique active site pocket and interaction sites for HIF-1 and VHL. *J. Biol. Chem.* **278,** 7558–7563.

Liu, C., Shi, Y., Han, Z., Pan, Y., Liu, N., Han, S., Chen, Y., Lan, M., Qiao, T., and Fan, D. (2003). Suppression of the dual-specificity phosphatase MKP-1 enhances HIF-1 trans-activation and increases expression of EPO. *Biochem. Biophys. Res. Commun.* **312**, 780–786.

Liu, L. Z., Hu, X. W., Xia, C., He, J., Zhou, Q., Shi, X., Fang, J., and Jiang, B. H. (2006). Reactive oxygen species regulate epidermal growth factor-induced vascular endothelial growth factor and hypoxia-inducible factor-1alpha expression through activation of AKT and P70S6K1 in human ovarian cancer cells. *Free Radic. Biol. Med.* **41**, 1521–1533.

Liu, Q., Berchner-Pfannschmidt, U., Moller, U., Brecht, M., Wotzlaw, C., Acker, H., Jungermann, K., and Kietzmann, T. (2004). A Fenton reaction at the endoplasmic reticulum is involved in the redox control of hypoxia-inducible gene expression. *Proc. Natl. Acad. Sci. USA* **101**, 4302–4307.

Liu, Y., Christou, H., Moritia, T., Laughner, E., Semenza, G. L., and Kourembanas, S. (1998). Carbon monoxide and nitric oxide suppress the hypoxic induction of vascular endothelial growth factor gene via the 5′ enhancer. *J. Biol. Chem.* **273**, 15257–15262.

Mahon, P. C., Hirota, K., and Semenza, G. L. (2001). FIH-1: A novel protein that interacts with HIF-1α and VHL to mediate repression of HIF-1 transcriptional activity. *Genes Dev.* **15**, 2675–2686.

Maranchie, J. K., and Zhan, Y. (2005). Nox4 is critical for hypoxia-inducible factor 2-alpha transcriptional activity in von Hippel-Lindau-deficient renal cell carcinoma. *Cancer Res.* **65**, 9190–9193.

Marchetti, S., Gimond, C., Roux, D., Gothie, E., Pouyssegur, J., and Pages, G. (2004). Inducible expression of a MAP kinase phosphatase-3-GFP chimera specifically blunts fibroblast growth and ras-dependent tumor formation in nude mice. *J. Cell Physiol.* **199**, 441–450.

Masson, N., Willam, C., Maxwell, P. H., Pugh, C. W., and Ratcliffe, P. J. (2001). Independent function of two destruction domains in hypoxia-inducible factor-alpha chains activated by prolyl hydroxylation. *EMBO J.* **20**, 5197–5206.

Mateo, J., Garcia-Lecea, M., Cadenas, S., Hernandez, C., and Moncada, S. (2003). Regulation of hypoxia-inducible factor-1alpha by nitric oxide through mitochondria-dependent and -independent pathways. *Biochem. J.* **376**, 537–544.

Maxwell, P. H., Wiesener, M. S., Chang, G. W., Clifford, S. C., Vaux, E. C., Cockman, M. E., Wykoff, C. C., Pugh, C. W., Maher, E. R., and Ratcliffe, P. J. (1999). The tumour suppressor protein VHL targets hypoxia-inducible factors for oxygen-dependent proteolysis. *Nature* **399**, 271–275.

Mazure, N. M., Chen, E. Y., Laderoute, K. R., and Giaccia, A. J. (1997). Induction of vascular endothelial growth factor by hypoxia is modulated by a phosphatidylinositol 3-kinase/Akt signaling pathway in Ha-ras-transformed cells through a hypoxia inducible factor-1 transcriptional element. *Blood* **90**, 3322–3331.

McNeill, L. A., Hewitson, K. S., Claridge, T. D., Seibel, J. F., Horsfall, L. E., and Schofield, C. J. (2002). Hypoxia-inducible factor asparaginyl hydroxylase (FIH-1) catalyses hydroxylation at the beta-carbon of asparagine-803. *Biochem. J.* **367**, 571–575.

Min, J. H., Yang, H., Ivan, M., Gertler, F., Kaelin, W. G., Jr., and Pavletich, N. P. (2002). Structure of an HIF-1α-pVHL complex: Hydroxyproline recognition in signaling. *Science* **296**, 1886–1889.

Murrant, C. L., and Reid, M. B. (2001). Detection of reactive oxygen and reactive nitrogen species in skeletal muscle. *Microsc. Res. Tech.* **55**, 236–248.

Oehme, F., Ellinghaus, P., Kolkhof, P., Smith, T. J., Ramakrishnan, S., Hutter, J., Schramm, M., and Flamme, I. (2002). Overexpression of PH-4, a novel putative proline 4-hydroxylase, modulates activity of hypoxia-inducible transcription factors. *Biochem. Biophys. Res. Commun.* **296**, 343–349.

Ohh, M., Park, C. W., Ivan, M., Hoffman, M. A., Kim, T. Y., Huang, L. E., Pavletich, N., Chau, V., and Kaelin, W. G. (2000). Ubiquitination of hypoxia-inducible factor requires direct binding to the beta-domain of the von Hippel-Lindau protein. *Nat. Cell Biol.* **2,** 423–427.

Page, E. L., Robitaille, G. A., Pouyssegur, J., and Richard, D. E. (2002). Induction of hypoxia-inducible factor-1alpha by transcriptional and translational mechanisms. *J. Biol. Chem.* **277,** 48403–48409.

Park, J. H., Kim, T. Y., Jong, H. S., Kim, T. Y., Chun, Y. S., Park, J. W., Lee, C. T., Jung, H. C., Kim, N. K., and Bang, Y. J. (2003). Gastric epithelial reactive oxygen species prevent normoxic degradation of hypoxia-inducible factor-1alpha in gastric cancer cells. *Clin. Cancer Res.* **9,** 433–440.

Pascual, O., Denavit-Saubie, M., Dumas, S., Kietzmann, T., Ghilini, G., Mallet, J., and Pequignot, J. M. (2001). Selective cardiorespiratory and catecholaminergic areas express the hypoxia-inducible factor-1alpha (HIF-1α) under *in vivo* hypoxia in rat brainstem. *Eur. J. Neurosci.* **14,** 1981–1991.

Pugh, C. W., O'Rourke, J. F., Nagao, M., Gleadle, J. M., and Ratcliffe, P. J. (1997). Activation of hypoxia-inducible factor-1: Definition of regulatory domains within the alpha subunit. *J. Biol. Chem.* **272,** 11205–11214.

Qian, D., Lin, H. Y., Wang, H. M., Zhang, X., Liu, D. L., Li, Q. L., and Zhu, C. (2004). Normoxic induction of the hypoxic-inducible factor-1 alpha by interleukin-1 beta involves the extracellular signal-regulated kinase 1/2 pathway in normal human cyto-trophoblast cells. *Biol. Reprod.* **70,** 1822–1827.

Quintero, M., Brennan, P. A., Thomas, G. J., and Moncada, S. (2006). Nitric Oxide Is a Factor in the Stabilization of Hypoxia-Inducible Factor-1α in Cancer: Role of Free Radical Formation. *Cancer Res.* **66,** 770–774.

Richard, D. E., Berra, E., and Pouyssegur, J. (2000). Nonhypoxic pathway mediates the induction of hypoxia-inducible factor 1alpha in vascular smooth muscle cells. *J. Biol. Chem.* **275,** 26765–26771.

Richard, D. E., Berra, E., Gothie, E., Roux, D., and Pouyssegur, J. (1999). p42/p44 mitogen-activated protein kinases phosphorylate hypoxia-inducible factor 1alpha (HIF-1α) and enhance the transcriptional activity of HIF-1. *J. Biol. Chem.* **274,** 32631–32637.

Robinson, K. M., Janes, M. S., Pehar, M., Monette, J. S., Ross, M. F., Hagen, T. M., Murphy, M. P., and Beckman, J. S. (2006). Selective fluorescent imaging of superoxide *in vivo* using ethidium-based probes. *Proc. Natl. Acad. Sci. USA* **103,** 15038–15043.

Rothe, G., and Valet, G. (1990). Flow cytometric analysis of respiratory burst activity in phagocytes with hydroethidine and 2′,7′-dichlorofluorescin. *J. Leukoc. Biol.* **47,** 440–448.

Salceda, S., and Caro, J. (1997). Hypoxia-inducible factor 1alpha (HIF-1α) protein is rapidly degraded by the ubiquitin-proteasome system under normoxic conditions. Its stabilization by hypoxia depends on redox-induced changes. *J. Biol. Chem.* **272,** 22642–22647.

Sandau, K. B., Zhou, J., Kietzmann, T., and Brune, B. (2001). Regulation of the hypoxia-inducible factor 1alpha by the inflammatory mediators nitric oxide and tumor necrosis factor-alpha in contrast to desferroxamine and phenylarsine oxide. *J. Biol. Chem.* **276,** 39805–39811.

Schoenfeld, A. R., Davidowitz, E. J., and Burk, R. D. (2001). Endoplasmic reticulum/cytosolic localization of von Hippel-Lindau gene products is mediated by a 64-amino acid region. *Int. J. Cancer* **91,** 457–467.

Semenza, G. L. (2006). Regulation of physiological responses to continuous and intermittent hypoxia by hypoxia-inducible factor 1. *Exp. Physiol.* **91,** 803–806.

Semenza, G. L., and Wang, G. L. (1992). A nuclear factor induced by hypoxia via *de novo* protein synthesis binds to the human erythropoietin gene enhancer at a site required for transcriptional activation. *Mol. Cell Biol.* **12,** 5447–5454.

Shatrov, V. A., Sumbayev, V. V., Zhou, J., and Brune, B. (2003). Oxidized low-density lipoprotein (oxLDL) triggers hypoxia-inducible factor-1alpha (HIF-1α) accumulation via redox-dependent mechanisms. *Blood* **101,** 4847–4849.

Shimmura, S., Tsubota, K., Oguchi, Y., Fukumura, D., Suematsu, M., and Tsuchiya, M. (1992). Oxiradical-dependent photoemission induced by a phacoemulsification probe. *Invest. Ophthalmol. Vis. Sci.* **33,** 2904–2907.

Sodhi, A., Montaner, S., Miyazaki, H., and Gutkind, J. S. (2001). MAPK and Akt act cooperatively but independently on hypoxia inducible factor-1alpha in rasV12 upregulation of VEGF. *Biochem. Biophys. Res. Commun.* **287,** 292–300.

Sogawa, K., Numayama-Tsuruta, K., Ema, M., Abe, M., Abe, H., and Fujii-Kuriyama, Y. (1998). Inhibition of hypoxia-inducible factor 1 activity by nitric oxide donors in hypoxia. *Proc. Natl. Acad. Sci. USA* **95,** 7368–7373.

Sowter, H. M., Raval, R. R., Moore, J. W., Ratcliffe, P. J., and Harris, A. L. (2003). Predominant role of hypoxia-inducible transcription factor (HIF)-1alpha versus HIF-2α in regulation of the transcriptional response to hypoxia. *Cancer Res.* **63,** 6130–6134.

Srinivas, V., Leshchinsky, I., Sang, N., King, M. P., Minchenko, A., and Caro, J. (2001). Oxygen sensing and HIF-1 activation does not require an active mitochondrial respiratory chain electron-transfer pathway. *J. Biol. Chem.* **276,** 21995–21998.

Stiehl, D. P., Jelkmann, W., Wenger, R. H., and Hellwig-Bürgel, T. (2002). Normoxic induction of the hypoxia-inducible factor 1alpha by insulin and interleukin-1beta involves the phosphatidylinositol 3-kinase pathway. *FEBS Lett.* **512,** 157–162.

Tacchini, L., Dansi, P., Matteucci, E., and Desiderio, M. A. (2001). Hepatocyte growth factor signalling stimulates hypoxia inducible factor-1 (HIF-1) activity in HepG2 hepatoma cells. *Carcinogenesis* **22,** 1363–1371.

Tacchini, L., De Ponti, C., Matteucci, E., Follis, R., and Desiderio, M. A. (2004). Hepatocyte growth factor-activated NF-κB regulates HIF-1 activity and ODC expression, implicated in survival, differently in different carcinoma cell lines. *Carcinogenesis* **25,** 2089–2100.

Tanimoto, K., Makino, Y., Pereira, T., and Poellinger, L. (2000). Mechanism of regulation of the hypoxia-inducible factor-1 alpha by the von Hippel-Lindau tumor suppressor protein. *EMBO J.* **19,** 4298–4309.

Tarpey, M. M., Wink, D. A., and Grisham, M. B. (2004). Methods for detection of reactive metabolites of oxygen and nitrogen: *In vitro* and *in vivo* considerations. *Am. J. Physiol. Regul. Integr. Comp. Physiol.* **286,** R431–R444.

Thornton, R. D., Lane, P., Borghaei, R. C., Pease, E. A., Caro, J., and Mochan, E. (2000). Interleukin 1 induces hypoxia-inducible factor 1 in human gingival and synovial fibroblasts. *Biochem. J.* **350,** 307–312.

Tonks, N. K. (2005). Redox redux: Revisiting PTPs and the control of cell signaling. *Cell* **121,** 667–670.

Treins, C., Giorgetti-Peraldi, S., Murdaca, J., Semenza, G. L., and Van Obberghen, E. (2002). Insulin stimulates hypoxia-inducible factor 1 through a phosphatidylinositol 3-kinase/target of rapamycin-dependent signaling pathway. *J. Biol. Chem.* **277,** 27975–27981.

Wakisaka, N., and Pagano, J. S. (2003). Epstein-Barr virus induces invasion and metastasis factors. *Anticancer Res.* **23,** 2133–2138.

Wakisaka, N., Kondo, S., Yoshizaki, T., Murono, S., Furukawa, M., and Pagano, J. S. (2004). Epstein-Barr Virus Latent Membrane Protein 1 Induces Synthesis of Hypoxia-Inducible Factor 1α. *Mol. Cell. Biol.* **24,** 5223–5234.

Walling, C. (1975). Fenton's reagent revisited. *Acc. Chem. Res.* **8,** 125–131.

Wang, G. L., and Semenza, G. L. (1995). Purification and characterization of hypoxia-inducible factor 1. *J. Biol. Chem.* **270,** 1230–1237.

Wang, G. L., Jiang, B. H., and Semenza, G. L. (1995a). Effect of altered redox states on expression and DNA-binding activity of hypoxia-inducible factor 1. *Biochem. Biophys. Res. Commun.* **212,** 550–556.

Wang, G. L., Jiang, B. H., and Semenza, G. L. (1995b). Effect of protein kinase and phosphatase inhibitors on expression of hypoxia-inducible factor 1. *Biochem. Biophys. Res. Commun.* **216,** 669–675.

Wang, G. L., Jiang, B. H., Rue, E. A., and Semenza, G. L. (1995c). Hypoxia-inducible factor 1 is a basic helix-loop-helix PAS heterodimer regulated by cellular O_2 tension. *Proc. Natl. Acad. Sci. USA* **92,** 5510–5514.

Wang, F. S., Wang, C. J., Chen, Y. J., Chang, P. R., Huang, Y. T., Sun, Y. C., Huang, H. C., Yang, Y. J., and Yang, K. D. (2004). Ras induction of superoxide activates ERK-dependent angiogenic transcription factor HIF-1α and VEGF-A expression in shock wave-stimulated osteoblasts. *J. Biol. Chem.* **279,** 10331–10337.

Wellman, T. L., Jenkins, J., Penar, P. L., Tranmer, B., Zahr, R., and Lounsbury, K. M. (2004). Nitric oxide and reactive oxygen species exert opposing effects on the stability of hypoxia-inducible factor-1α (HIF-1α) in explants of human pial arteries. *FASEB J.* **18,** 379–381.

Welsh, S. J., Bellamy, W. T., Briehl, M. M., and Powis, G. (2002). The redox protein thioredoxin-1 (Trx-1) increases hypoxia-inducible factor 1alpha protein expression: Trx-1 overexpression results in increased vascular endothelial growth factor production and enhanced tumor angiogenesis. *Cancer Res.* **62,** 5089–5095.

Wenger, R. H. (2002). Cellular adaptation to hypoxia: O_2-sensing protein hydroxylases, hypoxia-inducible transcription factors, and O_2-regulated gene expression. *FASEB J.* **16,** 1151–1162.

Wersto, R. P., Rosenthal, E. R., Crystal, R. G., and Spring, K. R. (1996). Uptake of fluorescent dyes associated with the functional expression of the cystic fibrosis transmembrane conductance regulator in epithelial cells. *Proc. Natl. Acad. Sci. USA* **93,** 1167–1172.

Yu, F., White, S. B., Zhao, Q., and Lee, F. S. (2001). HIF-1α binding to VHL is regulated by stimulus-sensitive proline hydroxylation. *Proc. Natl. Acad. Sci. USA* **98,** 9630–9635.

Zelko, I. N., and Folz, R. J. (2005). Extracellular superoxide dismutase functions as a major repressor of hypoxia-induced erythropoietin gene expression. *Endocrinology* **146,** 332–340.

Zelzer, E., Levy, Y., Kahana, C., Shilo, B. Z., Rubinstein, M., and Cohen, B. (1998). Insulin induces transcription of target genes through the hypoxia-inducible factor HIF-1α/ARNT. *EMBO J.* **17,** 5085–5094.

Zhao, H., Kalivendi, S., Zhang, H., Joseph, J., Nithipatikom, K., Vasquez-Vivar, J., and Kalyanaraman, B. (2003). Superoxide reacts with hydroethidine but forms a fluorescent product that is distinctly different from ethidium: Potential implications in intracellular fluorescence detection of superoxide. *Free Radic. Biol. Med.* **34,** 1359–1368.

Zhao, H., Joseph, J., Fales, H. M., Sokoloski, E. A., Levine, R. L., Vasquez-Vivar, J., and Kalyanaraman, B. (2005). Detection and characterization of the product of hydroethidine and intracellular superoxide by HPLC and limitations of fluorescence. *Proc. Natl. Acad. Sci. USA* **102,** 5727–5732.

Zhou, J., Callapina, M., Goodall, G. J., and Brune, B. (2004). Functional integrity of nuclear factor kappa B, phosphatidylinositol 3′-kinase, and mitogen-activated protein kinase signaling allows tumor necrosis factor alpha-evoked Bcl-2 expression to provoke internal ribosome entry site-dependent translation of hypoxia-inducible factor 1alpha. *Cancer Res.* **64,** 9041–9048.

Zhu, X. Y., Rodriguez-Porcel, M., Bentley, M. D., Chade, A. R., Sica, V., Napoli, C., Caplice, N., Ritman, E. L., Lerman, A., and Lerman, L. O. (2004). Antioxidant intervention attenuates myocardial neovascularization in hypercholesterolemia. *Circulation* **109,** 2109–2115.

CHAPTER TWENTY-THREE

GENETICS OF MITOCHONDRIAL ELECTRON TRANSPORT CHAIN IN REGULATING OXYGEN SENSING

Eric L. Bell* *and* Navdeep S. Chandel*,†

Contents

1. Introduction	448
2. Detecting HIF-1α Protein Levels	449
2.1. Background	449
2.2. Materials	450
2.3. Equipment	451
2.4. Methods	451
3. Detecting Intracellular ROS Levels	452
3.1. Background	452
3.2. Materials	453
3.3. Equipment	453
3.4. Methods	453
4. Method 1: Examining Hypoxic Stabilization of HIF-1α Protein in Cells Containing RNAI against the Rieske Fe-S Protein	454
4.1. Background	454
4.2. Materials	454
4.3. Equipment	455
4.4. Methods	455
5. Method 2: Examining the Role of ROS Generated from Mitochondrial Electron Transport in Hypoxic Stabilization of HIF-1α Protein	458
5.1. Background	458
5.2. Materials	458
5.3. Methods	458
6. Concluding Remarks	459
Acknowledgments	459
References	460

* Department of Medicine, Northwestern University Medical School, Chicago, Illinois
† Department of Cell and Molecular Biology, Northwestern University Medical School, Chicago, Illinois

Methods in Enzymology, Volume 43
ISSN 0076-6879, DOI: 10.1016/S0076-6879(07)35023-4

Abstract

Oxygen is the terminal electron acceptor in the mitochondrial electron transport chain and therefore is required for the generation of energy through oxidative phosphorylation.

In environments of decreased oxygen levels (hypoxia), organisms have developed an adaptive response through the activation of the hypoxia-inducible transcription factor (HIF) to maintain their energetic demand. In order to sense hypoxic environments, cells have developed oxygen-sensing machinery that allows for the activation of HIF. The mitochondrial electron transport chain is required for the oxygen-sensing pathway. This chapter outlines methods used to explore the role of the electron transport chain and a by-product of electron transport, reactive oxygen species, in oxygen sensing.

1. Introduction

Mammalian cells require energy in the form of ATP for survival. The majority of ATP is generated through oxidative phosphorylation (OXPHOS), thereby making oxygen necessary for the survival of mammalian cells. Accordingly, mammalian cells have developed the ability to sense decreases in oxygen levels within their environment in order to elicit an adaptive response to a potential decrease in energy supply. The adaptive response is mediated by activating a program of gene expression regulated by HIF. Activation of HIF is associated with numerous physiological processes, such as cellular metabolism, proliferation, cellular survival, and neo-vascularization (Semenza, 2003). HIF activation is also associated with pathophysiological processes, such as cancer and pulmonary hypertension (Semenza, 2000).

The mitochondrial electron transport chain is necessary for oxygen sensing; however, the exact mechanism for its involvement is unresolved. Currently, it is believed that electron transport contributes to oxygen sensing through its ability to generate reactive oxygen species or its ability to consume oxygen (Brunelle *et al.*, 2005; Chandel *et al.*, 1998; Doege *et al.*, 2005; Guzy *et al.*, 2005; Hagen *et al.*, 2003; Mansfield *et al.*, 2005). The latter model postulates that oxygen consumption by the mitochondria generates a gradient of oxygen within the cytosol of the cell. When cells become hypoxic, any available oxygen diffuses to the mitochondria, creating an environment within the cytosol that lacks oxygen, thereby inhibiting the activity of a class of enzymes that regulate the activation of HIF known as the prolyl hydroxylases (PHDs). These enzymes require oxygen to catalyze the hydroxylation of HIF–alpha (α) protein, which is required for the degradation of HIF-α protein. In hypoxic conditions, HIF-α is not hydroxylated and therefore is stabilized to initiate HIF transcriptional activity.

Recently, we published genetic data demonstrating that hypoxic stabilization of HIF-α protein and thus HIF activation are independent of OXPHOS, implying that mitochondrial oxygen consumption is dispensable for hypoxic activation of HIF (Brunelle *et al.*, 2005). Our model postulates that reactive oxygen species (ROS) generated from complex III of the mitochondria are responsible for HIF activation in hypoxic conditions. This chapter describes methods that can be used to investigate the role of ROS and electron transport through complex III in hypoxic stabilization of HIF-α protein.

2. DETECTING HIF-1α PROTEIN LEVELS

2.1. Background

The hypoxia-inducible factor is a heterodimer of HIF-α and of the aryl hydrocarbon nuclear translocator (ARNT, or HIF-beta [β]) (Wang and Semenza, 1995; Wang *et al.*, 1995). Both proteins are ubiquitously expressed; however, the stability of the proteins differ. ARNT is constitutively stabile, while HIF-α is liable depending on oxygen levels (Jiang *et al.*, 1996a). Under normal oxygen conditions, HIF-α is hydroxylated by a family of PHD enzymes (Bruick and McKnight, 2001; Epstein *et al.*, 2001). The PHDs require Fe^{++}, oxygen, and 2-oxoglutarate to catalyze the hydroxylation reaction. Specifically, HIF-α is hydroxylated on two proline residues (P402 and P564 for human HIF-1α), which are located within the oxygen-dependent degradation domain (ODD) (Masson *et al.*, 2001; Pugh *et al.*, 1997). Hydroxylated proline serves as a binding site for the von Hippel-Lindau protein (pVHL), the substrate recognition component of the VBC-CUL-2 E3 ubiquitin ligase complex (Ivan *et al.*, 2001; Iwai *et al.*, 1999; Jaakkola *et al.*, 2001). Once bound, pVHL tags HIF-α with ubiquitin, thereby targeting it for proteasomal degradation (Maxwell *et al.*, 1999; Salceda and Caro, 1997). In conditions of hypoxia, HIF-α is not hydroxylated and therefore is not targeted for degradation. Once stabilized, HIF-α dimerizes with ARNT and translocates to the nucleus to bind to HIF response elements located upstream of its target genes (Jiang *et al.*, 1996b). HIF target genes include vascular endothelial factor (VEGF) as well as numerous glycolytic genes (Mazure *et al.*, 1997). The importance of HIF-mediated transcription is highlighted by the data demonstrating that genetic knockout of either HIF-1α or ARNT is embryonic lethal at day E9.5 (Iyer *et al.*, 1998; Maltepe *et al.*, 1997). These embryos display vascular abnormalities and placental defects that do not allow the embryos to develop normally. HIF-α protein can be stabilized in normal oxygen conditions by inhibiting the enzymatic activity of the PHDs with the competitive inhibitor dimethyloxyglycine (DMOG). In hypoxic conditions, HIF-α

can be detected in nuclear extracts or whole cell lysates. Here, we describe methods for detecting HIF-α protein.

2.2. Materials

Phosphate-buffered saline (PBS) (Mediatech Inc., Herndon, VA)

10 × Lysis Buffer (for whole cell lysates) (Cell Signaling Technology, Danvers, MA)

Buffer A (for nuclear extracts): 10 mM HEPES-KOH, 1.5 mM MgCl$_2$, 10 mM KCl, 50 μg/ml phenylmethylsulfonyl fluoride (PMSF), 1 mM dithiothreitol (DTT), and cocktail of protease inhibitors (Roche Applied Science, Indianapolis, IN); sterilize for filtration and store at 4°

Buffer C (for nuclear extracts): 20 mM HEPES-KOH, 1.5 mM MgCl$_2$, 420 mM KCL, 25% (v/v) glycerol, 0.2 mM ethylenediaminetetraacetic acid (EDTA), 50 μg/ml PMSF, 1 mM DTT, and cocktail of protease inhibitors (Roche); sterilize for filtration and store at 4°

Cocktail of protease inhibitors (Roche Molecular Biochemicals/Boehringer Mannheim, Indianapolis, IN)

Bradford protein assay reagent (BioRad Laboratories, Hercules, CA)

Sample loading buffer (2×): 125 mM Tris base (pH 6.8), 4% (v/v) sodium dodecyl sulfate (SDS), 20% (v/v) glycerol, 200 mM DTT, 0.02% (w/v) bromophenol blue; store in aliquots at −20°

Running buffer: 25 mM Tris base (pH 8.3), 192 mM glycine, 0.1% (w/v) SDS

Transfer buffer: 25 mM Tris base (pH 8.3), 192 mM glycine, 20% (v/v) methanol

Ponceau S solution: 0.1% Ponceau S in 5% (v/v) acetic acid

Tris buffered saline (TBS): 100 mM Tris base (pH 7.5), 0.9% (w/v) NaCl

Tris buffered saline Tween 20 (TBS-T): 0.1% (v/v) Tween 20 in TBS

Blocking solution: 5% (w/v) nonfat milk powder in TBS-T; prepare fresh for each use

Anti–hypoxia-inducible factor 1α monoclonal antibody (clone 54; BD Biosciences, San Jose, CA)

Anti-RNA polymerase (Pol) II (clone C-21) (Santa Cruz Biotechnology, Santa Cruz, CA)

Monoclonal anti–a-tubulin (clone B-5–1-2 ascites fluid; Sigma-Aldrich, St. Louis, MO)

Anti-mouse immunoglobulin G (IgG) horseradish peroxidase (HRP) (Cell signaling)

Anti-rabbit IgG HRP (Cell signaling)

DMOG (Frontier Scientific, Logan, UT)

Hybond-ECL nitrocellulose (0.45 μm) (Amersham, Piscataway, NJ)

SuperSignal West Pico Chemiluminescent Substrate (Pierce, Rockford, IL)

Hyperfilm ECL (Amersham)

Tris-HCl Ready Gel Precast Gel, 7.5% resolving gel, 4% stacking gel (BioRad)

2.3. Equipment

Mini-PROTEAN 3 System (BioRad)
Trans-Blot SD Semi-Dry Electrophoretic Transfer Cell (BioRad)
INVIVO$_2$ hypoxic workstation (Biotrace International, Brigend, UK)

2.4. Methods

To pre-equilibrate media to hypoxic conditions, 100-mm cell culture dishes containing 10 ml of appropriate culture media are placed in either a cell culture incubator (5% CO_2) or in the hypoxic workstation (1.5% O_2, 5% CO_2, balanced with N_2) for a minimum of 8 h before experimentation. Cells at 70% confluence and cultured in 100-mm cell culture dishes are incubated in 10 ml of pre-equilibrated culture media in either hypoxic or normoxic conditions for 4 h. To prepare a positive control sample, cells are incubated with 1 mM DMOG for 4 h.

2.4.1. Nuclear extract preparation

After incubation, culture dishes are placed on ice and media is aspirated. Cells are washed once with 3 ml cold PBS and then scraped in 1 ml of cold buffer A. The cell suspension is then transferred to 1.5-ml microfuge tubes, briefly vortexed, and incubated on ice for 30 min. Cells are pelleted at 12,000 g for 1 min, and the supernatant is aspirated. Depending on size of the resulting pellet, nuclei are resuspended in 30 to 100 μl of cold buffer C (25 μl of buffer C per 1 × 10^6 cells), vortexed, and then incubated on ice for 30 min. After incubation, the soluble fraction is collected by spinning tubes for 1 min at 12,000 g, and the supernatant (nuclear extract) is transferred to new tubes and stored at −20° until analysis by gel electrophoresis.

2.4.2. Whole cell lysate preparation

After incubation, culture dishes are immediately placed on ice, and media is aspirated. Cells are washed with 3 ml cold PBS. Cells are scraped in 100 to 400 μl of 1× cold cell lysis buffer containing 100 μM PMSF (100 μl of 1× cell lysis buffer per 1 × 10^6 cells). Cell lysates are transferred to microfuge tubes and incubated on ice for 15 min. The lysates are stored at −20° until they are analyzed by gel electrophoresis.

2.4.3. Immunoblotting for HIF-1α protein levels

After thawing samples, tubes are centrifuged for 1 min at 12,000g to clear any cellular debris. Protein concentration is determined by the Bradford protein assay (BioRad). Nuclear extract (15–50 μg) or whole cell lysates

(50–100 μg) are added to 2× sample loading buffer and heated for 5 min at 85°. The samples are loaded onto a 7.5% SDS polyacrylamide gel, and the gel is run at 100 V until the dye front reaches the bottom of the gel (approximately 1 h). After electrophoresis, the proteins are transferred to a nitrocellulose membrane using the Trans-Blot SD Semi-Dry Electrophoretic Transfer Cell (BioRad) for 65 min at a constant voltage of 15 V (amps should not go above 0.4 Amps). To ensure protein transfer, the nitrocellulose membrane is stained with Ponceau S solution for 5 min on a rocking plate and then rinsed with ddH$_2$O until clear protein bands can be visualized. Once the membrane is rinsed, it is placed in blocking solution for 1 h at room temperature on a rocking plate. The blocking solution is then discarded, and the membrane is incubated in 10 ml of blocking solution containing the HIF-1α monoclonal antibody diluted 1:250. The membrane is incubated in this primary antibody solution overnight on a rocking plate at 4°. The next day, the primary antibody solution is discarded, and the membrane is washed 3× with 10 ml blocking solution on a rocking plate at room temperature. After washing, 10 ml of blocking solution containing the rabbit anti-mouse IgG HRP antibody (diluted 1:1000) is added to the membrane and allowed to incubate for 1 h at room temperature on a rocking plate. Subsequently, the secondary antibody solution is discarded, and the membrane is washed 2× with TBS-T for 30 min each time. Finally, the protein–antibody complex is visualized by adding the SuperSignal West Pico Chemiluminescent Substrate to the membrane for 5 min and then exposing the membrane to Hyperfilm ECL in a dark room.

2.4.4. Note

To control for equal loading of each sample, the membrane is incubated with blocking solution containing antibodies for a-tubulin (whole cell lysates) (diluted 1:2000) or RNA Pol II (nuclear extracts) (1:200). These proteins can be probed for on the same nitrocellulose membrane used to analyze HIF-1α protein levels by cutting the membranes in between the predicted molecular weight for the proteins.

3. Detecting Intracellular ROS Levels

3.1. Background

Reactive oxygen species are highly reactive and unstable, making them difficult to measure in biological systems. The evaluation of ROS in biological systems is determined using indirect measurements of the interaction between redox-sensitive molecules and ROS. Once modified, these redox-sensitive molecules emit luminescent or fluorescent signals, which allow for the relative quantification of the amount of ROS present within the system being tested. In order to detect intracellular ROS, the redox-sensitive molecule needs to be membrane-permeable and to preferably remain within the cell. Due to these requirements, we utilize the fluorescent probe 5-(and-6)-chloromethyl-2′,7′-dichlorodihydrofluorescein diacetate acetyl ester (CM-H$_2$DCFDA) to detect intracellular ROS. CM-H$_2$DCFDA is a cell-permeable nonfluorescent substrate. Once CM-H$_2$DCFDA crosses the membrane, it is deacetylated to CM-H$_2$DCFH within the cytosol, thus trapping it within the cell. ROS oxidize CM-H$_2$DCFDA into its fluorescent form CM-H$_2$DCF, which is readily detectable. This dye is also sensitive to reactive nitrogen species (RNS), such as nitric oxide and peroxynitrite. During hypoxia, an increase occurs in DCF fluorescence, which can be quenched by hydrogen peroxide scavengers. However, RNS scavengers do not suppress the oxidation of DCFH, indicating that oxidation of DCFH in hypoxia is primarily due to ROS and not RNS.

3.2. Materials

5-(and-6)-chloromethyl-2′,7′-dichlorodihydrofluorescein diacetate acetyl ester (CM-H$_2$DCFDA) (Molecular Probes, Invitrogen, Carlsbad, CA)
Minimal Essential Media a (MEMa) (Mediatech Inc., Herndon, VA)
PBS (Mediatech Inc)
Black 96-well microplates
Triton X-100

3.3. Equipment

INVIVO$_2$ hypoxic workstation (Biotrace International)
96-well fluorescence spectrometer (Molecular Devices, Sunnyvale, CA)

3.4. Methods

Cell culture dishes (60 mm) containing 4 ml MEMa supplemented with CM-H$_2$DCFDA (10 μM) are placed in either a cell culture incubator (5% CO$_2$) or in the hypoxic workstation (1.5% O$_2$, 5% CO$_2$, balanced with N$_2$) to pre-equilibrate for a minimum of 8 h. Aluminum foil is loosely wrapped around the dishes to prevent oxidation of the dye by light. Cells cultured to approximately 50% confluence in 60-mm cell culture dishes are incubated in 4 ml of the normoxic or hypoxic pre-equilibrated MEMa media supplemented with DCFH-DA (10 μM) for 4 h. The media is then removed, and cells are washed with normoxic or hypoxic pre-equilibrated PBS. Cells are lysed with 1 ml of 0.1% Triton X-100 within their respective incubators. Lysates are centrifuged for 1 min at 12,000g to remove any cellular debris. Cell lysates (200 μl) are placed in wells of the 96-well black microplate, and DCFH oxidation is measured using a fluorescence spectrometer at an excitation wavelength of 500 nm and an emission wavelength of 530 nm. To determine baseline fluorescence, a 200 -μl aliquot of 0.1% Triton X-100 is used as a blank. Values for the treated samples are normalized to the values obtained from the normoxic untreated samples.

4. METHOD 1: EXAMINING HYPOXIC STABILIZATION OF HIF-1α PROTEIN IN CELLS CONTAINING RNAI AGAINST THE RIESKE FE-S PROTEIN

4.1. Background

Previously, it has been demonstrated that hypoxic stabilization of HIF-1α protein requires ROS generated from the mitochondrial electron transport chain (Chandel *et al.*, 1998; Schroedl *et al.*, 2002). These studies used p^0 cells, which are devoid in mitochondrial DNA (mtDNA) and therefore lack a functional electron transport chain. In order to generate p^0 cells, cells are cultured in sublethal levels of ethidium bromide to inhibit the transcription and replication of mtDNA. Unfortunately, ethidium bromide can damage nuclear DNA. To rule out the possibility that mutation of nuclear DNA played a role in the data generated using p^0 cells, an approach that genetically inhibits ROS generation from the mitochondrial electron transport chain is necessary. Data generated with pharmacological inhibitors of electron transport indicate that mitochondrial complex III is the source of the ROS needed to stabilize HIF-1α protein. To genetically investigate the role of mitochondrial ROS, cells can be genetically modified to have reduced levels of the nuclear-encoded complex III protein subunit, Rieske Fe-S protein, with RNA interference techniques. In the absence of Rieske Fe-S protein, cells lack a functional complex III and therefore cannot generate ROS from this complex.

4.2. Materials

Human lung carcinoma cell line A549 (ATCC, Manassas, VA)
Dulbecco's modified essential medium (DMEM) with high glucose (4500 mg/l) (Mediatech Inc.)
Penicillin-streptomycin liquid (10,000 units/ml and 10,000 μg/m, respectively) (Mediatech Inc.)
Fetal bovine serum (FBS) (Gibco, Invitrogen, Carlsbad, CA)
1 *M* HEPES (Mediatech Inc.)
Uridine (Sigma)
Puromycin (Sigma)

Polybrene (Sigma)
Chloroquine (Sigma)
DMOG
pSIREN-RetroQ Vector (Clontech)
RetroPack PT67 cell line (Clontech)
Rieske Fe-S short hairpin RNA (shRNA) oligonucleotide (IDT DNA)
Target sequence 1 (top strand): 5′-AAGGUGCCUGACUUCUCUGAA-3′
Target sequence 2 (top strand): 5′-AAUGCCGUCACCCAGUUCGUU-3′
Control shRNA oligonucleotide (*Drosophila* HIF) (IDT DNA)
Target sequence (top strand): 5-CCUACAUCCCGAUCGAUGAUG-3
Anti–Rieske Fe-S Protein monoclonal antibody (Molecular Probes, Invitrogen, Carlsbad, CA)
*Trans*IT-LT1 transfection reagent (Mirus Bio, Madison, WI)
T4 ligase (New England Biolabs, Ipswich, MA)
Tris-EDTA (TE) 50 mM Tris-HCL (pH 7.8), 1 mM EDTA
BSA (Sigma)
One Shot TOP10 Chemically Competent cells (Invitrogen, Carlsbad, CA)
Luria broth (EMD Biosciences, San Diego, CA)
Luria broth/agar (EMD Biosciences)
Ampicillin (Sigma)
Qiaprep Spin Miniprep Kit (QIAGEN, Valencia, CA)
Endo-free Maxiprep Kit (QIAGEN)

4.3. Equipment

PTC-200 Thermal Cycler (MJ Research, Waltham, MA)
Orbital Shaker (Thermo Forma, Marietta, OH)
Incubator (37°)

4.4. Methods

Wild-type A549 are grown in high-glucose DMEM containing sodium pyruvate supplemented with 10% FBS, 100 units/ml penicillin, 100 μg/ml streptomycin, and 20 mM HEPES. For experiments analyzing HIF-1α protein levels, cells are cultured in 100-mm cell culture dishes; the cells are cultured in 60-mm cell culture dishes for experiments analyzing ROS levels. To stably knockdown the Rieske Fe-S protein, the targeting sequences are ligated into the retroviral shRNA pSiren construct. As a control, a sequence that is specific for *Drosophila melanogaster* HIF is used, since it has no target in mammalian cells. First, the oligonucleotides are resuspended in TE buffer to a concentration of 100 μM. Equal amounts of oligonucleotides for the top and bottom strands are mixed together for a final concentration of 50 μM of double-stranded oligonucleotide. The mixture is heated to 95° for 30 s in the thermal cycler to remove secondary structure. The mixture is then heated at 72° for 2 min, 37° for 2 min, and 25° for 2 min before placing on ice. The double-stranded oligonucleotide can be stored at −20° until ligating into pSiren. Prior to ligation, the double-stranded oligonucleotide is diluted to a concentration of 0.5 μM and then 1 μl is added to 2 μl Linearized pSIREN vector (25 ng/μl), 1.5 μl 10× T4 DNA Ligase Buffer, 0.5 μl BSA (10 mg/ml), 0.5 μl T4 DNA ligase (400 U/μl), and Nuclease-free H_2O to 15 μl. The ligation reaction is incubated for 3 h at room temperature. After 3 h, 2 μl of the ligation is used to transform One Shot TOP10 chemically competent cells for 30 min on ice. Cells are heat-shocked at 42° for 30 s and then placed on ice and supplemented with 250 μl of SOC media (comes with competent cells). The transformed cells are allowed to shake at 225 rpm in a 37° orbital shaker for 1 h, then plated on petri dishes coated with LB agar supplemented with 100 μg/ml of Ampicillin and incubated overnight at 37°. The next day, colonies are picked and propagated overnight in 3 ml of LB supplemented with 100 μg/ml ampicillin in the 37° orbital shaker. The following morning, cultures are centrifuged at 3000 rpm for 15 min to pellet the cells. Media is aspirated and the plasmid is purified using Qiaprep Spin Miniprep Kit. To check for insertion of the double-stranded oligonucleotide, a restriction site can be added to the end of the oligonucleotide, which is not present in the vector backbone. Then, a restriction digest can be performed on the plasmid purified with the miniprep kit to validate insertion. In addition, it is recommended that the plasmid be sequenced to confirm insertion of the double-stranded oligonucleotide. Once it been confirmed that the plasmid has the insert, a large-scale preparation is performed using the Qiaprep Maxi Prep Kit.

The retrovirus used to stably knock down the Rieske Fe-S protein and the control sequence is generated by transfecting the Rieske Fe-S pSiren plasmid or the dHIF pSiren plasmid into the PT67 packaging cell line. Prior to transfection, 40% confluent PT67 cells cultured in 100-mm cell culture dishes are incubated in 2.5 ml complete DMEM, which is supplemented with 25 uM chloroquine for 30 min to 2 h. *Trans*IT-LT1 Transfection Reagent (2.5 ul per 1 μg of plasmid) is added into 100 μl of serum-free media and allowed to incubate for 15 min at room temperature. Next, 10 to 15 ug of plasmid is added to the transfection mixture and allowed to incubate for 30 min at room temperature. Transfection reagent is then added dropwise to the PT67 cells. Twenty-four hours later, media is removed, cells are washed with 4 ml of PBS, and 4 ml of complete DMEM supplemented with 100 μg/ml uridine is added to the cells. A low volume of media is used to concentrate the virus and increase the titer.

It is necessary to add uridine (50 μg/ml) to the media since these cells are respiratorily deficient. After another 24 h, the 4 ml of media is collected and centrifuged for 5 min at 1000 rpm to pellet any PT67. After centrifugation, the media is filtered through a 0.45-um filter and supplemented with 8 μg/ml of polybrene. Subsequently, the virus-containing media is placed on either of the A549 cells, which are approximately 50% confluent in 100-mm cell culture dishes. The next day, the virus-containing media is aspirated, and fresh media containing uridine is added to the cells to promote replication. Two days later, cells stably expressing the shRNA against the Rieske Fe-S protein are selected for in complete DMEM containing uridine (50 ug/ml) and puromycin (1 ug/mL). As a control, cells that have not been infected should be placed in media containing puromycin. Once all the cells on the control plate are killed, it can be assumed that the remaining population in the infected plates is stably expressing the Rieske Fe-S shRNA or dHIF shRNA. To assess the efficiency of knockdown, immunoblotting for the Rieske Fe-S protein is performed (as previously described for HIF-1α protein) on lysates of both the dHIF and Rieske Fe-S stable cell lines cultured to 70% confluence in 100-mm cell culture dishes.

To determine whether the Rieske Fe-S protein is required for the hypoxic increase in ROS, the dHIF and Rieske Fe-S shRNA cells are incubated in normoxia (21% O_2) or hypoxia (1.5% O_2) in the presence of dichlorofluorescein diacetate (DCFH-DA). Intracellular ROS levels are determined as indicated in the protocol listed previously. The dHIF shRNA control cells demonstrate an increase in ROS in hypoxic conditions. This increase is not observed in the Rieske Fe-S cells, indicating the requirement for electron transport through complex III for hypoxic-induced increase in cytosolic ROS. To assess the impact of knocking down the Rieske Fe-S protein on the stabilization of HIF-1α protein, the dHIF and Rieske Fe-S stable cell lines were exposed to normoxia (21% O_2), hypoxia (1.5% O_2), or DMOG (1 mM) for 4 h. After incubation, HIF-1α protein levels were assessed by immunoblotting, as indicated in the protocol described earlier. Our experimental findings indicate that the Rieske Fe-S protein is necessary for hypoxic stabilization of HIF-1α protein. In the dHIF cells, HIF-1α protein was stabilized in hypoxic conditions as well as in the presence of DMOG. However, in the Rieske Fe-S cells, HIF-1α protein was not stabilized under hypoxic conditions while it was present in cells treated with DMOG, indicating that the role of the Riekse Fe-S protein is upstream of the PHDs. These data indicate that electron transport distal to complex III is required for hypoxic generation of ROS and stabilization of HIF-1α protein. However, just by examining this data, it is unclear whether the inability of the Rieske Fe-S shRNA cells to stabilize HIF-1α protein is due to a defect in ROS production or respiration.

5. Method 2: Examining the Role of ROS Generated from Mitochondrial Electron Transport in Hypoxic Stabilization of HIF-1α Protein

5.1. Background

Hypoxia increases the levels of ROS within the cytosol of the cell (Chandel *et al.*, 1998, 2000). Inhibiting the formation of mitochondrial ROS by impeding electron transport takes away hypoxic stabilization of HIF-1α protein. Inhibiting electron transport not only attenuates ROS production but also abolishes respiration. To determine whether it is ROS generated by the electron transport chain or the oxygen consumed by electron transport that is important for hypoxic stabilization of HIF-1α protein, we ectopically expressed different protein antioxidants and analyzed the stability of HIF-1α protein.

5.2. Materials

DMEM (Mediatech Inc.)
Anti-superoxide dismutase 1 (SOD1) antibody (BD Biosciences)
Anti-superoxide dismutase 2 (SOD2) antibody (BD Biosciences)
Anti-catalase antibody (Abcam, Cambridge, MA)
Anti-myc antibody (Invitrogen)
Adenovirus encoding SOD1 (The Gene Transfer Vector Core, University of Iowa)
Adenovirus encoding SOD2 (The Gene Transfer Vector Core, University of Iowa)
Adenovirus encoding catalase (The Gene Transfer Vector Core, University of Iowa)
Adenovirus encoding myc-tagged glutathione peroxidase (GPX) (The Gene Transfer Vector Core, University of Iowa)

5.3. Methods

Wild-type A549 and MDA-MB-435 cells are grown in high-glucose DMEM containing sodium pyruvate supplemented with 10% FBS, 100 units/ml penicillin, 100 μg/ml streptomycin, and 20 mM HEPES. To assess the effect of antioxidants on hypoxic HIF-1α protein stabilization, cells are cultured to 50% confluence in 100-mm cell culture dishes. Prior to adenoviral infection, 2 ml of non-supplemented DMEM is placed on cells. One plate of cells is used to count the exact number of cells present in order to calculate the amount of particle-forming units (pfu) added to each cell culture dish. Twenty particle-forming units per cell are added to each dish, and the cells are incubated for 2 h with gentle agitation every 30 min. Subsequently, 7 ml of complete media is added to the plates, and cells are incubated overnight. The following day, infected cell lines are exposed to normoxia (21% O_2) or hypoxia (1.5% O_2) for 4 h. Whole cell lysates are collected, as described in a previous section. The protein concentration of the lysates is determined, and 50 μg of lysate is used to analyze protein levels by immunoblotting. Membranes are probed for HIF-1α levels, SOD1 (1:500), SOD2 (1:1000), catalase (1:10,000), glutathione peroxidase (GPX 1:5000 of anti-myc antibody), or a-tubulin.

SOD1 and SOD2 quench superoxide to form hydrogen peroxide, while GPX and catalase convert hydrogen peroxide to water. We found that ectopic expression of SOD1 and SOD2 had no effect on hypoxic stabilization of HIF-1α protein levels. However, GPX and catalase attenuated hypoxic stabilization of HIF-1α protein. These data indicate that hydrogen peroxide, and not superoxide, is the ROS moiety that is responsible for hypoxic stabilization of HIF-1α. Moreover, these data also imply that the role of mitochondria in hypoxic stabilization is dependent on their production of ROS. These experiments do not rule out the possibility that the ability of mitochondria to consume oxygen plays a role in hypoxic stabilization of HIF-1α. Further experimental methods need to be designed to test this hypothesis.

 ## 6. CONCLUDING REMARKS

Current data indicate that mitochondria play a role in the hypoxic stabilization of HIF-1α protein; however, the mechanism is a topic for debate. It is believed that the mitochondria increase ROS in hypoxic conditions to inhibit the activity of the PHDs or generate a cytosolic oxygen gradient that would sequester oxygen from the PHDs, thereby inhibiting their enzymatic activity. The methods previously outlined are designed to genetically demonstrate the importance of mitochondrial electron transport distal to complex III in hypoxic stabilization of HIF-1α protein. To date, our experimental observations demonstrate that decreasing ROS generated by the mitochondria attenuates HIF-1α protein stabilization in hypoxia. Thus, ROS are crucial mediators of HIF activation; however, further experimentation needs to be performed to rule out a role of a cytosolic oxygen gradient in hypoxic activation of HIF.

ACKNOWLEDGMENTS

This work was supported in part by National Institutes of Health Grants (GM60472-07) to Navdeep S. Chandel. Eric Bell is supported by the American Heart Association Grant (0515563Z).

REFERENCES

Bruick, R. K., and McKnight, S. L. (2001). A conserved family of prolyl-4-hydroxylases that modify HIF. *Science* **294,** 1337–1340.

Brunelle, J. K., Bell, E. L., Quesada, N. M., Vercauteren, K., Tiranti, V., Zeviani, M., Scarpulla, R. C., and Chandel, N. S. (2005). Oxygen sensing requires mitochondrial ROS but not oxidative phosphorylation. *Cell Metab.* **1,** 409–414.

Chandel, N. S., Maltepe, E., Goldwasser, E., Mathieu, C. E., Simon, M. C., and Schumacker, P. T. (1998). Mitochondrial reactive oxygen species trigger hypoxia-induced transcription. *Proc. Natl. Acad. Sci. USA* **95,** 11715–11720.

Chandel, N. S., McClintock, D. S., Feliciano, C. E., Wood, T. M., Melendez, J. A., Rodriguez, A. M., and Schumacker, P. T. (2000). Reactive oxygen species generated at mitochondrial complex III stabilize hypoxia-inducible factor-1alpha during hypoxia: A mechanism of O_2 sensing. *J. Biol. Chem.* **275,** 25130–25138.

Doege, K., Heine, S., Jensen, I., Jelkmann, W., and Metzen, E. (2005). Inhibition of mitochondrial respiration elevates oxygen concentration but leaves regulation of hypoxia-inducible factor (HIF) intact. *Blood* **106,** 2311–2317.

Epstein, A. C., Gleadle, J. M., McNeill, L. A., Hewitson, K. S., O'Rourke, J., Mole, D. R., Mukherji, M., Metzen, E., Wilson, M. I., Dhanda, A., Tian, Y. M., and Masson, N. (2001). *C. elegans* EGL-9 and mammalian homologs define a family of dioxygenases that regulate HIF by prolyl hydroxylation. *Cell* **107,** 43–54.

Guzy, R. D., Hoyos, B., Robin, E., Chen, H., Liu, L., Mansfield, K. D., Simon, M. C., Hammerling, U., and Schumacker, P. T. (2005). Mitochondrial complex III is required for hypoxia-induced ROS production and cellular oxygen sensing. *Cell Metabolism* **1,** 401–408.

Hagen, T., Taylor, C. T., Lam, F., and Moncada, S. (2003). Redistribution of intracellular oxygen in hypoxia by nitric oxide: Effect on HIF-1α. *Science* **302,** 1975–1978.

Ivan, M., Kondo, K., Yang, H., Kim, W., Valiando, J., Ohh, M., Salic, A., Asara, J. M., Lane, W. S., and Kaelin, W. G., Jr. (2001). HIF-α targeted for VHL-mediated destruction by proline hydroxylation: Implications for O_2 sensing. *Science* **292,** 464–468.

Iwai, K., Yamanaka, K., Kamura, T., Minato, N., Conaway, R. C., Conaway, J. W., Klausner, R. D., and Pause, A. (1999). Identification of the von Hippel-Lindau tumor-suppressor protein as part of an active E3 ubiquitin ligase complex. *Proc. Natl. Acad. Sci. USA* **96,** 12436–12441.

Iyer, N. V., Kotch, L. E., Agani, F., Leung, S. W., Laughner, E., Wenger, R. H., Gassmann, M., Gearhart, J. D., Lawler, A. M., Yu, A. Y., and Semenza, G. L. (1998). Cellular and developmental control of O_2 homeostasis by hypoxia-inducible factor 1 alpha. *Genes Dev.* **12,** 149–162.

Jaakkola, P., Mole, D. R., Tian, Y. M., Wilson, M. I., Gielbert, J., Gaskell, S. J., Kriegsheim, A., Hebestreit, H. F., Mukherji, M., Schofield, C. J., Maxwell, P. H., Pugh, C. W., *et al.* (2001). Targeting of HIF-α to the von Hippel-Lindau ubiquitylation complex by O_2-regulated prolyl hydroxylation. *Science* **292,** 468–472.

Jiang, B. H., Semenza, G. L., Bauer, C., and Marti, H. H. (1996a). Hypoxia-inducible factor 1 levels vary exponentially over a physiologically relevant range of O_2 tension. *Am. J. Physiol.* **271,** C1172–C1180.

Jiang, B. H., Rue, E., Wang, G. L., Roe, R., and Semenza, G. L. (1996b). Dimerization, DNA binding, and transactivation properties of hypoxia-inducible factor 1. *J. Biol. Chem.* **271,** 17771–17778.

Maltepe, E., Schmidt, J. V., Baunoch, D., Bradfield, C. A., and Simon, M. C. (1997). Abnormal angiogenesis and responses to glucose and oxygen deprivation in mice lacking the protein ARNT. *Nature* **386,** 403–407.

Mansfield, K. D., Guzy, R. D., Pan, Y., Young, R. M., Cash, T. P., Schumacker, P. T., and Simon, M. C. (2005). Mitochondrial dysfunction resulting from loss of cytochrome c impairs cellular oxygen sensing and hypoxic HIF-α activation. *Cell Metabolism* **1,** 393–399.

Masson, N., Willam, C., Maxwell, P. H., Pugh, C. W., and Ratcliffe, P. J. (2001). Independent function of two destruction domains in hypoxia-inducible factor-alpha chains activated by prolyl hydroxylation. *EMBO J.* **20,** 5197–5206.

Maxwell, P. H., Wiesener, M. S., Chang, G. W., Clifford, S. C., Vaux, E. C., Cockman, M. E., Wykoff, C. C., Pugh, C. W., Maher, E. R., and Ratcliffe, P. J. (1999). The tumour suppressor protein VHL targets hypoxia-inducible factors for oxygen-dependent proteolysis. *Nature* **399,** 271–275.

Mazure, N. M., Chen, E. Y., Laderoute, K. R., and Giaccia, A. J. (1997). Induction of vascular endothelial growth factor by hypoxia is modulated by a phosphatidylinositol 3-kinase/Akt signaling pathway in Ha-ras-transformed cells through a hypoxia inducible factor-1 transcriptional element. *Blood* **90,** 3322–3331.

Pugh, C. W., O'Rourke, J. F., Nagao, M., Gleadle, J. M., and Ratcliffe, P. J. (1997). Activation of hypoxia-inducible factor-1: Definition of regulatory domains within the alpha subunit. *J. Biol. Chem.* **272,** 11205–11214.

Salceda, S., and Caro, J. (1997). Hypoxia-inducible factor 1alpha (HIF-1α) protein is rapidly degraded by the ubiquitin-proteasome system under normoxic conditions. Its stabilization by hypoxia depends on redox-induced changes. *J. Biol. Chem.* **272,** 22642–22647.

Schroedl, C., McClintock, D. S., Budinger, G. R., and Chandel, N. S. (2002). Hypoxic but not anoxic stabilization of HIF-1α requires mitochondrial reactive oxygen species. *Am. J. Physiol. Lung Cell Mol. Physiol.* **283,** L922–L931.

Semenza, G. L. (2000). HIF-1 and human disease: One highly involved factor. *Genes Dev.* **14,** 1983–1991.

Semenza, G. L. (2003). Targeting HIF-1 for cancer therapy. *Nat. Rev. Cancer* **3,** 721–732.

Wang, G. L., and Semenza, G. L. (1995). Purification and characterization of hypoxia-inducible factor 1. *J. Biol. Chem.* **270,** 1230–1237.

Wang, G. L., Jiang, B. H., Rue, E. A., and Semenza, G. L. (1995). Hypoxia-inducible factor 1 is a basic helix-loop-helix PAS heterodimer regulated by cellular O_2 tension. *Proc. Natl. Acad. Sci USA* **92,** 5510–5514.

CHAPTER TWENTY-FOUR

HYPOXIA-INDUCIBLE FACTOR-1α UNDER THE CONTROL OF NITRIC OXIDE

Bernhard Brüne *and* Jie Zhou

Contents

1. HIF-1 and Oxygen Sensing	464
2. Nitric Oxide: A Multifunctional Messenger	465
3. Accumulation of HIF-1α and Activation of HIF-1 by NO	467
4. Superoxide Stabilizes HIF-1α but Antagonizes NO Actions	470
5. Hypoxic Signal Transmission is Antagonized by NO	472
6. Summary and Conclusions	473
Acknowledgments	475
References	475

Abstract

Decreased oxygen availability evokes adaptive responses, which are primarily under the gene regulatory control of hypoxia-inducible factor 1 (HIF-1). HIF-1 is a heterodimer composed of the basic helix-loop-helix Per-ARNT-Sim (bHLH-PAS) protein HIF-1alpha (α) and the aryl hydrocarbon nuclear translocator (ARNT), also known as HIF-1beta (β). The HIF-1 transcriptional system senses decreased oxygen availability and transmits this signal into pathophysiological responses, such as angiogenesis, erythropoiesis, vasomotor control, an altered energy metabolism, and/or cell survival decisions. It is now appreciated that nitric oxide (NO) and/or derived reactive nitrogen species (RNS) participate in stability control of HIF-1α. Although initial observations showed that NO inhibits hypoxia-induced HIF-1α stabilization and HIF-1 transcriptional activation, later studies revealed that the exposure of cells from different species to chemically diverse NO donors, or conditions of endogenous NO formation, induced HIF-1α accumulation, HIF-1–DNA binding, and activation of downstream target gene expression under normoxic conditions. The opposing effects of NO under hypoxia versus normoxia are discussed based on direct and indirect reaction properties of NO, taking metal interactions as well as secondary reaction products, generated in the presence of oxygen or superoxide, into account. Considering HIF-1α as a target that is controlled by the bioavailability of NO

Institute of Biochemistry I/ZAFES, Johann Wolfgang Goethe-University, Frankfurt, Germany

Methods in Enzymology, Volume 435
ISSN 0076-6879, DOI: 10.1016/S0076-6879(07)35024-6

helps in the understanding of how signaling mechanisms are attributed to physiological and pathological transmission of NO actions with broad implications for medicine.

1. HIF-1 AND OXYGEN SENSING

Oxygen gradients are precisely controlled to meet requirements for oxidative phosphorylation, to generate ATP, and to prevent the potential risk of damage from excess oxygen associated with the formation of reactive oxygen species (ROS). To avoid the danger of metabolic demise from decreased oxygen availability, a hypoxia-responsive system senses a fall in tissue oxygen partial pressure and answers with a broad adaptive response to maintain and/or restore functionality. Pioneering work on the regulation of the gene encoding erythropoietin led to the discovery of HIF-1 (Semenza and Wang, 1992), with a conceptually attractive and "simple" picture of oxygen sensing being established today (Kaelin, 2002; Schofield and Ratcliffe, 2004; Semenza, 2002; Wenger, 2002). Essential regulatory elements comprise the oxygen-dependent transcription factor HIF-1 and those mechanisms that control posttranslational modifications of the HIF-1α subunit to affect protein stability and/or cofactor recruitment.

As a heterodimer, HIF-1 is composed of one of the three alpha subunits (HIF-1α, HIF-2α, or HIF-3α), and one beta subunit (Wang and Semenza, 1995). HIF-1β is constitutively expressed and identical to the heterodimerization partner of the dioxin receptor/aryl hydrocarbon receptor (AhR) known as ARNT. Under normoxia, the α subunits are recognized by an E3-ubiquitin ligase complex that contains, among others, the von Hippel Lindau protein (pVHL), followed by polyubiquitination and proteasomal degradation (Kallio et al., 1999; Salceda and Caro, 1997). This recognition explains a low, often undetectable protein amount of HIF-1α under normoxia. With the term "hypoxia-inducible," it became apparent that an oxygen-dependent enzymatic activity hydroxylates two proline residues (Pro402 and Pro564) within a domain of HIF-1α, known as the oxygen-dependent degradation domain (ODD) (Ivan et al., 2001; Jaakkola et al., 2001). Corresponding enzymes are known as orthologs of C. elegans Egl-9, designated prolyl-4-hydroxylase (P4H) domain-containing enzymes prolyl hydroxylases (PHDs) (Bruick and McKnight, 2001; Epstein et al., 2001; Oehme et al., 2002). PHDs require an active Fe^{2+} site for catalysis, with intracellular ascorbate acting as a natural antioxidant (Gerald et al., 2004; Semenza, 2004).

Hydroxylated prolyl residues of HIF-1α form two hydrogen bonds with pVHL side chains, allowing polyubiquitination and proteasomal degradation of HIF-1α (Kallio et al., 1999; Maxwell et al., 1999; Salceda and Caro, 1997).

Cells with a defective pVHL constitutively express HIF-1α and harbor a transcriptionally active HIF-1 complex, as known for manifestations of hereditary pVHL disease (Clifford *et al.*, 2001; Krieg *et al.*, 2000). Oxygen additionally affects DNA binding and transcriptional activation of HIF-1α by hydroxylating an asparagine (Asn803) residue within the C-terminal transactivation domain (CTAD) of the protein (Lando *et al.*, 2002). Hypoxia attenuates Pro564/402 as well as Asn803 modifications, resulting in protein stabilization and coactivator recruitment (p300/cAMP response element-binding protein [CBP]) and thus accounts for activation of HIF-1 (Kaelin, 2002).

The HIF-1α/HIF-1β complex binds to hypoxia-responsive element (HRE) sites with the core DNA sequence 5′-ACGTG-3′ to cause target gene activation. Examples comprise erythropoiesis and iron metabolism (e.g., erythropoietin or transferrin), glucose/energy metabolism (e.g., glucose transporters), cell proliferation/viability decisions (e.g., transforming growth factor β3), vascular development/remodeling, and/or vasomotor tone (e.g., vascular endothelial growth factor [VEGF] or inducible nitric oxide synthase [iNOS]) (Semenza, 2000; Wenger, 2002). The diversity of target genes links hypoxic signal transmission to fundamental physiological, pathological, and developmental processes. Therefore, it is without surprise that HIF-1 target genes are associated with disease states, such as ischemic cardiovascular disorders, pulmonary hypertension, pregnancy disorders, or cancer (Semenza, 2000; Wenger, 2002).

Interestingly enough, we learned that signals other than hypoxia participate in hypoxic signaling. This participation applies to several hormones, cytokines, growth factors, as well as NO and/or NO-derived species (RNS).

2. Nitric Oxide: A Multifunctional Messenger

The landmark discovery that NO is synthesized by mammalian cells was followed by studies demonstrating that this molecule plays a crucial role in the homeostatic regulation of vascular, immune, and neurological systems. NO is enzymatically produced by different NOS isoforms from the terminal guanido nitrogen of L-arginine. A rough classification discriminates inducible versus constitutively expressed NOS isoforms, which allows us to approximate a low versus high output system for endogenously generated NO (Kavya *et al.*, 2006; Stuehr *et al.*, 2004). NOS inhibitors are derivatives of the natural substrate L-arginine, such as N^G-monomethyl-L-arginine, and intervene pharmacologically with the NO-output/NO-signaling system (Moncada *et al.*, 1991). Once activated, NOS isoforms not only produce NO, the primary reaction product, but also those species resulting from oxidation, reduction, or adduction of NO in physiological milieus, thereby producing various nitrogen oxides, S-nitrosothiols, peroxynitrite (ONOO⁻), and transition metal adducts (Stamler *et al.*, 1992) (Fig. 24.1). Documentation of a role of NO in biology is often based on the use of NO donors, such as S-nitrosothiols (S-nitrosoglutathione), NONOates (Dea-NO, Deta-NO, spermine-NO), and transition metal nitrosyls (sodium nitroprusside).

The physiological chemistry of NO encompasses numerous partners and can be separated into direct and indirect effects (Grisham *et al.*, 1999). Direct effects are those reactions in which NO interacts directly with target molecules/proteins, such as soluble guanylyl cyclase. Indirects effects are those reactions mediated by NO-derived species, generated in the reactions of NO with molecular oxygen (O_2) or superoxide (O_2^-). It has been proposed that one of the primary reactions of NO *in vivo* is its reaction with superoxide (Beckman *et al.*, 1990) (Fig. 24.2). The diffusion-controlled reaction between NO and O_2^- produces a variety of reactive intermediates that can nitrate, nitrosate, or oxidize biological targets. Although nitrotyrosine can be formed during this reaction and often has been considered a biological footprint of ONOO⁻, an oxidative chemistry may prevail during decomposition, especially if the flux rate of one radical exceeds that of the other one (Thomas *et al.*, 2006). To comprehend NO chemistry, one needs to consider timing, location, and the rate of NO production, which determines the type of chemistry and dictates individual target interactions.

A growing body of evidence suggests that oxidative as well as nitrosative stress is sensed and closely associated with transcriptional regulation of

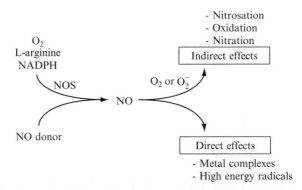

Figure 24.1 Nitric oxide (NO) formation and its biological entities. NO is generated by NO synthase (NOS) enzymes (utilizing oxygen, L-arginine, and nicotinamide adenine dinucleotide phosphate-oxidase [NADPH]) or by the breakdown of NO donors. Biological actions of NO comprise direct and indirect target interaction, with the latter ones requiring the presence of oxygen and/or superoxide. See text for further explanations.

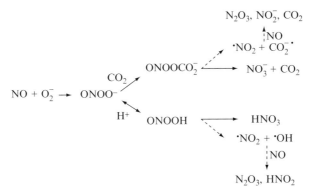

Figure 24.2 The interplay between nitric oxide (NO) and superoxide (O_2^-) and resulting chemical consequences. Peroxynitrite anion ($ONOO^-$) is presumably formed through the diffusion-limited reaction of NO with O_2^-. Peroxynitrous acid (ONOOH) decomposes to form nitrate (NO_3^-), free hydroxyl radical ($\bullet OH$), and nitrogen dioxide ($\bullet NO_2$). Compensating reactions with excess NO produces N_2O_3 and nitrite. In the presence of carbon dioxide (CO_2), the nitrosoperoxocarbonate anion ($ONOOCO_2^-$) forms, which may decompose to carbonate radical ($CO_3^- \bullet$) and $\bullet NO_2$. Further reactions in the presence of excess NO produce N_2O_3, nitrite, and carbon dioxide. Alternatively, $ONOOCO_2^-$ breaks down to nitrite and CO_2.

multiple target genes (Marshall *et al.*, 2000). Posttranslational modification of transcription factors by S-nitros(yl)ation, oxidation, glutathionylation, or phosphorylation may offer mechanisms by which cells sense redox changes to affect gene regulation. In addition, NO may modulate transcription/translation indirectly by affecting signaling pathways, such as mitogen-activated protein kinases, G-proteins, the Ras pathway, or phosphatidylinositol-3 kinase (PI3K) (Marshall *et al.*, 2000). Without surprise, the net effect of NO on gene regulation is variable and ranges from activation of transcription to its inhibition. This range may depend on NO-concentrations, the NO-related redox species, the redox state of the cell, the timing of NO delivery, and/or the presence of co-radical formation, such as superoxide. These particular features of the NO chemistry help to define the role of NO in (de)stabilizing HIF-1α under normoxia versus hypoxia.

3. Accumulation of HIF-1α and Activation of HIF-1 by NO

Initial studies in a number of cell systems, such as pig tubular LLC-PK_1 cells (Sandau *et al.*, 2000), human glioblastoma cells (Kimura *et al.*, 2000), or bovine pulmonary artery endothelial cells (Palmer *et al.*, 2000), showed accumulation of HIF-1α under the impact of NO. NO was provided by chemically distinct NO donors, such as S-nitrosoglutathione (GSNO, considered the most physiological NO donor), NOC-18 (Z-1-1 [2-aminoethyl-amino]diazen-1ium-1,2-diolate), NOC-5 (3-[hydroxy-1-(1-methylethyl)-2-nitrosohydazino]-1-propanamine), S-nitroso-N-acetyl-D, L-penicillamine (SNAP), or sodium nitroprusside (SNP) (Kimura *et al.*, 2000; Palmer *et al.*, 2000; Sandau *et al.*, 2000). These observations support the notion that the response was neither species nor cell-type specific and indeed resulted from decomposition of the NO donors. Time- and concentration-dependent effects of NO were established by using NO donors that show differences in their half lives (Sandau *et al.*, 2001). The fast release of NO from Dea-NO (half life $\sim$2 min) induced a rapid HIF-1α response, while the slow decomposing Deta-NO (half life $\sim$20 h) elicited delayed HIF-1α accumulation with an intermediate response elicited by spermineNONOate (half life $\sim$40 min). A cyclic guanosine monophosphate cGMP-mediated pathway was excluded by using a guanylyl cyclase inhibitor and the observation that lipophilic cGMP analogues did not reproduce HIF-1α accumulation (Kimura *et al.*, 2000; Palmer *et al.*, 2000). Initial studies looked into activation of the human VEGF promoter under normoxia by NO (Kimura *et al.*, 2000, 2001). Promoter deletion and mutation analysis suggested that the NO-responsive *cis*-elements were the HIF-1 binding site and an adjacent ancillary sequence that is located immediately downstream within the HRE. Accumulation of HIF-1α could not be mimicked by Angeli's salt, which argues against nitroxyl-mediated effects (Palmer *et al.*, 2000). However, considering that GSNO, a nitrosonium donor, provoked HIF-1α accumulation and that the GSNO effect could be reversed by dithiothreitol as well as acivicin (a proposed inhibitor of GSNO breakdown), the effect opened the possibility that S-nitrosation contributes to stabilizing HIF-1α (Palmer *et al.*, 2000). *In vitro*, NO^{+}-donating NO donors, such as GSNO and SNAP, provoked S-nitrosation of all 15 free thiol groups of purified HIF-1α, whereas spermine-NONOate, a NO radical–donating compound, did not share this behavior. However, spermine-NONOate, in the presence of O_2^- and generated by xanthine/xanthine oxidase, regained S-nitrosation, most likely via formation of a N_2O_3-like (and thus nitrosating) species (Sumbayev *et al.*, 2003). In renal carcinoma cell line (RCC4; constitutively expresses HIF-1α due to a defect in pVHL) and human embryonic kidney (HEK293) cells, GSNO or SNAP reproduced S-nitrosation of HIF-1α with the modification of three to four thiols, though only when S-nitrosation was followed by the biotin switch assay (Sumbayev *et al.*, 2003; Jaffrey *et al.*, 2001). However, so far, a causative role of S-nitrosation in stabilizing HIF-1α has not been established.

While NO donors serve as valuable tools for performing mechanistic studies, they cannot answer the question on intracellular NO concentrations needed to affect HIF-1 activity. One way to solve this problem is to overexpress iNOS to demonstrate the importance of NO as an intracellular messenger molecule (Sandau *et al.*, 2001). Expression of human iNOS caused nitrite accumulation in the cell supernatant and triggered HIF-1α accumulation in LLC-PK$_1$ cells, supporting the notion that NO indeed functions as an autocrine and/or paracrine factor. Moreover, a coculture or transwell system of macrophages and a detector cell line (e.g., tubular LLC-PK$_1$) established that lipopolysaccharide/interferon-γ–stimulated (and thus NO-producing macrophages, but not resting macrophages) elicited HIF-1α accumulation in the detector cell (Thomas *et al.*, 2004; Zhou *et al.*, 2003). Blocking NO production by pharmacologic interventions or scavenging NO with the use of carboxy-PTIO further points to NO as the relevant signaling molecule. These experiments underscored the ability of endogenously produced NO to stabilize HIF-1α.

Basic observations on a role of NO in eliciting HIF-1α have been confirmed in different settings. Cell density–induced activation of HRE demands production of NO, which acts as a diffusible paracrine factor secreted by densely cultured cells (Sheta *et al.*, 2001). In human prostate cells, NO activated HRE sites to confer a survival or growth advantage of these cells *in vivo*, and iNOS inhibitors blocked production of an angiogenic activity in thioglycolate-induced peritoneal macrophages (Xiong *et al.*, 1998). It is believed that VEGF is an important contributor to macrophage-dependent angiogenic activity and modulation of VEGF messenger RNA (mRNA) levels in RAW264.7 macrophages is regulated in part by the iNOS pathway. Attenuating iNOS allows formation of an antiangiogenic factor, which suggests NO as a player in the regulation of macrophage-dependent angiogenic activity *in vivo*, in wound repair, and possibly in solid tumor development (Xiong *et al.*, 1998).

Addressing molecular mechanisms of NO in stabilizing HIF-1α revealed decreased ubiquitination of HIF-1α and abrogated binding of pVHL to HIF-1α (Metzen *et al.*, 2003), which implied that NO directly attenuates hydroxylation of HIF-1α. Indeed, in an *in vitro* HIF-1α–pVHL capture assay, the NO donor GSNO dose-dependently attenuated PHD activity. These data suggest that hypoxia and NO employ overlapping, if not identical, signaling mechanisms to block PHD activity (Fig. 24.3). It is known that NO interacts with iron (II) in heme- or nonheme containing proteins (Grisham *et al.*, 1999), exemplified by spectroscopic studies when NO directly coordinates the ferrous iron in protocatechuate 4,5-dioxygenase, catechol 2,3-dioxygenase (Arciero *et al.*, 1985), or isopenicillin N synthase (Roach *et al.*, 1995). These enzymes coordinate Fe^{2+} in their catalytic site in a 2-histidine-1-carboxylate facial triad, which is the defining structural motif of mononuclear nonheme iron (II) enzymes (Hegg and Que, 1997).

Considering that PHDs are nonheme Fe^{2+}-containing enzymes, one may envision Fe^{2+}-coordination by NO to explain enzyme inhibition. These considerations point to a direct inhibition of PHD activity by NO and thus favors a direct NO action.

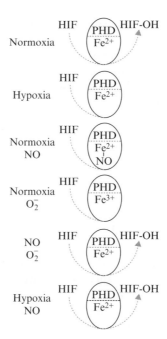

Figure 24.3 Regulation of prolyl hydroxylase (PHD) activity by hypoxia, nitric oxide (NO), and superoxide (O_2^-). PHD, containing a central Fe^{2+}, accounts for hydroxylation of hypoxia-inducible factor 1alpha (HIF-1α) under normoxia. Hypoxia (i.e., a lack of oxygen) inhibits PHD activity. NO and O_2^- may attenuate PHD activity by targeting Fe^{2+} or by promoting iron oxidation. PHD activity is regained under NO/O_2^- co-formation or when NO is supplied under hypoxia. See text for further explanations.

4. SUPEROXIDE STABILIZES HIF-1α BUT ANTAGONIZES NO ACTIONS

Superoxide formation affects the intracellular redox environment and may modulate NO signaling, suggesting that regulation of HIF-1α might be under the control of superoxide. One concept, although controversially discussed, proposes the generation of ROS by mitochondria under hypoxic conditions. Mitochondrial-produced ROS were considered to be both necessary and sufficient to stabilize HIF-1α (Chandel *et al.*, 1998). This idea was challenged because it heavily relied on pharmacological inhibition and because other groups reported that cells lacking functional mitochondria (p^0 cells) still stabilized HIF-1α in response to hypoxia (Srinivas *et al.*, 2001; Vaux *et al.*, 2001). However, more recently, genetic and pharmacologic inhibition of the electron transport precludes ROS formation, which impaired HIF induction by hypoxia but not anoxia (Brunelle *et al.*, 2005; Guzy *et al.*, 2005; Mansfield *et al.*, 2005), arguing that mitochondrial ROS alter the shape of the PHD dose-response curve. Mechanistically, inactivation of PHD activity via conversion of Fe^{2+} to Fe^{3+} or indirect regulation of hydroxylase activity via signaling pathways responsive to ROS has been predicted (see Fig. 24.3). Evidence exists that junD reduces ROS as part of a defense system against oxidative stress (Gerald et al., 2004). In junD-deficient cells, reduced PHD activity was assigned to oxidized Fe^{3+} and could be recovered by ascorbate substitution. In corroborating experiments using the pVHL–HIF-1α capture assay, it was shown that the redox cycler DMNQ (2,3-dimethoxy-1,4-naphthoquinone) dose-dependently produced a rise in ROS (O_2^- and/or H_2O_2 via superoxide dismutase-mediated conversion of superoxide), which correlated with HIF-1α accumulation and impaired PHD activity (Callapina et al., 2005; Kohl et al., 2006). One can propose that antioxidant systems that scavenge ROS might have a tonic influence to enhance PHD activity. In turn, ROS that oxidize PHD-bound iron might provoke HIF-1α stabilization, as seen with oxidants such as O_2^- or H_2O_2 (Chandel et al., 2000; Görlach et al., 2001; Kohl et al., 2006).

The role of ROS suggests that additional layers of complexity are about to be added to the emerging picture of oxygen sensing. This issue certainly applies to NO signaling considering the diffusion-controlled interaction between O_2^- and NO.

Potential sources of O_2^- are, among others, nicotinamide adenine dinucleotide phosphate-oxidase (NADPH) oxidases, xanthine oxidase, uncoupled NOS, and the mitochondrial electron transport chain. For mechanistic studies, a controlled formation of superoxide can be produced by xanthine/xanthine oxidase (Daiber et al., 2002) or redox cyclers, such as menadione or DMNQ (Gutierrez, 2000). Simultaneous generation of O_2^- and NO by xanthine/ xanthine oxidase and a NO donor caused an increase in oxidative intermediates, followed by dihydrorhodamine (DHR) oxidation, with a decrease in steady-state NO concentrations and a proportional reduction of NO-evoked HIF-1α stabilization (Thomas et al., 2006). Superoxide dismutase (SOD) restored NO responses. Intermediates formed from the reactions of O_2^- with NO were nontoxic and did not form 3-nitrotyrosine footprints. Basic observations are corroborated by using DMNQ to attenuate NO-elicited HIF-1α accumulation (Kohl et al., 2006). When HIF-1α was stabilized by 0.5 mM DETA-NONOate, low concentrations of DMNQ (<1 μM) revealed no effect, intermediate concentrations of DMNQ (1–40 μM) attenuated HIF-1α accumulation, and, with higher amounts of DMNQ (>60 μM), HIF-1α stability reappeared. The ability of ROS to attenuate NO-induced HIF-1α stabilization required proteasomal degradation and thus pointed to an intact hydroxylation system (see Fig. 24.3). In supporting studies, O_2^- generated by xanthine/xanthine oxidase impaired HIF-1α stability in cells exposed to hypoxia or NOC-18 (Wellman et al., 2004). Similar data from Hep3B cells showed that NO donors prevented hypoxic HIF-1α accumulation, an effect antagonized to some extent by glutathione analogues or peroxynitrite scavengers

(Agani *et al.*, 2002). These observations suggest that the ability of NO or O_2^- to stabilize HIF-1α depends, to some extent, on the formation of the corresponding co-signal. The primary consequence of O_2^- and NO cogeneration is a change in the cellular phenotype due to an altered bioavailability of NO or O_2^-. The rate of NO or O_2^- formation is critical in this scenario because O_2^- concentrations determine the steady-state concentrations of NO and vice versa. Thus, a greater rate of NO (O_2^-) production would be necessary in the presence of O_2^- (NO) to achieve the same HIF-1α response, as in the absence of O_2^- (NO). The unique features of these reactions are their simplicity. In this way, alterations in bioavailability of NO or O_2^- account for signaling rather than the resultant chemistry of newly formed, higher nitrogen oxides. These considerations may help to explain inconsistencies in the literature on the ability of NO or O_2^- to stabilize HIF-1α. Variables accounting for distinctions reside in the relative flux rates of NO or O_2^- and/or their co-formation.

5. Hypoxic Signal Transmission is Antagonized by NO

In 1998/1999, it was reported that carbon monoxide (CO) and NO inhibited accumulation of HIF-1α under hypoxia (Huang *et al.*, 1999; Liu *et al.*, 1998; Sogawa *et al.*, 1998) and transcriptional activation of HIF-1. For example, hypoxia-evoked luciferase reporter activity was reduced by single nucleotide polymorphism (SNP), GSNO, and 3-morpholinosydnonimine (Sogawa *et al.*, 1998). CO suppressed erythropoietin mRNA expression by hypoxia in a dose-dependent manner, though erythropoietin expression induced by either Co^{2+} or desferrioxamine was not affected (Huang *et al.*, 1999). Intriguingly, CO is chemically inert, undergoes no known chemical modification, and thus must be considered a rather selective ligand for Fe^{2+}-containing proteins. As pointed out, the chemistry of NO is unequally more complex, but shares with CO the ability to target Fe^{2+}-containing proteins. Therefore, the conjoint destabilization of HIF-1α and inhibition HIF-1 transactivation might be compatible with current concepts of HIF-1 stability regulation if CO and NO provoke HIF-1α degradation under hypoxia. Indeed, a more recent study proposes that activation of PHDs by NO derived from SNP in Hep3B cells occurs and requires a substantial amount of oxygen because reactivation is seen with 1% oxygen, but not in cells cultured under 0.4% oxygen (Wang *et al.*, 2002). HIF-1α–pVHL–binding assays confirmed that SNP efficiently recovered the activities of PHD, which were inhibited by Co^{2+} or desferrioxamine, suggesting that NO enhances the interaction between HIF-1α and pVHL through reactivation of PHDs.

The idea that PHD activity is regained under hypoxia/NO to provoke destruction of HIF-1α is in line with the observation by Huang et al. (1999) that the ODD of HIF-1α, which accounts for protein stability, is needed for reactive nitrogen intermediates (RNI) to reverse hypoxic HIF-1α stabilization (see Fig. 24.3). The concept was further strengthened by Hagen et al. (2003), who showed that expression of P402A/P564A-HIF-1α, a protein which lacks the hydroxylation motive and thus destabilization features, remains stable when subjected to DETA-NO and hypoxia (Hagen et al., 2003). Various inhibitors of the mitochondrial respiratory chain reproduced actions of the NO donor. Using a Renilla luciferase construct targeted to mitochondria as a monitor of available oxygen, it can be suggested that interfering with mitochondrial respiration under hypoxia redistributes oxygen towards nonrespiratory oxygen-dependent targets, such as PHD, so that they do not register hypoxia. This redistribution may account for CO as well, considering the potential inhibition of complex IV (i.e., cytochrome C oxidase). The ability of high NO concentrations to stabilize HIF-1α via direct inhibition of PHD was not impaired by inhibition of mitochondrial respiration (Mateo et al., 2003). Continuative examination implied that the role of mitochondrial respiratory chain inhibitors in destabilizing HIF-1α under intermediate hypoxia and pimonidazole staining—a marker of decreased oxygen availability—were eliminated in gas-permeable dishes (Doege et al., 2005). These data suggest that any interference with mitochondrial respiration may reverse an oxygen gradient in conventional cell culture dishes, thereby elevating the O_2 concentration in the medium, which then allows for HIF-1α degradation. Rather than redistributing intracellular oxygen, impaired mitochondrial respiration simply increases oxygen availability. In contemporary studies, a pVHL–HIF-1α capture assay proved that PHD activity was restored under hypoxia/DETA-NO, compared to hypoxia or low concentration of NO supplied individually (Callapina et al., 2005). HIF-1 itself may repress mitochondrial function and oxygen consumption by inhibiting pyruvate dehydrogenase, thereby preventing pyruvate to fuel the Krebs-cycle, which again will result in a relative increase in intracellular oxygen tension (Papandreou et al., 2006). Conclusively, the rate of oxygen consumption by mitochondria exerts a profound effect on the ability of cells to sense hypoxia and thus to stabilize HIF-1α. These considerations help to explain cell type–specific variations toward HIF-1α accumulation under hypoxia, depending on the respiratory capacity of individual cells.

6. Summary and Conclusions

It is becoming clear that HIF-1α not only is under the control of hypoxia, but also is affected by radical species, such as NO or O_2^-. NO has a dual role in stabilizing HIF-1α under normoxia, but destabilizing the protein under hypoxia. Superoxide accounts on one side for activation/stabilization of HIF-1, but on the other side antagonizes the action of NO. This juxtaposition implies fine tuning of HIF-1 responses within the triad hypoxia/NO/O_2^- (Fig. 24.4).

Assuming that radicals show the potential to mimic a kind of metabolic hypoxia requires us to define relevant (patho)physiological conditions. Taking into account that HIF-1 is relevant for evoking inflammatory reactions (at least in macrophages), it is suggested that radical formation and HIF-1 responses refer to inflammation. Inflammation may use the HIF-1 system to adapt to stress conditions to allow regeneration/healing simply by guaranteeing energy supply via angiogenesis and glucose uptake/metabolism. It will be the task of future workers to see whether NO, O_2^-, and hypoxia activate the same cluster of genes (i.e., HIF-1 targets), or whether cells respond with smaller or larger subsets of target genes in response to each agonist. The ability of NO to limit the protein amount of HIF-1α under hypoxia may point to a self-regulatory feedback loop, considering that iNOS contains an HRE site in its promoter region. The diffusion-controlled interaction of NO with O_2^- complicates the picture further, and it will be challenging to define the regulatory balance between NO and O_2^- in affecting HIF-1α. Modulation of HIF-1 responses in normoxia versus hypoxia by NO and O_2^- broadens the repertoire of functional consequences attributed to HIF-1. At the same time, it may help to better characterize the still-bewildering biological diversity of signaling mechanisms attributed to NO.

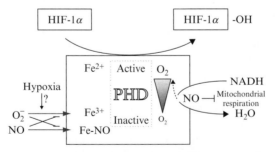

Figure 24.4 The oxygen sensor in the context of radical formation. An active prolyl hydroxylase (PHD) (Fe^{2+}, sufficient oxygen) hydroxylates hypoxia-inducible factor 1alpha (HIF-1α). Activity is attenuated at a low oxygen content, iron oxidation (Fe^{3+}), or the binding of nitric oxide (NO) to the central iron. Superoxide (O_2^-) formation and NO generation may account for PHD inhibition. NO signaling is interrupted by O_2^- and vice versa, with the further notion that hypoxia may be a sufficient signal to enhance O_2^- formation. Mitochondrial respiration, besides PHDs, consumes oxygen to produce H_2O. If NO blocks mitochondrial oxygen consumption, more oxygen remains available for PHD activity. See text for further explanations.

ACKNOWLEDGMENTS

We apologize to researchers whose primary observations, which form the basis for our current knowledge in this field, could not be cited due to space limitations or whose work has been acknowledged indirectly via reference to current reviews. Grants supporting our work over the past years came from Deutsche Forschungsgemeinschaft (BR999), Sander Foundation (2002.088.2), Deutsche Krebshilfe (10-2008), and European Community (PROLIGEN).

REFERENCES

Agani, F. H., Puchowicz, M., Chavez, J. C., Pichiule, P., and LaManna, J. (2002). Role of nitric oxide in the regulation of HIF-1α expression during hypoxia. *Am. J. Physiol.* **283,** C178–C186.

Arciero, D. M., Orville, A. M., and Lipscomb, J. D. (1985). [^{17}O]Water and nitric oxide binding by protocatechuate 4,5-dioxygenase and catechol 2,3-dioxygenase. Evidence for binding of exogenous ligands to the active site Fe^{2+} of extradiol dioxygenases. *J. Biol. Chem.* **260,** 14035–14044.

Beckman, J. S., Beckman, T. W., Chen, J., Marshall, P. A., and Freeman, B. A. (1990). Apparent hydroxyl radical production by peroxynitrite: Implications for endothelial injury from nitric oxide and superoxide. *Proc. Natl. Acad. Sci. USA* **87,** 1620–1624.

Bruick, R. K., and McKnight, S. L. (2001). A conserved family of prolyl-4-hydroxylases that modify HIF. *Science* **294,** 1337–1340.

Brunelle, J. K., Bell, E. L., Quesada, N. M., Vercauteren, K., Tiranti, V., Zeviani, M., Scarpulla, R. C., and Chandel, N. S. (2005). Oxygen sensing requires mitochondrial ROS but not oxidative phosphorylation. *Cell Metab.* **1,** 409–414.

Callapina, M., Zhou, J., Schmid, T., Kohl, R., and Brune, B. (2005). NO restores HIF-1α hydroxylation during hypoxia: Role of reactive oxygen species. *Free Radic. Biol. Med.* **39,** 925–936.

Chandel, N. S., Maltepe, E., Goldwasser, E., Mathieu, C. E., Simon, M. C., and Schumacker, P. T. (1998). Mitochondrial reactive oxygen species trigger hypoxia-induced transcription. *Proc. Natl. Acad. Sci. USA* **95,** 11715–11720.

Chandel, N. S., McClintock, D. S., Feliciano, C. E., Wood, T. M., Melendez, J. A., Rodriguez, A. M., and Schumacker, P. T. (2000). Reactive oxygen species generated at mitochondrial complex III stabilize hypoxia-inducible factor-1alpha during hypoxia: A mechanism of O$_2$ sensing. *J. Biol. Chem.* **275,** 25130–25138.

Clifford, S. C., Cockman, M. E., Smallwood, A. C., Mole, D. R., Woodward, E. R., Maxwell, P. H., Ratcliffe, P. J., and Maher, E. R. (2001). Contrasting effects on HIF-1α regulation by disease-causing pVHL mutations correlate with patterns of tumorigenesis in von Hippel-Lindau disease. *Hum. Mol. Genet.* **10,** 1029–1038.

Daiber, A., Frein, D., Namgaladze, D., and Ullrich, V. (2002). Oxidation and nitrosation in the nitrogen monoxide/superoxide system. *J. Biol. Chem.* **277,** 11882–11888.

Doege, K., Heine, S., Jensen, I., Jelkmann, W., and Metzen, E. (2005). Inhibition of mitochondrial respiration elevates oxygen concentration but leaves regulation of hypoxia-inducible factor (HIF) intact. *Blood* **106,** 2311–2317.

Epstein, A. C., Gleadle, J. M., McNeill, L. A., Hewitson, K. S., O'Rourke, J., Mole, D. R., Mukherji, M., Metzen, E., Wilson, M. I., Dhanda, A., Tian, Y. M., Masson, N., *et al.* (2001). *C. elegans* EGL-9 and mammalian homologs define a family of dioxygenases that regulate HIF by prolyl hydroxylation. *Cell* **107,** 43–54.

Gerald, D., Berra, E., Frapart, Y. M., Chan, D. A., Giaccia, A. J., Mansuy, D., Pouyssegur, J., Yaniv, M., and Mechta-Grigoriou, F. (2004). JunD reduces tumor angiogenesis by protecting cells from oxidative stress. *Cell* **118,** 781–794.

Görlach, A., Diebold, I., Schini-Kerth, V. B., Berchner-Pfannschmidt, U., Roth, U., Brandes, R. P., Kietzmann, T., and Busse, R. (2001). Thrombin activates the hypoxia-inducible factor-1 signaling pathway in vascular smooth muscle cells: Role of the p22(phox)-containing NADPH oxidase. *Circ. Res.* **89,** 47–54.

Grisham, M. B., Jourd'Heuil, D., and Wink, D. A. (1999). Nitric oxide. I. Physiological chemistry of nitric oxide and its metabolites: Implications in inflammation. *Am. J. Physiol.* **276,** G315–G321.

Gutierrez, P. L. (2000). The metabolism of quinone-containing alkylating agents: Free radical production and measurement. *Front. Biosci.* **5,** D629–D638.

Guzy, R. D., Hoyos, B., Robin, E., Chen, H., Liu, L., Mansfield, K. D., Simon, M. C., Hammerling, U., and Schumacker, P. T. (2005). Mitochondrial complex III is required for hypoxia-induced ROS production and cellular oxygen sensing. *Cell Metab.* **1,** 401–408.

Hagen, T., Taylor, C. T., Lam, F., and Moncada, S. (2003). Redistribution of intracellular oxygen in hypoxia by nitric oxide: Effect on HIF1α. *Science* **302,** 1975–1978.

Hegg, E. L., and Que, L., Jr. (1997). The 2-His-1-carboxylate facial triad—an emerging structural motif in mononuclear non-heme iron(II) enzymes. *Eur. J. Biochem.* **250,** 625–629.

Huang, L. E., Willmore, W. G., Gu, J., Goldberg, M. A., and Bunn, H. F. (1999). Inhibition of hypoxia-inducible factor 1 activation by carbon monoxide and nitric oxide. *J. Biol. Chem.* **274,** 9038–9044.

Ivan, M., Kondo, K., Yang, H., Kim, W., Valiando, J., Ohh, M., Salic, A., Asara, J. M., Lane, W. S., and Kaelin, W. G., Jr. (2001). HIF-α targeted for VHL-mediated destruction by proline hydroxylation: Implications for O_2 sensing. *Science* **292,** 464–468.

Jaakkola, P., Mole, D. R., Tian, Y. M., Wilson, M. I., Gielbert, J., Gaskell, S. J., von Kriegsheim, A., Hebestreit, H. F., Mukherji, M., Schofield, C. J., Maxwell, P. H., Pugh, C. W., *et al.* (2001). Targeting of HIF-α to the von Hippel-Lindau ubiquitylation complex by O_2-regulated prolyl hydroxylation. *Science* **292,** 468–472.

Jaffrey, S. R., Erdjument-Bromage, H., Ferris, C. D., Tempst, P., and Snyder, S. H. (2001). Protein S-nitrosylation: A physiological signal for neuronal nitric oxide. *Nat. Cell Biol.* **3,** 193–197.

Kaelin, W. G., Jr. (2002). How oxygen makes its presence felt. *Genes Dev.* **16,** 1441–1445.

Kallio, P. J., Wilson, W. J., O'Brien, S., Makino, Y., and Poellinger, L. (1999). Regulation of the hypoxia-inducible transcription factor 1alpha by the ubiquitin-proteasome pathway. *J. Biol. Chem.* **274,** 6519–6525.

Kavya, R., Saluja, R., Singh, S., and Dikshit, M. (2006). Nitric oxide synthase regulation and diversity: Implications in Parkinson's disease. *Nitric Oxide* **15,** 280–294.

Kimura, H., Weisz, A., Kurashima, Y., Hashimoto, K., Ogura, T., D'Acquisto, F., Addeo, R., Makuuchi, M., and Esumi, H. (2000). Hypoxia response element of the human vascular endothelial growth factor gene mediates transcriptional regulation by nitric oxide: Control of hypoxia-inducible factor-1 activity by nitric oxide. *Blood* **95,** 189–197.

Kimura, H., Weisz, A., Ogura, T., Hitomi, Y., Kurashima, Y., Hashimoto, K., D'Acquisto, F., Makuuchi, M., and Esumi, H. (2001). Identification of hypoxia-inducible factor 1 ancillary sequence and its function in vascular endothelial growth factor gene induction by hypoxia and nitric oxide. *J. Biol. Chem.* **276,** 2292–2298.

Kohl, R., Zhou, J., and Brune, B. (2006). Reactive oxygen species attenuate nitric-oxide-mediated hypoxia-inducible factor-1alpha stabilization. *Free Radic. Biol. Med.* **40,** 1430–1442.

Krieg, M., Haas, R., Brauch, H., Acker, T., Flamme, I., and Plate, K. H. (2000). Upregulation of hypoxia-inducible factors HIF-1α and HIF-2α under normoxic conditions in renal carcinoma cells by von Hippel-Lindau tumor suppressor gene loss of function. *Oncogene* **19,** 5435–5443.

Lando, D., Peet, D. J., Whelan, D. A., Gorman, J. J., and Whitelaw, M. L. (2002). Asparagine hydroxylation of the HIF transactivation domain: A hypoxic switch. *Science* **295,** 858–861.

Liu, Y., Christou, H., Morita, T., Laughner, E., Semenza, G. L., and Kourembanas, S. (1998). Carbon monoxide and nitric oxide suppress the hypoxic induction of vascular endothelial growth factor gene via the 5′ enhancer. *J. Biol. Chem.* **273,** 15257–15262.

Mansfield, K. D., Guzy, R. D., Pan, Y., Young, R. M., Cash, T. P., Schumacker, P. T., and Simon, M. C. (2005). Mitochondrial dysfunction resulting from loss of cytochrome c impairs cellular oxygen sensing and hypoxic HIF-α activation. *Cell Metab.* **1,** 393–399.

Marshall, H. E., Merchant, K., and Stamler, J. S. (2000). Nitrosation and oxidation in the regulation of gene expression. *FASEB J.* **14,** 1889–1900.

Mateo, J., Garcia-Lecea, M., Cadenas, S., Hernandez, C., and Moncada, S. (2003). Regulation of hypoxia-inducible factor-1alpha by nitric oxide through mitochondria-dependent and -independent pathways. *Biochem. J.* **376,** 537–544.

Maxwell, P. H., Wiesener, M. S., Chang, G.-W., Clifford, S. C., Vaux, E. C., Cockman, M. E., Wykoff, C. C., Pugh, C. W., Maher, E. R., and Ratcliffe, P. J. (1999). The tumour suppressor protein VHL tragets hypoxia-inducible factors for oxygen-dependent proteolysis. *Nature* **399,** 271–275.

Metzen, E., Zhou, J., Jelkmann, W., Fandrey, J., and Brüne, B. (2003). Nitric oxide impairs normoxic degradation of HIF-1α by inhibition of prolyl hydroxylases. *Mol. Biol. Cell* **14,** 3470–3481.

Moncada, S., Palmer, R. M., and Higgs, E. A. (1991). Nitric oxide: Physiology, pathophysiology, and pharmacology. *Pharmacol. Rev.* **43,** 109–142.

Oehme, F., Ellinghaus, P., Kolkhof, P., Smith, T. J., Ramakrishnan, S., Hutter, J., Schramm, M., and Flamme, I. (2002). Overexpression of PH-4, a novel putative proline 4-hydroxylase, modulates activity of hypoxia-inducible transcription factors. *Biochem. Biophys. Res. Commun.* **296,** 343–349.

Palmer, L. A., Gaston, B., and Johns, R. A. (2000). Normoxic stabilization of hypoxia-inducible factor-1 expression and activity: Redox-dependent effect of nitrogen oxides. *Mol. Pharmacol.* **58,** 1197–1203.

Papandreou, I., Cairns, R. A., Fontana, L., Lim, A. L., and Denko, N. C. (2006). HIF-1 mediates adaptation to hypoxia by actively downregulating mitochondrial oxygen consumption. *Cell Metab.* **3,** 187–197.

Roach, P. L., Clifton, I. J., Fulop, V., Harlos, K., Barton, G. J., Hajdu, J., Andersson, I., Schofield, C. J., and Baldwin, J. E. (1995). Crystal structure of isopenicillin N synthase is the first from a new structural family of enzymes. *Nature* **375,** 700–704.

Salceda, S., and Caro, J. (1997). Hypoxia-inducible factor 1alpha (HIF-1α) protein is rapidly degraded by the ubiquitin-proteasome system under normoxic conditions. *J. Biol. Chem.* **272,** 22642–22647.

Sandau, K. B., Faus, H. G., and Brüne, B. (2000). Induction of hypoxia-inducible-factor 1 by nitric oxide is mediated via the PI3K pathway. *Biochem. Biophys. Res. Commun.* **278,** 263–267.

Sandau, K. B., Fandrey, J., and Brüne, B. (2001). Accumulation of HIF-1α under the influence of nitric oxide. *Blood* **97,** 1009–1015.

Schofield, C. J., and Ratcliffe, P. J. (2004). Oxygen sensing by HIF hydroxylases. *Nat. Rev. Mol. Cell Biol.* **5,** 343–354.

Semenza, G. L. (2000). HIF-1 and human disease: One highly involved factor. *Genes Dev.* **14,** 1983–1991.

Semenza, G. L. (2002). HIF-1 and tumor progression: Pathophysiology and therapeutics. *Trends Mol. Med.* **8,** S62–S67.

Semenza, G. L. (2004). Hydroxylation of HIF-1: Oxygen sensing at the molecular level. *Physiology (Bethesda)* **19,** 176–182.

Semenza, G. L., and Wang, G. L. (1992). A nuclear factor induced by hypoxia via *de novo* protein synthesis binds to the human erythropoietin gene enhancer at a site required for transcriptional activation. *Mol. Cell Biol.* **12,** 5447–5454.

Sheta, E. A., Trout, H., Gildea, J. J., Harding, M. A., and Theodorescu, D. (2001). Cell density mediated pericellular hypoxia leads to induction of HIF-1α via nitric oxide and Ras/MAP kinase mediated signaling pathways. *Oncogene* **20,** 7624–7634.

Sogawa, K., Numayama-Tsuruta, K., Ema, M., Abe, M., Abe, H., and Fujii-Kuriyama, Y. (1998). Inhibition of hypoxia-inducible factor 1 activity by nitric oxide donors in hypoxia. *Proc. Natl. Acad. Sci. USA* **95,** 7368–7373.

Srinivas, V., Leshchinsky, I., Sang, N., King, M. P., Minchenko, A., and Caro, J. (2001). Oxygen sensing and HIF-1 activation does not require an active mitochondrial respiratory chain electron-transfer pathway. *J. Biol. Chem.* **276,** 21995–21998.

Stamler, J. S., Singel, D. J., and Loscalzo, J. (1992). Biochemistry of nitric oxide and its redox-activated forms. *Science* **258,** 1898–1902.

Stuehr, D. J., Santolini, J., Wang, Z. Q., Wei, C. C., and Adak, S. (2004). Update on mechanism and catalytic regulation in the NO synthases. *J. Biol. Chem.* **279,** 36167–36170.

Sumbayev, V. V., Budde, A., Zhou, J., and Brüne, B. (2003). HIF-1α protein as a target for S-nitrosation. *FEBS Lett.* **535,** 106–112.

Thomas, D. D., Espey, M. G., Ridnour, L. A., Hofseth, L. J., Mancardi, D., Harris, C. C., and Wink, D. A. (2004). Hypoxic inducible factor 1alpha, extracellular signal-regulated kinase, and p53 are regulated by distinct threshold concentrations of nitric oxide. *Proc. Natl. Acad. Sci. USA* **101,** 8894–8899.

Thomas, D. D., Ridnour, L. A., Espey, M. G., Donzelli, S., Ambs, S., Hussain, S. P., Harris, C. C., Degraff, W., Roberts, D. D., Mitchell, J. B., and Wink, D. A. (2006). Superoxide fluxes limit nitric oxide-induced signaling. *J. Biol. Chem.* **281,** 25984–25993.

Vaux, E. C., Metzen, E., Yeates, K. M., and Ratcliffe, P. J. (2001). Regulation of hypoxia-inducible factor is preserved in the absence of a functioning mitochondrial respiratory chain. *Blood* **98,** 296–302.

Wang, G. L., and Semenza, G. L. (1995). Purification and characterization of hypoxia-inducible factor 1. *J. Biol. Chem.* **270,** 1230–1237.

Wang, F., Sekine, H., Kikuchi, Y., Takasaki, C., Miura, C., Heiwa, O., Shuin, T., Fujii-Kuriyama, Y., and Sogawa, K. (2002). HIF-1α–prolyl hydroxylase: Molecular target of nitric oxide in the hypoxic signal transduction pathway. *Biochem. Biophys. Res. Commun.* **295,** 657–662.

Wellman, T. L., Jenkins, J., Penar, P. L., Tranmer, B., Zahr, R., and Lounsbury, K. M. (2004). Nitric oxide and reactive oxygen species exert opposing effects on the stability of hypoxia inducible factor-1alpha (HIF-1α) in explants of human pial arteries. *FASEB J.* **18,** 379–381.

Wenger, R. H. (2002). Cellular adaptation to hypoxia: O_2-sensing protein hydroxylases, hypoxia-inducible transcription factors, and O_2-regulated gene expression. *FASEB J.* **16,** 1151–1162.

Xiong, M., Elson, G., Legarda, D., and Leibovich, S. J. (1998). Production of vascular endothelial growth factor by murine macrophages: Regulation by hypoxia, lactate, and the inducible nitric oxide synthase pathway. *Am. J. Pathol.* **153,** 587–598.

Zhou, J., Fandrey, J., Schumann, J., Tiegs, G., and Brüne, B. (2003). NO and TNF-α released from activated macrophages stabilize HIF-1α in resting tubular LLC-PK1 cells. *Am. J. Physiol. Cell. Physiol.* **284,** C439–C446.

CHAPTER TWENTY-FIVE

Hypoxic Regulation of NF-κB Signaling

Eoin P. Cummins*, Katrina M. Comerford*, Carsten Scholz†,
Ulrike Bruning†, *and* Cormac T. Taylor*

Contents

1. Background	480
2. Treatment Protocols for Cellular Hypoxia Studies	481
2.1. Exposure of cells to ambient atmospheric hypoxia	482
2.2. Inhibition of hydroxylases in cultured cells	482
2.3. Manipulation of NF-κB signaling in cultured cells	483
3. Measurement of NF-κB Activity in Cultured Cells	484
3.1. NF-κB–luciferase promoter-reporter assay	484
3.2. NF-κB DNA–binding assay	485
3.3. Immunoblotting analysis	488
3.4. Peptide pull-down assay	490
4. Summary/Conclusions	491
References	491

Abstract

Hypoxia and inflammation are coincidental events in an array of diseased tissues, including chronically inflamed sites (e.g., inflammatory bowel disease, rheumatoid arthritis), growing tumors, myocardial infarcts, atherosclerotic plaques, healing wounds, and sites of bacterial infection (Murdoch *et al.*, 2005). An understanding of how hypoxia modulates the inflammatory response is critical in developing our fundamental understanding of inflammatory disease and identifying new windows of therapeutic opportunity. Nuclear factor-κB (NF-κB) is a master transcriptional regulator of inflammatory and antiapoptotic gene expression, the activation of which has significant implications in disease development. Recent work has uncovered mechanisms by which hypoxia modulates the activation of NF-κB in cells through decreased oxygen-dependent suppression of the key regulators of this pathway. This work has implicated a novel role for proline and asparagine

* UCD Conway Institute, University College Dublin, Dublin, Ireland
† University of Lübeck, Lübeck, Germany

Methods in Enzymology, Volume 435
ISSN 0076-6879, DOI: 10.1016/S0076-6879(07)35025-8

hydroxylases in the modulation of NF-κB activity. Here, we describe methodologies used to demonstrate and interrogate hypoxic induction of the NF-κB pathway.

1. BACKGROUND

Over the course of evolution, we have developed the ability to adapt to hypoxia through the induction of a specific transcriptional pathway governed by the hypoxia inducible factor 1 (HIF-1), a transcription factor that regulates the expression of genes promoting angiogenesis, vasodilatation, glycolysis, and erythropoiesis (Schofield and Ratcliffe, 2005). The induction of such genes leads to increased tissue perfusion and anaerobic metabolism, thus maintaining ATP levels and forming a critical adaptive pathway to deal with the hypoxic threat. As well as a primary role in this adaptive response, HIF-1 also appears to play an important role in the control of inflammatory processes. For example, HIF-1 controls the expression of a number of proinflammatory genes (e.g., interleukin [IL]-8, adenosine 2B receptor) (Kim *et al.*, 2006; Kong *et al.*, 2006) and is itself under the control of a number of inflammatory mediators, such as cytokines (e.g., tumor necrosis factor alpha [TNFα], IL-1) (Zhou and Brune, 2006) and nitric oxide (Hagen *et al.*, 2003). HIF-1 also plays an important role in macrophage survival after migration from the bloodstream to sites of inflammation through the enhanced glycolytic maintenance of ATP levels (Cramer *et al.*, 2003). Furthermore, hypoxia also activates NF-κB, a critical transcription factor in the promotion and progression of inflammatory and antiapoptotic processes, (Cummins and Tayor, 2005). However, the mechanism by which hypoxia activates NF-κB and thus contributes to inflammation via this pathway remains poorly understood. A greater understanding of the oxygen-sensing and signaling pathways leading to the activation of these important transcriptional regulators is critical in the identification of new windows of therapeutic opportunity in a diverse range of pathological processes where hypoxia and inflammation coexist.

The mechanism by which HIF-1 is activated in hypoxia is relatively well understood (Schofield and Ratcliffe, 2005). HIF-1α is constitutively synthesized at a high level in normoxia, but its level is repressed by members of the 2-oxoglutarate–dependent dioxygenase superfamily, namely the prolyl hydroxylases (PHDs). Three PHD isoforms that regulate HIF-dependent transcriptional activity have been described to date (PHD1, PHD2, and PHD3) (Berra *et al.*, 2006). Oxygen-dependent modification of specific proline residues within consensus LxxLAP motifs (Pro402 and Pro564) in the oxygen-dependent degradation domain of HIF-1α by these enzymes, primarily the PHD2 isoform, results in targeting of HIF-1α for ubiquitylation via an E3 ligase complex initiated by the binding of the von Hippel Lindau protein (pVHL) and subsequent proteasomal degradation. A further hydroxylation of Asn803 in the transactivation domain of HIF-1α by factor-inhibiting HIF (FIH), an asparagine hydroxylase, represents a second mechanism of oxygen–dependent repression through inhibition of transactivation (Lando *et al.*, 2002). Similar mechanisms exist for HIF-2α (Haase, 2006). The hypoxic sensitivity of the HIF pathway is achieved by the absolute requirement of hydroxylases for molecular oxygen as a cosubstrate (with iron and the Krebs cycle intermediate 2-oxoglutarate). Inhibition of this pathway in hypoxia with the resultant stabilization and transactivation of HIF-α subunits represents a paradigm for oxygen sensing and hypoxia-responsive alterations in gene expression.

The pathways leading to NF-κB activation have been best characterized for their role in inflammation in response to a host of proinflammatory ligands (e.g., TNFα, IL-1, lipopolysaccharide, lymphotoxin beta [LTβ], viral infection, CD40 ligand engagement) (Karin and Greten, 2005). A complex sequence of events resulting from receptor occupation by these ligands triggers cascades involving a diverse array of adaptor molecules and signaling enzymes, which are stimulus specific. However, a convergence point exists for these signals at the level of the IκB kinases (IKKα, -β, and -γ), which form the IKK complex.

A substantial amount of evidence now exists that hypoxia activates NF-κB–dependent gene transcription and increases the sensitivity of this pathway to activation by proinflammatory stimuli, such as cytokines (Cummins and Taylor, 2005; Cummins *et al.*, 2006). However, the underlying mechanisms have remained elusive. We have hypothesized that hypoxia modulates NF-κB–dependent signaling through decreased hydroxylation of key regulators of this pathway, namely the IKKs. In recent studies (Cummins *et al.*, 2006), we have hypothesized that IKKβ (or an upstream regulator) is hydroxylated by PHD1 and PHD2 in normoxia, an event that results in suppressed IKKβ enzymatic activity leading to negative modulation of NF-κB signaling. Reversal of this hydroxylation in hypoxia would result in increased basal as well as cytokine-stimulated IKKβ activity, an event that has significant implications for NF-κB–dependent expression of proinflammatory and antiapoptotic genes. Cockman *et al.* (2006) have also recently demonstrated a role for FIH in the hydroxylation of members of the NF-κB pathway. Thus, we propose that the suppression of hydroxylation of components of the NF-κB pathway in hypoxia represents a potentially critical point of crosstalk between hypoxic and inflammatory signaling. In this methodological review, we will outline a number of techniques that may be used to interrogate the activation or inhibition of NF-κB–dependent signaling in hypoxia.

2. TREATMENT PROTOCOLS FOR CELLULAR HYPOXIA STUDIES

A number of approaches may be used to expose cultured cells to ambient atmospheric hypoxia or conditions that mimic hypoxia, including pharmacologic inhibition and genetic knockdown of hydroxylases. NF-κB signaling in cells may be altered using a number of approaches, including cytokine treatment and overexpression of wild-type and mutant components of the NF-κB signaling pathway. These treatments will be discussed in detail in this section.

2.1. Exposure of cells to ambient atmospheric hypoxia

For these studies, cells are grown to confluence in normal tissue culture conditions and placed in a hypoxia chamber, which allows the manipulation of cells and preparation of cell extracts while maintaining a hypoxic environment (Coy Laboratories, Grass Lake, MI). This manipulation is important to exclude possible complicating responses associated with reoxygenation. Typically, normobaric atmospheric oxygen levels are maintained at 1% (5% CO_2, balanced N_2, 37°, humidified). Fluorescence oximetry measurements demonstrate that medium PO_2 values equilibrate with atmospheric values within approximately 20 to 30 min. Medium PO_2 values at the level of the cell monolayer are generally significantly lower than those in the atmosphere due to the presence of an oxygen gradient through the medium layer generated by cellular oxygen consumption. In studies of acute hypoxia, medium is pre-equilibrated overnight in the hypoxia chamber, and this hypoxic medium is put on the cells in the hypoxia chamber at the beginning of the experiment, thus achieving "instantaneous hypoxia." Medium PO_2 measurements after equilibration and at time of harvest are determined in the hypoxia chamber using fluorescent quenching oximetry (Oxford Optronix, Oxford, UK) to ensure ambient hypoxia is achieved and maintained.

2.2. Inhibition of hydroxylases in cultured cells

A pharmacologic approach can be adopted to investigate the involvement of hydroxylases in hypoxia-induced NF-κB activation. Dimethyloxalyl glycine (DMOG) (Cayman Chemicals, Ann Arbor, MI) is a cell-permeable 2-oxoglutarate analogue and is thus a nonselective inhibitor of the 2-oxoglutarate–dependent dioxygenase family of enzymes, including PHD1, -2, and -3 and the asparagine hydroxylase, FIH. Cells are grown to confluence on 100-mm petri dishes and treated with vehicle (0.1% dimethyl sulfoxide [DMSO]) or 1 mM DMOG. Cells are maintained in ambient normoxia for 24 h. Following exposure to DMOG, various procedures (outlined later) may be utilized to investigate the impact on NF-κB signaling. An alternative and more specific method for the removal of hydroxylases from cultured cells involves the use of small interfering RNA (siRNA) directed against specific hydroxylase isoforms. RNA interference by siRNA against PHD isoforms is described in detail in a separate chapter in this issue (see Chapter 6).

2.3. Manipulation of NF-κB signaling in cultured cells

The NF-κB pathway is classically responsive to activation through a number of proinflammatory ligand receptors. A prototypic activator of NF-κB is the proinflammatory cytokine TNFα. HeLa cells in culture demonstrate a robust activation of the NF-κB pathway upon stimulation with TNFα (0–10 ng/ml). The sensitivity of the cellular response to TNFα may be manipulated by overexpressing various components of the NF-κB pathway in cells and investigating the impact of this on the basal and stimulated NF-κB response. Having identified a role for PHDs in the regulation of hypoxia-induced NF-κB activation, we sought to investigate the role of specific proline residues (P191 in IKKβ and P190 in IKKα) within LxxLAP motifs of IKKs. We hypothesized that mutation of these proline residues to alanine (which cannot act as hydroxyl-acceptors) would result in altered sensitivity of the NF-κB pathway to TNFα. To do this, we first generated constructs allowing the transient overexpression of wild-type and mutated IKK isoforms in HeLa cells.

In vitro site-directed mutagenesis is a polymerase chain reaction (PCR)-based technique that permits alteration of vector sequences in order to interrogate the functionality of specific residues on recombinant proteins. The Quikchange (Stratagene, La Jolla, CA) site-directed mutagenesis kit was used for these studies. This strategy permits oligo-mediated introduction of site-specific mutations onto the double-stranded plasmid DNA. Template pCMV6-XL5 IKKβ DNA was obtained from OriGene (Rockville, MD). Plasmids were reconstituted in ultrapure dH₂O and used to transform DH5α-competent *Escherichia coli* (Invitrogen, Carlsbad, CA), and plasmid isolation using a Maxi Prep kit (QIAGEN, Valencia, CA) was carried out according to the manufacturer's instructions.

The strategy aimed to design primers that would mutate the proline residue within the LxxLAP consensus sites of IKKα and IKKβ. Conversion of proline to alanine requires a single-base change of CCA to GCA, and alanine was thus chosen as the amino acid to which the proline would be converted. The DNA sequence 25 bases upstream and downstream of the critical proline were analyzed, and mutagenic primers were designed and synthesized (Sigma-Genosys/Sigma-Aldrich, St. Louis, MO). The primers used were as follows:

IKKα:
Forward *5′ GCAGTATCTGGCCgCAGAGCTCTTTGAGAATAAGC3′*
Reverse: *5′ GCTTATTCTCAAAGAGCTCTGcGGCCAGATACTGC 3′*
IKKβ:
Forward: *5′ CCCTGCAGTACCTGGCCgCAGAGCTACTGGAGC3′*
Reverse: *5′ GCTCCAGTAGCTCTGcGGCCAGGTACTGCAGGG 3′*
(Lower case letters denote the target of the mutagenesis reaction)

A PCR reaction was set up using primers to mutate the proline residues within the LxxLAP consensus sequence of IKKβ. This action was done by passing the vector DNA through a series of thermal cycles (95° for 30 s; 16 cycles of 95° for 30 s, 55° for 30 s, and 68° for 8 min, 8 s [IKKα] or 7 min, 27 s [IKKβ] [1 min/kb]). The reaction volume was 50 μl. Parental DNA (25 ng) was first denatured, mutagenic primers annealed to the target site, and a specific PfuTurbo DNA polymerase extended the mutagenic primers, thus generating double-stranded DNA (dsDNA) molecules with the mutation incorporated within one of the strands. The PCR products were then treated for 90 min with DpnI restriction endonuclease that digests methylated, non-mutated, parental DNA. Reaction mixture (5 μl) was subsequently transformed into 50 μl One-Shot supercompetent cells, where the single-stranded DNA is converted into duplex form *in vivo*. The transformants were plated onto LB agar plates with 100 μg/ml ampicillin and incubated at 37° overnight. Colonies were selected from the agar plates, and each colony was grown up in an overnight culture of 5 ml LB/AMP (1% Tryptone [w/v], 0.5% yeast extract [w/v], 1% NaCl [w/v] and 100 μg/ml ampicillin) on a shaking platform at 225 rpm at 37°. One milliliter from the overnight prep was then used, and the DNA was subsequently isolated according to the manufacturer's instructions using a MiniPrep Kit (QIAGEN). The isolated DNA was submitted for sequencing (MWG, Ebersberg, Germany) with the following sequencing primers:

Forward primer: VP1.5   5′ GGACTTTCCAAAATGTCG 3′
Reverse primer: XL39   5′ ATTAGGACAAGGCTGGTGGG 3′.
Incorporation of the mutation was confirmed following sequencing.

Using these constructs, the role of P190/191 in IKKα/β may be investigated using the approaches outlined next.

3. MEASUREMENT OF NF-κB ACTIVITY IN CULTURED CELLS

3.1. NF-κB–luciferase promoter-reporter assay

A straightforward mechanism to determine NF-κB activity in cells exposed to hypoxia involves the introduction of artificial DNA constructs into cells, which encode the gene for firefly luciferase under the control of a con-catemer of NF-κB responsive elements. In a number of studies, transient transfection of plasmid DNA was performed in HeLa cells using FuGENE 6 Transfection reagent (Roche Applied Science, Indianapolis, IN). Prior to transfection, cells were grown to approximately 70% confluence in complete minimum essential medium (Sigma, St. Louis, MO) containing 2 m*M* glutamine, 100 U/ml penicillin, 1× nonessential amino acids, and 10% fetal calf serum (Gibco, Invitrogen, Carlsbad, CA). For each 60-mm petri dish to be transfected, 50 μl of Optimem-1 serum-free media (Gibco), 3 μl of FuGENE 6, and 1 μg of NF-κB–Luciferase plasmid DNA (pNFκB-Luc; Cis-Reporting Systems, Stratagene, Cedar Creek, TX) were mixed gently and incubated for 10 min in a sterile 1.5-ml Eppendorf tube. The mixture was added dropwise to cells in 1 to 2 ml of fresh complete media. Cells were maintained overnight at 37° (5% CO_2). The following morning, the media was replaced prior to treatment.

In an experiment to determine the role of the putative hydroxyl receptor residue Pro190 of IKKα, control vector pcDNA3.1, wild-type IKKα, and P190A mutant IKKα constructs were co-transfected with the NF-κB–luciferase reporter construct into separate plates of cells. Following transfection, cells were treated with 0, 0.01, or 0.1 ng/ml recombinant human TNFα (Sigma-Aldrich, St. Louis, MO) for 24 h. Media was aspirated, and cells were washed once with phosphate buffered saline (PBS). PBS was aspirated, and 200 μl of 1× luciferase lysis buffer (Promega, Madison, WI) was added to the plate. Cells were incubated in lysis buffer for ~5 min, following which cells were scraped. Lysates were transferred to a 1.5-ml Eppendorf tube. Tubes containing lysate were clarified by centrifugation at room temperature at 14,000 rpm for 6 min.

Lysates and reagents were allowed to equilibrate to room temperature. In order to assess levels of luciferase expression, 20 μl of lysate was mixed vigorously in a plastic test tube using a pipette for 5 s with 100 μl of luciferase substrate solution (Promega). The test tube was placed in a desktop luminometer (Junior LB 9509, Berthold Technologies, Oak Ridge, TN), and light units were read over a 20-s period. The cumulative light units were displayed; the readout is relative light units (RLU). RLU readings are normalized according to sample protein content or Renilla-Luciferase controls, as desired. Coexpression of wild-type IKKα caused an increase in basal and TNF-stimulated NF-κB activity when compared to pcDNA control–transfected cells, whereas the P190A mutant IKKα-transfected cells had diminished responses, suggesting a positive regulatory role for Pro190 in regulating IKKα-dependent NF-κB activity (Fig. 25.1).

3.2. NF-κB DNA–binding assay

In order to examine the nature and kinetics of hypoxia-induced NF-κB activation upstream of transcriptional activity, we examined DNA binding of the principal transactivating member of the NF-κB family, p65. The TransAM^{TM} nuclear binding assay (Active Motif, Carlsbad, CA) is an enzyme-linked immunosorbent assay (ELISA)-based technique used to detect and quantify transcription factor activation in a 96-well plate. HeLa cells were grown to confluence on 100-mm petri dishes. Cells were treated with pre-equilibrated normoxic or hypoxic media and maintained in

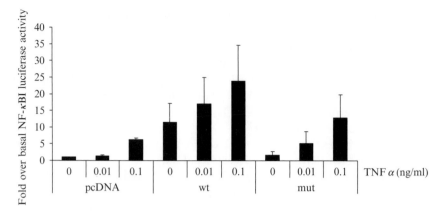

Figure 25.1 Basal and tumor necrosis factor alpha (TNFα)-stimulated NF-κB activity was measured in cells overexpressing control vector pcDNA3.1, wild-type IκB kinase alpha (IKKα) (wt), and P190A mutant IKKα (mt) constructs. Overexpression of wt increased basal and stimulated–NF-κB activity, while overexpression of mt suppressed basal- and stimulated–NF-κB activity. Results shown are foldover basal normalized pcDNA3.1 control values ± SEM for $n = 4$ independent experiments.

ambient normoxia or hypoxia for 4 h. Hypoxia-treated cells were washed, scraped, and maintained on ice prior to removal from the hypoxia chamber to avoid potentially confounding reoxygenation-dependent events.

Nuclear lysates for use in TransAM^{TM} nuclear binding assay were prepared. The media was aspirated and cells placed on ice. Cells were washed with 5 ml of ice-cold PBS containing phosphatase inhibitors (PIBs) (0.5 ml of PIB [125 mM NaF, 250 mM β-glycerophosphate, 25 mM NaVO$_3$] to 10 ml of 1× PBS). PIB was aspirated, and 3 ml of fresh, ice-cold PIB was added to the plate. Cells were disrupted by scraping using a cell scraper and deposited in ice-chilled 15-ml falcon tubes. Cells were then centrifuged for 5 min at 1000 rpm. Supernatant was discarded.

The cell pellet was resuspended in 500 μl of buffer A (10 mM HEPES [pH 8.0], 1.5 mM MgCl$_2$, 10 mM KCl, 200 mM sucrose, 0.5 mM dithiothreitol [DTT], and 0.25% NP-40 [IPEGAL] plus protease inhibitor cocktail [Sigma]). Cells were incubated for 10 min on ice. Lysate was triturated three to four times to disrupt any clumps of cells, collected in a 1.5-ml Eppendorf tube, and stored on ice. Lysate was then centrifuged at 12,000 g for 1 min at 4°. The supernatant is the cytosolic extract. The pellet was then resuspended in 70 μl of complete lysis buffer (1 ml lysis buffer AM2, 5 μl 1 M DTT, 10 μl protease inhibitor) and rocked gently on ice for 30 min. Samples were briefly vortexed before centrifugation at 14,000g for 10 min at 4°. The supernatant is the nuclear extract. Protein determination was carried out on the extracts (BioRad Laboratories, Hercules, CA). Protein-normalized nuclear lysates (20 μl) were then combined with complete binding buffer AM3 and incubated with an immobilized oligonucleotide corresponding to the consensus p65 NF-κB transcription factor–binding site in the wells of the 96-well plate. The nuclear lysate was incubated with the oligonucleotide at room temperature (rt) for 1 h on a shaker platform (100 rpm). Lysate was removed, and wells were washed three times with 200 μl of 1× wash buffer AM2. Anti-p65 NF-κB primary antibody (100 μl; 1:1000 dilution in 1× antibody binding buffer AM2) was then added to each of the wells. This antibody has specificity for an epitope on the transcription factor, which is only exposed when the transcription factor is activated and bound to its target DNA. The antibody solution was incubated in the wells at rt for 1 h without rotation. Antibody solution was then removed, and the wells were washed three times with 200 μl of 1× wash buffer AM2. Anti-rabbit horseradish peroxidase (HRP)-conjugated immunoglobulin G (IgG) (100 μl; 1:1000 dilution in 1× antibody binding buffer AM2) was then added to each of the wells. The antibody solution was incubated in the wells at rt for 1 h without rotation in the dark. Antibody solution was then removed, and the wells were washed four times with 200 μl of 1× wash buffer AM2. Developing solution (100 μl) was added to each of the wells and incubated at rt in the dark for 2 to 12 min. The colorimetric reaction was then terminated using 100 μl stop solution prior to color saturation. Optical density was measured on a microplate spectrophotometer at 450 nm with an optional reference wavelength at 595 nm.

We carried out TransAM^{TM} DNA-binding assays under the following conditions. HeLa cells were grown to confluence on 100-mm petri dishes. Cells were then maintained in normoxia (21% O_2, 37°, 5% CO_2) in a humidified environment for 1 h or exposed to pre-equilibrated hypoxic media and maintained in hypoxia (1% O_2, 37°, 5% CO_2) for 1 h. After 1 h, both sets of samples were treated with 0.0, 0.01, or 0.1 ng/ml TNFα for a further 1 h in their respective conditions. We found hypoxia to enhance TNF-dependent NF-κB signaling (Cummins *et al.*, 2006). In an additional study using the same system, we demonstrate in this chapter that pharmacologic inhibition of PHDs using the pan-hydroxylase inhibitor DMOG (1 mM) also causes a robust activation of NF-κB signaling (Fig. 25.2).

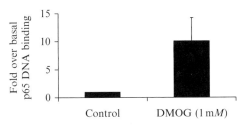

Figure 25.2 TransAM^{TM} DNA binding assay of HeLa cells exposed to vehicle (dimethyl sulfoxide [DMSO]) or the pan-hydroxylase inhibitor dimethyloxalyl glycine (DMOG) (1 mM) for 24 h. Results shown are foldover basal (vehicle) optical density (450 nm ref 595 nm) $\pm$ SEM for $n = 5$ independent experiments.

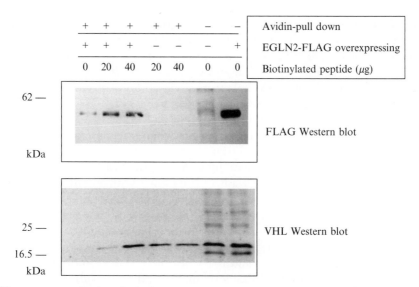

+	+	+	+	+	−	−	Avidin-pull down
+	+	+	−	−	−	+	EGLN2-FLAG overexpressing
0	20	40	20	40	0	0	Biotinylated peptide (μg)

Figure 25.3 Peptide pull-down assay. Whole cell lysates from normal HeLa cells and HeLa cells transiently overexpressing EGLN2-FLAG were incubated with 0, 20, or 40 μg IκB kinase beta (IKKβ)-like biotinylated peptide (Biotin-GSLATSFVGTLQY-LAPELLEQQKY-COOH). Peptide was pulled down using monomeric avidin-agarose, and immunoblotting was carried out using (A) anti-FLAG antibody to detect hydroxylase association with the IKKβ-like peptide or (B) anti–von Hippel-Lindau (VHL) antibody to detect VHL association with the IKKβ-like peptide. In both cases, whole cell lysates that were not exposed to peptide or avidin-agarose from normal HeLa cells and HeLa cells transiently overexpressing EGLN2-FLAG were included as appropriate controls.

3.3. Immunoblotting analysis

In order to examine the kinetics of hypoxia-induced NF-κB signaling upstream of p65 nuclear translocation and nuclear binding, the following experiment was carried out. Hela cells were grown to confluence on petri dishes, exposed to pre-equilibrated hypoxic media, and maintained in hypoxia for 0, 5, 15, 30, 60, or 120 min. Western blotting analysis of IκBα, IKKα, and IKKβ was carried out to examine the expression levels and phosphorylation status of these upstream NF-κB regulatory proteins. Cells were exposed to hypoxia, as previously described, and cytosolic extracts were prepared as follows. The culture media was aspirated from the petri dishes and cells placed on ice. Cells were then washed briefly in ice-cold PBS. The PBS was then aspirated, and 200 μl of cytosolic-extract lysis buffer—buffer A (10 mM HEPES [pH 8.0], 1.5 mM MgCl$_2$, 10 mM KCl, 200 mM sucrose, 0.5 mM DTT, and 0.25% NP-40 [IPEGAL] plus protease inhibitor cocktail) was added to the petri dish (100 mm). Cells were incubated for 10 min on ice before being scraped with a cell scraper. Lysate was triturated three to four times to disrupt any clumps of cells, collected in a 1.5-ml Eppendorf tube, and stored on ice. Lysate was then centrifuged at 12,000g for 1 min at 4°. The supernatant is the cytosolic extract. Cell lysate was stored at −20°. Hypoxia-treated cells were washed, scraped, lysed, and maintained on ice prior to removal from the hypoxia chamber to avoid the potentially confounding reintroduction of oxygen to the samples. Protein determination and normalization was carried out on the extracts. Cytosolic extracts were prepared for electrophoresis in 5× Laemmli buffer (250 mM Tris-HCL [pH 6.8], 10% sodium dodecyl sulfate [SDS], 0.05% bromophenol blue, 25% glycerol, 6% β-mercaptoethanol). Protein samples were denatured by boiling for 10 min and were loaded into pre-cleaned wells using a long-tip pipette. Proteins were separated on a vertical BioRad (Hercules, CA) dual-slab minigel system using a 5% acrylamide stacking gel and a 7.5% acrylamide resolving gel. Electrophoresis was carried out at a voltage of 100 V for 15 min and 120 V until the dye-front reached the bottom of the gel.

Following SDS polyacrylamide gel electrophoresis (SDS-PAGE), gels were removed from the slab minigel system and placed in chilled transfer buffer (48 mM Tris, 39 mM glycine, 20% [v/v] methanol). Fiber pads, filter paper, and nitrocellulose membrane (BioRad) were also moistened in chilled transfer buffer. The fiber pads, filter paper, nitrocellulose membrane, and gels from electrophoresis step were arranged in a "sandwich" configuration. The proteins within the gels were transferred onto the nitrocellulose membrane using a tank transfer system (BioRad) for 80 min at 100 V (constant voltage). Following transfer, nitrocellulose membranes were stained with a Ponceau S solution (Sigma) to confirm successful transfer of proteins and to assess equal loading of the proteins. Ponceau S solution was washed off initially in distilled H$_2$O to reveal protein bands and subsequently in Tris-buffered saline Tween (TBST) (50 mM Tris, 150 mM NaCl, 0.1% Tween 20) for 5 min to clean the membrane fully of residual Ponceau S solution.

Nonspecific protein sites on the membranes were saturated using the appropriate blocking buffer (5% BSA [w/v] [Sigma] TBST or 5% dried milk [w/v] [Fluka] TBST) for 1 h at rt or overnight at 4°. After blocking, the blots were incubated with primary antibody diluted in appropriate blocking buffer. Total IKKα and IKKβ as well as phospho-IKKα(S180)/IKKβ(S181) (Cell Signaling Technology, Danvers, MA) antibodies were diluted 1:1000 in 5% BSA in TBST. Phospho-IκBα(S32/36) (Cell Signaling Technology) antibody was diluted 1:1000 in 5% dried milk (w/v) (Fluka) TBST. Blots were incubated overnight at 4° or at rt for 2 h with mild agitation. Following incubation with the primary antibody, blots were washed for 15 min to 1 h in TBST with at least three changes of wash buffer on an orbital shaker at rt. After washing, the blots were incubated with the appropriate HRP-conjugated secondary antibody (Cell Signaling Technology) and diluted in the appropriate blocking buffer for 60 min with mild agitation at rt. Following incubation with secondary antibody, blots were washed for 15 min to 1 h in TBST with at least three changes of wash buffer on an orbital shaker at rt. The protein bands on the nitrocellulose membrane were visualized using enhanced chemiluminescence (ECL) solution (Santa Cruz Biotechnology, Santa Cruz, CA).

3.4. Peptide pull-down assay

Our work to date has implicated a role for hydroxylases in regulating NF-κB signaling. Mutation of the proline residue (P191) within the LxxLAP motif of IKKβ resulted in altered sensitivity of the kinase to hypoxia. Taken together, this data suggested that the PHDs were acting on and modulating IKKβ through this motif. To test this hypothesis further, we examined whether a biotinylated synthetic peptide analogous to that surrounding the LxxLAP motif of IKKβ could interact with a hydroxylase *in vitro*. Furthermore, we wished to examine whether this peptide could interact with the VHL tumor suppressor protein *in vitro* in an analogous way to that of the HIF-1α LxxLAP motif.

A synthetic IKKβ-like peptide was designed and synthesized (SYN-BIOSCI, Dublin, CA). This peptide was synthetically N-terminally biotinylated and spanned amino acids 176 to 199 of IKKβ (includes Serines 177/181, the LxxLAP motif, and downstream leucine residues). The peptide sequence is identical to that of wild-type IKKβ with the exception of a single cysteine to alanine mutation to avoid spurious oxidation reactions. An approach similar to this previously and successfully has been used to identify hydroxylation of an HIF-1–like peptide (Ivan *et al.*, 2001).

Peptide sequence: Biotin-GSLATSFVGTLQYLAPELLEQQKY-COOH

First, a FLAG-tagged hydroxylase was overexpressed in HeLa cells by transient transfection of pEGLN2 (EGLN2 is also known as PHD1)-FLAG (a generous gift from Dr. William Kaelin, Harvard Medical School, Boston, MA) plasmid DNA, as previously outlined. HeLa cells were grown to 70% confluence on 100-mm petri dishes (four in total for transfection, three were untransfected). pEGLN2-FLAG plasmid DNA (2 μg) was transiently transfected into the cells. Post-transfection (24–48 h), the culture media was aspirated from the petri dishes and cells placed on ice. Cells were then washed briefly in ice-cold PBS. The PBS was aspirated, and 200 μl NETN lysis buffer (20 mM Tris-HCl [pH 8.0], 100 mM NaCl, 1 mM ethylenediaminetetraacetic acid [EDTA], 0.5% NP-40) was added to the dishes for 5 min. Cells were scraped using a cell scraper, and lysate was triturated three to four times to disrupt any clumps of cells, collected in a 1.5-ml Eppendorf tube, and stored on ice. Lysate was then centrifuged at 12,000g for 5 min at 4°. The supernatant was collected. Whole cell lysates corresponding to cells overexpressing EGLN2-FLAG (PHD1-FLAG) and control HeLa whole cell lysates were protein-normalized and divided into aliquots of

500 μg total protein in 1.5-ml Eppendorf tubes. Lysates were then brought to a common volume of 500 μl with NETN lysis buffer. Biotinylated IKKβ-like peptide (0, 20, or 40 μg) was incubated with 500 μg control HeLa or EGLN2-FLAG–overexpressing whole cell lysate at rt for 2 h with rotation (end over end). Pre-washed SoftLink avidin resin (20 μl) (Promega, Madison, WI) was then added to the tubes at rt for 1 h with rotation (end over end) to pull down the biotinylated peptide and any associated proteins. Samples were then centrifuged for 60 s at 4900 g to pellet the avidin resin. The avidin resin was washed three times with 750 μl NETN lysis buffer. Sixty microliters of 1× Laemmli-reducing sample buffer was added to each of the tubes. Samples were boiled on a heating block for 10 min and subsequently separated by SDS-PAGE, as previously described. Immunoblotting was carried out as described previously using small pore-diameter nitrocellulose (0.2 μm) (BioRad). Association of the tagged hydroxylase with the IKKβ-like biotinylated peptide was examined using an anti-FLAG antibody (Sigma) (1:1000 5% BSA in TBST). Association of VHL with the IKKβ-like biotinylated peptide was examined using an anti-VHL antibody (BD Biosciences, San Jose, CA) (1:1000 5% dried milk [Fluka] in TBST). Whole cell extracts from EGLN2-FLAG overexpressing cells and HeLa cells were run alongside the pull-down extracts as controls.

4. SUMMARY/CONCLUSIONS

Tissue hypoxia is an important environmental feature in a number of inflammatory disease states where the activation of NF-κB has significant implications for disease progression. Recent work has begun to shed light on the mechanism by which hypoxia activates this important event. Using the techniques outlined in this chapter, we can gain greater insight into the mechanisms underlying hypoxic activation of NF-κB.

REFERENCES

Berra, E., Ginouves, A., and Pouyssegur, J. (2006). The hypoxia-inducible factor hydroxylases bring fresh air into hypoxia signaling. *EMBO Rep.* **7,** 41–45.

Cockman, M. E., Lancaster, D. E., Stolze, I. P., Hewitson, K. S., McDonough, M. A., Coleman, M. L., Coles, C. H., Yu, X., Hay, R. T., Ley, S. C., Pugh, C. W., Oldham, N. J., *et al.* (2006). Posttranslational hydroxylation of ankyrin repeats in IκB proteins by the hypoxia-inducible factor (HIF) asparaginyl hydroxylase, factor inhibiting HIF (FIH). *Proc. Natl. Acad. Sci. USA* **103,** 14767–14772.

Cramer, T., Yamanishi, Y., Clausen, B. E., Forster, I., Pawlinski, R., Mackman, N., Haase, V. H., Jaenisch, R., Corr, M., Nizet, V., Firestein, G. S., Gerber, H. P., *et al.* (2003). HIF-1α is essential for myeloid cell-mediated inflammation. *Cell* **112,** 645–657.

Cummins, E. P., and Taylor, C. T. (2005). Hypoxia-responsive transcription factors. *Pflugers Arch.* **450,** 363–371.

Cummins, E. P., Berra, E., Comerford, K. M., Ginouves, A., Fitzgerald, K. T., Seeballuck, F., Godson, C., Nielsen, J. E., Moynagh, P., Pouyssegur, J., and Taylor, C. T. (2006). Prolyl hydroxylase-1 negatively regulates IκB kinase-β, giving insight into hypoxia-induced NF-κB activity. *Proc. Natl. Acad. Sci. USA* **103,** 18154–18159.

Haase, V. H. (2006). Hypoxia-inducible factors in the kidney. *Am. J. Physiol. Renal Physiol.* **291,** F271–F281.

Hagen, T., Taylor, C. T., Lam, F., and Moncada, S. (2003). Redistribution of intracellular oxygen in hypoxia by nitric oxide: Effect on HIF-1α. *Science* **302,** 1975–1978.

Ivan, M., Kondo, K., Yang, H., Kim, W., Valiando, J., Ohh, M., Salic, A., Asara, J. M., Lane, W. S., and Kaelin, W. G. (2001). HIFα targeted for VHL-mediated destruction by proline hydroxylation: Implications for oxygen sensing. *Science* **292,** 449–451.

Karin, M., and Greten, F. R. (2005). NF-κB: Linking inflammation and immunity to cancer development and progression. *Nat. Rev. Immunol.* **5,** 749–759.

Kim, K. S., Rajagopal, V., Gonsalves, C., Johnson, C., and Kalra, V. K. (2006). A novel role of hypoxia-inducible factor in cobalt chloride- and hypoxia-mediated expression of IL-8 chemokine in human endothelial cells. *J. Immunol.* **177,** 7211–7224.

Kong, T., Westerman, K. A., Faigle, M., Eltzschig, H. K., and Colgan, S. P. (2006). HIF-dependent induction of adenosine A2B receptor in hypoxia. *FASEB J.* **20,** 2242–2250.

Lando, D., Peet, D. J., Whelan, D. A., Gorman, J. J., and Whitelaw, M. L. (2002). Asparagine hydroxylation of the HIF transactivation domain: A hypoxic switch. *Science* **295,** 858–861.

Murdoch, C., Muthana, M., and Lewis, C. E. (2005). Hypoxia regulates macrophage functions in inflammation. *J. Immunol.* **175,** 6257–6263.

Schofield, C. J., and Ratcliffe, P. J. (2005). Signalling hypoxia by HIF hydroxylases. *Biochem. Biophys. Res. Commun.* **338,** 617–626.

Zhou, J., and Brune, B. (2006). Cytokines and hormones in the regulation of hypoxia inducible factor-1alpha (HIF-1α). *Cardiovasc. Hematol. Agents Med. Chem.* **4,** 189–197.

Author Index

A

Aarbodem, Y., 348
Abai, A. M., 185
Abashkin, Y. G., 182
Abbott, B., 89, 91, 100
Abboud, H. E., 427
Abe, H., 425, 428, 472
Abe, M., 472
Abeloff, M. D., 348
Abraham, R. T., 327, 330
Abramovitch, R., 98, 332
Acar, N., 231
Acker, H., 19, 308, 325, 413, 423, 428, 429, 430
Acker, T., 91, 98, 201, 413, 465
Acquaviva, A. M., 158
Adachi, T., 135, 231, 409, 426, 427
Adak, S., 465
Adam, M., 305
Adams, M. D., 126
Addeo, R., 467, 468
Addison, C. L., 281
Adelman, D. M., 89, 91, 100, 133, 204, 213, 214
Adler, H. S., 375
Aebersold, D. M., 348
Aebi, M., 188
Aebischer, P., 186
Affolter, M., 127, 128
Agami, R., 109, 115
Agani, F., 89, 90, 100, 133, 135, 203, 224, 227, 327, 332, 348, 350, 449
Agani, F. H., 330, 412, 472
Aguet, M., 93
Aherne, W., 330, 391
Ahmet, I., 148
Ahn, B. S., 335
Ahn, C. H., 187
Ahn, D., 148
Ahn, M. Y., 114
Aiello, L. P., 206, 207
Airley, R., 312
Aitola, M., 19
Akazawa, R., 280
Akinc, A., 118
Akiyama, N., 312
Alam, J., 231, 236, 325
Alberch, J., 186
Albina, J. E., 407
Albitar, M., 160

Alcaide, M., 46, 112
Ali, M., 160
Ali, M. A., 330
Ali, R. R., 184, 185
Alitalo, K., 201, 237
Alkayed, N. J., 233
Allegrini, P. R., 148
Allen, D., 231
Allen, W. E., 223, 224, 332
Allshire, R. C., 63
Alon, N., 376
Alonzo, E. J., 209
Alt, F. W., 213
Altieri, A. S., 10
Altman, R., 324, 325
Aman, P., 362
Amanatides, P. G., 126
Amann, K., 130, 229, 235
Ambs, S., 466, 471
Ameln, H., 236
Ameri, K., 279, 281, 312, 406
Amersi, F., 231
Ames, C., 98, 100, 233
Amezcua, C. A., 6, 7, 9, 20
Amidi, M., 423, 425
Aminova, L., 230, 233
An, J., 279
Anagnostou, A., 207
Anaya, J. M., 365, 366
Anderson, D., 310
Anderson, K. C., 281
Anderson, M. G., 126
Anderson, W. F., 277
Andersson, I., 28, 469
Andre, H., 237
Andreeva, N., 425
Andrew, D. J., 126
Ang, S. O., 184, 372
Angelidakis, D., 348
Anlezark, G. M., 310
Annex, B. H., 151
Antonarakis, S. E., 158, 160, 168, 173, 181, 332
Antonini, J., 327, 426, 430
Apasov, S. G., 406, 407
Aplin, R. T., 63, 77
Appelhoff, R. J., 27, 46, 63, 79, 111, 115, 413
Aprelikova, O., 46, 116, 230
Aragones, J., 98
Arai, F., 212

Arai, M., 281
Arai, T., 409
Araki, K., 202
Arany, Z., 223, 408, 425, 428
Aravind, L., 6
Arbeit, J., 4
Arbeit, J. M., 89, 96, 97, 98, 100, 204, 237, 276
Arciero, D. M., 469
Arenberg, D. A., 357
Arend, M., 130, 229, 235
Arenzana, N., 280
Arison, R. N., 314
Armour, S. M., 47
Armstrong, J., 406
Armutcu, F., 236
Arnold, S. M., 278, 300
Arnould, T., 409
Arquier, N., 131
Arsham, A. M., 91, 254, 276, 377
Asara, J., 372
Asara, J. M., 4, 26, 27, 36, 88, 89, 110, 324, 327,
    425, 449, 464, 490
Ascensao, J. L., 159
Ascher, P. W., 314
Ash, R. C., 159
Ashburner, M., 125
Ashcroft, M., 4, 328, 330, 377, 391
Askham, Z., 184, 185, 312
Asman, M., 209, 332
Aster, J. C., 69
Atadja, P., 392
Ataka, K., 185
Athauda, G., 374
Atwell, G. J., 310, 311
Aulmann, M., 148
Aust, C., 184
Austin, E. A., 312
Avery, R. L., 206
Ayoub, I. A., 230, 233
Azpiazu, N., 135

                          B

Baader, E., 229
Baba, M., 374, 375, 377
Babior, B. M., 423
Bach, F. H., 231, 236
Bacher, M., 357
Bachmann, S., 160, 181, 225, 227, 229, 230, 232,
    233, 235, 236, 237
Bachmann, U., 231
Bachtiary, B., 348
Back, T. C., 377
Bacon, N. C., 129
Bacon, N. C. M., 27
Baddeley, J., 310
Badrawi, S. M., 173, 188

Bae, J. M., 330
Bae, M. H., 114
Bae, M. K., 114
Baek, J. H., 46, 114, 348, 349, 351, 373
Baer, A. N., 407
Bagci, C., 236
Bagg, A., 185, 186
Baglioni, C., 109
Bahlmann, F. H., 231
Bailey, D. M., 158
Baker, J. E., 231
Bakker, A., 406
Balamurugan, K., 43
Balanos, G. M., 372
Baldwin, J. E., 28, 30, 469
Ballschmieter, P., 374
Balrak, W., 79, 332, 378
Bang, B. K., 148
Bang, Y. J., 426
Banisadre, M., 159
Banning, A., 424
Baranova, O., 181
Barat, A., 237
Bardor, M., 189
Bardos, J. I., 328
Bariety, J., 160
Baringhaus, K. H., 229
Barnhart, K. M., 185
Barr, E. W., 71
Barrett, J. C., 46, 116, 230
Barry, R., 374
Barry, R. E., 376
Bartel, D. P., 109
Barth, S., 43, 47, 54
Bartholomew, A., 186
Bartlett, S. M., 27, 129
Barton, A., 365, 366
Barton, G. J., 7, 8, 469
Bartram, M., 374
Bartz, K. V., 172
Bartz, S., 378
Bartz, S. R., 112, 230
Basheer, S., 190
Bassler, B. L., 19
Bastidas, N., 201, 325
Batchelder, R. M., 308
Batmunkh, C., 182
Bauer, C., 154, 160, 182, 224, 227, 228, 233, 449
Baugh, J. A., 364, 365
Baunoch, D., 89, 91, 204, 227, 332, 449
Bax, A., 10
Bayko, L., 327, 328
Beach, D., 109
Bean, J. M., 305
Bean, M., 229, 230, 234
Beart, P. M., 232
Beaty, T., 93
Bechmann, I., 225, 227, 229

Beck, H., 91, 97, 201
Beck, I., 88, 100, 424
Becker, A., 305
Beckman, B. S., 172, 325
Beckman, J. S., 435, 466
Beckman, T. W., 466
Beckmann, J. S., 188
Bedogni, B., 325, 326, 364
Beer, K. T., 348
Begent, R. H., 315
Behar, K. L., 134
Behrens, J., 348
Behrooz, A., 327, 328
Beischlag, T. V., 19
Beitner-Johnson, D., 135
BelAiba, R. S., 410, 413, 423, 424, 425, 426, 427, 430
Belawat, P., 135
Belda, E., 237
Beliard, R., 189
Bell, E. L., 447, 448, 449, 470
Bell, J. C., 249, 276, 277, 279, 281
Bellamy, W. T., 330, 428
Bellorin-Font, E., 209
Belozerov, V. E., 329
Bemis, L., 114, 362, 363
Benizri, E., 27, 46, 111, 112, 230, 331, 376
Bennett, J., 117
Bennett, M., 95, 182, 211, 212, 225, 235
Bennett, M. J., 90, 95, 182, 211, 212, 227, 332
Bennewith, K. L., 248, 325, 326
Benov, L., 425
Bensinger, S. J., 88, 100
Bentley, M. D., 426
Bento, M. C., 372
Bentzen, S. M., 305
Benvin, S., 109
Ben-yosef, Y., 201
Benzing, T., 374
Berchner-Pfannschmidt, U., 19, 47, 405, 410, 412, 413, 426, 427, 428, 429, 430, 471
Berenfeld, L., 407
Berg-Dixon, S., 348, 350
Bergeron, M., 232, 233
Bergers, G., 99, 378
Berglund, H., 7
Bergman, B., 309
Berkenstam, A., 209, 332
Bermudes, D., 314
Bernabeu, C., 182
Bernards, R., 109
Bernasconi, S., 201
Bernaudin, M., 232
Berndt, A., 7, 12, 19
Bernhagen, J., 362, 363
Bernhard, E. J., 330
Bernhardt, W., 27, 47, 226, 229
Bernhardt, W. M., 130, 221, 225, 229, 230, 235

Bernstein, E., 109
Berra, E., 27, 46, 47, 107, 111, 112, 135, 230, 327, 328, 330, 331, 376, 407, 409, 413, 414, 426, 427, 429, 430, 464, 471, 480, 481, 487
Berrigan, D., 124
Bert, A. G., 410
Bertilsson, G., 209, 332
Bertolotti, A., 279, 280, 282
Bertschinger, M., 188
Berx, G., 373
Besuchet, S. N., 188
Bethge, L., 32, 64, 71, 72, 75
Bettan, M., 185
Betten, M., 373
Beuzard, Y., 185, 186
Bhattacharya, S., 4, 30, 44, 62, 63, 72, 89, 115, 223, 226, 328, 378, 408, 425, 429
Bhoumik, A., 46
Bi, M., 252, 254, 275, 279, 281, 283, 287
Bianchi, M., 154
Bianchi, R., 154
Bibby, M. C., 310
Bickel, C., 423
Biju, M. P., 95, 100, 181, 182, 224, 227, 234, 236, 407
Bilton, R., 114
Bindra, R. S., 325
Binley, K., 184, 185, 312
Birchmeier, W., 348
Birle, D. C., 330
Birmingham, A., 330, 392
Birner, P., 348
Bisgrove, B. W., 373
Biswas, S. K., 201
Bitterman, H., 201
Blacklow, S. C., 69
Blagosklonny, M. V., 182
Blais, J., 252, 254, 279, 281, 283, 287
Blais, J. D., 249, 276, 277, 279, 281
Blanca, I., 209
Blanchard, K. L., 158
Blancher, C., 330
Blankenship, C., 372, 376
Block, K., 427
Blouin, C. C., 408, 409, 428
Blouw, B., 98, 99, 100, 233, 378
Blumenfeld, N., 186
Blydt-Hansen, T. D., 231
Bobrowicz, P., 188
Bocca, S. N., 27, 129, 130, 131, 132
Boddy, J. L., 365
Boehm, S. M., 231
Boelens, R., 7, 12, 20
Bohensky, J., 408
Bohl, D., 185, 186
Bohn, M., 184
Bohrer, W., 160
Bokoch, G. M., 423, 424

Boland, M. P., 310
Bolli, R., 222
Bollinger, J. M., Jr., 71
Bolontrade, M., 97
Bonanou, S., 8
Bondesson, M., 70, 76
Bondke, A., 148, 154
Bondurant, M. C., 160, 181
Bondy, S. C., 434
Bonello, S., 410, 413, 423, 425, 426, 427, 430
Bonifacino, J. S., 372
Bonizalzi, M. E., 229
Bonner, W. M., 19
Bono, F., 98, 332
Bonomi, C. A., 313, 388, 389
Bonventre, J. V., 222
Booker, G. W., 64, 70, 75, 82
Booth, G., 90, 91
Borgel, S. D., 313, 388, 389
Borghaei, R. C., 428
Borkhardt, A., 367
Borriello, A., 372
Bos, R., 348
Bosch, A., 185
Bosch, M., 186
Bosch-Marce, M., 93
Bosenberg, M., 89, 94, 97, 375, 377
Bosze, J., 99
Botella, L. M., 182
Bottaro, D. P., 374
Bouchier-Hayes, D., 183
Boulahbel, H., 47
Bourel, D., 189
Boven, E., 426
Boviatsis, E., 348
Bowtell, D. D., 46
Boxer, G. E., 314
Boyer, M., 309
Boyer, N., 51, 53, 230
Bozoukova, V., 185
Bracken, C. P., 79, 332, 378, 414
Bradfield, C. A., 7, 89, 91, 92, 204, 227, 332, 377, 449
Bradshaw, J., 109
Brahimi-Horn, M. C., 114, 115
Bramlage, B., 118
Brand, A. H., 126
Brandes, R. P., 410, 412, 423, 425, 426, 427, 429, 430, 471
Brant, A., 357
Brantjes, H., 115
Brauch, H., 465
Braun, R. D., 300
Braunschweig, T., 313, 388, 389
Brecht, M., 428, 429
Brechtel, K., 425
Brehn, J. K., 314
Breier, G., 89, 93, 201

Breitenecker, G., 348
Breiter, N., 184
Brenner, M., 183
Brenner, M. C., 229, 230, 234
Bridgewater, J. A., 315
Briehl, M. M., 330, 428
Brigelius-Flohe, R., 424
Briggs, J., 180
Brill-Almon, E., 187
Brines, M., 159, 165, 187, 231, 232
Britti, D., 230
Brizel, D. M., 300, 305
Brogiolo, W., 135
Brooks, B. A., 5, 14, 424
Brooks, D. J., 182
Brooks, J. T., 372
Broome, P. M., 364
Broome Powell, M., 325, 326
Brorson, J. R., 91
Brower, C. S., 377
Brown, E. J., 95, 182
Brown, J., 98, 100
Brown, J. A., 281
Brown, J. M., 4, 277, 281, 297, 298, 299, 300, 302, 303, 304, 305, 306, 307, 308, 309, 312, 314, 315
Brown, P., 231
Brown, T. G., 232
Browne, G. J., 255
Browne, J. K., 173, 188
Brudler, R., 6
Brueckl, C., 430
Bruegge, K., 182
Brugarolas, J., 136, 254
Brugarolas, J. B., 327
Bruick, R. K., 3, 4, 6, 8, 10, 13, 14, 18, 27, 29, 44, 46, 47, 62, 63, 67, 68, 75, 88, 89, 110, 114, 116, 129, 131, 182, 226, 230, 324, 328, 378, 393, 395, 413, 425, 429, 449, 464
Brummelkamp, T. R., 109, 115
Brune, B., 47, 231, 327, 407, 410, 412, 426, 428, 429, 463, 467, 468, 469, 471, 473, 480
Brunelle, J. K., 448, 449
Bruneval, P., 160
Bruning, U., 479
Brunnelle, J. K., 470
Brunner, H., 362, 363
Brusselmans, K., 90, 91, 93, 97, 98, 201, 332
Bryant, B., 314
Bucala, R., 357, 362, 364
Buchner, C., 185
Buck, L. T., 248, 276
Buck, M. R., 47
Buck, M. R. G., 28, 35
Buckley, A., 364
Buckley, M. R., 89, 94, 97, 375, 377
Budd, S. L., 435
Budde, A., 468

Budinger, G. R., 454
Budka, H., 348
Buechler, P., 311, 348, 349
Buelow, R., 231
Buju, M. P., 88, 100
Bullock, A. N., 4, 30, 44, 62, 63, 72, 89, 226, 328, 378, 425, 429
Bumcrot, D., 118
Bump, E. A., 302
Bunger, M. K., 92
Bungert, J., 172
Bunn, H. F., 129, 131, 158, 172, 182, 191, 223, 408, 425, 428, 472, 473
Bunz, F., 332
Burchard, J., 112
Burd, R., 307
Burger, A., 98
Burger-Kentischer, A., 362, 363
Burghard, H., 229
Burk, R., 372, 373, 376
Burk, R. D., 374, 377, 429
Burke, B., 312, 406
Burkhardt, T., 227, 228
Burley, S. K., 9, 19
Burlington, H., 160
Burn, T. C., 63
Burri, P., 348
Burry, T. N., 349, 373
Burt, S. K., 182
Bush, A., 230, 233
Busse, R., 410, 412, 423, 425, 426, 427, 429, 430, 471
Busuttil, R. W., 160, 231
Butler, D., 28, 29, 32, 33, 35, 39
Butler, T. W., 309
Butman, J. A., 237, 372
Buttrick, P. M., 229, 230, 234
Buyaner, D., 186
Byrd, R. A., 10
Byrne, A., 364
Byrne, T., 188

C

Cachau, R. E., 182
Cadenas, S., 473
Cadwell, C. C., 213
Cai, H., 423
Cai, Z., 234
Cairns, R. A., 112, 473
Calandra, T., 362, 363
Calcedo, R., 185, 186
Caldas, G., 372
Caldwell, C., 407
Caldwell, C. C., 406, 407
Calfon, M., 279, 280
Callaghan, M. J., 201, 325
Callapina, M., 412, 428, 471, 473
Calvani, M., 329, 330, 391, 393, 395

Calzada, M. J., 373
Camenisch, G., 43, 44, 46, 47, 54, 112, 182, 230, 423, 425
Camenisch, I., 182
Camilleri, J. P., 160
Campean, V., 130, 229, 235
Campo, L., 365
Campochiaro, P. A., 206, 207
Campos, M. E., 32
Candinas, D., 227, 228
Cantera, R., 127
Cao, H., 277, 281, 307
Cao, R., 209, 332
Cao, Y., 209, 304, 307, 332
Capla, J. M., 201, 325
Caplice, N., 426
Capobianco, A., 231, 232
Carcagno, C. M., 188
Card, P. B., 6, 8, 10, 13, 18, 19, 393, 395
Cardellina, J., 20, 329, 330, 391, 395, 397
Cardellina, J. H., 391, 393, 395, 399
Cardellina, J. H. II, 329, 330
Cardenas-Navia, L. I., 248
Cardona, A., 185
Cardone, L., 7, 12, 19
Carducci, M. A., 392
Carey, R. W., 314
Carlini, R. G., 209
Carmeliet, P., 89, 90, 93, 98, 201, 205, 213, 332, 407
Caro, J., 88, 100, 160, 181, 327, 330, 392, 408, 424, 425, 426, 428, 449, 464, 470
Carrero, P., 5, 425, 428
Carroll, K., 180
Carroll, V., 330, 377, 391
Carskadon, S., 357
Carter, B. D., 375, 376
Carter, J. J., 63
Carter, J. P., 313, 388, 389
Carver-Moore, K., 89, 93, 100, 204
Casadevall, N., 160, 185, 186, 187
Casagrande, S., 231, 232
Casanova, J., 126
Cash, T. P., 250, 251, 254, 255, 276, 279, 448, 470
Castellanos, M. C., 112, 349, 373, 374
Castilho, R. F., 435
Castilla, M. A., 237
Castranova, V., 327, 426, 430
Cathew, R., 126
Caudy, A. A., 109
Caughey, G. E., 406
Cavallaro, U., 348
Cavet, G., 112
Celniker, S. E., 126
Centanin, L., 27, 129, 130, 131, 132, 133, 136, 137
Cepek, L., 184

Ceradini, D. J., 201, 325
Cerami, A., 159, 165, 187
Cerniglia, G. J., 330
Ceylan, H., 236
Chachami, G., 8
Chade, A. R., 426
Chait, B. T., 6, 9
Chamovitz, D. A., 362, 364
Chan, D. A., 46, 47, 114, 133, 323, 324, 327, 328, 329, 356, 362, 363, 413, 429, 464, 471
Chan, M. K., 6
Chandel, N. S., 327, 426, 447, 448, 449, 454, 458, 470, 471
Chandra, A., 349, 373, 374
Chandramouli, G. V., 46, 116, 230
Chandramouli, N., 72
Chang, G. W., 4, 129, 183, 226, 327, 425, 449, 464
Chang, K., 109
Chang, K. J., 236
Chang, K. T., 372
Chang, P. R., 430
Chang, Y. S., 148
Chaplin, D. J., 298, 300, 304
Chapman, J. D., 4
Chapman-Smith, A., 5, 8
Chappell, J. B., 435
Charbonneau, M., 412
Charnock, J. M., 28, 30
Chatterjee, P. K., 230
Chau, N. M., 330, 391
Chau, V., 327, 328, 425
Chauhan, V., 277, 278, 281
Chavez, J. C., 181, 230, 233, 472
Chelmicki, T., 427
Chen, C. F., 236, 327, 328
Chen, E., 135, 327, 328, 330
Chen, E. Y., 277, 327, 328, 330, 430, 449
Chen, F., 332
Chen, H., 89, 93, 100, 184, 204, 207, 327, 372, 448, 470
Chen, J., 213, 466
Chen, K. K., 173, 188
Chen, L., 116
Chen, P., 407
Chen, Q., 134
Chen, T., 72
Chen, W., 128
Chen, X., 114, 277, 280, 362, 363
Chen, Y., 19, 281, 430
Chen, Y. J., 430
Cheng, L. S., 376
Chenuaud, P., 185, 186
Cherel, Y., 185, 186
Cherny, R. A., 230, 233
Chesney, J., 357
Cheung, C., 148
Chi, J. T., 325, 326

Chi, S. M., 158, 160, 173, 181, 332
Chiang, G. G., 327, 330
Chiarugi, P., 430
Chien, W., 29, 30
Chiles, K., 135, 330
Chilvers, E. R., 409
Chirat, F., 189
Chisholm, V., 180
Chitnis, S., 364
Chittick, P., 126
Chiu, D., 94
Cho, E. J., 376
Cho, W. K., 377
Cho, Y. S., 313, 392, 426, 427
Chodosh, L. A., 204, 332, 424
Choi, A. M., 231, 236
Choi, B. H., 187
Choi, B. K., 188
Choi, B. S., 148
Choi, E., 330
Choi, Y., 426, 427
Chong, W., 201
Chou, A., 330
Chou, T. B., 126
Chow, V. W., 349, 373
Chowdhury, R., 27, 28, 226
Choyke, P. L., 237, 372
Chrast, R., 168
Christie, J. M., 7
Christofori, G., 348
Christou, H., 472
Chu, C. K., 63
Chu, F. F., 5, 14, 424
Chun, Y. S., 313, 330, 392, 426, 427
Chung, H. S., 413
Chung, P., 280
Chung, T. D., 72
Chung, W. J., 426, 427
Chung, Y. L., 374, 375
Chung-Faye, G., 310
Churchill-Davidson, I., 298
Cirillo, P., 372
Clackson, T., 185
Claffey, K., 375
Clamp, M. E., 7, 8
Claridge, T. D., 28, 35, 36, 47, 63, 328, 356, 429
Clark, J., 310
Clark, P., 309
Clark, R. A., 427
Clark, S. G., 280
Clausen, B. E., 94, 100, 213, 227, 406, 480
Clauss, M., 201
Cleaver, O., 201
Cleland, L. G., 406
Clendenin, C., 92
Clevers, H., 115

Clifford, S. C., 4, 129, 183, 226, 327, 375, 378, 425, 449, 464, 465
Clifton, I. J., 27, 28, 30, 32, 33, 35, 39, 47, 63, 226, 469
Clore, G. M., 10
Clowes, A. W., 186
Clowes, M. M., 186
Cobb, L. M., 310
Cockman, M. E., 4, 55, 64, 65, 79, 129, 183, 226, 230, 327, 375, 378, 425, 449, 464, 465, 481
Coffer, P., 375
Cohen, B., 135, 425
Cohen, H., 375
Cohen, M. V., 234
Cohen, S. L., 6, 9
Cohn, Z. A., 431
Cointe, D., 189
Coito, A. J., 231
Cole, R. N., 114
Cole, S., 310
Coleman, M. L., 55, 64, 65, 79, 230, 481
Coles, B., 310
Coles, C. H., 55, 64, 65, 79, 230, 481
Colgan, S. P., 96, 308, 325, 406, 407, 412, 480
Collen, D., 90, 93
Collingridge, D. R., 304
Collins, I., 330, 391
Collins, M. K., 315
Collis, S. J., 392
Comerford, K. M., 308, 325, 414, 479, 481, 487
Comoglio, P. M., 374
Compernolle, V., 90, 91, 93, 201
Conaway, J. W., 4, 44, 226, 229, 234, 377, 425, 429, 449
Conaway, R. C., 4, 44, 226, 229, 234, 377, 425, 429, 449
Condliffe, A. M., 409
Condorelli, F., 201
Conger, A. D., 298, 300
Conklin, D. S., 117
Conley, L. H., 5, 14, 424
Connolly, E., 250, 251, 254, 255, 262
Connor, J. R., 230, 233
Connors, T. A., 310
Conrad, P. W., 135
Conti, C. J., 97
Cook-Johnson, R. J., 406
Cooley, L., 126
Cope, G. A., 363
Copeland, N. G., 159, 162, 168
Copeland, R. A., 72
Corke, K. P., 312, 406, 407
Corle, C., 99, 325, 326
Cornell-Kennon, S., 313, 330, 331, 392, 397, 398
Cornesse, Y., 425
Corr, M., 94, 100, 213, 227, 406, 480
Corry, J., 310
Corum, L., 327, 426, 430

Costa, M., 182, 327
Costas, M., 27, 29, 30, 38
Cotes, P. M., 173
Coumailleau, P., 425, 428
Court, D. L., 159, 162, 168
Covell, D. G., 386
Covello, K. L., 133, 377
Cowburn, A. S., 409
Cowden, K. D., 92
Cowen, R. L., 312
Cowfer, D., 180
Crabtree, H. G., 300
Cramer, T., 94, 95, 100, 213, 227, 406, 408, 409, 480
Cramer, W., 300
Craven, C. J., 7, 12
Creighton-Gutteridge, M., 329, 330
Crespo, P., 375
Crews, S. T., 5, 127, 128, 129
Crimin, K., 114
Criscuolo, M. E., 188
Cronkite, E. P., 160
Crouzet, J., 185
Crowder, R., 376
Crystal, R. G., 184, 436
Cucciolla, V., 372
Cuevas, Y., 46, 112
Cuff, J. A., 7, 8
Cui, J., 98, 100, 233
Cui, L., 234
Culcasi, M., 437
Cullen, B. R., 112
Cummins, E. P., 413, 414, 479, 481, 487
Cummins, J. M., 332
Cunliffe, C. J., 35
Cunningham, J. M., 158
Cunningham, M. R., 63
Currens, M. J., 386
Currie, M. J., 201
Curth, K., 135
Curtin, P. T., 181
Cutrone, E. C., 80
Cuttitta, F., 329, 330, 392
Cuzzocrea, S., 230

D

Dachs, G. U., 276, 304, 312
D'Acquisto, F., 467, 468
Dadak, A., 96, 204, 233, 332
Dahm, K., 302
Daiber, A., 471
Dale, D. C., 186
Dalgard, C. L., 47
Dalle, B., 186
Damaert, A., 201
D'Amato, R., 205, 206
Damdimopoulos, A. E., 19

Dame, C., 172
Damert, A., 89
Dames, S. A., 64, 397
Dammacco, F., 207
D'Amore, P. A., 98, 205, 206
D'Andrea, R. J., 211
Danel, C., 184
Dang, C. V., 120
Dang, D. T., 332
Dang, L. H., 332
D'Angelo, G., 51, 53, 131, 230, 327
Dann, C. E. III, 4, 63
Danos, O., 184, 186
Dansi, P., 426
Dapp, C., 236
Darnell, J. E., Jr., 386
Das, D. K., 234
Da Silva, J. L., 160
Datta, K., 115
Datta, V., 95, 408
Davidowitz, E. J., 372, 373, 376, 377, 429
Davidson, G., 327
Davidson, R. C., 188
Davidson, T. L., 327
Davies, B., 158
Davis, D. A., 332
Davis, R. J., 231, 236
Davis, S., 205
Davis, T. O., 314
Davis-Smyth, T., 201
Day, D., 312
Dayan, F., 115
De, B., 231
de Alwis, M., 184, 185
Deans, R., 186
De Castro, M., 186
Deccache, Y., 327
de Ceaurriz, J., 185
Dechant, G., 375
Decker, J., 376
Declercq, C., 89, 93
Dedner, D., 184
Degenkolbe, D. M., 311
De Giacomo, A. F., 281
Déglon, N., 186
Degraff, W., 466, 471
De Groot, K., 231
De Guzman, R. N., 64, 397
Dehghani, F., 423, 425
Dehne, M. G., 32
Deisenhofer, J., 63
Dekanty, A., 123, 129, 131, 132, 135, 136, 138
de la Monte, S. M., 63
de la Vega, M. C., 254
Della, N., 206
Dell-Era, P., 207
del Peso, L., 46, 112
De Marzo, A. M., 311, 348, 349

Demasi, M., 406
Demazy, C., 327
Dembrowski, C., 184
Demiryurek, A. T., 236
Demol, H., 46
Denavit-Saubie, M., 425
Dendorfer, A., 425
Denekamp, J., 304, 305
Deng, C., 281
Deng, X., 426
Deng, X. W., 362, 364
Denham, J., 310
Denko, N. C., 112, 277, 281, 307, 324, 325, 327, 328, 473
Denny, W. A., 307, 308, 311, 312, 329, 330
de Noronha, R. G., 330
Depinho, R. A., 89, 183
De Ponti, C., 428
de Ramon, E., 365, 366
Derewenda, Z., 7
DeRisi, J., 280
Derix, N. M., 7, 12
Derouazi, M., 188
Dervan, P. B., 395
Dery, M. A., 114
Desbaillets, I., 227, 228, 304
Deshaies, R. J., 363
Desiderio, M. A., 426, 428
Dessypris, E. N., 407
Detwiller, K. Y., 98
Deudero, J. J., 237
Devor, D. E., 89, 92, 100
de Vries, S. C., 6
Dewerchin, M., 90, 91, 97, 98, 201, 332
Dewhirst, M. W., 117, 300, 304, 305, 307, 325
de Wit, C., 148, 149, 150
Dhanda, A., 4, 27, 29, 30, 36, 44, 46, 88, 89, 110, 112, 131, 226, 229, 324, 429, 449
Dhar, D. K., 367
Diaz, S., 189
Dicker, A. P., 307
Dickson, B., 127
Didrickson, S., 97, 100
Diebold, I., 410, 412, 426, 427, 429, 430, 471
Diehl, A., 251, 252, 253, 256, 276, 277, 278, 279, 281, 282, 283, 284
Diehl, R. E., 63
Diemer, K., 423
Dietz, R., 225, 230, 235
Diez, R., 114
Diez-Juan, A., 98
Digicaylioglu, M., 148, 232
Dikalov, S. I., 427, 437
Dikshit, M., 465
Dillon, M. E., 124
Dinchuk, J. E., 63
Dinenno, F. A., 229

Ding, K., 13, 14, 18, 90, 95, 182, 211, 212, 227, 332, 393
Ding, M., 327, 426, 430
Dinh, D. H., 117
Di Toro, D. M., 327
Dixon, R., 6
Dixon, R. A., 63
Djonov, V., 348
Djordjevic, T., 410, 413, 423, 424, 425, 426, 427, 430
Doe, B. G., 160
Doe, C. Q., 375
Doedens, A., 87, 95, 408
Doege, K., 46, 112, 184, 230, 448, 473
Doherty, A., 302
Doi, M., 7, 12, 19
Doi, T., 181
Doll, C. J., 248, 276
Dominguez, J., 209
Dominiak, P., 425
Donald, S. P., 47, 182, 230
Dong, Z., 236
Doni, A., 201
Donn, R., 365, 366
Donnelly, S. C., 364, 365
Donzelli, S., 466, 471
Dor, Y., 91, 98, 201, 332
Dorie, M. J., 309
Dorner, A. J., 279
Dornhofer, N., 325, 326
Dorrington, K. L., 182
Dos-Santos, B., 182
Dossenbach, C., 127
Dostie, J., 249, 250, 252, 253, 254, 255, 262, 265, 276, 278
Double, J. A., 310
Doucette, D., 372
Dowbenko, D., 212
Dowd, M., 89, 93, 100, 204
Doweiko, J., 172
Downes, M., 310
Downey, H. F., 223, 234
Downey, J. M., 234
Downing, M. R., 173
Dragatsis, I., 98, 100, 233
Dragsten, P. R., 182
Dranitzki-Elhalel, M., 187
Dreier, B., 348
Driver, S. E., 108
Drysdale, L., 91, 201
Duan, L. J., 46, 89, 92
Duarte, C. A., 162
Dubois, C. M., 412
Ducker, K., 127
Duckert, T., 231
Dumas, S., 425
Dumont, V., 327

Dunst, J., 305
Dunwoodie, S. L., 63
Duplan, E., 51, 53, 131, 230
Dupraz, P., 186
Durand, R. E., 248, 298, 300, 307
Düx, P. E., 20
Duyndam, M. C., 426
Dvorak, H., 375
Dyson, H. J., 64, 397
Dzau, V. J., 230
Dzus, A. L., 407

E

Eberhardt, C., 89, 93
Ebert, B. L., 89, 94, 97, 128, 182, 375, 377
Ebert, M., 298, 300
Ebner, A., 128
Eckardt, K. U., 27, 47, 116, 160, 181, 191, 221, 223, 224, 225, 226, 227, 229, 230, 231, 234, 235, 236, 237, 332
Ecker, J. R., 362, 364
Eckhardt, K., 47, 54, 182, 237
Eckner, R., 223, 408
Edery, I., 5, 14
Edgar, B. A., 136, 137, 138
Edge, R., 281
Edmunds, S. J., 310, 311
Egawa, K., 330
Egginton, S., 236
Egido, J., 237
Egrie, J. C., 173, 188
Ehrenreich, H., 184
Ehrismann, D., 27, 28, 33, 34, 35, 37, 46, 47, 62, 63, 64, 72, 77
Ekatodramis, D., 148
El Assal, O. N., 367
Elbashir, S. M., 109
Elkins, J. M., 4, 30, 44, 62, 63, 64, 72, 75, 77, 89, 226, 328, 378, 425, 429
Ellinghaus, P., 425, 464
Elliot, S. G., 188
Elliott, S., 188
Ellis, H. M., 168
Ellis, L. M., 135, 426
Ellisen, L. W., 136
Ellison, J. A., 98, 100, 233
Ellison, L. O., 309
Elliston, K. O., 63
Elson, D. A., 89, 96, 97, 98, 100, 204, 237, 276
Elson, G., 469
Elson, L. A., 310
Eltzschig, H. K., 407, 480
Elwell, J. H., 314
Ema, M., 7, 8, 17, 89, 166, 201, 425, 428, 472
Emi, M., 281
Emmert-Buck, M. R., 89, 92, 100

Endler, A., 116
Endre, Z., 148
Engel, J. D., 159, 172
Engelbart, K., 314
Engelman, R. M., 234
Engle, J. M., 32
Enholm, B., 201
Epp, E. R., 306
Eppler, S., 180
Epstein, A. C., 4, 27, 29, 30, 36, 44, 46, 88, 89,
    110, 112, 131, 226, 229, 324, 429, 449, 464
Erbel, P. J., 6, 8, 10, 13, 14, 18, 19, 393, 395
Erdag, S., 184
Erdjument-Bromage, H., 46, 468
Erdmann, V. A., 116, 181, 234, 332
Eremina, V., 237
Eriksson, U., 201
Erler, J. T., 325, 326
Ernest, I., 409, 414
Escuin, D., 313, 330, 392
Esko, J., 96, 204, 209, 211
Espey, M. G., 466, 469, 471
Espinosa, M. B., 414
Esteban, M. A., 349, 373, 374
Esumi, H., 467, 468
Etages, S. D., 114
Ettenger, R., 231
Evans, A. J., 349, 373
Evans, C. A., 126
Evans, J. W., 309
Evans, S. M., 4, 325, 330
Evert, G., 98

F

Fabbro, D., 207
Fahner, J., 237
Fahrig, M., 89, 93
Faigle, M., 480
Fales, H. M., 435
Falls, T., 281
Fan, D., 430
Fan, F., 135, 426
Fan, J., 376
Fandrey, J., 19, 47, 148, 172, 180, 182, 327,
    405, 407, 408, 409, 410, 412, 413, 416,
    428, 468, 469
Fang, H. M., 88, 100
Fang, J., 63, 330, 426
Farahi, N., 409
Farber, M., 406
Farese, R. V., 375, 376
Farkas, M. E., 395
Fath, D., 330, 392
Faus, H. G., 467, 468
Fawcett, T. W., 279
Fedele, A. O., 79, 332, 378
Federle, M. J., 19

Fedoruk, M. N., 118
Feeney, E. P., 19
Feijoo-Cuaresma, M., 373
Feldman, D., 275, 277, 281, 377
Feldser, D., 135, 330, 348, 350
Feliciano, C. E., 327, 426, 458, 471
Fels, D., 252, 254, 277, 278, 279, 281, 283, 287
Feng, L. L., 63
Ferguson, D. J., 160
Ferguson, T., 182
Fernandez, J. R., 162
Fernandez, R., 135
Fernando, N. T., 98
Ferrara, N., 89, 93, 96, 99, 100, 201, 204, 207,
    209, 211, 213, 237
Ferraro, M., 372
Ferre-D'Amare, A. R., 9, 19
Ferreira, G., 135, 348, 350
Ferreira, V., 89, 93
Ferrell, J. E., Jr., 109
Ferrer, K., 186
Ferreyra, R., 377
Ferriero, D. M., 232, 233
Ferris, C. D., 468
Ferry, D., 310, 311
Fesik, S. W., 98
Fielitz, J., 225, 230, 235
Figueroa, Y. G., 325
Filipenko, V., 249, 276, 277, 279
Filmus, J., 327, 328
Finkel, T., 423
Finkelmeier, D., 362, 363
Finkielstein, C. V., 114, 362, 363
Finlay, M., 7, 8
Fire, A., 108
Firestein, G. S., 94, 100, 213, 227, 406, 480
Firth, J. D., 182
Fish, J. E., 349, 373
Fisher, B. J., 235
Fisher, J. W., 160
Fisher, R., 20, 310, 391, 395, 397
Fisher, R. J., 329, 330, 391, 393, 395
Fisher, T. S., 114
Fitzgerald, K. T., 414, 481, 487
Flaction, R., 188
Flaherty, K. R., 357
Flamme, I., 43, 47, 48, 54, 89, 200, 201, 425,
    464, 465
Flanigan, C. C., 314
Flashman, E., 27, 28, 33, 34, 35, 37, 46, 47, 62,
    64, 72, 77, 226
Flavell, R. A., 231, 236
Fleischmann, W., 406
Flemming, B., 154
Fliegauf, M., 374
Flieger, O., 362, 363
Flippin, L., 89, 183

Fliser, D., 231
Flores, A., 46
Fluck, M., 236
Flugel, D., 410, 413, 426
Flynn, D. C., 327, 426, 430
Focht, R. J., 63
Foley, E. D., 206, 207
Folkman, J., 205
Follis, R., 428
Folz, R. J., 427
Fong, A., 183
Fong, G.-H., 46, 89, 91, 92, 201
Fontana, L., 112, 204, 324, 325, 332, 473
Fontijn, D., 426
Forster, I., 94, 100, 213, 227, 406, 480
Fosnot, J., 117
Foster, C. A., 298
Foster, J. L., 304
Fourney, P., 183
Fournier, J. G., 160
Fowler, A. A. III, 235
Fowler, J. F., 304
Fox, G. M., 173, 188
Fox, M. E., 314
Fox, S. B., 201, 365
Fox-Talbot, K., 234
France, D. S., 313, 330, 331, 392, 397, 398
Franchini, M., 183
Franco, D., 90
Francoijs, K. J., 6
Franklin, T. J., 35
Frapart, Y. M., 47, 413, 429, 464, 471
Frappell, P. B., 46
Frasch, M., 135
Fratelli, M., 154, 231, 232
Frazier, M. R., 124, 136
Frede, S., 405, 407, 408, 409, 410, 428
Freedman, B. D., 95
Freedman, S. J., 313, 330, 331, 392, 397, 398
Freeman, B. A., 424, 466
Freeman, B. D., 407
Freeman, R. S., 375, 376
Freemont, A. J., 407
Freezer, L. J., 314
Frei, C., 136, 137, 138
Frei, U., 116, 181, 225, 226, 227, 229, 230,
    232, 233, 234, 235, 236, 237, 332
Frein, D., 471
Freitag, P., 407, 408, 409, 410, 428
Frejaville, C., 437
Frelin, C., 51, 53, 131, 134, 230
Frew, I. J., 46
Fricke, A., 302
Fridovich, I., 425
Friedlos, F., 310
Friedman, P. A., 63
Frietsch, T., 148

Fritsch, E. F., 173
Frohlich, R., 425
Frohlich, T., 89
Frolich, T., 200
Frydman, J., 377
Fuchs, J., 425
Fuchs, M., 32
Fuchs, U., 367
Fujii-Kuriyama, Y., 5, 7, 8, 17, 89, 199,
    425, 428, 472
Fujimoto, T., 281
Fujita, T., 236
Fukuda, K., 135, 409, 426, 427
Fukuda, R., 135, 426, 427, 430
Fukuda, S., 234
Fukumoto, A., 362
Fukumura, D., 98, 332, 433
Fukushima, Y., 212
Fulop, V., 469
Furie, M. B., 201
Furst, R., 430
Furuta, G. T., 96, 412

G

Gabay, L., 127
Gagneux, P., 189
Galban, S., 376
Gale, N. W., 205
Galiano, R. D., 201, 325
Galic, M., 135
Galis, Z. S., 427
Galle, R. F., 126
Gallo, R. L., 95, 408
Galluzzo, M., 374
Galson, D. L., 158, 182
Gambhir, S. S., 160
Gantier, M., 364
Gao, G., 185, 186
Gao, J., 166
Gao, N., 327, 426, 430
Garci, J. A., 211, 212
Garcia, J. A., 13, 14, 18, 90, 95, 182, 211, 212,
    225, 227, 235, 332, 393
Garcia-Lecea, M., 473
Gardner, K. H., 3, 6, 7, 8, 9, 10, 12, 13, 14, 18,
    19, 20, 393, 395
Gardner, L. B., 332
Garner, C. D., 28, 30
Garrard, S., 7
Garsky, V. M., 63
Gascon, A. R., 186
Gaskell, R. M., 35
Gaskell, S. J., 4, 26, 27, 36, 88, 89, 110, 131,
    183, 226, 229, 324, 327, 372, 425, 429,
    449, 464
Gasparian, L., 115

Gassmann, M., 89, 90, 96, 98, 100, 125, 130, 133, 148, 154, 182, 203, 204, 224, 227, 228, 231, 233, 236, 276, 308, 325, 332, 348, 408, 423, 425, 449
Gaston, B., 467, 468
Gatter, K. C., 79, 223, 224, 311, 332, 348
Gatterbauer, B., 348
Gatti, R. A., 372
Gatzemeier, U., 309
Gaur, A., 90, 95, 211, 212, 227, 332
Gavin, B. J., 92, 378
Gaw, A. F., 281
Gawlitzek, M., 180
Gazit, G., 277
Ge, W., 109
Gearhart, J. D., 89, 90, 100, 133, 158, 160, 203, 224, 227, 332, 348, 449
Geenen, D., 229, 230, 234
Geiger, G., 362, 363
Gejyo, F., 185
Gembruch, U., 148
Genius, J., 416
Genn, D. N., 33, 34, 37, 46, 62, 64, 72, 77
Georgatsou, E., 8
George, R. A., 126
Georgescu, M. M., 135, 330
Geraghty, J., 183
Gerald, D., 47, 413, 429, 464, 471
Gerber, H.-P., 93, 94, 96, 99, 100, 201, 204, 209, 211, 213, 227, 237, 406, 480
Gergerliouglu, H. S., 236
Gericke, D., 314
Gertler, F., 425
Gertsenstein, M., 89, 93, 133
Gervais, M. L., 349, 373, 376
Gervasi, D. C., 4, 44, 226, 229, 234, 429
Gerwien, J., 154
Gerzelis, I., 348
Getzoff, E. D., 6
Ghabrial, A., 126
Ghenoiu, C., 374
Ghezzi, P., 154
Ghikonti, I., 348
Ghilini, G., 425
Giaccia, A. J., 4, 20, 46, 47, 114, 133, 135, 204, 277, 281, 307, 312, 313, 314, 315, 323, 324, 325, 326, 327, 328, 329, 330, 332, 356, 362, 363, 364, 386, 413, 429, 430, 449, 464, 471
Giannakakou, P., 313, 330, 392
Giannoudis, A., 407
Gibert, P., 124
Gibson, C., 98, 100, 233
Gidday, J. M., 222, 232, 233
Gielbert, J., 4, 26, 27, 36, 88, 89, 110, 131, 183, 226, 229, 324, 327, 372, 425, 429, 449, 464
Gilchrist, G. W., 124
Gildea, J. J., 469
Giles, R. H., 376

Gill, V. J., 314
Gilles-Gonzalez, M. A., 6
Gillespie, D., 98, 100
Gimond, C., 430
Gingras, A. C., 251
Ginouves, A., 27, 46, 111, 112, 230, 331, 376, 414, 480, 481, 487
Ginzinger, D. G., 89, 97, 237
Giordano, F. J., 95, 96, 204, 209, 211, 233, 236
Giordano, S., 374
Giorgetti-Peraldi, S., 135, 330, 425
Giudice, L. C., 327
Giuliani, R., 207
Glacet, A., 189
Glass, G. A., 158
Glazer, L., 127
Glazer, P. M., 325
Gleadle, J. M., 4, 27, 29, 30, 36, 44, 46, 62, 63, 72, 79, 88, 89, 110, 111, 112, 115, 129, 130, 131, 132, 182, 226, 229, 276, 324, 328, 378, 413, 425, 429, 449, 464
Gleiter, C., 231, 232
Glickman, J., 89, 94, 97, 100, 375, 377
Glimcher, L. H., 277, 281, 307
Gnant, M., 348
Gnarra, J. R., 89, 92, 100, 374
Gobe, G., 148
Goberdhan, D. C., 135
Gocayne, J. D., 126
Goda, N., 231
Godson, C., 414, 481, 487
Godzik, C. A., 431
Goemans, C., 231, 232
Gohil, M., 204
Goldberg, M. A., 158, 172, 223, 408, 472, 473
Goldfarb, M., 181
Golding, M. C., 109
Goldman, L. A., 80
Goldwasser, E., 160, 173, 407, 448, 454, 458, 470
Golic, K. G., 125
Goligorsky, M. S., 160
Gomez, L. M., 365, 366
Gondi, C. S., 117
Gong, C., 374
Gong, W., 6
Gong, Y., 330
Gonsalves, C., 481
Gonter, J., 32
Gonzalez, F. J., 133
Gonzalez, C., 6
Gonzalez-Escribano, M. F., 365, 366
Gonzalez-Gay, M. A., 365, 366
Gonzalez-Pacheco, F. R., 237
Goodall, G. J., 410, 428
Goodman, E. C., 135
Gordan, J. D., 377
Gordeuk, V. R., 184, 237, 372
Gorin, Y., 427

Görlach, A., 410, 412, 413, 421, 423, 424, 425, 426, 427, 429, 430, 471
Gorman, J. J., 4, 26, 27, 44, 62, 63, 65, 67, 68, 70, 72, 75, 114, 226, 328, 378, 413, 425, 429, 465, 481
Gorospe, M., 376
Gorr, T. A., 125, 129, 130, 131, 408
Gorselink, M., 97
Gothie, E., 135, 328, 409, 430
Gottlieb, E., 47
Gottschalk, A. R., 135, 327, 328, 330
Gozal, D., 236
Graber, S. E., 181
Gradin, K., 5, 70, 76, 325, 326, 364
Graeber, T. G., 307
Grafe, M., 232, 233
Granatino, N., 27, 28, 30, 38, 63
Grandmont, S., 412
Grant, R., 185, 186
Grant, R. L., 185
Grasso, G., 154
Gray, L. H., 298, 300
Greco, O., 312
Green, M. R., 112
Green, S. R., 109
Greenspan, R. J., 126
Gregersen, P. K., 364
Greijer, A. E., 348
Greiner, R. H., 348
Grell, M., 362, 363
Grepin, R., 410
Greten, F. R., 409, 481
Griendling, K. K., 423
Griethe, W., 181, 225, 230, 232, 233, 237
Griffiths, L., 312
Grimm, C., 231, 233
Grinberg, A., 89, 92, 100, 407
Grisham, M. B., 434, 466, 469
Groemping, Y., 423
Grolleau, A., 255
Gronenborn, A. M., 10
Groppe, J., 127
Gross, J., 425
Groszer, M., 233
Groulx, I., 132
Gruber, M., 95, 182
Gu, H., 213, 407
Gu, J., 129, 425, 472, 473
Gu, Y., 310, 311
Gu, Y. Z., 7, 89, 92, 332
Guan, Y., 184, 237, 372
Guenzler-Pukall, V., 183
Guidi, G. C., 183
Guillemin, K., 126, 127
Gujrati, M., 117
Gunaratnam, L., 229
Gunji, Y., 201
Gunningham, S. P., 201

Gunnison, K. M., 278
Gunzler, V., 4, 27, 33, 36, 39, 44, 46, 47, 51, 62, 64, 71, 72, 73, 89, 130, 183, 226, 229, 234, 235, 429
Guo, C., 377
Guo, M., 378
Guo, R., 27
Guo, S., 109
Gupta, A. K., 330
Gurel, A., 236
Gurtner, G. C., 201, 325
Gustafsson, J. A., 19
Gustafsson, M. V., 70, 76
Gustafsson, T., 236
Gutierrez, P. L., 471
Gutkind, J. S., 330, 375, 433
Guyton, K. Z., 279
Guzy, R. D., 448, 470
Gygi, S. P., 249

H

Ha, Y., 187
Haas, R., 465
Haase, V. H., 88, 92, 94, 95, 96, 100, 133, 181, 182, 204, 213, 224, 227, 234, 236, 237, 332, 377, 378, 406, 407, 412, 480, 481
Haas-Kogan, D., 135, 327, 328, 330
Habelhah, H., 46
Haberberger, T., 4, 44, 226, 229, 234, 429
Hacker, D. L., 188
Hackett, S. F., 206
Hacohen, N., 126, 127
Hada, A., 281
Haddad, G. G., 129, 134
Haddad, J. J., 426
Hadsell, D., 96
Hafen, E., 127, 135, 136
Hagen, T., 47, 412, 413, 435, 448, 476, 480
Hagensen, M., 46
Hahn, S. M., 330
Hai, T., 279, 281
Haigh, J., 237
Hainfellner, J. A., 348
Hajdu, J., 28, 469
Hall, E. J., 301, 302
Haller, H., 231
Hamada, J., 425
Hamamori, Y., 186
Hamilton, K. E., 407
Hamilton, S. R., 188
Hammarback, J. A., 280
Hammarstrom, M., 7
Hammer, R. E., 89, 91, 100, 201, 202, 204, 332
Hammer, S., 325, 326
Hammerling, U., 448, 470
Hammond, E., 277, 281, 307
Hammond, S. M., 109

Hampton-Smith, R. J., 61
Han, C., 201, 365
Han, J., 135
Han, J. Y., 330
Han, S., 430
Han, Z., 430
Hanahan, D., 348
Hanauske, A., 229
Hanauske-Abel, H. M., 39, 47, 71, 229
Hancock, J. T., 423
Hankinson, O., 5, 14, 19, 89, 91, 100, 276, 424
Hannon, G. J., 109, 117
Hanrahan, C., 135, 330
Hanrahan, C. F., 348
Hao, B., 6
Haque, D., 425
Haque, M., 332
Hara, S., 425
Hara, T., 28, 30
Harati, M., 187
Harborth, J., 109, 118
Hard, T., 7
Harding, H. P., 249, 252, 254, 276, 277, 279,
    280, 281, 282, 283, 287
Harding, M. A., 469
Hardy, W., 186
Harlos, K., 28, 30, 36, 356, 469
Harmey, J., 183
Haroon, Z. A., 325
Harpal, K., 89, 93
Harper, S. M., 6, 7, 9, 12, 20
Harrelson, J. M., 305
Harris, A. L., 4, 46, 47, 79, 111, 116, 201,
    223, 224, 276, 279, 281, 304, 311, 312, 325,
    326, 330, 332, 348, 365, 372, 377, 392,
    411, 425, 426, 429
Harris, C. C., 466, 469, 471
Harris, P. A., 310
Harris, S. R., 304
Harrison, C., 46
Harrison, D. G., 423, 427
Harrison, J. F., 136
Hart, M. L., 230
Harten, S. K., 349, 373, 374
Hartikka, J., 185
Harwood, S., 231
Hasebe, Y., 330
Hasegawa, J., 148
Hashimoto, K., 467, 468
Hasle, H., 237
Hasselblatt, M., 184
Hattab, M., 114
Hattori, K., 374
Hausinger, R. P., 71
Hausmaninger, H., 348
Hausser, A., 362, 363
Hawley, R. S., 125
Hay, M. P., 307, 308

Hay, R. T., 55, 64, 65, 79, 230, 481
Hayashi, N., 159
Hayashi, S., 127
Hayes, L., 114
Haze, K., 280
He, H., 327, 430
He, J., 426
He, Z., 378
Heard, J. M., 185, 186
Hebestreit, H. F., 4, 26, 27, 36, 88, 89, 110,
    131, 183, 226, 229, 324, 327, 372, 425,
    429, 449, 464
Hedley, D. W., 330
Hedlund, B. E., 182
Hefti, M. H., 6
Hegg, E. L., 469
Hegge, J. O., 185
Heidbreder, M., 425
Heikkinen, P., 46
Heilmaier, C., 407, 410
Heine, S., 448, 473
Heinicke, K., 148
Heinrich, D., 327
Heinz, G., 184
Heiwa, O., 472
Helenius, A., 188
Hellgren, N., 7
Hellings, P., 97
Hellingwerf, K. J., 7, 12, 20
Hellwig, T., 182, 407
Hellwig-Bürgel, T., 46, 112, 172, 181, 182, 230,
    407, 410, 426
Helsby, N. A., 311, 312
Helton, R., 98, 100, 233
Hempelmann, G., 32
Hendershot, L. M., 278, 280, 282
Henderson, D. W., 375
Henderson, L. M., 435
Henderson, N. L., 63
Hendriks, J., 7, 12
Henk, J. M., 298
Hennig, S., 7, 12, 19
Henri, A., 186
Henry, W. L., Jr., 407
Heppner, F., 314
Herbert, J. M., 98, 332
Hergovich, A., 374, 376
Herkert, O., 424
Hernandez, C., 473
Hernandez, R. M., 186
Herrera, A. M., 162
Hershey, J. W., 251
Hertel, B., 231
Herweijer, H., 185
Heryet, A., 160
Hescheler, J., 308, 325
Hess, J., 410, 413, 423, 424, 425, 426, 427, 430
Heumann, R., 231, 232

Hewick, R., 173
Hewitson, K. S., 4, 25, 27, 28, 29, 30, 32, 33, 34, 35, 36, 37, 38, 39, 44, 46, 47, 55, 62, 63, 64, 65, 71, 72, 75, 77, 79, 88, 89, 110, 112, 131, 226, 229, 230, 324, 328, 378, 425, 429, 449, 464, 481
Hewitt, S. M., 313, 388, 389
Heyes, J. A., 118
Heyman, S. N., 181
Heymans, S., 97
Hickey, P. R., 407
Hickey, R. P., 95, 96, 233, 236
Hicks, K. O., 307, 310, 311, 312
Higgins, D. F., 377
Higgs, E. A., 465
Higuchi, M., 159, 161, 163, 165, 166, 168
Higuchi, N., 185
Hill, C. M., 160
Hill, P., 349, 373, 374
Hill, R. P., 305, 376
Hill, S. A., 300, 304
Hillan, K. J., 89, 93, 100, 204
Hilliard, G., 182
Hilton, D. A., 311, 348, 349
Hinnebusch, A. G., 252
Hino, S. I., 280
Hinton, D. P., 10
Hinze-Selch, D., 184
Hira, E., 367
Hirahara, H., 185
Hirahra, H., 185
Hirai, S., 375
Hiraoka, M., 312
Hirata, M., 185
Hirota, K., 44, 62, 65, 89, 115, 135, 184, 328, 372, 378, 409, 413, 425, 426, 427, 428, 429
Hirsh, V., 309
Hirsila, M., 27, 33, 36, 39, 46, 47, 51, 62, 64, 71, 72, 73, 75
Hirst, D. G., 312
Hirst, V. K., 308
Hishikawa, Y., 367
Hite, K., 329, 330
Hitomi, Y., 468
Hizawa, N., 365
Ho, V., 46, 89, 92
Hobbs, S., 67
Hochachka, P. W., 248, 276
Hockel, M., 275, 348
Hocking, D. J., 309
Hodgkiss, R. J., 301
Hoet, P., 91, 201
Hofer, T., 327
Hoffart, L. M., 71
Hoffman, E. C., 5, 14, 424
Hoffman, M., 375, 378
Hoffman, M. A., 327, 328, 425

Hofmann, F., 207
Hofseth, L. J., 469
Hogan, M. C., 95, 236
Hogenesch, J. B., 7, 89, 92, 332
Hogg, A., 311, 312
Hohne, M., 374
Holash, J., 205
Holbrook, N. J., 279
Holcik, M., 252, 281
Holland, J. F., 314
Hollander, A. P., 407
Holley, C. L., 249
Hollien, J., 280
Hollingshead, M., 313, 388, 389
Hollis, G. F., 72
Holme, E., 72
Holstege, F. C., 115
Holt, R. A., 126
Holtermann, G., 423
Holthusen, H., 300
Hon, W. C., 36, 356
Honda, Y., 201, 206
Honess, D., 304, 312
Hong, K., 300
Hong, S. B., 377
Hong, S. H., 413
Hong, W. K., 330
Hoover, P., 427
Hopfer, H., 46
Hopfer, U., 46
Hopkins, D., 188
Hoppeler, H., 236
Horner, J. W., 89, 183
Hornsey, S., 298, 300
Horsfall, L. E., 63, 328, 429
Horstrup, J. H., 181, 225, 227, 229, 232, 233, 235, 236
Horvat, R., 348
Ho-Schleyer, S. C., 284, 289
Hoskin, P. J., 300
Hoskins, R. A., 126
Hosoya-Ohmura, S., 161, 163, 165
Houck, D. R., 182
Housman, D. E., 307
Howell, J. J., 254, 276
Howlett, R. A., 95, 236
Hoyos, B., 448, 470
Hsieh, C. C., 306
Hsu, R., 300, 307
Hsu, T., 131
Hu, C. J., 95, 182, 204, 332, 377, 424
Hu, G., 109
Hu, N., 252, 254, 279, 281, 283, 287
Hu, S., 128
Hu, X. W., 426
Huang, C., 327
Huang, E., 4
Huang, H., 135

Huang, H. C., 430
Huang, L. E., 19, 46, 89, 97, 129, 182, 223, 237, 325, 326, 327, 328, 332, 408, 412, 425, 428, 472, 473
Huang, L. H., 72
Huang, Q., 117
Huang, S. P., 357
Huang, X., 187, 229, 230, 234, 357
Huang, Y., 96, 204, 209, 211, 233, 327
Huang, Y. T., 430
Huang, Z. J., 5, 14
Huasinger, R. P., 27, 29, 30, 38
Hubbard, S. R., 280
Huber, B. E., 312
Huber, R. M., 63
Hudson, C. C., 327, 330
Hudson, K. M., 324, 325
Huetter, J., 48
Huey, R. B., 124
Hughes, C. M., 312
Hughson, F. M., 19
Hughson, M. D., 378
Hulscher, T. M., 426
Hung, G., 277
Hunt, J. W., 302
Hunter, T., 330
Hurley, R. L., 136
Hurtado-Ziola, N., 95, 408
Huso, D. L., 93
Hussain, S., 310
Hussain, S. P., 466, 471
Hussels, C. S., 172
Hutter, J., 425, 464
Hwang, K. S., 377
Hwu, P., 314

I

Iaina, A., 187
Iannettoni, M. D., 357
Ibrahim, H., 325
Ichikawa, F., 185
Ignacy, W., 183
Iino, N., 185
Ikeya, T., 135
Iliopoulos, O., 92, 372, 373, 376, 378
Im, D. S., 377
Imagawa, S., 159, 161, 162, 163, 165, 166, 168, 171, 172, 181
Imazeki, I., 185
Imazeki, N., 281
Imura, N., 425
Inagi, R., 236
Ingelfinger, J. R., 236
Iqball, S., 184, 185, 312
Irisarri, M., 27, 129, 130, 131, 132, 135, 136, 138
Iruela-Arispe, M. L., 166

Irwin, M. S., 373, 376
Isaac, D. D., 126
Isaacs, J. S., 5, 329, 330, 374, 375, 392, 411
Isaacs, W. B., 311, 348, 349
Ischiropoulos, H., 434, 435
Ishida, J., 171
Ishiguro, H., 171
Ishikawa, M., 231
Ishizaka, N., 236
Ismail-Beigi, F., 327, 328
Ito, W., 171
Ito, Y., 166, 185, 433
Ivan, M., 4, 26, 27, 36, 44, 88, 89, 110, 226, 229, 234, 324, 327, 328, 372, 375, 378, 425, 429, 449, 464, 490
Ivanov, S., 47
Ivashchenko, Y., 225, 230, 237
Iwai, K., 375, 449
Iwakoshi, N. N., 281
Iwamoto, T., 223, 234
Iwanaga, T., 168
Iyer, N. V., 89, 90, 93, 100, 133, 135, 203, 204, 224, 227, 332, 348, 350, 449
Iyer, S., 231

J

Jaakkola, P., 4, 26, 27, 36, 46, 88, 89, 110, 131, 183, 226, 229, 324, 327, 372, 425, 429, 449, 464
Jablonski, K., 46
Jacks, T., 307
Jackson, A. L., 112
Jackson, M., 310
Jackson, S. P., 302
Jacob, S., 184
Jacobs, C. S., 395
Jacobs, K., 173
Jacobsson, L., 362
Jaenisch, R., 94, 100, 158, 165, 213, 227, 377, 406, 480
Jaffar, M., 312
Jaffrey, S. R., 468
Jahn, H., 184
Jain, R. K., 205
Jain, S., 92
Jakesz, R., 348
Jalaguier, S., 19
James, M. J., 406
Janes, M. S., 435
Jansson, E., 236
Jarecki, J., 128, 130
Jarman, M., 310
Jarrott, B., 232
Jedlickova, K., 237, 372
Jeffers, M., 374
Jefferson, A. B., 135, 327, 328, 330

Jefferson, L. S., 256
Jeffrey, P. D., 19
Jeffrey, S. S., 325, 326
Jeffs, L. B., 118
Jelinek, J., 184, 372
Jelkmann, W., 46, 47, 112, 148, 158, 159, 160, 172, 173, 179, 180, 181, 182, 183, 184, 187, 189, 190, 191, 229, 230, 407, 410, 412, 423, 426, 448, 469, 473
Jenkins, J., 426, 427, 471
Jenkins, N. A., 159, 162, 168
Jennings, R. B., 222, 237
Jensen, I., 448, 473
Jensen, M. P., 27, 29, 30, 38
Jensen, R. L., 98, 100
Jeong, D. G., 63, 429, 430
Jeong, H. J., 413
Jeong, J. W., 114
Jeschke, G., 28, 35, 47
Jewell, U. R., 148, 154
Jewett, M. A., 373
Jia, S., 63
Jiang, B. H., 4, 5, 13, 88, 89, 112, 129, 181, 204, 224, 327, 330, 348, 356, 386, 387, 393, 395, 424, 425, 426, 428, 430, 449
Jiang, C., 223, 408
Jiang, G., 330
Jiang, H. Q., 27
Jiang, L., 127
Jiang, Y., 188, 330
Jiang, Z., 330
Jimenez, S., 237
Jin, S., 70, 76
Jirstrom, K., 348
Jitrapakdee, S., 67
Johannes, F. J., 362, 363
Johannes, G. J., 249, 250, 251
John, M., 118
Johns, R. A., 467, 468
Johnson, C., 427, 481
Johnson, D. W., 148
Johnson, E., 128, 130
Johnson, E. A., 134
Johnson, M. H., 160
Johnson, M. S., 313, 330, 392
Johnson, N., 409
Johnson, R. S., 4, 20, 87, 88, 89, 90, 92, 94, 95, 96, 97, 98, 99, 100, 133, 135, 158, 159, 173, 181, 182, 204, 209, 211, 213, 224, 225, 227, 233, 234, 236, 237, 276, 313, 324, 325, 326, 327, 328, 329, 330, 332, 377, 378, 386, 406, 407, 408, 409, 412
Johnson, W. A., 126
Johnston, R. B., Jr., 431
Johren, O., 425
Joiner, M. C., 305
Jokilehto, T., 46

Jones, D., 185, 235
Jones, D. T., 392
Jones, E. Y., 36, 356
Jones, J. G., 99
Jones, J. T., 109
Jones, N. M., 232
Jones, O. T., 423
Jones, R. G., 250, 251, 254, 255, 276, 279
Jones, R. W., 88, 100, 181
Jones, S. A., 423
Jones, S. S., 173
Jong, H. S., 426
Jonghaus, W., 48
Jorcano, J. L., 97
Jordan, R. B., 34
Jorgensen, R., 109
Jorieux, S., 189
Joseph, J., 434, 435
Joseph, S. K., 95, 407
Joseph-Silverstein, J., 201
Jothy, S., 373, 378
Joukov, V., 201
Jourd'Heuil, D., 466, 469
Jousse, C., 254, 279
Joyce, M. R., 183
Ju, T. C., 233
Judge, A. D., 118
Judware, R. R., 414
Jumbe, N. L., 184
Jung, C. R., 377
Jung, H. C., 426
Jung, J. Y., 148
Jung, Y. D., 135, 426
Jung, Y. J., 5, 329, 330, 374, 375, 392, 411
Jungermann, K., 135, 425, 428, 429, 430
Jurgensen, J. S., 130, 181, 225, 227, 229, 230, 232, 233, 235, 236, 237

K

Kaanders, J. H., 325
Kachhap, S. K., 392
Kadoya, T., 46
Kaelin, W. G., Jr., 4, 26, 27, 36, 44, 88, 89, 92, 94, 97, 99, 110, 136, 183, 226, 229, 230, 234, 313, 324, 327, 328, 331, 356, 371, 372, 373, 374, 375, 376, 377, 378, 395, 425, 429, 449, 464, 465, 490
Kaelin, W. J., 372, 374, 375, 378
Kagaya, Y., 154, 166
Kageyama, R., 133
Kahana, C., 135, 425
Kahn, C. R., 95, 236
Kajimura, M., 231
Kalivendi, S., 434, 435
Kallio, P. J., 5, 425, 464
Kallman, R. F., 305

Kallunki, P., 154
Kalra, K. L., 349
Kalra, V. K., 481
Kaluz, S., 391
Kaluzova, M., 391
Kalyanaraman, B., 434, 435
Kamada, M., 375
Kameda, S., 185
Kamura, T., 377, 425, 449
Kan, O., 312
Kaneko, S., 375
Kanemoto, S., 280
Kang, C., 12
Kanga, S., 365
Kanopka, A., 332
Kantsevoy, S. V., 332
Kanwar, J. R., 313
Kany, S., 130, 229, 235
Kaper, F., 135, 251, 254, 255, 327, 328, 330
Kaplan, E., 187
Kappel, A., 89, 201
Kaptein, R., 7, 12, 20
Kapurniotu, A., 362, 363
Karaczyn, A., 47
Karakuzu, O., 6, 8, 10, 13, 18, 393, 395
Karhausen, J., 96, 308, 325, 406, 407, 412
Karibe, A., 166
Karin, M., 409, 481
Karoui, H., 437
Kaska, D. D., 229
Kasprzak, K. S., 47, 182, 230
Kastriotis, I., 348
Kasuno, K., 135
Kato, H., 171, 231
Kato, J. Y., 362
Kato, S., 374
Katori, M., 231
Katschinski, D. M., 5, 43, 47, 54, 148, 182,
        237, 327
Kaufman, B., 4
Kaufman, R. J., 173, 249, 250, 252, 253, 254,
        255, 262, 265, 276, 278, 279, 280, 281,
        283, 287
Kaugman, R. J., 278
Kaule, G., 51
Kaur, B., 313, 392
Kavantzas, N., 348
Kavya, R., 465
Kawachi, H., 185
Kawauchi, S., 166, 201
Kay, L. E., 10
Kay, M. A., 117
Ke, B., 231
Kedes, L., 186
Keenan, S., 231, 232
Kehler, J., 377
Keifer, P., 10
Keilty, G., 312

Keith, B., 89, 91, 95, 100, 181, 182, 204, 224,
        227, 234, 250, 251, 254, 255, 276, 279,
        332, 377, 424
Keller, G. A., 93
Keller, S., 231
Kelley, J. A., 63
Kelley, R., 126
Kelly, B., 46, 114, 348, 350, 430
Kelly, R. C., 19
Kelm, M., 148, 149, 150, 234
Kemphues, K. J., 109
Kendrick, K. F., 95
Kenny, L., 310
Kerbel, R. S., 327, 328, 373, 378
Kerensky, R. A., 222
Kershaw, N. J., 27, 30, 63
Keshet, E., 91, 201
Kessimian, L., 207
Kewley, R. J., 64, 70, 75, 82
Key, M. E., 349
Khan, A. H., 310
Khatri, J. J., 427
Khomenko, T., 426
Khwaja, F., 330
Kieckens, L., 89, 93
Kiemer, A. K., 430
Kieran, M. W., 161, 165
Kieser, A., 375
Kiessling, I., 148
Kieswich, J., 231
Kietzmann, T., 135, 410, 412, 413, 421, 423,
        424, 425, 426, 427, 428, 429, 430, 471
Kikkawa, Y., 237
Kikuchi, H., 185
Kikuchi, Y., 8, 472
Kim, C. H., 426, 427
Kim, C. M., 375, 377
Kim, E., 330
Kim, H., 376
Kim, H. H., 426, 427
Kim, H. M., 413
Kim, H. Y., 185
Kim, J., 148, 151, 313, 392
Kim, J. M., 187, 377
Kim, J. W., 120
Kim, K. S., 481
Kim, K. W., 114
Kim, M. J., 95, 236
Kim, M. S., 313, 330, 392, 426, 427
Kim, N. K., 426
Kim, S. H., 114
Kim, S. J., 63, 413, 429, 430
Kim, S. K., 135
Kim, S. T., 185, 376
Kim, S. W., 187, 426, 427
Kim, T. Y., 327, 328, 425, 426
Kim, W., 4, 26, 27, 36, 88, 89, 110, 324, 327,
        356, 372, 425, 449, 464, 490

Kim, W. H., 377
Kim, W. Y., 89, 94, 97, 183, 349, 356, 372, 373, 374, 375, 377
Kim, Y. S., 148
Kimball, S. R., 256
Kimura, C., 433
Kimura, H., 300, 412, 467, 468
Kimura, S., 202
King, G. L., 207
King, M. P., 426, 470
Kingsman, S., 184, 185, 312
Kirkpatrick, D. L., 330, 392
Kirkpatrick, J. P., 117
Kirkpatrick, L., 330, 392
Kirschner, K. M., 149, 172
Kiryu, J., 184
Kisanuki, Y. Y., 204
Kishida, T., 374, 375
Kita, M., 184
Kita, T., 409
Kivinen, L., 201
Kivirikko, K. I., 27, 30, 33, 36, 39, 46, 47, 51, 62, 64, 71, 72, 73, 75, 229
Kizaka-Kondoh, S., 135, 426, 427
Klagsbrun, M., 205
Klambt, C., 127
Klaus, S., 130, 183, 229, 235
Klausmeyer, P., 399
Klausner, R., 374
Klausner, R. D., 89, 92, 100, 372, 449
Klco, J. M., 313, 331, 372, 375, 378, 395
Klebanov, B., 407
Kleemann, R., 362, 363
Klefstrom, J., 201
Klein, B. E., 205
Klein, F., 308, 325
Klein, R., 205
Kleinman, M. E., 201, 325
Kley, N., 92, 372, 376, 378
Kling, P. J., 182
Klitzman, B., 300
Knaus, U. G., 423, 424
Knerlich, F., 184
Knick, V. C., 312
Knight, M., 97, 100
Knight, M. C., 133
Knocke, T. H., 348
Knolle-Veentjer, S., 184
Knopfle, G., 172
Knowles, H. J., 47, 411, 426, 429
Knox, R. J., 310, 315
Kobata, A., 188
Kobata, T., 213, 407
Kobayashi, C., 425
Kobayashi, S. V., 112
Kobayashi, T., 97, 100, 133, 168, 184
Koch, C. J., 98, 302, 307, 325, 330, 332

Koch, W. J., 151
Kochling, J., 181
Kodama, T., 181
Koditz, J., 46, 47, 112, 230
Koehne, P., 226, 230
Koeleman, B. P., 365, 366
Koga, S., 201
Koh, M. Y., 183
Kohin, S., 236
Kohl, N. E., 63
Kohl, R., 412, 429, 471, 473
Kohn, E. C., 348
Koivunen, P., 27, 33, 36, 39, 46, 47, 51, 62, 64, 71, 72, 73, 75
Kojima, H., 213, 406, 407
Kojima, I., 236
Kokame, K., 280
Kolch, W., 375
Kolkhof, P., 425, 464
Komatsu, N., 154
Kon, T., 117
Kondo, K., 4, 26, 27, 36, 44, 88, 89, 110, 226, 229, 234, 324, 327, 372, 425, 429, 449, 464, 490
Kondo, M., 362
Kondo, S., 63, 280
Kondo, Y., 425
Kondoh, N., 281
Kong, C., 325, 326
Kong, D., 20, 329, 330, 391, 393, 395, 397
Kong, G., 114
Kong, T., 407, 480
Kong, X., 330, 392
Konstantinidou, A. E., 348
Koochekpour, S., 374
Koong, A. C., 135, 275, 277, 278, 281, 307, 325, 327, 328, 330, 357, 360, 362, 363, 364
Kooy, N. W., 435
Koritzinsky, M., 247, 249, 250, 251, 252, 253, 254, 255, 256, 262, 265, 276, 277, 278, 279, 281, 282, 283, 284, 287
Korkolopoulou, P., 348
Koromilas, A., 251, 252, 253, 256, 276, 277, 278, 279, 281, 282, 283, 284
Korpisalo, P., 237
Koshiba, M., 407
Koshiji, M., 325, 326
Kostas, S. A., 108
Kostyk, S. K., 205, 206
Kotch, L. E., 89, 90, 100, 133, 203, 204, 224, 227, 332, 348, 449
Kotenko, S. V., 80
Kotsis, F., 374
Koumenis, C., 249, 250, 251, 252, 253, 254, 255, 256, 262, 265, 275, 276, 277, 278, 279, 281, 282, 283, 284
Kourembanas, S., 472
Koury, M. J., 158, 159, 160, 163, 181, 412

Koury, S. T., 158, 160, 181
Kovacs, M. S., 309
Kowalski, J., 93
Koyama, S., 206
Koyama, T., 433
Kozak, K. R., 89, 91, 100
Kozak, M., 268
Kraggerud, S. M., 250, 251
Krajewski, J., 182
Krantz, S. B., 407
Krasnow, M. A., 126, 127, 128, 130
Krause, C. D., 80
Krawczyk, M., 148
Krebs, C., 71
Kreiss, P., 185
Krek, W., 325, 374, 376
Kremmer, E., 375
Krempler, A., 302
Krieg, A. J., 323, 335
Krieg, M., 465
Krieg, S. A., 335
Kriegsheim, A., 4, 88, 89, 110, 131, 183, 226,
    229, 324, 327, 372, 449
Krishnamachary, B., 46, 114, 347, 348, 349, 350,
    351, 373
Krissansen, G. W., 313
Kronblad, A., 348
Krotz, F., 430
Krull, M., 32
Krystal, G. W., 330
Kubista, E., 348
Kubota, Y., 374
Kubowitz, F., 406
Kuebler, W. M., 430
Kugler, J., 226, 230
Kuhn, K. M., 266
Kuhn, W., 374
Kuhne, M., 302
Kukreja, R. C., 234, 235
Kulkarni, A. R., 201, 325
Kumamoto, K., 325, 326
Kumar, G. K., 93
Kumar, V., 201
Kume, A., 185
Kung, A. L., 313, 330, 331, 392, 397, 398
Kunii, H., 154
Kunnapuu, J., 237
Kupiec-Weglinski, J. W., 231
Kurashima, Y., 467, 468
Kurimoto, M., 184
Kurnick, J. E., 407
Kurokawa, H., 11, 12
Kuroki, A., 117
Kurosawa, Y., 171
Kurreck, J., 116, 181, 234, 332
Kurtz, A., 160
Kurzeja, R. J., 27, 28, 226
Kuschinsky, W., 148

Kuwabara, K., 201
Kvietikova, I., 154, 423, 425
Kwack, K., 426, 427
Kwak, N., 207
Kwan, J. T., 189
Kwasny, W., 348
Kwok, S. F., 362, 364

L

Labosky, P. A., 377
Labugger, R., 148, 149, 150
Lacombe, C., 160
Laderoute, K. R., 327, 328, 330, 430, 449
La Ferla, K., 172, 181
Lahat, N., 201
Lahousse, S. A., 63
Lai, I. R., 236
Lai, P. H., 173, 188
Laiho, M., 201
Laissue, J., 348
Lajoie, G., 237
Lakatta, E. G., 148
Lakka, S. S., 117
Lal, A., 325
Lam, F., 47, 448, 476, 480
Lam, K., 118
LaManna, J. C., 230, 233, 472
Lamb, D., 310
Lamb, R., 365, 366
Lambeth, J. D., 423
Lambin, P., 249, 250, 252, 253, 254, 255, 262,
    265, 276, 278
Lan, L., 330
Lan, M., 430
Lancaster, D. E., 55, 63, 64, 65, 77, 79, 230, 481
Land, S. C., 248, 276, 426
Landazuri, M. O., 46, 112, 373
Landberg, G., 348
Lando, D., 4, 26, 27, 44, 62, 63, 65, 67, 68, 70,
    72, 75, 114, 226, 328, 378, 413, 425, 429,
    465, 481
Landry, A. L., 231
Lane, P., 428
Lane, W. S., 4, 26, 27, 36, 88, 89, 110, 324, 327,
    372, 425, 449, 464, 490
Lang, K. J., 249, 250, 255, 268
Langer, J. A., 80
Langsetmo, I., 230, 233
Langsetmo Parobok, I., 183
Lansford, R., 213
Lanzen, J., 300
Lapetina, E. G., 414
Lappin, T. R., 46, 160, 372
Larcher, F., 97
Larcher, T., 185, 186
Larsen, A. K., 154
Lartigau, E., 305

LaRusch, J., 348, 350
Lasko, P., 135
Lasky, L. A., 212
Lasne, F., 185
Lassman, C., 231
Lau, K. W., 372, 377
Lau, S., 186
Laude, K. M., 427
Laughner, E., 89, 90, 100, 133, 135, 203, 204, 224, 227, 311, 330, 332, 348, 349, 449, 472
Lauricella, R., 437
LaVallee, T. M., 313, 330, 392
Lavista-Llanos, S., 27, 129, 130, 131, 132, 135, 136, 138
Law, S., 357
Lawler, A. M., 89, 90, 100, 133, 203, 224, 227, 332, 348, 449
Layfield, L. J., 305
Layton, D. M., 372
Lazaris, A. Ch., 348
Lazo, J. S., 308
Le, A. D., 98, 100
Le, L., 5, 327
Le, N. T., 230
Le, Q. T., 277, 281, 307, 325, 326
Leach, K. M., 172
Leahy, M., 300
Leavy, J., 201
LeBal, C. P., 434
Lebherz, C., 185, 186
Lechpammer, M., 372
Lee, A., 6
Lee, A. C., 118
Lee, A. H., 277, 281, 307
Lee, A. S., 277, 278
Lee, B. R., 413
Lee, C., 63, 429, 430
Lee, C. T., 426
Lee, C. Y., 375
Lee, D. S., 11, 12
Lee, E., 429, 430
Lee, E. C., 159, 162, 168
Lee, E. M., 232
Lee, E. S., 207
Lee, F. S., 26, 27, 46, 324, 327, 372, 429
Lee, H. J., 28, 30
Lee, H. Y., 330
Lee, J., 212, 313, 330, 331, 392, 397, 398
Lee, J. C., 313, 392
Lee, K., 280
Lee, K. J., 114
Lee, M., 187
Lee, M. O., 114
Lee, P. Y., 357
Lee, S., 132, 229, 374, 375, 376, 411
Lee, S. C., 223, 234
Lee, S. E., 426, 427
Lee, S. H., 166

Lee, S. K., 429, 430
Lee, S. M., 63, 376, 429, 430
Lee, W. H., 19
Lee, W. W., 308
Lee, Z. H., 426, 427
Leedham, D. L., 372
Lefebvre, V. H., 250, 251
Le Gall, M., 409
Legarda, D., 469
Le Hir, M., 160
Lehmann, M., 231
Lehnert, K., 313
Lei, K., 136
Leibovich, S. J., 469
Leigh, J. S., 95
Leigh, S., 183
Leist, M., 154
Lemieux, C., 109
Lemmon, M. J., 308, 309, 314
Lemmon, M. L., 314
Le Moigne, F., 437
Lendahl, U., 70, 76
Lendeckel, W., 109
Leng, L., 355
Leonard, S. S., 327, 426, 430
Lerman, A., 426
Lerman, L. O., 426
Leroy, Y., 189
Leshchinsky, I., 408, 426, 470
Lessner, S. M., 427
Leung, E., 313
Leung, S. W., 89, 90, 100, 129, 133, 203, 224, 227, 332, 348, 425, 449
Levine, J. P., 201, 325
Levine, R. L., 435
Levinson, R., 207
Levy, Y., 135, 425
Lewczuk, P., 184, 231, 232
Lewis, C. E., 279, 281, 312, 406, 407, 479
Lewis, D. L., 185
Lewis, I. D., 211
Lewis, J., 27, 28, 226
Lewis, J. O., 5, 129
Lewis, S. E., 126
Ley, S. C., 55, 64, 65, 79, 230, 481
Li, B., 112, 114
Li, C., 148
Li, C. Y., 117, 304, 307
Li, D. M., 135
Li, F., 117
Li, H., 188
Li, J., 64, 77, 115, 278, 327
Li, J. L., 372, 377
Li, L., 98, 212, 230, 364, 378
Li, M., 6
Li, M. Z., 109
Li, P., 99, 100
Li, P. W., 126

Li, Q. L., 430
Li, V., 27, 28, 226
Li, W., 437
Liang, D., 330, 392
Liang, J., 279
Liao, D., 96, 99, 325, 326
Liberman, Y., 187
Lieb, M. E., 27
Lienard, B. M., 27, 28, 32, 33, 35, 39, 226
Lieubeau-Teillet, B., 373, 378
Lim, A. L., 112, 473
Lim, K. C., 172
Lim, M., 311, 348, 349
Lim, S. W., 148
Lin, A., 183
Lin, C. H., 173, 188
Lin, E. Y., 99
Lin, F. K., 173, 188
Lin, H., 160
Lin, H. Y., 430
Lin, J., 181
Lin, J. T., 357
Lin, M. T., 357
Lin, X., 364
Lin, Z., 330, 392
Lindberg, R. L., 148, 149, 150
Linden, T., 47, 54, 182, 237
Lindstedt, S., 72
Linehan, W. M., 46, 89, 92, 100, 230, 372, 374, 375, 377
Ling, F. C., 308, 325
Linggi, M. S., 375, 376
Link, J., 63
Linke, S., 61, 64, 70, 75, 77, 79, 82, 332, 378
Linsley, P. S., 112
Liotta, L. A., 348
Lippi, G., 183
Lippin, Y., 187
Lipscomb, E. A., 376
Lipscomb, J. D., 469
Lipson, S. A., 98, 100
Lipton, S. A., 148, 232, 233
Lisy, K., 79, 332, 378
Lisztwan, J., 325, 374, 376
Little, E., 277
Litz, J., 330
Liu, C., 372, 430
Liu, D., 96, 233
Liu, D. L., 430
Liu, E., 184, 237, 372
Liu, G., 134
Liu, K. J., 327, 426, 430
Liu, L., 250, 251, 254, 255, 276, 279, 425, 448, 470
Liu, L. Z., 426
Liu, M., 233, 327, 330
Liu, M. Y., 327
Liu, N., 430

Liu, Q., 181, 182, 224, 227, 234, 428, 429
Liu, S., 378
Liu, S. C., 315
Liu, X., 166
Liu, Y., 412, 472
Liu, Y. V., 114
Livingston, D. M., 223, 313, 331, 408, 425, 428
Llimargas, M., 126, 127
Lloyd, M. D., 28, 29, 30
Lo, J., 89, 90, 98, 100, 211, 225, 227, 332
Lobe, C. G., 377
Lodish, H. F., 158, 165
Lofgren, J., 180
Lolkema, M. P., 376
Lonergan, K. M., 92, 378
Loomis, D. C., 327, 330
Lopez, A. D., 222
Lopez de Silanes, I., 376
Lopez-Nevot, M. A., 365, 366
Lorz, C., 237
Loscalzo, J., 466
Louis, C. A., 407
Louis, D. N., 92, 378
Louis, N. A., 308, 325, 407
Lounsbury, K. M., 426, 427, 471
Low, K. B., 314
Lowe, S. W., 307
Loya, F., 160
Lu, H., 47, 234
Lu, L., 89, 93, 100, 204
Lu, P. D., 254, 279
Lu, Z., 330
Ludolph, B., 234
Lukashev, D., 406, 407
Lukyanov, Y., 330
Luna, J. D., 206
Lundkvist, J., 70, 76
Lupu, F., 91, 201
Luschnig, S., 126
Lutwyche, J. K., 5, 8
Lutz, M. S., 374
Lyapina, S., 363
Lyden, D., 205
Lygate, C., 27, 47, 226, 229
Lyn, P., 201

M

Ma, E., 129, 134
Ma, J., 231
Ma, M. C., 236
Ma, W., 377
Ma, Y., 172, 278
Mabjeesh, N. J., 313, 330, 392
Macann, A., 310
MacDougall, I. C., 191, 231
Machulik, A., 425
Macino, G., 109

Mack, F., 92, 100, 378
Mackay, A., 186
MacKenzie, E. D., 47
Mackinnon, R., 6
Mackman, N., 94, 100, 213, 227, 406, 480
Madan, A., 181
Maddrell, S. H., 136
Maeda, M., 330
Magagnin, M. G., 249, 250, 252, 253, 254, 255, 262, 265, 276, 277, 278
Magid, R., 427
Maguire, A. M., 117
Mahajan, A., 64, 77
Mahajan, R., 185
Maher, E. R., 4, 129, 183, 226, 327, 375, 378, 425, 449, 464, 465
Maher, G. W., 377
Mahmud, N., 186
Mahon, P. C., 44, 46, 62, 65, 89, 114, 115, 328, 378, 413, 425, 429
Maity, A., 327, 328, 330
Majamaa, K., 39, 47, 71
Makino, N., 231
Makino, Y., 5, 209, 236, 332, 406, 411, 425, 464
Makuuchi, M., 467, 468
Malkin, D., 376
Mallet, J., 425
Mallet, R. T., 223, 234
Malmgren, R. A., 314
Malone, P. R., 365
Maltepe, E., 89, 91, 133, 204, 213, 214, 227, 332, 448, 449, 454, 458, 470
Manalo, D. J., 234
Mancada, S., 476
Mancardi, D., 469
Manche, L., 109
Mandai, M., 206
Mandriota, S., 181, 225, 227, 229, 235, 236, 372, 377
Manikas, K., 348
Manning, B. D., 136
Manning, G., 126, 127
Mansfield, K. D., 47, 448, 470
Mansuy, D., 47, 413, 429, 464, 471
Mansy, S. S., 6
Mantovani, A., 201
Mao, J., 378
Mao, M., 112, 230, 378
Maranchie, J. K., 46, 230, 372, 427
Marchetti, S., 430
Marciniak, S. J., 254, 279
Marck, B. T., 90, 95, 211, 212, 227, 332
Mareel, M. M., 348
Marincola, F. M., 314
Marsden, P. A., 373
Marshall, H. E., 467
Marshall, P. A., 466
Marth, J. D., 94

Marti, H. H., 224, 449
Martin, F., 47, 54, 173, 188
Martin, J., 315, 365, 366
Martin, L., 184, 185, 312
Martin, M. E., 254
Martindale, J. L., 279, 376
Martinelli, V., 372
Martinet, D., 188
Martinez, A., 329, 330, 392
Martinez, J., 159, 162, 168
Martinez-Yamout, M., 64, 397
Martin-Puig, S., 112
Martins, P. N., 231
Maruyama, H., 185
Maruyama, K., 231
Marxsen, J. H., 46, 112, 230
Mason, S. D., 95, 236
Mason, Y. M., 44, 46
Masson, N., 4, 26, 27, 29, 30, 36, 46, 88, 89, 110, 112, 131, 226, 229, 324, 327, 328, 425, 429, 449, 464
Massoud, T. F., 160
Mastrofrancesco, B., 407
Masuda, S., 184
Mateo, J., 412, 473
Mathews, M. B., 109
Mathieu, C. E., 448, 454, 458, 470
Mathioudakis, N., 33, 34, 37, 46, 62, 64, 72, 77
Matsubara, O., 281
Matsuda, Y., 7, 17, 89
Matsui, M., 362, 364
Matsui, S., 184
Matsui, T., 280
Matsumoto, A. M., 89, 90, 91, 95, 100, 151, 201, 202, 211, 212, 227, 332
Matsumoto, K., 181
Matsumoto, M., 201, 236
Matsumura, M., 201
Matsuoka, T., 171
Matteucci, E., 426, 428
Mauchline, M. L., 314
Maulik, N., 234
Maxwell, A. P., 160, 372, 377
Maxwell, P. H., 4, 14, 26, 27, 36, 46, 47, 88, 89, 93, 98, 110, 129, 131, 160, 181, 182, 183, 184, 185, 186, 223, 225, 226, 227, 229, 232, 233, 235, 236, 276, 311, 312, 324, 325, 327, 328, 332, 348, 349, 356, 372, 373, 374, 375, 378, 425, 429, 449, 464, 465
Maynard, M. A., 349, 373
Mayser, H., 233
Mazan-Mamczarz, K., 376
Mazure, N. M., 114, 115, 327, 328, 330, 430, 449
Mazurek, B., 425
Mazzone, M., 374
McCaffrey, A. P., 117
McCarthy, H. O., 312
McClintock, D. S., 327, 426, 454, 458, 471

McCloud, T. G., 399
McClure, B., 310
McCormick, T. S., 414
McCullough, B., 185, 186
McDonald, D. M., 89, 97, 237
McDonald, E., 330, 391
McDonough, M. A., 27, 28, 30, 32, 33, 35, 38, 39, 55, 63, 64, 65, 77, 79, 226, 230, 481
McErlane, V., 312
McFate, T., 47
McGary, E. C., 172
McGlave, P. B., 159
McGowan, F. X., 407
McKeown, S. R., 310, 312
McKnight, S. L., 4, 7, 27, 29, 88, 89, 90, 91, 100, 110, 129, 131, 201, 202, 226, 324, 332, 449, 464
McLaughlin, P. J., 63
McLellan, A., 205, 206
McMahon, S., 412
McManaman, J., 96
McMullin, M. F., 46, 372
McNamara, C. J., 372
McNeill, L. A., 4, 27, 28, 29, 30, 32, 33, 35, 36, 39, 44, 46, 47, 62, 63, 64, 71, 72, 75, 77, 88, 89, 110, 112, 131, 226, 229, 324, 328, 378, 425, 429, 449, 464
McNulty, W., 95, 96, 98, 100, 204, 236, 276
McWilliams, R., 93
Means, R. T., Jr., 407
Mechta-Grigoriou, F., 47, 413, 429, 464, 471
Meek, J., 185
Mehn, M. P., 27, 29, 30, 38
Mehranpour, P., 437
Mehta, S., 180
Meinhardt, A., 357
Mei-Zhahav, C., 187
Mekhail, K., 229
Melendez, J. A., 327, 426, 458, 471
Melillo, G., 20, 201, 313, 329, 330, 357, 385, 386, 387, 388, 389, 390, 391, 392, 393, 395, 397, 399
Melinek, J., 231
Meller, B., 148
Mello, C. C., 108
Melo, L. G., 230
Melo, M. E., 188
Melton, D. A., 201
Melton, R. G., 310
Memmert, K., 313, 330, 331, 392, 397, 398
Mendez, F., 373
Menges, T., 32
Menne, J., 231
Mennini, T., 231, 232
Menzies, K., 27
Merchant, K., 467
Merino, M., 374, 375
Merrick, W. C., 251, 252
Merrifield, R. B., 72
Messadi, D. V., 98, 100
Metz, C., 357, 362
Metzen, E., 4, 19, 27, 29, 30, 36, 44, 46, 47, 88, 89, 110, 112, 131, 182, 184, 226, 229, 230, 324, 407, 412, 429, 448, 449, 464, 469, 470, 473
Metzger, R. J., 128
Metzstein, M. M., 126
Meuse, L., 117
Meyer, T., 109
Miasnikova, G. Y., 184, 237, 372
Michael, N. P., 315
Michalak, M., 280
Michel, G., 5, 327, 409, 414
Michelson, K. S., 4, 44, 226, 229, 234, 429
Michieli, P., 374
Michiels, C., 5, 327, 409, 413, 414, 425, 426, 427
Migliaccio, C., 372
Mikami, B., 11, 12
Milanini, J., 409
Milkiewicz, M., 236
Miller, A., 201
Miller, C. N., 278
Miller, J. J., 236
Millhorn, D. E., 135, 182
Mimnaugh, E. G., 329, 330, 392
Mimura, J., 8, 425, 428
Mimuro, J., 185
Min, J. H., 425
Minato, N., 375, 449
Minchenko, A., 426, 470
Minegishi, N., 166
Miner, J. H., 237
Minet, E., 5, 327, 409, 413, 414
Minks, M. A., 109
Minor, D. L., Jr., 64
Minton, N. P., 314, 315
Miron, M., 135
Mischak, H., 375
Mischke, R., 362, 363
Mitchell, J. B., 466, 471
Mitchell, R. A., 355, 357, 360, 362, 363, 364
Mitchley, B. C. V., 310
Mitsock, L. D., 171
Mitsuhashi, Y., 327, 328
Miura, C., 472
Miura, Y., 172
Miyamoto, N., 206
Miyamoto, T., 212
Miyata, T., 236
Miyazaki, H., 330, 433
Miyazaki, J., 185, 204
Mizrachi, S., 187
Mizukami, H., 185
Mlodzik, M., 135
Moch, H., 325
Mochan, E., 428

Modaressi, S., 425
Modi, N., 180
Modrich, P., 325, 326
Moe, O., 95, 182, 212
Moeller, B. J., 304, 307
Mohr, A., 184
Mohr, C., 27, 28, 226
Mohyeldin, A., 47
Mole, D. R., 4, 26, 27, 29, 30, 36, 44, 46, 88, 89,
    110, 112, 131, 183, 184, 226, 229, 324, 327,
    372, 374, 375, 378, 411, 425, 429, 449,
    464, 465
Molineaux, C. J., 183
Moller, U., 428, 429
Molls, M., 305
Mommers, E. C., 348
Moncada, S., 47, 448, 465, 473, 480
Mondotte, J. A., 27
Monette, J. S., 435
Monks, A., 329, 330, 386, 391, 393, 395
Montagne, J., 135
Montalto, M. C., 308, 325
Montaner, S., 330, 433
Monteiro, J., 364
Montgomery, M. K., 108
Moon, C., 148
Moons, L., 97, 98
Moore, J., 377
Moore, J. W., 116, 330, 425
Moore, M. W., 89, 93, 100, 204
Moorman, A., 90
Morais Cabral, J. H., 6
Moran, S. M., 7, 89, 92, 332
Morello, F., 230
Moreno-Murciano, M. P., 98
Morgillo, F., 330
Mori, K., 277, 278, 280, 281, 307
Mori, S., 160
Morikawa, K., 280
Morimoto, C., 406, 411
Morishita, Y., 160
Morita, M., 166, 201
Morita, T., 472
Morris, M. A., 95, 211, 212, 225, 235
Morton, K. E., 314
Mosavi, L. K., 64
Mosca, J., 186
Moschella, M. C., 27
Möse, G., 314
Möse, J. R., 314
Mosher, J. T., 129
Moss, S. E., 205
Mota-Philipe, H., 231
Motohashi, H., 159, 161, 163
Mottet, D., 5, 327, 409, 413, 414
Mottram, J. C., 300
Moulder, J. E., 302, 303
Moullier, P., 185, 186

Moynagh, P., 408, 414, 481, 487
Muchmore, E., 189
Mufson, A., 173
Muhling, J., 32
Mukai, H. Y., 159, 161, 162, 163, 165, 166,
    168, 171
Mukherji, M., 4, 26, 27, 29, 30, 36, 44, 46, 88, 89,
    110, 112, 131, 183, 226, 229, 324, 327, 372,
    425, 429, 449, 464
Mukhopadhyay, C. K., 47, 54
Mukhopadhyay, D., 115, 375
Mukuria, C. J., 189
Mulder, F. A. A., 20
Muller, R. U., 374
Muller, W. J., 99
Muncan, V., 115
Murakami, T., 184, 280
Muranyi, A., 63
Murdaca, J., 135, 330, 425
Murdoch, C., 406, 407, 479
Murillas, R., 97
Murphy, M. P., 435
Murrant, C. L., 431, 432
Murray, C. J., 222
Murray, D. M., 312
Murray, G. I., 310, 312
Murray, M., 312
Murray, P. I., 310
Murry, C. E., 222, 237
Musacchio, A., 162
Muschen, M., 308, 325
Mushinski, J. F., 375
Muthana, M., 406, 479
Muthukrishnan, E., 183
Muzzopappa, M., 27, 129, 130, 131, 132
Myers, J. N., 330
Myers, J. W., 109
Myint, H., 310, 311
Myllyharju, J., 27, 30, 33, 36, 39, 46, 47, 51, 62,
    64, 71, 72, 73, 75
Myllyla, R., 39, 47, 71, 229

N

Naczki, C., 251, 252, 253, 254, 256, 276, 277,
    278, 279, 281, 282, 283, 284, 287
Naegeli, H. U., 313, 330, 331, 392, 397, 398
Naffakh, N., 184, 186
Nafz, B., 154
Nagahata, T., 281
Nagai, H., 281
Nagano, M., 199
Nagao, M., 128, 129, 168, 425, 449
Nagasawa, H., 348, 349, 351, 373
Nagasawa, T., 162, 171, 181
Nagashima, Y., 374
Nagasue, N., 367
Naglich, J., 372, 376

Nagy, A., 46, 89, 92, 133, 237
Naiki, M., 189
Nairz, K., 135
Nakagawa, A., 362
Nakahata, T., 159, 161, 163, 165, 166, 168
Nakaigawa, N., 374
Nakajima, O., 166, 201
Nakamura, E., 372, 375, 376
Nakamura, H., 406, 411
Nakano, M., 166
Nakano, Y., 181
Nakayama, K., 46
Nam, Y., 69
Nambu, J. R., 5, 128, 129
Namgaladze, D., 471
Namiuchi, S., 154
Nangaku, M., 236
Napier, M., 315
Napoli, C., 109, 426
Naranjo, S., 46, 112
Naranjo-Suarez, S., 373
Narouz-Ott, L., 48
Natarajan, R., 235
Naylor, S., 184, 185, 312
Nebuloni, M., 201
Neckers, L., 5, 329, 330, 374, 375, 392, 411
Neeman, M., 98, 332
Neff, N. T., 63
Neff, T. B., 183
Negishi, M., 280
Neichi, T., 185
Neiditch, M. B., 19
Neil, D. A., 227, 228
Neil, L. C., 7, 12, 20
Neill, S. D., 173
Nejfelt, M. K., 158, 160, 173, 181, 332
Nelson, D., 277, 281, 307
Nelson, M., 186
Nemery, B., 91, 201
Nemukhin, A. V., 182
Neter, E., 314
Nett, J. H., 188
Neufeld, T. P., 135
Neumann, A. K., 88, 95, 100, 407
Neumayer, H. H., 183
Neville, M., 96
Newcomb, E. W., 330
Newman, D. J., 330, 392
Ng, J. C., 135
Ngo, F. Q., 306
Nguyen, K., 423, 425
Ni, R., 27
Nichol, L., 27
Nicholls, D. G., 435
Nicholls, L. G., 160, 276
Nichols, L., 47, 226, 229
Nicholson, A. C., 330

Nickols, N. G., 395
Nico, B., 207
Nicolau, M., 325, 326
Nicosia, R. F., 201
Nielsen, J., 154
Nielsen, J. E., 414, 481, 487
Nieminen-Kelha, M., 231
Niname, N., 327
Nishi, K., 409
Nishihira, J., 362, 365
Nishihiri, J., 362
Nishikawa, Y., 185
Nishimura, M., 365
Nishimura, S., 159
Nithipatikom, K., 434, 435
Nitschke, R., 374
Niwa, H., 202
Niwa, M., 280
Nizet, V., 94, 95, 100, 213, 227, 406, 408, 480
Nobili, B., 372
Noble, M. A., 185
Noda, M., 312
Noguchi, A., 189
Noguchi, C. T., 159
Nohara, R., 409
Noiri, E., 236
Noiseux, N., 230
Noll, G., 148, 149, 150
Nomura, S., 213, 407
Nordenskjold, B., 348
Nordsmark, M., 305
Nose, K., 330
Novoa, I., 252, 254, 279, 281, 283, 287
Noyszewski, E. A., 95
Numayama-Tsuruta, K., 472
Nusse, R., 135
Nuyens, D., 97
Nwogu, J. I., 229, 230, 234
Nyborg, J., 252
Nystrom, G., 129

O

Oakeley, E. J., 325
Obara, N., 157, 159, 161, 162, 163, 171, 181
Obeid, A., 300
Oberhuber, G., 348
Obermair, A., 348
O'Brien, S., 5, 425, 428, 464
O'Brien, T. S., 349, 373, 374
Ockaili, R., 235
O'Connell, F., 89, 183
O'Connor, P. M., 171
Oda, H., 127
Oda, S., 409
Oda, T., 409
Odh, G., 362

O'donnell, J. L., 183
Oehme, F., 43, 47, 48, 54, 425, 464
Ogata, M., 280
Ogawa, K., 281
Ogawa, S., 201
Oguchi, Y., 433
Ogura, T., 182, 185, 467, 468
Oh, H., 201, 206
Oh, J., 46, 114
Oh, S. H., 330
Oh, T. K., 185
Ohashi, H., 184
Ohh, M., 4, 26, 27, 36, 88, 89, 92, 110, 324, 327, 328, 372, 373, 375, 376, 378, 425, 449, 464, 490
Ohkotin, D., 372
Ohmine, K., 162, 171
Ohneda, K., 161, 163, 165, 199, 212
Ohneda, O., 159, 161, 163, 165, 166, 168, 171, 201, 212
Ohno, S., 375
Ohnuma, K., 406, 411
Ohsawa, S., 8
Ohse, T., 236
Ohshiro, T., 129
Ohta, A., 407
Ohta, J., 154, 159, 166
Oikawa, M., 154
Oike, M., 433
Ojima, T., 184
Okada, N., 14
Okada, T., 185, 280, 281
Okamoto, K., 5, 236, 406, 411, 425, 428
Okamoto, N., 206, 207
Okhotin, D., 237
Oksanen, S., 183
Oktay, Y., 90, 95, 182, 211, 212, 225, 227, 235, 332
Okuda, H., 375
Okuyama, H., 348, 349, 351, 373
Oldenburg, K. R., 72
Oldham, N. J., 4, 27, 28, 30, 32, 33, 35, 39, 44, 47, 55, 62, 63, 64, 65, 72, 79, 89, 226, 230, 328, 378, 425, 429, 481
Oldham, S., 129, 131, 132, 135, 136, 138
Olenyuk, B. Z., 395
Olfert, I. M., 95, 236
Olive, P. L., 298, 300
Oliver, D. O., 173
Olivero, W. C., 117
Olmos, G., 112
Olofsson, B., 201
Olsen, B. R., 133
Olson, J. J., 330
Olver, I., 309
Olver, R. E., 426
Omran, H., 374

Onda, M., 281
Ong, E. T., 300
Ono, M., 166
Ono, T., 367
Onodera, K., 159
Oosthuyse, B., 97
Orban, P. C., 94
Orive, G., 186
Orkin, S. H., 161, 165
O'Rourke, J., 4, 27, 29, 30, 36, 44, 46, 88, 89, 110, 112, 129, 131, 226, 229, 276, 324, 425, 429, 449, 464
Orringer, M. B., 357
Ortego-Centeno, N., 365, 366
Orville, A. M., 469
Osborne, W. R., 186
O'Shea, K. S., 89, 93, 100, 204
Osmanian, C., 307
Osmond, M. K., 160
Ossent, P., 148
Otani, A., 201, 206
Otani, H., 234
Otterbein, L. E., 231, 236
Otterness, D. M., 327, 330
Otto, T., 19, 407, 410
Ouyang, L., 98, 100, 233
Overgaard, J., 305
Overgaard, M., 305
Oving, I., 115
Ozaki, H., 206, 207
Ozaki, K., 206
Ozawa, K., 162, 171, 185
Ozer, A., 46, 116

P

Paavonen, K., 201
Pachori, A. S., 230
Pack, S., 377
Paddison, P. J., 109
Pagano, J. S., 428
Page, E. L., 408, 409, 428
Pagel, H., 47, 182, 237, 407
Pages, G., 330, 409, 430
Paik, D., 148
Paine-Murrieta, G., 330, 392
Pajusola, K., 237
Pak, B., 135, 348, 350
Pal, S., 375
Palmer, C., 96
Palmer, D., 310
Palmer, L. A., 467, 468
Palmer, R. M., 465
Pan, X., 159, 161, 163
Pan, Y., 47, 430, 448, 470
Papac, D., 180
Papahadjopoulos, D., 300
Papandreou, I., 112, 473

Paraskeva, E., 8
Paricio, N., 135
Parisi, G., 372
Park, C. W., 327, 328, 425
Park, E. J., 20, 329, 330, 391, 393, 395, 397
Park, F., 185
Park, H., 12, 187, 426, 427, 429, 430
Park, H. C., 187
Park, J., 201
Park, J. H., 426
Park, J. W., 313, 330, 392, 426, 427
Park, S. K., 94, 204, 332
Park, S. R., 187
Parker, A. R., 302
Parsa, C. J., 151
Partridge, L., 124
Pascal, L. S., 151
Pascual, O., 425
Pasquali, C., 231, 232
Passaniti, A., 327
Pastore, Y., 237
Pat, B., 148
Patel, J. H., 100
Patel, N. S., 230, 231
Patil, C. K., 280
Patil, S., 186
Patsouris, E., 348
Patterson, A. V., 310, 311, 312
Patterson, L. H., 310, 312
Patton, S. M., 230, 233
Paules, R., 279
Pause, A., 449
Pavletich, N., 327, 328, 425
Pavlopoulos, P. M., 348
Pawelek, J. M., 314
Pawling, J., 89, 93
Pawlinski, R., 94, 100, 213, 227, 406, 480
Payen, E., 185, 186
Pearson, M., 46, 114
Pease, E. A., 428
Peck, L. S., 136
Pedraz, J. L., 186
Peduto, G., 186
Peet, D. J., 4, 26, 27, 44, 61, 62, 63, 64, 65, 67,
    68, 70, 72, 75, 77, 79, 82, 114, 226, 328,
    332, 378, 413, 414, 425, 429, 465, 481
Pehar, M., 435
Pei, Y., 112
Pellequer, J. L., 6
Pelto-Huikko, M., 19
Penar, P. L., 426, 427, 471
Peng, J., 91, 201
Peng, T., 357
Peng, Y. J., 93
Peng, Z. Y., 64
Pennacchietti, S., 374
Penniment, M., 310

Pequignot, J. M., 425
Percy, M. J., 46, 372
Perdiki, M., 348
Pereira, T., 70, 76, 237, 425
Perkins, A. C., 161, 165
Perkins, G. L., 126
Perrakis, A., 28
Perrimon, N., 126
Perrotta, S., 372
Persengiev, S. P., 112
Peruzzi, B., 374
Pescador, N., 46, 112
Peters, H., 304, 325
Peters, L., 310
Petit, E., 232
Petit, V., 128
Petrat, K., 308, 325
Petrofski, J. A., 151
Petroski, C. J., 63
Petry, A., 423
Petry, E., 301
Pettersen, E. O., 249, 250
Peyssonnaux, C., 95, 408, 409
Pfander, D., 133
Pham, T. T., 117
Phang, J. M., 47, 182, 230
Philipp, S., 225, 230, 232, 233, 234, 235
Philipsen, S., 161, 163, 165
Piantadosi, C. A., 160
Pichiule, P., 181, 472
Piecha, G., 183
Pierce, E. A., 206, 207
Pietri, S., 437
Pigott, K. H., 300
Pili, R., 392
Pilz, B., 225, 230, 235
Pinedo, H. M., 348, 426
Pinsky, D. J., 201
Pippard, M. J., 173
Pisarczyk, M. J., 407
Plaisance, S., 91, 201
Plank, C., 348
Plant, B. J., 364
Plate, K. H., 465
Plotkin, M. D., 160
Podjarny, E., 187
Poellinger, L., 5, 70, 76, 158, 159, 173, 209, 236,
    237, 325, 326, 327, 332, 364, 406, 411, 425,
    428, 464
Pogrebniak, A., 423, 430
Pollard, J. W., 99, 407
Pollefeyt, S., 89, 93
Poloni, M., 96, 98, 100, 204, 276
Polyakova, L. A., 237, 372
Pommier, Y., 313, 388, 391, 392
Pompeani, A. J., 19
Ponce, S., 186

Pon Fong, M. T., 115
Ponting, C. P., 6
Pontoglio, M., 182
Poon, G. K., 315
Popko, B.1, 279
Pore, N., 327, 328, 330
Porteous, D. D., 160
Porter, F. D., 89, 92, 100
Poser, W., 184
Post, D. E., 313, 392
Potter, C. J., 135
Potter, R., 348
Poulsen, M., 310
Pouyssegur, J., 27, 46, 47, 107, 111, 112, 114,
    115, 135, 230, 327, 328, 330, 331, 376, 407,
    409, 413, 414, 426, 427, 428, 429, 430, 464,
    471, 480, 481, 487
Powell, D. R., 327
Powell, M. E., 300
Powell-Braxton, L., 89, 93, 100, 204
Powell-Coffman, J. A., 27
Powis, G., 183, 330, 392, 428
Prabhakar, N. R., 93
Pradet-Balade, B., 266
Pratschke, J., 231
Pratt, R. D., 189
Pratt, R. E., 230
Prchal, J. F., 237
Prchal, J. T., 237, 372
Prendergast, G. C., 19
Presta, M., 207
Pribluda, V. S., 313, 330, 392
Price, J. C., 71
Print, C., 409
Prodinger, A., 348
Prosnitz, L. R., 305
Prost, M., 205
Proud, C. G., 255
Provost, N., 185, 186
Pruijn, F. B., 307
Prywes, R., 280
Puchowicz, M., 472
Pugh, C. W., 4, 14, 26, 27, 30, 36, 44, 46, 47,
    55, 62, 63, 64, 65, 72, 75, 77, 79, 88,
    89, 100, 110, 111, 128, 129, 131, 160,
    181, 183, 223, 224, 225, 226, 227, 229, 230,
    236, 276, 311, 324, 325, 327, 328, 332,
    348, 356, 372, 377, 378, 392, 425, 429, 449,
    464, 481
Pullen, S. M., 311, 312
Punzo, F., 372
Putz, V., 308, 325
Puype, M., 46
Pyronnet, S., 249, 250, 252, 253, 254, 255, 262,
    265, 276, 278
Pyrzynska, B., 330

Q

Qadri, F., 425
Qi, L., 114, 362, 363
Qi, R. C., 72
Qian, D., 430
Qian, D. Z., 392
Qiao, T., 430
Quaggin, S. E., 237
Quan, G. H., 185
Quaschning, T., 148, 149, 150
Que, L., Jr., 27, 29, 30, 38, 469
Quesada, N. M., 448, 449, 470
Quintero, M., 412

R

Rabbani, Z. N., 117
Rabinowitz, J. E., 185, 186
Rabow, A. A., 386
Radcliffe, L. A., 92
Raden, D., 278
Radimerski, T., 135
Radtke, F., 93
Raes, M., 5, 327, 409, 413, 414
Rafii, S., 205
Raftery, M., 231
Ragel, B. T., 98, 100
Ragione, F. D., 372
Rago, C., 332
Raida, M., 279, 281
Rainey, K., 348, 349, 351, 373
Rajagopal, V., 481
Rajasekhar, V. K., 249, 255
Rak, J., 327, 328, 373, 378
Raleigh, J., 252, 254, 279, 281, 283, 287
Ramakrishnan, D., 93
Ramakrishnan, K., 425
Ramakrishnan, M., 277
Ramakrishnan, S., 464
Raman, C. S., 11, 12
Ramaswamy, S., 28
Ramesh, N., 186
Ramirez, J. R., 182
Ramirez, S., 88, 100
Ramírez-Bergeron, D. L., 204
Rangell, L., 93
Rankin, E. B., 100, 181, 182, 224, 227, 234,
    377, 378
Rao, J. S., 117
Rao, J. Y., 98, 100
Rapisarda, A., 201, 313, 329, 330, 387, 388, 389,
    390, 391, 392
Rapka, M., 374
Rastani, S., 251, 252, 253, 256, 276, 277, 278,
    279, 281, 282, 283, 284

Ratcliffe, P. J., 4, 14, 25, 26, 27, 30, 33, 34, 36, 37, 44, 46, 47, 62, 63, 64, 71, 72, 75, 77, 79, 88, 89, 98, 100, 111, 115, 116, 128, 129, 130, 131, 132, 133, 136, 137, 148, 160, 181, 182, 183, 186, 223, 224, 225, 226, 227, 229, 235, 236, 276, 311, 312, 324, 325, 327, 328, 332, 348, 356, 372, 374, 375, 377, 378, 411, 413, 425, 426, 429, 449, 464, 465, 470, 480
Raught, B., 252
Raval, R. R., 46, 47, 111, 116, 349, 372, 373, 374, 377, 425, 426, 429
Raychaudhuri, S., 324, 325
Recio, M. J., 302
Reddy, A., 327
Regan, M., 89, 94, 97, 375, 377
Regitz-Zagrosek, V., 225, 230, 235
Regulier, E., 186
Reich, S. J., 117
Reichman-Fried, M., 127
Reichner, J. S., 407
Reid, C., 173
Reid, M. B., 431, 432
Reiling, J. H., 136
Reilly, M., 232
Reimann, C., 172, 181
Reimer, K. A., 222, 237
Reincke, U., 160
Reinhard, C., 135
Reinke, P., 181
Reis, C., 302
Reiter, J. F., 373
Remacle, J., 5, 409, 414
Reme, C. E., 233
Ren, X., 182
Ren, Y., 182, 357
Rendon, B. E., 357, 360, 362, 363, 364
Restifo, N. P., 314
Reutzel-Selke, A., 231
Rey, A., 309
Reyes, H., 5, 14, 424
Reynolds, M., 47
Rha, J., 181, 182, 224, 227, 234
Rheinlander, C., 425
Rhoads, R. E., 72
Ria, R., 207
Riballo, E., 302
Ribati, D., 207
Ribeiro, C., 128
Rich, I. N., 159
Richard, D. E., 114, 135, 327, 328, 407, 408, 409, 412, 426, 427, 428, 430
Richardson, D. R., 230
Richardson, J. A., 90, 95, 182, 204, 211, 212, 227, 332
Richardson, K. S., 362
Richardson, R. S., 95
Rickles, R. J., 109
Riddle, L., 207

Ridnour, L. A., 466, 469, 471
Riedel, J., 98
Rief, N., 302
Riel, R. U., 151
Riggins, G. J., 325
Rigoutsos, I., 63
Rinsch, C., 186
Rintelen, F., 135
Riordan, M. V., 4, 30, 44, 62, 63, 72, 89, 226, 328, 378, 425, 429
Rios, S., 188
Risau, W., 89, 200, 201
Rischin, D., 310
Riss, J., 46, 230, 372
Ristimäki, A., 201
Ritman, E. L., 426
Rittinger, K., 423
Ritzen, M., 184
Rivera, V. M., 185
Roach, P. L., 469
Robbins, P. A., 182, 372
Roberts, A. M., 349, 373
Roberts, D. D., 466, 471
Roberts, I. J., 301
Roberts, J. J., 310
Roberts, J. R., 327, 426, 430
Roberts, R. A., 182
Roberts, W. K., 306
Robertson, N., 330
Robin, E., 448, 470
Robinson, B. A., 201
Robinson, K. J., 375
Robinson, K. M., 435
Robitaille, G. A., 409, 428
Robson, T., 312
Roche, O., 349, 373
Rock, S., 127
Rockwell, S., 302, 303
Rodriguez, A. M., 327, 426, 458, 471
Rodriguez, E. R., 234
Rodriguez-Porcel, M., 426
Roe, J. S., 376
Roe, R., 5, 129, 387, 393, 395, 425, 449
Roecker, A. J., 330
Roessler, L., 374
Roger, T., 362, 363
Rogers, P., 330, 391
Roh, B. L., 160
Roland, I., 5, 409, 414
Rologis, D., 348
Romano, N., 109
Romero, N. M., 123
Romero-Ramirez, L., 277, 278, 281, 307
Ron, D., 249, 254, 276, 277, 279, 280, 282
Ronai, Z., 46
Roncali, L., 207
Rondon, I. J., 172
Rong, Y., 427

Rönicke, V., 201
Rorsman, H., 362
Rosbash, M., 5, 14
Rose, D. W., 19
Rosen, S., 181
Rosenberger, C., 181, 225, 227, 229, 232, 233, 235, 236
Rosenfeld, M. G., 19
Rosengren, E., 362
Rosenthal, E. R., 436
Ross, M. F., 435
Ross, W. C. J., 310
Rossi, F., 372
Rossi, J. J., 117
Rossler, K., 348
Roth, P. H., 88, 100
Roth, U., 135, 410, 412, 425, 426, 427, 429, 430, 471
Rothe, G., 434
Rothstein, M., 209
Rotoli, B., 372
Roux, D., 27, 46, 111, 112, 115, 135, 230, 328, 331, 376, 409, 430
Rouyer-Fessard, P., 185, 186
Roy, B., 277
Roy, J., 134
Royall, J. A., 435
Rozamus, L. W., 185
Ruas, J. L., 70, 76
Rubin, G. M., 126
Rubinstein, M., 135, 425
Rubinstenn, G., 20
Rudat, V., 305
Rudersdorf, R., 173
Rudge, J. S., 205
Rue, E. A., 4, 5, 13, 88, 89, 112, 181, 204, 348, 356, 386, 387, 393, 395, 424, 425, 428, 430, 449
Rulicke, T., 148, 149, 150
Rulifson, E. J., 135
Runge, A., 204
Rupar, M. J., 63
Ruparelia, K., 310
Ruschitzka, F. T., 148, 149, 150
Russell, D. W., 7, 89, 90, 91, 100, 201, 202, 332
Russell, R. C., 349, 373
Russo, D. M., 27, 129, 130, 131, 132
Rust, R. T., 135
Rustenbeck, H.-H., 184
Rutkowski, D. T., 278
Rutkowski, K., 407
Rutter, J., 6, 7, 9, 20
Ryan, A. M., 93
Ryan, H. E., 89, 90, 96, 97, 98, 100, 135, 204, 211, 225, 227, 276, 327, 328, 330, 332
Ryden, L., 348
Ryeom, S. W., 98
Ryu, S. E., 63, 429, 430

S

Saarinen, T., 10
Sabio, J. M., 365, 366
Sablotzki, A., 32
Saccani, A., 201
Saccani, S., 201
Sadek, C. M., 19
Sadighi Akha, A. A., 278
Sadlon, T. J., 211
Sadri, N., 279
Safran, M., 89, 94, 97, 183, 375, 377
Sagami, I., 11, 12
Saigo, K., 129
Saito, A., 280
Saito, K., 160
Saito, T., 171
Saitoh, K., 375
Sajan, M. P., 375, 376
Sakai, N., 375
Sakamoto, F., 185
Sakamoto, M., 133
Sakata, Y., 185
Sakkas, D., 348
Sakuma, M., 154
Salceda, S., 327, 425, 449, 464
Salic, A., 4, 26, 27, 36, 88, 89, 110, 324, 327, 372, 425, 449, 464, 490
Salloum, F. N., 234, 235
Salnikow, K., 47, 182, 230
Saluja, R., 465
Salvetti, A., 185
Samakovlis, C., 126, 127, 128
Samal, B., 186
Samardzija, M., 233
Samonigg, H., 348
Samons, M., 19
Samoylenko, A., 135, 425, 430
Sampson, B. A., 153
Samulski, R. J., 185, 186
Sanchez, E., 365, 366
Sanchez-Elsner, T., 182
Sandau, K. B., 327, 407, 412, 429, 467, 468, 469
Sander, F., 5, 14, 424
Sandor, Z., 426
Sang, N., 330, 392, 408, 426, 470
Santolini, J., 465
Sanz-Rodriguez, F., 182
Saravanan, A., 349, 373
Sarmiere, P., 376
Sartorelli, A. C., 308
Sasai, K., 312
Sasaki, F., 362
Sasaki, H., 234
Sasaki, R., 160, 168, 187
Sasazuki, T., 327, 328
Sassone-Corse, P., 7, 12, 19
Sato, J. D., 98, 100

Sato, S., 425
Sato, T., 160
Sato, Y., 362
Satoh, K., 166
Satterlee, J. D., 12
Sauer, B., 94
Sauer, H., 308, 325
Saunders, M., 300, 312
Sausville, E. A., 313, 330, 386, 387, 390, 392
Savelkouls, K., 249, 250, 252, 253, 254, 255, 262, 265, 276, 278
Savino, C., 154
Sawdey, M., 185
Sawitzki, B., 231
Scalzitti, J. M., 348
Scandurro, A. B., 325
Scarpulla, R. C., 448, 449, 470
Scharf, O., 4
Schau, M., 129, 425
Scheel, J. R., 98, 100, 233
Schefold, J. C., 226, 230
Scheid, A., 47, 154, 182, 237
Schelter, J., 112
Scherer, S. E., 126
Scherman, D., 185
Schermer, B., 374
Scheuermann, T. H., 3
Scheuner, D., 252, 254, 279, 281, 283, 287
Schiche, A., 225, 230, 235
Schindl, M., 348
Schindler, C., 135, 325, 327, 328, 330
Schindler, S. G., 5, 327
Schini-Kerth, V. B., 410, 412, 426, 427, 429, 430, 471
Schioppa, T., 201
Schipani, E., 97, 100, 133
Schlabach, M. R., 109
Schlaifer, J. D., 222
Schlemminger, I., 30, 63, 64, 75, 77
Schlessinger, J., 135
Schley, G., 148
Schlisio, S., 375, 376
Schmedt, C., 109
Schmid, T., 410, 412, 471, 473
Schmidbauer, G., 231
Schmid-Schonbein, H., 148
Schmidt, J. V., 89, 91, 204, 227, 332, 449
Schmidt, l., 375, 377
Schmidt, L. S., 377
Schmidt, R., 183
Schnee, T., 330
Schneider, B. L., 186
Schnell, M. A., 185
Schnurr, K., 424
Schoen, F. J., 153
Schoenfeld, A., 372, 373, 376
Schoenfeld, A. R., 377, 429

Schofield, C. J., 4, 25, 26, 27, 28, 29, 30, 32, 33, 34, 35, 36, 37, 38, 39, 44, 46, 47, 62, 63, 64, 71, 72, 75, 77, 88, 89, 110, 131, 148, 183, 226, 229, 324, 327, 328, 356, 372, 378, 425, 429, 449, 464, 469, 480
Scholz, C., 479
Scholz, H., 148, 149, 154, 172
Scholze, C. K., 225, 227, 229
Schoppmann, S. F., 348
Schramm, M., 425, 464
Schramm, U., 148
Schroder, M., 278
Schroedl, C., 454
Schulz, R., 234
Schulze, S., 7, 12, 19
Schumacher, D., 148
Schumacker, P. T., 327, 426, 448, 454, 458, 470, 471
Schumann, J., 410, 412, 469
Schuster, S. J., 160
Schwartzentruber, D. J., 314
Scigalla, P., 183
Scortegagna, M., 90, 95, 182, 211, 212, 225, 227, 235, 332
Scott, H. S., 168
Scott, O. C., 298, 300
Scott, P. A., 201
Scudiero, D. A., 313, 330, 386, 387, 390, 392, 399
Scully, S. P., 305
Seagroves, T. N., 96, 99, 325, 326
Searle, P., 310
Searls, T., 29
Sebestyen, M. G., 185
Secomb, T. W., 300, 307
Sedelnikova, O. A., 19
Sedlacek, R., 306
Seeballuck, F., 414, 481, 487
Seehra, J., 173
Seeley, T., 47, 183
Seeliger, M., 231, 233
Segal, N. H., 98
Seibel, J. F., 63, 64, 75, 77, 328, 429
Seidl, J., 229
Seigneuric, R., 249, 250, 252, 253, 254, 255, 262, 265, 276, 278
Seitz, T., 375
Seizinger, B., 372, 376
Sekine, H., 472
Sekizuka, E., 231
Selak, M. A., 47
Selby, M., 313, 330, 387, 390, 392
Sellers, W. R., 327
Semenza, G. L., 4, 5, 13, 15, 17, 20, 26, 44, 46, 62, 65, 88, 89, 90, 91, 93, 100, 112, 114, 115, 120, 129, 133, 135, 158, 159, 160, 173, 181, 182, 203, 204, 206, 213, 223, 224, 227, 230, 232, 233, 234, 248, 276, 311, 312, 324,

327, 328, 329, 330, 332, 347, 348, 349, 350, 351, 356, 372, 373, 378, 386, 387, 393, 395, 407, 408, 409, 413, 424, 425, 426, 427, 428, 429, 430, 448, 449, 464, 465, 472
Semizarov, D., 98
Seo, M. S., 207
Sergueeva, A. I., 184, 237, 372
Serino, G., 363
Sethuraman, N., 188
Setoguchi, Y., 184
Seymour, L., 310
Shah, S., 230, 233
Sham, J. S., 93
Shaner, D., 32
Shanks, J. H., 160
Shannon, A. M., 183
Shapiro, D. J., 335
Shapiro, S., 201
Sharma, N., 377
Sharma, S. D., 93
Sharma, S. K., 315
Sharp, F. R., 232, 233
Sharp, P. A., 109
Sharples, E. J., 230, 231
Shatrov, V. A., 426
Shaw, S., 148, 149, 150
Sheaff, M., 231
Sheffield, P., 7
Sheldon, P. W., 304
Shelley, C. S., 407
Shelton, J. M., 90, 95, 182, 211, 212, 227, 332
Shen, J., 280
Shen, L., 327, 430
Shen, N., 185
Shen, X., 280
Shen, Y., 98
Sheridan, M. R., 304
Sherry, R. M., 314
Sherwood, R. F., 310
Sheta, E. A., 469
Sheth, N., 109
Shevchenko, A., 363
Shi, G., 281
Shi, S. R., 349
Shi, X., 327, 426, 430
Shi, Y., 279, 430
Shiba, N., 154
Shibahara, S., 166, 201
Shibanuma, M., 330
Shibasaki, F., 116
Shibata, T., 312
Shilo, B. Z., 27, 126, 127, 129, 135, 425
Shimizu, R., 163
Shimizu, T., 11, 12
Shimmura, S., 433
Shimoda, L. A., 93
Shin, S. J., 148
Shina, A., 181

Shiosaka, S., 280
Shirasawa, S., 327, 328
Shizukuda, Y., 223, 234
Shoemaker, A., 98
Shoemaker, C., 171, 173
Shoemaker, R. H., 20, 313, 329, 330, 386, 387, 388, 389, 390, 391, 392, 393, 395, 397, 399
Shrigley, R. J., 126
Shuda, M., 281
Shuin, T., 375, 472
Shun, C. T., 357
Shvarts, A., 348
Siatskas, M., 186
Sica, V., 426
Siddiq, A., 230, 233
Siddiqui, A. S., 7, 8
Siddiqui, N., 372
Siim, B. G., 4, 20, 309, 313, 324, 329, 386
Silman, A., 365, 366
Silva, J. M., 109
Silver, G., 306
Simon, M. C., 47, 89, 91, 92, 95, 100, 133, 181, 182, 204, 213, 214, 224, 227, 234, 250, 251, 254, 255, 276, 279, 332, 377, 378, 424, 448, 449, 454, 458, 470
Simons, J. W., 135, 311, 313, 330, 348, 349, 392
Simos, G., 8
Sinclair, A. M., 188
Singel, D. J., 466
Singh, B. K., 32
Singh, S., 465
Singla, V., 373
Sinowaitz, F., 160
Siolas, D., 109
Siren, A. L., 231, 232
Sitkovsky, M. V., 213, 406, 407
Skals, M., 46
Skinner, H. D., 327
Smalling, R., 173, 188
Smallwood, A., 375, 378, 465
Smith, G. C., 302
Smith, L. E., 205, 206, 207
Smith, T. G., 372
Smith, T. J., 425, 464
Snijckers, C. M., 376
Snove, O., Jr., 117
Snyder, S. H., 468
Soares, A. S., 28, 32, 33, 35, 39
Soares, M., 231, 236
Sobolewski, A., 409
Socolovsky, M., 94, 100, 377
Sodhi, A., 330, 433
Sodoyer, R., 187
Sofras, F., 348
Sogawa, K., 5, 7, 8, 17, 89, 412, 425, 428, 472
Sohn, T. K., 114
Soilleux, E. J., 79
Sokoloski, E. A., 435

Sola, M. C., 172
Solano, R., 362, 364
Solomon, S. D., 230
Solway, K. E., 232
Sonenberg, N., 135, 251, 252, 253, 256, 276, 277,
        278, 279, 281, 282, 283, 284
Song, E. J., 114
Song, H., 99, 378
Sonnenfeld, M., 129
Sood, M., 237
Sood, R., 279
Sordet, O., 313, 388, 391, 392
Sorokina, I., 4, 44, 226, 229, 234, 429
Soronen, J., 237
Soto, C., 237
Soucy, G. M., 408, 409, 428
Southan, C., 310
Southard, D., 231
Southon, E., 377
Soutoglou, E., 182
Sowter, H. M., 116, 279, 281, 372, 377, 425
Spandau, J. M., 231
Sparrow, D. B., 63
Spearman, H., 184, 185, 312
Spence, S., 377
Spielmann, P., 43, 46, 47, 112, 182, 230
Spinowitz, B. S., 189
Spradling, A. C., 126
Spring, K. R., 436
Springer, C. J., 315
Srinivas, V., 327, 408, 426, 470
Srivastava, G., 357
Stabinsky, Z., 173, 188
Stadheim, T. A., 188
Stadler, P., 305
Stahl, G. L., 230, 407
Stahl, R. A., 46
Stahl, S., 129
Staller, P., 325
Stallmach, T., 148, 149, 150
Stallock, J. P., 135
Stamler, J. S., 151, 466, 467
Stanbridge, E. J., 391
Stary, C. M., 236
Staver, M., 98
Stawicki, S., 184
St Croix, B., 325
Stearman, R., 374
Stein, I., 255, 268
Steiner, M., 207
Stemmer-Rachamimov, A. O., 92, 378
Stenberg, Y., 63
Steneberg, P., 127
Stenflo, J., 63, 72
Stengel, P., 46
Stenmark, K., 114, 362, 363
Stephen, A. G., 329, 330, 391, 393, 395
Stephenson, R. C., 183

Stern, A. M., 63, 72
Stetler-Stevenson, W., 374
Stevens, M. F. G., 392
Stewart, K., 231
Stewart, V., 213
Stickle, N. H., 376
Stiefel, M., 184
Stiehl, D. P., 43, 44, 46, 47, 112, 230,
        408, 410, 426
Stobbe, C. C., 302
Stock, F., 314
Stocker, H., 135
Stockmann, C., 172, 181, 408, 409, 428
Stockton, D. W., 184, 372
Stojdl, D. F., 279
Stojkoski, C., 64, 70, 75, 82
Stokoe, D., 135, 327, 328, 330
Stolze, I. P., 55, 63, 64, 65, 79, 115, 230,
        413, 481
Storkebaum, E., 97
Stramm, L., 279
Stratford, I. J., 276, 304, 310, 312
Straub, J. A., 376
Strausberg, R. L., 325
Stribbling, S. M., 315
Strickland, T. W., 188
Stroka, D. M., 182, 227, 228
Stuart, D. I., 36, 356
Stuehr, D. J., 465
Sturgeon, C., 186
Su, W., 182
Suda, T., 212
Suematsu, M., 231, 433
Suen, L. F., 327
Sufan, R. I., 349, 373
Sugawa, M., 185
Sugawara, J., 327
Sugaya, T., 171
Suggs, S., 173, 188
Sugi, M., 154
Sugimoto, H., 362
Sugiyama, F., 171
Suit, H. D., 306
Sukhu, L., 185
Suliman, H. B., 160
Sulitkova, J., 325
Sullivan, R., 205, 206
Sumbayev, V. V., 328, 426, 468
Sun, B., 362
Sun, B. K., 148
Sun, H., 135
Sun, X., 93, 313
Sun, Y., 134
Sun, Y. C., 430
Sundberg, C. J., 236
Sung, H. J., 427
Suquet, C., 12
Sutherland, D. C., 126, 127, 128

Sutphin, P. D., 114, 324, 325, 327, 328, 329, 356, 362, 363
Suwabe, N., 159, 161, 163
Suzuki, E., 189
Suzuki, M., 362
Suzuki, N., 157, 159, 161, 162, 163, 165, 166, 168, 171, 181, 201
Suzuki, T., 160
Suzuki, Y., 237
Suzuma, I., 184, 206
Suzuma, K., 184, 201, 206
Svensson, K., 209, 332
Swartz, G. M., 313, 330, 392
Swem, D. L., 19
Swiersz, L., 325
Swing, D. A., 159, 162, 168
Sylvester, J. T., 93
Symmons, D., 365, 366
Szabo, S., 426
Szamosi, I., 32
Sztejnberg, L., 425

T

Tabarini, D., 135
Tacchini, L., 426, 428
Tackett, S., 116
Tada, H., 154, 166
Taetle, R., 182
Taghavi, P., 348, 350
Taher, M., 234
Tait, A. S., 47
Takabuchi, S., 135, 409, 426, 427
Takagi, H., 201, 206, 207
Takahama, Y., 133
Takahashi, D., 365
Takahashi, J. S., 92
Takahashi, S., 159, 161, 163, 165, 166, 168, 181, 201
Takaki, Y., 375
Takaku, F., 160
Takano-Shimizu, T., 63
Takasaki, C., 5, 8, 472
Takeda, H., 46, 89, 92
Takeda, K., 46, 89, 92
Takeichi, M., 127
Takeuchi, M., 188
Takeuchi, T., 63
Talan, M. I., 148
Talbot, D., 309
Talbot, N. P., 372
Talianidis, I., 182
Talks, K. L., 223, 224, 311, 332, 348
Tam, J. P., 72
Tamai, M., 166, 201
Tamourin, P., 160
Tan, C., 330
Tan, C. C., 88, 100, 160, 181

Tanaka, H., 5, 209, 236, 332, 406, 411, 425, 428
Tanaka, I., 362
Tanaka, K., 281
Tanaka, T., 221, 236
Tanaka-Matakatsu, M., 127
Tang, N., 92, 96, 204, 209, 211, 312, 378, 406
Tang, W., 117
Tang, Y., 232
Tangvoranuntakul, P., 189
Taniguchi, M., 280
Taniguchi, S., 202
Tanii, I., 280
Tanimoto, K., 171, 425
Tanner, E. A., 313, 330, 331, 392, 397, 398
Tannock, I. F., 307
Tao, L. H., 231, 236
Tao, W., 135
Tarnawski, A. S., 426
Tarpey, M. M., 434
Taubman, M. B., 27
Taya, S., 7, 17, 89
Taylor, B. L., 5, 6, 20
Taylor, C. T., 47, 413, 414, 448, 476, 479, 480, 481, 487
Taylor, E., 160
Taylor, G., 374
Taylor, M. S., 111
Taylor, R. T., 19
Tazuke, S. I., 327
Tazzyman, D., 312, 406
Tchernyshyov, I., 120
Teh, B. T., 373
Teicher, B. A., 308
Telfer, B. A., 304, 312
Temes, E., 112, 373
Tempst, P., 46, 468
Tendler, D. S., 182
Teng, C. M., 330
Teng, Q., 330
Tepper, O. M., 201, 325
Terris, D. J., 305, 325
Tessarollo, L., 159, 162, 168, 377
Theilmeier, G., 97
Theodorakis, E. A., 95, 408
Theodorescu, D., 469
Theodoropoulos, V. E., 348
Theres, H., 148, 154
Therond, P. P., 131
Thiel, M., 407
Thiele, E. H., 314
Thiemermann, C., 230, 231
Thiery, J. P., 348
Thoma, C. R., 376
Thomas, D. D., 466, 469, 471
Thomas, G., 135
Thomas, J. D., 249, 250, 251
Thomas, T., 5
Thomas-Tsagli, E., 348

Thomlinson, R. H., 298
Thompson, A., 28
Thompson, C. B., 47, 250, 251, 254, 255, 276, 279
Thompson, R. B., 151
Thomson, W., 365, 366
Thornton, R. D., 428
Thrasher, A. J., 184, 185
Thulasiraman, V., 377
Thurmond, F., 90, 95, 211, 212, 227, 332
Thurston, G., 89, 97, 237
Thymara, I., 348
Tian, H., 7, 89, 90, 91, 100, 201, 202, 332
Tian, J., 186
Tian, X., 63
Tian, Y. M., 4, 26, 27, 29, 30, 36, 44, 46, 47, 62, 63, 72, 79, 88, 89, 110, 111, 112, 115, 131, 183, 226, 229, 324, 327, 328, 372, 378, 413, 425, 429, 449, 464
Tiegs, G., 410, 412, 469
Tihan, T., 99
Till, J. H., 280
Tiranti, V., 448, 449, 470
Tirasophon, W., 279, 280
Tjwa, M., 91, 93, 98, 201
To, K. K., 19, 46, 325, 326
Todo, S., 362
Tolentino, M. J., 117
Tom, S., 180
Tomaszewski, J. E., 96, 100, 378, 412
Tomita, K., 166, 201
Tomita, S., 133
Tomita, T., 129, 131
Tomoda, K., 362
Tonks, N. K., 430
Topalian, S. L., 314
Topol, I. A., 182
Torchia, D. A., 10
Torcia, G., 201
Tordo, P., 437
Toro, J., 374, 375
Torres-Cabala, C., 374, 375, 377
Toso, J. F., 314
Tran, M. G., 349, 372, 373, 374, 377
Tranmer, B., 426, 427, 471
Travers, K. J., 284, 289
Treacy, M., 372
Treins, C., 135, 330, 425
Treisman, R., 127
Trepel, J., 374, 375, 411
Trewick, S. C., 63
Trinh, Q. T., 312
Trinidad, B., 47, 412, 413
Tröger, J., 47, 54
Trotter, M. J., 300
Trottier, E., 114
Trout, H., 469

Tsai, M. D., 64, 77
Tschank, G., 229
Tsoukala, V., 348
Tsubota, K., 433
Tsuchiya, M., 433
Tsuchiya, T., 182
Tsuge, T., 362, 363, 364
Tsukamoto, S., 161, 163, 165
Tu, B. P., 277, 284, 289
Tuccio, B., 437
Tuckerman, J. R., 36, 46
Tullius, S. G., 231
Turcotte, S., 323
Turka, L. A., 88, 95, 100, 407
Turley, H., 46, 63, 79, 111, 115, 223, 224, 311, 332, 348, 365, 413
Turner, K., 201
Turtle, E., 183
Tuschl, T., 109, 112

U

Uchida, K., 116
Udenfriend, S., 72
Udono, T., 166, 201
Ueda, N., 133
Uemura, T., 127
Ueno, M., 133
Uezono, S., 407
Ullrich, V., 231, 471
Unger, T. L., 181, 182, 224, 227, 234
Unterman, T. G., 135
Urabe, A., 160
Urabe, M., 185
Uranchimeg, B., 201, 313, 329, 330, 387, 388, 389, 390, 391, 392
Urano, F., 280
Urbanke, C., 7, 12, 19
Urquilla, P., 183

V

Vacca, A., 207
Vadas, M. A., 410
Vago, L., 201
Vainchenker, W., 186
Valegard, K., 28, 30
Valentine, S., 310, 311
Valet, G., 434
Valiando, J., 4, 26, 27, 36, 88, 89, 110, 324, 327, 372, 425, 449, 464, 490
van Berkel, M., 348
Vandekerckhove, J., 46
van Den Berg, S., 7
van den Beucken, T., 249, 250, 252, 253, 254, 255, 262, 265, 276, 277, 278
Vandenhoeck, A., 89, 93
van der Groep, P., 348

van der Kogel, A. J., 304, 312, 325
van der Wall, E., 348
Vandewalle, C., 373
van de Wetering, M., 115
Vande Woude, G., 374
van Diest, P. J., 348
Van Dorpe, J., 97
VanDusen, W. J., 63
Van Dyke, M. M., 395
van Leenen, D., 115
Van Meir, E. G., 313, 329, 392
Van Obberghen, E., 135, 330, 425
Van Roy, F. M., 348
van Scheltinga, A. C., 28, 30
van Tinteren, H., 348
Van Zijl, P., 314
Varela, E., 182
Varet, B., 160
varez Arroyo, M. V., 237
Vargas, S. O., 153
Varia, M., 252, 254, 279, 281, 283, 287
Varki, A., 189
Varki, N., 189
Vasquez-Vivar, J., 434, 435
Vasselli, J. R., 46, 230, 372, 374
Vattem, K. M., 254, 279
Vaupel, P., 275, 348
Vaux, E. C., 4, 129, 183, 226, 327, 425,
    449, 464, 470
Vaxillaire, M., 182
Vazquez, F., 327
Verbert, A., 189
Vercauteren, K., 448, 449, 470
Verheul, H. M. W., 392
Verhey, L., 306
Verma, A., 47
Vervoort, J., 6
Vesey, D. A., 148
Vessella, J. A., 407
Vidal, J. A., 188
Vieira, A., 313, 330, 331, 392, 397, 398
Vigne, P., 51, 53, 131, 134, 230
Viklund, J. A. C., 28, 30
Villa, P., 154
Villabona, C., 414
Villar, D., 46, 112
Villeval, J. L., 186
Vinals, F., 409
Viollet, B., 182
Vleugel, M. M., 348
Vocke, C., 375, 377
Voest, E. E., 376
Vogel, J., 148
Vogel, K. S., 375
Vogel, O., 148
Vogt, L., 423, 425
Vogt, P. K., 330

Vojnovic, B., 304
Volitakis, I., 230, 233
Vollmar, A. M., 430
Volmat, V., 27, 46, 111, 112, 230, 331, 376
Voloshin, Y., 237, 372
Voncken, J. W., 249, 250, 252, 253, 254, 255,
    262, 265, 276, 278
von Kriegsheim, A., 26, 27, 36, 425, 429, 464
von Pawel, J., 309
von Reutern, M., 89
von Roemeling, R., 309
Vortmeyer, A. O., 377
Voss, A. K., 5
Vuister, G. W., 20
Vujaskovic, Z., 117
Vuorikoski, S., 237

W

Wachsberger, P., 307
Wagner, I., 225, 230, 237
Wagner, J. R., 89, 92, 100
Wagner, K., 148, 149, 150, 187
Wagner, K. D., 148, 154
Wagner, K. F., 148, 327
Wagner, K. U., 96
Wagner, N., 148, 154
Wagner, P. D., 95, 236
Wagner, T., 46
Wakabayashi, T., 160
Wakasugi, S., 202
Wakatsuki, T., 281
Wakisaka, N., 411, 428
Walisser, J. A., 377
Walker, C. L., 327
Wallace, J. C., 67
Wallace, T. J., 308, 325
Wallach, T., 172
Walling, C., 424
Walmsley, S. R., 409
Walter, G., 314
Walter, P., 280
Walther, M., 375, 377
Walton, G. B., 151
Walz, G., 374
Wands, J. R., 63
Wang, B., 313, 330, 331, 392, 397, 398
Wang, C. J., 430
Wang, D., 313
Wang, E., 115
Wang, F. S., 430, 472
Wang, G. L., 4, 5, 13, 15, 17, 88, 89, 100, 112,
    159, 173, 181, 182, 204, 223, 230, 332, 348,
    356, 386, 387, 393, 395, 408, 424, 425, 428,
    430, 449, 464
Wang, H., 117
Wang, H. M., 430

Wang, J., 236, 313, 330, 331
Wang, L., 96, 204, 209, 211
Wang, L. Y., 204, 332, 424
Wang, M. H., 236
Wang, P., 374
Wang, Q. P., 63
Wang, S., 313, 331
Wang, S. S., 437
Wang, V., 332
Wang, W., 376
Wang, X., 255, 280
Wang, Y., 236, 280
Wang, Z. Q., 465
Wanger, R. H., 46, 47
Wanner, R. M., 182
Wappner, P., 27, 123, 125, 127, 128, 129, 130,
    131, 132, 133, 135, 136, 137, 138
Ward, G., 300
Ward, H. P., 407
Ward, J. M., 89, 92, 100
Ward, M., 129
Warnecke, C., 116, 130, 181, 221, 225,
    226, 227, 229, 230, 232, 233, 234,
    235, 237, 332
Wartenberg, M., 308, 325
Wartman, L., 7, 89, 332
Wary, K., 281
Wasley, L. C., 279
Watanabe, D., 184
Watanabe, J., 166
Watanabe, M., 11, 12
Watanabe, Y., 63
Watson, D. G., 47
Watson, I. R., 376
Watts, M. E., 301
Weber, K., 109
Wee, S., 356
Wei, C. C., 465
Wei, G., 234
Wei, N., 363
Wei, W., 375, 376
Weidemann, A., 130, 225, 229, 230,
    235, 237
Weikert, S., 180
Weinberg, R. A., 348
Weiner, D. J., 185, 186
Weinmann, R., 88, 100, 424
Weisinger, J. R., 209
Weiss, S., 32
Weissman, J. S., 277, 280, 284, 289
Weisz, A., 467, 468
Weizman, I., 126, 129
Wek, R. C., 254, 279
Weldon, C. W., 325
Welford, R. W., 4, 27, 28, 30, 38, 44, 62, 63, 72,
    89, 226, 328, 378, 425, 429
Welford, S. M., 325, 326, 364
Welinhinda, A. A., 279

Wellman, M., 412
Wellman, T. L., 426, 427, 471
Wellmann, S., 148
Wells, M., 312, 406
Welm, B., 96
Welsh, S. J., 183, 330, 392, 428
Wendling, F., 160
Weng, A. P., 69
Wenger, R. H., 5, 26, 43, 44, 47, 54, 89,
    90, 100, 112, 133, 148, 149, 150, 154,
    164, 182, 203, 224, 227, 228, 230, 232, 237,
    327, 332, 348, 407, 410, 423, 424, 425, 426,
    449, 465
Wenzel, A., 231, 233
Weppler, S. A., 249, 250, 252, 253, 254, 255,
    262, 265, 276, 278
Wersto, R. P., 436
Wesolowski, E., 205, 206
West, D. K., 109
Westerman, K. A., 480
Westphal, H., 89, 92, 100
Whaley, J., 372, 376
Whaley, J. M., 92, 378
Whang, H. Y., 314
Whang, K., 98, 100
Wharton, K. A., Jr., 5, 129
Whelan, D. A., 4, 26, 27, 44, 62, 63, 65, 67, 68,
    70, 72, 75, 114, 226, 328, 378, 413, 425,
    429, 465, 481
Whillans, D. W., 302
Whisson, M. E., 310
White, E. S., 357
White, S. B., 26, 27, 324, 327, 372, 429
Whitelaw, M. L., 4, 5, 8, 26, 27, 44, 62, 63, 64,
    65, 67, 68, 70, 72, 75, 79, 82, 114, 226, 328,
    332, 378, 413, 414, 425, 429, 465, 481
Whitney, K. D., 99
Wicht, M., 188
Wiecek, A., 183
Wiegand, S. J., 205
Wiener, C. M., 90, 91, 93
Wierzbicki, A. S., 30
Wiesener, M., 47, 116, 181, 225, 226, 229, 230,
    234, 237, 332
Wiesener, M. S., 4, 14, 27, 93, 129, 130, 181,
    183, 221, 223, 224, 225, 226, 227, 229, 232,
    233, 235, 236, 327, 332, 425, 449, 464
Wiessner, C., 148
Wigfield, S., 392
Wilk, R., 126, 129
Willam, C., 26, 27, 47, 130, 221, 223, 224, 225,
    226, 229, 230, 235, 237, 324, 327, 328, 332,
    425, 449
Willard, M. T., 313, 330, 392
Willenbrock, R., 225, 230, 232, 233, 235
Williams, K. J., 304, 312
Williams, P., 427
Williams, R., 330, 392

Williams, R. R., 330, 392
Williams, S. C., 204
Willmore, W. G., 472, 473
Wilms, S., 184
Wilsbacher, L. D., 92
Wilson, C., 135
Wilson, J. M., 185, 186
Wilson, L. A., 373
Wilson, M. I., 4, 26, 27, 29, 30, 36, 44, 46, 88, 89,
    110, 112, 131, 183, 226, 229, 324, 327, 356,
    372, 425, 429, 449, 464
Wilson, W. J., 332, 425, 464
Wilson, W. R., 300, 303, 305, 307, 308, 310,
    311, 312
Winearls, C. G., 173
Wink, D. A., 434, 466, 469, 471
Winner, M., 355, 357, 360,
    362, 363, 364
Wirbelauer, C., 376
Wirthner, R., 43, 46, 47,
    112, 230
Wischnewski, H., 188
Witte, L., 201
Witters, L. A., 136
Wloch, M. K., 185
Wolf, C. R., 312
Wolf, D. A., 356, 363
Wolf, E., 7, 12, 19
Wolf, G., 46
Wolfe, F., 364
Wolff, J. A., 185
Wolff, M., 417
Wong, J., 357
Wood, J. M., 207
Wood, M., 46, 116, 230
Wood, S. M., 14, 223, 224, 332
Wood, T. M., 327, 426, 458, 471
Woodcock, M., 301
Woods, H. A., 136
Woodward, E., 375, 378, 465
Workman, P., 330, 391
World Health Organization, 188
Worthington, J., 365, 366
Wotzlaw, C., 19, 428, 429, 430
Wouters, B. G., 247, 249, 251, 252, 253, 256,
    276, 277, 278, 279, 281, 282, 283, 284, 302,
    305, 306, 309
Wouters, M. A., 63
Wright, B. D., 93
Wright, P. E., 64, 397
Wu, H., 158, 165, 166
Wu, J., 278
Wu, L. C., 46, 116
Wu, M. S., 357
Wurm, F. M., 188
Wykoff, C. C., 4, 63, 79, 115, 129, 183, 226, 325,
    327, 413, 425, 449, 464

Wyman, R. J., 134
Wynn, R., 63
Wysk, M., 231, 236

X

Xenaki, D., 304
Xi, L., 234
Xia, C., 426
Xiong, M., 469
Xu, J. H., 34
Xu, L., 47
Xu, S., 108
Xu, T., 134, 135
Xu, X. J., 63

Y

Yagami, K., 171
Yague, S., 237
Yakkundi, A., 312
Yalcin, A., 109
Yamac, H., 47, 412, 413
Yamada, E., 207
Yamada, H., 206, 207, 409
Yamaguchi, E., 365
Yamaguchi, T., 231
Yamamoto, A., 280
Yamamoto, K., 280
Yamamoto, M., 157, 159, 161, 162, 163, 165,
    166, 168, 171, 172, 181, 281
Yamamoto, N., 281
Yamamura, K., 202
Yamanaka, K., 449
Yamanaka, S., 362
Yamanishi, Y., 94, 100, 213, 227, 406, 480
Yamanoi, A., 367
Yamao, H., 154
Yamashita, T., 166, 201
Yamaura, G., 234
Yan, L.-J., 90, 95, 211, 212, 227, 332
Yan, Q., 230, 378
Yanagi, H., 280
Yanagisawa, M., 204
Yanase, H., 168
Yancopoulos, G. D., 205
Yanez-Mo, M., 373
Yang, C. W., 148
Yang, D. I., 233
Yang, H., 4, 44, 88, 89, 110, 226, 229, 234, 324,
    327, 372, 375, 376, 378, 425, 429, 449,
    464, 490
Yang, H. F., 26, 27, 36
Yang, J., 3, 13, 14, 18, 95, 393, 407
Yang, J. C., 314
Yang, K. D., 430
Yang, S., 310, 311
Yang, X., 117

Yang, Y., 160, 375, 377
Yang, Y. J., 430
Yang, Y. T., 233
Yaniv, M., 47, 182, 413, 429, 464, 471
Yao, M., 374, 375
Yaoita, E., 185
Yaqoob, M. M., 230, 231
Yarchoan, R., 332
Yasinska, I. M., 328
Yauch, R. L., 92, 378
Ye, J., 275
Yeates, K. M., 14, 470
Yee, J., 357
Yeh, J. L., 96, 233
Yeh, P., 327, 328
Yen, S. E., 324, 327, 328, 356
Yeo, E. J., 313, 330, 392
Yeowell, D., 183
Yian, Y. M., 30, 449
Yildiz, Ö., 7, 12, 19
Yim, S., 429, 430
Yin, C., 234
Yin, T., 212
Yla-Herttuala, S., 237
Yodoi, J., 135, 425, 428
Yokoo, T., 171
Yokotani, N., 7, 17, 89
Yomogida, K., 159
Yoneda, T., 427
Yoneda-Kato, N., 362
Yoo, J., 377
Yoo, M. A., 114
Yoo, S. J., 413
Yoo, Y. G., 114
Yoon, D., 19
Yoon, S. S., 98
Yoshida, E. S., 376
Yoshida, H., 277, 280, 281, 307
Yoshinaga, K., 280
Yost, H. J., 373
Youn, H. D., 376
Young, F., 213
Young, L. H., 96, 233
Young, R. M., 448, 470
Yu, A. Y., 89, 90, 93, 100, 133, 203, 206, 224, 227, 232, 233, 332, 348, 449
Yu, D., 159, 162, 168
Yu, F., 26, 27, 324, 327, 372, 429
Yu, H., 377
Yu, X., 55, 64, 65, 79, 230, 481
Yuan, G., 93
Yuan, Y., 182
Yui, M., 154
Yujnovsky, I., 7, 12, 19
Yun, C., 279
Yun, J. K., 414
Yuncu, M., 236
Yura, T., 280

Z

Zabludoff, S. D., 313, 330, 331, 392, 397, 398
Zaborowska, Z., 116, 181, 234, 332
Zack, D. J., 206
Zafari, A. M., 437
Zagzag, D., 311, 330, 348, 349, 351, 373
Zahler, S., 430
Zahr, R., 426, 427, 471
Zähringer, C., 425, 426, 427
Zalek, J., 313, 388, 389
Zamore, P. D., 109
Zanjani, E. D., 159, 160
Zappia, V., 372
Zbar, B., 374, 375, 377
Zeira, E., 187
Zeitlin, S., 98, 100, 233
Zelko, I. N., 427
Zelzer, E., 129, 133, 135, 425
Zeman, E. M., 308
Zeng, H., 279, 280
Zentgraf, H., 374
Zeviani, M., 448, 449, 470
Zhan, Y., 427
Zhang, G. J., 395
Zhang, H., 114, 135, 330, 434, 435
Zhang, J. H., 72
Zhang, L., 3, 13, 14, 18, 91, 201, 230, 393
Zhang, L. P., 327
Zhang, Q., 95, 98, 100, 182, 212
Zhang, S. X., 236
Zhang, X., 117, 430
Zhang, X. G., 327, 430
Zhang, Y., 185, 186, 254, 279, 280, 282
Zhang, Z., 327, 330, 426, 430
Zhang, Z. F., 98, 100
Zhang, Z. H., 28, 30
Zhao, H., 281, 434, 435
Zhao, Q., 26, 27, 46, 324, 327, 372, 429
Zhao, Y. G., 36
Zheng, J. Z., 129, 327, 330, 425, 426, 430
Zheng, X., 70, 76
Zhitkovich, A., 47, 182, 230
Zhong, H., 135, 311, 313, 330, 348, 349, 392
Zhong, X. S., 327
Zhou, C., 356, 363
Zhou, J., 47, 407, 410, 412, 426, 428, 429, 463, 468, 469, 471, 473, 480
Zhou, Q., 426
Zhou, W., 330
Zhu, C., 430
Zhu, L., 99, 100, 234
Zhu, X., 112
Zhu, X. H., 327
Zhu, X. Y., 426
Zhulin, I. B., 5, 6, 20
Ziff, E. B., 19

Zimmer, M., 372
Zimmermann, T. S., 118
Zolotarjova, N. I., 63
Zoltick, P. W., 185
Zon, L. I., 161, 165

Zondlo, J., 27, 28, 226
Zünd, G., 407
Zundel, W., 114, 135, 327, 328, 330, 355, 357, 360, 362, 363, 364
Zweier, J. L., 234

Subject Index

A

Akt, reactive oxygen species and regulation of hypoxia-inducible factor activity, 430

Angiogenesis, hypoxia-inducible factor transgenic mouse studies
HIF-1α null endothelial cells, 210–211
HIF-2α knockdown mice, 205–207, 209

Apoptosis, von Hippel–Lindau protein and neuronal apoptosis regulation, 375–376

AQN4, *see* Banoxantrone

ARD1, *see* Arrest-defective-1

Arrest-defective-1, RNA interference effects on HIF-α stability, 114

Aryl hydrocarbon nuclear receptor translocator, *see* Hypoxia-inducible factor; Per/ARNT/Sim domains

Asparaginyl hydroxylases, *see* Factor-inhibiting hypoxia-inducible factor; β-Hydroxylase

ATF4
detection by immunohistochemistry in hypoxic areas, 286–287
unfolded protein response and PERK-eIF2α-ATF4 arm, 278–279

ATF6
cleavage assay, 285–286
nuclear translocation assay, 285–286
unfolded protein response pathway, 280–281

B

Bacteria artificial chromosome transgenic reporter mice, *see* Erythropoietin

BAH, *see* β-Hydroxylase

Banoxantrone, hypoxic cytotoxicity in cancer, 310

Brain, hypoxia-inducible factor activation and protection, 232–233

C

E-Cadherin, von Hippel–Lindau protein and expression, 373

Cancer
hypoxia-inducible factor expression
assays
chromatin immunoprecipitation, 334–335
human exonic evidence-based oligonucleotide microarrays, 335–337

Northern blot, 333
quantitative reverse transcriptase-polymerase chain reaction, 334
Western blot, 334
experimental induction, 326
inhibitors, 329–331
knockout mouse studies, 332–333
posttranslational modifications, 328
regulation, 324–325
reporter assays, 333
responsive elements, 333
RNA interference, 331–332
stabilizers, 327–328
target genes and tumor promotion, 324–326

hypoxia-inducible factor tumor biology
astrocytoma, 99
mammary cancer, 99
von Hippel–Lindau disease, 99–100
xenografts, 98

hypoxia-inducible factor-1α expression and invasion/metastasis
immunohistochemistry, 349–350
invasion assay, 350–351
overview, 347–349
quantitative reverse transcriptase-polymerase chain reaction, 351–352
transepithelial resistance measurement of cell–cell adhesion, 351

migration inhibitory factor and tumor expression
immunohistochemistry, 365
plasma analysis and genomic DNA extraction, 365–366
polymorphisms and genotyping, 364–366

tumor hypoxia
acute versus chronic hypoxia, 298–300
consequences for cancer treatment
chemotherapy, 307–308
overview, 300–301, 324–325
radiation therapy, 302–307
dynamics, 300
exploitation in cancer treatment
cytotoxins, 308–311
gene therapy, 311–312
hypoxia-inducible factor-1 targeting, 312–314
obligate anaerobes, 314–315
oxygen tension measurement, 299–300

Cancer (*cont.*)
  unfolded protein response, hypoxia, and tumor
      formation, 275–276, 281–282
Carbon monoxide, hypoxia-inducible factor
      activation, 229
Chemotherapy, hypoxia consequences for cancer
      treatment, 307–308
Chemotin, hypoxia-inducible factor-1
      inhibition, 329
Chromatin immunoprecipitation, hypoxia-
      inducible factor expression analysis in
      tumors, 334–335
Clostridium, tumor necrosis targeting, 314–315
Coimmunoprecipitation
  factor-inhibiting hypoxia-inducible factor and
      binding proteins, 79–82
  hypoxia-inducible factor subunits, 15–17
Confocal microscopy, ATF6 nuclear translocation
      assay, 285–286
COP9 signalosome
  function, 356–357
  migration inhibitory factor function
      coimmunoprecipitation, 363
      CSN5-dependent activities, 364
      deneddylation assay, 363–364
      overview, 362–363
  subunits, 356
CSN, *see* COP9 signalosome
Cytochrome C, reduction assay for reactive
      oxygen species, 431–432

D

Desferrioxamine, hypoxia-inducible factor
      stabilization, 230, 326
Dichlorofluorescein, reactive oxygen species assay
      in cells, 434, 452–454
Dihydroethidine, reactive oxygen species assay in
      cells, 434–435
Dihydrorhodamine, reactive oxygen species assay
      in cells, 435–436
Double-stranded RNA-like endoplasmic
      reticulum resident kinase, *see* PERK
Drosophila melanogaster, hypoxia response
  advantages as model system, 124–126
  hypoxia-inducible factors
      Fatiga prolyl hydroxylase
        development studies of mutants,
            132–133
        overview, 131
      green fluorescent protein reporters, 130
      growth control and cell size determination
            role, 136–138
      Sima
        development studies of mutants,
            132–133
        HIF-1α homology, 128–129
        oxygen regulation, 131–132

phosphatidylinositol 3-kinase regulation,
            135–136
      target of rapamycin regulation, 135–136
      Tango and HIF-1β homology, 135
  oxygen starvation adaptation, 134
  prospects for study, 138–1939
  respiratory system overview, 126–128

E

Echinomycin, hypoxia-inducible factor-1
      inhibition, 329, 395
eEF2, phosphorylation and translation regulation
      in hypoxia, 255
eIF2α
  phosphorylation and translation regulation in
      hypoxia
      gene-specific effects, 253–254
      global effects, 252–253
  unfolded protein response and PERK-eIF2α-
      ATF4 arm, 278–279
eIF4F, complex availability and translation
      regulation in hypoxia, 254–255
Electrophoretic mobility shift assay, hypoxia-
      inducible factor, 17–18
ELISA, *see* Enzyme-linked immunosorbent assay
EMSA, *see* Electrophoretic mobility shift assay
Endoplasmic reticulum stress, *see* Unfolded
      protein response
Enzyme-linked immunosorbent assay, high
      throughput screening of hypoxia-inducible
      factor-1 inhibitors, 394–397
EPO, *see* Erythropoietin
Erythropoietin
  bacteria artificial chromosome transgenic
      reporter mice
      bacteria artificial chromosome library and
            recombination, 162
      erythropoietic and non-erythropoietic
            function analysis, 165–166
      essential *cis*-elements for *in vivo* expression,
            171–173
      genetic manipulation of mouse lines, 161
      green fluorescence protein expression under
            erythropoietin gene regulatory region,
            166–169
      hypoxic chamber, 161
      immunohistochemistry, 162–163
      luciferase reporter monitoring *in vivo*,
            163, 165
      overview, 159, 161
      renal erythropoietin-producing cell
            identification, 169–171
  expression regulation
      divalent transition metal effects, 182–183
      hypoxia response, 148, 158, 180–181
      hypoxia-inducible factor, 181–182
      hypoxia-responsive elements, 158–159, 181

iron chelator effects, 182
transcription factors, 181
gene therapy
    clinical prospects, 187
    immune response in *ex vivo* approaches,
        186–187
    vectors, 184–186
hypoxic preconditioning response, 231
neuroprotection, 232
recombinant proteins
    commercial preparations, 188–189
    gene activation technology, 189–190
    indications, 190
    posttranslational modifications and
        expression systems, 187–188
    structure, 180
tissue distribution, 158–160, 180
transgenic mouse studies in permanent
        coronary artery ligation model
    animals and surgery, 149–150
    cardiac protection, 153
    erythropoietin plasma levels, 151
    erythropoietin receptor expression in
        myocardium, 151
    immunohistochemical analysis, 150
    infarct size determination, 150
    WT1 as marker in ischemic injury, 153–154

F

Factor-inhibiting hypoxia-inducible factor
    assay
        carbon dioxide capture radioassay
            background calculations, 74
            incubation conditions, 72–74
            kinetic parameters, 74–75
            principles, 71–72
            sensitivity, 75–76
        protein–protein interaction assays
            coimmunoprecipitation, 79–82
            overview, 76
            pull-down assay, 76–79
        substrate preparation, 69–71
    hypoxia-inducible factor regulation in
        inflammation, 413
    oxygen sensitivity, 62–63
    protein–protein interactions, 63–64
    purification
        Escherichia coli expression system, 67
        histidine-tagged protein, 66
        human embryonic kidney cell transfection
            system, 67–68
        maltose-binding protein fusion protein,
            65–66
        quantification and purity assessment, 68–69
    structure, 63–64
    substrate specificity, 64
Fibronectin, von Hippel–Lindau protein
    interaction, 378

FIH-1, *see* Factor-inhibiting hypoxia-inducible
    factor
Fluorescence resonance energy transfer, Per/
    ARNT/Sim domains in hypoxia-inducible
    factor, 19
FRET, *see* Fluorescence resonance energy transfer

G

GATA-2
    erythropoietin expression regulation, 181
    inhibitors, 191
Geldanamycin, hypoxia-inducible factor-1
    inhibition, 329
Gene therapy, hypoxia-selective gene therapy for
    cancer treatment, 311–312
Gene therapy, *see* Erythropoietin

H

HDAC1, *see* Histone deacetylase 1
Heart, hypoxia-inducible factor activation and
    protection, 233–235
Heat shock proteins, hypoxia-inducible factor
    regulation in inflammation, 414
HEEBO, *see* Human exonic evidence-based
    oligonucleotide microarray
Hematopoiesis, hypoxia-inducible factor
        transgenic mouse studies
    HIF-1α null mice, 213
    HIF-1β/ARNT null mice, 213–214
    HIF-2α knockdown mice, 212–213
    HIF-2α null mice, 211–212
Heme oxygenase-1, hypoxic preconditioning
    response, 231
HIF, *see* Hypoxia-inducible factor
High throughput screening, hypoxia-inducible
        factor-1 inhibitors
    cell-based screens, 387–390
    DNA-binding inhibitors in cell-free assays,
        393–395
    enzyme-linked immunosorbent assay, 394–397
    protein–protein interaction inhibitors in
        cell-free assays, 395–397
    small molecule inhibitor findings, 392–395
    transcription inhibitors, 397–398
    validation, 390–392
Histone deacetylase 1, RNA interference effects
    on HIF-α stability, 114
HTS, *see* High throughput screening
Human exonic evidence-based oligonucleotide
    microarray, hypoxia-inducible factor
    expression analysis in tumors, 335–337
Hydrogen peroxide, *see also* Reactive oxygen
        species
    hypoxia-inducible factor regulation in
        inflammation, 413
β-Hydroxylase, structure, 63
Hydroxyl radical, *see* Reactive oxygen species

Hypoxia-inducible factor
  asparaginyl hydroxylation, see Factor-inhibiting
      hypoxia-inducible factor
  Drosophila studies, see Drosophila melanogaster,
      hypoxia response
  electrophoretic mobility shift assay, 17–18
  erythropoietin target gene, see Erythropoietin
  heterodimers, 4, 26, 44, 88
  induction, 4
  inflammatory response, see Inflammation
  inhibitor screening for HIF-1
    bioassay-directed isolation of natural
        product inhibitors, 398–399
    high throughput screening
      cell-based screens, 387–390
      DNA-binding inhibitors in cell-free
          assays, 393–395
      enzyme-linked immunosorbent assay,
          394–397
      protein–protein interaction inhibitors in
          cell-free assays, 395–397
      small molecule inhibitor findings,
          392–395
      transcription inhibitors, 397–398
      validation, 390–392
    overview, 386
  nitric oxide regulation, 469–470
  nitric oxide regulation, see Nitric oxide
  nuclear lysate preparation and extraction, 15
  oxygen sensitivity, 424–425, 464–465
  PAS domains, see Per/ARNT/Sim domains
  preconditioning studies, see Preconditioning
  prolyl hydroxylation, see Hypoxia-inducible
      factor prolyl hydroxylase
  reporter assay, 13–14
  RNA interference
    factor-inhibiting hypoxia-inducible factor
        knockdown, 114–116
    HIF-1/HIF-2 target gene specificity, 116
    HIF-α stability studies
      acetylation/deacetylation enzyme
          knockdown, 114
      prolyl hydroxylase knockdown,
          110–112, 114
    therapeutic prospects, 116–118
  stabilizing compounds, 183
  transgenic mouse systems
    angiogenesis studies
      HIF-1α null endothelial cells, 210–211
      HIF-2α knockdown mice, 205–207, 209
    conditional knockouts with loxP system, 94
    hematopoiesis studies
      HIF-1α null mice, 213
      HIF-1β/ARNT null mice, 213–214
      HIF-2α knockdown mice, 212–213
      HIF-2α null mice, 211–212
    HIF-α subunit
      heterozygous knockouts, 93

      knockouts, 90–91
      targeted knockouts, 94–98
      HIF-β/ARNT subunit knockouts, 91–92
    historical perspective, 89–90
    physiological effects
      cardiovascular system, 96
      chondrocytes, 96–97
      colon, 96
      hematopoiesis, 95
      liver, 94
      lymphocytes, 95
      mammary gland, 96
      myeloid lineage, 94–95
      nervous system, 97–98
      skeletal muscle, 95
      skin, 97
    prolyl hydroxylase domain enzyme
        knockouts, 92–93
    tumor biology
      astrocytoma, 99
      mammary cancer, 99
      von Hippel–Lindau disease, 99–100
      xenografts, 98
    vascular endothelial growth factor
        knockouts, 93
    vasculogenesis studies
      HIF-1α null endothelial cells, 204
      HIF-1α null mice, 203–204
      HIF-1β/ARNT null mice, 204–205
      HIF-2α knockdown mice, 201–202
      HIF-2α null mice, 201
      von Hippel–Lindau protein knockouts,
          92, 94
  tumor biology, see Cancer
  Western blot of HIF-1α
    electrophoresis and blotting, 451–452
    lysate preparation, 451
    materials, 450–451
    nuclear extract preparation, 451
    overview, 449–450
Hypoxia-inducible factor prolyl hydroxylase
  ascorbate reduction, 47
  assays
    carbon dioxide capture radioassay, 30–32
    classification, 29
    comparison of formats, 38–39
    fluorescence assay, 32–33
    inhibitor testing, 32
    mass spectrometry identification of modified
        protein, 35–36
    oxygen consumption assay, 33–35
    succinate radioassay
      extract requirements, 53–54
      incubation conditions, 35, 51–53
      protein versus peptide target assay, 55
      thin-layer chromatography for substrate
          purity assessment, 54–55
    von Hippel–Lindau protein binding

capture assay, 36–37
HIF-α peptide binding, 48–50
Fatiga in *Drosophila*
development studies of mutants, 132–133
overview, 131
hypoxia induction, 44, 46
inhibitor development, 47, 229–230
knockout mice, 46, 92–93
oxygen affinity and inhibition, 46, 88, 448
prolyl hydroxylase domain family of enzymes,
27, 44
protein–protein interactions, 46
purification of recombinant proteins
PHD2 from *Escherichia coli*, 28–29
baculovirus–insect cell expression system, 48
reactive oxygen species inhibition, 47
substrate modification sites, 27
substrate specificity, 27
surface plasmon resonance of enzyme–substrate
interactions, 37–38
tissue distribution, 46–47
Hypoxia translation regulation
global changes
genetic background effects, 250–251
kinetics and oxygen dependency, 250
mechanisms
eEF2 phosphorylation, 255
eIF2α phosphorylation
eIF4F complex availability, 254–255
gene-specific effects, 253–254
global effects, 252–253
overview, 251–252
overview, 248–249
polysome assay
fractionation, detection, and collection, 270
gene-specific translation efficiency
mechanistic parameters, 265–266
messenger RNA translation percentage
and recruitment, 265
normalization, 263, 265
ribosome density, 265–266
RNA isolation, 263
translation efficiency, 266
genome-wide analysis of translation
efficiency, 266–267
lysate preparation, 269–270
principles, 256–258
quantitative analysis of polysome profiles,
258, 260
ribosomal RNA fraction in polysomes,
260–261
ribosomes per messenger RNA, 261–262
RNA isolation from gradient fractions, 270
sucrose gradient preparation, 269
protein synthesis assays
gel electrophoresis, 268–269
gene-specific effects, 256
global effects, 256

metabolic radiolabeling, 268
scintillation counting, 269
reporter assays, 267–268
Hypoxic preconditioning, *see* Preconditioning

I

IKK, *see* Nuclear factor-κB
IL-1, *see* Interleukin-1
Inflammation
hypoxia and immune response, 406–407
mediators and hypoxia-inducible factor
regulation
activity regulation, 413–414
overview, 406–408
posttranslational regulation, 411–413
prospects for study, 414–415
transcriptional regulation, 408–410
translational regulation, 410–411
Interleukin-1, hypoxia-inducible factor
regulation in inflammation, 410
Invasion, *see* Cancer
IRE1, unfolded protein response and
IRE1-XBP1 arm, 279–280
Ischemic preconditioning. *see* Preconditioning

K

Kidney, hypoxia-inducible factor activation and
protection, 235–236

L

Lung, hypoxia-inducible factor activation and
protection, 236

M

Macrophage migration inhibitory factor,
see Migration inhibitory factor
MAPK, *see* Mitogen-activated protein kinase
Mass spectrometry, hypoxia-inducible factor
prolyl hydroxylase assay, 35–36
c-Met, von Hippel–Lindau protein crosstalk,
374–375
Metastasis, *see* Cancer
2-Methoxyestradiol, hypoxia-inducible factor-1α
inhibition, 313
MIF, *see* Migration inhibitory factor
Migration inhibitory factor
COP9 signalosome function
coimmunoprecipitation, 363
CSN5-dependent activities, 364
deneddylation assay, 363–364
overview, 362–363
hypoxia induction, 364
RNA interference
knockdown assessment
catalytic activity assay, 361–362

Migration inhibitory factor (*cont.*)
    reverse transcriptase–polymerase chain
        reaction, 359–360
    Western blot, 360–361
    short hairpin RNA transfection, 357–358
    tumor expression
        immunohistochemistry, 365
        plasma analysis and genomic DNA
            extraction, 365–366
        polymorphisms and genotyping,
            364–366
Misomidazole, radiosensitization, 304
Mitogen-activated protein kinase, reactive
    oxygen species and regulation of hypoxia-
    inducible factor activity, 430
MS, *see* Mass spectrometry
Muscle, hypoxia-inducible factor activation and
    protection, 236
Myocardial infarction, erythropoietin transgenic
    mouse studies in permanent coronary artery
    ligation model
    animals and surgery, 149–150
    cardiac protection, 153
    erythropoietin
        plasma levels, 151
        receptor expression in myocardium, 151
    immunohistochemical analysis, 150
    infarct size determination, 150
    WT1 as marker in ischemic injury,
        153–154

N

NF-κB, *see* Nuclear factor-κB
Nitric oxide
    HIF-1α regulation
        accumulation of protein, 467–468
        activation, 468–469
        hypoxic signal transmission antagonism by
            nitric oxide, 472–473
        prospects for study, 473–474
        superoxide stabilization and antagonism of
            nitric oxide actions, 470–472
    hypoxia-inducible factor prolyl hydroxylase
        regulation, 469–470
    hypoxia-inducible factor regulation in
        inflammation, 412–413
Nitroimidazole, radiosensitization, 304
NMR, *see* Nuclear magnetic resonance
NO, *see* Nitric oxide
Northern blot
    formation and signaling, 465–467
    hypoxia-inducible factor expression analysis in
        tumors, 333
Nuclear factor-κB
    cell hypoxia studies
        DNA-binding assay, 485–487
        hydroxylase inhibitors, 482
        hypoxia chamber culture, 482

IKK
    IKKβ pull-down assay, 490–491
    site-directed mutagenesis for signaling
        manipulation, 483–484
    reporter assay for nuclear factor-κB
        expression, 484–485
    Western blot, 488–490
    hypoxia activation, 480–481
    hypoxia-inducible factor regulation in
        inflammation, 409, 413
Nuclear magnetic resonance, Per/ARNT/Sim
    domain
    structure studies, 10–11
    protein–protein interaction analysis, 11–13

O

2-Oxoglutarate
    hypoxia-inducible factor stabilizing analogs,
        183–184
    thin-layer chromatography for substrate purity
        assessment, 54–55

P

p53, von Hippel–Lindau protein interactions, 376
PAS domains, *see* Per/ARNT/Sim domains
Per/ARNT/Sim domains
    coimmunoprecipitation of subunits, 15–17
    delineation, 5–7
    fluorescence resonance energy transfer, 19
    hypoxia-inducible factor
        function, 5
        mutation studies, 13–14
    limited proteolysis experiments, 8–10
    nuclear magnetic resonance
        protein–protein interaction analysis, 11–13
        structure studies, 10–11
    protein distribution, 5
    recombinant protein expression in *Escherichia*
        coli, 7–8
    structure, 6–7, 18–19
PERK
    phosphorylation assay, 282–285
    unfolded protein response and PERK-eIF2α-
        ATF4 arm, 278–279
PHD, *see* Hypoxia-inducible factor prolyl
    hydroxylase
Phosphatidylinositol 3-kinase
    reactive oxygen species and regulation of
        hypoxia-inducible factor activity, 430
    Sima regulation, 135–136
PI3K, *see* Phosphatidylinositol 3-kinase
PKC, *see* Protein kinase C
Polymerase chain reaction
    hypoxia-inducible factor expression analysis in
        tumors with quantitative reverse
        transcriptase-polymerase chain reaction,
        334, 351–352

migration inhibitory factor knockdown
        assessment with reverse transcriptase-
        polymerase chain reaction, 359–360
Polysome assay, hypoxia effects on translation
    fractionation, detection, and collection, 270
    gene-specific translation efficiency
        mechanistic parameters, 265–266
        messenger RNA translation percentage and
            recruitment, 265
        normalization, 263, 265
        ribosome density, 265–266
        RNA isolation, 263
        translation efficiency, 266
    genome-wide analysis of translation efficiency,
        266–267
    lysate preparation, 269–270
    principles, 256–258
    quantitative analysis of polysome profiles,
        258, 260
    ribosomal RNA fraction in polysomes, 260–261
    ribosomes per messenger RNA, 261–262
    RNA isolation from gradient fractions, 270
    sucrose gradient preparation, 269
PR104, hypoxic cytotoxicity in cancer, 310–311
Preconditioning
    clinical application of ischemic
        preconditioning, 222–223, 238
    definition, 222
    hypoxia-inducible factors
        activation and protection
            lung protection, 236
            muscle protection, 236
            myocardial protection, 233–235
            nephroprotection, 235–236
            neuroprotection, 232–233
        activation strategies
            carbon monoxide, 229
            hypoxia-inducible factor prolyl
                hydroxylase inhibition, 229–230
            hypoxic hypoxia, 227–229
        chronic hypoxic/ischemic disease response,
            236–237
        degradation, 225–226
        hypoxia response, 223–224
        structure, 225
        target genes, 226–227, 231
    hypoxic preconditioning and gene induction
        erythropoietin, 231
        heme oxygenase-1, 231
Primary cilium, von Hippel–Lindau protein and
    disease, 373–374
PRK-like endoplasmic reticulum resident kinase,
    see PERK
Prolyl hydroxylase, see Hypoxia-inducible factor
    prolyl hydroxylase
Protein kinase C, factor-inhibiting hypoxia-
    inducible factor regulation, 115
PX-478, hypoxia-inducible factor-1 inhibition, 392

R

Radiation therapy
    fractionated radiation and reoxygenation, 305
    hypoxia consequences for cancer treatment,
        303–307
    oxygen enhancement ratio, 302
    radiosensitization, 304–305
Ras, hypoxia-inducible factor interactions, 328
Reactive oxygen species
    assays
        chemiluminescence assay, 432–433
        cytochrome C reduction assay, 431–432
        electron paramagnetic resonance, 436–437
        fluorecent dyes for intracellular detection
            dichlorofluorescein, 434, 452–454
            dihydroethidine, 434–435
            dihydrorhodamine, 435–436
            tissue measurements, 436
    cellular sources, 422
    hypoxia-inducible factor prolyl hydroxylase
        inhibition, 47
    hypoxia-inducible factor regulation
        direct regulation, 428
        overview, 425–427
        signaling
            phosphorylative signaling, 429–430
            prolyl hydroxylases, 428–429
            transcriptional regulation, 427–428
    hypoxic stabilization studies of HIF-1α
        mitochondrial electron transport-generated
            reactive oxygen species, 458–459
        Rieske iron–sulfur protein RNA
            interference
                materials, 454–455
                overview, 454
                transfection, 455–457
    signaling function, 423–424
Retinopathy of prematurity, HIF-2α knockdown
    mouse studies, 205–206
RNA interference
    historical perspective, 108–109
    hypoxia signaling pathway studies
        factor-inhibiting hypoxia-inducible factor
            knockdown, 114–116
        HIF-1/HIF-2 target gene specificity, 116
        HIF-α stability studies
            acetylation/deacetylation enzyme
                knockdown, 114
            prolyl hydroxylase knockdown,
                110–112, 114
        therapeutic prospects, 116–118
    hypoxia-inducible factor targeting in cancer,
        331–332
    migration inhibitory factor
        knockdown assessment
            catalytic activity assay, 361–362
            reverse transcriptase-polymerase chain
                reaction, 359–360

RNA interference (*cont.*)
 Western blot, 360–361
 short hairpin RNA transfection, 357–358
 principles, 108–109
 Rieske iron–sulfur protein knockdown and
  hypoxic stabilization of HIF-1α
  materials, 454–455
  overview, 454
  transfection, 455–457
ROP, *see* Retinopathy of prematurity
ROS, *see* Reactive oxygen species

S

Sima
 development studies of mutants, 132–133
 hypoxia-inducible factor-1α homology,
  128–129
 oxygen regulation, 131–132
 phosphatidylinositol 3-kinase regulation,
  135–136
 target of rapamycin regulation, 135–136
SPR, *see* Surface plasmon resonance
Stress erythropoiesis, 158
Superoxide, *see* Reactive oxygen species
Surface plasmon resonance, hypoxia-inducible
 factor prolyl hydroxylase–substrate
 interactions, 37–38

T

Tango, hypoxia-inducible factor-1β
 homology, 135
Target of rapamycin, Sima regulation, 135–136
Taxol, hypoxia-inducible factor-1α
 inhibition, 313
TGF-β, *see* Transforming growth factor-β
Thin-layer chromatography, 2-oxoglutarate
 substrate purity assessment, 54–55
Tirapazamine, hypoxic cytotoxicity in cancer,
 308–310
TLC, *see* Thin-layer chromatography
Topotecan, hypoxia-inducible factor-1
 inhibition, 329, 392
TOR, *see* Target of rapamycin
TPZ, *see* Tirapazamine
Transforming growth factor-β, hypoxia-inducible
 factor regulation in inflammation, 412
Translation, *see* Hypoxia translation regulation
Tumor, *see* Cancer

U

Unfolded protein response
 ATF4 detection by immunohistochemistry in
  hypoxic areas, 286–287
 ATF6
  cleavage assay, 285–286

 nuclear translocation assay, 285–286
  pathway, 280–281
 causes, 277–278
 cellular adaptation to hypoxia, 276–277
 IRE1-XBP1 arm, 279–280
 PERK phosphorylation assay, 282–285
 PERK-eIF2α-ATF4 arm, 278–279
 tumor hypoxia and formation, 275–276,
  281–282
 XBP1 slicing assay, 287–289
UPR, *see* Unfolded protein response

V

Vascular endothelial growth factor (VEGF)
 knockout mice, 93
 retinal neovascularization studies, 207, 209
Vasculogenesis, hypoxia-inducible factor
 transgenic mouse studies
 HIF-1α null endothelial cells, 204
 HIF-1α null mice, 203–204
 HIF-1β/ARNT null mice, 204–205
 HIF-2α knockdown mice, 201–202
 HIF-2α null mice, 201
VHL, *see* von Hippel–Lindau protein
Vincristine, hypoxia-inducible factor-1α
 inhibition, 313
von Hippel–Lindau disease
 features, 372
 mouse models, 377–378
von Hippel–Lindau protein
 epithelial differentiation regulation
 E-cadherin, 373
  primary cilium, 373–374
 hypoxia-inducible factor prolyl hydroxylase
  binding
  capture assay, 36–37
  HIF-α peptide binding, 48–50
  physiological function, 356
 hypoxia-inducible factor regulation
 overview, 372–373
 inflammation, 411–412
 IKKβ pull-down assay, 490–491
 knockout mice, 92, 94, 99–100
 c-Met crosstalk, 374–375
 neuronal apoptosis regulation, 375–376
 p53 interactions, 376
 phosphorylation, 376–377
 polyubiquitinylation, 377

W

Western blot
 HIF-1α
  electrophoresis and blotting, 451–452
  lysate preparation, 451
  materials, 450–451

nuclear extract preparation, 451
overview, 449–450
hypoxia-inducible factor expression analysis in
    tumors, 334
migration inhibitory factor knockdown
    assessment, 360–361
nuclear factor-κB, 488–490
PERK phosphorylation assay, 282–285
WT1
hypoxia-inducible factor-1 regulation, 148
ischemic injury marker, 153–154
knockout mice, 149

X

XBP1
slicing assay, 287–289
unfolded protein response and IRE1-XBP1
    arm, 279–280, 279–280

Y

YC-1, hypoxia-inducible factor-1 inhibition,
    313–314, 392

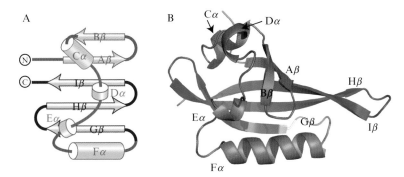

Thomas H. Scheuermann, *et al.*, Figure 1.1 Per/ARNT/Sim (PAS) domain topology and structure. (A) Schematic of a typical arrangement of secondary structure elements in PAS domains. (B) Ribbon diagram of the solution structure of the hypoxia-inducible factors (HIF)-β PAS-B domain (Card *et al.*, 2005). In both panels, the PAS motif/S1 box (red) and PAC motif/S2 box (blue) sequence homology elements are indicated, and secondary structure elements are labeled using the nomenclature initially established for FixL (Gong *et al.*, 1998).

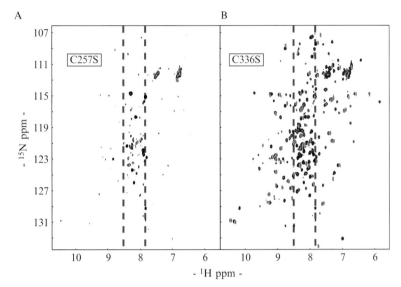

Thomas H. Scheuermann, *et al.*, Figure 1.3 Use of nuclear magnetic resonance (NMR) spectroscopy to validate the structures of recombinant proteins. Displayed are ^{15}N/^{1}H heteronuclear spin quantum coherence (HSQC) spectra collected from two different Gβ1-HIF-2α PAS-B fusion proteins bearing single amino acid substitutions in the Per/ARNT/Sim (PAS) domain. (A) Hypoxia-inducible factors (HIF)-2α PAS-B (C257S). In this spectrum, the majority of peaks are found within the ^{1}H ~7.9- to 8.5-ppm (denoted by blue lines) boundaries associated with poorly folded or disordered protein elements. The poor signal intensity of this spectrum also implies that this construct is aggregation prone, a function of NMR spectroscopy's sensitivity to macromolecular size. (B) HIF-2α PAS-B (C336S). A ^{15}N/^{1}H HSQC spectrum of this construct indicates that it is a well-folded protein, as demonstrated by the presence of a large number of well-dispersed peaks showing similar line widths. The spectrum collected from the Gβ1-fused mutant PAS domain is given in black, while a reference Gβ1-only spectrum is given in red. This example underscores the importance of confirming the folded state of recombinantly expressed proteins. While neither of these conservative substitutions was expected to disrupt the HIF-2α PAS-B structure, qualitative analysis of these spectra provided early insight regarding the viability of these reagents for subsequent studies.

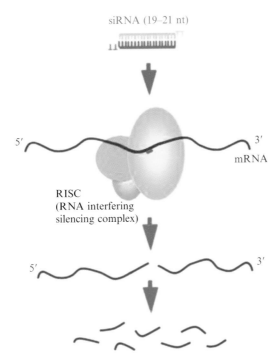

Edurne Berra and Jacques Pouysségur, Figure 6.1 Schema of the evolutionary conserved RNA interference (RNAi) process. Chemically synthesized, enzymatically produced, or shRNAs and shRNA-miRs–derived small interfering RNAs (siRNAs) trigger RNAi silencing. One strain of the siRNA (those containing the 5′ extremity are more thermolabile) is incorporated into a protein complex called RNA interfering silencing complex (RISC). This complex "guides" the siRNA to the target messenger RNA (mRNA), then binds to and cleaves it in the middle of the complementary region. Following this first cleavage, several nonspecific nucleases are able to achieve complete mRNA degradation leading to gene silencing.

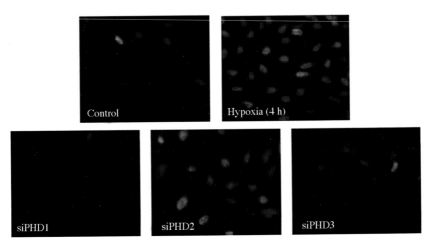

Edurne Berra and Jacques Pouysségur, Figure 6.2 Prolyl hydroxylase domain (PHD) 2 is the oxygen sensor setting low, steady-state levels of hypoxia-inducible transcription factor (HIF)-1α in normoxia. HeLa cells were transfected with the corresponding small interfering RNAs (siRNAs; 20 nM), and 48 h later, the expression of HIF-1α was analyzed by IF. Cells were incubated in hypoxia (1% O_2 for 4 h) as a positive control.

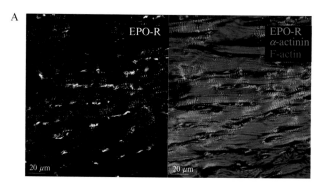

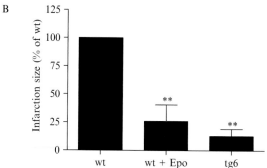

Giovanni G. Camici, Figure 8.1 (A) Immunohistochemical analysis reveals abundant expression of erythropoietin receptor (EPO-R) in wild-type (wt) mice myocardium (left panel). According to the expression pattern of the cardiomyocyte-specific marker alpha (α)-actinin (right panel, red) and myofibril marker F-actin (right panel, blue), EPO-R (right panel, green) localizes in the membrane of cardiomyocytes. (B) Administration of recombinant EPO (rhEPO) (wt + EPO) as well as constitutively overexpressed EPO (tg6) drastically reduce infarction size upon left anterior descending (LAD) ligation ($n = 8$; $P < 0.05$ for all groups versus wt). Infarct size is represented as percentage of wt mice $\pm$ SEM (Standard Error Mean).

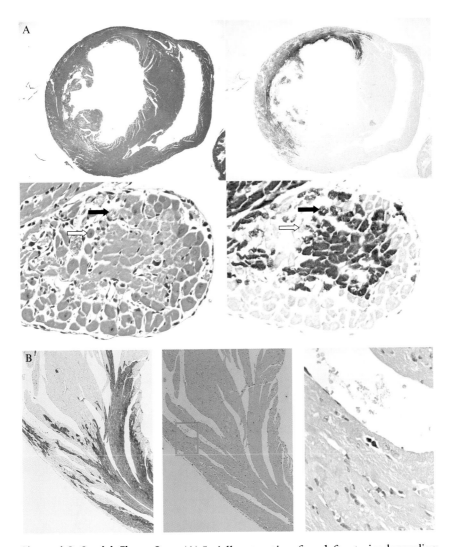

Giovanni G. Camici, Figure 8.2 (A) Serially cut sections from left anterior descending (LAD) ligated hearts of wild-type (wt) mice were stained with hematoxylin and eosin (H&E) (left panel) and Wilms' tumor suppressor gene (WT1; right panel). Upper and lower panel correspond to a 20- and 60-fold magnification, respectively. Black and white arrows indicate necrotic and viable cardiomyocytes, respectively. (B) To exclude nonspecific binding of WT1 (left panel), parallel sections from infarcted hearts were stained with CD45 antibody. No unspecific staining was observed (middle panel). Magnified (60x) selection (red box) in the right panel shows specific staining of CD45-positive lymphocytes.

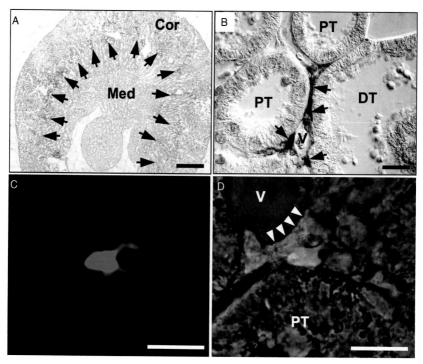

Norio Suzuki et al., Figure 9.4 The renal erythropoietin (EPo)–producing (REP) cells. (A, B) Anti–green fluorescent protein (GFP) immunohistochemistry of the kidney section from the anemic *wt-Epo–GFP* transgenic mouse. The renal EPo-producing cells are stained black (arrows). The cells are surrounded by the renal tubules (DT, distal tube; PT, proximal tube) in deeper regions of the outer medulla (Med). Cor and V indicate the renal cortex and vessel, respectively. (C, D) Fluorescent image of REP cell in the anemic kidney of *wt-Epo–GFP* transgenic mouse. REP cells display a unique stellar shape, with the projections extending in various directions. (D) Anti-PECAM1-PE immunofluorescence and phase contrast images are merged with the GFP image. Arrowheads indicate a PECAM1-positive vascular endothelial cell associating with REP cells. Scale bars indicate 1 mM (A) and 10 μM (B–D).

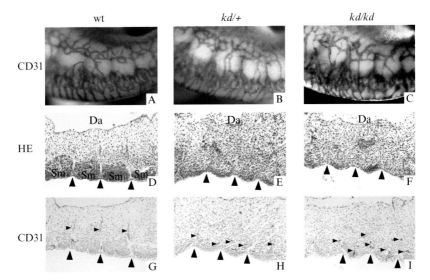

Osamu Ohneda *et al.,* Figure 11.3 Neovascularization in wild-type (wt), hypoxia-inducible factor (*HIF*)-2α$^{kd/+}$, *HIF*-2α$^{kd/kd}$ embryos. Whole mount embryos (A, B, C) (E10.5) and sectioned samples (G, H, I) were immunostained with an anti–CD31 antibody. Blue-color staining represents positive for CD31. Sections were also stained by hematoxylene and eosin (H & E) in D, E, and F. There is altered CD31 staining, vessel formation, and somite structure in the *HIF-2α* mutant embryos. Large arrowheads indicate the intersomite regions, and small arrowheads indicate vessels. Da: dosal aorta, Sm: somite.

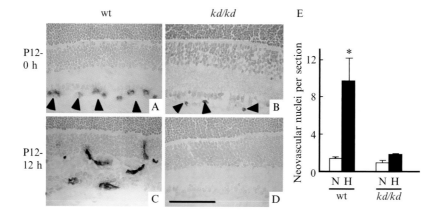

Osamu Ohneda *et al.,* **Figure 11.5** Retinal neovascularization in a mouse model of retinopathy of prematurity (ROP). Retinas from wild-type (wt) (A, C) and hypoxia-inducible factor (HIF)-$2\alpha^{kd/kd}$ (B, D) mice at P12–0 h and 12 h after shifting from hyperoxia to normoxia were subjected to immunostaining with anti-CD34 antibody (A–D). Arrowheads indicate neovascular buds detected in the ganglion cell layer (GCL) (A, B). Note the elongation of the neovascular buds and vessel infiltration into the inner nuclear layer (INL) of *wt* mice (C); neovascularizatin was altered in the retina of HIF-$2\alpha^{kd/kd}$ mice (D). The number of neovascular nuclei in the retinas of wt and HIF-$2\alpha^{kd/kd}$ (kd/kd) mice was counted, and the average number of nuclei per section is shown. Bar indicates 100 μm. $^{*}P < 0.01$.

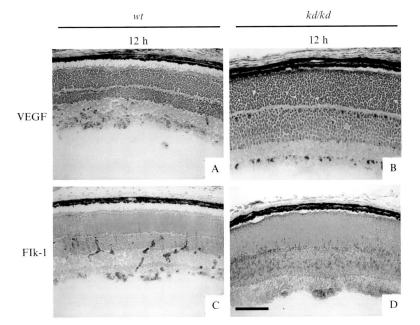

Osamu Ohneda *et al.*, Figure 11.7 Immunohistochemical staining for angiogenic factors in the retinas of wild-type (*wt*) and mutant mice in the retinopathy of prematurity (ROP) model. Frozen retinal sections from *wt* (A, C) and hypoxia-inducible factor (*HIF*)-$2\alpha^{kd/kd}$ (kd/kd), (B, D) mice were immunostained with anti-vascular endothelial growth factor (VEGF) (A, B), and Flk-1 (C, D) antibodies at P12–12 h time point. Bar indicates 100 μM.

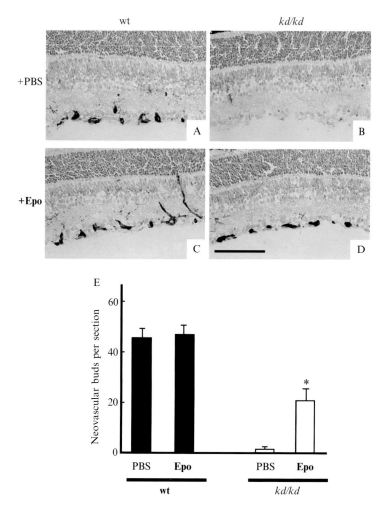

Osamu Ohneda *et al.*, Figure 11.8   Effects of EPO on retinal neovascularization. Wild-type (*wt*) (A, C) and hypoxia-inducible factor (*HIF*)-2α$^{kd/kd}$ (kd/kd) (B, D) mice were examined for the formation of neovascular buds following the administration of phosphate-buffered saline (PBS) (A, B) or EPO (C, D) at P12–24 h. CD34-positive neovascular buds were scored under the microscope (E). $^*P < 0.01$. Bar indicates 100 μm.

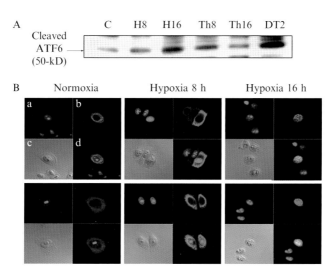

Constantinos Koumenis, *et al.,* **Figure 14.2** (A) Hypoxia induces proteolytic cleavage and activation of ATF6. Production of the 50-kD proteolytic fragment of ATF6 is increased under prolonged hypoxia. HeLa cells were transfected with FLAG-tagged ATF6 and exposed to hypoxia ($\leq 0.02\%$ O_2) for 8 and 16 h, the endoplasmic reticulum (ER) stress activators thapsigargin for 8 or 16 h, and the dithiothreitol (DTT) for 2 h. Cell lysates were immunoblotted with an anti-FLAG antibody. (B) The N-terminal, 50-kD–cleaved ATF6 fragment translocates to the nucleus following prolonged hypoxia. Confocal microscopy was performed on fixed cells using an anti-FLAG fluorescein isothiocyanate (FITC)-labeled antibody (Panel b) and counterstaining nuclei with propidium iodide (Panel a). (Panel c) Differential interference contrast (DIC) microscopy. (Panel d) Combined image of Panels a and b.

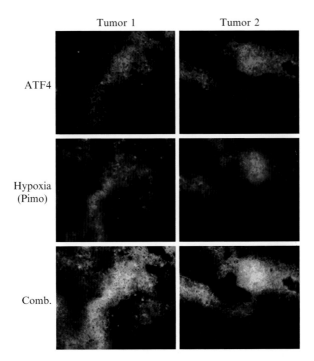

Constantinos Koumenis, *et al.*, Figure 14.3 Immunofluorescence of ATF4 and hypoxia in two primary human cervical tumors. Sections were stained with an anti-ATF4 (top) polyclonal antibody followed by a fluorescein isothiocyanate (FITC)-conjugated secondary antibody and with a Cy-3–labelled anti-pimonidazole antibody (middle). Bottom panels are combined images. Adapted with permission from Bi *et al.* (2005).

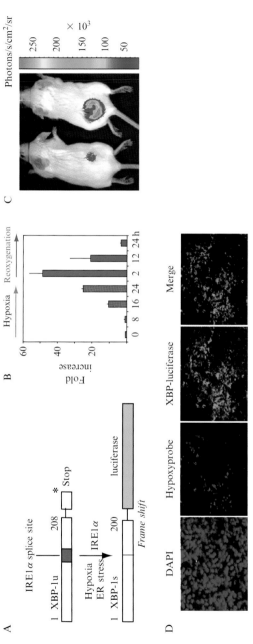

Constantinos Koumenis, *et al.*, Figure 14.4 Hypoxia activates XBP-1 splicing (A,B). XBP-1 splicing occurs within solid tumors throughout their growth (C,D).

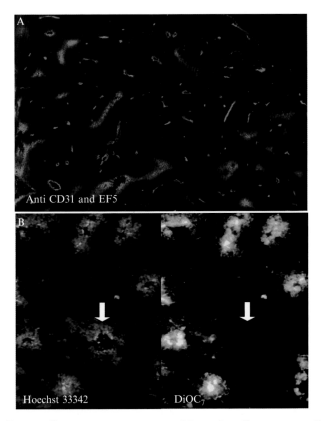

J. Martin Brown, Figure 15.1 Two types of hypoxia affect tumor cells: chronic hypoxia (A), which is a consequence of the fact that the distance between the cells and the vasculature is greater than the diffusion distance of oxygen through respiring tissue, and acute hypoxia (B), which is caused by fluctuating blood flow in the tumor. (A) Chronic hypoxic tumor cells (anti-EF5, red) in this human soft tissue sarcoma are seen in regions distant from blood vessels (anti-CD31, green). Hypoxic cells were labeled with a fluorescent monoclonal antibody against EF5, a nitroimidazole that is metabolized selectively in hypoxic cells and then covalently binds to them. Photograph courtesy of Drs. Cameron Koch and Sydney Evans, University of Philadelphia; adapted from Brown (2002). (B) Acute hypoxia. Visualization of a specific tumor area from a mouse injected with Hoechst 3342 (blue) 20 m in prior to injection with $DiOC_7$ (green). Each dye labels the cells immediately surrounding open blood vessels. In these photographs, the two dyes label the same vessels except for the one marked with an arrow, which has shut down in the 20-min period between injection of the first and second dyes. Photographs courtesy of Dr. Andrew Minchinton, BC Cancer Center, Vancouver; adapted from Brown (2002).

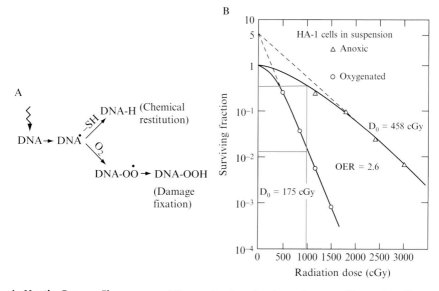

J. Martin Brown, Figure 15.2 The mechanism for the resistance of hypoxic cells to radiation and a typical radiation survival curve of cells irradiated *in vitro* under either hypoxic or aerobic conditions. (A) Radiation produces free radicals in cellular molecules, including DNA, which is the target for radiation killing. Molecular oxygen has a high affinity for this radical and binds covalently to it, producing a peroxy radical, which "fixes" the damage leading to DNA strand breaks and base damage. In the absence of oxygen, cellular nonprotein sulfhydryls (of which glutathione is the most important) can donate an H atom to the radical, thereby restoring it to its undamaged form. (B) Radiation survival curves of Chinese hamster HA-1 cells determined by clonogenic assay. The cells were irradiated under either hypoxic or aerobic conditions and plated for colony formation. All colonies of more than 50 cells after 12 days of incubation at 37° were counted as indicating surviving cells. The red lines show that, at a dose of 1000 cGy (10 Gy), survival under hypoxic conditions is approximately 50% and under aerobic conditions is approximately 1%.

A Oxic cell

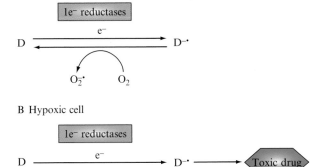

B Hypoxic cell

J. Martin Brown, Figure 15.6 Typical mechanism by which prodrugs act as hypoxic-selective cytotoxins. The nontoxic prodrug must be a substrate for intracellular one-electron reductases, such as cytochrome P450 reductase, which adds an electron to the prodrug, thereby converting it to a free radical. In oxic cells, the unpaired electron in the prodrug radical is rapidly transferred to molecular oxygen, forming superoxide and regenerating the initial prodrug. This futile redox cycle prevents buildup of the prodrug radical when O_2 is present. Hypoxia-selective cell killing is achieved if the prodrug radical that accumulates in hypoxic cells is more cytotoxic than the superoxide formed in oxic cells. In principle, the prodrug radical could itself be the cytotoxin; however, more commonly, it undergoes further reactions to form the ultimate toxic species. From Brown and Wilson (2004), with permission.

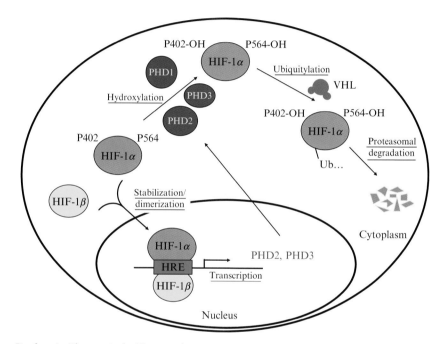

Denise A. Chan, *et al.*, Figure 16.1 Regulation of hypoxia–inducible factors. Under normal oxygenated conditions, a family of prolyl hydroxylases (PHDs) hydroxylate proline 402 or proline 564 of hypoxia-inducible transcription factor alpha (HIF-1α). Hydroxylation of either residue results in recognition by the von Hippel-Lindau (VHL)/ elongin B/elongin C complex, which functions as an E3 ubiquitin ligase, targeting HIF-1α to the proteasome for degradation. Conversely, under hypoxic conditions, HIF-1α is not hydroxylated and is able to dimerize with HIF-1β to drive transcription of target genes. Two of the target genes are PHD2 and PHD3, which feedback to regulate HIF-1α.

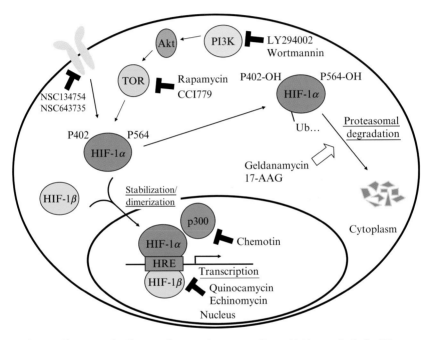

Denise A. Chan, _et al._, Figure 16.2 Pathways implicated in hypoxia-inducible transcription factor (HIF) inhibition. Several HIF inhibitors and their targets are shown in this schematic.